STUDENT SOLUT

to accompany

CALCULUS
ONE AND SEVERAL VARIABLES

EIGHTH EDITION

SATURNINO SALAS

EINAR HILLE

GARRET ETGEN
University of Houston

PREPARED BY

BRADLEY E. GARNER
University of Houston - Clear Lake

CARRIE J. GARNER

JOHN WILEY & SONS, INC.
New York • Chichester • Weinheim • Brisbane • Singapore • Toronto

COVER PHOTO   © Robert Shafer/Tony Stone Images

ISBN 0-471-32959-2

Printed in the United States of America

10 9 8 7 6

Printed and bound by Victor Graphics, Inc.

CONTENTS

CHAPTER 1

SECTION 1.2

1. rational, complex

3. rational, complex

5. integer, rational, complex

7. integer, rational, complex

9. integer, rational, complex

11. $\dfrac{3}{4} = 0.75$

13. $\sqrt{2} > 1.414$

15. $-\dfrac{2}{7} < -0.28517$

17. $|6| = 6$

19. $|3 - 7| = 4$

21. $|-5| + |-8| = 13$

23. $|5 - \sqrt{5}| = 5 - \sqrt{5}$

25.

27.

29.

31.

33.

35.

37.

39.

41. bounded, lower bound 0, upper bound 4

43. not bounded

45. not bounded

47. bounded above, upper bound $\sqrt{2}$

49. $x^2 - 10x + 25 = (x - 5)^2$

51. $8x^6 + 64 = 8(x^2 + 2)(x^4 - 2x^2 + 4)$

53. $4x^2 + 12x + 9 = (2x + 3)^2$

55. $x^2 - x - 2 = (x - 2)(x + 1) = 0; \quad x = 2, -1$

57. $x^2 - 6x + 9 = (x - 3)^2; \quad x = 3$

59. $x^2 - 2x + 2 = 0; \quad$ no real zeros

61. $5! = 120$

63. $\dfrac{8!}{3!5!} = \dfrac{8 \cdot 7 \cdot 6 \cdot 5 \cdot 4 \cdot 3 \cdot 2 \cdot 1}{3 \cdot 2 \cdot 1 \cdot 5 \cdot 4 \cdot 3 \cdot 2 \cdot 1} = 56$

65. $\dfrac{7!}{0!7!} = \dfrac{7!}{1 \cdot 7!} = 1$

67. Let r be a rational number and s an irrational number. Suppose $r + s$ is rational. Then $(r + s) - r = s$ is rational which contradicts the fact that s is irrational.

69. The product of a rational and an irrational number may either be rational or irrational; $0 \cdot \sqrt{2} = 0$ is rational, $1 \cdot \sqrt{2} = \sqrt{2}$ is irrational.

71. Suppose that $\sqrt{2} = p/q$ where p and q are integers and $q \neq 0$. Assume that p and q have no common factors (other than ± 1). Then $p^2 = 2q^2$ and p^2 is even. This implies that $p = 2r$ is even. Thus $2q^2 = 4r^2$ which implies that q^2 is even, and hence q is even. It now follows that p and q are both even and contradicts the assumption that p and q have no common factors.

73. Let x be the length of a rectangle that has perimeter P. Then the width y of the rectangle is given by $y = \frac{1}{2}(P - x)$ and the area is

$$A = x\left(\frac{1}{2}P - x\right) = \left(\frac{P}{4}\right)^2 - \left(x - \frac{P}{4}\right)^2.$$

It now follows that the area is a maximum when $x = P/4$. Since $y = P/4$ when $x = P/4$, the rectangle of perimeter P having the largest area is a square.

PROJECT   1.2

1. $\dfrac{p}{q}$ terminates when q is of the form $2^m 5^n$ for m, n nonnegative integers.

3. (a) $x = 13.201201\cdots$, $1000x = 13201.201201\cdots$. Therefore, $999x = 13188$ and $x = \dfrac{13188}{999}$.

(b) $2.777\cdots = \dfrac{25}{9}$.

(c) $x = 0.2323\cdots$, $100x = 23.2323\cdots$. Therefore, $99x = 23$ and $x = \dfrac{23}{99}$.

(d) $4.16\overline{3} = \dfrac{3477}{900}$

Let $x = 0.999\cdots$, Then $10x - x = 9.\overline{9} - 0.\overline{9}$, so $x = \dfrac{9}{9} = 1$

SECTION 1.3

1. $2 + 3x < 5$

$3x < 3$

$x < 1$

Ans: $(-\infty, 1)$

3. $16x + 64 \leq 16$

$16x \leq -48$

$x \leq -3$

Ans: $(-\infty, -3]$

5. $\frac{1}{2}(1 + x) < \frac{1}{3}(1 - x)$

$3(1 + x) < 2(1 - x)$

$3 + 3x < 2 - 2x$

$5x < -1$

$x < -\frac{1}{5}$

Ans: $(-\infty, -\frac{1}{5})$

7. $x^2 - 1 < 0$

$(x + 1)(x - 1) < 0$

Ans: $(-1, 1)$

9. $x(x - 1)(x - 2) > 0$

Ans: $(0, 1) \cup (2, \infty)$

11. $x^3 - 2x^2 + x \geq 0$

$x(x - 1)^2 \geq 0$

Ans: $[0, \infty)$

13.

$$\frac{1}{x} < x$$

$$x - \frac{1}{x} > 0$$

$$\frac{x^2 - 1}{x} > 0$$

$$x(x-1)(x+1) > 0 \quad \text{(by 1.3.1)}$$

$$(x+1)\,x\,(x-1) > 0$$

Ans: $(-1,0) \cup (1,\infty)$

15.

$$\frac{x}{x-5} > \frac{1}{4}$$

$$\frac{x}{x-5} - \frac{1}{4} > 0$$

$$\frac{4x - (x-5)}{4(x-5)} > 0$$

$$\frac{3x+5}{4(x-5)} > 0$$

$$4(x-5)(3x+5) > 0 \quad \text{(by 1.3.1)}$$

$$(3x+5)(x-5) > 0$$

Ans: $\left(-\infty, -\dfrac{5}{3}\right) \cup (5,\infty)$

17.

$$\frac{x^2 - 9}{x+1} > 0$$

$$(x+1)(x-3)(x+3) > 0 \quad \text{(by 1.4.1)}$$

$$(x+3)(x+1)(x-3) > 0$$

Ans : $(-3,-1) \cup (3,\infty)$

19.

$$x^3(x-2)(x+3)^2 < 0$$

$$(x+3)^2 x\,(x-2) < 0$$

Ans: $(0,2)$

21.

$$x^2(x-2)(x+6) > 0$$

$$(x+6)\,x^2(x-2) > 0$$

Ans: $(-\infty, -6) \cup (2,\infty)$

23.

$$\frac{1}{x-1} + \frac{4}{x-6} > 0$$

$$\frac{x-6+4(x-1)}{(x-1)(x-6)} > 0$$

$$\frac{5x-10}{(x-1)(x-6)} > 0$$

$$5(x-2)(x-1)(x-6) > 0$$

$$(x-1)(x-2)(x-6) > 0$$

Ans: $(1,2) \cup (6,\infty)$

25.

$$\frac{2x-6}{x^2-6x+5} < 0$$

$$2(x-3)(x-1)(x-5) < 0$$

$$(x-1)(x-3)(x-5) < 0$$

Ans: $(-\infty, 1) \cup (3,5)$

27. $(-2,2)$

29. $(-\infty,-3) \cup (3,\infty)$

31. $\left(\frac{3}{2}, \frac{5}{2}\right)$

33. $(-1,0) \cup (0,1)$

35. $\left(\frac{3}{2},2\right) \cup \left(2,\frac{5}{2}\right)$

37. $(-5,3) \cup (3,11))$

39. $\left(-\frac{5}{8}, -\frac{3}{8}\right)$

41. $(-\infty,-4) \cup (-1,\infty)$

43. $\left(-\infty, -\frac{8}{5}\right) \cup (2,\infty)$

45. $|x-0| < 3$ or $|x| < 3$

47. $|x-2| < 5$

49. $|x-(-2)| < 5$ or $|x+2| < 5$

51. $|x - 2| < A \implies 2|x - 2| = |2x - 4| < 2A \implies |2x - 4| < 3$

provided that $0 < A \leq \frac{3}{2}$

53. $|x + 1| < 2 \implies 3|x + 1| = |3x + 3| < 6 \implies |3x + 3| < A$

provided that $A \geq 6$

55. $x < \sqrt{x} < 1 < \dfrac{1}{\sqrt{x}} < \dfrac{1}{x}$

57. If a and b have the same sign, then $ab > 0$. Suppose that $a < b$. Then $a - b < 0$ and

$$\frac{1}{b} - \frac{1}{a} = \frac{a - b}{ab} < 0.$$

Thus, $(1/b) < (1/a)$.

59. With $a \geq 0$ and $b \geq 0$

$$b \geq a \implies b - a = (\sqrt{b} + \sqrt{a})(\sqrt{b} - \sqrt{a}) \geq 0 \implies \sqrt{b} - \sqrt{a} \geq 0 \implies \sqrt{b} \geq \sqrt{a}.$$

61. By the hint

$$\big|\,|a| - |b|\,\big|^2 = (|a| - |b|)^2 = |a|^2 - 2|a|\,|b| + |b|^2 = a^2 - 2|ab| + b^2$$

$$\leq a^2 - 2ab + b^2 = (a - b)^2.$$

$$(ab \leq |ab|)$$

Taking the square root of the extremes, we have

$$\big|\,|a| - |b|\,\big| \leq \sqrt{(a - b)^2} = |a - b|.$$

63. With $0 \leq a \leq b$

$$a(1 + b) = a + ab \leq b + ab = b(1 + a).$$

Division by $(1 + a)(1 + b)$ gives

$$\frac{a}{1 + a} \leq \frac{b}{1 + b}.$$

65. Suppose that $a < b$. Then

$$a = \frac{a + a}{2} \leq \frac{a + b}{2} \leq \frac{b + b}{2} = b.$$

$\dfrac{a + b}{2}$ is the midpoint of the line segment $\overline{ab}$.

SECTION 1.4

1. $d(P_0, P_1) = \sqrt{(6-5)^2 + (-3-0)^2} = \sqrt{1+9} = \sqrt{10}$

3. $d(P_0, P_1) = \sqrt{[5-(-3)]^2 + (-2-2)^2} = \sqrt{64+16} = 4\sqrt{5}$

5. $d(P_0, P_1) = \sqrt{(2-2)^2 + (-2-4)^2} = \sqrt{0+36} = 6$

7. $\left(\dfrac{2+6}{2}, \dfrac{4+8}{2}\right) = (4,6)$ **9.** $\left(\dfrac{2+7}{2}, \dfrac{-3-3}{2}\right) = (\tfrac{9}{2}, -3)$

11. $\left(\dfrac{\sqrt{3}+0}{2}, \dfrac{0+\sqrt{3}}{2}\right) = \tfrac{1}{2}(\sqrt{3}, \sqrt{3})$ **13.** $m = \dfrac{5-1}{(-2)-4} = \dfrac{4}{-6} = -\dfrac{2}{3}$

15. $m = \dfrac{b-a}{a-b} = -1$ **17.** $m = \dfrac{0-y_0}{x_0-0} = -\dfrac{y_0}{x_0}$

19.   Equation is in the form $y = mx + b$. Slope is 2; y-intercept is -4.

21.   Write equation as $y = \tfrac{1}{3}x + 2$. Slope is $\tfrac{1}{3}$; y-intercept is 2.

23.   Write equation as $y = \tfrac{7}{3}x + \tfrac{4}{3}$. Slope is $\tfrac{7}{3}$; y-intercept is $\tfrac{4}{3}$.

25. $y = 5x + 2$ **27.** $y = -5x + 2$ **29.** $y = 3$ **31.** $x = -3$

33.   Every line parallel to the x-axis has an equation of the form $y = a$ constant. In this case $y = 7$.

35.   The line $3y - 2x + 6 = 0$ has slope $\tfrac{2}{3}$. Every line parallel to it has that same slope. The line through $P(2, 7)$ with slope $\tfrac{2}{3}$ has equation $y - 7 = \tfrac{2}{3}(x - 2)$, which reduces to $3y - 2x - 17 = 0$.

37.   The line $3y - 2x + 6 = 0$ has slope $\tfrac{2}{3}$. Every line perpendicular to it has slope $-\tfrac{3}{2}$. The line through $P(2, 7)$ with slope $-\tfrac{3}{2}$ has equation $y - 7 = -\tfrac{3}{2}(x - 2)$, which reduces to $2y + 3x - 20 = 0$.

39. $\left(\tfrac{1}{2}\sqrt{2}, \tfrac{1}{2}\sqrt{2}\right), \left(-\tfrac{1}{2}\sqrt{2}, -\tfrac{1}{2}\sqrt{2}\right)$ [Substitute $y = x$ into $x^2 + y^2 = 1$.]

41. $(3, 4)$ [Write $4x + 3y = 24$ as $y = \tfrac{4}{3}(6 - x)$ and substitute into $x^2 + y^2 = 25$.]

43. $(1, 1)$ **45.** $\left(-\tfrac{2}{23}, \tfrac{38}{23}\right)$

47.   We select the side joining $A(1, -2)$ and $B(-1, 3)$ as the base of the triangle.

length of side AB : $\sqrt{29}$

equation of line through A and B : $5x + 2y - 1 = 0$

length of altitude from vertex $C(2, 4)$ to side AB : $\dfrac{|5(2) + 2(4) - 1|}{\sqrt{29}} = \dfrac{17}{\sqrt{29}}$

area of triangle: $\dfrac{1}{2}\left(\sqrt{29}\right)\left(\dfrac{17}{\sqrt{29}}\right) = \dfrac{17}{2}$

49. $(y + 1)^2 = x + 1$;   parabola, vertex at $(-1, 1)$

51. $2(x - 2)^2 + 3(y + 1)^2 = 6$,   or $\dfrac{(x - 2)^2}{3} + \dfrac{(y + 1)^2}{2} = 1$;   ellipse, center at $(2, -1)$

53. $(y - 2)^2 - 4(x - 1)^2 = 4$,   or $\dfrac{(y - 2)^2}{4} - \dfrac{(x - 1)^2}{1} = 1$;   hyperbola, center at $(1, 2)$

55. $4(x - 3)^2 - (y + 2)^2 = 16$,   or $\dfrac{(x - 3)^2}{4} - \dfrac{y + 2)^2}{16} = 1$;   hyperbola, center at $(3, -2)$

57.   Substitute $y = m(x - 5) + 12$ into $x^2 + y^2 = 169$ and you get a quadratic in x that involves m. That quadratic has a unique solution iff $m = -\frac{5}{12}$. (A quadratic $ax^2 + bx + c = 0$ has a unique solution iff $b^2 - 4ac = 0$. This is clear from the general quadratic formula.)

59.   The slope of the line through the center of the circle and the point P is $\dfrac{3 - (-1)}{-1 - 1} = -2$, so the slope of the tangent line is $\frac{1}{2}$. The equation for the tangent line to the circle at P is

$$(y + 1) = \frac{1}{2}(x - 1), \quad \text{or} \quad x - 2y - 3 = 0.$$

61.   midpoint of line segment $\overline{PQ}$: $\left(\dfrac{5}{2}, \dfrac{5}{2}\right)$

slope of line segment $\overline{PQ}$: $\dfrac{13}{3}$

equation of the perpendicular bisector: $y - \dfrac{5}{2} = -\left(\dfrac{3}{13}\right)\left(x - \dfrac{5}{2}\right)$ or $3x + 13y - 40 = 0$

63. $d\,(P_0, P_1) = \sqrt{(-2 - 1)^2 + (5 - 3)^2} = \sqrt{13}$, $d\,(P_0, P_2) = \sqrt{[-2 - (-1)]^2 + (5 - 0)^2} = \sqrt{26}$,

$d\,(P_1, P_2) = \sqrt{[1 - (-1)]^2 + (3 - 0)^2} = \sqrt{13}$.

Since $d\,(P_0, P_1) = d\,(P_1, P_2)$,   the triangle is isosceles.

Since $[d\,(P_0, P_1)]^2 + [d\,(P_1, P_2)]^2 = [d\,(P_0, P_2)]^2$,   the triangle is a right triangle.

65. $d\,(P_0, P_1) = \sqrt{(3 - 1)^2 + (4 - 1)^2} = \sqrt{13}$, $d\,(P_0, P_2) = \sqrt{[3 - (-2)]^2 + (4 - 3)^2} = \sqrt{26}$,

$d\,(P_1, P_2) = \sqrt{[1 - (-2)]^2 + (1 - 3)^2} = \sqrt{13}$.

Since $d\,(P_0, P_1) = d\,(P_1, P_2)$,   the triangle is isosceles.

Since $[d\,(P_0, P_1)]^2 + [d\,(P_1, P_2)]^2 = [d\,(P_0, P_2)]^2$,   the triangle is a right triangle.

67.   The coordinates of M are $\left(\dfrac{a}{2}, \dfrac{b}{2}\right)$;   and

$d\,(M, (0, b)) = d\,(M, (0, a)) = d\,(M, (0, 0)) = \frac{1}{2}\sqrt{a^2 + b^2}$.

69.   Denote the points $(0, 1)$, $(3, 4)$ and $(-1, 6)$ by A, B and C, respectively. The midpoints of the line segments $\overline{AB}$, $\overline{AC}$, and $\overline{BC}$ are $P\,(2, 2)$, $Q\,(0, 3)$ and $R\,(1, 5)$, respectively.
An equation for the line through A and R is: $x = 1$.
An equation for the line through B and Q is: $y = \frac{1}{3}x + 3$.
An equation for the line through C and P is: $y - 2 = -\frac{4}{3}(x - 2)$.
These three lines intersect at the point $(1, \frac{10}{3})$.

71. Let $A(0,0)$ and $B(a,0)$, $a > 0$, be adjacent vertices of a parallelogram. If $C(b,c)$ is the vertex opposite B, then the vertex D opposite A has coordinates $(a + b, c)$ [see the figure].

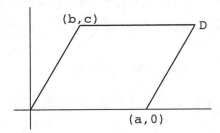

The line through A and D has equation: $y = \dfrac{c}{a + b}\, x.$

The line through B and C has equation: $y = -\dfrac{c}{a - b}\,(x - a).$

These lines intersect at the point $\left(\dfrac{a + b}{2}, \dfrac{c}{2}\right)$ which is the midpoint of each of the line segments $\overline{AD}$ and $\overline{BC}$.

73. Since the relation between F and C is linear, $F = mC + b$ for some constants m and C. Setting $C = 0$ and $F = 32$ gives $b = 32$. Thus $F = mC + 32$. Now, letting $C = 100$ and $F = 212$ gives $m = (212 - 32)/100 = 9/5$. Therefore

$$F = \frac{9}{5} C + 32$$

The Fahrenheit and Centigrade temperatures are equal when

$$C = F = \frac{9}{5} C + 32$$

which implies $C = F = -40°.$

SECTION 1.5

1.   (a) $f(0) = 2(0)^2 - 3(0) + 2 = 2$ (b) $f(1) = 2(1)^2 - 3(1) + 2 = 1$

     (c) $f(-2) = 2(-2)^2 - 3(-2) + 2 = 16$ (d) $f(\frac{3}{2}) = 2(3/2)^2 - 3(3/2) + 2 = 2$

3.   (a) $f(0) = \sqrt{0^2 + 2 \cdot 0} = 0$ (b) $f(1) = \sqrt{1^2 + 2 \cdot 1} = \sqrt{3}$

     (c) $f(-2) = \sqrt{(-2)^2 + 2(-2)} = 0$ (d) $f(\frac{3}{2}) = \sqrt{(3/2)^2 + 2(3/2)} = \frac{1}{2}\sqrt{21}$

5.   (a) $f(0) = \dfrac{2 \cdot 0}{|0 + 2| + 0^2} = 0$ (b) $f(1) = \dfrac{2 \cdot 1}{|1 + 2| + 1^2} = \dfrac{1}{2}$

     (c) $f(-2) = \dfrac{2 \cdot (-2)}{|-2 + 2| + (-2)^2} = -1$ (d) $f(\frac{3}{2}) = \dfrac{2 \cdot (3/2)}{|(3/2) + 2| + (3/2)^2} = \dfrac{12}{23}$

7.   (a) $f(-x) = (-x)^2 - 2(-x) = x^2 + 2x$ (b) $f(1/x) = (1/x)^2 - 2(1/x) = \dfrac{1 - 2x}{x^2}$

(c) $f(a + b) = (a + b)^2 - 2(a + b) = a^2 + 2ab + b^2 - 2a - 2b$

9. (a) $f(-x) = \sqrt{1 + (-x)^2} = \sqrt{1 + x^2}$ (b) $f(1/x) = \sqrt{1 + (1/x)^2} = |x|/\sqrt{1 + x^2}$

(c) $f(a + b) = \sqrt{1 + (a + b)^2} = \sqrt{a^2 + 2ab + b^2 + 1}$

11. (a) $f(a + h) = 2(a + h)^2 - 3(a + h) = 2a^2 + 4ah + 2h^2 - 3a - 3h$

(b) $\dfrac{f(a + h) - f(a)}{h} = \dfrac{[2(a + h)^2 - 3(a + h)] - [2a^2 - 3a]}{h} = \dfrac{4ah + 2h^2 - 3h}{h} = 4a + 2h - 3$

13. $x = 1, 3$ **15.** $x = -2$ **17.** $x = -3, 3$

19. $\text{dom}(f) = (-\infty, \infty);$ $\text{range}(f) = [0, \infty)$

21. $\text{dom}(f) = (-\infty, \infty);$ $\text{range}(f) = (-\infty, \infty)$

23. $\text{dom}(f) = (-\infty, 0) \cup (0, \infty);$ $\text{range}(f) = (0, \infty)$

25. $\text{dom}(f) = (-\infty, 1];$ $\text{range}(f) = [0, \infty)$ **27.** $\text{dom}(f) = (-\infty, 7];$ $\text{range}(f) = [-1, \infty)$

29. $\text{dom}(f) = (-\infty, 2);$ $\text{range}(f) = (0, \infty)$ **31.** horizontal line one unit above the x-axis.

33. line through the origin with slope 2.   **35.** line through $(0, 2)$ with slope $\frac{1}{2}$.

37. upper semicircle of radius 2 centered at the origin.

39. $\text{dom}(f) = (-\infty, \infty)$ **41.** $\text{dom}(f) = (-\infty, 0) \cup (0, \infty);$ $\text{range}(f) = \{-1, 1\}$

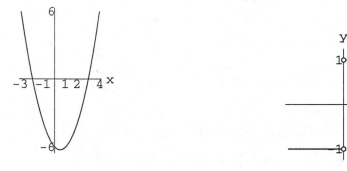

43. $\text{dom}(f) = [0, \infty);$

range$(f) = [1, \infty).$

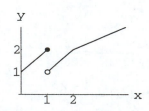

45.   The curve is the graph of a function: domain $[-2, 2]$,   range $[-2, 2]$.

47.   The curve is not the graph of a function; it fails the *vertical line test.*

49.   odd: $f(-x) = (-x)^3 = -x^3 = -f(x)$.

51.   neither even nor odd: $g(-x) = -x(-x - 1) = x(x + 1)$; $g(-x) \neq g(x)$ and $g(-x) \neq -g(x)$ **53.**   even.

55.   (a)

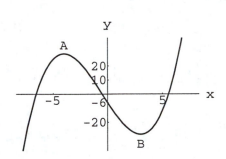

57. $-5 \leq x \leq 8$, $0 \leq y \leq 100$

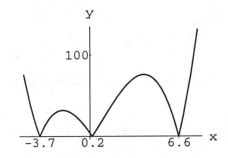

(b) $-6.566, -0.493, 5.559$

(c) $A(-4, 28.667)$, $B(3, -28.500)$

59. $A = \dfrac{C^2}{4\pi}$,   where C is the circumference; dom $(A) = [0, \infty)$

61. $V = s^{3/2}$, where s is the area of a face; dom $V = [0, \infty)$

63. $S = 3d^2$, where d is the diagonal of a face; dom $(S) = [0, \infty)$

65. $A = \dfrac{\sqrt{3}}{4} x^2$, where x is the length of a side; dom $(A) = [0, \infty)$

67.   Let y be the length of the rectangle. Then
$$x + 2y + \frac{\pi x}{2} = 15 \quad \text{and} \quad y = \frac{15}{2} - \frac{2 + \pi}{4} x, \qquad 0 < x < \frac{30}{2 + \pi}$$
The area $A = xy + \frac{1}{2}\pi (x/2)^2 = \left(\frac{15}{2} - \frac{2 + \pi}{4} x\right) x + \frac{1}{8}\pi x^2 = \frac{15}{2} x - \frac{x^2}{2}\frac{\pi}{8} x^2 \quad 0 < x < \frac{30}{2 + \pi}$.

69.   Let y be the length of the beam. Then $y = \sqrt{d^2 - x^2}$, $0 < x < d$.

The cross-sectional area $A = x\sqrt{d^2 - x^2}$.

71.   The coordinates x and y are related by the equation $y = -\dfrac{b}{a}(x - a)$, $0 \leq x \leq a$.

The area A of the rectangle is given by $A = xy = x\left[-\dfrac{b}{a}(x - a)\right] = bx - \dfrac{b}{a} x^2$, $0 \leq x \leq a$.

73.   Let P be the perimeter of the square. Then the edge length of the square is $P/4$ and the area of the square is $A_s = (P/4)^2 = P^2/16$. Now, the circumference of the circle is $28 - P$ which implies that the radius is $\dfrac{1}{2\pi}(28 - \pi)$. Thus, the area of the circle is $A_c = \pi \left[\dfrac{1}{2\pi}(28 - P)\right]^2 = \dfrac{1}{4\pi}(28 - P)^2$. and the total area is $A_s + A_c = \dfrac{P^2}{16} + \dfrac{1}{4\pi}(28 - P)^2, \quad 0 \le P \le 28$.

75.   Set length plus girth equal to 108.   Then $l = 108 - 2\pi r$, and $V = (108 - 2\pi r)\pi r^2$.

SECTION 1.6

1.   polynomial, degree 0   **3.**   rational function   **5.**   neither

7.   neither   **9.**   neither

11. $\operatorname{dom}(f) = (-\infty, \infty)$ **13.** $\operatorname{dom}(f) = (-\infty, \infty)$ **15.** $\operatorname{dom}(f) = \{x : x \ne \pm 2\}$

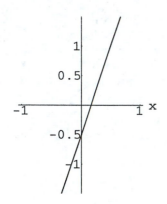   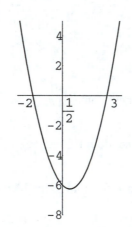   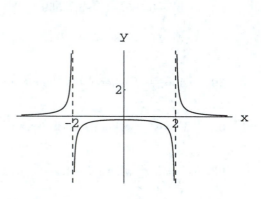

17. $225\left(\dfrac{\pi}{180}\right) = \dfrac{5\pi}{4}$ **19.** $(-300)\left(\dfrac{\pi}{180}\right) = -\dfrac{5\pi}{3}$ **21.** $15\left(\dfrac{\pi}{180}\right) = \dfrac{\pi}{12}$

23. $\left(-\dfrac{3\pi}{2}\right)\left(\dfrac{180}{\pi}\right) = -270°$ **25.** $\left(\dfrac{5\pi}{3}\right)\left(\dfrac{180}{\pi}\right) = 300°$ **27.** $2\left(\dfrac{180}{\pi}\right) \cong 114.59°$

29. $\sin x = \tfrac{1}{2}; \quad x = \pi/6, \ 5\pi/6$ **31.** $\tan(x/2) = 1; \quad x = \pi/2$

33. $\cos x = \sqrt{2}/2; \quad x = \pi/4, \ 7\pi/4$ **35.** $\cos 2x = 0; \quad x = \pi/4, \ 3\pi/4, \ 5\pi/4, \ 7\pi/4$

37. $\sin 51° \cong 0.7772$ **39.** $\sin(2.352) \cong 0.7101$

41. $\tan 72.4° \cong 3.1524$ **43.** $\tan(11.249) \cong -3.8611$

45. $\sec(4.360) \cong -2.8974$ **47.** $\sin x = 0.5231; \quad x = 0.5505, \ \pi - 0.5505$

49. $\tan x = 6.7192; \quad x = 1.4231, \ \pi + 1.4231$ **51.** $\sec x = -4.4073; \quad x = 1.7997, \ \pi + 1.7997$

53. $\operatorname{dom}(f) = (-\infty, \infty); \operatorname{range}(f) = [0, 1]$ **55.** $\operatorname{dom}(f) = (-\infty, \infty); \operatorname{range}(f) = [-2, 2]$

57. $\operatorname{dom}(f) = \left(k\pi - \dfrac{\pi}{2}, k\pi + \dfrac{\pi}{2}\right), \quad k = 0, \pm 1, \pm 2, \cdots; \quad \operatorname{range}(f) = [1, \infty)$

59.

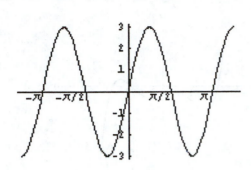

61.

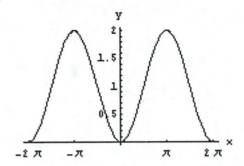

63.

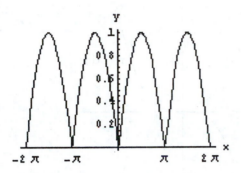

65. odd   **67.** even   **69.** odd

71. Assume that $\theta_2 > \theta_1$. Let $m_1 = \tan\theta_1$, $m_2 = \tan\theta_2$.. The angle α between l_1 and l_2 is the smaller of $\theta_2 - \theta_1$ and $180° - [\theta_2 - \theta_1]$. In the first case

$$\tan\alpha = \tan[\theta_2 - \theta_1] = \frac{\tan\theta_2 - \tan\theta_1}{1 + \tan\theta_2 \tan\theta_1} = \frac{m_2 - m_1}{1 + m_2 m_1} > 0$$

In the second case, $\tan\alpha = \tan[180° - (\theta_2 - \theta_1)] = -\tan(\theta_2 - \theta_1) = -\dfrac{m_2 - m_1}{1 + m_2 m_1} > 0$

Thus $\tan\alpha = \left|\dfrac{m_2 - m_1}{1 + m_2 m_1}\right|$

73. $\left(\frac{23}{37}, \frac{116}{37}\right)$; $\alpha \cong 73°$ $[m_1 = -3 = \tan\theta_1, \ \theta_1 \cong 108°; \ m_2 = \frac{7}{10} = \tan\theta_2, \ \theta_2 \cong 35°]$

75. $\left(-\frac{17}{13}, -\frac{2}{13}\right)$; $\alpha \cong 82°$ $[m_1 = \frac{5}{6} = \tan\theta_1, \ \theta_1 \cong 40°; \ m_2 = -\frac{8}{5} = \tan\theta_2, \ \theta_2 \cong 122°]$

77. $h = b\sin A = a\sin B$ (see figure)

so $\dfrac{\sin A}{a} = \dfrac{\sin B}{b}$

Similarly, $\dfrac{\sin A}{a} = \dfrac{\sin C}{c}$

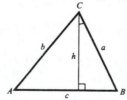

79. $A = \frac{1}{2}ah = \frac{1}{2}a^2\sin\theta$

(see figure)

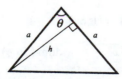

81.   (a)

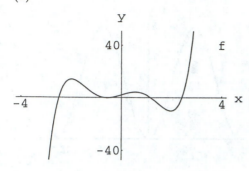

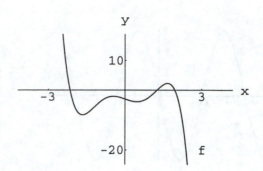

83.   (b)

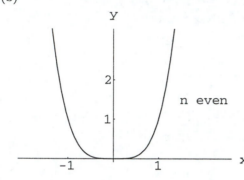

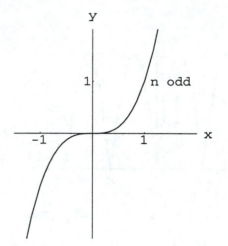

(c) $f_k(x) \geq f_{k+1}(x)$ on $[0,1]$; $f_{k+1}(x) \geq f_k(x)$ on $[1,\infty)$

SECTION 1.7

1. $(f+g)(2) = f(2) + g(2) = 3 + \frac{9}{2} = \frac{15}{2}$ **3.** $(f \cdot g)(-2) = f(-2)g(-2) = 15 \cdot \frac{7}{2} = \frac{105}{2}$

5. $(2f - 3g)(\frac{1}{2}) = 2f(\frac{1}{2}) - 3g(\frac{1}{2}) = 2 \cdot 0 - 3 \cdot \frac{9}{4} = -\frac{27}{4}$

7. $(f \circ g)(1) = f[g(1)] = f(2) = 3$

9. $(f+g)(x) = f(x) + g(x) = x - 1$; $\mathrm{dom}\,(f+g) = (-\infty, \infty)$

$(f-g)(x) = f(x) - g(x) = 3x - 5$; $\mathrm{dom}\,(f-g) = (-\infty, \infty)$

$(f \cdot g)(x) = f(x)g(x) = -2x^2 + 7x - 6$; $\mathrm{dom}\,(f \cdot g) = (-\infty, \infty)$

$(f/g)(x) = \dfrac{2x - 3}{2 - x}$; $\mathrm{dom}\,(f/g) = \{x : x \neq 2\}$

11. $(f+g)(x) = x + \sqrt{x-1} - \sqrt{x+1}$; $\mathrm{dom}\,(f+g) = [1, \infty)$

$(f-g)(x) = \sqrt{x-1} + \sqrt{x+1} - x$; $\mathrm{dom}\,(f-g) = [1, \infty)$

$(f \cdot g)(x) = \sqrt{x-1}\,(x - \sqrt{x+1}) = x\sqrt{x-1} - \sqrt{x^2 - 1}$; $\mathrm{dom}\,(f \cdot g) = [1, \infty)$

$$(f/g)(x) = \frac{\sqrt{x-1}}{x - \sqrt{x+1}}; \quad \text{dom}\,(f/g) = \{x : x \geq 1 \ \text{and} \ x \neq \tfrac{1}{2}(1+\sqrt{5})\}$$

13.   (a) $(6f + 3g)(x) = 6(x + 1/\sqrt{x}) + 3(\sqrt{x} - 2/\sqrt{x}) = 6x + 3\sqrt{x}; \quad x > 0$

(b) $(f - g)(x) = x + 1/\sqrt{x} - (\sqrt{x} - 2/\sqrt{x}) = x + 3/\sqrt{x} - \sqrt{x}; \quad x > 0$

(c) $(f/g)(x) = \dfrac{x\sqrt{x} + 1}{x - 2}; \quad x > 0, \ x \neq 2$

15.

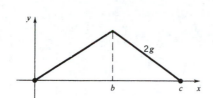

17.

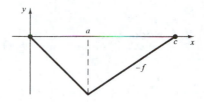

19.

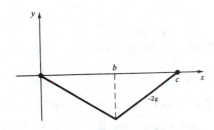

21.

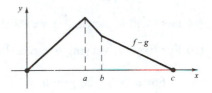

23. $(f \circ g)(x) = 2x^2 + 5; \ \text{dom}\,(f \circ g) = (-\infty, \infty)$ **25.** $(f \circ g)(x) = \sqrt{x^2 + 5}; \ \text{dom}\,(f \circ g) = (-\infty, \infty)$

27. $(f \circ g)(x) = \dfrac{x}{x - 2}; \ \text{dom}\,(f \circ g) = \{x : \ x \neq 0, 2\}$

29. $(f \circ g)(x) = \sqrt{1 - \cos^2 2x} = |\sin 2x|; \ \text{dom}\,(f \circ g) = (-\infty, \infty)$

31. $(f \circ g \circ h)(x) = 4\,[\,g(h(x))\,] = 4\,[\,h(x) - 1\,] = 4(x^2 - 1); \quad \text{dom}\,(f \circ g \circ h) = (-\infty, \infty)$

33. $(f \circ g \circ h)(x) = \dfrac{1}{g(h(x))} = \dfrac{1}{1/[2h(x) + 1]} = 2h(x) + 1 = 2x^2 + 1; \quad \text{dom}\,f \circ g \circ h) = (-\infty, \infty)$

35.   Take $f(x) = \dfrac{1}{x}$ since $\dfrac{1 + x^4}{1 + x^2} = F(x) = f(g(x)) = f\left(\dfrac{1 + x^2}{1 + x^4}\right).$

37.   Take $f(x) = 2\sin x$ since $2\sin 3x = F(x) = f(g(x)) = f(3x).$

39.   Take $g(x) = \left(1 - \dfrac{1}{x^4}\right)^{2/3}$ since $\left(1 - \dfrac{1}{x^4}\right)^2 = F(x) = f(g(x)) = [\,g(x)\,]^3.$

41.   Take $g(x) = 2x^3 - 1 \ (\text{or} \ -(2x^3 - 1))$ since $(2x^3 - 1)^2 + 1 = F(x) = f(g(x)) = [\,g(x)\,]^2 + 1.$

43. $(f \circ g)(x) = f(g(x)) = \sqrt{g(x)} = \sqrt{x^2} = |x|;$

$$(g \circ f)(x) = g(f(x)) = [f(x)]^2 = [\sqrt{x}]^2 = x, \quad x \geq 0$$

45. $(f \circ g)(x) = f(g(x)) = 1 - \sin^2 x = \cos x;$ $\quad (g \circ f)(x) = g(f(x)) = \sin f(x) = \sin(1 - x^2)$

47. $(f \circ g)(x) = f(g(x)) = (x - 1) + 1 = x;$ $\quad (g \circ f)(x) = g(f(x)) = \sqrt[3]{(x^3 + 1) - 1} = x$

49. fg is even since $(fg)(-x) = f(-x)g(-x) = f(x)g(x) = (fg)(x).$

51. (a) If f is even, then

$$f(x) = \begin{cases} -x, & -1 \leq x < 0 \\ 1, & x < -1. \end{cases}$$

(b) If f is odd, then

$$f(x) = \begin{cases} x, & -1 \leq x < 0 \\ -1, & x < -1. \end{cases}$$

53. $g(-x) = f(-x) + f[-(-x)] = f(-x) + f(x) = g(x)$

55. $f(x) = \dfrac{1}{2}\underbrace{[f(x) + f(-x)]}_{\text{even}} + \dfrac{1}{2}\underbrace{[f(x) - f(-x)]}_{\text{odd}}$

57. (a) For fixed b, varying a varies the x-coordinate of the vertex of the parabola.

(b) For fixed a, varying b varies the y-coordinate of the parabola

59. (a) For $a > 0$, the graph of $f(x - a)$ is the graph of f shifted horizontally a units to the right; for $a < 0$, the graph of $f(x - a)$ is the graph of f shifted horizontally $|a|$ units to the left.

(b) For $b > 1$, the graph of $f(bx)$ is the graph of f compressed horizontally; for $0 < b < 1$, the graph of $f(bx)$ is the graph of f stretched horizontally; for $-1 < b < 0$, the graph of $f(bx)$ is the graph of f stretched horizontally and reflected in the y-axis; for $b < -1$, the graph of $f(bx)$ is the graph of f compressed horizontally and reflected in the y-axis.

(c) The graph of $f(x) + c$ is the graph of f shifted c units up if $c > 0$ and shifted $|c|$ units down if $c < 0$.

61. (a) For $A > 0$, the graph of Af is the graph of f scaled vertically by the factor A; for $A < 0$, the graph of Af is the graph of f scaled vertically by the factor $|A|$ and then reflected in the x-axis.

(b) See Exercise 59(b).

PROJECT 1.7

1. (a) $(f \circ g)(x) = 3[\frac{1}{3}(x + 5)] - 5 = x$ and $(g \circ f)(x) = \frac{1}{3}[(3x - 5) + 5] = x$

(b) $(\sqrt[3]{x})^3 = x = \sqrt[3]{x^3}$

(c) $(f \circ g)(x) = \dfrac{1}{\dfrac{1 - x}{x} + 1} = \dfrac{x}{(1 - x) + x} = x$ and $(g \circ f)(x) = \dfrac{1 - \dfrac{1}{x + 1}}{\dfrac{1}{x + 1}} = (x + 1) - 1 = x$

SECTION 1.8

1. Let S be the set of integers for which the statement is true. Since $2(1) \leq 2^1$, S contains 1. Assume now that $k \in S$. This tells us that $2k \leq 2^k$, and thus

$$2(k+1) = 2k + 2 \leq 2^k + 2 \leq 2^k + 2^k = 2(2^k) = 2^{k+1}.$$

$$(k \geq 1)$$

This places $k+1$ in S.

We have shown that

$$1 \in S \quad \text{and that} \quad k \in S \quad \text{implies} \quad k+1 \in S.$$

It follows that S contains all the positive integers.

3. Let S be the set of integers for which the statement is true. Since $(1)(2) = 2$ is divisible by $2, 1 \in S$.

Assume now that $k \in S$. This tells us that $k(k+1)$ is divisible by 2 and therefore

$$(k+1)(k+2) = k(k+1) + 2(k+1)$$

is also divisible by 2. This places $k+1 \in S$.

We have shown that

$$1 \in S \text{ and that } k \in S \text{ implies } k+1 \in S.$$

It follows that S contains all the positive integers.

5. Use

$$1^2 + 2^2 + \cdots + k^2 + (k+1)^2 = \tfrac{1}{6}k(k+1)(2k+1) + (k+1)^2$$

$$= \tfrac{1}{6}(k+1)[k(2k+1) + 6(k+1)]$$

$$= \tfrac{1}{6}(k+1)(2k^2 + 7k + 6)$$

$$= \tfrac{1}{6}(k+1)(k+2)(2k+3)$$

$$= \tfrac{1}{6}(k+1)[(k+1) + 1][2(k+1) + 1].$$

7. By Exercise 6 and Example 1

$$1^3 + 2^3 + \cdots + (n-1)^3 = [\tfrac{1}{2}(n-1)n]^2 = \tfrac{1}{4}(n-1)^2 n^2 < \tfrac{1}{4}n^4$$

and

$$1^3 + 2^3 + \cdots + n^3 = [\tfrac{1}{2}n(n+1)]^2 = \tfrac{1}{4}n^2(n+1)^2 > \tfrac{1}{4}n^4.$$

9. Use

$$\frac{1}{\sqrt{1}} + \frac{1}{\sqrt{2}} + \frac{1}{\sqrt{3}} + \cdots + \frac{1}{\sqrt{n}} + \frac{1}{\sqrt{n+1}}$$

$$> \sqrt{n} + \frac{1}{\sqrt{n+1}+\sqrt{n}}\left(\frac{\sqrt{n+1}-\sqrt{n}}{\sqrt{n+1}-\sqrt{n}}\right) = \sqrt{n+1}.$$

11. Let S be the set of integers for which the statement is true. Since

$$3^{2(1)+1} + 2^{1+2} = 27 + 8 = 35$$

is divisible by 7, we see that $1 \in S$.

Assume now that $k \in S$. This tells us that

$$3^{2k+1} + 2^{k+2} \text{ is divisible by 7.}$$

It follows that

$$3^{2(k+1)+1} + 2^{(k+1)+2} = 3^2 \cdot 3^{2k+1} + 2 \cdot 2^{k+2}$$

$$= 9 \cdot 3^{2k+1} + 2 \cdot 2^{k+2}$$

$$= 7 \cdot 3^{2k+1} + 2(3^{2k+1} + 2^{k+2})$$

is also divisible by 7. This places $k + 1 \in S$.

We have shown that

$$1 \in S \qquad \text{and that} \qquad k \in S \quad \text{implies} \quad k + 1 \in S.$$

It follows that S contains all the positive integers.

13. For all positive integers $n \geq 2$,

$$\left(1 - \frac{1}{2}\right)\left(1 - \frac{1}{3}\right) \cdots \left(1 - \frac{1}{n}\right) = \frac{1}{n}.$$

To see this, let S be the set of integers n for which the formula holds. Since $1 - \frac{1}{2} = \frac{1}{2}$, $\quad 2 \in S$. Suppose now that $k \in S$. This tells us that

$$\left(1 - \frac{1}{2}\right)\left(1 - \frac{1}{3}\right) \cdots \left(1 - \frac{1}{k}\right) = \frac{1}{k}$$

and therefore that

$$\left(1 - \frac{1}{2}\right)\left(1 - \frac{1}{3}\right) \cdots \left(1 - \frac{1}{k}\right)\left(1 - \frac{1}{k+1}\right) = \frac{1}{k}\left(1 - \frac{1}{k+1}\right) = \frac{1}{k}\left(\frac{k}{k+1}\right) = \frac{1}{k+1}.$$

This places $k + 1 \in S$ and verifies the formula for $n \geq 2$.

15.   From the figure, observe that adding a vertex V_{N+1} to an N-sided polygon increases the number of diagonals by $(N-2)+1 = N-1$.   Then use the identity

$$\tfrac{1}{2}N(N-3) + (N-1) = \tfrac{1}{2}(N+1)(N+1-3).$$

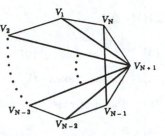

17.   To go from k to $k+1$, take $A = \{a_1, \cdots, a_{k+1}\}$ and $B = \{a_1, \cdots, a_k\}$. Assume that B has 2^k subsets: $B_1, B_2, \cdots B_{2^k}$. The subsets of A are then $B_1, B_2, \cdots, B_{2^k}$ together with

$$B_1 \cup \{a_{k+1}\}, \; B_2 \cup \{a_{k+1}\}, \cdots, B_{2^k} \cup \{a_{k+1}\}.$$

This gives $2(2^k) = 2^{k+1}$ subsets for A.

CHAPTER 2

SECTION 2.1

1. (a) 2 (b) −1 (c) does not exist (d) −3

3. (a) does not exist (b) −3 (c) does not exist (d) −3

5. (a) does not exist (b) does not exist (c) does not exist (d) 1

7. (a) 2 (b) 2 (c) 2 (d) −1

9. (a) 0 (b) 0 (c) 0 (d) 0

11. $c = 0, 6$ **13.** −1 **15.** 12 **17.** 1

19. $\frac{3}{2}$ **21.** does not exist

23. $\lim\limits_{x \to 3} \dfrac{2x - 6}{x - 3} = \lim\limits_{x \to 3} 2 = 2$

25. $\lim\limits_{x \to 3} \dfrac{x - 3}{x^2 - 6x + 9} = \lim\limits_{x \to 3} \dfrac{x - 3}{(x - 3)^2} = \lim\limits_{x \to 3} \dfrac{1}{x - 3};$ does not exist

27. $\lim\limits_{x \to 2} \dfrac{x - 2}{x^2 - 3x + 2} = \lim\limits_{x \to 2} \dfrac{x - 2}{(x - 1)(x - 2)} = \lim\limits_{x \to 2} \dfrac{1}{x - 1} = 1$

29. does not exist **31.** $\lim\limits_{x \to 0} \dfrac{2x - 5x^2}{x} = \lim\limits_{x \to 0} (2 - 5x) = 2$

33. $\lim\limits_{x \to 1} \dfrac{x^2 - 1}{x - 1} = \lim\limits_{x \to 1} \dfrac{(x - 1)(x + 1)}{x - 1} = \lim\limits_{x \to 1} (x + 1) = 2$

35. 0 **37.** 1 **39.** 16

41. does not exist **43.** does not exist **45.** 4

47.
$$\lim\limits_{x \to 1} \frac{\sqrt{x^2 + 1} - \sqrt{2}}{x - 1} = \lim\limits_{x \to 1} \frac{(\sqrt{x^2 + 1} - \sqrt{2})(\sqrt{x^2 + 1} + \sqrt{2})}{(x - 1)(\sqrt{x^2 + 1} + \sqrt{2})}$$

$$= \lim\limits_{x \to 1} \frac{x^2 - 1}{(x - 1)(\sqrt{x^2 + 1} + \sqrt{2})} = \lim\limits_{x \to 1} \frac{x + 1}{\sqrt{x^2 + 1} + \sqrt{2}} = \frac{2}{2\sqrt{2}} = \frac{1}{\sqrt{2}}$$

49. $f(x) = x^2, \quad c = 2, \quad f(2) = 4$

$$\frac{f(2 + h) - f(2)}{h} = \frac{(2 + h)^2) - 4}{h} = \frac{4 + 4h + h^2 - 4}{h} = 4 + h$$

$$\lim\limits_{h \to 0} \frac{f(2 + h) - f(2)}{h} = \lim\limits_{h \to 0} (4 + h) = 4$$

tangent line: $y - 4 = 4(x - 2)$ or $y = 4x - 4$

51. $f(x) = 1 - 2x + x^2, \quad c = -1, \quad f(-1) = 4$

$$\frac{f(-1+h) - f(-1)}{h} = \frac{1 - 2(-1+h) + (-1+h)^2 - 4}{h} = \frac{4 - 4h + h^2 - 4}{h} = -4 + h$$

$$\lim_{h \to 0} \frac{f(-1+h) - f(-1)}{h} = \lim_{h \to 0} (-4 + h) = -4$$

tangent line: $\quad y - 4 = -4(x+1) \quad$ or $\quad y = -4x$

53. $f(x) = \sqrt{x}, \quad c = 1 \quad f(1) = 1$

$$\frac{f(1+h) - f(1)}{h} = \frac{\sqrt{1+h} - 1}{h} = \frac{\sqrt{1+h} - 1}{h} \cdot \frac{\sqrt{1+h} + 1}{\sqrt{1+h} + 1} = \frac{h}{h(1+h) + 1} = \frac{1}{\sqrt{1+h} + 1}$$

$$\lim_{h \to 0} \frac{f(1+h) - f(1)}{h} = \lim_{h \to 0} \frac{1}{\sqrt{1+h} + 1} = \frac{1}{2}$$

tangent line: $\quad y - 1 = \frac{1}{2}(x - 1) \quad$ or $\quad y = \frac{1}{2}x + \frac{1}{2}$

55. $f(x) = \sqrt[3]{x}, \quad c = 0, \quad f(0) = 0$

$$\frac{f(0+h) - f(0)}{h} = \frac{\sqrt[3]{h}}{h} = \frac{1}{h^{2/3}}$$

$$\lim_{h \to 0} \frac{f(0+h) - f(0)}{h} = \lim_{h \to 0} \frac{1}{h^{2/3}} \quad \text{does not exist}$$

57. (a)

$$f(1/\pi) = f(2/\pi) = f(3/\pi) = f(4/\pi) = 0$$

$$f\left(\frac{1}{\pi/2}\right) = f\left(\frac{1}{5\pi/2}\right) = f\left(\frac{1}{9\pi/2}\right) = 1$$

$$f\left(\frac{1}{3\pi/2}\right) = f\left(\frac{1}{7\pi/2}\right) = f\left(\frac{1}{11\pi/2}\right) = -1$$

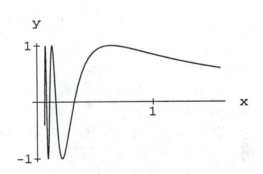

(b) neither limit exists

(c)

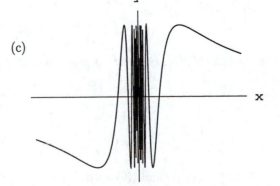

59. 2   **61.** $\frac{3}{2}$ **63.** 2.7182817

SECTION 2.2

1. $\frac{1}{2}$

3. $\lim\limits_{x \to 0} \dfrac{x(1+x)}{2x^2} = \lim\limits_{x \to 0} \dfrac{1+x}{2x};$ does not exist

5. $\lim\limits_{x \to 1} \dfrac{x^4 - 1}{x - 1} = \lim\limits_{x \to 1} (x^3 + x^2 + x + 1)$ does not exist

9. -1 **11.**   does not exist   **13.** 0

15. $\lim\limits_{x \to 2^+} f(x) = \lim\limits_{x \to 2^+} (x^2 - x) = 2$ **17.** 1

19. 1   **21.** δ_1 and δ_2 **23.** $\frac{1}{2}\epsilon$ **25.** 2ϵ

27. Since
$$|(2x - 5) - 3| = |2x - 8| = 2|x - 4|,$$
we can take $\delta = \frac{1}{2}\epsilon$:
$$\text{if}\quad 0 < |x - 4| < \tfrac{1}{2}\epsilon\quad \text{then,}\quad |(2x - 5) - 3| = 2|x - 4| < \epsilon.$$

29. Since
$$|(6x - 7) - 11| = |6x - 18| = 6|x - 3|,$$
we can take $\delta = \frac{1}{6}\epsilon$:
$$\text{if}\quad 0 < |x - 3| < \tfrac{1}{6}\epsilon\quad \text{then}\quad |(6x - 7) - 11| = 6|x - 3| < \epsilon.$$

31. Since
$$\big||1 - 3x| - 5\big| = \big||3x - 1| - 5\big| \le |3x - 6| = 3|x - 2|,$$
we can take $\delta = \frac{1}{3}\epsilon$:
$$\text{if}\quad 0 < |x - 2| < \tfrac{1}{3}\epsilon\quad \text{then}\quad \big||1 - 3x| - 5\big| \le 3|x - 2| < \epsilon.$$

33. Statements (b), (e), (g), and (i) are necessarily true.

35. (i) $\lim\limits_{x \to 3} \dfrac{1}{x - 1} = \dfrac{1}{2}$ (ii) $\lim\limits_{h \to 0} \dfrac{1}{(3 + h) - 1} = \dfrac{1}{2}$

  (iii) $\lim\limits_{x \to 3} \left(\dfrac{1}{x - 1} - \dfrac{1}{2} \right) = 0$ (iv) $\lim\limits_{x \to 3} \left| \dfrac{1}{x - 1} - \dfrac{1}{2} \right| = 0$

37. By (2.2.5) parts (i) and (iv) with $L = 0$

39.   Let $\epsilon > 0$. If
$$\lim_{x \to c} f(x) = L,$$
then there must exist $\delta > 0$ such that

(*)   if $0 < |x - c| < \delta$ then $|f(x) - L| < \epsilon.$

Suppose now that
$$0 < |h| < \delta.$$

Then
$$0 < |(c + h) - c| < \delta$$
and thus by (*)
$$|f(c + h) - L| < \epsilon.$$

This proves that

if $\lim_{x \to c} f(x) = L$ then $\lim_{h \to 0} f(c + h) = L.$

If, on the other hand,
$$\lim_{h \to 0} f(c + h) = L,$$
then there must exist $\delta > 0$ such that

(**)   if $0 < |h| < \delta$ then $|f(c + h) - L| < \epsilon.$

Suppose now that
$$0 < |x - c| < \delta.$$

Then by (**)
$$|f(c + (x - c)) - L| < \epsilon.$$

More simply stated,
$$|f(x) - L| < \epsilon.$$

This proves that

if $\lim_{h \to 0} f(c + h) = L$ then $\lim_{x \to c} f(x) = L.$

41.   Set $\delta = \epsilon \sqrt{c}$. By the hint,
$$\text{if} \quad 0 < |x - c| < \epsilon \sqrt{c} \quad \text{then} \quad |\sqrt{x} - \sqrt{c}| < \frac{1}{\sqrt{c}} |x - c| < \epsilon.$$
(b)   Set $\delta = \epsilon^2$. If $0 < x < \epsilon^2$,   then $|\sqrt{x} - 0| = \sqrt{x} < \epsilon.$

43.   Take $\delta = $ minimum of 1 and $\epsilon/7$. If $0 < |x - 1| < \delta$,   then $0 < x < 2$
and $|x - 1| < \epsilon/7$. Therefore
$$|x^3 - 1| = |x^2 + x + 1| |x - 1| < 7|x - 1| < 7(\epsilon/7) = \epsilon.$$

45.   Set $\delta = \epsilon^2$.   If $3 - \epsilon^2 < x < 3$,   then $-\epsilon^2 < x - 3$, $0 < 3 - x < \epsilon^2$
and therefore $|\sqrt{3 - x} - 0| < \epsilon.$

47.   Suppose, on the contrary, that $\lim_{x \to c} f(x) = L$ for some particular c. Taking $\epsilon = \frac{1}{2}$, there must exist
$\delta > 0$ such that

$$\text{if} \quad 0 < |x - c| < \delta, \quad \text{then} \quad |f(x) - L| < \tfrac{1}{2}.$$

Let x_1 be a rational number satisfying $0 < |x_1 - c| < \delta$ and x_2 an irrational number satisfying $0 < |x_2 - c| < \delta$. (That such numbers exist follows from the fact that every interval contains both rational and irrational numbers.) Now $f(x_1) = L$ and $f(x_2) = 0$. Thus we must have both

$$|1 - L| < \tfrac{1}{2} \quad \text{and} \quad |0 - L| < \tfrac{1}{2}.$$

From the first inequality we conclude that $L > \tfrac{1}{2}$. From the second, we conclude that $L < \tfrac{1}{2}$. Clearly no such number L exists.

49.   We begin by assuming that $\lim\limits_{x \to c^+} f(x) = L$ and showing that

$$\lim_{h \to 0} f(c + |h|) = L.$$

Let $\epsilon > 0$. Since $\lim\limits_{x \to c^+} f(x) = L$, there exists $\delta > 0$ such that

(*)   if $\quad c < x < c + \delta \quad$ then $\quad |f(x) - L| < \epsilon.$

Suppose now that $0 < |h| < \delta$. Then $c < c + |h| < c + \delta$ and, by (*),

$$|f(c + |h|) - L| < \epsilon.$$

Thus $\quad \lim\limits_{h \to 0} f(c + |h|) = L.$

Conversely we now assume that $\lim\limits_{h \to 0} f(c + |h|) = L$. Then for $\epsilon > 0$ there exists $\delta > 0$ such that

(**)   if $\quad 0 < |h| < \delta \quad$ then $\quad |f(c + |h|) - L| < \epsilon.$

Suppose now that $c < x < c + \delta$. Then $0 < x - c < \delta$ so that, by (**),

$$|f(c + (x - c)) - L| = |f(x) - L| < \epsilon.$$

Thus $\quad \lim\limits_{x \to c^+} f(x) = L.$

51.   (a) Let $\epsilon = L$. Since $\lim\limits_{x \to c} f(x) = L$, there exists $\delta > 0$ such that if $0 < |x - c| < \delta$ then

$$L - f(x) \leq |L - f(x)| = |f(x) - L| < L$$

Therefore, $f(x) > L - L = 0$ for all $x \in (c - \delta, c + \delta)$; take $\gamma = \delta$.

(b) Let $\epsilon = -L$ and repeat the argument in part (a).

53.   (a) Let $\lim\limits_{x \to c} f(x) = L$ and $\lim\limits_{x \to c} g(x) = M$, and let $\epsilon > 0$. There exist positive numbers δ_1 and δ_2 such that

$$|f(x) - L| < \epsilon/2 \quad \text{if} \quad 0 < |x - c| < \delta_1$$

and

$$|g(x) - M| < \epsilon/2 \quad \text{if} \quad 0 < |x - c| < \delta_2$$

Let $\delta = \min(\delta_1, \delta_2)$. Then

$$M - L = M - g(x) + g(x) - f(x) + f(x) - L \geq [M - g(x)] + [f(x) - L]$$

$$\geq -\epsilon/2 - \epsilon/2 = -\epsilon$$

for all x such that $0 < |x - c| < \delta$. Since ϵ is arbitrary, it follows that $M \geq L$.

(b) No. For example, if $f(x) = x^2$ and $g(x) = |x|$ on $(-1, 1)$, then $f(x) < g(x)$ on $(-1, 1)$ except at $x = 0$, but $\lim_{x \to 0} x^2 = \lim_{x \to 0} |x| = 0$.

55. $f(x) = 2x^2 - 3x, \quad a = 2, \quad f(2) = 2$

$$\lim_{x \to a} \frac{f(x) - f(a)}{x - a} = \lim_{x \to a} \frac{2x^2 - 3x - 2}{x - 2}$$

$$= \lim_{x \to 2} \frac{(x - 2)(2x + 1)}{x - 2} = \lim_{x \to 2} (2x + 1) = 5$$

tangent line: $\quad y - 2 = 5(x - 2) \quad$ or $\quad y = 5x - 8$.

57. $f(x) = \sqrt{x}, \quad a = 4, \quad f(4) = 2$

$$\lim_{x \to a} \frac{f(x) - f(a)}{x - a} = \lim_{x \to 4} \frac{\sqrt{x} - 2}{x - 4} = \lim_{x \to 4} \frac{\sqrt{x} - 2}{x - 2} \cdot \frac{\sqrt{x} + 2}{\sqrt{x} + 2}$$

$$= \lim_{x \to 4} \frac{x - 4}{(x - 4)(\sqrt{x} + 2)} = \lim_{x \to 4} \frac{1}{\sqrt{x} + 2} = \frac{1}{4}$$

tangent line: $\quad y - 2 = \frac{1}{4}(x - 4) \quad$ or $\quad y = \frac{1}{4}x + 1$.

59. $\lim_{h \to 0} \dfrac{f(x + h) - f(x)}{h} = \lim_{h \to 0} \dfrac{[5(x + h) + 2] - (5x + 2)}{h} = \lim_{h \to 0} \dfrac{5h}{h} = 5$

61. $\lim_{h \to 0} \dfrac{f(x + h) - f(x)}{h} = \lim_{h \to 0} \dfrac{[4(x + h) + 5(x + h)^2] - (4x + 5x^2)}{h}$

$$= \lim_{h \to 0} \frac{4x + 4h + 5x^2 + 10xh + 5h^2 - 4x - 5x^2}{h}$$

$$= \lim_{h \to 0} \frac{h(4 + 10x) + 5h^2}{h} = \lim_{h \to 0} (4 + 10x + 5h) = 4 + 10x$$

63. $\lim_{h \to 0} \dfrac{f(x + h) - f(x)}{h} = \lim_{h \to 0} \dfrac{\dfrac{1}{x + h + 1} - \dfrac{1}{x + 1}}{h} = \lim_{h \to 0} - \dfrac{h}{h(x + h + 1)(x + 1)}$

$$= -\lim_{h \to 0} \frac{1}{(x + h + 1)(x + 1)} = -\frac{1}{(x + 1)^2}$$

65. $\lim\limits_{x \to \frac{1}{3}} \dfrac{\cos \pi x - \cos(\pi/3)}{x - \frac{1}{3}} \cong -2.72070$ \qquad **67.** $\lim\limits_{x \to 0} \dfrac{3^x - 3^0}{x} \cong 1.09861$

PROJECT 2.2

1. For $\epsilon = 0.25$, take $\delta = 0.1$. \qquad For $\epsilon = 0.10$, take $\delta = 0.04$.

3. For $\epsilon = 0.25$, take $\delta = 0.23$. \qquad For $\epsilon = 0.1$, take $\delta = 0.14$.

SECTION 2.3

1. (a) 3   (b) 4   (c) −2   (d) 0   (e) does not exist   (f) $\frac{1}{3}$

3. $\lim\limits_{x\to4}\left(\frac{1}{x}-\frac{1}{4}\right)\left(\frac{1}{x-4}\right)=\lim\limits_{x\to4}\left(\frac{4-x}{4x}\right)\left(\frac{1}{x-4}\right)=\lim\limits_{x\to4}\frac{-1}{4x}=-\frac{1}{16}$;   Theorem 2.3.2 does not apply

   since $\lim\limits_{x\to4}\frac{1}{x-4}$ does not exist.

5. 3   7. −3   9. 5

11. does not exist   13. −1   15. does not exist

17. $\lim\limits_{h\to0}h\left(1+\frac{1}{h}\right)=\lim\limits_{h\to0}(h+1)=1$ 19. $\lim\limits_{x\to2}\frac{x^2-4}{x-2}=\lim\limits_{x\to2}\frac{x+2}{1}=4$

21. $\lim\limits_{x\to4}\frac{\sqrt{x}-2}{x-4}=\lim\limits_{x\to4}\frac{\sqrt{x}-2}{x-4}\cdot\frac{\sqrt{x}+2}{\sqrt{x}+2}=\lim\limits_{x\to4}\frac{x-4}{(x-4)(\sqrt{x}+2)}=\frac{1}{4}$

23. $\lim\limits_{x\to1}\frac{x^2-x-6}{(x+2)^2}=\lim\limits_{x\to1}\frac{(x+2)(x-3)}{(x+2)^2}=\lim\limits_{x\to1}\frac{x-3}{x+2}=-\frac{2}{3}$

25. $\lim\limits_{h\to0}\frac{1-1/h^2}{1-1/h}=\lim\limits_{h\to0}\frac{h^2-1}{h^2-h}=\lim\limits_{h\to0}\frac{(h+1)(h-1)}{h(h-1)}=\lim\limits_{h\to0}\frac{h+1}{h}$;   does not exist

27. $\lim\limits_{h\to0}\frac{1-1/h}{1+1/h}=\lim\limits_{h\to0}\frac{h-1}{h+1}=-1$

29. $\lim\limits_{t\to-1}\frac{t^2+6t+5}{t^2+3t+2}=\lim\limits_{t\to-1}\frac{(t+1)(t+5)}{(t+1)(t+2)}=\lim\limits_{t\to-1}\frac{t+5}{t+2}=4$

31. $\lim\limits_{t\to0}\frac{t+a/t}{t+b/t}=\lim\limits_{t\to0}\frac{t^2+a}{t^2+b}=\frac{a}{b}$

33. $\lim\limits_{x\to1}\frac{x^5-1}{x^4-1}=\lim\limits_{x\to1}\frac{(x-1)(x^4+x^3+x^2+x+1)}{(x-1)(x^3+x^2+x+1)}=\lim\limits_{x\to1}\frac{x^4+x^3+x^2+x+1}{x^3+x^2+x+1}=\frac{5}{4}$

35. $\lim\limits_{h\to0}h\left(1+\frac{1}{h^2}\right)=\lim\limits_{h\to0}\frac{h^2+1}{h}$;   does not exist

37. $\lim\limits_{x\to-4}\left(\frac{2x}{x+4}+\frac{8}{x+4}\right)=\lim\limits_{x\to-4}\frac{2x+8}{x+4}=\lim\limits_{x\to-4}2=2$

39. (a) $\lim\limits_{x\to4}\left(\frac{1}{x}-\frac{1}{4}\right)=\lim\limits_{x\to4}\frac{4-x}{4x}=0$

    (b) $\lim\limits_{x\to4}\left[\left(\frac{1}{x}-\frac{1}{4}\right)\left(\frac{1}{x-4}\right)\right]=\lim\limits_{x\to4}\left[\left(\frac{4-x}{4x}\right)\left(\frac{1}{x-4}\right)\right]=\lim\limits_{x\to4}\left(-\frac{1}{4x}\right)=-\frac{1}{16}$

    (c) $\lim\limits_{x\to4}\left[\left(\frac{1}{x}-\frac{1}{4}\right)(x-2)\right]=\lim\limits_{x\to4}\frac{(4-x)(x-2)}{4x}=0$

    (d) $\lim\limits_{x\to4}\left[\left(\frac{1}{x}-\frac{1}{4}\right)\left(\frac{1}{x-4}\right)^2\right]=\lim\limits_{x\to4}\frac{4-x}{4x(x-4)^2}=\lim\limits_{x\to4}\frac{1}{4x(4-x)}$;   does not exist

41. (a) $\displaystyle\lim_{x\to 4}\frac{f(x)-f(4)}{x-4}=\lim_{x\to 4}\frac{(x^2-4x)-(0)}{x-4}=\lim_{x\to 4}x=4$

(b) $\displaystyle\lim_{x\to 1}\frac{f(x)-f(1)}{x-1}=\lim_{x\to 1}\frac{x^2-4x+3}{x-1}=\lim_{x\to 1}\frac{(x-1)(x-3)}{x-1}=\lim_{x\to 1}(x-3)=-2$

(c) $\displaystyle\lim_{x\to 3}\frac{f(x)-f(1)}{x-3}=\lim_{x\to 3}\frac{x^2-4x+3}{x-3}=\lim_{x\to 3}\frac{(x-1)(x-3)}{x-3}=\lim_{x\to 3}(x-1)=2$

(d) $\displaystyle\lim_{x\to 3}\frac{f(x)-f(2)}{x-3}=\lim_{x\to 3}\frac{x^2-4x+4}{x-3};$ does not exist

43. $f(x)=1/x,\quad g(x)=-1/x$ with $c=0$

45.   True. Let $\displaystyle\lim_{x\to c}[f(x)+g(x)]=L.$ If $\displaystyle\lim_{x\to c}g(x)=M$ exists, then $\displaystyle\lim_{x\to c}f(x)=\lim_{x\to c}[f(x)+g(x)-g(x)]=$
$L-M$ also exists. This contradicts the fact that $\displaystyle\lim_{x\to c}f(x)$ does not exist.

47.   True. If $\displaystyle\lim_{x\to c}\sqrt{f(x)}=L$ exists, then $\displaystyle\lim_{x\to c}\sqrt{f(x)}\sqrt{f(x)}=L^2$ also exists.

49.   False;   for example set $f(x)=x$ and $c=0$

51.   False;   for example, set $f(x)=1-x^2,\ g(x)=1+x^2$, and $c=0$.

53.   If $\displaystyle\lim_{x\to c}f(x)=L$ and $\displaystyle\lim_{x\to c}g(x)=L,$ then

$\displaystyle\lim_{x\to c}h(x)=\lim_{x\to c}\tfrac{1}{2}\{[f(x)+g(x)]-|f(x)-g(x)|\}$

$\displaystyle=\lim_{x\to c}\tfrac{1}{2}[f(x)+g(x)]-\lim_{x\to c}\tfrac{1}{f}(x)-g(x)|$

$\displaystyle=\tfrac{1}{2}(L+L)-\tfrac{1}{2}(L-L)=L.$

A similar argument works for H.

55.   (a) Suppose on the contrary that $\displaystyle\lim_{x\to c}g(x)$ does exist. Let $\displaystyle L=\lim_{x\to c}g(x).$ Then

$$\lim_{x\to c}f(x)g(x)=\lim_{x\to c}f(x)\cdot\lim_{x\to c}g(x)=0\cdot L=0.$$

This contradicts the fact that $\displaystyle\lim_{x\to c}f(x)g(x)=1$

(b) $\displaystyle\lim_{x\to c}g(x)$ exists since $\displaystyle\lim_{x\to c}g(x)=\lim_{x\to c}\frac{f(x)g(x)}{f(x)}=\frac{1}{L}.$

57.   (a) $\displaystyle\lim_{h\to 0}\frac{f(x+h)-f(x)}{h}=\lim_{h\to 0}\frac{x+h-x}{h}=1$

(b) $\displaystyle\lim_{h\to 0}\frac{f(x+h)-f(x)}{h}=\lim_{h\to 0}\frac{(x+h)^2-x^2}{h}=\lim_{h\to 0}\frac{x^2+2xh+h^2-x^2}{h}$

$$= \lim_{h \to 0} (2x + h) = 2x$$

(c) $\displaystyle \lim_{h \to 0} \frac{f(x+h) - f(x)}{h} = \lim_{h \to 0} \frac{(x+h)^3 - x^3}{h} = \lim_{h \to 0} \frac{x^3 + 3x^2 h + 3xh^2 + h^3 - x^3}{h}$

$$= \lim_{h \to 0} (3x^2 + 3xh + h^2) = 3x^2$$

(d) $\displaystyle \lim_{h \to 0} \frac{f(x+h) - f(x)}{h} = \lim_{h \to 0} \frac{(x+h)^4 - x^4}{h} = \lim_{h \to 0} \frac{x^4 + 4x^3 h + 6x^2 h^2 + 4xh^3 + h^4 - x^4}{h}$

$$= \lim_{h \to 0} (4x^3 + 6x^2 h + 4xh^2 + h^3) = 4x^3$$

(e) $\displaystyle \lim_{h \to 0} \frac{f(x+h) - f(x)}{h} = \lim_{h \to 0} \frac{(x+h)^n - x^n}{h} = nx^{n-1}$ for any positive integer n.

SECTION 2.4

1.   (a) f is discontinuous at $x = -3,\ 0,\ 2,\ 6$

(b)   at -3, neither; f is continuous from the right at 0; at 2 and 6, neither

3.   continuous   5.   continuous   7.   continuous

9.   removable discontinuity   11.   jump discontinuity   13.   continuous

15.   jump discontinuity

17.   19.

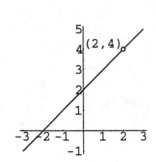

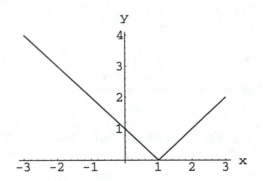

removable discontinuity at 2   no discontinuities

21.

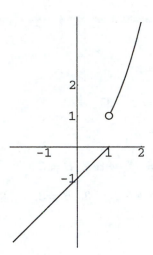

jump discontinuity at 1

23.

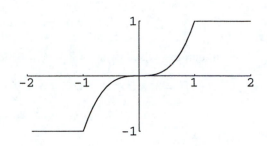

no discontinuities

25.

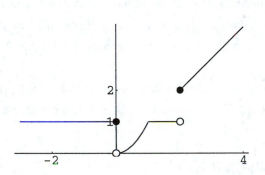

jump discontinuities at 0 and 2

27.

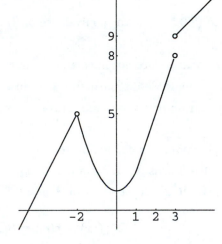

removable discontinuity at -2

jump discontinuity at 3

29.

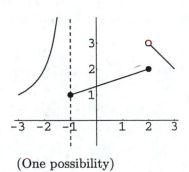

(One possibility)

31. $f(1) = 2$

33. impossible; $\lim\limits_{x \to 1^-} f(x) = -1$; $\lim\limits_{x \to 1^+} f(x) = 1$

35.   Since $\lim\limits_{x\to 1^-} f(x) = 1$ and $\lim\limits_{x\to 1^+} f(x) = A - 3 = f(1)$, take $A = 4$.

37.   The function f is continuous at $x = 1$ iff

$$f(1) = \lim\limits_{x\to 1^-} f(x) = A - B \quad \text{and} \quad \lim\limits_{x\to 1^+} f(x) = 3$$

are equal;   that is, $A - B = 3$. The function f is discontinuous at $x = 2$ iff

$$\lim\limits_{x\to 2^-} f(x) = 6 \quad \text{and} \quad \lim\limits_{x\to 2^+} f(x) = f(2) = 4B - A$$

are unequal;   that is, iff $4B - A \neq 6$. More simply we have $A - B = 3$ with $B \neq 3$:

$$A - B = 3,\ 4B - A \neq 6 \quad \Longrightarrow \quad A - B = 3,\ 3B - 3 \neq 6 \quad \Longrightarrow \quad A - B = 3,\ B \neq 3.$$

39. $f(5) = \frac{1}{6}$ 　　　　　　　　　　　　　　　　　　**41.** $f(5) = \frac{1}{3}$

43.   nowhere;   see Figure 2.1.8

45. $x = 0$, $x = 2$,   and all non-integral values of x

47.   Refer to (2.2.5). Use the equivalence of (i) and (ii) setting $L = f(c)$.

49.   Suppose that g does not have a non-removable discontinuity at c. Then either g is continuous at c or it has a removable discontinuity at c. In either case, $\lim g(x)$ as $x \to c$ exists. Since $g(x) = f(x)$ except at a finite set of points $x_1, x_2, \ldots, x_n$, $\lim f(x)$ exists as $x \to c$ by Exercise 54, Section 2.3.

51.   By implication, f is defined on $(c - p, c + p)$. The given inequality implies that $B \geq 0$. If $B = 0$, then $f \equiv f(c)$ is a constant function and hence is continuous. Now assume that $B > 0$. Let $\epsilon > 0$ and let $\delta = \min\{\epsilon/B, p\}$. If $|x - c| < \delta$ then $x \in (c - p, c + p)$ and

$$|f(x) - f(c)| \leq B|x - c| < B \cdot \delta \leq B \cdot \frac{\epsilon}{B} = \epsilon$$

Thus, f is continuous at c.

53. $$\lim\limits_{h\to 0} [f(c + h) - f(c)] = \lim\limits_{h\to 0} \left[\frac{f(c + h) - f(c)}{h} \cdot h \right] = \lim\limits_{h\to 0} \left[\frac{f(c + h) - f(c)}{h} \right] \cdot \lim\limits_{h\to 0} h = L \cdot 0 = 0.$$
Therefore f is continuous at c by Exercise 47.

SECTION 2.5

1. $\lim\limits_{x\to 0} \dfrac{\sin 3x}{x} = \lim\limits_{x\to 0} 3 \left(\dfrac{\sin 3x}{3x} \right) = 3(1) = 3$ 　　　　**3.** $\lim\limits_{x\to 0} \dfrac{3x}{\sin 5x} = \lim\limits_{x\to 0} \dfrac{3}{5} \left(\dfrac{5x}{\sin 5x} \right) = \dfrac{3}{5}(1) = \dfrac{3}{5}$

5. $\lim\limits_{x\to 0} \dfrac{\sin 4x}{\sin 2x} = \lim\limits_{x\to 0} \dfrac{4x}{2x} \cdot \dfrac{\sin 4x}{4x} \cdot \dfrac{2x}{\sin 2x} = 2(1)(1) = 2$

7. $\lim\limits_{x\to 0} \dfrac{\sin x^2}{x} = \lim\limits_{x\to 0} x \left(\dfrac{\sin x^2}{x^2} \right) = \lim\limits_{x\to 0} x \cdot \lim\limits_{x\to 0} \dfrac{\sin x^2}{x^2} = 0(1) = 0$

9. $\lim\limits_{x\to 0} \dfrac{\sin x}{x^2} = \lim\limits_{x\to 0} \dfrac{(\sin x)/x}{x}$;   does not exist

11. $\displaystyle\lim_{x\to 0}\frac{\sin^2 3x}{5x^2}=\lim_{x\to 0}\frac{9}{5}\left(\frac{\sin 3x}{3x}\right)^2=\frac{9}{5}(1)=\frac{9}{5}$

13. $\displaystyle\lim_{x\to 0}\frac{2x}{\tan 3x}=\lim_{x\to 0}\frac{2x\cos 3x}{\sin 3x}=\lim_{x\to 0}\frac{2}{3}\left(\frac{3x}{\sin 3x}\right)\cos 3x=\frac{2}{3}(1)(1)=\frac{2}{3}$

15. $\displaystyle\lim_{x\to 0}x\csc x=\lim_{x\to 0}\frac{x}{\sin x}=1$

17. $\displaystyle\lim_{x\to 0}\frac{x^2}{1-\cos 2x}=\lim_{x\to 0}\frac{x^2}{1-\cos 2x}\cdot\left(\frac{1+\cos 2x}{1+\cos 2x}\right)=\lim_{x\to 0}\frac{x^2(1+\cos 2x)}{\sin^2 2x}$

$$=\lim_{x\to 0}\frac{1}{4}\left(\frac{2x}{\sin 2x}\right)^2(1+\cos 2x)=\frac{1}{4}(1)(2)=\frac{1}{2}$$

19. $\displaystyle\lim_{x\to 0}\frac{1-\sec^2 2x}{x^2}=\lim_{x\to 0}\frac{-\tan^2 2x}{x^2}=\lim_{x\to 0}\frac{-\sin^2 2x}{x^2\cos^2 2x}=\lim_{x\to 0}\left[-4\left(\frac{\sin 2x}{2x}\right)^2\frac{1}{\cos^2 2x}\right]=-4$

21. $\displaystyle\lim_{x\to 0}\frac{2x^2+x}{\sin x}=\lim_{x\to 0}(2x+1)\frac{x}{\sin x}=1$

23. $\displaystyle\lim_{x\to 0}\frac{\tan 3x}{2x^2+5x}=\lim_{x\to 0}\frac{1}{x(2x+5)}\frac{\sin 3x}{\cos 3x}=\lim_{x\to 0}\frac{3}{2x+5}\left(\frac{\sin 3x}{3x}\right)\frac{1}{\cos 3x}=\frac{3}{5}(1)(1)=\frac{3}{5}$

25. $\displaystyle\lim_{x\to 0}\frac{\sec x-1}{x\sec x}=\lim_{x\to 0}\frac{\dfrac{1}{\cos x}-1}{x\left(\dfrac{1}{\cos x}\right)}=\lim_{x\to 0}\frac{1-\cos x}{x}=0$

27. $\dfrac{2\sqrt{2}}{\pi}$

29. $\displaystyle\lim_{x\to \pi/2}\frac{\cos x}{x-\pi/2}=\lim_{h\to 0}\frac{\cos(h+\pi/2)}{h}=\lim_{h\to 0}\frac{-\sin h}{h}=-1$

$$h=x-\pi/2\qquad\qquad\cos(h+\pi/2)=\cos h\cos\pi/2-\sin h\sin\pi/2$$

31. $\displaystyle\lim_{x\to \pi/4}\frac{\sin(x+\pi/4)-1}{x-\pi/4}=\lim_{h\to 0}\frac{\sin(h+\pi/2)-1}{h}=\lim_{h\to 0}\frac{\cos h-1}{h}=0$

$$h=x-\pi/4$$

33. Equivalently we will show that $\displaystyle\lim_{h\to 0}\cos(c+h)=\cos c$. The identity

$$\cos(c+h)=\cos c\cos h-\sin c\sin h$$

gives

$$\lim_{h\to 0}\cos(c+h)=\cos c\left(\lim_{h\to 0}\cos h\right)-\sin c\left(\lim_{h\to 0}\sin h\right)=(\cos c)(1)-(\sin c)(0)=\cos c.$$

35. $f(x)=\sin x;\quad a=\pi/4$

$$\lim_{h\to 0}\frac{f(a+h)-f(a)}{h}=\lim_{h\to 0}\frac{\sin\left(\frac{\pi}{4}+h\right)-\sin\left(\frac{\pi}{4}\right)}{h}=\lim_{h\to 0}\frac{\sin(\pi/4)\cos h+\cos(\pi/4)\sin h-\sin(\pi/4)}{h}$$

$$= \lim_{h \to 0} \frac{-\sin(\pi/4)(1 - \cos h) + \cos(\pi/4)\sin h}{h}$$

$$= -\sin(\pi/4) \lim_{h \to 0} \frac{1 - \cos h}{h} + \cos(\pi/4) \lim_{h \to 0} \frac{\sin h}{h} = \cos(\pi/4) = \frac{\sqrt{2}}{2}$$

tangent line: $y - \dfrac{\sqrt{2}}{2} = \dfrac{\sqrt{2}}{2}\left(x - \dfrac{\pi}{4}\right)$

37. $f(x) = \cos 2x; \quad a = \pi/6$

$$\lim_{h \to 0} \frac{f(a+h) - f(a)}{h} = \lim_{h \to 0} \frac{\cos 2\left(\frac{\pi}{6} + h\right) - \cos\left(2\frac{\pi}{6}\right)}{h} = \lim_{h \to 0} \frac{\cos(2h + \pi/3) - \cos(\pi/3)}{h}$$

$$= \lim_{h \to 0} \frac{\cos(\pi/3)\cos 2h - \sin(\pi/3)\sin 2h - \cos(\pi/3)}{h}$$

$$= -\cos(\pi/3) \lim_{h \to 0} 2\frac{1 - \cos 2h}{h} - \sin(\pi/3) \lim_{h \to 0} 2\frac{\sin 2h}{h}$$

$$= -\cos(\pi/3) \cdot 2 \cdot 0 - \sin(\pi/3) \cdot 2 \cdot 1 = -\sqrt{3}$$

tangent line: $y - \dfrac{1}{2} = -\sqrt{3}\left(x - \dfrac{\pi}{6}\right)$

39.   For $x \neq 0$, $|x \sin(1/x)| = |x| \, |\sin(1/x)| \leq |x|$. Thus,

$$-|x| \leq |x \sin(1/x)| \leq |x|$$

Since $\lim_{x \to 0} (-|x|) = \lim_{x \to 0} |x| = 0$, the result follows by the pinching theorem.

41.   For x close to 1(radian), $0 < \sin x \leq 1$. Thus,

$$0 < |x - 1| \sin x \leq |x - 1|$$

and the result follows by the pinching theorem.

43.   Suppose that there is a number B such that $|f(x)| \leq B$ for all $x \neq 0$. Then $|x f(x)| \leq B|x|$ and

$$-B|x| \leq x f(x) \leq B|x|$$

The result follows by the pinching theorem.

45.   Suppose that there is a number B such that $\left|\dfrac{f(x) - L}{x - c}\right| \leq B$ for $x \neq c$. Then

$$0 \leq |f(x) - L| = \left|(x - c)\frac{f(x) - L}{x - c}\right| \leq B|x - c|$$

By the pinching theorem, $\lim_{x \to c} |f(x) - L| = 0$ which implies $\lim_{x \to c} f(x) = L$.

SECTION 2.6

1.   Let $f(x) = 2x^3 - 4x^2 + 5x - 4$. Then f is continuous on $[1, 2]$ and $f(1) = -1 < 0$. $f(2) = 6 > 0$.

Thus by the intermediate-value theorem, there is a c in $[1, 2]$ such that $f(c) = 0$.

3. Let $f(x) = \sin x + 2\cos x - x^2$ Then f is continuous on $[0, \frac{\pi}{2}]$ and $f(0) = 2 > 0$, $\quad f(\frac{\pi}{2}) = 1 - \frac{\pi^2}{4} < 0$.

   Thus by the intermediate-value theorem, there is a c in $[0, \frac{\pi}{2}]$ such that $f(c) = 0$.

5. Let $f(x) = x^2 - 2 + \frac{1}{2x}$. Then f is continuous on $[\frac{1}{4}, 1]$ and $f(\frac{1}{4}) = \frac{1}{16} > 0$, $\quad f(1) = -\frac{1}{2} < 0$.

   Thus by the intermediate-value theorem, there is a c in $[\frac{1}{4}, 1]$ such that $f(c) = 0$.

7. Let $f(x) = x^3 - \sqrt{x+2}$. Then f is continuous on $[1, 2]$ and $f(1) = 1 - \sqrt{3} < 0$, $\quad f(2) = 6 > 0$.

   Thus by the intermediate-value theorem, there is a c in $[1, 2]$ such that $f(c) = 0$.

   i.e. $c^3 = \sqrt{c+2}$.

9. Let $R(x) = (x-2)^2(10-2x)$.   Then $R(x) = 0$ has solutions at $x = 2$ and $x = 5$.

   Thus the intervals of interest are $(-\infty, 2), (2, 5)$ and $(5, \infty)$.

   By inspection, $R(x) > 0$ on $(-\infty, 2) \cup (2, 5)$.

11. Let $R(x) = x^3 - 2x^2 + x$.   Then $R(x) = 0$ has solutions at $x = 0$ and $x = 1$.

    Thus the intervals of interest are $(-\infty, 0], [0, 1]$ and $[1, \infty)$.

    By inspection $R(x) \le 0$ on $(-\infty, 0] \cup 1$.

13. Let $R(x) = \frac{1}{x-1} + \frac{4}{x-6}$.   Then $R(x) = 0$ has a solutions at $x = 2$ and is undefined at $x = 1$ and $x = 6$.

    Thus the intervals of interest are $(-\infty, 1), (1, 2), (2, 6)$ and $(6, \infty)$.

    By inspection $R(x) > 0$ on $(1, 2) \cup (6, \infty)$.

15. $f(x)$ is continuous on $[0, 1]$. $f(0) = 0 < 1$ and $f(1) = 4 > 1$.

    Thus by the intermediate value theorem there is a c in $[0, 1]$ such that $f(c) = 1$.

17. Let $f(x) = x^3 - 4x + 2$.   Then $f(x)$ is continuous on $[-3, 3]$.

    Checking the integer values on this interval,

    $$f(-3) = -13 < 0, \quad f(-2) = 2 > 0, \quad f(0) = 2 > 0, \quad f(1) = -1 < 0, \text{ and } f(2) = 2 > 0.$$

    Thus by the intermediate value theorem there are roots in $(-3, -2), (0, 1)$ and $(1, 2)$.

19.

21.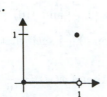

23.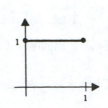

25.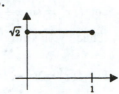

27. Impossible

29.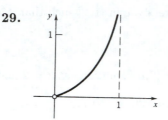

31. Set $g(x) = x - f(x)$. Since g is continuous on $[0, 1]$ and $g(0) \leq 0 \leq g(1)$, there exists c in $[0, 1]$ such that $g(c) = c - f(c) = 0$.

33. Since f is bounded on $(-p, p)$, it follows from Exercise 43, Section 2.5, that $\lim_{x \to 0} x f(x) = 0$. Thus,

$$\lim_{x \to 0} g(x) = \lim_{x \to 0} x f(x) = 0 = g(0)$$

which implies that g is continuous at 0.

35. The cubic polynomial $P(x) = x^3 + ax^2 + bx + c$ is continuous on $(-\infty, \infty)$.. Writing P as

$$P(x) = x^3 \left(1 + \frac{a}{x} + \frac{b}{x^2} + \frac{c}{x^3} \right) \quad x \neq 0$$

it follows that $P(x) < 0$ for large negative values of x and $P(x) > 0$ for large positive values of x. Thus there exists a negative number N such that $P(x) < 0$ for $x < N$, and a positive number M such that $P(x) > 0$ for $x > M$. By the intermediate-value theorem, P has a zero in $[N, M]$.

37. Let $A(r)$ denote the area of a circle with radius r, $r \in [0, 10]$. Then $A(r) = \pi r^2$ is continuous on $[0, 10]$, and $A(0) = 0$ and $A(10) = 100\pi \cong 314$. Since $0 < 250 < 314$ it follows from the intermediate value theorem that there exists a number $c \in (0, 10)$ such that $A(c) = 250$.

39. Inscribe a rectangle in a circle of radius R and introduce a coordinate system as shown in the figure. Then the area of the rectangle is given by

$$A(x) = 4x\sqrt{R^2 - x^2}, \quad x \in [0, R].$$

Since A is continuous on $[0, R]$, A has a maximum value.

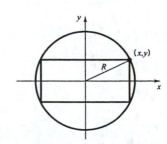

41. $f(-3) = -9$, $f(-2) = 5$; $f(0) = 3$, $f(1) = -1$; $f(1) = -1$, $f(2) = 1$ Thus, f has a zero in

$(-3, -2,)$ in $(0, 1)$ and in $(1, 2)$.

$r_1 = -2.4909$, $r_2 = 0.6566$, and $r_3 = 1.8343$

43. $f(-2) = -5.6814$, $f(-1) = 1.1829$; $f(0) = 0.5$, $f(1) = -0.1829$; $f(1) = -0.1829$, $f(2) = 6.681$

Thus, f has a zero in $(-2, -1)$, in $(0, 1)$ and in $(1, 2)$.

$r_1 = -1.3482$, $r_2 = 0.2620$, and $r_3 = 1.0816$

45. f is bounded.

$\max(f) = 1$ $[f(1) = 1]$

$\min(f) = -1$ $[f(-1) = -1]$

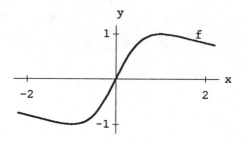

47. f is bounded.

$\max(f) = 0.5$

$\min(f) \cong 0.3540$

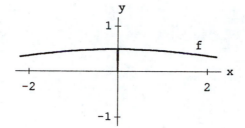

PROJECT 2.6

1. $\dfrac{2 - 1}{2^n} < 0.01$ $\implies$ $2^n > 100$ $\implies$ $n > 6$.

Thus, minimum number of iterations required is $n = 7$.

$$\sqrt{2} \simeq \frac{181}{128} \simeq 1.4140625.$$

$\dfrac{2 - 1}{2^n} < 0.0001$ $\implies$ $2^n > 10,000$ $\implies$ $n > 13$.

Thus, minimum number of iterations required is $n = 14$.

3. For $f(x) = x^2 - 2$, the first three iterations are:

$$c_1 = \frac{4}{3} \simeq 1.333\ldots$$

$$c_2 = \frac{7}{5} = 1.4$$

$$c_3 = \frac{24}{17} \simeq 1.41176$$

For $f(x) = x^3 + x - 9$, the first three iterations are:

$$c_1 = \frac{15}{8} = 1.875$$

$$c_2 \simeq 1.918471$$

$$c_3 \simeq 1.920112$$

These approximations appear to converge more rapidly than the approximations obtained by the bisection method.

CHAPTER 3

SECTION 3.1

1. $f'(x) = \lim\limits_{h \to 0} \dfrac{f(x+h) - f(x)}{h} = \lim\limits_{h \to 0} \dfrac{4 - 4}{h} = \lim\limits_{h \to 0} 0 = 0$

3. $f'(x) = \lim\limits_{h \to 0} \dfrac{f(x+h) - f(x)}{h} = \lim\limits_{h \to 0} \dfrac{[2 - 3(x+h)] - [2 - 3x]}{h} = \lim\limits_{h \to 0} \dfrac{-3h}{h} = \lim\limits_{h \to 0} -3 = -3$

5. $f'(x) = \lim\limits_{h \to 0} \dfrac{f(x+h) - f(x)}{h} = \lim\limits_{h \to 0} \dfrac{[5(x+h) - (x+h)^2] - (5x - x^2)}{h}$

$\qquad = \lim\limits_{h \to 0} \dfrac{5h - 2xh - h^2}{h} = \lim\limits_{h \to 0} (5 - 2x - h) = 5 - 2x$

7. $f'(x) = \lim\limits_{h \to 0} \dfrac{f(x+h) - f(x)}{h} = \lim\limits_{h \to 0} \dfrac{(x+h)^4 - x^4}{h}$

$\qquad = \lim\limits_{h \to 0} \dfrac{(x^4 + 4x^3h + 6x^2h^2 + 4xh^3 + h^4) - x^4}{h}$

$\qquad = \lim\limits_{h \to 0} (4x^3 + 6x^2h + 4xh^2 + h^3) = 4x^3$

9. $f'(x) = \lim\limits_{h \to 0} \dfrac{f(x+h) - f(x)}{h} = \lim\limits_{h \to 0} \dfrac{\sqrt{x+h-1} - \sqrt{x-1}}{h}$

$\qquad = \lim\limits_{h \to 0} \dfrac{(x+h-1) - (x-1)}{h(\sqrt{x+h-1} + \sqrt{x-1})} = \lim\limits_{h \to 0} \dfrac{1}{\sqrt{x+h-1} + \sqrt{x-1}} = \dfrac{1}{2\sqrt{x-1}}$

11. $f'(x) = \lim\limits_{h \to 0} \dfrac{f(x+h) - f(x)}{h} = \lim\limits_{h \to 0} \dfrac{\dfrac{1}{(x+h)^2} - \dfrac{1}{x^2}}{h}$

$\qquad = \lim\limits_{h \to 0} \dfrac{x^2 - (x^2 + 2hx + h^2)}{hx^2(x+h)^2} = \lim\limits_{h \to 0} \dfrac{-2x - h}{x^2(x+h)^2} = -\dfrac{2}{x^3}$

13. $f'(2) = \lim\limits_{h \to 0} \dfrac{f(2+h) - f(2)}{h} = \lim\limits_{h \to 0} \dfrac{(3h - 1)^2 - 1}{h}$

$\qquad = \lim\limits_{h \to 0} \dfrac{9h^2 - 6h}{h} = \lim\limits_{h \to 0} (9h - 6) = -6$

15. $f'(2) = \lim\limits_{h \to 0} \dfrac{f(2+h) - f(2)}{h} = \lim\limits_{h \to 0} \dfrac{\dfrac{9}{6+h} - \dfrac{3}{2}}{h}$

$\qquad = \lim\limits_{h \to 0} \dfrac{18 - 3(6+h)}{2h(6+h)} = \lim\limits_{h \to 0} \dfrac{-3}{2(6+h)} = -\dfrac{1}{4}$

17. $f'(2) = \lim\limits_{h \to 0} \dfrac{f(2+h) - f(2)}{h} = \lim\limits_{h \to 0} \dfrac{(2 + h + \sqrt{4+2h}) - 4}{h}$

$\qquad = \lim\limits_{h \to 0} \left(1 + \dfrac{\sqrt{4+2h} - 2}{h}\right) = \lim\limits_{h \to 0} 1 + \dfrac{(4+2h) - 4}{h(\sqrt{4+2h} + 2)}$

$$= \lim_{h \to 0} 1 + \frac{2}{\sqrt{4 + 2h} + 2} = \frac{3}{2}$$

19. Slope of tangent at $(2, 4)$ is $f'(2) = 4$.   Tangent $y - 4 = 4(x - 2)$;

normal $y - 4 = -\frac{1}{4}(x - 2)$.

21. Slope of tangent at $(4, 4)$ is $f'(4) = -3$.   Tangent $y - 4 = -3(x - 4)$;

normal $y - 4 = \frac{1}{3}(x - 4)$.

23. Slope of tangent at $(-2, \frac{1}{4})$ is $\frac{1}{4}$.   Tangent $y - \frac{1}{4} = \frac{1}{4}(x + 2)$;

normal $y - \frac{1}{4} = -4(x + 2)$.

25. (a) f is not continuous at $c = -1$ and $c = 1$; f has a removable discontinuity at $c = -1$

and a jump discontinuity at $c = 1$.

(b) f is continuous but not differentiable at $c = 0$ and $c = 3$.

27.   at $x = -1$ **29.**   at $x = 0$ **31.**   at $x = 1$

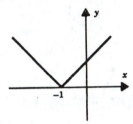 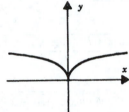 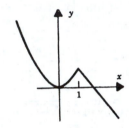

33. $f'(1) = 4$

$$\lim_{h \to 0^-} \frac{f(1 + h) - f(1)}{h} = \lim_{h \to 0^-} \frac{4(1 + h) - 4}{h} = 4$$

$$\lim_{h \to 0^+} \frac{f(1 + h) - f(1)}{h} = \lim_{h \to 0^+} \frac{2(1 + h)^2 + 2 - 4}{h} = 4$$

35. $f'(-1)$ does not exist

$$\lim_{h \to 0^-} \frac{f(-1 + h) - f(-1)}{h} = \lim_{h \to 0^-} \frac{h - 0}{h} = 1$$

$$\lim_{h \to 0^+} \frac{f(-1 + h) - f(-1)}{h} = \lim_{h \to 0^+} \frac{h^2 - 0}{h} = 0$$

37.

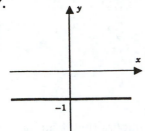

39.

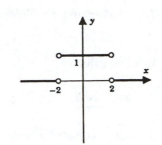

41.

43. $f(x) = x^2; \quad c = 1$

45. $f(x) = \sqrt{x}; \quad c = 4$

47. $f(x) = \cos x; \quad c = \pi$

49. Since $f(1) = 1$ and $\lim\limits_{x \to 1^+} f(x) = 2$, f is not continuous at 1. Therefore, by (3.1.4),

f is not differentiable at 1.

51. (a) $f'(x) = \begin{cases} 2(x+1), & x < 0 \\ 2(x-1), & x > 0 \end{cases}$

(b) $\lim\limits_{h \to 0^-} \dfrac{f(0+h) - f(0)}{h} = \lim\limits_{h \to 0^-} \dfrac{(h+1)^2 - 1}{h} = \lim\limits_{h \to 0^-} (h+2) = 2,$

$\lim\limits_{h \to 0^+} \dfrac{f(0+h) - f(0)}{h} = \lim\limits_{h \to 0^+} \dfrac{(h-1)^2 - 1}{h} = \lim\limits_{h \to 0^+} (h-2) = -2.$

53. $f(x) = c$, c any constant

55. $f(x) = |x+1|;$ or $f(x) = \begin{cases} 0, & x \neq -1 \\ 1, & x = -1 \end{cases}$

57. $f(x) = 2x + 5$

59. (a) $\lim\limits_{x \to 2^+} f(x) = \lim\limits_{x \to 2^-} f(x) = f(2) = 2$ Thus, f is continuous at $x = 2$.

(b) $f'_-(2) = \lim\limits_{h \to 0^-} \dfrac{f(2+h) - f(2)}{h} = \lim\limits_{h \to 0^-} \dfrac{(2+h)^2 - (2+h) - 2}{h} = 3$

$f'_+(2) = \lim\limits_{h \to 0^+} \dfrac{f(2+h) - f(2)}{h} = \lim\limits_{h \to 0^+} \dfrac{2(2+h) - 2 - 2}{h} = 2$

(c) No, since $f'_-(2) \neq f'_+(2)$.

61. (a) $f'(x) = \lim\limits_{h \to 0} \dfrac{f(x+h) - f(x)}{h} = \lim\limits_{h \to 0} \dfrac{\sqrt{1 - (x+h)} - \sqrt{1 - x}}{h}$

$= \lim\limits_{h \to 0} \dfrac{-h}{h \left(\sqrt{1 - (x+h)} + \sqrt{1 - x} \right)} = \dfrac{-1}{2\sqrt{1 - x}}$

(b) $f'_+(0) = \lim\limits_{h \to 0^+} \dfrac{\sqrt{1 - h} - 1}{h} = -\dfrac{1}{2}$

(c) $f'_-(1) = \lim\limits_{h \to 0^-} \dfrac{\sqrt{1 - (1+h)}}{h} = \lim\limits_{h \to 0^-} \dfrac{\sqrt{-h}}{h} = \lim\limits_{h \to 0^-} \dfrac{-1}{\sqrt{-h}}$ does not exist.

63. Suppose $f'_-(c) = \lim\limits_{h \to 0^-} \dfrac{f(c+h) - f(c)}{h} = L = \lim\limits_{h \to 0^+} \dfrac{f(c+h) - f(c)}{h} = f'_+(c).$

Then $\lim\limits_{h\to 0}\dfrac{f(c+h)-f(c)}{h}=L$ exists and f is differentiable at c.

65.   (a) Since $|\sin(1/x)|\le 1$ it follows that

$$-x\le f(x)\le x\qquad\text{and}\qquad -x^2\le g(x)\le x^2$$

Thus $\lim\limits_{x\to 0}f(x)=f(0)=0$ and $\lim\limits_{x\to 0}g(x)=g(0)=0,$ which implies that f and g

are continuous at 0.

(b) $\lim\limits_{h\to 0}\dfrac{h\sin(1/h)-0}{h}=\lim\limits_{h\to 0}\sin(1/h)$ does not exist.

(c) $\lim\limits_{h\to 0}\dfrac{h^2\sin(1/h)-0}{h}=\lim\limits_{h\to 0}h\sin(1/h)=0.$ Thus g is differentiable at 0 and $g'(0)=0.$

67. $f'(1)=\lim\limits_{h\to 0}\dfrac{f(1+h)-f(1)}{h}=\lim\limits_{h\to 0}\dfrac{[(1+h)^2-3(1+h)]-(-2)]}{h}=\lim\limits_{h\to 0}\dfrac{-h+h^2}{h}=-1$

$f'(1)=\lim\limits_{x\to 1}\dfrac{f(x)-f(1)}{x-1}=\lim\limits_{x\to 1}\dfrac{(x^2-3x)-(-2)}{x-1}=\lim\limits_{x\to 1}\dfrac{(x-2)(x-1)}{x-1}=\lim\limits_{x\to 1}(x-2)=-1$

69. $f'(-1)=\lim\limits_{h\to 0}\dfrac{f(-1+h)-f(-1)}{h}=\lim\limits_{h\to 0}\dfrac{(-1+h)^{1/3}+1}{h}$

$\qquad =\lim\limits_{h\to 0}\dfrac{(-1+h)^{1/3}+1}{h}\cdot\dfrac{(-1+h)^{2/3}-(-1+h)^{1/3}+1}{(-1+h)^{2/3}-(-1+h)^{1/3}+1}$

$\qquad =\lim\limits_{h\to 0}\dfrac{h}{h\left((-1+h)^{2/3}-(-1+h)^{1/3}+1\right)}=\dfrac{1}{3}$

$f'(-1)=\lim\limits_{x\to -1}\dfrac{f(x)-f(-1)}{x-(-1)}=\lim\limits_{x\to -1}\dfrac{x^{1/3}+1}{x+1}=\lim\limits_{x\to -1}\dfrac{x^{1/3}+1}{x+1}\cdot\dfrac{x^{2/3}-x^{1/3}+1}{x^{2/3}-x^{1/3}+1}$

$\qquad =\lim\limits_{x\to -1}\dfrac{x+1}{(x+1)(x^{2/3}-x^{1/3}+1)}=\dfrac{1}{3}$

71.   (a) $D=\dfrac{(2+h)^{5/2}-2^{5/2}}{h}\qquad -1\le h\le 1$

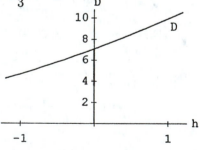

(b) $f'(2)\cong 7.071$

(c) $D(0.001)\cong 7.074,\quad D(-0.001)\cong 7.068$

73.   (a) Let $f(x)=4x-x^3.$

Then $f'(x)=4-3x^2;\quad f'(3/2)=\frac{11}{4}$

$T(x)=-\frac{11}{4}\left(x-\frac{3}{2}\right)+\frac{21}{8}$

(c) $(1.453, 1.547)$

(b)

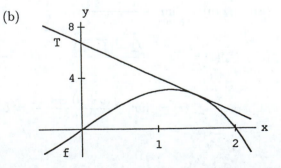

SECTION 3.2

1. $F'(x) = -1$ **3.** $F'(x) = 55x^4 - 18x^2$ **5.** $F'(x) = 2ax + b$

7. $F'(x) = 2x^{-3}$ **9.** $G'(x) = (x^2 - 1)(1) + (x - 3)(2x) = 3x^2 - 6x - 1$

11. $G'(x) = \dfrac{(1-x)(3x^2) - x^3(-1)}{(1-x)^2} = \dfrac{3x^2 - 2x^3}{(1-x)^2}$

13. $G'(x) = \dfrac{(2x+3)(2x) - (x^2-1)(2)}{(2x+3)^2} = \dfrac{2(x^2 + 3x + 1)}{(2x+3)^2}$

15. $G'(x) = (x-1)(1) + (x-2)(1) = 2x - 3$

17. $G'(x) = \dfrac{(x-2)(1/x^2) - (6 - 1/x)(1)}{(x-2)^2} = \dfrac{-2(3x^2 - x + 1)}{x^2(x-2)^2}$

19. $G'(x) = (9x^8 - 8x^9)\left(1 - \dfrac{1}{x^2}\right) + \left(x + \dfrac{1}{x}\right)(72x^7 - 72x^8) = -80x^9 + 81x^8 - 64x^7 + 63x^6$

21. $f'(x) = -x(x-2)^{-2}$; $f'(0) = -\frac{1}{4}$, $f'(1) = -1$

23. $f'(x) = \dfrac{(1+x^2)(-2x) - (1-x^2)(2x)}{(1+x^2)^2} = \dfrac{-4x}{(1+x^2)^2}$; $f'(0) = 0$, $f'(1) = -1$

25. $f'(x) = \dfrac{(cx+d)a - (ax+b)c}{(cx+d)^2} = \dfrac{ad - bc}{(cx+d)^2}$, $f'(0) = \dfrac{ad - bc}{d^2}$, $f'(1) = \dfrac{ad - bc}{(c+d)^2}$

27. $f'(x) = xh'(x) + h(x)$; $f'(0) = 0h'(0) + h(0) = 0(2) + 3 = 3$

29. $f'(x) = h'(x) + \dfrac{h'(x)}{[h(x)]^2}$, $f'(0) = h'(0) + \dfrac{h'(0)}{[h(0)]^2} = 2 + \dfrac{2}{3^2} = \dfrac{20}{9}$

31. $f'(x) = \dfrac{(x+2)(1) - x(1)}{(x+2)^2} = \dfrac{2}{(x+2)^2}$,

slope of tangent at $(-4, 2) : f'(-4) = \frac{1}{2}$,

equation for tangent: $y - 2 = \frac{1}{2}(x + 4)$

33. $f'(x) = (x^2 - 3)(5 - 3x^2) + (5x - x^3)(2x)$;

slope of tangent at $(1, -8) : f'(1) = (-2)(2) + (4)(2) = 4$,

equation for tangent: $y + 8 = 4(x - 1)$

35. $f'(x) = (x-2)(2x-1) + (x^2 - x - 11)(1) = 3(x-3)(x+1)$;

$f'(x) = 0$ at $x = -1, 3$; $(-1, 27)$, $(3, -5)$

37. $f'(x) = \dfrac{(x^2+1)(5) - 5x(2x)}{(x^2+1)^2} = \dfrac{5(1 - x^2)}{(x^2+1)^2}$, $f'(x) = 0$ at $x = \pm 1$; $(-1, -5/2)$, $(1, 5/2)$

39. $f'(x) = 1 - 8/x^3$, $f'(x) = 0$ at $x = 2$; $(2, 3)$

41. slope of line 4; slope of tangent $f'(x) = -2x$; $-2x = 4$ at $x = -2$; $(-2, -10)$

43. slope of line $-1/5$; slope of tangent $3x^2 - 2x$;

perpendicular when $3x^2 - 2x = 5$; $x = -1, 5/3$; $(-1, -2)$, $\left(\frac{5}{3}, \frac{50}{27}\right)$

45. $f(x) = x^3 + x^2 + x + C$

47. $f(x) = \dfrac{2x^3}{3} - \dfrac{3x^2}{2} + \dfrac{1}{x} + C$

49.

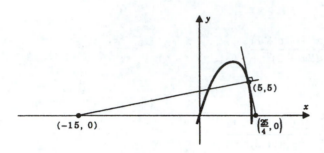

slope of tangent at $(5, 5)$ is $f'(5) = -4$

tangent $y - 5 = -4(x - 5)$ intersects

x-axis at $\left(\frac{25}{4}, 0\right)$

normal $y - 5 = \frac{1}{4}(x - 5)$ intersects

x-axis at $(-15, 0)$

area of triangle is

$$\tfrac{1}{2}(5)\left(15 + \tfrac{25}{4}\right) = \tfrac{425}{8}$$

51. If the point $(1, 3)$ lies on the graph, we have $f(1) = 3$ and thus

$(*)$ $\qquad\qquad\qquad\qquad\qquad A + B + C = 3.$

If the line $4x + y = 8$ (slope -4) is tangent to the graph at $(2, 0)$, then

$f(2) = 0$ and $f'(2) = -4$. Thus,

$(**)$ $\qquad\qquad\qquad 4A + 2B + C = 0 \quad$ and $\quad 4A + B = -4.$

Solving the equations in $(*)$ and $(**)$, we find that $A = -1, \quad B = 0, \quad C = 4.$

53. Let $f(x) = ax^2 + bx + c$. Then $f'(x) = 2ax + b$ and $f'(x) = 0$ at $x = -b/2a$.

55. Let $f(x) = x^3 - x$. The secant line through $(-1, f(-1)) = (-1, 0)$ and $(2, f(2)) = (2, 6)$ has slope

$m = \dfrac{6 - 0}{2 - (-1)} = 2.$ Now, $f'(x) = 3x^2 - 1$ and $3c^2 - 1 = 2$ implies $c = -1, 1.$

57. Let $f(x) = 1/x$, $x > 0$. Then $f'(x) = -1/x^2$. An equation for the tangent line to the graph of f at the point $(a, f(a))$, $a > 0$, is $y = (-1/a^2)x + 2/a$. The y-intercept is $2/a$ and the x-intercept is $2a$. The area of the triangle formed by this line and the coordinate axes is: $A = \frac{1}{2}(2/a)(2a) = 2$ square units.

59. Let (x, y) be the point on the graph that the tangent line passes through. $f'(x) = 3x^2 - 1$, so $x^3 - x - 2 = (3x^2 - 1)(x + 2).$ Thus $x = 0$ or $x = -3.$ The lines are $y = -x$ and $y + 24 = 26(x + 3).$

61. Since f and $f + g$ are differentiable, $g = (f + g) - f$ is differentiable. The functions $f(x) = |x|$ and $g(x) = -|x|$ are not differentiable at $x = 0$ yet their sum $f(x) + g(x) \equiv 0$ is differentiable for all x.

63. Since

$$\left(\frac{f}{g}\right)(x) = \frac{f(x)}{g(x)} = f(x) \cdot \frac{1}{g(x)},$$

it follows from the product and reciprocal rules that

$$\left(\frac{f}{g}\right)'(x) = \left(f \cdot \frac{1}{g}\right)'(x) = f(x)\left(-\frac{g'(x)}{[g(x)]^2}\right) + f'(x) \cdot \frac{1}{g(x)} = \frac{g(x)f'(x) - f(x)g'(x)}{[g(x)]^2}.$$

65. $F'(x) = 2x\left(1 + \frac{1}{x}\right)(2x^3 - x + 1) + (x^2 + 1)\left(\frac{-1}{x^2}\right)(2x^3 - x + 1) + (x^2 + 1)\left(1 + \frac{1}{x}\right)(6x^2 - 1)$

67. $g(x) = [f(x)]^2 = f(x) \cdot f(x)$

$g'(x) = f(x)f'(x) + f(x)f'(x) = 2f(x)f'(x)$

69. $g'(x) = 3(x^3 - 2x^2 + x + 2)^2(3x^2 - 4x + 1)$

71. (a) $f'_+(-1) = \lim\limits_{h \to 0^+} \dfrac{f(-1+h) - f(-1)}{h} = \lim\limits_{h \to 0^+} \dfrac{(-1+h)^2 - 4(-1+h) + 2 - 7}{h}$

$$= \lim\limits_{h \to 0^+} \frac{-6h + h^2}{h} = -6$$

$f'_-(3) = \lim\limits_{h \to 0^-} \dfrac{f(3+h) - f(3)}{h} = \lim\limits_{h \to 0^-} \dfrac{(3+h)^2 - 4(3+h) + 2 + 1}{h}$

$$= \lim\limits_{h \to 0^-} \frac{2h + h^2}{h} = 2$$

(b) If $g(x) = x^2 - 4x + 2$ then $g'(x) = 2x - 4$, and $g'(-1) = -6$, $g'(3) = 2$.

73. We want f to be continuous at $x = 2$. That is, we want

$$\lim\limits_{x \to 2^-} f(x) = f(2) = \lim\limits_{x \to 2^+} f(x).$$

This gives

(1) $\qquad\qquad\qquad\qquad 8A + 2B + 2 = 4B - A.$

We also want

$$\lim\limits_{x \to 2^-} f'(x) = \lim\limits_{x \to 2^+} f'(x).$$

This gives

(2) $\qquad\qquad\qquad\qquad 12A + B = 4B.$

Equations (1) and (2) together imply that $A = -2$ and $B = -8$.

75. (a) $\dfrac{\sin(0 + 0.001) - \sin 0}{0.001} \cong 0.99999 \qquad \dfrac{\sin(0 - 0.001) - \sin 0}{-0.001} \cong 0.99999$

$\dfrac{\sin[(\pi/6) + 0.001] - \sin(\pi/6)}{0.001} \cong 0.86578 \qquad \dfrac{\sin[(\pi/6) - 0.001] - \sin(\pi/6)}{-0.001} \cong 0.86628$

$\dfrac{\sin[(\pi/4) + 0.001] - \sin(\pi/4)}{0.001} \cong 0.70675 \qquad \dfrac{\sin[(\pi/4) - 0.001] - \sin(\pi/4)}{-0.001} \cong 0.70746$

$$\frac{\sin[(\pi/3) + 0.001] - \sin(\pi/3)}{0.001} \cong 0.49957 \qquad \frac{\sin[(\pi/3) - 0.001] - \sin(\pi/3)}{-0.001} \cong 0.50043$$

$$\frac{\sin[(\pi/2) + 0.001] - \sin(\pi/2)}{0.001} \cong -0.0005 \qquad \frac{\sin[(\pi/2) - 0.001] - \sin(\pi/2)}{-0.001} \cong 0.0005$$

(b) $\cos 0 = 1, \quad \cos(\pi/6) \cong 0.866025, \quad \cos(\pi/4) \cong 0.707107, \quad \cos(\pi/3) = 0.5, \quad \cos(\pi/2) = 0$

(c)   If $f(x) = \sin x$ then $f'(x) = \cos x$.

77.   (a) $\dfrac{2^{0+0.001} - 2^0}{0.001} \cong 0.69339 \qquad \dfrac{2^{0-0.001} - 2^0}{-0.001} \cong 0.69291$

$\dfrac{2^{1+0.001} - 2^1}{0.001} \cong 1.38678 \qquad \dfrac{2^{1-0.001} - 2^1}{-0.001} \cong 1.38581$

$\dfrac{2^{2+0.001} - 2^2}{0.001} \cong 2.77355 \qquad \dfrac{2^{2-0.001} - 2^2}{-0.001} \cong 2.77163$

$\dfrac{2^{3+0.001} - 2^3}{0.001} \cong 5.54710 \qquad \dfrac{2^{3-0.001} - 2^0}{-0.001} \cong 5.54326$

(b) $\dfrac{f'(x)}{f(x)} \cong 0.693$ \qquad (c)   If $f(x) = 2^x$ then $f'(x) = 2^x K$, where $K \cong 0.693$.

SECTION 3.3

1. $\dfrac{dy}{dx} = 12x^3 - 2x$ \hspace{2cm} **3.** $\dfrac{dy}{dx} = 1 + \dfrac{1}{x^2}$

5. $\dfrac{dy}{dx} = \dfrac{(1+x^2)(1) - x(2x)}{(1+x^2)^2} = \dfrac{1 - x^2}{(1+x^2)^2}$ \hspace{1cm} **7.** $\dfrac{dy}{dx} = \dfrac{(1-x)2x - x^2(-1)}{(1-x)^2} = \dfrac{x(2-x)}{(1-x)^2}$

9. $\dfrac{dy}{dx} = \dfrac{(x^3-1)3x^2 - (x^3+1)3x^2}{(x^3-1)^2} = \dfrac{-6x^2}{(x^3-1)^2}$ \hspace{0.5cm} **11.** $\dfrac{d}{dx}(2x-5) = 2$

13. $\dfrac{d}{dx}[(3x^2 - x^{-1})(2x+5)] = (3x^2 - x^{-1})2 + (2x+5)(6x + x^{-2}) = 18x^2 + 30x + 5x^{-2}$

15. $\dfrac{d}{dt}\left(\dfrac{t^4}{2t^3-1}\right) = \dfrac{(2t^3-1)4t^3 - t^4(6t^2)}{(2t^3-1)^2} = \dfrac{2t^3(t^3-2)}{(2t^3-1)^2}$

17. $\dfrac{d}{du}\left(\dfrac{2u}{1-2u}\right) = \dfrac{(1-2u)2 - 2u(-2)}{(1-2u)^2} = \dfrac{2}{(1-2u)^2}$

19. $\dfrac{d}{du}\left(\dfrac{u}{u-1} - \dfrac{u}{u+1}\right) = \dfrac{(u-1)(1) - u}{(u-1)^2} - \dfrac{(u+1)(1) - u}{(u+1)^2}$

$$= -\dfrac{1}{(u-1)^2} - \dfrac{1}{(u+1)^2} = -\dfrac{2(1+u^2)}{(u^2-1)^2}$$

21. $\dfrac{d}{dx}\left(\dfrac{x^3 + x^2 + x + 1}{x^3 - x^2 + x - 1}\right) = \dfrac{(x^3 - x^2 + x - 1)(3x^2 + 2x + 1) - (x^3 + x^2 + x + 1)(3x^2 - 2x + 1)}{(x^3 - x^2 + x - 1)^2}$

$$= \frac{-2(x^4 + 2x^2 + 1)}{(x^2 + 1)^2(x - 1)^2} = \frac{-2}{(x - 1)^2}$$

23. $\dfrac{dy}{dx} = (x + 1)\dfrac{d}{dx}\left[(x + 2)(x + 3)\right] + (x + 2)(x + 3)\dfrac{d}{dx}(x + 1)$

$$= (x + 1)(2x + 5) + (x + 2)(x + 3)$$

At $x = 2$, $\dfrac{dy}{dx} = (3)(9) + (4)(5) = 47$.

25. $\dfrac{dy}{dx} = \dfrac{(x + 2)\dfrac{d}{dx}\left[(x - 1)(x - 2)\right] - (x - 1)(x - 2)(1)}{(x + 2)^2}$

$$= \dfrac{(x + 2)(2x - 3) - (x - 1)(x - 2)}{(x + 2)^2}$$

At $x = 2$, $\quad \dfrac{dy}{dx} = \dfrac{4(1) - 1(0)}{16} = \dfrac{1}{4}$.

27. $f'(x) = 21x^2 - 30x^4$, $f''(x) = 42x - 120x^3$ **29.** $f'(x) = 1 + 3x^{-2}$, $f''(x) = -6x^{-3}$

31. $f(x) = 2x^2 - 2x^{-2} - 3$, $\quad f'(x) = 4x + 4x^{-3}$, $\quad f''(x) = 4 - 12x^{-4}$

33. $\dfrac{dy}{dx} = x^2 + x + 1$ **35.** $\dfrac{dy}{dx} = 8x - 20$ **37.** $\dfrac{dy}{dx} = 3x^2 + 3x^{-4}$

$\quad\;\; \dfrac{d^2y}{dx^2} = 2x + 1$ $\qquad\qquad\quad \dfrac{d^2y}{dx^2} = 8$ $\qquad\qquad\quad\; \dfrac{d^2y}{dx^2} = 6x - 12x^{-5}$

$\quad\;\; \dfrac{d^3y}{dx^3} = 2$ $\qquad\qquad\qquad\;\; \dfrac{d^3y}{dx^3} = 0$ $\qquad\qquad\qquad\; \dfrac{d^3y}{dx^3} = 6 + 60x^{-6}$

39. $\dfrac{d}{dx}\left[x\dfrac{d}{dx}(x - x^2)\right] = \dfrac{d}{dx}\left[x(1 - 2x)\right] = \dfrac{d}{dx}\left[x - 2x^2\right] = 1 - 4x$

41. $\dfrac{d^4}{dx^4}[3x - x^4] = \dfrac{d^3}{dx^3}[3 - 4x^3] = \dfrac{d^2}{dx^2}[-12x^2] = \dfrac{d}{dx}[-24x] = -24$

43. $\dfrac{d^2}{dx^2}\left[(1 + 2x)\dfrac{d^2}{dx^2}(5 - x^3)\right] = \dfrac{d^2}{dx^2}\left[(1 + 2x)(-6x)\right] = \dfrac{d^2}{dx^2}\left[-6x - 12x^2\right] = -24$

45. $y = x^4 - \dfrac{x^3}{3} + 2x^3 + C$ **47.** $y = x^5 - \dfrac{1}{x^4} + C$

49. Let $p(x) = ax^2 + bx + c$. Then $p'(x) = 2ax + b$ and $p''(x) = 2a$. Now

$$p''(1) = 2a = 4 \Longrightarrow a = 2; \quad p'(1) = 2(2)(1) + b = -2 \Longrightarrow b = -6;$$

$$p(1) = 2(1)^2 - 6(1) + c = 3 \Longrightarrow c = 7$$

Thus $p(x) = 2x^2 - 6x + 7$.

51. (a) If $k = n$, $\quad f^{(n)}(x) = n!$ $\qquad\qquad$ (b) If $k > n$, $\quad f^n(x) = 0$.

(c) If $k < n$, $f^{(n)}(x) = n(n-1)(n-2)\cdots(n-k+1)x^{n-k}$.

53.   Let $f(x) = \begin{cases} x^2 & x \geq 0 \\ 0 & x \leq 0 \end{cases}$

(a) $f'_+(0) = \lim\limits_{h \to 0+} \dfrac{f(0+h) - f(0)}{h} = \lim\limits_{h \to 0+} \dfrac{h^2 - 0}{h} = 0$ and

$f'_-(0) = \lim\limits_{h \to 0-} \dfrac{f(0+h) - f(0)}{h} = \lim\limits_{h \to 0-} \dfrac{0}{h} = 0$

Therefore, f is differentiable at 0 and $f'(0) = 0$.

(b) $f'(x) = \begin{cases} 2x & x \geq 0 \\ 0 & x \leq 0 \end{cases}$

(c) $f''_+(0) = \lim\limits_{h \to 0+} \dfrac{f'(0+h) - f'(0)}{h} = \lim\limits_{h \to 0+} \dfrac{2h - 0}{h} = 2$ and

$f''_-(0) = \lim\limits_{h \to 0-} \dfrac{f'(0+h) - f'(0)}{h} = \lim\limits_{h \to 0-} \dfrac{0}{h} = 0$

Since $f''_+(0) \neq f''_-(0)$, $f''(0)$ does not exist.

(d)

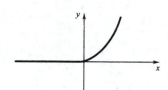

55.   It suffices to give a single counterexample. For instance, if

$f(x) = g(x) = x$, then $(fg)(x) = x^2$ so that $(fg)''(x) = 2$ but

$f(x)g''(x) + f''(x)g(x) = x \cdot 0 + 0 \cdot x = 0$.

57. $f''(x) = 6x$;   (a) $x = 0$ (b) $x > 0$ (c) $x < 0$

59. $f''(x) = 12x^2 + 12x - 24$;   (a) $x = -2,\ 1$ (b) $x < -2,\quad x > 1$ (c) $-2 < x < 1$

61.   The result is true for $n = 1$:

$$\frac{d^1 y}{lx^1} = \frac{dy}{dx} = -x^{-2} = (-1)^1 1!\, x^{-1-1}.$$

If the result is true for $n = k$:

$$\frac{d^k y}{dx^k} = (-1)^k k! \, x^{-k-1}$$

then the result is true for $n = k + 1$:

$$\frac{d^{k+1} y}{dx^{k+1}} = \frac{d}{dx}\left[\frac{d^k y}{dx^k}\right] = \frac{d}{dx}\left[(-1)^k k! \, x^{-(k+1)}\right] = (-1)^{(k+1)}(k+1)! \, x^{-(k+1)-1}.$$

63. $\dfrac{d}{dx}(uvw) = uv\dfrac{dw}{dx} + uw\dfrac{dv}{dx} + vw\dfrac{du}{dx}$

65.   (a) $f(x) = x^3 + x^2 - 4x + 1; \quad f'(x) = 3x^2 + 2x - 4.$

(b)

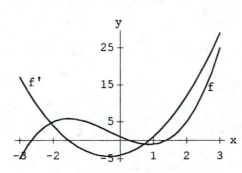

(c) The graph is "falling" when $f'(x) < 0$;

   The graph is "rising" when $f'(x) > 0.$

67.   (a)   Let $f(x) = \frac{1}{2}x^3 - 3x^2 + 4x + 1.$ (b)

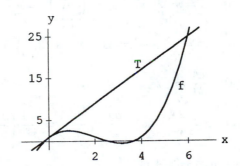

Then $f'(x) = \frac{3}{2}x^2 - 6x + 4$ and

$f'(0) = 4.$

Tangent line at $x = 0: \quad y = 4x + 1$

(c)   Solving $\frac{1}{2}x^3 - 3x^2 + 4x + 1 = 4x + 1$ for x gives $x = 6;$ the graph and the tangent line

intersect at $(6, 25).$

PROJECT 3.3

1. $g(x) = f^4(x) = f(x)f^3(x)$

   $g'(x) = f(x)[f^3(x)]' + f^3(x)f'(x) = f(x)3f^2(x)f'(x) + f^3(x)f'(x) = 4f^3(x)f'(x).$

3. We know the result holds for all positive integers. If $k = 0$, then $g(x) = [f(x)] = 1$ (provided $f(x) \neq 0$) and $g'(x) = 0 = 0[f(x)]^{-1}$. If k is a negative integer, then

$$g(x) = \frac{1}{f^n(x)}, \quad (f(x) \neq 0)$$

where $n = -k$ is a positive integer. Thus

$$g'(x) = \frac{-nf^{n-1}(x)f'(x)}{f^{2n}(x)} \text{ (reciprocal rule) } = \frac{-nf'(x)}{f^{n+1}} = -nf^{-n-1}(x)f'(x) = kf^{k-1}(x)f'(x).$$

Thus the result holds for all integers n.

SECTION 3.4

1. $A = \pi r^2$, $\dfrac{dA}{dr} = 2\pi r$. When $r = 2$, $\dfrac{dA}{dr} = 4\pi$. **3.** $A = \dfrac{1}{2}z^2$, $\dfrac{dA}{dz} = z$. When $z = 4$, $\dfrac{dA}{dz} = 4$.

5. $y = \dfrac{1}{x(1+x)}$, $\dfrac{dy}{dx} = \dfrac{-(2x+1)}{x^2(1+x)^2}$. At $x = 2$, $\dfrac{dy}{dx} = -\dfrac{5}{36}$.

7. $V = \dfrac{4}{3}\pi r^3$, $\dfrac{dV}{dr} = 4\pi r^2 = $ the surface area of the ball.

9. $y = 2x^2 + x - 1$, $\dfrac{dy}{dx} = 4x + 1$; $\dfrac{dy}{dx} = 4$ at $x = \frac{3}{4}$. Therefore $x_0 = \frac{3}{4}$.

11. (a) $w = s\sqrt{2}$, $V = s^3 = \left(\dfrac{w}{\sqrt{2}}\right)^3 = \dfrac{\sqrt{2}}{4}w^3$, $\dfrac{dV}{dw} = \dfrac{3\sqrt{2}}{4}w^2$.

(b) $z^2 = s^2 + w^2 = 3s^2$, $z = s\sqrt{3}$. $V = s^3 = \left(\dfrac{z}{\sqrt{3}}\right)^3 = \dfrac{\sqrt{3}}{9}z^3$, $\dfrac{dV}{dz} = \dfrac{\sqrt{3}}{3}z^2$.

13. (a) $\dfrac{dA}{d\theta} = \dfrac{1}{2}r^2$ (b) $\dfrac{dA}{dr} = r\theta$

(c) $\theta = \dfrac{2A}{r^2}$ so $\dfrac{d\theta}{dr} = \dfrac{-4A}{r^3} = \dfrac{-4}{r^3}\left(\dfrac{1}{2}r^2\theta\right) = \dfrac{-2\theta}{r}$

15. $y = ax^2 + bx + c$, $\qquad\qquad\qquad\qquad z = bx^2 + ax + c$.

$\dfrac{dy}{dx} = 2ax + b$, $\qquad\qquad\qquad\qquad \dfrac{dz}{dx} = 2bx + a$.

$\dfrac{dy}{dx} = \dfrac{dz}{dx}$ iff $2ax + b = 2bx + a$. With $a \neq b$, this occurs only at $x = \dfrac{1}{2}$.

17. $x(5) = -6$; $v(t) = 3 - 2t$ so $v(5) = -7$ and speed $= 7$; $a(t) = -2$ so $a(5) = -2$.

19. $x(1) = 6$; $v(t) = -18/(t+2)^2$ so $v(1) = -2$ and speed $= 2$,

$a(t) = 36/(t+2)^3$ so $a(1) = 4/3$.

21. $x(1) = 0$, $v(t) = 4t^3 + 18t^2 + 6t - 10$ so $v(1) = 18$ and speed $= 18$,

$a(t) = 12t^2 + 36t + 6$ so $a(1) = 54$.

23. $v(t) = 3t^2 - 6t + 3 = 3(t-1)^2 \geq 0$; the object never changes direction.

25. $v(t) = 1 - \dfrac{5}{(t+2)^2}$;   the object changes direction (from left to right) at $t = -2 + \sqrt{5}$.

27. A **29.** A **31.** A and B **33.** A **35.** A and C

37. The object is moving right when $v(t) > 0$. Here,

$$v(t) = 4t^3 - 36t^2 + 56t = 4t(t-2)(t-7); \quad v(t) > 0 \text{ when } 0 < t < 2 \text{ and } 7 < t.$$

39. The object is speeding up when $v(t)$ and $a(t)$ have the same sign.

$v(t) = 5t^3(4-t)$ sign of $v(t)$:

$a(t) = 20t^2(3-t)$ sign of $a(t)$:

Thus, $0 < t < 3$ and $4 < t$.

41. The object is moving left and slowing down when $v(t) < 0$ and $a(t) > 0$.

$v(t) = 3(t-5)(t+1)$ sign of $v(t)$:

$a(t) = 6(t-2)$ sign of $a(t)$:

Thus, $2 < t < 5$.

43. The object is moving right and speeding up when $v(t) > 0$ and $a(t) > 0$.

$v(t) = 4t(t-2)(t-4)$ sign of $v(t)$:

$a(t) = 4(3t^2 - 12t + 8)$ sign of $a(t)$:

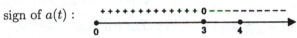

Thus, $0 < t < 2 - \tfrac{2}{3}\sqrt{3}$ and $4 < t$.

45. Since $v_0 = 0$ the equation of motion is

$$y(t) = -16t^2 + y_0.$$

We want to find y_0 so that $y(6) = 0$. From

$$0 = -16(6)^2 + y_0$$

we get $y_0 = 576$ feet.

47. The object's height and velocity at time t are given by

$$y(t) = -\frac{1}{2}gt^2 + v_0 t \quad \text{and} \quad v(t) = -gt + v_0$$

Since the object's velocity at its maximum height is 0, it takes v_0/g seconds to reach

maximum height, and

$$y(v_0/g) = -\tfrac{1}{2}g(v_0/g)^2 + v_0(v_0/g) = v_0^2/2g \quad \text{or} \quad v_0^2/19.6 \text{ (meters)}$$

49. At time t, the object's height is $y(t) = -\frac{1}{2}gt^2 + v_0 t + y_0$, and its velocity is $v(t) = -gt + v_0$. Suppose that $y(t_1) = y(t_2)$, $t_1 \neq t_2$. Then

$$-\tfrac{1}{2}\,gt_1^2 + v_0 t_1 + y_0 = -\tfrac{1}{2}\,gt_2^2 + v_0 t_2 + y_0$$

$$\tfrac{1}{2}\,g(t_2^2 - t_1^2) = v_0(t_2 - t_1)$$

$$gt_2 + gt_1) = 2v_0$$

From this equation, we get $-(-gt_1 + v_0) = -gt_2 + v_0$ and so $|v(t_1)| = |v(t_2)|$.

51.   In the equation

$$y(t) = -16t^2 + v_0 t + y_0$$

we take $v_0 = -80$ and $y_0 = 224$. The ball first strikes the ground when

$$-16t^2 - 80t + 224 = 0;$$

that is, at $t = 2$. Since

$$v(t) = y'(t) = -32t - 80,$$

we have $v(2) = -144$ so that the speed of the ball the first time it strikes the ground is 144 ft/sec. Thus, the speed of the ball the third time it strikes the ground is $\tfrac{1}{4}\left[\tfrac{1}{4}(144)\right] = 9$ ft/sec.

53.   The equation is $y(t) = -16t^2 + 32t$.   (Here $y_0 = 0$ and $v_0 = 32$.)

(a)   We solve $y(t) = 0$ to find that the stone strikes the ground at $t = 2$ seconds.

(b)   The stone attains its maximum height when $v(t) = 0$. Solving

$$v(t) = -32t + 32 = 0, \quad \text{we get } t = 1 \quad \text{and, thus, the maximum height is } y(1) = 16 \text{ feet.}$$

(c)   We want to choose v_0 in

$$y(t) = -16t^2 + v_0 t$$

so that $y(t_0) = 36$ when $v(t_0) = 0$ for some time t_0.

From $v(t) = -32t + v_0 = 0$ we get $t_0 = v_0/32$ so that

$$-16\left(\frac{v_0}{32}\right)^2 + v_0\left(\frac{v_0}{32}\right) = 36, \quad \text{or} \quad \frac{v_0^2}{64} = 36.$$

Thus, $v_0 = 48$ ft/sec.

55.   For all three parts of the problem the basic equation is

$$y(t) = -16t^2 + v_0 t + y_0$$

with

(∗) $y(t_0) = 100 \quad \text{and} \quad y(t_0 + 2) = 16$

for some time $t_0 > 0$.

We are asked to find y_0 for a given value of v_0.

From $(*)$ we get

$$16 - 100 = y(t_0 + 2) - y(t_0)$$

$$= [-16(t_0 + 2)^2 + v_0(t_0 + 2) + y_0] - [-16t_0{}^2 + v_0 t_0 + y_0]$$

$$= -64t_0 - 64 + 2v_0$$

so that

$$t_0 = \tfrac{1}{32}(v_0 + 10).$$

Substituting this result in the basic equation and noting that $y(t_0) = 100$, we have

$$-16\left(\frac{v_0 + 10}{32}\right)^2 + v_0\left(\frac{v_0 + 10}{32}\right) + y_0 = 100$$

and therefore

$(**)$ $$y_0 = 100 - \frac{v_0{}^2}{64} + \frac{25}{16}.$$

We use $(**)$ to find the answer to each part of the problem.

   (a) $v_0 = 0$ so $y_0 = \frac{1625}{16}$ ft   (b) $v_0 = -5$ so $y_0 = \frac{6475}{64}$ ft   (c) $v_0 = 10$ so $y_0 = 100$ ft

57.   Let $y/_0 > 0$ be the initial height. The equation of motion becomes:

$$0 = -16(8)^2 + 5(8) + y_0, \quad \text{so } y_0 = 984 \text{ ft.}$$

59. $C(x) = 200 + 0.02x + 0.0001x^2, \quad C'(x) = 0.02 + 0.002x$

Marginal cost at $x = 100$ units: $C'(100) = 0.04$

Actual cost of 101st unit: $C(101) - C(100) = 0.0401$

61. $C(x) = 200 + 0.01x + \dfrac{100}{x}, \quad C'(x) = 0.01 - \dfrac{100}{x^2}$

Marginal cost at $x = 100$ units: $C'(100) = 0$

Actual cost of producing the 101st unit: $C(101) - C(100) = 0$

63. $C(x) = 1000 + 25x - \dfrac{x^2}{10}, \quad C'(x) = 25 - \dfrac{x}{5}$

Marginal cost of producing 10 motors: $C'(10) = 23$

Actual cost of producing the 10th motor: $C(11) - C(10) = 22.90$

65.   (a) Profit function: $P(x) = R(x) - C(x) = 20x - \dfrac{x^2}{50} - (4x + 1400) = 16x - \dfrac{x^2}{50} - 1400.$

      Break-even points: $16x - \dfrac{x^2}{50} - 1400 = 0$ so $x^2 - 800x + 70{,}000 = 0$

      Thus $x = 100$, or $x = 700$ units.

   (b) $P'(x) = 16 - \dfrac{x}{25}; \quad P'(x) = 0 \implies x = 400$ units.

(c)

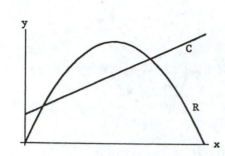

67. (a) $v(t) = 3t^2 - 14t + 10, \ 0 \le t \le 5$

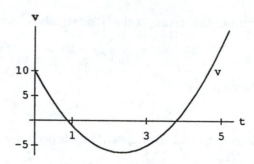

(b) The object is moving to the right when $0 < t < 0.88$ and when $3.79 < t < 5$.

The object is moving to the left when $0.88 < t < 3.79$

(c)

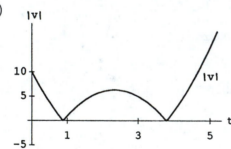

The object stops at times $t \cong 0.88$ and $t \cong 3.79$.

The maximum speed is $v \cong 6.33$ at $t \cong 2.33$.

(d) $a(t) = 6t - 14$

The object is speeding up when $v(t)$ and $a(t)$ have the same sign: $0.88 < t < 2.33$ and $3.79 < t < 5$.

The object is slowing down when $v(t)$ and $a(t)$ have opposite sign: $0 < t < 0.88$ and $2.33 < t < 3.79$.

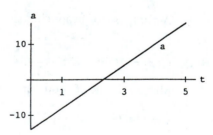

SECTION 3.5

1. $f(x) = x^4 + 2x^2 + 1, \quad f'(x) = 4x^3 + 4x = 4x(x^2 + 1)$

$f(x) = (x^2 + 1)^2, \quad f'(x) = 2(x^2 + 1)(2x) = 4x(x^2 + 1)$

3. $f(x) = 8x^3 + 12x^2 + 6x + 1, \quad f'(x) = 24x^2 + 24x + 6 = 6(2x + 1)^2$

$$f(x) = (2x+1)^3, \quad f'(x) = 3(2x+1)^2(2) = 6(2x+1)^2$$

5. $\quad f(x) = x^2 + 2 + x^{-2}, \quad f'(x) = 2x - 2x^{-3} = 2x(1 - x^{-4})$

$\qquad f(x) = (x + x^{-1})^2, \quad f'(x) = 2(x + x^{-1})(1 - x^{-2}) = 2x(1 + x^{-2})(1 - x^{-2}) = 2x(1 - x^{-4})$

7. $\quad f'(x) \quad = -1(1 - 2x)^{-2} \dfrac{d}{dx}(1 - 2x) = 2(1 - 2x)^{-2}$

9. $\quad f'(x) \quad = 20(x^5 - x^{10})^{19} \dfrac{d}{dx}(x^5 - x^{10}) = 20(x^5 - x^{10})^{19}(5x^4 - 10x^9)$

11. $\quad f'(x) \quad = 4\left(x - \dfrac{1}{x}\right)^3 \dfrac{d}{dx}\left(x - \dfrac{1}{x}\right) = 4\left(x - \dfrac{1}{x}\right)^3\left(1 + \dfrac{1}{x^2}\right)$

13. $\quad f'(x) \quad = 4(x - x^3 - x^5)^3 \dfrac{d}{dx}(x - x^3 - x^5) = 4(x - x^3 - x^5)^3(1 - 3x^2 - 5x^4)$

15. $\quad f'(t) \quad = 100(t^2 - 1)^{99} \dfrac{d}{dt}(t^2 - 1) = 200t(t^2 - 1)^{99}$

17. $\quad f'(t) \quad = 4(t^{-1} + t^{-2})^3 \dfrac{d}{dt}(t^{-1} + t^{-2}) = 4(t^{-1} + t^{-2})^3(-t^{-2} - 2t^{-3})$

19. $\quad f'(x) \quad = 4\left(\dfrac{3x}{x^2 + 1}\right)^3 \dfrac{d}{dx}\left(\dfrac{3x}{x^2 + 1}\right) = 4\left(\dfrac{3x}{x^2 + 1}\right)^3\left[\dfrac{(x^2 + 1)3 - 3x(2x)}{(x^2 + 1)^2}\right] = \dfrac{324x^3(1 - x^2)}{(x^2 + 1)^5}$

21. $\quad f'(x) \quad = -\left(\dfrac{x^3}{3} + \dfrac{x^2}{2} + \dfrac{x}{1}\right)^{-2} \dfrac{d}{dx}\left(\dfrac{x^3}{3} + \dfrac{x^2}{2} + \dfrac{x}{1}\right) = -\left(\dfrac{x^3}{3} + \dfrac{x^2}{2} + x\right)^{-2}(x^2 + x + 1)$

23. $\quad \dfrac{dy}{dx} = \dfrac{dy}{du}\dfrac{du}{dx} = \dfrac{-2u}{(1 + u^2)^2} \cdot (2)$

$\qquad$ At $x = 0$, we have $u = 1$ and thus $\quad \dfrac{dy}{dx} = \dfrac{-4}{4} = -1$.

25. $\quad \dfrac{dy}{dx} = \dfrac{dy}{du}\dfrac{du}{dx} = \dfrac{(1 - 4u)2 - 2u(-4)}{(1 - 4u)^2} \cdot 4(5x^2 + 1)^3(10x) = \dfrac{2}{(1 - 4u)^2} \cdot 40x(5x^2 + 1)^3$

$\qquad$ At $x = 0$, we have $u = 1$ and thus $\dfrac{dy}{dx} = \dfrac{2}{9}(0) = 0$.

27. $\quad \dfrac{dy}{dt} = \dfrac{dy}{du}\dfrac{du}{dx}\dfrac{dx}{dt} = \dfrac{(1 + u^2)(-7) - (1 - 7u)(2u)}{(1 + u^2)^2}(2x)(2)$

$\qquad\qquad = \dfrac{7u^2 - 2u - 7}{(1 + u^2)^2}(4x) = \dfrac{4x(7x^4 + 12x - 2)}{(x^4 + 2x^2 + 2)^2} = \dfrac{4(2t - 5)[7(2t - 5)^4 + 12(2t - 5)^2 - 2]}{[(2t - 5)^4 + 2(2t - 5)^2 + 2]^2}$

29. $\quad \dfrac{dy}{dx} = \dfrac{dy}{ds}\dfrac{ds}{dt}\dfrac{dt}{dx} = 2(s + 3) \cdot \dfrac{1}{2\sqrt{t - 3}} \cdot (2x)$

$\qquad$ At $x = 2$, we have $t = 4$ so that $s = 1$ and thus $\dfrac{dy}{dx} = 2(4)\dfrac{1}{2 \cdot 1}(4) = 16$.

31. $\quad (f \circ g)'(0) = f'(g(0))g'(0) = f'(2)g'(0) = (1)(1) = 1$

33. $(f \circ g)'(2) = f'(g(2))g'(2) = f'(2)g'(2) = (1)(1) = 1$

35. $(g \circ f)'(1) = g'(f(1))f'(1) = g'(0)f'(1) = (1)(1) = 1$

37. $(f \circ h)'(0) = f'(h(0))h'(0) = f'(1)h'(0) = (1)(2) = 2$

39. $(g \circ f \circ h)'(2) = g'(f(h(2)))\ f'(h(2))h'(2) = g'(1)f'(0)h'(2) = (0)(2)(2) = 0$

41. $f'(x) = 4(x^3 + x)^3(3x^2 + 1)$

$f''(x) = 3(4)(x^3 + x)^2(3x^2 + 1)^2 + 4(x^3 + x)^3(6x) = 12(x^3 + x)^2[(3x^2 + 1)^2 + 2x(x^3 + x)]$

43. $f'(x) = 3\left(\dfrac{x}{1 - x}\right)^2 \cdot \dfrac{1}{(1 - x)^2} = \dfrac{3x^2}{(1 - x)^4}$

$f''(x) = \dfrac{6x(1 - x)^4 - 3x^2(4)(1 - x)^3(-1)}{(1 - x)^8} = \dfrac{6x(1 + x)}{(1 - x)^5}$

45. $2xf'(x^2 + 1)$ **47.** $2f(x)f'(x)$

49. $f'(x) = -4x(1 + x^2)^{-3}$;   (a) $x = 0$ (b) $x < 0$ (c) $x > 0$

51. $f'(x) = \dfrac{1 - x^2}{(1 + x^2)^2}$;   (a) $x = \pm 1$ (b) $-1 < x < 1$ (c) $x < -1,\ \ x > 1$

53. $v(t) = 5(t + 1)(t - 9)^2(t - 3)$;   the object changes direction (from left to right) at $t = 3$.

55. $v(t) = 12t^3(t^2 - 12)^3(t^2 - 4)$;   the object changes direction (from right to left) at $t = 2$

and (from left to right) at $t = 2\sqrt{3}$.

57. $\dfrac{n!}{(1 - x)^{n+1}}$ **59.** $n!\,b^n$

61. $y = (x^2 + 1)^3 + C$ **63.** $y = (x^3 - 2)^2 + C$

65. $L'(x) = \dfrac{1}{x^2 + 1} \cdot 2x = \dfrac{2x}{x^2 + 1}$

67. $T'(x) = 2f(x) \cdot f'(x) + 2g(x) \cdot g'(x) = 2f(x) \cdot g(x) - 2g(x) \cdot f(x) = 0$

69.   Suppose $p(x) = (x - a)^2 q(x)$, where $q(a) \neq 0$. Then

$p'(x) = 2(x - a)q(x) + (x - a)^2 q'(x)$ and $p''(x) = 2q(x) + 4(x - a)q'(x) + (x - a)^2 q''(x)$,

and it follows that $p(a) = p'(a) = 0$, and $p''(a) \neq 0$.

Now suppose that $p(a) = p'(a) = 0$ and $p''(a) \neq 0$.

$$p(a) = 0 \quad \Longrightarrow \quad p(x) = (x - a)g(x) \quad \text{for some polynomial } g.$$

Then $p'(x) = y(x) + (x - a)g'(x)$ and

$$p'(a) = 0 \implies g(a) = 0 \text{ and so} g(x) = (x - a)q(x) \text{for some polynomial} q.$$

Therefore, $p(x) = (x - a)^2 q(x)$. Finally, $p''(a) \neq 0$ implies $q(a) \neq 0$.

71. Let p be a polynomial function of degree n. The number a is a root of p of multiplicity k, $(k < n)$ if and only if $p(a) = p'(a) = \cdots = p^{(k-1)}(a) = 0$ and $p^{(k)}(a) \neq 0$.

73. $\dfrac{dy}{dt} = \dfrac{dy}{dx} \cdot \dfrac{dx}{dt} = (3x^2 - 3)(4t - 1)$

At $t = 2$, $x(2) = 8$ and $\dfrac{dy}{dt} = [3(8)^2 - 3][4(2) - 1] = 1323$.

75. $V = \frac{4}{3}\pi r^3$ and $\dfrac{dr}{dt} = 2$ cm/sec. By the chain rule, $\dfrac{dV}{dt} = \dfrac{dV}{dr}\dfrac{dr}{dt} = 4\pi r^2 \dfrac{dr}{dt} = 8\pi r^2$.

At the instant the radius is 10 centimeters, the volume is increasing at the rate

$$\dfrac{dV}{dt} = 8\pi(10)^2 = 800\pi \ \text{cm}^3/\text{sec}.$$

77. $KE = \frac{1}{2}mv^2$; $\dfrac{d(KE)}{dt} = \dfrac{d(KE)}{dv} \cdot \dfrac{dv}{dt} = mv\dfrac{dv}{dt}$.

SECTION 3.6

1. $\dfrac{dy}{dx} = -3\sin x - 4\sec x \tan x$

3. $\dfrac{dy}{dx} = 3x^2 \csc x - x^3 \csc x \cot x$

5. $\dfrac{dy}{dt} = -2\cos t \sin t$

7. $\dfrac{dy}{du} = 4\sin^3\sqrt{u}\,\dfrac{d}{du}(\sin\sqrt{u}) = 4\sin^3\sqrt{u}\,\cos\sqrt{u}\,\dfrac{d}{du}(\sqrt{u}) = 2u^{-1/2}\sin^3\sqrt{u}\,\cos\sqrt{u}$

9. $\dfrac{dy}{dx} = \sec^2 x^2 \dfrac{d}{dx}(x^2) = 2x\sec^2 x^2$

11. $\dfrac{dy}{dx} = 4[x + \cot\pi x]^3[1 - \pi\csc^2\pi x]$

13. $\dfrac{dy}{dx} = \cos x, \ \dfrac{d^2y}{dx^2} = -\sin x$

15. $\dfrac{dy}{dx} = \dfrac{(1 + \sin x)(-\sin x) - \cos x\,(\cos x)}{(1 + \sin x)^2} = \dfrac{-\sin x - (\sin^2 x + \cos^2 x)}{(1 + \sin x)^2} = -(1 + \sin x)^{-1}$

$\dfrac{d^2y}{dx^2} = (1 + \sin x)^{-2}\dfrac{d}{dx}(1 + \sin x) = \cos x\,(1 + \sin x)^{-2}$

17. $\dfrac{dy}{du} = 3\cos^2 2u\,\dfrac{d}{du}(\cos 2u) = -6\cos^2 2u \sin 2u$

$\dfrac{d^2y}{du^2} = -6[\cos^2 2u\,\dfrac{d}{du}(\sin 2u) + \sin 2u\,\dfrac{d}{du}(\cos^2 2u)]$

$\quad = -6[2\cos^3 2u + \sin 2u\,(-4\cos 2u \sin 2u)] = 12\cos 2u\,[2\sin^2 2u - \cos^2 2u]$

19. $\dfrac{dy}{dt} = 2\sec^2 2t, \ \dfrac{d^2y}{dt^2} = 4\sec 2t\,\dfrac{d}{dt}(\sec 2t) = 8\sec^2 2t \tan 2t$

21. $\dfrac{dy}{dx} = x^2(3\cos 3x) + 2x\sin 3x$

$$\frac{d^2y}{dx^2} = [x^2(-9\sin 3x) + 2x(3\cos 3x)] + [2x(3\cos 3x) + 2(\sin 3x)]$$

$$= (2 - 9x^2)\sin 3x + 12x\cos 3x$$

23. $y = \sin^2 x + \cos^2 x = 1$ so $\dfrac{dy}{dx} = \dfrac{d^2y}{dx^2} = 0$

25. $\dfrac{d^4}{dx^4}(\sin x) = \dfrac{d^3}{dx^3}(\cos x) = \dfrac{d^2}{dx^2}(-\sin x) = \dfrac{d}{dx}(-\cos x) = \sin x$

27. $\dfrac{d}{dt}\left[t^2\dfrac{d^2}{dt^2}(t\cos 3t)\right] = \dfrac{d}{dt}\left[t^2\dfrac{d}{dt}(\cos 3t - 3t\sin 3t)\right]$

$$= \frac{d}{dt}[t^2(-3\sin 3t - 3\sin 3t - 9t\cos 3t)] = \frac{d}{dt}[-6t^2\sin 3t - 9t^3\cos 3t]$$

$$= (-18t^2\cos 3t - 12t\sin 3t) + (27t^3\sin 3t - 27t^2\cos 3t) = (27t^3 - 12t)\sin 3t - 45t^2\cos 3t$$

29. $\dfrac{d}{dx}[f(\sin 3x)] = f'(\sin 3x)\dfrac{d}{dx}(\sin 3x) = 3\cos 3x\, f'(\sin 3x)$

31. $\dfrac{dy}{dx} = \cos x;$ slope of tangent at $(0,0)$ is 1; tangent: $y = x.$

33. $\dfrac{dy}{dx} = -\csc^2 x;$ slope of tangent at $(\frac{\pi}{6}, \sqrt{3})$ is $-4,$ an equation for
tangent: $y - \sqrt{3} = -4\left(x - \dfrac{\pi}{6}\right).$

35. $\dfrac{dy}{dx} = \sec x \tan x,$ slope of tangent at $(\frac{\pi}{4}, \sqrt{2})$ is $\sqrt{2},$ an equation for
tangent is $y - \sqrt{2} = \sqrt{2}\left(x - \dfrac{\pi}{4}\right).$

37. $\dfrac{dy}{dx} = -\sin x;$ $x = \pi$

39. $\dfrac{dy}{dx} = \cos x - \sqrt{3}\sin x;$ $\dfrac{dy}{dx} = 0$ gives $\tan x = \dfrac{1}{\sqrt{3}};$ $x = \dfrac{\pi}{6}, \dfrac{7\pi}{6}$

41. $\dfrac{dy}{dx} = 2\sin x \cos x = \sin 2x;$ $x = \dfrac{\pi}{2}, \pi, \dfrac{3\pi}{2}$

43. $\dfrac{dy}{dx} = \sec^2 x - 2;$ $\dfrac{dy}{dx} = 0$ gives $\sec x = \pm\sqrt{2};$ $x = \dfrac{\pi}{4}, \dfrac{3\pi}{4}, \dfrac{5\pi}{4}, \dfrac{7\pi}{4}$

45. $\dfrac{dy}{dx} = 2\sec x \tan x + \sec^2 x;$ since $\sec x$ is never zero, $\dfrac{dy}{dx} = 0$ gives

$2\tan x + \sec x = 0$ so that $\sin x = -1/2;$ $x = \dfrac{7\pi}{6}, \dfrac{11\pi}{6}$

47.   We want $v(t) > 0$ and $a(t) > 0.$

$v(t) = 3\cos 3t$ 　　　　　　　　　　 sign of $v(t)$:

$a(t) = -9\sin 3t$ 　　　　　　　　　　 sign of $a(t)$:

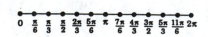

Thus, $\pi < t < \dfrac{2\pi}{3},$ $\dfrac{7\pi}{6} < t < \dfrac{4\pi}{3},$ $\dfrac{11\pi}{6} < t < 2\pi.$

49.   We want $v(t) > 0$ and $a(t) > 0.$

$v(t) = \cos t + \sin t$

sign of $v(t)$:

$a(t) = -\sin t + \cos t$

sign of $a(t)$:

Thus, $0 < t < \dfrac{\pi}{4}$ and $\dfrac{7\pi}{4} < t < 2\pi.$

51.   We want $v(t) > 0$ and $a(t) > 0.$

$v(t) = 1 - 2\sin t$

sign of $v(t)$:

$a(t) = -2\cos t$

sign of $a(t)$:

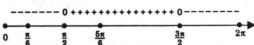

Thus, $\dfrac{5\pi}{6} < t < \dfrac{3\pi}{2}.$

53.   (a) $\dfrac{dy}{dt} = \dfrac{dy}{du}\dfrac{du}{dx}\dfrac{dx}{dt} = (2u)(\sec x \tan x)\pi = 2\pi \sec^2 \pi t \tan \pi t$

(b) $y = \sec^2 \pi t - 1, \quad \dfrac{dy}{dt} = 2\sec \pi t\,(\sec \pi t \tan \pi t)\pi = 2\pi \sec^2 \pi t \tan \pi t$

55.   (a) $\dfrac{dy}{dt} = \dfrac{dy}{du}\dfrac{du}{dx}\dfrac{dx}{dt} = 4\left[\dfrac{1}{2}(1-u)\right]^3\left(-\dfrac{1}{2}\right)(-\sin x)(2) = 4\left[\dfrac{1}{2}(1-\cos 2t)\right]^3 \sin 2t$

$$= 4\sin^6 t\,(2\sin t \cos t) = 8\sin^7 t \cos t$$

(b) $y = \left[\dfrac{1}{2}(1-\cos 2t)\right]^4 = \sin^8 t, \quad \dfrac{dy}{dt} = 8\sin^7 t \cos t$

57. $\dfrac{d^n}{dx^n}(\cos x) = \begin{cases} (-1)^{(n+1)/2}\sin x, & n \text{ odd} \\ (-1)^{n/2}\cos x, & n \text{ even} \end{cases}$

59. $\dfrac{d}{dx}(\cos x) = \dfrac{d}{dx}\left(\sin(\dfrac{\pi}{2}x)\right) = -\cos(\dfrac{\pi}{2}x) = -\sin x.$

61. $f'(0) = \lim_{h\to 0}\dfrac{\sin(0+h) - \sin 0}{h} = \lim_{h\to 0}\dfrac{\sin h}{h} = \lim_{x\to 0}\dfrac{\sin x}{x}$

63. $f(x) = 2\sin x + 3\cos x + C$ **65.** $f(x) = \sin 2x + \sec x + C$

67. $f(x) = \sin(x^2) + \cos 2x + C$

69.   (a) $f'(x) = \sin(1/x) + x\cos(1/x)(-1/x^2) = \sin(1/x) - (1/x)\cos(1/x)$

$$g'(x) = 2x \sin(1/x) + x^2 \cos(1/x)(-1/x^2) = 2x \sin(1/x) - \cos(1/x)$$

(b) $\lim\limits_{x \to 0} g'(x) = \lim\limits_{x \to 0} [2x \sin(1/x) - \cos(1/x)] = -\lim\limits_{x \to 0} \cos(1/x)$ does not exist

71.   (a) Continuity:

$$\lim_{x \to (2\pi/3)-} \sin x = \frac{\sqrt{3}}{2}, \quad \lim_{x \to (2\pi/3)-} (ax + b) = \frac{2\pi a}{3} + b; \quad \text{thus} \quad \frac{2\pi a}{3} + b = \frac{\sqrt{3}}{2}$$

Differentiability:

$$\lim_{x \to (2\pi/3)+} \cos x = -\frac{1}{2}, \quad \lim_{x \to (2\pi/3)+} (a) = a; \quad \text{thus} \quad a = -\frac{1}{2}$$

Therefore, f is differentiable at $2\pi/3$ if $a = -\frac{1}{2}$ and $b = \frac{1}{2}\sqrt{3} + \frac{1}{3}\pi$

(b)

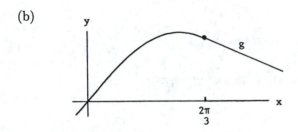

73.   Let $y(t) = A \sin \omega t + B \cos \omega t$. Then

$$y'(t) = \omega A \cos \omega t - \omega B \sin \omega t \quad \text{and} \quad y''(t) = -\omega^2 A \sin \omega t - \omega^2 B \cos \omega t$$

Thus,

$$\frac{d^2 y}{dt^2} + \omega^2 y = 0.$$

75. $A = \frac{1}{2} c^2 \sin x; \quad \dfrac{dA}{dx} = \frac{1}{2} c^2 \cos x$

77.   (a)

θ	5	1	0.1	0.01	0.001
$\dfrac{\sin \theta}{\theta}$	0.01743	0.01745	0.01745	0.01745	0.01745

(b) $\dfrac{\pi}{180} \cong 0.01745$

79.   (a)

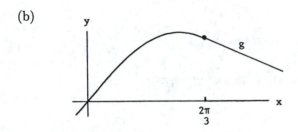

(b) $f(x) = 0$ at $x = 0$ and $x \cong 0.81$

(c) $f'(x) = 0$ at $x \cong -1.25, \quad x \cong -0.68,$ and $x \cong 0.43$

81.   (a)   (b)   (c)

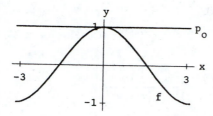
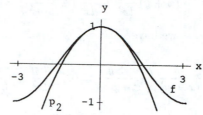
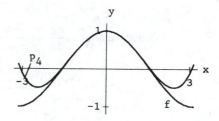

SECTION 3.7

1. $x^2 + y^2 = 4$ **3.** $4x^2 + 9y^2 = 36$

$$2x + 2y\frac{dy}{dx} = 0$$ $$8x + 18y\frac{dy}{dx} = 0$$

$$\frac{dy}{dx} = \frac{-x}{y}$$ $$\frac{dy}{dx} = \frac{-4x}{9y}$$

5. $x^4 + 4x^3y + y^4 = 1$

$$4x^3 + 12x^2y + 4x^3\frac{dy}{dx} + 4y^3\frac{dy}{dx} = 0$$

$$\frac{dy}{dx} = -\frac{x^3 + 3x^2y}{x^3 + y^3}$$

7. $(x - y)^2 - y = 0$

$$2(x - y)\left(1 - \frac{dy}{dx}\right) - \frac{dy}{dx} = 0$$

$$\frac{dy}{dx} = \frac{2(x - y)}{2(x - y) + 1}$$

9. $\sin(x + y) = xy$

$$\cos(x + y)\left(1 + \frac{dy}{dx}\right) = x\frac{dy}{dx} + y$$

$$\frac{dy}{dx} = \frac{y - \cos(x + y)}{\cos(x + y) - x}$$

11. $y^2 + 2xy = 16$

$$2y\frac{dy}{dx} + 2x\frac{dy}{dx} + 2y = 0$$

$$(x + y)\frac{dy}{dx} + y = 0.$$

Differentiating a second time, we have

$$(x+y)\frac{d^2y}{dx^2} + \frac{dy}{dx}\left(2 + \frac{dy}{dx}\right) = 0.$$

Substituting $\quad \dfrac{dy}{dx} = \dfrac{-y}{x+y}, \quad$ we have

$$(x+y)\frac{d^2y}{dx^2} - \frac{y}{(x+y)}\left(\frac{2x+y}{x+y}\right) = 0, \quad \frac{d^2y}{dx^2} = \frac{2xy+y^2}{(x+y)^3} = \frac{16}{(x+y)^3}.$$

13.
$$y^2 + xy - x^2 = 9$$

$$2y\frac{dy}{dx} + x\frac{dy}{dx} + y - 2x = 0.$$

Differentiating a second time, we have

$$\left[2\left(\frac{dy}{dx}\right)^2 + 2y\frac{d^2y}{dx^2}\right] + \left[x\frac{d^2y}{dx^2} + \frac{dy}{dx}\right] + \frac{dy}{dx} - 2 = 0$$

$$(2y+x)\frac{d^2y}{dx^2} + 2\left[\left(\frac{dy}{dx}\right)^2 + \frac{dy}{dx} - 1\right] = 0.$$

Substituting $\quad \dfrac{dy}{dx} = \dfrac{2x-y}{2y+x}, \quad$ we have

$$(2y+x)\frac{d^2y}{dx^2} + 2\left[\frac{(2x-y)^2 + (2x-y)(2y+x) - (2y+x)^2}{(2y+x)^2}\right] = 0$$

$$\frac{d^2y}{dx^2} = \frac{10(y^2+xy-x^2)}{(2y+x)^3} = \frac{90}{(2y+x)^3}.$$

15.
$$4\tan y = x^3$$

$$4\sec^2 y\,\frac{dy}{dx} = 3x^2$$

$$\frac{dy}{dx} = \frac{3}{4}x^2\cos^2 y$$

$$\frac{d^2y}{dx^2} = \frac{3}{2}x\cos^2 y + \frac{3}{4}x^2\left(2\cos y(-\sin y)\frac{dy}{dx}\right)$$

$$= \frac{3}{2}x\cos^2 y - \frac{9}{8}x^4\sin y\cos^3 y$$

17. $\quad x^2 - 4y^2 = 9, \quad 2x - 8y\dfrac{dy}{dx} = 0.$

At $(5, 2)$, we get $\quad \dfrac{dy}{dx} = \dfrac{5}{8}. \quad$ Then,

$$2 - 8\left[y\frac{d^2y}{dx^2} + \left(\frac{dy}{dx}\right)^2\right] = 0.$$

At $(5, 2)$ we get

$$2 - 8\left[2\frac{d^2y}{dx^2} + \frac{25}{64}\right] = 0 \quad \text{so that} \quad \frac{d^2y}{dx^2} = -\frac{9}{128}.$$

19. $\cos(x + 2y) = 0 \qquad -\sin(x + 2y)\left(1 + 2\frac{dy}{dx}\right) = 0.$

At $(\pi/6, \pi/6)$,   we get $\frac{dy}{dx} = -1/2.$ Then,

$$-\cos(x + 2y)\left(1 + 2\frac{dy}{dx}\right)^2 - \sin(x + 2y)\left(2\frac{d^2y}{dx^2}\right) = 0.$$

At $(\pi/6, \pi/6)$,   we get

$$-\cos\frac{\pi}{2}(0)^2 - \sin\frac{\pi}{2}\left(2\frac{d^2y}{dx^2}\right) = 0 \quad \text{so that} \quad \frac{d^2y}{dx^2} = 0.$$

21. $\qquad 2x + 3y = 5$

$$2 + 3\frac{dy}{dx} = 0$$

slope of tangent at $(-2, 3)$: $-2/3$

tangent: $y - 3 = -\frac{2}{3}(x + 2)$

normal: $y - 3 = \frac{3}{2}(x + 2)$

23. $\qquad x^2 + xy + 2y^2 = 28$

$$2x + x\frac{dy}{dx} + y + 4y\frac{dy}{dx} = 0$$

slope of tangent at $(-2, -3)$: $-1/2$

tangent: $y + 3 = -\frac{1}{2}(x + 2)$

normal: $y + 3 = 2(x + 2)$

25.
$$x = \cos y$$

$$1 = -\sin y\frac{dy}{dx}$$

slope of tangent at $\left(\frac{1}{2}, \frac{\pi}{3}\right)$: $\frac{-2}{\sqrt{3}}$

tangent: $y - \frac{\pi}{3} = -\frac{2}{\sqrt{3}}\left(x - \frac{1}{2}\right)$

normal: $y - \frac{\pi}{3} = \frac{\sqrt{3}}{2}\left(x - \frac{1}{2}\right)$

27. $\frac{dy}{dx} = \frac{1}{2}(x^3 + 1)^{-1/2}\frac{d}{dx}(x^3 + 1) = \frac{3}{2}x^2(x^3 + 1)^{-1/2}$

29. $\frac{dy}{dx} = x\left(\frac{1}{2}(x^2 + 1)^{-1/2}(2x)\right) + (x^2 + 1)^{1/2} = (1 + 2x^2)(x^2 + 1)^{-1/2}$

31. $\frac{dy}{dx} = \frac{1}{4}(2x^2 + 1)^{-3/4}\frac{d}{dx}(2x^2 + 1) = x(2x^2 + 1)^{-3/4}$

33. $\dfrac{dy}{dx} = \sqrt{2-x^2}\left[\dfrac{-x}{\sqrt{3-x^2}}\right] + \sqrt{3-x^2}\left[\dfrac{-x}{\sqrt{2-x^2}}\right] = \dfrac{x(2x^2-5)}{\sqrt{2-x^2}\,\sqrt{3-x^2}}$

35. $\dfrac{d}{dx}\left(\sqrt{x} + \dfrac{1}{\sqrt{x}}\right) = \dfrac{d}{dx}\left(x^{1/2} + x^{-1/2}\right) = \dfrac{1}{2}x^{-1/2} - \dfrac{1}{2}x^{-3/2} = \dfrac{1}{2}x^{-3/2}(x-1)$

37. $\dfrac{d}{dx}\left(\dfrac{x}{\sqrt{x^2+1}}\right) = \dfrac{d}{dx}\left(x(x^2+1)^{-1/2}\right)$

$$= x\left(-\dfrac{1}{2}(x^2+1)^{-3/2}(2x)\right) + (x^2+1)^{-1/2} = (x^2+1)^{-3/2}$$

39. $\dfrac{d}{dx}\left(x^{1/3} + x^{-1/3}\right) = \dfrac{1}{3}x^{-2/3} - \dfrac{1}{3}x^{-4/3} = \dfrac{1}{3}x^{-4/3}(x^{2/3} - 1)$

41. (a)   (b)   (c)

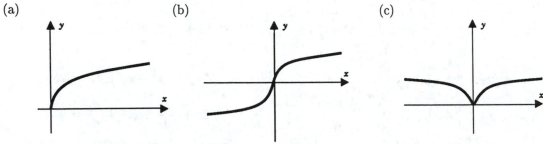

43. $y = (a+bx)^{1/3};\quad \dfrac{dy}{dx} = \dfrac{b}{3}(a+bx)^{-2/3};\quad \dfrac{d^2y}{dx^2} = \dfrac{-2b^2}{9}(a+bx)^{-5/3}$

45. $y = \sqrt{x}\,\tan\sqrt{x};\quad \dfrac{dy}{dx} = \dfrac{1}{2\sqrt{x}}\tan\sqrt{x} + \sqrt{x}\,\sec^2\sqrt{x}\left(\dfrac{1}{2\sqrt{x}}\right) = \dfrac{1}{2\sqrt{x}}\tan\sqrt{x} + \dfrac{1}{2}\sec^2\sqrt{x}$

$\dfrac{d^2y}{dx^2} = \dfrac{2\sqrt{x}\,\sec^2\sqrt{x}\,(1/2\sqrt{x}) - \tan\sqrt{x}\,(1/\sqrt{x})}{4x} + \sec\sqrt{x}\,\sec\sqrt{x}\,\tan\sqrt{x}\,(1/2\sqrt{x})$

$= \dfrac{\sqrt{x}\,\sec^2\sqrt{x} - \tan\sqrt{x} + 2x\,\sec^2\sqrt{x}\,\tan\sqrt{x}}{4x\sqrt{x}}$

47. Differentiation of $x^2 + y^2 = r^2$ gives $2x + 2y\dfrac{dy}{dx} = 0$ so that the slope of the normal line is

$$\dfrac{-1}{dy/dx} = \dfrac{y}{x}\quad (x \neq 0).$$

Let (x_0, y_0) be a point on the circle. Clearly, if $x_0 = 0$, the normal line, $x = 0$, passes through the origin. If $x_0 \neq 0$, the normal line is

$$y - y_0 = \dfrac{y_0}{x_0}(x - x_0),\quad \text{which simplifies to}\quad y = \dfrac{y_0}{x_0}x,$$

a line through the origin.

49. For the parabola $y^2 = 2px + p^2$, we have $2y\dfrac{dy}{dx} = 2p$ and the slope of a tangent is given by $m_1 = p/y$.

For the parabola $y^2 = p^2 - 2px$, we obtain $m_2 = -p/y$ as the slope of a tangent. The parabolas intersect at the points $(0, \pm p)$. At each of these points $m_1 m_2 = -1$; the parabolas intersect at right angles.

51. For $y = x^2$ we have $m_1 = \dfrac{dy}{dx} = 2x$; for $x = y^3$ we have $3y^2 \dfrac{dy}{dx} = 1$ or $m_2 = \dfrac{dy}{dx} = 1/3y^2$.

At $(1, 1)$, $m_1 = 2$, $m_2 = 1/3$ and
$$\tan \alpha = \left| \frac{m_1 - m_2}{1 - m_1 m_2} \right| = \left| \frac{2 - (1/3)}{1 + 2(1/3)} \right| = 1 \quad \Rightarrow \quad \alpha = \frac{\pi}{4}$$

At $(0, 0)$, $m_1 = 0$ and m_2 is undefined. Thus $\alpha = \pi/2$.

53. The hyperbola and the ellipse intersect at the points $(\pm 3, \pm 2)$. For the hyperbola, $\dfrac{dy}{dx} = \dfrac{x}{y}$ and for the ellipse $\dfrac{dy}{dx} = -\dfrac{4x}{9y}$. The product of these slopes is $-\dfrac{4x^2}{9y^2}$. This product is -1 at each of the points of intersection. Therefore the hyperbola and ellipse are orthogonal.

55. For the circles, $\dfrac{dy}{dx} = -\dfrac{x}{y}$, $y \neq 0$, and for the straight lines, $\dfrac{dy}{dx} = m = \dfrac{y}{x}$, $x \neq 0$. Since the product of the slopes is -1, it follows that the two families are orthogonal trajectories.

57. The line $x + 2y + 3 = 0$ has slope $m = -1/2$. Thus, a line perpendicular to this line will have slope 2. A tangent line to the ellipse $4x^2 + y^2 = 72$ has slope $m = \dfrac{dy}{dx} = -\dfrac{4x}{y}$. Setting $-\dfrac{4x}{y} = 2$ gives $y = -2x$. Substituting into the equation for the ellipse, we have
$$4x^2 + 4x^2 = 72 \quad \Rightarrow \quad 8x^2 = 72 \quad \Rightarrow \quad x = \pm 3$$

It now follows that $y = \mp 6$ and the equations of the tangents are:

at $(3, -6)$: $\quad y + 6 = 2(x - 3)$ or $y = 2x - 12$;

at $(-3, 6)$: $\quad y - 6 = 2(x + 3)$ or $y = 2x + 12$.

59. Differentiate the equation $(x^2 + y^2)^2 = x^2 - y^2$ implicitly with respect to x:
$$2(x^2 + y^2)\left(2x + 2y \frac{dy}{dx}\right) = 2x - 2y \frac{dy}{dx}$$

Now set $dy/dx = 0$. This gives
$$2x(x^2 + y^2) = x$$
$$x^2 + y^2 = \frac{1}{2} \quad (x \neq 0)$$

Substituting this result into the original equation, we get
$$x^2 - y^2 = \frac{1}{4}$$

Now

$$\begin{matrix} x^2 + y^2 = 1/2 \\ x^2 - y^2 = 1/4 \end{matrix} \Rightarrow x = \pm\frac{\sqrt{6}}{4}, \quad y = \pm\frac{\sqrt{2}}{4}$$

Thus, the points on the curve at which the tangent line is horizontal are:

$$(\sqrt{6}/4, \sqrt{2}/4), \quad (\sqrt{6}/4, -\sqrt{2}/4), \quad (-\sqrt{6}/4, \sqrt{2}/4), \quad (-\sqrt{6}/4, -\sqrt{2}/4).$$

61. Differentiate the equation $x^{1/2} + y^{1/2} = c^{1/2}$ implicitly with respect to x :

$$\frac{1}{2}x^{-1/2} + \frac{1}{2}y^{-1/2}\frac{dy}{dx} = 0 \quad \text{which implies} \quad \frac{dy}{dx} = -\left(\frac{y}{x}\right)^{1/2}$$

An equation for the tangent line to the graph at the point (x_0, y_0) is

$$y - y_0 = -\left(\frac{y_0}{x_0}\right)^{1/2}(x - x_0)$$

The x- and y-intercepts of this line are

$$a = (x_0 y_0)^{1/2} + x_0 \quad \text{and} \quad b = (x_0 y_0)^{1/2} + y_0 \quad \text{respectively.}$$

Now

$$a + b = 2(x_0 y_0)^{1/2} + x_0 + y_0 = \left(x_0^{1/2} + y_0^{1/2}\right)^2 = c.$$

63. (a)

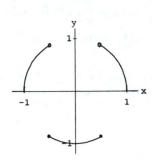

(b) $2x + 2y\dfrac{dy}{dx} = 0 \quad$ and $\quad \dfrac{dy}{dx} = -\dfrac{x}{y}$

At $(-\sqrt{3}/2, 1/2)$, $\quad \dfrac{dy}{dx} = \sqrt{3}.$

At $(\sqrt{3}/2, 1/2)$, $\quad \dfrac{dy}{dx} = -\sqrt{3}.$

At $(0, -1)$, $\quad \dfrac{dy}{dx} = 0.$

(c) $y = -\sqrt{1 - x^2}$ for $-\frac{1}{2} \le x \le \frac{1}{2}$

$$y'_+(-1/2) = \lim_{h \to 0^+} \frac{y(-\frac{1}{2}+h) - y(-\frac{1}{2})}{h} = \lim_{h \to 0^+} \frac{-\sqrt{1 - (-\frac{1}{2}+h)^2} + \frac{\sqrt{3}}{2}}{h}$$

$$= \lim_{h \to 0^+} \frac{-\sqrt{3 + 4h - 4h^2} + \sqrt{3}}{2h} = -\frac{1}{\sqrt{3}}$$

$$y'_-(1/2) = \lim_{h \to 0^-} \frac{y(\frac{1}{2}+h) - y(\frac{1}{2})}{h} = \lim_{h \to 0^-} \frac{-\sqrt{1 - (\frac{1}{2}+h)^2} - \frac{\sqrt{3}}{2}}{h}$$

$$= \lim_{h \to 0^-} \frac{-\sqrt{3 - 4h - 4h^2} + \sqrt{3}}{2h} = \frac{1}{\sqrt{3}}$$

65. By numerical work, $f'(16) \cong 0.375;$ from (3.7.1)

$$f'(x) = \frac{3}{4}x^{-1/4}, \quad \text{and} \quad f'(16) = \tfrac{3}{8} = 0.375.$$

67. $x = t, \quad y = \sqrt{4 - t^2}$ $x = t, \quad y = -\sqrt{4 - t^2}$

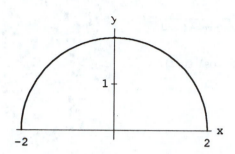   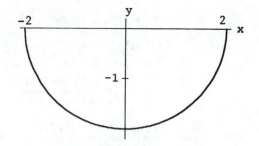

69.   (a) The graph of $x^4 = x^2 - y^2$ is:

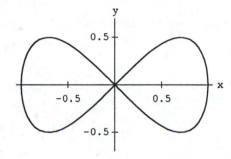

(b) Differentiate the equation $x^4 = x^2 - y^2$ implicitly with respect to x :

$$4x^3 = 2x - 2y\,\frac{dy}{dx}$$

Now set $dy/dx = 0$. This gives $4x^3 = 2x$ which implies $x = \pm\frac{\sqrt{2}}{2}$.

SECTION 3.8

1. $x + 2y = 2, \quad \dfrac{dx}{dt} + 2\dfrac{dy}{dt} = 0$

(a) If $\dfrac{dx}{dt} = 4,$ then $\dfrac{dy}{dt} = -2$ units/sec.   (b) If $\dfrac{dy}{dt} = -2,$ then $\dfrac{dx}{dt} = 4$ units/sec.

3. $y^2 = 4(x + 2), \quad 2y\dfrac{dy}{dt} = 4\dfrac{dx}{dt}$ and $\dfrac{dx}{dt} = \frac{1}{2}\,y\dfrac{dy}{dt}$

At the point $(7, 6),$ $\dfrac{dy}{dt} = 3.$ Therefore $\dfrac{dx}{dt} = \frac{1}{2} \cdot 6 \cdot 3 = 9$ units/sec.

5.   Let $s = \sqrt{x^2 + y^2}$ denote the distance to the origin at time t. Since $x = 4\cos t$ and $y = 2\sin t,$

we have

$$s(t) = \sqrt{16\cos^2 t + 4\sin^2 t} = \sqrt{12\cos^2 t + 4}$$

$$\frac{ds}{dt} = \frac{1}{2}(12\cos^2 t + 4)^{-1/2}(-24\cos t \sin t)$$

$$= \frac{-12\cos t \sin t}{\sqrt{12\cos^2 t + 4}}$$

At $t = \pi/4$, $\dfrac{ds}{dt} = \dfrac{-12\cos(\pi/4)\sin(\pi/4)}{\sqrt{12\cos^2(\pi/4) + 4}} = -\frac{3}{5}\sqrt{10}.$

7.

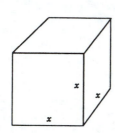

Find $\dfrac{dx}{dt}$ and $\dfrac{dS}{dt}$ when $V = 27\text{m}^3$ given that $\dfrac{dV}{dt} = -2\text{m}^3/\text{min}.$

(*) $V = x^3$, $S = 6x^2$

Differentiation of equations (*) gives

$$\frac{dV}{dt} = 3x^2 \frac{dx}{dt} \quad \text{and} \quad \frac{dS}{dt} = 12x \frac{dx}{dt}.$$

When $V = 27$, $x = 3$. Substituting $x = 3$ and $dV/dt = -2$, we get

$$-2 = 27\frac{dx}{dt} \quad \text{so that} \quad \frac{dx}{dt} = -2/27 \quad \text{and} \quad \frac{dS}{dt} = 12(3)\left(\frac{-2}{27}\right) = -8/3.$$

The rate of change of an edge is $-2/27$ m/min;   the rate of change of the surface area is $-8/3$ m²/min.

9.

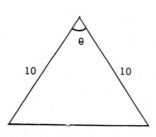

(a) $A = \frac{1}{2} \cdot 10 \cdot 10 \cdot \sin\theta = 50\sin\theta$ (see the figure)

(b) $\dfrac{d\theta}{dt} = 10° = \frac{10}{360}(2\pi) = \dfrac{\pi}{18}$ radians

$\dfrac{dA}{dt} = 50\cos\theta \dfrac{d\theta}{dt}$

At the instant $\theta = 60° = \pi/3$ radians, $\dfrac{dA}{dt} = 50\cos(\pi/3)\dfrac{\pi}{18} \cong 4.36$ cm²/min

(c) $\dfrac{dA}{d\theta} = 50\cos\theta = 0 \;\Rightarrow \theta = \pi/2$;   the triangle has maximum area when $\theta = \pi/2$.

11.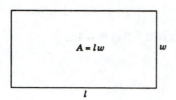

We will find the values of l for which $\dfrac{dA}{dt} < 0$

given that $\dfrac{dl}{dt} = 1$ cm/sec and

$P = 2(l + w) = 24.$

We combine $A = lw$ and $l + w = 12$ to write $A = 12l - l^2$. Differentiating with respect to t, we have

$$\frac{dA}{dt} = 12\frac{dl}{dt} - 2l\frac{dl}{dt} = 2(6 - l)\frac{dl}{dt}.$$

Since $\dfrac{dl}{dt} = 1$, $\dfrac{dA}{dt} < 0$ for $l > 6$. The area of the rectangle starts to decrease when the length is 6 cm.

13.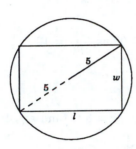

Find $\dfrac{dA}{dt}$ when $l = 6$ in.

given that $\dfrac{dl}{dt} = -2$ in./sec.

By the Pythagorean theorem

$$l^2 + w^2 = 100.$$

Also, $A = lw$. Thus, $A = l\sqrt{100 - l^2}$. Differentiation with respect to t gives

$$\frac{dA}{dt} = l\left(\frac{-l}{\sqrt{100 - l^2}}\right)\frac{dl}{dt} + \sqrt{100 - l^2}\,\frac{dl}{dt}.$$

Substituting $l = 6$ and $dl/dt = -2$, we get

$$\frac{dA}{dt} = 6\left(\frac{-6}{8}\right)(-2) + (8)(-2) = -7.$$

The area is decreasing at the rate of 7 in.^{2}/sec.

15.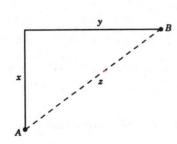

Compare $\dfrac{dy}{dt}$ to $\dfrac{dx}{dt} = -13$ mph

given that $z = 16$ and $\dfrac{dz}{dt} = -17$

when $x = y$.

By the Pythagorean theorem $x^2 + y^2 = z^2$. Thus,

$$2x\frac{dx}{dt} + 2y\frac{dy}{dt} = 2z\frac{dz}{dt}.$$

Since $x = y$ when $z = 16$, we have $x = y = 8\sqrt{2}$ and

$$2(8\sqrt{2})(-13) + 2(8\sqrt{2})\frac{dy}{dt} = 2(16)(-17).$$

Solving for dy/dt, we get

$$-13\sqrt{2} + \sqrt{2}\frac{dy}{dt} = -34 \quad \text{or} \quad \frac{dy}{dt} = \frac{1}{\sqrt{2}}(13\sqrt{2} - 34) \cong -11.$$

Thus, boat A wins the race.

17. We want to find $\dfrac{dA}{dt}$ when $\dfrac{dx}{dt} = 2$ and $x = 12$.

$A = \dfrac{1}{2}x\sqrt{169 - x^2}$, so $\dfrac{dA}{dt} = \left[\dfrac{1}{2}\sqrt{169 - x^2} - \dfrac{x^2}{2\sqrt{169 - x^2}}\right]\dfrac{dx}{dt}$

When $\dfrac{dx}{dt} = 2$ and $x = 12$, $\dfrac{dA}{dt} = -\dfrac{119}{5}\,ft^2/\text{sec}$.

19. We want to find dV/dt when $V = 1000$ ft^3 and $P = 5$ lb/in.^{2} given that $dP/dt = -0.05$ lb/in.^{2}/hr.

Differentiating $PV = C$ with respect to t, we get

$$P\frac{dV}{dt} + V\frac{dP}{dt} = 0 \quad \text{so that} \quad 5\frac{dV}{dt} + 1000(-0.05) = 0. \quad \text{Thus,} \quad \frac{dV}{dt} = 10.$$

The volume increases at the rate of 10 ft^3/hr.

21.

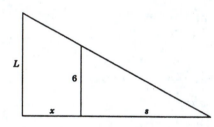

Find $\dfrac{ds}{dt}$ when $x = 3$ ft (and $s = 4$ ft)

given that $\dfrac{dx}{dt} = 400$ ft/min.

By similar triangles

$$\frac{L}{x + s} = \frac{6}{s}.$$

Substitution of $x = 3$ and $s = 4$ gives us $\dfrac{L}{7} = \dfrac{6}{4}$ so that the lamp post is

$L = 10.5$ ft tall. Rewriting

$$\frac{10.5}{x + s} = \frac{6}{s} \quad \text{as} \quad s = \frac{4}{3}x$$

and differentiating with respect to t, we find that

$$\frac{ds}{dt} = \frac{4}{3}\frac{dx}{dt} = \frac{1600}{3}.$$

The shadow lengthens at the rate of $1600/3$ ft/min.

23. Let $W(t) = 150\left(1 + \frac{1}{4000}r\right)^{-2}$. We want to find dW/dt when $r = 400$ given that

$dr/dt = 10$ mi/sec. Differentiating with respect to t, we get

$$\frac{dW}{dt} = -300\left(1 + \frac{1}{4000}r\right)^{-3}\left(\frac{1}{4000}\right)\frac{dr}{dt}$$

Now set $r = 400$ and $dr/dt = 10$. Then

$$\frac{dW}{dt} = -300\left(1 + \frac{400}{4000}\right)^{-3}\frac{10}{4000} \cong 0.5634\text{lbs/sec}$$

25.

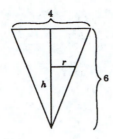

Find $\dfrac{dh}{dt}$ when $h = 3$ in.

given that $\dfrac{dV}{dt} = -\dfrac{1}{2}$ cu in./min.

By similar triangles

$$r = \tfrac{1}{3}h.$$

Thus $V = \tfrac{1}{3}\pi r^2 h = \tfrac{1}{27}\pi h^3$. Differentiating with respect to t , we get

$$\frac{dV}{dt} = \frac{1}{9}\pi h^2 \frac{dh}{dt}.$$

When $h = 3$,

$$-\frac{1}{2} = \frac{1}{9}\pi(9)\frac{dh}{dt} \quad \text{and} \quad \frac{dh}{dt} = -\frac{1}{2\pi}.$$

The water level is dropping at the rate of $1/2\pi$ inches per minute.

27. $\dfrac{dV}{dt} = 4\pi r^2 \dfrac{dr}{dt}$ and $\dfrac{dSA}{dt} = 8\pi r \dfrac{dr}{dt}.$ Thus when $\dfrac{dSA}{dt} = 4$ and $\dfrac{dr}{dt} = 0.1$

we get $r = 5\pi$ and $\dfrac{dV}{dt} = 10\pi^3$ cubic cm/min.

29.

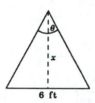

6 ft

Find $\dfrac{d\theta}{dt}$ when $x = 4$ ft

given that $\dfrac{dx}{dt} = 2$ in./min.

$(*) \qquad \tan\dfrac{\theta}{2} = \dfrac{3}{x}$

Differentiation of $(*)$ with respect to t gives

$$\frac{1}{2}\sec^2\frac{\theta}{2}\frac{d\theta}{dt} = -\frac{3}{x^2}\frac{dx}{dt} \quad \text{or} \quad \frac{d\theta}{dt} = -\frac{6}{x^2}\cos^2\frac{\theta}{2}\frac{dx}{dt}.$$

Note that $dx/dt = 2$ in./min=1/6 ft/min. When $x = 4$, we have $\cos\theta/2 = 4/5$ and thus

$$\frac{d\theta}{dt} = -\frac{6}{16}\left(\frac{4}{5}\right)^2\left(\frac{1}{6}\right) = -\frac{1}{25}.$$

The vertex angle decreases at the rate of 0.04 rad/min.

31.

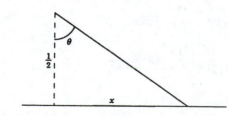

$\dfrac{1}{2}$

x

Find $\dfrac{dx}{dt}$ when $x = 1$ mi

given that $\dfrac{d\theta}{dt} = 2\pi$ rad/min.

$(*) \qquad \tan\theta = \dfrac{x}{1/2} = 2x$

Differentiation of $(*)$ with respect to t gives

$$\sec^2\theta\,\frac{d\theta}{dt} = 2\frac{dx}{dt}.$$

When $x = 1$, we get $\sec\theta = \sqrt{5}$ and thus $\dfrac{dx}{dt} = 5\pi$.

The light is traveling at 5π mi/min.

33.   We have $\tan\theta = \dfrac{x}{40}$, so $\sec^2\theta\dfrac{d\theta}{dt} = \dfrac{1}{40}\dfrac{dx}{dt}$, and $\dfrac{dx}{dt} = 4$.

At $t = 15, x = 60$ and $\sec\theta = \dfrac{\sqrt{5200}}{40}$, so $\dfrac{d\theta}{dt} = \dfrac{2}{65}$ rad/sec.

35.   We have $\sin\theta = \dfrac{4}{5}$ so $\tan\theta = \dfrac{4}{3} = \dfrac{x}{h}$.   Thus $x = \dfrac{4}{3}h$. $V = 12\left(\dfrac{3+2x+3}{2}h\right) = 36h + 16h^2$.

Thus $\dfrac{dv}{dt} = (36 + 32h)\dfrac{dh}{dt}$, so at $\dfrac{dV}{dt} = 10$ and $h = 2$, $\dfrac{dh}{dt} = \dfrac{1}{10}$ ft/min.

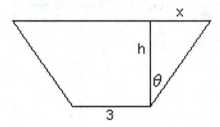

37.

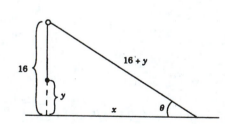

Find $\dfrac{d\theta}{dt}$ when $y = 4$ ft

given that $\dfrac{dx}{dt} = 3$ ft/sec.

$\tan\theta = \dfrac{16}{x}$, $x^2 + (16)^2 = (16 + y)^2$

Differentiating $\tan\theta = 16/x$ with respect to t, we obtain

$$\sec^2\theta\frac{d\theta}{dt} = \frac{-16}{x^2}\frac{dx}{dt}   \text{and thus}   \frac{d\theta}{dt} = \frac{-16}{x^2}\cos^2\theta\frac{dx}{dt}.$$

From $x^2 + (16)^2 = (16 + y)^2$ we conclude that $x = 12$, when $y = 4$. Thus

$$\cos\theta = \frac{x}{16 + y} = \frac{12}{20} = \frac{3}{5}   \text{and}   \frac{d\theta}{dt} = \frac{-16}{(12)^2}\left(\frac{3}{5}\right)^2(3) = \frac{-3}{25}.$$

The angle decreases at the rate of 0.12 rad/sec.

39.

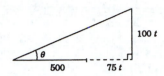

Find $\dfrac{d\theta}{dt}$ when $t = 6$ min.

$$\tan\theta = \frac{100t}{500 + 75t} = \frac{4t}{20 + 3t}$$

Differentiation with respect to t gives

$$\sec^2\theta\,\frac{d\theta}{dt} = \frac{(20 + 3t)4 - 4t(3)}{(20 + 3t)^2} = \frac{80}{(20 + 3t)^2}.$$

When $t = 6$

$$\tan\theta = \tfrac{24}{38} = \tfrac{12}{19} \quad \text{and} \quad \sec^2\theta = 1 + \left(\tfrac{12}{19}\right)^2 = \tfrac{505}{361}$$

so that

$$\frac{d\theta}{dt} = \frac{80}{(20 + 3t)^2} \cdot \frac{1}{\sec^2\theta} = \frac{80}{(38)^2} \cdot \frac{361}{505} = \frac{4}{101}.$$

The angle increases at the rate of 4/101 rad/min.

PROJECT 3.8

1. length of arc $= r\theta$, speed $= \dfrac{d}{dt}[r\theta] = r\dfrac{d\theta}{dt} = r\omega$

3. We know that $d\theta/dt = \omega$ and, at time t, $\theta = \theta_0$. Therefore $\theta = \omega t + \theta_0$. It follows that

$$x(t) = r\cos(\omega t + \theta_0) \quad \text{and} \quad y(t) = r\sin(\omega t + \theta_0).$$

$$x(t) = r\cos(\omega t + \theta_0), \quad y(t) = r\sin(\omega t + \theta_0)$$

$$v(t) = x'(t) = -r\omega\sin(\omega t + \theta_0) = -\omega\,y(t)$$

$$a(t) = -r\omega^2\cos(\omega t + \theta_0) = -\omega^2\,x(t)$$

$$v(t) = y'(t) = r\omega\cos(\omega t + \theta_0) = \omega\,x(t)$$

$$a(t) = -r\omega^2\sin(\omega t + \theta_0) = -\omega^2\,y(t)$$

5. From Exercise 4, $\dfrac{dA_T}{dt} = \tfrac{1}{2}r^2\omega\cos\theta$ and $\dfrac{dA_S}{dt} = \tfrac{1}{2}r^2\omega - \tfrac{1}{2}r^2\omega\cos\theta$

Now,

$$\frac{1}{2}r^2\omega\cos\theta = \frac{1}{2}r^2\omega - \frac{1}{2}r^2\omega\cos\theta \quad\Longrightarrow\quad \cos\theta = \frac{1}{2} \quad\Longrightarrow\quad \theta = \frac{\pi}{3}.$$

SECTION 3.9

1.
$$\Delta V = (x+h)^3 - x^3$$
$$= (x^3 + 3x^2h + 3xh^2 + h^3) - x^3$$
$$= 3x^2h + 3xh^2 + h^3,$$
$$dV = 3x^2h,$$
$$\Delta V - dV = 3xh^2 + h^3 \quad \text{(see figure)}$$

3. $f(x) = x^{1/3}, \quad x = 1000, \quad h = 10, \quad f'(x) = \frac{1}{3}x^{-2/3}$

$$\sqrt[3]{1010} = f(x+h) \cong f(x) + hf'(x) = \sqrt[3]{1000} + 10\left[\frac{1}{3}(1000)^{-2/3}\right] = 10\frac{1}{30}$$

5. $f(x) = x^{1/4}, \quad x = 16, \quad h = -1, \quad f'(x) = \frac{1}{4}x^{-3/4}$

$$(15)^{1/4} = f(x+h) \cong f(x) + hf'(x) = (16)^{1/4} + (-1)\left[\frac{1}{4}(16)^{-3/4}\right] = 1\frac{31}{32}$$

7. $f(x) = x^{1/5}, \quad x = 32, \quad h = -2, \quad f'(x) = \frac{1}{5}x^{-4/5}$

$$(30)^{1/5} = f(x+h) \cong f(x) + hf'(x) = (32)^{1/5} + (-2)\left[\frac{1}{5}(32)^{-4/5}\right] = 1.975$$

9. $f(x) = x^{3/5}, \quad x = 32, \quad h = 1, \quad f'(x) = \frac{3}{5}x^{-2/5}$

$$(33)^{3/5} = f(x+h) \cong f(x) + hf'(x) = (32)^{3/5} + (1)\left[\frac{3}{5}(32)^{-2/5}\right] = 8.15$$

11. $f(x) = \sin x, \quad x = \frac{\pi}{4}, \quad h = \frac{\pi}{180}, \quad f'(x) = \cos x$

$$\sin 46° = f(x+h) \cong f(x) + hf'(x) = \sin\frac{\pi}{4} + \frac{\pi}{180}\cos\frac{\pi}{4} = \frac{\sqrt{2}}{2}\left(1 + \frac{\pi}{180}\right) \cong 0.719$$

13. $f(x) = \tan x, \quad x = \frac{\pi}{6}, \quad h = \frac{-\pi}{90}, \quad f'(x) = \sec^2 x$

$$\tan 28° = f(x+h) \cong f(x) + hf'(x) = \tan\frac{\pi}{6} + \left(\frac{-\pi}{90}\right)\left(\frac{4}{3}\right) = \frac{\sqrt{3}}{3} - \frac{2\pi}{135} \cong 0.531$$

15. $f(2.8) \cong f(3) + (-0.2)f'(3) = 2 + (-0.2)(2) = 1.6$

17. $V(x) = \pi x^2 h; \quad \text{volume} = V(r+t) - V(r) \cong tV'(r) = 2\pi rht$

19. $V(x) = x^3, \quad V'(x) = 3x^2, \quad \Delta V \cong dV = V'(10)h = 300h$

$$|dV| \le 3 \quad\Longrightarrow\quad |300h| \le 3 \quad\Longrightarrow\quad |h| \le 0.01, \quad \text{error} \le 0.01 \text{ feet}$$

21. $V(r) = \frac{2}{3}\pi r^3$ and $dr = 0.01$.

$$V(r + 0.01) - V(r) \cong V'(r)(0.01) = 2\pi r^2(0.01)$$

$$= 2\pi(600)^2(0.01)   \text{(50 ft = 600 in)}$$

$$= 22619.5 \text{ in}^3 \quad \text{or} \quad 98 \text{ gallons (approx.)}$$

23. $P = 2\pi\sqrt{\dfrac{L}{g}}$ implies $P^2 = 4\pi^2 \dfrac{L}{g}$

Differentiating with respect to t, we have

$$2P\frac{dP}{dt} = \frac{4\pi^2}{g}\cdot\frac{dL}{dt} = \frac{P^2}{L}\cdot\frac{dL}{dt} \quad \text{since} \quad \frac{P^2}{L} = \frac{4\pi^2}{g}.$$

Thus $\dfrac{dP}{P} = \dfrac{1}{2}\cdot\dfrac{dL}{L}$

25. $L = 3.26$ ft, $P = 2$ sec, and $dL = 0.01$ ft

$$\frac{dP}{P} = \frac{1}{2}\cdot\frac{dL}{L}$$

$$dP = \frac{1}{2}\cdot\frac{dL}{L}\cdot P = \frac{1}{2}\cdot\frac{0.01}{3.26}\cdot 2 \quad dP \cong 0.00307 \text{ sec}$$

27. $A(x) = \dfrac{1}{4}\pi x^2$, $dA = \dfrac{1}{2}\pi x h$, $\dfrac{dA}{A} = 2\dfrac{h}{x}$

$$\frac{dA}{A} \le 0.01 \quad \Longleftrightarrow \quad 2\frac{h}{x} \le 0.01 \quad \Longleftrightarrow \quad \frac{h}{x} \le 0.005 \quad \text{within } \frac{1}{2}\%$$

29. (a) $x_{n+1} = \dfrac{1}{2}x_n + 12\left(\dfrac{1}{x_n}\right)$ (b) $x_4 \cong 4.89898$

31. (a) $x_{n+1} = \dfrac{2}{3}x_n + \dfrac{25}{3}\left(\dfrac{1}{x_n}\right)^2$ (b) $x_4 \cong 2.92402$

33. (a) $x_{n+1} = \dfrac{x_n \sin x_n + \cos x_n}{\sin x_n + 1}$ (b) $x_4 \cong 0.73909$

35. (a) $x_{n+1} = \dfrac{2x_n \cos x_n - 2\sin x_n}{2\cos x_n - 1}$ (b) $x_4 \cong 1.89549$

37.   Let $f(x) = x^{1/3}$. Then $f'(x) = \frac{1}{3}x^{-2/3}$. The Newton-Raphson method applied to

this function gives:

$$x_{n+1} = x_n - \frac{f(x_n)}{f'(x_n)} = x_n - \frac{x_n^{1/3}}{\frac{1}{3}x_n^{-2/3}} = -2x_n.$$

Choose any $x_1 \ne 0$. Then $x_2 = -2x_1$, $x_3 = -2x_2 = 4x_1$, $\cdots$,

$$x_n = -2x_{n-1} = (-1)^{n-1}2^n x_1, \quad \cdots.$$

39.   (a) Let $f(x) = x^4 - 2x^2 - \frac{17}{16}$. Then $f'(x) = 4x^3 - 4x$. The Newton-Raphson method

applied to this function gives:

$$x_{n+1} = x_n - \frac{x_n^4 - 2x_n^2 - \frac{17}{16}}{4x_n^3 - 4x_n}$$

If $x_1 = \frac{1}{2}$, then $x_2 = -\frac{1}{2}$, $x_3 = \frac{1}{2}$, $\cdots x_n = (-1)^{n-1}\frac{1}{2}$, $\cdots$.

(b) $x_1 = 2$, $x_2 = 1.71094$, $x_3 = 1.58569$, $x_4 = 1.56165$; $f(x_4) = 0.00748$

41.   (a) Let $f(x) = x^k - a$. Then $f'(x) = kx^{k-1}$. The Newton-Raphson method applied to

this function gives:

$$x_{n+1} = x_n - \frac{x_n^k - a}{kx_n^{k-1}} = x_n - \frac{1}{k}x_n + \frac{1}{k}\frac{a}{x_n^{k-1}}$$

$$= \frac{1}{k}\left[(k-1)x_n + \frac{a}{x_n^{k-1}}\right]$$

(b) Let $a = 23$, $k = 3$ and $x_1 = 3$. Then

$$x_1 = 3, \quad x_2 = 2.85185, \quad x_3 = 2.84389, \quad x_4 = 2.84382; \quad f(x_4) = -0.00114$$

43.   (a) and (b)

45. $\lim\limits_{h \to 0} \dfrac{g_1(h) + g_2(h)}{h} = \lim\limits_{h \to 0} \dfrac{g_1(h)}{h} + \lim\limits_{h \to 0} \dfrac{g_2(h)}{h} = 0 + 0 = 0$

$\lim\limits_{h \to 0} \dfrac{g_1(h)g_2(h)}{h} = \lim\limits_{h \to 0} h \dfrac{g_1(h)g_2(h)}{h^2} = \left(\lim\limits_{h \to 0} h\right)\left(\lim\limits_{h \to 0} \dfrac{g_1(h)}{h}\right)\left(\lim\limits_{h \to 0} \dfrac{g_2(h)}{h}\right) = (0)(0)(0) = 0$

PROJECT 3.9

1.   Let $F(x) = f(x) - x$. Then F is continuous on $[a, b]$, and $F(b) = f(b) - b \le 0$. If either $f(a) - a = 0$ or $f(b) - b = 0$, then a and/or b is a fixed point. If $f(a) - a \ne 0$ and $f(b) - b \ne 0$, Then $F(a) > 0$ and $F(b) < 0$, and, by the intermediate value theorem, there exists at least one $c \in (a, b)$ such that $F(c) = f(c) - c = 0$. The number c is a fixed point of f.

CHAPTER 4

SECTION 4.1

1. f is differentiable on $(0,1)$, continuous on $[0,1]$; and $f(0) = f(1) = 0$.

   $$f'(c) = 3c^2 - 1; \quad 3c^2 - 1 = 0 \Longrightarrow c = \frac{\sqrt{3}}{3} \quad \left(-\frac{\sqrt{3}}{3} \notin (0,1)\right)$$

3. f is differentiable on $(0, 2\pi)$, continuous on $[0, 2\pi]$; and $f(0) = f(2\pi) = 0$.

   $$f'(c) = 2\cos 2c; \quad 2\cos 2c = 0 \Longrightarrow 2c = \frac{\pi}{2} + n\pi, \quad \text{and} \quad c = \frac{\pi}{4} + \frac{n\pi}{2}, \quad n = 0, \pm 1, \pm 2 \ldots$$

   Thus, $c = \dfrac{\pi}{4}, \dfrac{3\pi}{4}, \dfrac{5\pi}{4}, \dfrac{7\pi}{4}$

5. $f'(c) = 2c, \quad \dfrac{f(b) - f(a)}{b - a} = \dfrac{4 - 1}{2 - 1} = 3; \quad 2c = 3 \implies c = 3/2$

7. $f'(c) = 3c^2, \quad \dfrac{f(b) - f(a)}{b - a} = \dfrac{27 - 1}{3 - 1} = 13; \quad 3c^2 = 13 \implies c = \dfrac{1}{3}\sqrt{39} \quad \left(-\dfrac{1}{3}\sqrt{39} \text{ is not in } [a, b]\right)$

9. $f'(c) = \dfrac{-c}{\sqrt{1 - c^2}}, \quad \dfrac{f(b) - f(a)}{b - a} = \dfrac{0 - 1}{1 - 0} = -1; \quad \dfrac{-c}{\sqrt{1 - c^2}} = -1 \implies c = \dfrac{1}{2}\sqrt{2}$

   $(-\dfrac{1}{2}\sqrt{2} \text{ is not in } [a, b])$

11. f is continuous on $[-1, 1]$, differentiable on $(-1, 1)$ and $f(-1) = f(1) = 0$.

    $$f'(x) = \frac{-x(5 - x^2)}{(3 + x^2)^2\sqrt{1 - x^2}}, \quad f'(c) = 0 \text{ for } c \text{ in } (-1, 1) \text{ implies } c = 0.$$

13. No. By the mean-value theorem there exists at least one number $c \in (0, 2)$ such that

    $$f'(c) = \frac{f(2) - f(0)}{2 - 0} = \frac{3}{2} > 1.$$

15. f is everywhere continuous and everywhere differentiable except possibly at $x = -1$.

    f is continuous at $x = -1$: as you can check,

    $$\lim_{x \to -1^-} f(x) = 0, \quad \lim_{x \to -1^+} f(x) = 0, \quad \text{and} \quad f(-1) = 0.$$

    f is differentiable at $x = -1$ and $f'(-1) = 2$: as you can check,

    $$\lim_{h \to 0^-} \frac{f(-1 + h) - f(-1)}{h} = 2 \quad \text{and} \quad \lim_{h \to 0^+} \frac{f(-1 + h) - f(-1)}{h} = 2.$$

    Thus f satisfies the conditions of the mean-value theorem on every closed interval $[a, b]$.

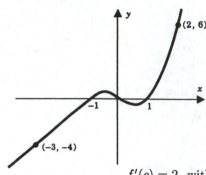

$$f'(x) = \begin{cases} 2, & x \le -1 \\ 3x^2 - 1, & x > -1 \end{cases}$$

$$\frac{f(2) - f(-3)}{2 - (-3)} = \frac{6 - (-4)}{2 - (-3)} = 2.$$

$f'(c) = 2$ with $c \in (-3, 2)$ iff $c = 1$ or $-3 < c \le -1$.

17. Let $f(x) = Ax^2 + Bx + C$. Then $f'(x) = 2Ax + B$. By the mean-value theorem

$$f'(c) = \frac{f(b) - f(a)}{b - a} = \frac{(Ab^2 + Bb + C) - (Aa^2 + Ba + C)}{b - a}$$
$$= \frac{A(b^2 - a^2) + B(b - a)}{b - a} = A(b + a) + B$$

Therefore, we have

$$2Ac + B = A(b + a) + B \implies c = \frac{a + b}{2}$$

19. $\dfrac{f(1) - f(-1)}{1 - (-1)} = 0$ and $f'(x)$ is never zero. This result does not violate the mean-value theorem

since f is not differentiable at 0; the theorem does not apply.

21. Set $P(x) = 6x^4 - 7x + 1$. If there existed three numbers $a < b < c$ at which $P(x) = 0$, then by Rolle's theorem $P'(x)$ would have to be zero for some x in (a, b) and also for some x in (b, c). This is not the case: $P'(x) = 24x^3 - 7$ is zero only at $x = (7/24)^{1/3}$.

23. Set $P(x) = x^3 + 9x^2 + 33x - 8$. Note that $P(0) < 0$ and $P(1) > 0$. Thus, by the intermediate-value theorem, there exists some number c between 0 and 1 at which $P(x) = 0$. If the equation $P(x) = 0$ had an additional real root, then by Rolle's theorem there would have to be some real number at which $P'(x) = 0$. This is not the case: $P'(x) = 3x^2 + 18x + 33$ is never zero since the discriminant $b^2 - 4ac = (18)^2 - 12(33) < 0$.

25. Let c and d be two consecutive roots of the equation $P'(x) = 0$. The equation $P(x) = 0$ cannot have two or more roots between c and d for then, by Rolle's theorem, $P'(x)$ would have to be zero somewhere between these two roots and thus between c and d. In this case c and d would no longer be consecutive roots of $P'(x) = 0$.

27. Suppose that f has two fixed points $a, b \in I$, with $a < b$. Let $g(x) = f(x) - x$. Then $g(a) = f(a) - a = 0$ and $g(b) = f(b) - b = 0$. Since f is differentiable on I, we can conclude that g is differentiable on (a, b) and continuous on $[a, b]$. By Rolle's theorem, there exists a number $c \in (a, b)$ such that $g'(c) = f'(c) - 1 = 0$ or $f'(c) = 1$. This contradicts the assumption that $f'(x) < 1$ on I.

29. (a) $f'(x) = 3x^2 - 3 > 0$ for all x in $(-1, 1)$. Also, f is differentiable on $(-1, 1)$ and continuous on

$[-1, 1]$. Thus there cannot be a and b in $(-1, 1)$ such that $f(a) = f(b) = 0$, or they would contradict Rolle's theorem.

(b) When $f(x) = 0$, $b = 3x - x^3 = x(3 - x^2)$. When x is in $(-1, 1)$, then $|x(3 - x^2)| < 2$.
    Thus $|b| < 2$.

31.   For $p(x) = x^n + ax + b$, $p'(x) = nx^{n-1} + a$, which has at most one real zero for n even $\left(x = -\dfrac{a}{n}^{\frac{1}{n-1}} \right)$.
If there were more than two distinct real roots of $p(x)$, then by Rolle's theorem there would be more than one zero of $p'(x)$. Thus there are at most two distinct real roots of $p(x)$.

33.   If $x_1 = x_2$, then $|f(x_1) - f(x_2)|$ and $|x_1 - x_2|$ are both 0 and the inequality holds. If $x_1 \neq x_2$, then by the mean-value theorem

$$\frac{f(x_1) - f(x_2)}{x_1 - x_2} = f'(c)$$

for some number c between x_1 and x_2. Since $|f'(c)| \leq 1$:

$$\left| \frac{f(x_1) - f(x_2)}{x_1 - x_2} \right| \leq 1 \quad \text{and thus} \quad |f(x_1) - f(x_2)| \leq |x_1 - x_2|.$$

35.   Set, for instance, $f(x) = \begin{cases} 1, & a < x < b \\ 0, & x = a, b \end{cases}$

37.   (a) By the mean-value theorem, there exists a number $c \in (a, b)$ such that $f(b) - f(a) = f'(c)(b - a)$.
If $f'(x) \leq M$ for all $x \in (a, b)$, then it follows that

$$f(b) \leq f(a) + M(b - a)$$

(b) If $f'(x) \geq m$ for all $x \in (a, b)$, then it follows that

$$f(b) \geq f(a) + m(b - a)$$

(c) If $|f'(x)| \leq L$ on (a, b), then $-L \leq f'(x) \leq L$ on (a, b) and the result follows from parts (a) and (b).

39.   Let $f(x) = \cos x$ and $g(x) = \sin x$ on $I = (-\infty, \infty)$. Then

$$f(x)g'(x) - g(x)f'(x) = \cos^2 x + \sin^2 x = 1 \text{ for all } x \in I$$

The result follows from Exercise 38.

41. $$f'(x_0) = \lim_{h \to 0} \frac{f(x_0 + h) - f(x_0)}{h} = \lim_{h \to 0} \frac{f'(x_0 + \theta h)h}{h} = \lim_{h \to 0} f'(x_0 + \theta h)$$

$$\text{(by the hint)}$$

$$= \lim_{x \to x_0} f'(x) = L$$

$$\text{(by 2.2.5)}$$

43.   Using the hint, F is continuous on $[a, b]$, differentiable on (a, b), and $F(a) = F(b)$. Thus by Exercise

42, there is a c in (a, b) such that $F'(c) = 0$.

Thus $[f(b) - f(a)]g'(c) - [g(b) - g(a)]f'(c) = 0$ and $\dfrac{f(b) - f(a)}{g(b) - g(a)} = \dfrac{f'(c)}{g'(c)}$.

45.   Let $f_1(t)$ and $f_2(t)$ be the positions of the cars at time t. Consider $f(t) = f_1(t) - f_2(t)$. Let T be the time the cars finish the race. Then $f(t)$ satisfies the hypothesis of Exercise 42, so there is a c in $(0, T)$ such that $f'(c) = 0$. Hence $f_1'(t) = f_2'(t)$, so the cars had the same velocity at time c.

47.   Let $s(t)$ denote the distance that the car has traveled in t seconds since applying the brakes, $0 \le t \le 6$. Then $s(0) = 0$ and $s(6) = 280$. Assume that s is differentiable on $(0, 6)$ and continuous on $[0, 6]$. Then, by the mean-value theorem, there exists a time $c \in (0, 6)$ such that

$$s'(c) = v(c) = \frac{s(6) - s(0)}{6 - 0} = \frac{280}{6} \cong 46.67 \text{ ft/sec}$$

Now $v(0) \ge v(c) = 46.7$ ft/sec. Thus, the driver must have been exceeding the speed limit (44 ft/sec) at the instant he applied his brakes.

49.   Let $f(x) = \sqrt{x}$. Then $f'(x) = \dfrac{1}{2\sqrt{x}}$. Using Exercise 48, we have

$$\sqrt{65} = \sqrt{64 + 1} = f(64 + 1) \cong f(64) + f'(64)(1) = \sqrt{64} + \frac{1}{2\sqrt{64}} = 8.0625$$

51.   (a)   Let $f(x) = 1 + 4x - 2\cos x$, $x \in I = (-\infty, \infty)$. If f had two (or more) zeros on I, then, by Rolle's theorem, f' would have to have a zero on I. But, $f'(x) = 4 + 2\sin x > 0$ on I. Thus f has at most one zero on I.

      (b) $f(0) = -1$ and $f(1) \cong 3.92$. Thus f has a zero in $(0, 1)$.

      (b) $x_{n+1} = x_n - \dfrac{1 + 4x_n - 2\cos x_n}{4 + 2\sin x_n}$; $x_1 = 0$, $x_2 = 0.25$, $x_3 \cong 0.2361$

53. $f(x) = 1 - x^3 - \cos(\pi x/2)$ is differentiable on $(0, 1)$, continuous on $[0, 1]$, and $f(0) = f(1) = 0$.

$f'(x) = -3x^2 + \dfrac{\pi}{2}\sin(\pi x/2)$

$f'(c) = 0$ at $c \cong 0.676$

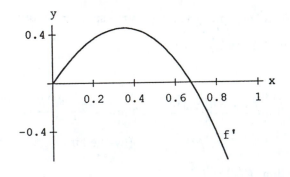

55. $f(x) = x^4 - 7x^2 + 2; \quad f'(x) = 4x^3 - 14x$

$$g(x) = 4x^3 - 14x - \frac{f(3) - f(1)}{3 - 1} = 4x^3 - 14x - 12$$

$g(c) = 0$ at $c \cong 2.205$

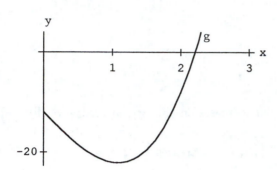
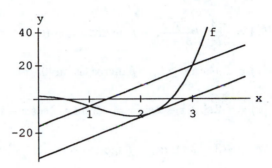

SECTION 4.2

1. $f'(x) = 3x^2 - 3 = 3\left(x^2 - 1\right) = 3(x + 1)(x - 1)$

f increases on $(-\infty, -1]$ and $[1, \infty)$, decreases on $[-1, 1]$

3. $f'(x) = 1 - \dfrac{1}{x^2} = \dfrac{x^2 - 1}{x^2} = \dfrac{(x + 1)(x - 1)}{x^2}$

f increases on $(-\infty, -1]$ and $[1, \infty)$, decreases on $[-1, 0)$ and $(0, 1]$ (f is not defined at 0)

5. $f'(x) = 3x^2 + 4x^3 = x^2(3 + 4x)$

f increases on $\left[-\frac{3}{4}, \infty\right)$, decreases on $\left(-\infty, -\frac{3}{4}\right]$

7. $f'(x) = 4(x + 1)^3$

f increases on $[-1, \infty)$, decreases on $(-\infty, -1]$

9. $f(x) = \begin{cases} \dfrac{1}{2 - x}, & x < 2 \\ \dfrac{1}{x - 2}, & x > 2 \end{cases}$ $f'(x) = \begin{cases} \dfrac{1}{(2 - x)^2}, & x < 2 \\ \dfrac{-1}{(x - 2)^2}, & x > 2 \end{cases}$

f increases on $(-\infty, 2)$, decreases on $(2, \infty)$ (f is not defined at 2)

11. $f'(x) = -\dfrac{4x}{\left(x^2 - 1\right)^2}$

f increases on $(-\infty, -1)$ and $(-1, 0]$, decreases on $[0, 1)$ and $(1, \infty)$ (f is not defined at ± 1)

13. $f(x) = \begin{cases} x^2 - 5, & x < -\sqrt{5} \\ -\left(x^2 - 5\right), & -\sqrt{5} \le x \le \sqrt{5} \\ x^2 - 5, & \sqrt{5} < x \end{cases}$ $f'(x) = \begin{cases} 2x, & x < -\sqrt{5} \\ -2x, & -\sqrt{5} < x < \sqrt{5} \\ 2x, & \sqrt{5} < x \end{cases}$

f increases on $[-\sqrt{5}, 0]$ and $[\sqrt{5}, \infty)$, decreases on $(-\infty, -\sqrt{5}]$ and $[0, \sqrt{5}]$

15. $f'(x) = \dfrac{2}{(x + 1)^2};$ f increases on $(-\infty, -1)$ and $(-1, \infty)$ (f is not defined at -1)

17. $f'(x) = -\dfrac{7\left(1-\sqrt{x}\right)^6}{\sqrt{x}\left(1+\sqrt{x}\right)^8}$; f decreases on $[0,\infty)$

19. $f'(x) = \dfrac{x}{\left(2+x^2\right)^2}\sqrt{\dfrac{2+x^2}{1+x^2}}$ f increases on $[0,\infty)$,   decreases on $(-\infty, 0\,]$

21. $f'(x) = \dfrac{-3}{2x^2}\sqrt{\dfrac{x}{3-x}}$; f decreases on $(0, 3]$

23. $f'(x) = 1 + \sin x \geq 0$; f increases on $[0, 2\pi]$

25. $f'(x) = -2\sin 2x - 2\sin x = -2\sin x\,(2\cos x + 1)$; f increases on $\left[\frac{2}{3}\pi, \pi\right]$, decreases on $\left[0, \frac{2}{3}\pi\right]$

27. $f'(x) = \sqrt{3} + 2\sin 2x$; f increases on $\left[0, \frac{2}{3}\pi\right]$ and $\left[\frac{5}{6}\pi, \pi\right]$, decreases on $\left[\frac{2}{3}\pi, \frac{5}{6}\pi\right]$

29. $\dfrac{d}{dx}\left(\dfrac{x^3}{3} - x\right) = f'(x)$ $\implies$ $f(x) = \dfrac{x^3}{3} - x + C$

   $f(1) = 2$ $\implies$ $2 = \frac{1}{3} - 1 + C$, so $C = \frac{8}{3}$. Thus, $f(x) = \frac{1}{3}x^3 - x + \frac{8}{3}$.

31. $\dfrac{d}{dx}\left(x^5 + x^4 + x^3 + x^2 + x\right) = f'(x)$ $\implies$ $f(x) = x^5 + x^4 + x^3 + x + C$

   $f(0) = 5$ $\implies$ $5 = 0 + C$, so $C = 5$. Thus, $f(x) = x^5 + x^4 + x^3 + x^2 + x + 5$.

33. $\dfrac{d}{dx}\left(\dfrac{3}{4}x^{4/3} - \dfrac{2}{3}x^{3/2}\right) = f'(x)$ $\implies$ $f(x) = \dfrac{3}{4}x^{4/3} - \dfrac{2}{3}x^{3/2} + C$

   $f(0) = 1$ $\implies$ $1 = 0 + C$, so $C = 1$. Thus, $f(x) = \frac{3}{4}x^{4/3} - \frac{2}{3}x^{3/2} + 1,\ x \geq 0$.

35. $\dfrac{d}{dx}\left(2x - \cos x\right) = f'(x)$ $\implies$ $f(x) = 2x - \cos x + C$

   $f(0) = 3$ $\implies$ $3 = 0 - 1 + C$, so $C = 4$. Thus, $f(x) = 2x - \cos x + 4$.

37. $f'(x) = \begin{cases} 1, & x < -3 \\ -1, & -3 < x < -1 \\ 1, & -1 < x < 1 \\ -2, & 1 < x \end{cases}$

   f increases on $(-\infty, -3)$ and $[-1, 1]$;

   decreases on $[-3, -1]$ and $[1, \infty)$

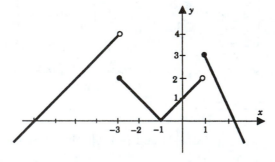

39. $f'(x) = \begin{cases} -2x, & x < 1 \\ -2, & 1 < x < 3 \\ 3, & 3 < x \end{cases}$

   f increases on $(-\infty, 0]$ and $[3, \infty)$;

   decreases on $[0, 1)$ and $[1, 3]$

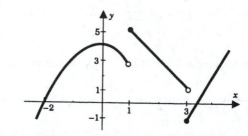

41.

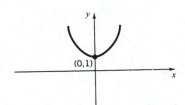

43.

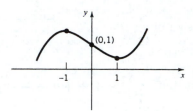

45.

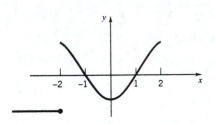

47.

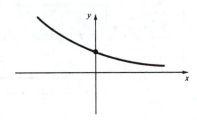

49.   Not possible; f is increasing, so $f(2)$ must be greater than $f(-1)$.

51.   Let $x(t) = t^3 - 6t^2 + 9t + 2$. Then

$$v(t) = 3t^2 - 12t + 9 = 3(t-1)(t-3)$$

sign of v :

$$a(t) = 6t - 12 = 6(t-2)$$

sign of a :

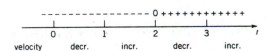

sign of v : $+ + + 0 - - - - - - 0 + + +$

sign of a : $- - - - - - 0 + + + + + +$

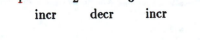

speed   decr   incr   decr   incr

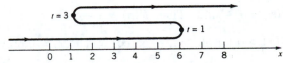

53.   Let $x(t) = 2\sin 3t, \quad t \in [0, \pi]$. Then

$$v(t) = 6\cos 3t$$

sign of v :

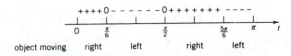

$$a(t) = -18\sin 3t$$

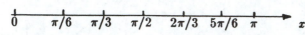

sign of a :

velocity   decr.   incr.   decr.   incr.   decr.   incr.

sign of v : $+ + + 0 - - - - \; 0 + + + + \; 0 - -$

sign of a : $- - \; - - \; -0 + + + + \; 0 - - - -$

$$0 \qquad \pi/6 \quad \pi/3 \quad \pi/2 \quad 2\pi/3 \; 5\pi/6 \; \pi \qquad x$$

decr   **incr**   **decr**   **incr**   **decr**   **incr**

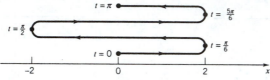

55.   (a) $M \leq L \leq N$ (b)   none   (c) $M = L = N$

57.   Suppose there is a c in (a, b) such that $f'(c) < 0$. Then by theorem 4.1.2, $f(c + h) < f(c)$ for h sufficiently small. This contradicts the fact that f is increasing on (a, b). Thus $f'(c) \geq 0$ for all c in (a, b).

59.   (a) $f'(x) = 2\sec x(\sec x \tan x) = 2\sec^2 x \tan x$ and $g'(x) = 2\tan x \sec^2 x$.

   Therefore, $f'(x) = g'(x)$ for all $x \in I$.

   (b) Evaluating $\sec^2 x - \tan^2 x = C$ at $x = 0$ gives $C = 1$.

61.   Let f and g be functions such that $f'(x) = -g(x)$ and $g'(x) = f(x)$. Then:

   (a) Differentiating $f^2(x) + g^2(x)$ with respect to x, we have

$$2f(x)f'(x) + 2g(x)g'(x) = -2f(x)g(x) + 2g(x)f(x) = 0.$$

   Thus, $f^2(x) + g^2(x) = C$ (constant).

   (b) $f(a) = 1$ and $g(a) = 0$ implies $C = 1$. The functions $f(x) = \cos(x - a)$, $g(x) = \sin(x - a)$

   have these properties.

63.   Let $f(x) = x - \sin x$. Then $f'(x) = 1 - \cos x$.

   (a) $f'(x) \geq 0$ for all $x \in (-\infty, \infty)$ and $f'(x) = 0$ only at $x = \dfrac{\pi}{2} + n\pi$, $n = 0, \pm 1, \pm 2, \ldots$

   It follows from Theorem 4.2.3 that f is increasing on $(-\infty, \infty)$.

   (b) Since f is increasing on $(-\infty, \infty)$ and $f(0) = 0 - \sin 0 = 0$, we have:

$$f(x) > 0 \;\; \text{for all} \;\; x > 0 \Rightarrow x > \sin x \;\; \text{on} \;\; (0, \infty);$$

$$f(x) < 0 \;\; \text{for all} \;\; x < 0 \Rightarrow x < \sin x \;\; \text{on} \;\; (-\infty, 0).$$

65. Let $f(x) = \tan x$ and $g(x) = x$ for $x \in [0, \pi/2)$. Then $f(0) = g(0) = 0$ and $f'(x) = \sec^2 x > g'(x) = 1$ for $x \in (0, \pi/2)$. Thus, $\tan x > x$ for $x \in (0, \pi/2)$ by Exercise 64(a).

67. Choose an integer $n > 1$. Let $f(x) = (1+x)^n$ and $g(x) = 1 + nx$, $x > 0$. Then, $f(0) = g(0) = 1$ and $f'(x) = n(1+x)^{n-1} > g'(x) = n$ since $(1+x)^{n-1} > 1$ for $x > 0$. The result follows from Exercise 64(a).

69. $4° \cong 0.06981$ radians. By Exercises 63 and 68,
$$0.6981 - \frac{(0.6981)^3}{6} = 0.06975 < \sin 4° < 0.6981$$

71. Let $f(x) = 3x^4 - 10x^3 - 4x^2 + 10x + 9$, $x \in [-2, 5]$. Then $f'(x) = 12x^3 - 30x^2 - 8x + 10$.

$f'(x) = 0$ at $x \cong -0.633$, 0.5, 2.633

f is decreasing on $[-2, -0.633]$

and $[0.5, 2.633]$

f is increasing on $[-0.633, 0.5]$

and $[2.633, 5]$

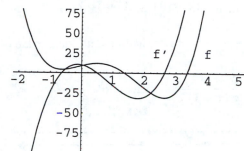

73. Let $f(x) = x \cos x - 3 \sin 2x$, $x \in [0, 6]$. Then $f'(x) = \cos x - x \sin x - 6 \cos 2x$.

$f'(x) = 0$ at $x \cong 0.770$, 2.155, 3.798, 5.812

f is decreasing on $[0, 0.770]$, $[2.155, 3.798]$

and $[5.812, 6]$

f is increasing on $[0.770, 2.155]$

and $[3.798, 5.812]$

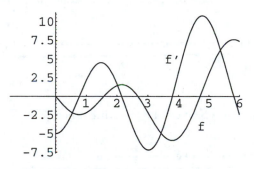

PROJECT 4.2

1.
$$\frac{d}{dt}\left[mgy + \tfrac{1}{2}mv^2\right] = mg\frac{dy}{dt} + \frac{1}{2}m\frac{d}{dt}(v^2)$$
$$= mgv + \frac{1}{2}m\left[2v\frac{dv}{dt}\right]$$
$$= mgv + mv\frac{dv}{dt}$$
$$= mgv + mv(-g) \quad (\text{since} \quad dv/dt = a = -g)$$
$$= mgv - mgv = 0$$

3. $y(t) = \tfrac{1}{2}gt^2 + y_0 \implies gt = \sqrt{2g(y_0 - y)}$

$v(t) = y'(t) = -gt$ Therefore, $|v(t)| = \sqrt{2g(y_0 - y)}$.

SECTION 4.3

1. $f'(x) = 3x^2 + 3 > 0;$ no critical nos, no local extreme values

3. $f'(x) = 1 - \dfrac{1}{x^2};$ critical nos $-1, 1$

$f''(x) = \dfrac{2}{x^3},\ f''(-1) = -2,\ f''(1) = 2$ $f(-1) = -2$ local max, $f(1) = 2$ local min

5. $f'(x) = 2x - 3x^2 = x(2 - 3x);$ critical nos $0, \frac{2}{3}$

$f''(x) = 2 - 6x;$ $f''(0) = 2,\ f''\left(\frac{2}{3}\right) = -2$

$f(0) = 0$ local min, $f\left(\frac{2}{3}\right) = \frac{4}{27}$ local max

7. $f'(x) = \dfrac{2}{(1-x)^2};$ no critical nos, no local extreme values

9. $f'(x) = -\dfrac{2(2x+1)}{x^2(x+1)^2};$ critical no $-\dfrac{1}{2}$

$f\left(-\frac{1}{2}\right) = -8$ local max

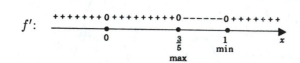

11. $f'(x) = x^2(5x - 3)(x - 1);$ critical nos $0, \frac{3}{5}, 1$

$f\left(\dfrac{3}{5}\right) = \dfrac{2^2 3^3}{5^5}$ local max

$f(1) = 0$ local min

no local extreme at 0

13. $f'(x) = (5 - 8x)(x - 1)^2;$ critical nos $\frac{5}{8}, 1$

$f\left(\dfrac{5}{8}\right) = \dfrac{27}{2048}$ local max

no local extreme at 1

15. $f'(x) = \dfrac{x(2 + x)}{(1 + x)^2};$ critical nos $-2, 0$

$f(-2) = -4$ local max

$f(0) = 0$ local min

17. $f'(x) = \begin{cases} 2x + 1, & x < -2, x > 1 \\ -(2x + 1), & -2 < x < 1 \end{cases};$ critical nos $-2, -\frac{1}{2}, 1$

$f(-2) = 0$ local min

$f\left(-\frac{1}{2}\right) = \frac{9}{4}$ local max

$f(1) = 0$ local min

19. $f'(x) = \frac{1}{3}x(7x + 12)(x + 2)^{-2/3}$;   critical nos $-2, -\frac{12}{7}, 0$

f': +++++++ dne +++ 0 ----------- 0 ++++

-2 $-\frac{12}{7}$ 0

no extreme   max   min

$f\left(-\frac{12}{7}\right) = \frac{144}{49}\left(\frac{2}{7}\right)^{1/3}$ local max

$f(0) = 0$ local min

21. $f(x) = \begin{cases} 2 - 3x, & x \le -\frac{1}{2} \\ x + 4, & -\frac{1}{2} < x < 3 \\ 3x - 2, & 3 \le x \end{cases}$ $f'(x) = \begin{cases} -3, & x < -\frac{1}{2} \\ 1, & -\frac{1}{2} < x < 3 \\ 3, & 3 < x \end{cases}$

critical nos $-\frac{1}{2}, 3$

f': -------- dne ++++++++ dne ++++++++

$-\frac{1}{2}$ 3

min   no extreme

$f\left(-\frac{1}{2}\right) = \frac{7}{2}$ local min

no local extreme at 3

23. $f'(x) = \frac{2}{3}x^{-4/3}(x - 1)$;   critical nos $0, -1$

f': --------   -------- 0 +++++++++

0 1

no extreme   min

$f(1) = 3$ local min

no local extreme at 0

25. $f'(x) = \cos x - \sin x$;   critical nos $\frac{1}{4}\pi, \frac{5}{4}\pi$

$f''(x) = -\sin x - \cos x$, $f''\left(\frac{1}{4}\pi\right) = -\sqrt{2}$, $f''\left(\frac{5}{4}\pi\right) = \sqrt{2}$

$f\left(\frac{1}{4}\pi\right) = \sqrt{2}$ local max, $f\left(\frac{5}{4}\pi\right) = -\sqrt{2}$ local min

27. $f'(x) = \cos x \left(2\sin x - \sqrt{3}\right)$;   critical nos $\frac{1}{2}\pi, \frac{1}{3}\pi, \frac{2}{3}\pi$

f': -------- 0 +++++ 0 ----- 0 +++++++++

0 $\frac{\pi}{3}$ $\frac{\pi}{2}$ $\frac{2\pi}{3}$ π

min   max   min

$f\left(\frac{1}{3}\pi\right) = f\left(\frac{2}{3}\pi\right) = -\frac{3}{4}$ local mins

$f\left(\frac{1}{2}\pi\right) = 1 - \sqrt{3}$ local max

29. $f'(x) = \cos^2 x - \sin^2 x - 3\cos x + 2 = (2\cos x - 1)(\cos x - 1)$

critical pts $\frac{1}{3}\pi, \frac{5}{3}\pi$

f': ---- 0+ +++++++++++++++++++++ 0 -----

0 $\frac{\pi}{3}$ $\frac{5\pi}{3}$ 2π

min   min

$f\left(\frac{1}{3}\pi\right) = \frac{2}{3}\pi - \frac{5}{4}\sqrt{3}$ local min

$f\left(\frac{5}{3}\pi\right) = \frac{10}{3}\pi + \frac{5}{4}\sqrt{3}$ local max

31. (i) f increases on $(c - \delta, c]$ and decreases on $[c, c + \delta)$.

(ii) f decreases on $(c - \delta, c]$ and increases on $[c, c + \delta)$.

(iii)   If $f'(x) > 0$ on $(c - \delta, c) \cup (c, c + \delta)$, then, since f is continuous at c, f increases on $(c - \delta, c]$ and also on $[c, c + \delta)$. Therefore, in this case, f increases on $(c - \delta, c + \delta)$. A similar argument shows that, if $f'(x) < 0$ on $(c - \delta, c) \cup (c, c + \delta)$, then f decreases on $(c - \delta, c + \delta)$.

33. Solving $f'(x) = 2ax + b = 0$ gives a critical point at $x = -\dfrac{b}{2a}$.   Since $f''(x) = 2a$, f has a local maximum at $-\dfrac{b}{2a}$ if $a < 0$ and a local minimum at $-\dfrac{b}{2a}$ if $a > 0$.

35.
$$P(x) = x^4 - 8x^3 + 22x^2 - 24x + 4$$
$$P'(x) = 4x^3 - 24x^2 + 44x - 24$$
$$P''(x) = 12x^2 - 48x + 44$$

Since $P'(1) = 0$, $P'(x)$ is divisible by $x - 1$. Division by $x - 1$ gives
$$P'(x) = (x - 1)\left(4x^2 - 20x + 24\right) = 4(x - 1)(x - 2)(x - 3).$$
The critical pts are 1, 2, 3. Since
$$P''(1) > 0, \quad P''(2) < 0, \quad P''(3) > 0,$$
$P(1) = -5$ is a local min, $P(2) = -4$ is a local max, and $P(3) = -5$ is a local min.

Since $P'(x) < 0$ for $x < 0$, P decreases on $(-\infty, 0]$. Since $P(0) > 0$, P does not take on the value 0 on $(-\infty, 0]$.

Since $P(0) > 0$ and $P(1) < 0$, P takes on the value 0 at least once on $(0, 1)$. Since $P'(x) < 0$ on $(0, 1)$, P decreases on $[0, 1]$. It follows that P takes on the value zero only once on $[0, 1]$.

Since $P'(x) > 0$ on $(1, 2)$ and $P'(x) < 0$ on $(2, 3)$, P increases on $[1, 2]$ and decreases on $[2, 3]$. Since $P(1)$, $P(2)$, $P(3)$ are all negative, P cannot take on the value 0 between 1 and 3.

Since $P(3) < 0$ and $P(100) > 0$, P takes on the value 0 at least once on $(3, 100)$. Since $P'(x) > 0$ on $(3, 100)$, P increases on $[3, 100]$. It follows that P takes on the value zero only once on $[3, 100]$.

Since $P'(x) > 0$ on $(100, \infty)$, P increases on $[100, \infty)$. Since $P(100) > 0$, P does not take on the value 0 on $[100, \infty)$.

37. (a)   (b)

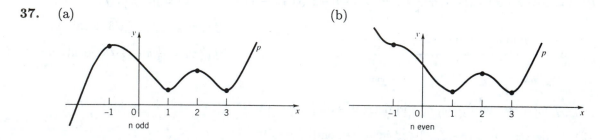

39. Let $f(x) = \dfrac{ax}{x^2 + b^2}$. Then $f'(x) = \dfrac{a\left(b^2 - x^2\right)}{\left(b^2 + x^2\right)^2}$. Now

$$f'(0) = \frac{a}{b^2} = 1 \Rightarrow a = b^2 \text{ and } f'(x) = \frac{b^2 \left(b^2 - x^2\right)}{\left(b^2 + x^2\right)^2}$$

$$f'(-2) = \frac{b^2 \left(b^2 - 4\right)}{\left(b^2 + 4\right)^2} = 0 \Rightarrow b = \pm 2$$

Thus, $a = 4$ and $b = \pm 2$.

41. Let δ be any positive number and consider f on the interval $(-\delta, \delta)$. Let n be a positive integer such that

$$0 < \frac{1}{\frac{\pi}{2} + 2n\pi} < \delta \text{ and } 0 < \frac{1}{\frac{-\pi}{2} + 2n\pi} < \delta.$$

Then

$$f\left(\frac{1}{\frac{\pi}{2} + 2n\pi}\right) > 0 \text{ and } f\left(\frac{1}{\frac{-\pi}{2} + 2n\pi}\right) < 0.$$

Thus f takes on both positive and negative values in every interval centered at 0 and it follows that f cannot have a local maximum or minimum at 0.

43. The function $D(x) = \sqrt{x^2 + [f(x)]^2}$ gives the distance from the origin to the point $(x, f(x))$ on the graph of f. Since the graph of f does not pass through the origin,

$$D'(x) = \frac{x + f(x)f'(x)}{\sqrt{x^2 + [f(x)]^2}}$$

is defined for all $x \in \mathrm{dom}\,(f)$. Suppose that D has a local extreme value at c. Then

$$D'(c) = \frac{c + f(c)f'(c)}{\sqrt{c^2 + [f(c)]^2}} = 0 \Rightarrow c + f(c)f'(c) = 0 \text{ and } f'(c) = -\frac{c}{f(c)}$$

Suppose that $c \neq 0$. The slope of the line through $(0,0)$ and $(c, f(c))$ is given by $m_1 = \dfrac{f(c)}{c}$ and the slope of the tangent line to the graph of f at $x = c$ is given by $m_2 = f'(c) = -\dfrac{c}{f(c)}$. Since $m_1 m_2 = -1$, these two lines are perpendicular. If $c = 0$, then the tangent line to the graph of f is horizontal and the line through $(0,0)$ and $(0, f(0))$ is vertical.

45. (a) Let $f(x) = x^4 - 7x^2 - 8x - 3$. Then $f'(x) = 4x^3 - 14x - 8$ and $f''(x) = 12x^2 - 14$. Since $f'(2) = -4 < 0$ and $f'(3) = 58 > 0$, f' has at least one zero in $(2, 3)$. Since $f''(x) > 0$ for $x \in (2, 3)$, f' is increasing on this interval and so it has exactly one zero. Thus, f has exactly one critical number c in $(2, 3)$.

     (b) $c \cong 2.1091$; f has a local minimum at c.

47. (a) Let $f(x) = \sin x + \dfrac{x^2}{2} - 2x$. Then $f'(x) = \cos x + x - 2$ and $f''(x) = -\sin x + 1$. Since $f'(2) = -0.4161 < 0$ and $f'(3) = 0.01 > 0$, f' has at least one zero in $(2, 3)$. Since $f''(x) > 0$ for $x \in (2, 3)$, f' is increasing on this interval and so it has exactly one zero. Thus, f has exactly one critical number

c in $(2, 3)$.

(b) $x_{n+1} = x_n - \dfrac{\cos x_n + x_n - 2}{-\sin x_n + 1}$; $x_1 = 3$, $x_2 = 2.9883$, $x_3 = 2.9883$. Thus $c \cong 2.9883$;

   f has a local minimum at c.

49. (a)

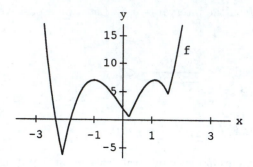

critical numbers: $x_1 \cong -2.085$, $x_2 \cong -1$, $x_3 \cong 0.207$, $x_4 \cong 1.096$, $x_5 = 1.544$

local extreme values: $f(-2.085) \cong -6.255$, $f(-1) = 7$, $f(0.207) \cong 0.621$, $f(1.096) \cong 7.097$,

$f(1.544) \cong 4.635$

(b) f is increasing on $[-2.085, -1]$, $[0.207, 1.096]$, and $[1.544, 4]$

   f is decreasing on $[-4, -2.085]$, $[-1, 0.207]$, and $[1.096, 1.544]$

51.

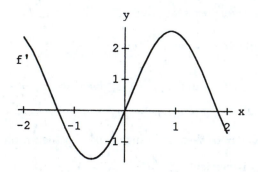

critical numbers of f: $x_1 \cong -1.326$, $x_2 = 0$, $x_3 \cong 1.816$

$f''(-1.326) \cong -4 < 0$ $\Rightarrow$ f has a local maximum at $x = -1.326$

$f''(0) = 4 > 0$ $\Rightarrow$ f has a local minimum at $x = 0$

$f''(1.816) \cong -4$ $\Rightarrow$ f has a local maximum at $x = 1.816$

SECTION 4.4

1. $f'(x) = \frac{1}{2}(x+2)^{-1/2}$, $x > -2$;

++++++++
f' : ●————————→
         -2 x
         min

critical no -2;

$f(-2) = 0$ endpt and abs min;   as $x \to \infty$, $f(x) \to \infty$;   so no abs max

3. $f'(x) = 2x - 4$, $x \in (0,3)$;

f' :

$$---------0++++++$$
$$\begin{array}{ccc} 0 & 2 & 0 \\ \textbf{max} & \textbf{min} & \textbf{max} \end{array}$$

critical nos $0, 2, 3$;

$f(0) = 1$ endpt and abs max, $f(2) = -3$ local and abs min, $f(3) = -2$ endpt max

5. $f'(x) = 2x - \dfrac{1}{x^2} = \dfrac{2x^3 - 1}{x^2}$, $x \neq 0$; $f'(x) = 0$ at $x = 2^{-1/3}$

critical no $2^{-1/3}$; $f''(x) = 2 + \dfrac{2}{x^3}$, $f''\left(2^{-1/3}\right) = 6$

$f\left(2^{-1/3}\right) = 2^{-2/3} + 2^{1/3} = 2^{-2/3} + 2 \cdot 2^{-2/3} = 3 \cdot 2^{-2/3}$ local min

7. $f'(x) = \dfrac{2x^3 - 1}{x^2}$, $x \in \left(\dfrac{1}{10}, 2\right)$;

f' :

$$------0++++++++++$$
$$\begin{array}{ccc} \frac{1}{10} & 2^{-\frac{1}{3}} & 2 \\ \textbf{max} & \textbf{min} & \textbf{max} \end{array}$$

critical nos $\frac{1}{10}, 2^{-1/3}, 2$;

$f\left(\frac{1}{10}\right) = 10\frac{1}{100}$ endpt and abs max, $f\left(2^{-1/3}\right) = 3 \cdot 2^{-2/3}$ local and abs min,

$f(2) = 4\frac{1}{2}$ endpt max

9. $f'(x) = 2x - 3$, $x \in (0,2)$;

f'

$$------------------0+++++++$$
$$\begin{array}{ccc} 0 & \frac{3}{2} & 2 \\ \textbf{max} & \textbf{min} & \textbf{max} \end{array}$$

critical nos $0, \frac{3}{2}, 2$;

$f(0) = 2$ endpt and abs max, $f\left(\frac{3}{2}\right) = -\frac{1}{4}$ local and abs min,

$f(2) = 0$ endpt max

11. $f'(x) = \dfrac{(2-x)(2+x)}{(4+x^2)^2}$, $x \in (-3,1)$;

f' :

$$------0+++++++++++++++$$
$$\begin{array}{ccc} -3 & -2 & 1 \end{array}$$

critical nos $-3, -2, 1$;

$f(-3) = -\frac{3}{13}$ endpt max, $f(-2) = -\frac{1}{4}$ local and abs min, $f(1) = \frac{1}{5}$ endpt and abs max

13. $f'(x) = 2\left(x - \sqrt{x}\right)\left(1 - \dfrac{1}{2\sqrt{x}}\right)$, $x > 0$;

f' :

$$++++++0-------------0++++$$
$$\begin{array}{ccc} 0 & \frac{1}{4} & 1 \end{array} \quad x$$

critical nos $0, \frac{1}{4}, 1$;

$f(0) = 0$ endpt and abs min, $f\left(\frac{1}{4}\right) = \frac{1}{16}$ local max, $f(1) = 0$ local and abs min;

as $x \to \infty$, $f(x) \to \infty$;   so no abs max

15. $f'(x) = \dfrac{3(2-x)}{2\sqrt{3-x}}$, $x < 3$

f' :

$$+++++++++++++++0------$$
$$\begin{array}{cc} 2 & 3 \end{array}$$

critical nos $2, 3$;

$f(2) = 2$ local and abs max, $f(3) = 0$ endpt min;

as $x \to -\infty$, $f(x) \to -\infty$;   so no abs min

17. $f'(x) = -\frac{1}{3}(x-1)^{-2/3}, \quad x \neq 1;$

$f':$

$$\text{- - - - - - - - dne - - - - - - - - - - - -}$$
$$1 \quad\quad x$$

critical no 1;

no local extremes; $\begin{array}{l} \text{as} \quad x \to \infty, \quad f(x) \to -\infty \\ \text{as} \quad x \to -\infty, \quad f(x) \to \infty \end{array} \Bigg\}$ no abs extremes

19. $f'(x) = \sin x \left(2\cos x + \sqrt{3}\right), x \in (0, \pi);$

$f':$

$$\text{+ + + + + + + + + + + + + + + + + 0 - - - - -}$$
$$0 \quad\quad \frac{5\pi}{6} \quad\quad \pi$$

critical nos $0, \frac{5}{6}\pi, \pi$;

$f(0) = -\sqrt{3}$ endpt and abs min, $f\left(\frac{5}{6}\pi\right) = \frac{7}{4}$ local and abs max, $f(\pi) = \sqrt{3}$ endpt min

21. $f'(x) = -3\sin x \left(2\cos^2 x + 1\right) < 0, \quad x \in (0, \pi);$ critical nos $0, \pi$;

$f(0) = 5$ endpt and abs max, $f(\pi) = -5$ endpt and abs min

23. $f'(x) = \sec^2 x - 1 \geq 0, \quad x \in \left(-\frac{1}{3}\pi, \frac{1}{2}\pi\right);$ critical nos $-\frac{1}{3}\pi, 0;$

$f\left(-\frac{1}{3}\pi\right) = \frac{1}{3}\pi - \sqrt{3}$ endpt and abs min, no abs max

25.

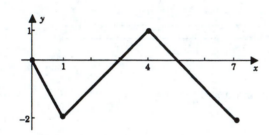

$$f'(x) = \begin{cases} -2, & 0 < x < 1 \\ 1, & 1 < x < 4 \\ -1, & 4 < x < 7 \end{cases}$$

critical nos 0, 1, 4, 7

$f(0) = 0$ endpt max, $f(1) = -2$ local and abs min,

$f(4) = 1$ local and absolute max, $f(7) = -2$ endpt and abs min

27.

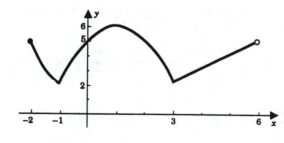

$$f'(x) = \begin{cases} 2x, & -2 < x < -1 \\ 2 - 2x, & -1 < x < 3 \\ 1, & 3 < x < 6 \end{cases}$$

critical nos $-2, -1, 1, 3$

$f(-2) = 5$ endpt max, $f(-1) = 2$ local and abs min,

$f(1) = 6$ local and abs max, $f(3) = 2$ local and abs min

29.

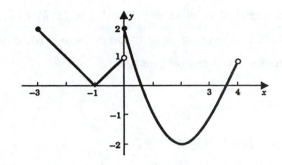

$$f'(x) = \begin{cases} -1, & -3 < x < -1 \\ 1, & -1 < x < 0 \\ 2x - 4, & 0 < x < 3 \\ 2, & 3 \le x < 4 \end{cases}$$

critical nos $-3, -1, 0, 2$

$f(-3) = 2$ endpt and abs max, $f(-1) = 0$ local min,

$f(0) = 2$ local and abs max, $f(2) = -2$ local and abs min

31.

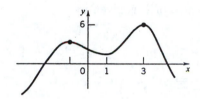

33. $f(-3) = 0$ and $f'(x) > 0$ on $(-3, -1)$

$\Rightarrow f(-1) > 0.$

$f(3) = 0$ and $f'(x) > 0$ on $(1, 3) \Rightarrow f(1) < 0.$

It now follows that f has a zero on $(-1, 1)$,

contradicting the fact that $f(x) \ne 0$ for

$x \in (-3, 3).$

35.   Let $p(x) = x^3 + ax^2 + bx + c$. Then $p'(x) = 3x^2 + 2ax + b$ is a quadratic with discriminant $\Delta = 4a^2 - 12b = 4(a^2 - 3b)$. If $a^2 \le 3b$, then $\Delta \le 0$. This implies that $p'(x)$ does not change sign on $(-\infty, \infty)$ and hence p is either increasing on $(-\infty, \infty)$ (if $a \le 0$) or decreasing ($a \ge 0$). In either case, p has no extreme values. On the other hand, if $a^2 - 3b > 0$, then $\Delta > 0$ and p' has two real zeros, c_1 and c_2, from which it follows that p has extreme values at c_1 and c_2. Thus, if p has no extreme values, then we must have $a^2 - 3b \le 0$.

37.   By contradiction. If f is continuous at c, then, by the first-derivative test (4.3.4), $f(c)$ is not a local maximum.

39. If f is not differentiable on (a, b), then f has a critical point at each point c in (a, b) where $f'(c)$ does not exist. If f is differentiable on (a, b), then by the mean-value theorem there exists c in (a, b) where $f'(c) = [f(b) - f(a)]/(b - a) = 0$. This means c is a critical point of f.

41. Let M be a positive number. Then

$$P(x) - M \geq a_n x^n - \left(|a_{n-1}|x^{n-1} + \cdots + |a_1|x + |a_0| + M\right) \quad \text{for} \quad x > 0$$

$$\geq a_n x^n - \left(|a_{n-1}| + \cdots + |a_1| + |a_0| + M\right) \quad \text{for} \quad x > 1$$

It now follows that

$$P(x) - M \geq 0 \quad \text{for} \quad x \geq K = \left(\frac{|a_{n-1}| + \cdots + |a_1| + |a_0| + M}{a_n}\right)^{1/n} + 1.$$

43. Let R be a rectangle with its diagonals having length c, and let x be the length of one of its sides. Then the length of the other side is $y = \sqrt{c^2 - x^2}$ and the area of R is given by

$$A(x) = x\sqrt{c^2 - x^2}$$

Now

$$A'(x) = \sqrt{c^2 - x^2} - \frac{x^2}{\sqrt{c^2 - x^2}}$$

$$= \frac{c^2 - 2x^2}{\sqrt{c^2 - x^2}},$$

and

$$A'(x) = 0 \implies x = \frac{\sqrt{2}}{2}c$$

It is easy to verify that A has a maximum at $x = \frac{\sqrt{2}}{2}c$. Since $y = \frac{\sqrt{2}}{2}c$ when $x = \frac{\sqrt{2}}{2}c$, it follows that the rectangle of maximum area is a square

45. Setting $R'(x) = \frac{v^2 \cos 2x}{16} = 0$, gives $x = \frac{\pi}{4}$. Since $R''(\frac{\pi}{4}) = -\frac{v^2}{8} < 0$, $x = \frac{\pi}{4}$ is a maximum.

47.

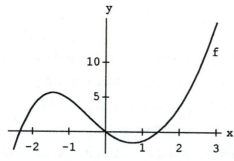

critical nos: $x_1 = -1.452$, $x_2 = 0.760$

$f(-1.452)$ local maximum

$f(0.727)$ local minimum

$f(3)$ absolute maximum

$f(-2.5)$ absolute minimum

49.

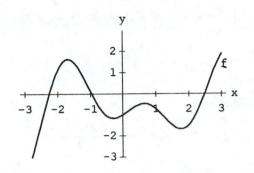

critical numbers: $x_1 = -1.683$, $x_2 = -0.284$,

$x_3 = 0.645$, $x_4 = 1.760$

$f(-1.683), f(0.645)$ local maxima

$f(-0.284), f(1.760)$ local minima

$f(\pi)$ absolute maximum

$f(-\pi)$ absolute minimum

SECTION 4.5

1. Set $P = xy$ and $y = 40 - x$. We want to maximize

$$P(x) = x(40 - x), \quad 0 < x < 40.$$

$$P'(x) = 40 - 2x, \quad P'(x) = 0 \implies x = 20.$$

Since P increases on $(0, 20]$ and decreases on $[20, 40)$, the abs max of P occurs when $x = 20$. Then, $y = 20$ and $xy = 400$.

The maximal value of xy is 400.

3.

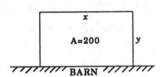

Minimize P

$$P = x + 2y, \quad 200 = xy, \quad y = 200/x$$

$$P(x) = x + \frac{400}{x}, \quad 0 < x.$$

$$P'(x) = 1 - \frac{400}{x^2}, \quad P'(x) = 0 \implies x = 20.$$

Since P decreases on $(0, 20]$ and increases on $[20, \infty)$, the abs min of P occurs when $x = 20$.

To minimize the fencing, make the garden 20 ft (parallel to barn) by 10 ft.

5.

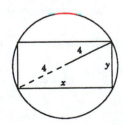

Maximize A

$$A = xy, \quad x^2 + y^2 = 8^2, \quad y = \sqrt{64 - x^2}$$

$$A(x) = x\sqrt{64 - x^2}, \quad 0 < x < 8.$$

$$A'(x) = \sqrt{64 - x^2} + x\left(\frac{-x}{\sqrt{64 - x^2}}\right) = \frac{64 - 2x^2}{\sqrt{64 - x^2}}, \quad A'(x) = 0 \implies x = 4\sqrt{2}.$$

Since A increases on $(0, 4\sqrt{2}\,]$ and decreases on $[4\sqrt{2}, 8)$, the abs max of A occurs when $x = 4\sqrt{2}$. Then, $y = 4\sqrt{2}$ and $xy = 32$.

The maximal area is 32.

7.

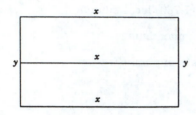

<u>Maximize A</u>

$$A = xy, \quad 2y + 3x = 600, \quad y = \frac{600 - 3x}{2}$$

$A(x) = x\left(300 - \frac{3}{2}x\right), \quad 0 < x < 200.$

$A'(x) = 300 - 3x, \quad A'(x) = 0 \implies x = 100.$

Since A increases on $(0, 100]$ and decreases on $[100, 200)$, the abs max of A occurs when $x = 100$. Then, $y = 150.$

The playground of greatest area measures 100 ft by 150 ft. (The fence divider is 100 ft long.)

9.

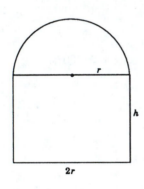

<u>Maximize L</u>

To account for the semi-circular portion admitting less light per square foot, we multiply its area by $1/3$.

$$L = 2rh + \frac{1}{3}\left(\frac{\pi r^2}{2}\right),$$

$$2r + 2h + \pi r = 24, \quad h = \tfrac{1}{2}(24 - 2r - \pi r)$$

$$L = 2r\left(\frac{24 - 2r - \pi r}{2}\right) + \frac{1}{6}\pi r^2$$

$$L(r) = 24r - \left(2 + \frac{5}{6}\pi\right)r^2, \quad 0 < r < \frac{30}{2 + \pi}.$$

$$L'(r) = 24 - \left(4 + \frac{5}{3}\pi\right)r, \quad L'(r) = 0 \implies r = \frac{72}{12 + 5\pi}.$$

Since $L''(r) < 0$ for all r in the domain of L, the local max at $r = 72/(12 + 5\pi)$ is the abs max.

For the window that admits the most light, take the radius of the semicircle as $\dfrac{72}{12 + 5\pi} \cong 2.6$ ft and the height of the rectangular portion as $\dfrac{72 + 24\pi}{12 + 5\pi} \cong 5.32$ ft.

11.

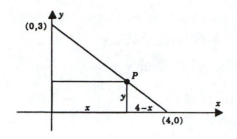

Maximize A

$$A = xy, \quad \frac{3}{4} = \frac{y}{4-x} \qquad \text{(similar triangles)}$$

$$y = \tfrac{3}{4}(4-x)$$

$$A(x) = \frac{3x}{4}(4-x), \quad 0 < x < 4.$$

$$A'(x) = 3 - \frac{3x}{2}, \quad A'(x) = 0 \implies x = 2.$$

Since A increases on $(0, 2]$ and decreases on $[2, 4)$, the abs max of A occurs when $x = 2$.

To maximize the area of the rectangle, take P as the point $\left(2, \tfrac{3}{2}\right)$.

13.

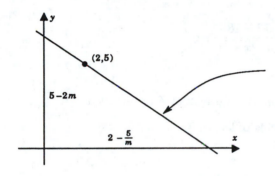

Minimize A

$$A = \tfrac{1}{2}(x\text{-intercept})\,(y\text{-intercept})$$

Equation of line: $y - 5 = m(x - 2)$

x-intercept: $2 - \dfrac{5}{m}$

y-intercept: $5 - 2m$

$$A = \frac{1}{2}\left(2 - \frac{5}{m}\right)(5 - 2m) = 10 - 2m - \frac{25}{2m}$$

$$A(m) = 10 - 2m - \frac{25}{2m}, \quad m < 0.$$

$$A'(m) = -2 + \frac{25}{2m^2}, \quad A'(m) = 0 \implies m = -\frac{5}{2}.$$

Since $A''(m) = -25/m^3 > 0$ for $m < 0$, the local min at $m = -5/2$ is the abs min.

The triangle of minimal area is formed by the line of slope $-5/2$.

15.

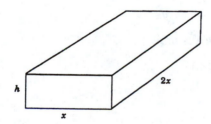

Maximize V

$$V = 2x^2 h, \quad 2\left(2x^2 + xh + 2xh\right) = 100, \quad h = \frac{50 - 2x^2}{3x}$$

$$V = 2x^2\left(\frac{50 - 2x^2}{3x}\right)$$

$$V(x) = \tfrac{100}{3}x - \tfrac{4}{3}x^3, \quad 0 < x < 5.$$

$V'(x) = \frac{100}{3} - 4x^2, \quad V'(x) = 0 \implies x = \frac{5}{3}\sqrt{3}.$

Since $V''(x) = -8x < 0$ on $(0,5)$, the local max at $x = \frac{5}{3}\sqrt{3}$ is the abs max.

The base of the box of greatest volume measures $\frac{5}{3}\sqrt{3}$ in. by $\frac{10}{3}\sqrt{3}$ in.

17.

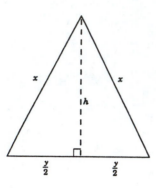

Maximize A

$A = \frac{1}{2}hy$

$2x + y = 12 \implies y = 12 - 2x$

Pythagorean Theorem:

$h^2 + \left(\frac{y}{2}\right)^2 = x^2 \implies h = \sqrt{x^2 - \left(\frac{y}{2}\right)^2}$

Thus, $h = \sqrt{x^2 - (6-x)^2} = \sqrt{12x - 36}.$

$A(x) = (6-x)\sqrt{12x - 36}, \quad 3 < x < 6.$

$A'(x) = -\sqrt{12x - 36} + (6-x)\left(\frac{6}{\sqrt{12x - 36}}\right) = \frac{72 - 18x}{\sqrt{12x - 36}},$

$A'(x) = 0 \implies x = 4.$

Since A increases on $(3,4]$ and decreases on $[4,6)$, the abs max of A occurs at $x = 4$.

The triangle of maximal area is equilateral with side of length 4.

19.

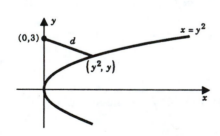

Minimize d

$d = \sqrt{(y^2 - 0)^2 + (y - 3)^2}$

The square-root function is increasing;

d is minimal when $D = d^2$ is minimal.

$D(y) = y^4 + (y-3)^2, \quad y \text{ real.}$

$D'(y) = 4y^3 + 2(y-3) = (y-1)\left(4y^2 + 4y + 6\right), \quad D'(y) = 0 \text{ at } y = 1.$

Since $D''(y) = 12y^2 + 2 > 0$, the local min at $y = 1$ is the abs min.

The point $(1,1)$ is the point on the parabola closest to $(0,3)$.

21.

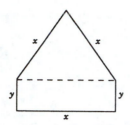

<u>Maximize A</u>

$$A = xy + \frac{\sqrt{3}}{4}x^2, \quad 30 = 3x + 2y, \quad y = \frac{30 - 3x}{2}$$

$A(x) = 15x - \frac{3}{2}x^2 + \frac{\sqrt{3}}{4}x^2, \quad 0 < x < 10.$

$A'(x) = 15 - 3x + \frac{\sqrt{3}}{2}x, \quad A'(x) = 0 \implies x = \frac{30}{6 - \sqrt{3}} = \frac{10}{11}\left(6 + \sqrt{3}\right).$

Since $A''(x) = -3 + \frac{\sqrt{3}}{2} < 0$ on $(0, 10)$, the local max at $x = \frac{10}{11}\left(6 + \sqrt{3}\right)$ is the abs max.

The pentagon of greatest area is composed of an equilateral triangle with side $\frac{10}{11}\left(6 + \sqrt{3}\right) \cong 7.03$ in. and rectangle with height $\frac{15}{11}\left(5 - \sqrt{3}\right) \cong 4.46$ in.

23.

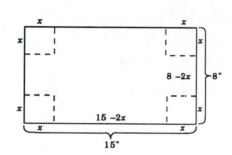

<u>Maximize V</u>

$$V = x(8 - 2x)(15 - 2x)$$

$\left.\begin{array}{r} x > 0 \\ 8 - 2x > 0 \\ 15 - 2x > 0 \end{array}\right\} \implies 0 < x < 4$

$V(x) = 120x - 46x^2 + 4x^3, \quad 0 < x < 4.$

$V'(x) = 120 - 92x + 12x^2 = 4(3x - 5)(x - 6), \quad V'(x) = 0$ at $x = \frac{5}{3}$.

Since V increases on $\left(0, \frac{5}{3}\right]$ and decreases on $\left[\frac{5}{3}, 4\right)$, the abs max of V occurs when $x = \frac{5}{3}$.

The box of maximal volume is made by cutting out squares $5/3$ inches on a side.

25.

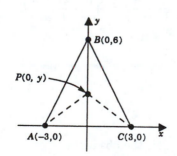

<u>Minimize $\overline{AP} + \overline{BP} + \overline{CP} = S$</u>

length $AP = \sqrt{9 + y^2}$

length $BP = 6 - y$

length $CP = \sqrt{9 + y^2}$

$S(y) = 6 - y + 2\sqrt{9 + y^2}, \quad 0 \le y \le 6.$

$S'(y) = -1 + \frac{2y}{\sqrt{9 + y^2}}, \quad S'(y) = 0 \implies y = \sqrt{3}.$

Since

$$S(0) = 12, \quad S\left(\sqrt{3}\right) = 6 + 3\sqrt{3} \cong 11.2, \quad \text{and} \quad S(6) = 6\sqrt{5} \cong 13.4,$$

the abs min of S occurs when $y = \sqrt{3}$.

To minimize the sum of the distances, take P as the point $\left(0, \sqrt{3}\right)$.

27.

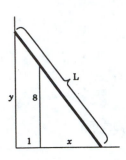

Minimize L

$L^2 = y^2 + (x+1)^2$.

By similar triangles $\dfrac{y}{x+1} = \dfrac{8}{x}$, $y = \dfrac{8}{x}(x+1)$.

$L^2 = \left[\left(\dfrac{8}{x}\right)(x+1)\right]^2 + (x+1)^2 = (x+1)^2\left(\dfrac{64}{x^2}+1\right)$

Since L is minimal when L^2 is minimal, we consider the function

$$f(x) = (x+1)^2\left(\dfrac{64}{x^2}+1\right), \quad x > 0.$$

$$f'(x) = 2(x+1)\left(\dfrac{64}{x^2}+1\right) + (x+1)^2\left(\dfrac{-128}{x^3}\right)$$

$$= \dfrac{2(x+1)}{x^3}\left[x^3 - 64\right], \quad f'(x) = 0 \implies x = 4.$$

Since f decreases on $(0,4]$ and increases on $[4,\infty)$, the abs min of f occurs when $x = 4$.

The shortest ladder is $5\sqrt{5}$ ft long.

29.

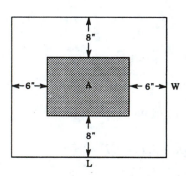

Maximize A

(We use feet rather than inches to reduce arithmetic.)

$A = (L-1)\left(W - \frac{4}{3}\right)$

$LW = 27 \implies W = \dfrac{27}{L}$

$A = (L-1)\left(\dfrac{27}{L} - \dfrac{4}{3}\right) = \dfrac{85}{3} - \dfrac{27}{L} - \dfrac{4}{3}L$

$A(L) = \dfrac{85}{3} - \dfrac{27}{L} - \dfrac{4}{3}L, \quad 1 < L < \dfrac{81}{4}$.

$A'(L) = \dfrac{27}{L^2} - \dfrac{4}{3}, \quad A'(L) = 0 \implies L = \dfrac{9}{2}$.

Since $A'(L) = -54/L^3 < 0$ for $1 < L < \frac{81}{4}$, the max at $L = \frac{9}{2}$ is the abs max.

The banner has length $9/2$ ft $= 54$ in. and height 6 ft $= 72$ in.

31.

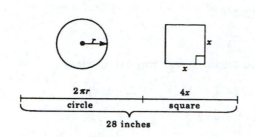

Find the extreme values of A

$$A = \pi r^2 + x^2$$

$$2\pi r + 4x = 28 \quad \Longrightarrow \quad x = 7 - \tfrac{1}{2}\pi r.$$

$$A(r) = \pi r^2 + \left(7 - \frac{1}{2}\pi r\right)^2, \quad 0 \le r \le \frac{14}{\pi}.$$

Note: the endpoints of the domain correspond to the instances when the string is not cut: $r = 0$ when no circle is formed, $r = 14/\pi$ when no square is formed.

$$A'(r) = 2\pi r - \pi\left(7 - \frac{1}{2}\pi r\right), \quad A'(r) = 0 \quad \Longrightarrow \quad r = \frac{14}{4 + \pi}.$$

Since $A''(r) = 2\pi + \pi^2/2 > 0$ on $(0, 14/\pi)$, the abs min of A occurs when $r = 14/(4 + \pi)$ and the abs max of A occurs at one of the endpts: $A(0) = 49$, $A(14/\pi) = 196/\pi > 49$.

(a) To maximize the sum of the two areas, use all of the string to form the circle.

(b) To minimize the sum of the two areas, use $2\pi r = 28\pi/(4 + \pi) \cong 12.32$ inches of string for the circle.

33.

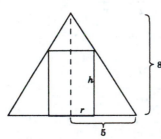

Maximize V
$V = \pi r^2 h$
By similar triangles
$$\frac{8}{5} = \frac{h}{5 - r} \quad \text{or} \quad h = \frac{8}{5}(5 - r).$$

$$V(r) = \frac{8\pi}{5}r^2(5 - r), \quad 0 < r < 5.$$

$$V'(r) = \frac{8\pi}{5}\left(10r - 3r^2\right), \quad V'(r) = 0 \quad \Longrightarrow \quad r = 10/3.$$

Since V increases on $(0, 10/3]$ and decreases on $[10/3, 5)$, the abs max of V occurs when $r = 10/3$.

The cylinder with maximal volume has radius $10/3$ and height $8/3$.

35.

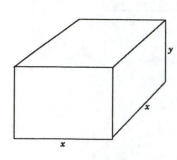

<u>Minimize C</u>

In dollars,

$C =$ cost base $+$ cost top $+$ cost sides

$\quad = .35\left(x^2\right) + .15\left(x^2\right) + .20(4xy)$

$\quad = \frac{1}{2}x^2 + \frac{4}{5}xy$

Volume $= x^2y = 1250 \quad y = \dfrac{1250}{x^2}$

$C(x) = \dfrac{1}{2}x^2 + \dfrac{1000}{x}, \quad x > 0.$

$C'(x) = x - \dfrac{1000}{x^2}, \quad C'(x) = 0 \quad \Longrightarrow \quad x = 10.$

Since $C''(x) = 1 + 2000/x^3 > 0$ for $x > 0$, the local min of C at $x = 10$ is the abs min.

The least expensive box is 12.5 ft tall with a square base 10 ft on a side.

37.

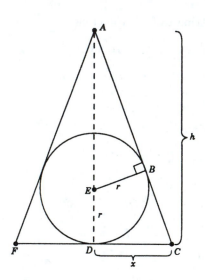

<u>Minimize A</u>

$A = \frac{1}{2}(h)(2x) = hx$

Triangles ADC and ABE are similar:

$\dfrac{AD}{DC} = \dfrac{AB}{BE} \quad$ or $\quad \dfrac{h}{x} = \dfrac{AB}{r}.$

Pythagorean Theorem:

$r^2 + (AB)^2 = (h - r)^2.$

Thus

$r^2 + \left(\dfrac{hr}{x}\right)^2 = (h - r)^2.$

Solving this equation for h we find that

$h = \dfrac{2x^2r}{x^2 - r^2}.$

$A(x) = \dfrac{2x^3r}{x^2 - r^2}, \quad x > r.$

$A'(x) = \dfrac{\left(x^2 - r^2\right)\left(6x^2r\right) - 2x^3r(2x)}{\left(x^2 - r^2\right)^2} = \dfrac{2x^2r\left(x^2 - 3r^2\right)}{\left(x^2 - r^2\right)^2},$

$A'(x) = 0 \quad \Longrightarrow \quad x = r\sqrt{3}.$

Since A decreases on $\left(r, r\sqrt{3}\,\right]$ and increases on $\left[r\sqrt{3}, \infty\right)$, the local min at $x = r\sqrt{3}$ is the abs min of A. When $x = r\sqrt{3}$, we get $h = 3r$ so that $FC = 2r\sqrt{3}$ and $AF = FC = \sqrt{h^2 + x^2} = 2r\sqrt{3}.$

The triangle of least area is equilateral with side of length $2r\sqrt{3}$.

39.

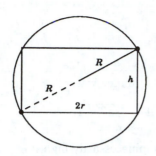

<u>Maximize V</u>

$V = \pi r^2 h$

By the Pythagorean Theorem,

$(2r)^2 + h^2 = (2R)^2$

so

$h = 2\sqrt{R^2 - r^2}.$

$V(r) = 2\pi r^2 \sqrt{R^2 - r^2}, \quad 0 < r < R.$

$V'(r) = 2\pi \left[2r\sqrt{R^2 - r^2} - \dfrac{r^3}{\sqrt{R^2 - r^2}} \right] = \dfrac{2\pi r \left(2R^2 - 3r^2 \right)}{\sqrt{R^2 - r^2}}$

$V'(r) = 0 \quad \Longrightarrow \quad r = \tfrac{1}{3} R\sqrt{6}.$

Since V increases on $\left(0, \tfrac{1}{3} R\sqrt{6}\right]$ and decreases on $\left[\tfrac{1}{3} R\sqrt{6}, R\right)$, the local max at $r = \tfrac{1}{3} R\sqrt{6}$ is the abs max.

The cylinder of maximal volume has base radius $\tfrac{1}{3} R\sqrt{6}$ and height $\tfrac{2}{3} R\sqrt{3}$.

41.

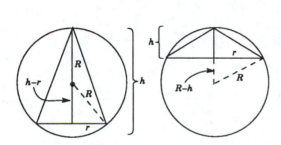

<u>Maximize V</u>

$V = \tfrac{1}{3}\pi r^2 h$

Pythagorean Theorem

Case 1 : $\quad (h - R)^2 + r^2 = R^2$

Case 2 : $\quad (R - h)^2 + r^2 = R^2$

Case 1 : $h \geq R$ Case 2 : $h \leq R$

In both cases

$r^2 = R^2 - (R - h)^2 = 2hR - h^2.$

$V(h) = \tfrac{1}{3}\pi \left(2h^2 R - h^3 \right), \quad 0 < h < 2R.$

$V'(h) = \dfrac{1}{3}\pi \left(4hR - 3h^2 \right), \quad V'(h) = 0 \quad \text{at} \quad h = \dfrac{4R}{3}.$

Since V increases on $\left(0, \tfrac{4}{3} R\right]$ and decreases on $\left[\tfrac{4}{3} R, 2R\right)$, the local max at $h = \tfrac{4}{3} R$ is the abs max.

The cone of maximal volume has height $\tfrac{4}{3} R$ and radius $\tfrac{2}{3} R\sqrt{2}$.

43.

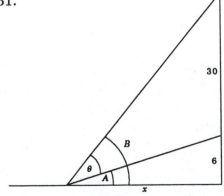

Minimize C

In units of $\$10,000$,

$$C = \frac{\text{cost of cable}}{\text{underground}} + \frac{\text{cost of cable}}{\text{under water}}$$

$$= 3(4-x) + 5\sqrt{x^2+1}.$$

Clearly, the cost is unnecessarily high if

$$x > 4 \quad \text{or} \quad x < 0.$$

$C(x) = 12 - 3x + 5\sqrt{x^2+1}, \ \ 0 \le x \le 4.$

$C'(x) = -3 + \dfrac{5x}{\sqrt{x^2+1}}, \ \ C'(x) = 0 \ \implies \ x = 3/4.$

Since the domain of C is closed, the abs min can be identified by evaluating C at each critical point:

$$C(0) = 17, \quad C\left(\tfrac{3}{4}\right) = 16, \quad C(4) = 5\sqrt{17} \cong 20.6.$$

The minimum cost is $\$160,000$.

45. $P'(\theta) = \dfrac{-mW(m\cos\theta - \sin\theta)}{(m\sin\theta + \cos\theta)^2}; \quad P$ is minimized when $\tan\theta = m.$

47. Minimize $I = \dfrac{a}{x^2} + \dfrac{b}{(s-x)^2}.$

$I'(x) = -\dfrac{2a}{x^3} + \dfrac{2b}{(s-x)^3}, \quad I'(x) = 0 \implies x = \dfrac{a^{\frac{1}{3}}s}{a^{\frac{1}{3}} + b^{\frac{1}{3}}}.$

49. The slope of the line through (a,b) and $(x, f(x))$ is $\dfrac{f(x) - b}{x - a}.$

Let $D(x) = [x-a]^2 + [b - f(x)]^2.$ Then $D'(x) = 0$

$\implies \quad 2[x-a] + 2[b - f(x)]f''(x) = 0$

$\implies \quad f'(x) = -\dfrac{x-a}{b - f(x)}.$

51.

Maximize θ

Since the tangent function increases on $[0, \pi/2)$, we can maximize θ by maximizing $\tan\theta$.

$$\tan\theta = \tan(B - A)$$

$$= \frac{\tan B - \tan A}{1 + \tan B \tan A}$$

$$= \frac{36/x - 6/x}{1 + (36/x)(6/x)} = \frac{30x}{x^2 + 216}.$$

Thus, we consider

$$f(x) = \frac{30x}{x^2 + 216}, \quad x \geq 0.$$

$$f'(x) = \frac{(x^2 + 216)\,30 - 30x(2x)}{(x^2 + 216)^2} = \frac{30\left(216 - x^2\right)}{(x^2 + 216)^2},$$

$$f'(x) = 0 \quad \Longrightarrow \quad x = 6\sqrt{6}.$$

Since f increases on $[0, 6\sqrt{6}]$ and decreases on $[6\sqrt{6}, \infty)$, the local max at $x = 6\sqrt{6}$ is the abs max. The observer should sit $6\sqrt{6}$ ft from the screen.

53.   Let x be the number of customers and P the net profit in dollars. Then $0 \leq x \leq 250$ and

$$P(x) = \begin{cases} 12x, & 0 \leq x \leq 50 \\ [12 - 0.06(x - 50)]x, & 50 < x \leq 250; \end{cases}$$

$$P'(x) = \begin{cases} 12, & 0 \leq x \leq 50 \\ 62x - 0.06x^2, & 50 < x \leq 250. \end{cases}$$

The critical numbers are $x = 0$, $x = 50$, $x = 125$, and $x = 250$. From $P(0) = 0$, $P(50) = 600$, $P(125) = 937.50$, and $P(250) = 0$, we conclude that the net profit is maximized by servicing 125 customers.

55. $A'(x) = \dfrac{xC'(x) - C(x)}{x^2}, \quad A'(x) = 0 \quad \Longrightarrow \quad C'(x) = \dfrac{C(x)}{x}.$

57. $y = mx - \dfrac{1}{800}(m^2 + 1)x^2.$ When $y = 0$, $\quad x = \dfrac{800m}{m^2 + 1}.$

Differentiating x with respect to m, $\quad x' = \dfrac{800 - 800m^2}{(m^2 + 1)^2} = 0$

$\Longrightarrow \quad m = 1.$

59.   Driving at x mph, the trip takes $\dfrac{300}{x}$ hours and uses $\left(2 + \dfrac{1}{600}x^2\right)\dfrac{300}{x}$ gallons of fuel.

Thus the expences are $\quad E = 1.35\left(2 + \dfrac{1}{600}x^2\right)\dfrac{300}{x} - 13\left(\dfrac{300}{x}\right) = 0.675x - \dfrac{3090}{x}.$

Differentiating, $\quad E' = 0.675 + \dfrac{3090}{x^2}$, which never equals zero.

Thus the minimal expenses occur at the endpoints:

$$x = 35 \quad \text{or} \quad x = 55.$$

Evaluating E at these points shows that the mnimal expenses occur when the truck is driven at 55 mph.

PROJECT 4.5

1. Distance over water: $\sqrt{36 + x^2}$.

Distance over land: $12 - x$.

Total energy: $E(x) = W\sqrt{36 + x^2} + L(12 - x)$.

3. (a) $W = kL$, $k > 1$, so $E(x) = kL\sqrt{36 + x^2} + L(12 - x)$, for $0 \leq x \leq 12$.

$E'(x) = \dfrac{kLx}{\sqrt{36 + x^2}} - L = 0 \implies x = \dfrac{6}{\sqrt{k^2 - 1}}.$

$E'(x) < 0$ on $(0, \dfrac{6}{\sqrt{k^2 - 1}})$ and $E'(x) > 0$ on $(\dfrac{6}{\sqrt{k^2 - 1}}, 12)$, so E has an absolute minimum

at $\dfrac{6}{\sqrt{k^2 - 1}}$.

(b) As k increases, x decreases.

As $k \to 1^+$, x increases.

(c) $x = 12 \implies k = \dfrac{\sqrt{5}}{2} \simeq 1.12.$

(d) No

SECTION 4.6

1. (a) f is increasing on $[a, b]$, $[d, n]$; f is decreasing on $[b, d]$, $[n, p]$.

(b) The graph of f is concave up on (c, k), (l, m);

The graph of f is concave down on (a, c), (k, l), (m, p).

The x-coordinates of the points of inflection are: $x = c$, $x = k$, $x = l$, $x = m$.

3. $f'(x) = -x^{-2}$, $f''(x) = 2x^{-3}$;

concave down on $(-\infty, 0)$, concave up on $(0, \infty)$; no pts of inflection

5. $f'(x) = 3x^2 - 3$, $f''(x) = 6x$;

concave down on $(-\infty, 0)$, concave up on $(0, \infty)$; pt of inflection $(0, 2)$

7. $f'(x) = x^3 - x$, $f''(x) = 3x^2 - 1$;

concave up on $\left(-\infty, -\frac{1}{3}\sqrt{3}\right)$ and $\left(\frac{1}{3}\sqrt{3}, \infty\right)$, concave down on $\left(-\frac{1}{3}\sqrt{3}, \frac{1}{3}\sqrt{3}\right)$;

pts of inflection $\left(-\frac{1}{3}\sqrt{3}, -\frac{5}{36}\right)$ and $\left(\frac{1}{3}\sqrt{3}, -\frac{5}{36}\right)$

9. $f'(x) = -\dfrac{x^2 + 1}{(x^2 - 1)^2}$, $f''(x) = \dfrac{2x\left(x^2 + 3\right)}{(x^2 - 1)^3}$;

concave down on $(-\infty, -1)$ and $(0, 1)$, concave up on $(-1, 0)$ and on $(1, \infty)$;

pt of inflection $(0, 0)$

11. $f'(x) = 4x^3 - 4x, \quad f''(x) = 12x^2 - 4;$

concave up on $\left(-\infty, -\frac{1}{3}\sqrt{3}\right)$ and $\left(\frac{1}{3}\sqrt{3}, \infty\right)$, concave down on $\left(-\frac{1}{3}\sqrt{3}, \frac{1}{3}\sqrt{3}\right)$;

pts of inflection $\left(-\frac{1}{3}\sqrt{3}, \frac{4}{9}\right)$ and $\left(\frac{1}{3}\sqrt{3}, \frac{4}{9}\right)$

13. $f'(x) = \dfrac{-1}{\sqrt{x}\left(1 + \sqrt{x}\right)^2}, \quad f''(x) = \dfrac{1 + 3\sqrt{x}}{2x\sqrt{x}\left(1 + \sqrt{x}\right)^3};$

concave up on $(0, \infty);$ no pts of inflection

15. $f'(x) = \frac{5}{3}(x + 2)^{2/3}, \quad f''(x) = \frac{10}{9}(x + 2)^{-1/3};$

concave down on $(-\infty, -2)$, concave up on $(-2, \infty);$ pt of inflection $(-2, 0)$

17. $f'(x) = 2\sin x \cos x = \sin 2x, \quad f''(x) = 2\cos 2x;$

concave up on $\left(0, \frac{1}{4}\pi\right)$ and $\left(\frac{3}{4}\pi, \pi\right)$, concave down on $\left(\frac{1}{4}\pi, \frac{3}{4}\pi\right)$;

pts of inflection $\left(\frac{1}{4}\pi, \frac{1}{2}\right)$ and $\left(\frac{3}{4}\pi, \frac{1}{2}\right)$

19. $f'(x) = 2x + 2\cos 2x, \quad f''(x) = 2 - 4\sin 2x;$

concave up on $\left(0, \frac{1}{12}\pi\right)$ and on $\left(\frac{5}{12}\pi, \pi\right)$, concave down on $\left(\frac{1}{12}\pi, \frac{5}{12}\pi\right)$;

pts of inflection $\left(\dfrac{1}{12}\pi, \dfrac{72 + \pi^2}{144}\right)$ and $\left(\dfrac{5}{12}\pi, \dfrac{72 + 25\pi^2}{144}\right)$

21. $f(x) = x^3 - 9x$

(a) $f'(x) = 3x^2 - 9 = 3(x^2 - 3)$

$f'(x) \geq 0 \Rightarrow x \leq -\sqrt{3}$ or $x \geq \sqrt{3};$

$f'(x) \leq 0 \Rightarrow -\sqrt{3} \leq x \leq \sqrt{3}.$

Thus, f is increasing on $(-\infty, -\sqrt{3}] \cup [\sqrt{3}, \infty)$

and decreasing on $[-\sqrt{3}, \sqrt{3}].$

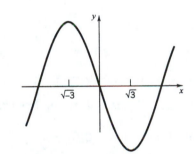

(b) $f(-\sqrt{3}) \cong 10.39$ is a local maximum;

$f(\sqrt{3}) \cong -10.39$ is a local minimum.

(c) $f''(x) = 6x;$

The graph of f is concave up on $(0, \infty)$ and concave down on $(-\infty, 0).$

(d) point of inflection: $(0, 0)$

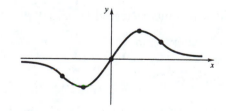

23. $f(x) = \dfrac{2x}{x^2 + 1}$

(a) $f'(x) = -\dfrac{2(x + 1)(x - 1)}{(x^2 + 1)^2}$

$f'(x) \geq 0 \Rightarrow -1 \leq x \leq 1;$

$f'(x) \leq 0 \implies x \leq -1$ or $x \geq 1$.

Thus, f is increasing on $[-1, 1]$;

and decreasing on $(-\infty, -1] \cup [1, \infty)$.

(b) $f(-1) = -1$ is a local minimum;

$f(1) = 1$ is a local maximum.

(c) $f''(x) = \dfrac{4x(x + \sqrt{3})(x - \sqrt{3})}{(x^2 + 1)^3}$

$f''(x) > 0 \implies x \leq -\sqrt{3}$ or $x \geq \sqrt{3}$;

$f''(x) < 0 \implies -\sqrt{3} < x < \sqrt{3}$.

The graph of f is concave up on $(-\sqrt{3}, 0) \cup (\sqrt{3}, \infty)$ and concave down

on $(-\infty, -\sqrt{3}) \cup (0, \sqrt{3})$.

(d) points of inflection: $(-\sqrt{3}, -\sqrt{3}/2)$, $(0,0)$, $(\sqrt{3}, \sqrt{3}/2)$

25. $f(x) = x + \sin x$, $x \in [-\pi, \pi]$

(a) $f'(x) = 1 + \cos x$

$f'(x) > 0$ on $(-\pi, \pi)$

Thus, f is increasing on $[-\pi, \pi]$.

(b) No local extrema

(c) $f''(x) = -\sin x$

$f''(x) > 0$ for $x \in (-\pi, 0)$;

$f''(x) < 0$ for $x \in (0, \pi)$.

The graph of f is concave up on $(\pi, 0)$ and concave down on $(0, \pi)$.

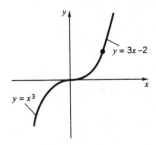

(d) point of inflection: $(0,0)$

27. $f(x) = \begin{cases} x^3, & x < 1 \\ 3x - 2, & x \geq 1. \end{cases}$

(a) $f'(x) = \begin{cases} 3x^2, & x < 1 \\ 3, & x \geq 1; \end{cases}$

$f'(x) > 0$ on $(-\infty, 0) \cup (0, \infty)$

Thus, f is increasing on $(-\infty, \infty)$.

(b) No local extrema

(c) $f''(x) = \begin{cases} 6x, & x < 1 \\ 0, & x \geq 1; \end{cases}$

$f''(x) > 0$ for $x \in (0, 1)$; $f''(x) < 0$ for $x \in (-\infty, 0)$.

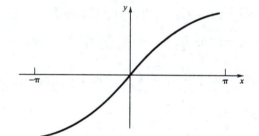

Thus, the graph of f is concave up on $(0,1)$ and concave down on $(-\infty, 0)$.

The graph of f is a straight line for $x \geq 1$.

(d) point of inflection: $(0,0)$

29.

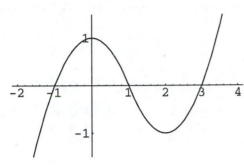

31.

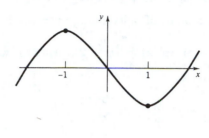

33. Since $f''(x) = 6x - 2(a+b+c)$, set $d = \frac{1}{3}(a+b+c)$. Note that $f''(d) = 0$ and that f is concave down on $(-\infty, d)$ and concave up on (d, ∞); $(d, f(d))$ is a point of inflection.

35. Since $(-1, 1)$ lies on the graph, $1 = -a + b$.

Since $f''(x)$ exists for all x and there is a pt of inflection at $x = \frac{1}{3}$, we must have $f''\left(\frac{1}{3}\right) = 0$.

Therefore

$$0 = 2a + 2b.$$

Solving these two equations, we find $a = -\frac{1}{2}$ and $b = \frac{1}{2}$.

Verification: the function

$$f(x) = -\frac{1}{2}x^3 + \frac{1}{2}x^2$$

has second derivative $f''(x) = -3x + 1$. This does change sign at $x = \frac{1}{3}$.

37. First, we require that $\left(\frac{1}{6}\pi, 5\right)$ lie on the curve:

$$5 = \frac{1}{2}A + B.$$

Next we require that $\dfrac{d^2y}{dx^2} = -4A \cos 2x - 9B \sin 3x$ be zero (and change sign) at $x = \frac{1}{6}\pi$:

$$0 = -2A - 9B.$$

Solving these two equations, we find $A = 18$, $B = -4$.

Verification: the function

$$f(x) = 18\cos 2x - 4\sin 3x$$

has second derivative $f''(x) = -72\cos 2x + 36\sin 3x$. This does change sign at $x = \frac{1}{6}\pi$.

39. Let $f'(x) = 3x^2 - 6x + 3$. Then we must have $f(x) = x^3 - 3x^2 + 3x + c$ for some constant c. Note that $f''(x) = 6x - 6$ and $f''(1) = 0$. Since $(1, -2)$ is a point of inflection of the graph of f, $(1, -2)$

must lie on the graph. Therefore,

$$1^3 - 3(1)^2 + 3(1) + c = -2 \quad \text{which implies} \quad c = -3$$

and so $f(x) = x^3 - 3x^2 + 3x - 3$.

41. (a) $p''(x) = 6x + 2a$ is negative for $x < -a/3$, and positive for $x > -a/3$. Therefore, the graph of p has a point of inflection at $x = -a/3$.

(b) $p'(x) = 3x^2 + 2ax + b$ has two real zeros iff $a^2 > 3b$.

43. (a)

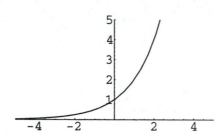

(b) No.   If $f''(x) < 0$ and $f'(x) < 0$ for all x, then

$f(x) < f'(0)x + f(0)$ on $(0, \infty)$, which implies that

$f(x) \to -\infty$ as $x \to \infty$.

45.

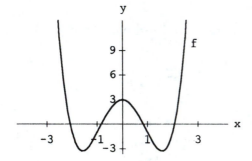

(a) concave up on $(-4, -0.913) \cup (0.913, 4)$

concave down on $(-0.913, 0.913)$

(b) pts of inflection at $x = -0.913, \quad 0.913$

47.

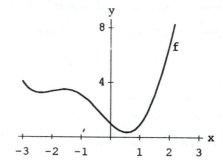

(a) concave up on: $(-\pi, -1.996) \cup (-.0345, 2.550)$

concave down on $(-1.996, -0.345) \cup (2.550, \pi)$

(b) pts of inflection at $x = -1.996, \ -0.345, \ 2,550$

49.

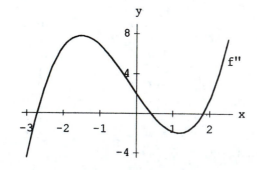

(a) concave up on $(-2.726, 0.402) \cup (1.823, 2.5)$

concave down on $(-3, -2.726) \cup (0.402, 1.823)$

(b) pts of inflection at $x = -2.726, 0.402, 1.823$.

SECTION 4.7

1. (a) ∞ (b) $-\infty$ (c) ∞ (d) 1

    (e) 0   (f) $x = -1$, $x = 1$ (g) $y = 0$, $y = 1$

3. vertical: $x = \frac{1}{3}$; horizontal: $y = \frac{1}{3}$ 5. vertical: $x = 2$; horizontal: none

7. vertical: $x = \pm 3$; horizontal: $y = 0$ 9. vertical: $x = -\frac{4}{3}$; horizontal: $y = \frac{4}{9}$

11. vertical: $x = \frac{5}{2}$; horizontal: $y = 0$ 13. vertical: none; horizontal: $y = \pm\frac{3}{2}$

15. vertical: $x = 1$; horizontal: $y = 0$ 17. vertical: none; horizontal: $y = 0$

19. vertical: $x = \left(2n + \frac{1}{2}\right)\pi$; horizontal: none   21. $f'(x) = \frac{4}{3}(x+3)^{1/3}$; neither

23. $f'(x) = -\frac{4}{5}(2-x)^{-1/5}$; cusp   25. $f'(x) = \frac{6}{5}x^{-2/5}\left(1 - x^{3/5}\right)$; tangent

27. $f(-2)$ undefined; neither

29. $f'(x) = \begin{cases} \frac{1}{2}(x-1)^{-1/2}, & x > 1 \\ -\frac{1}{2}(1-x)^{-1/2}, & x < 1; \end{cases}$ cusp

31. $f'(x) = \begin{cases} \frac{1}{3}(x+8)^{-23}, & x > -8 \\ -\frac{1}{3}(x+8)^{-2/3}, & x < -8; \end{cases}$ cusp

33. f not continuous at 0; neither

35.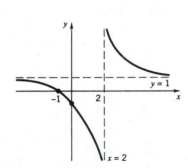

37.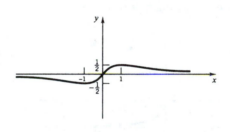

39. $f(x) = x - 3x^{1/3}$

(a) $f'(x) = 1 - \dfrac{1}{x^{2/3}}$

f is increasing on $(-\infty, -1] \cup [1, \infty)$

f is decreasing on $[-1, 1]$

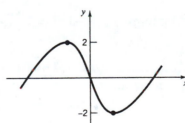

(b) $f''(x) = \frac{2}{3}x^{-5/3}$

concave up on $(0, \infty)$; concave down on $(-\infty, 0)$

vertical tangent at $(0, 0)$

41. $f(x) = \frac{3}{5} x^{5/3} - 3x^{2/3}$

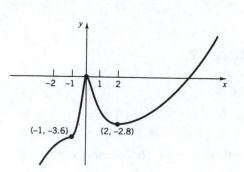

(a) $f'(x) = x^{2/3} - 2x^{-1/3} = \dfrac{x - 2}{x^{1/3}}$

f is increasing on $(-\infty, 0] \cup [2, \infty)$

f is decreasing on $[0, 2]$

(b) $f''(x) = \frac{2}{3} x^{-1/3} + \frac{2}{3} x^{-4/3} = \dfrac{2x + 2}{3x^{4/3}}$

concave up on $(-1, \infty)$; concave down on $(-\infty, -1)$

vertical cusp at $(0, 0)$

43.

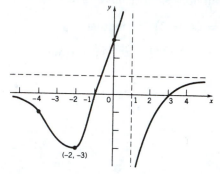

vertical asymptote: $x = 1$

horizontal asymptotes: $y = 0$, $y = 2$

no vertical tangents or cusps

45.

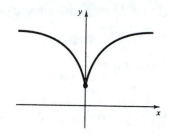

no asymptotes

vertical cusp at $(0, 1)$

47.

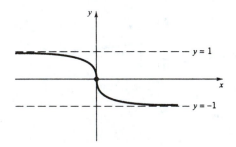

horizontal asymptotes: $y = -1$, $y = 1$

vertical tangent at $(0, 0)$

49. (a) p odd; (b) p even.

51.

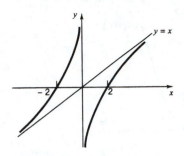

vertical asymptote: $x = 0$

oblique asymptote: $y = x$

53.

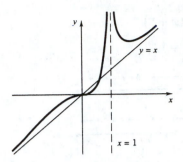

vertical asymptote: $x = 1$

oblique asymptote: $y = x$

SECTION 4.8

[Rough sketches; not scale drawings]

1. $f(x) = (x - 2)^2$

$f'(x) = 2(x - 2)$

$f''(x) = 2$

f':   ----------0+++++++++
              2

f'':   +++++++++++++++++++++

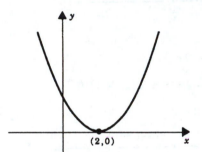

3. $f(x) = x^3 - 2x^2 + x + 1$

$f'(x) = (3x - 1)(x - 1)$

$f''(x) = 6x - 4$

f':   ++++++0--------0++++++++
                        1

f'':   ----------0++++++++++++
              $\frac{2}{3}$

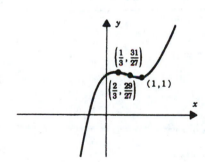

5. $f(x) = x^3 + 6x^2, \quad x \in [-4, 4]$

$f'(x) = 3x(x + 4)$

$f''(x) = 6x + 12$

f':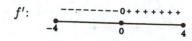

f'':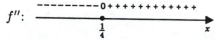

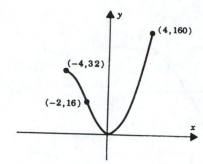

7. $f(x) = \frac{2}{3}x^3 - \frac{1}{2}x^2 - 10x - 1$

$f'(x) = (2x - 5)(x + 2)$

$f''(x) = 4x - 1$

f':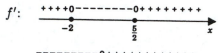

f'':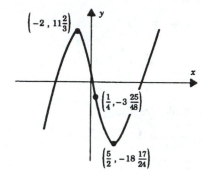

9. $f(x) = x^2 + 2x^{-1}$

$f'(x) = 2x - 2x^{-2} = 2\left(x^3 - 1\right)/x^2$

$f''(x) = 2 + 4x^{-3}$

f':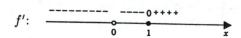

f'':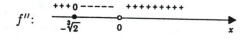

asymptote: $x = 0$

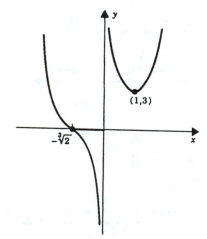

11. $f(x) = (x - 4)/x^2$

$f'(x) = (8 - x)/x^3$

$f''(x) = (2x - 24)/x^4$

f':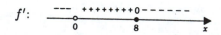

f'':

asymptotes: $x = 0, \; y = 0$

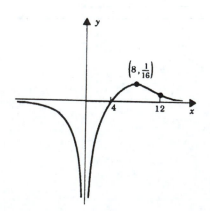

13. $f(x) = 2x^{1/2} - x, \quad x \in [0,4]$

$f'(x) = x^{-1/2}\left(1 - x^{1/2}\right)$

$f''(x) = -\frac{1}{2}x^{-3/2}$

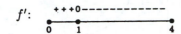

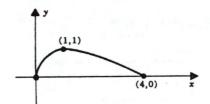

15. $f(x) = 2 + (x+1)^{6/5}$

$f'(x) = \frac{6}{5}(x+1)^{1/5}$

$f''(x) = \frac{6}{25}(x+1)^{-4/5}$

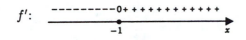

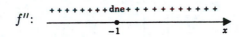

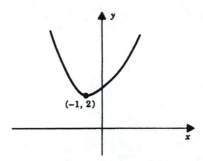

17. $f(x) = 3x^5 + 5x^3$

$f'(x) = 15x^2\left(x^2 + 1\right)$

$f''(x) = 30x\left(2x^2 + 1\right)$

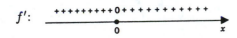

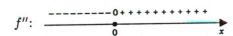

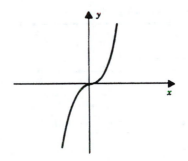

19. $f(x) = 1 + (x-2)^{5/3}$

$f'(x) = \frac{5}{3}(x-2)^{2/3}$

$f''(x) = \frac{10}{9}(x-2)^{-1/3}$

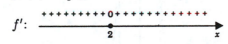

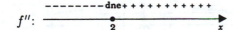

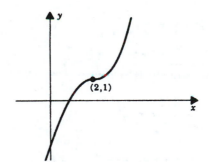

21. $f(x) = \dfrac{2x}{4x - 3}$

$f'(x) = -6(4x - 3)^{-2}$

$f''(x) = 48(4x - 3)^{-3}$

$f':$

$f'':$

asymptotes: $x = 3/4, \; y = 1/2$

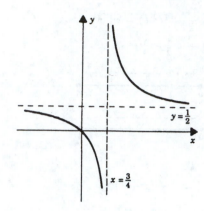

23. $f(x) = \dfrac{x}{(x + 3)^2}$

$f'(x) = \dfrac{3 - x}{(x + 3)^3}$

$f''(x) = \dfrac{2x - 12}{(x + 3)^4}$

$f':$

$f'':$

asymptotes: $x = -3, \; y = 0$

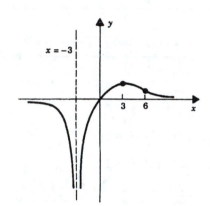

25. $f(x) = \dfrac{x^2}{x^2 - 4}$

$f'(x) = \dfrac{-8x}{\left(x^2 - 4\right)^2}$

$f''(x) = \dfrac{8\left(3x^2 + 4\right)}{\left(x^2 - 4\right)^3}$

$f':$

$f'':$

asymptotes: $x = -2, \; x = 2, \; y = 1$

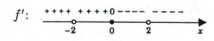

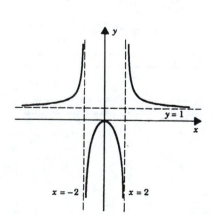

27. $f(x) = x(1-x)^{1/2}$

$f'(x) = \frac{1}{2}(1-x)^{-1/2}(2-3x)$

$f''(x) = \frac{1}{4}(1-x)^{-3/2}(3x-4)$

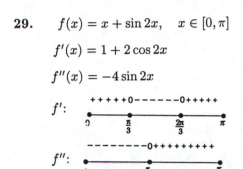

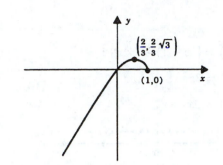

29. $f(x) = x + \sin 2x, \quad x \in [0, \pi]$

$f'(x) = 1 + 2\cos 2x$

$f''(x) = -4\sin 2x$

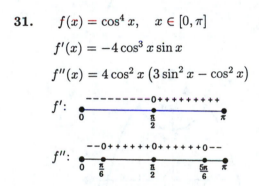

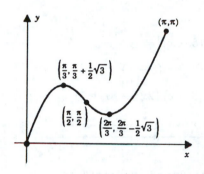

31. $f(x) = \cos^4 x, \quad x \in [0, \pi]$

$f'(x) = -4\cos^3 x \sin x$

$f''(x) = 4\cos^2 x \left(3\sin^2 x - \cos^2 x\right)$

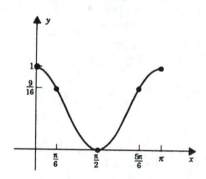

33. $f(x) = 2\sin^3 x + 3\sin x, \quad x \in [0, \pi]$

$f'(x) = 3\cos x \left(2\sin^2 x + 1\right)$

$f''(x) = 9\sin x \left(1 - 2\sin^2 x\right)$

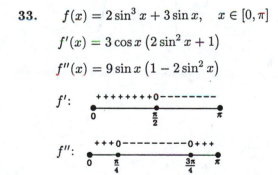

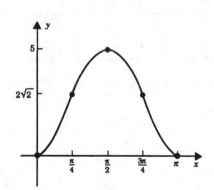

35. $f(x) = [(x+1) - 1]^3 + 1$

$f'(x) = 3x^2$

$f''(x) = 6x$

f':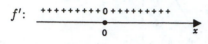

f'':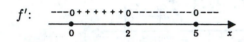

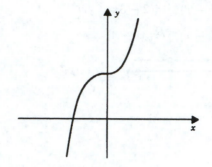

37. $f(x) = x^2(5-x)^3$

$f'(x) = 5x(2-x)(5-x)^2$

$f''(x) = 10(5-x)\left(2x^2 - 8x + 5\right)$

f':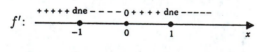

f'':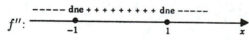

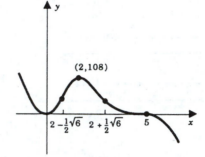

39. $f(x) = \begin{cases} 4 - x^2, & |x| > 1 \\ x^2 + 2, & -1 \le x \le 1 \end{cases}$

$f'(x) = \begin{cases} -2x, & |x| > 1 \\ 2x, & -1 < x < 1 \end{cases}$

$f''(x) = \begin{cases} -2, & |x| > 1 \\ 2, & -1 < x < 1 \end{cases}$

f':
f'':

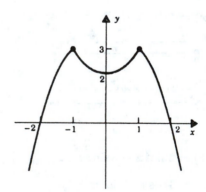

41. $f(x) = x(x-1)^{1/5}$

$f'(x) = \frac{1}{5}(x-1)^{-4/5}(6x-5)$

$f''(x) = \frac{2}{25}(x-1)^{-9/5}(3x-5)$

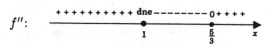

f':

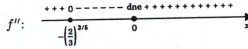

f'':

vertical tangent at $(1,0)$

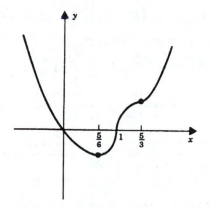

43. $f(x) = x^2 - 6x^{1/3}$

$f'(x) = 2x^{-2/3}\left(x^{5/3}-1\right)$

$f''(x) = \frac{2}{3}x^{-5/3}\left(3x^{5/3}+2\right)$

f':

f'':

vertical tangent at $(0,0)$

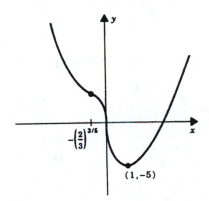

45. $f(x) = \left(\dfrac{x}{x-2}\right)^{1/2}; \quad x \le 0, \; x > 2$

$f'(x) = -\left(\dfrac{x}{x-2}\right)^{-1/2}(x-2)^{-2}$

$f''(x) = (2x-1)\left(\dfrac{x}{x-2}\right)^{-3/2}(x-2)^{-4}$

f':

f'':

asymptotes: $x = 2, \; y = 1$

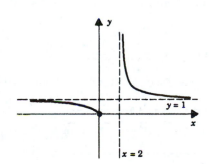
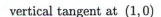

47. $f(x) = x^2 \left(x^2 - 2\right)^{-1/2}, \quad |x| > \sqrt{2}$

$f'(x) = x \left(x^2 - 4\right) \left(x^2 - 2\right)^{-3/2}$

$f''(x) = 2 \left(x^2 + 4\right) \left(x^2 - 2\right)^{-5/2}$

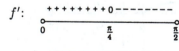

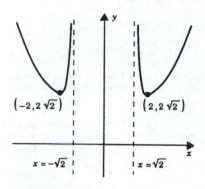

asymptotes: $\quad x = -\sqrt{2}, \; x = \sqrt{2}$

49. $f(x) = 2\sin 3x, \quad x \in [0, \pi]$

$f'(x) = 6\cos 3x$

$f''(x) = -18\sin 3x$

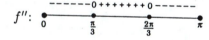

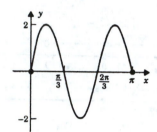

51. $f(x) = 2\tan x - \sec^2 x, \quad x \in (0, \pi/2)$

$\qquad = -(1 - \tan x)^2$

$f'(x) = 2\sec^2 x \, (1 - \tan x)$

$f''(x) = -2\sec^2 x \, \left(3\tan^2 x - 2\tan x + 1\right)$

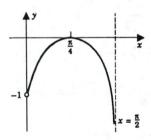

asymptote: $\quad x = \frac{1}{2}\pi$

53. $f(x) = \dfrac{\sin x}{1 - \sin x}, \quad x \in (-\pi, \pi)$

$f'(x) = \dfrac{\cos x}{(1 - \sin x)^2}$

$f''(x) = \dfrac{1 - \sin x + \cos^2 x}{(1 - \sin x)^3}$

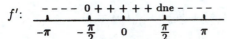

f':

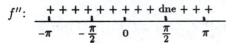

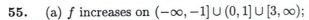

f'':

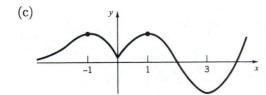

asymptote: $x = \frac{1}{2}\pi$

55.   (a) f increases on $(-\infty, -1] \cup (0, 1] \cup [3, \infty)$;

f decreases on $[-1, 0) \cup [1, 3]$;   critical numbers: $x = -1, 0, 1, 3$.

(b)

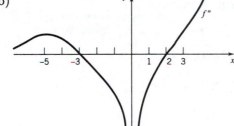

concave up on $(-\infty, -3) \cup (2, \infty)$

concave down on $(-3, 0) \cup (0, 2)$.

(c)

The graph does not necessarily have

a horizontal asymptote.

57.   Solve the equation

$$\frac{x^2}{a^2} - \frac{y^2}{b^2} = 1$$

for y :

$$y^2 = \frac{b^2(x^2 - a^2)}{a^2} \quad \text{and}$$

$$y = \pm\frac{b}{a}\sqrt{x^2 - a^2} = \pm\frac{b}{a}x\sqrt{1 - \frac{a^2}{x^2}}$$

Now, for $|x|$ large, $y \cong \pm\dfrac{b}{a}x$.

CHAPTER 5

SECTION 5.1

1. $L_f(P) = 0(\frac{1}{4}) + \frac{1}{2}(\frac{1}{4}) + 1(\frac{1}{2}) = \frac{5}{8},$ $\qquad$ $U_f(P) = \frac{1}{2}(\frac{1}{4}) + 1(\frac{1}{4}) + 2(\frac{1}{2}) = \frac{11}{8}$

3. $L_f(P) = \frac{1}{4}(\frac{1}{2}) + \frac{1}{16}(\frac{1}{4}) + 0(\frac{1}{4}) = \frac{9}{64},$ $\qquad$ $U_f(P) = 1(\frac{1}{2}) + \frac{1}{4}(\frac{1}{4}) + \frac{1}{16}(\frac{1}{4}) = \frac{37}{64}$

5. $L_f(P) = 1(\frac{1}{2}) + \frac{9}{8}(\frac{1}{2}) = \frac{17}{16},$ $\qquad$ $U_f(P) = \frac{9}{8}(\frac{1}{2}) + 2(\frac{1}{2}) = \frac{25}{16}$

7. $L_f(P) = \frac{1}{16}(\frac{3}{4}) + 0(\frac{1}{2}) + \frac{1}{16}(\frac{1}{4}) + \frac{1}{4}(\frac{1}{2}) = \frac{3}{16},$ $\quad$ $U_f(P) = 1(\frac{3}{4}) + \frac{1}{16}(\frac{1}{2}) + \frac{1}{4}(\frac{1}{4}) + 1(\frac{1}{2}) = \frac{43}{32}$

9. $L_f(P) = 0\left(\frac{\pi}{6}\right) + \frac{1}{2}\left(\frac{\pi}{3}\right) + 0\left(\frac{\pi}{2}\right) = \frac{\pi}{6},$ $\qquad$ $U_f(P) = \frac{1}{2}\left(\frac{\pi}{6}\right) + 1\left(\frac{\pi}{3}\right) + 1\left(\frac{\pi}{2}\right) = \frac{11\pi}{12}$

11. $\quad$ (a) $\quad L_f(P) \le U_f(P) \quad$ but $\quad 3 \not\le 2.$

$\quad$ (b) $\quad L_f(P) \le \displaystyle\int_{-1}^{1} f(x)\,dx \le U_f(P) \quad$ but $\quad 3 \not\le 2 \le 6.$

$\quad$ (c) $\quad L_f(P) \le \displaystyle\int_{-1}^{1} f(x)\,dx \le U_f(P) \quad$ but $\quad 3 \le 10 \not\le 6.$

13. $\quad$ (a) $\quad L_f(P) = -3x_1(x_1 - x_0) - 3x_2(x_2 - x_1) - \cdots - 3x_n(x_n - x_{n-1}),$

$\qquad\quad U_f(P) = -3x_0(x_1 - x_0) - 3x_1(x_2 - x_1) - \cdots - 3x_{n-1}(x_n - x_{n-1})$

$\quad$ (b) $\quad$ For each index i

$$-3x_i \le -\tfrac{3}{2}(x_i + x_{i-1}) \le -3x_{i-1}.$$

Multiplying by $\Delta x_i = x_i - x_{i-1}$ gives

$$-3x_i\,\Delta x_i \le -\tfrac{3}{2}(x_i^2 - x_{i-1}^2) \le -3x_{i-1}\,\Delta x_i.$$

Summing from $i = 1$ to $i = n$, we find that

$$L_f(P) \le -\tfrac{3}{2}(x_1^2 - x_0^2) - \cdots - \tfrac{3}{2}(x_n^2 - x_{n-1}^2) \le U_f(P).$$

The middle sum collapses to

$$-\tfrac{3}{2}(x_n^2 - x_0^2) = -\tfrac{3}{2}(b^2 - a^2).$$

Thus

$$L_f(P) \le -\frac{3}{2}(b^2 - a^2) \le U_f(P) \quad \text{so that} \quad \int_a^b -3x\,dx = -\frac{3}{2}(b^2 - a^2).$$

15.

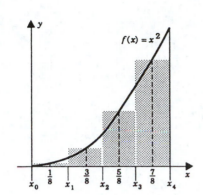

17.   (a) $\Delta x_1 = \Delta x_2 = \frac{1}{8}$, $\Delta x_3 = \Delta x_4 = \Delta x_5 = \frac{1}{4}$

(b) $\|P\| = \frac{1}{4}$

(c) $m_1 = 0$, $m_2 = \frac{1}{4}$, $m_3 = \frac{1}{2}$, $m_4 = 1$, $m_5 = \frac{3}{2}$

(d) $f(x_1^*) = \frac{1}{8}$, $f(x_2^*) = \frac{3}{8}$, $f(x_3^*) = \frac{3}{4}$, $f(x_4^*) = \frac{5}{4}$,

   $f(x_5^*) = \frac{3}{2}$

(e) $M_1 = \frac{1}{4}$, $M_2 = \frac{1}{2}$, $M_3 = 1$, $M_4 = \frac{3}{2}$, $M_5 = 2$

(f) $L_f(P) = \frac{25}{32}$ (g) $S^*(P) = \frac{15}{16}$

(h) $U_f(P) = \frac{39}{32}$ (i) $\displaystyle\int_a^b f(x)\,dx = 1$

19. $L_f(P) = x_0{}^3(x_1 - x_0) + x_1{}^3(x_2 - x_1) + \cdots + x_{n-1}^3(x_n - x_{n-1})$

   $U_f(P) = x_1{}^3(x_1 - x_0) + x_2{}^3(x_2 - x_1) + \cdots + x_n{}^3(x_n - x_{n-1})$

For each index i

$$x_{i-1}^3 \le \tfrac{1}{4}\left(x_i{}^3 + x_i{}^2 x_{i-1} + x_i x_{i-1}^2 + x_{i-1}^3\right) \le x_i{}^3$$

and thus by the hint

$$x_{i-1}^3(x_i - x_{i-1}) \le \tfrac{1}{4}\left(x_i{}^4 - x_{i-1}^4\right) \le x_i{}^3(x_i - x_{i-1}).$$

Adding up these inequalities, we find that

$$L_f(P) \le \tfrac{1}{4}\left(x_n{}^4 - x_0{}^4\right) \le U_f(P).$$

Since $x_n = 1$ and $x_0 = 0$, the middle term is $\dfrac{1}{4}$: $\displaystyle\int_0^1 x^3\,dx = \dfrac{1}{4}.$

21.   Let $P = \{x_0, x_1, x_2, \ldots, x_n\}$ be a regular partition of $[a, b]$ and let $\Delta x = (b - a)/n$.

Since f is increasing on $[a, b]$,

$$L_f(P) = f(x_0)\Delta x + f(x_1)\Delta x + \cdots + f(x_{n-1})\Delta x$$

and

$$U_f(P) = f(x_1)\Delta x + f(x_2)\Delta x + \cdots + f(x_n)\Delta x.$$

Now,

$$U_f(P) - L_f(P) = [f(x_n) - f(x_0)]\Delta x = [f(b) - f(a)]\Delta x.$$

23.   Necessarily holds: $L_g(P) \le \int_a^b g(x)\,dx < \int_a^b f(x)\,dx \le U_f(P).$

25. Necessarily holds: $L_g(P) \leq \int_a^b g(x)\, dx < \int_a^b f(x)\, dx$

27. Necessarily holds: $U_f(P) \geq \int_a^b f(x)\, dx > \int_a^b g(x)\, dx$

29. (a) By Definition 5.1.5, $L_f(P) \leq I \leq U_f(P)$. Subtracting

$L_f(P)$ from these inequalities gives

$$0 \leq I - L_f(P) \leq U_f(P) - L_f(P)$$

(b) From Definition 5.2.3,

$$L_f(P) - U_f(P) \leq I - U_f(P) \leq 0$$

Now multiply by -1 and the result follows.

31. (a) $f'(x) = \dfrac{x}{\sqrt{1+x^2}} > 0$ for $x \in (0, 2)$. Thus, f is increasing on $[0, 2]$.

(b) Let $P = \{x_0, x_1, \ldots, x_n\}$ be a regular partition of $[0, 2]$ and let $\Delta x = 2/n$

By Exercise 30 (a),

$$\int_0^2 f(x)\, dx - L_f(P) \leq |f(2) - f(0)|\,\frac{2}{n} = \frac{2(\sqrt{5}-1)}{n} \cong \frac{2.47}{n}$$

It now follows that $\int_0^2 f(x)\, dx - L_f(P) < 0.1$ if $n > 25$.

(c) $\int_0^2 f(x)\, dx \cong 2.96$

33. Let S be the set of positive integers for which the statement is true. Since $1 = \dfrac{1(2)}{2} = 1$, $1 \in S$.
Assume that $k \in S$. Then

$$1 + 2 + \cdots + k + k + 1 = (1 + 2 + \cdots + k) + k + 1 = \frac{k(k+1)}{2} + k + 1$$

$$= \frac{(k+1)(k+2)}{2}$$

Thus, $k + 1 \in S$ and so S is the set of positive integers.

35. Let $f(x) = x$ and let $P = \{x_0, x_1, x_2, \ldots, x_n\}$ be a regular partition of $[0, b]$. Then $\Delta x = b/n$
and $x_i = \dfrac{ib}{n}$, $i = 0, 1, 2, \ldots, n$.

(a) Since f is increasing on $[0, b]$,

$$L_f(P) = \left[f(0) + f\left(\frac{b}{n}\right) + f\left(\frac{2b}{n}\right) + \cdots + f\left(\frac{(n-1)b}{n}\right) \right]\frac{b}{n}$$

$$= \left[0 + \frac{b}{n} + \frac{2b}{n} + \cdots + \frac{(n-1)b}{n} \right]\frac{b}{n}$$

$$= \frac{b^2}{n^2}[1 + 2 + \cdots + (n-1)]$$

(b)
$$U_f(P) = \left[f\left(\frac{b}{n}\right) + f\left(\frac{2b}{n}\right) + \cdots + f\left(\frac{(n-1)b}{n}\right) + f(b) \right] \frac{b}{n}$$

$$= \left[\frac{b}{n} + \frac{2b}{n} + \cdots + \frac{(n-1)b}{n} + b \right] \frac{b}{n}$$

$$= \frac{b^2}{n^2} \left[1 + 2 + \cdots + (n-1) + n \right]$$

(c) By Exercise 33,

$$L_f(P) = \frac{b^2}{n^2} \cdot \frac{(n-1)n}{2} \quad \text{and} \quad U_f(P) = \frac{b^2}{n^2} \cdot \frac{n(n+1)}{2}.$$

As $n \to \infty$, $L_f(P) \to \frac{b^2}{2}$ and $U_f(P) \to \frac{b^2}{2}$. Therefore, $\int_0^2 x\,dx = \frac{b^2}{2}$.

37. (a) $\dfrac{1}{n^2}(1 + 2 + \cdots + n) = \dfrac{1}{n^2}\left[\dfrac{n(n+1)}{2}\right] = \dfrac{1}{2} + \dfrac{1}{2n}$

(b) $S_n^* = \dfrac{1}{2} + \dfrac{1}{2n}$, $\displaystyle\int_0^1 x\,dx = \left[\dfrac{1}{2}x^2\right]_0^1 = \dfrac{1}{2}$

$$\left| S_n^* - \int_0^1 x\,dx \right| = \frac{1}{2n} < \frac{1}{n} < \epsilon \quad \text{if} \quad n > \frac{1}{\epsilon}$$

39. (a) $\dfrac{1}{n^4}(1^3 + 2^3 + \cdots + n^3) = \dfrac{1}{n^4}\left[\dfrac{n^2(n+1)^2}{4}\right] = \dfrac{1}{4} + \dfrac{1}{2n} + \dfrac{1}{4n^2}$

(b) $S_n^* = \dfrac{1}{4} + \dfrac{1}{2n} + \dfrac{1}{4n^2}$, $\displaystyle\int_0^1 x^3\,dx = \left[\dfrac{1}{4}x^4\right]_0^1 = \dfrac{1}{4}$

$$\left| S_n^* - \int_0^1 x^3\,dx \right| = \frac{1}{2n} + \frac{1}{4n^2} < \frac{1}{n} < \epsilon \quad \text{if} \quad n > \frac{1}{\epsilon}$$

41. Let P be an arbitrary partition of $[0, 4]$. Since each $m_i = 2$ and each $M_i \geq 2$,

$$L_g(P) = 2\Delta x_1 + \cdots + 2\Delta x_n = 2(\Delta x_1 + \cdots + \Delta x_n) = 2 \cdot 4 = 8$$

and

$$U_g(P) \geq 2\Delta x_1 + \cdots + 2\Delta x_n = 2(\Delta x_1 + \cdots + \Delta x_n) = 2 \cdot 4 = 8.$$

Thus

$$L_g(P) \leq 8 \leq U_g(P) \quad \text{for all partitions } P \text{ of } [0, 4].$$

Uniqueness: Suppose that
(*) $L_g(P) \leq I \leq U_g(P)$ for all partitions P of $[0,4]$.
Since $L_g(P) = 8$ for all P, I is at least 8. Suppose now that $I > 8$ and choose a partition P of $[0,4]$
with max $\Delta x_i < \frac{1}{5}(I-8)$ and

$$0 = x_1 < \cdots < x_{i-1} < 3 < x_i < \cdots < x_n = 4.$$

Then

$$U_g(P) = 2\Delta x_1 + \cdots + 2\Delta x_{i-1} + 7\Delta x_i + 2\Delta x_{i+1} + \cdots + 2\Delta x_n$$

$$= 2(\Delta x_1 + \cdots + \Delta x_n) + 5\Delta x_i$$

$$= 8 + 5\Delta x_i < 8 + \tfrac{5}{5}(I - 8) = I$$

and I does not satisfy ($*$). This contradiction proves that I is not greater than 8 and therefore $I = 8$.

43.   Let $P = \{x_0, x_1, x_2, \ldots, x_n\}$ be any partition of $[2, 10]$.

(a) Since each subinterval $[x_{i-1}, x_i]$ contains both rational and irrational numbers, $m_i = 4$ and $M_i = 7$. Thus,

$$L_f(P) = 4\Delta x_1 + 4\Delta x_2 + \cdots + 4\Delta x_n = 4(\Delta x_1 + \Delta x_2 + \cdots + \Delta x_n) = 4(10 - 2) = 32$$

and

$$U_f(P) = 7\Delta x_1 + 7\Delta x_2 + \cdots + 7\Delta x_n = 7(\Delta x_1 + \Delta x_2 + \cdots + \Delta x_n) = 7(10 - 2) = 56$$

Therefore, $L_f(P) \leq 40 \leq U_f(P)$.

(b) Every number $I \in [32, 56]$ satisfies the inequalities

$$L_f(P) \leq I \leq U_f(P) \quad \text{for all partitions } P$$

(c) See part (a).

45.   (a) $L_f(P) \cong 0.6105$, $U_f(P) \cong 0.7105$

(b) $\dfrac{1}{2}[L_f(P) + U_f(P)] \cong 0.6605$ (c) $S^*(P) \cong 0.6684$

PROJECT 5.1

1.   (a) If the object traveled at its minimum speed m_i on $[t_{i-1}, t_i]$, then it would travel a distance of $m_i \Delta t_i$ units; if the object traveled at its maximum speed M_i on $[t_{i-1}, t_i]$, then it would travel a distance of $M_i \Delta t_i$ units. Thus, the actual distance traveled, s_i, must be somewhere inbetween. i.e.
$$m_i \Delta t_i \leq s_i \leq M_i \Delta t_i, \quad i = 1, 2, \ldots, n$$
Adding these inequalities, we get
$$m_1 \Delta t_1 + m_2 \Delta t_2 + \cdots + m_n \Delta t_n \leq D \leq M_1 \Delta t_1 + M_2 \Delta t_2 + \cdots + M_n \Delta t_n,$$
where $D = s_1 + s_2 + \cdots + s_n$.   Since the result holds for any partition P,   we have
$$D = \int_a^b s(t)\, dt.$$

(b) Since $m_i \leq s(t_i^*0 \leq M_i$ for each i,
$$m_1 \Delta t_1 + m_2 \Delta t_2 + \cdots + m_n \Delta t_n \leq S^*(P) \leq M_1 \Delta t_1 + M_2 \Delta t_2 + \cdots + M_n \Delta t_n,$$
Thus $\displaystyle \lim_{\|P\| \to 0} S*(P) = \int_a^b s(t)\, dt.$

3.   Distance $= \displaystyle\int_0^3 s(t)\, dt = \int_0^3 (t^2 + t + 1)\, dt = \left[\dfrac{t^3}{3} + \dfrac{t^2}{2} + t\right]_0^3 = \dfrac{33}{2}$ units.

SECTION 5.2

1.   (a) $\displaystyle \int_0^5 f(x)\, dx = \int_0^2 f(x)\, dx + \int_2^5 f(x)\, dx = 4 + 1 = 5$

(b) $\displaystyle \int_1^2 f(x)\, dx = \int_0^2 f(x)\, dx - \int_0^1 f(x)\, dx = 4 - 6 = -2$

(c) $\displaystyle\int_1^5 f(x)\,dx = \int_0^5 f(x)\,dx - \int_0^1 f(x)\,dx = 5 - 6 = -1$

(d) 0   (e) $\displaystyle\int_2^0 f(x)\,dx = -\int_0^2 f(x)\,dx = -4$

(f) $\displaystyle\int_5^1 f(x)\,dx = -\int_1^5 f(x)\,dx = 1$

3. With $P = \left\{1, \dfrac{3}{2}, 2\right\}$ and $f(x) = \dfrac{1}{x}$, we have

$$0.5 < \frac{7}{12} = L_f(P) \le \int_1^2 \frac{dx}{x} \le U_f(P) = \frac{5}{6} < 1.$$

5. (a) $F(0) = 0$ (b) $F'(x) = x\sqrt{x+1}$ (c) $F'(2) = 2\sqrt{3}$

(d) $F(2) = \displaystyle\int_0^2 t\sqrt{t+1}\,dt$ (e) $-F(x) = \displaystyle\int_x^0 t\sqrt{t+1}\,dt$

7. (a) $F'(x) = \dfrac{1}{x} > 0$ for $x > 0$.   (b) $F''(x) = -\dfrac{1}{x^2} < 0$ for $x > 0$.

   Thus, F is increasing on $(0, \infty)$;   The graph of F is concave down on $(0, \infty)$;

   there are no critical numbers.   there are no points of inflection.

(c)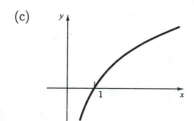

9. $F'(x) = \dfrac{1}{x^2 + 9}$;   (a) $\dfrac{1}{10}$ (b) $\dfrac{1}{9}$ (c) $\dfrac{4}{37}$ (d) $\dfrac{-2x}{(x^2+9)^2}$

11. $F'(x) = -x\sqrt{x^2+1}$;   (a) $\sqrt{2}$ (b) 0 (c) $-\tfrac{1}{4}\sqrt{5}$ (d) $-\left(\sqrt{x^2+1} - \dfrac{x^2}{\sqrt{x^2+1}}\right)$

13. $F'(x) = \cos \pi x$;   (a) -1 (b) 1 (c) 0 (d) $-\pi \sin \pi x$

15. (a) Since $P_1 \subseteq P_2$, $U_f(P_2) \le U_f(P_1)$ but $5 \not\le 4$.

   (b) Since $P_1 \subseteq P_2$, $L_f(P_1) \le L_f(P_2)$ but $5 \not\le 4$.

17. Let $u = x^3$. Then $F(u) = \displaystyle\int_1^u t\cos t\,dt$ and

$$\frac{dF}{dx} = \frac{dF}{du}\frac{du}{dx} = u\cos u\,(3x^2) = 3x^5 \cos x^3.$$

19. $F(x) = \displaystyle\int_{x^2}^1 (t - \sin^2 t)\,dt = -\int_1^{x^2} (t - \sin^2 t)\,dt$. Let $u = x^2$. Then

$$\frac{dF}{dx} = \frac{dF}{du}\frac{du}{dx} = -(u - \sin^2 u)(2x) = 2x\left[\sin^2(x^2) - x^2\right].$$

21. (a) $F(0) = 0$

(b) $F'(0) = 2 + \dfrac{\sin 2(0)}{1 + 0^2} = 2$

(c) $F''(0) = \dfrac{(1+0)^2 2\cos 2(0) - \sin 2(0)(2)(0)}{(1+0)^2} = 2$

23. $F'(x) = \dfrac{x-1}{1+x^2} = 0 \implies x = 1$ is a critical number.

$F''(x) = \dfrac{(1+x^2) - 2x(x-1)}{(1+x^2)^2}$, so $F''(1) = \dfrac{1}{2} > 0$ means $x = 1$ is a local minimum.

25. (a)

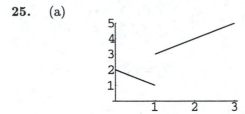

(b)

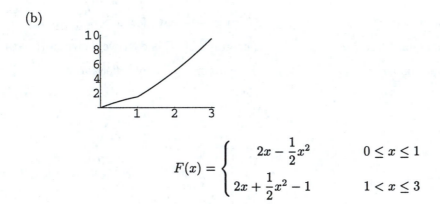

$$F(x) = \begin{cases} 2x - \dfrac{1}{2}x^2 & 0 \le x \le 1 \\[2mm] 2x + \dfrac{1}{2}x^2 - 1 & 1 < x \le 3 \end{cases}$$

(c) f is discontinuous at $x = 1$, F is continuous but not differentiable at $x = 1$.

27. $f(x) = \dfrac{d}{dx}\left(\dfrac{2x}{4+x^2}\right) = \dfrac{8 - 2x^2}{(4+x^2)^2}$

(a) $f(0) = \dfrac{1}{2}$ (b) $f(x) = 0$ at $x = -2, 2$

29. By the hint $\dfrac{F(b) - F(a)}{b - a} = F'(c)$ for some c in (a, b). The result follows by observing that

$$F(b) = \int_a^b f(t)\,dt, \quad F(a) = 0, \quad \text{and} \quad F'(c) = f(c).$$

31. Set $G(x) = \displaystyle\int_a^x f(t)\,dt$. Then $F(x) = \displaystyle\int_c^a f(t)\,dt + G(x)$. First, note that $\displaystyle\int_c^a f(t)\,dt$

is a constant. By (5.2.5) G, and thus F, is continuous on $[a, b]$, is differentiable

on (a, b), and $F'(x) = G'(x) = f(x)$ for all x in (a, b).

33. (a) $F(x) = -\int_0^c f(t)\,dt + \int_0^x f(t)\,dt$ and $G(x) = -\int_0^d f(t)\,dt + \int_0^x f(t)\,dt$.

Thus $F'(x) = f(x)dx + G'(x)$, so by theorem 4.2.4, F and G differ by a constant.

(b) $F(x) - G(x) = -\int_0^c f(t)\,dt + \int_0^d f(t)\,dt$

$$= -\int_0^c f(t)\,dt + \int_0^c f(t)\,dt + \int_c^d f(t)\,dt \;=\; \int_c^d f(t)\,dt.$$

PROJECT 5.2

1.   (a) Dom $(F) = (-\infty, \infty)$

   (b) By Theorem 5.2.5, F is continuous on (∞, ∞).

   (c)
$$F(-x) = \int_0^{-x} \sin^2(t^2)\,dt$$

   Set $u = -t$. Then $du = -dt$, and $u = 0$ at $t = 0$, $u = x$ at $t = -x$. Then

$$F(-x) = \int_0^x \sin^2(-u)^2\,(-du) = -\int_0^x \sin^2(u^2)\,du = -F(x).$$

3. $F''(x) = 2\sin(x^2)\cos(x^2)2x = 2x\sin(2x^2)$ $F''(x) = 0$ when $x^2 = \dfrac{n}{2}\pi$, $n = 0, 1, 2, \ldots$. That is, when $x = \pm\sqrt{n\pi/2}$. Since F'' changes signs at each of these values, the points $\left(\pm\sqrt{n\pi/2},\, F\left(\pm\sqrt{n\pi/2}\right)\right)$ are points of inflection. The graph of F is concave up on $\left(0, \sqrt{\pi/2}\right)$, concave down on $\left(\sqrt{\pi/2}, \sqrt{\pi}\right)$, and so on.

SECTION 5.3

1. $\displaystyle\int_0^1 (2x - 3)\,dx = [x^2 - 3x]_0^1 = (-2) - (0) = -2$

3. $\displaystyle\int_{-1}^0 5x^4\,dx = [x^5]_{-1}^0 = (0) - (-1) = 1$

5. $\displaystyle\int_1^4 2\sqrt{x}\,dx = 2\int_1^4 x^{1/2}\,dx = 2\left[\frac{2}{3}x^{3/2}\right]_1^4 = \frac{4}{3}\left[x^{3/2}\right]_1^4 = \frac{4}{3}(8 - 1) = \frac{28}{3}$

7. $\displaystyle\int_1^5 2\sqrt{x - 1}\,dx = \int_1^5 2(x - 1)^{1/2}\,dx = \left[\frac{4}{3}(x - 1)^{3/2}\right]_1^5 = \frac{4}{3}[4^{3/2} - 0] = \frac{32}{3}$

9. $\displaystyle\int_{-2}^0 (x + 1)(x - 2)\,dx = \int_{-2}^0 (x^2 - x - 2)\,dx = \left[\frac{x^3}{3} - \frac{x^2}{2} - 2x\right]_{-2}^0 = \left[0 - \left(\frac{-8}{3} - 2 + 4\right)\right] = \frac{2}{3}$

11. $\displaystyle\int_1^2 \left(3t + \frac{4}{t^2}\right) dt = \int_1^2 (3t + 4t^{-2})\,dt = \left[\frac{3}{2}t^2 - 4t^{-1}\right]_1^2 = \left[(6 - 2) - \left(\frac{3}{2} - 4\right)\right] = \frac{13}{2}$

13. $\displaystyle\int_0^1 (x^{3/2} - x^{1/2})\,dx = \left[\frac{2}{5}x^{5/2} - \frac{2}{3}x^{3/2}\right]_0^1 = \left[\left(\frac{2}{5} - \frac{2}{3}\right) - 0\right] = -\frac{4}{15}$

15. $\displaystyle\int_0^1 (x + 1)^{17}\,dx = \left[\frac{1}{18}(x + 1)^{18}\right]_0^1 = \frac{1}{18}(2^{18} - 1)$

17. $\displaystyle\int_0^a (\sqrt{a} - \sqrt{x})^2\,dx = \int_0^a (a - 2\sqrt{a}\,x^{1/2} + x)\,dx = \left[ax - \frac{4}{3}\sqrt{a}\,x^{3/2} + \frac{x^2}{2}\right]_0^a = a^2 - \frac{4}{3}a^2 + \frac{a^2}{2} = \frac{1}{6}a^2$

19. $\displaystyle\int_1^2 \frac{6 - t}{t^3}\,dt = \int_1^2 (6t^{-3} - t^{-2})\,dt = [-3t^{-2} + t^{-1}]_1^2 = \left[\frac{-3}{4} + \frac{1}{2}\right] - [-3 + 1] = \frac{7}{4}$

21. $\int_0^1 x^2(x-1)\,dx = \int_0^1 (x^3 - x^2)\,dx = \left[\dfrac{x^4}{4} - \dfrac{x^3}{3}\right]_0^1 = -\dfrac{1}{12}$

23. $\int_1^2 2x(x^2 + 1)\,dx = \int_1^2 (2x^3 + 2x)\,dx = \left[\dfrac{x^4}{2} + x^2\right]_1^2 = 12 - \dfrac{3}{2} = \dfrac{21}{2}$

25. $\int_0^{\pi/2} \cos x\,dx = [\sin x]_0^{\pi/2} = 1$

27. $\int_0^{\pi/4} 2\sec^2 x\,dx = 2\,[\tan x]_0^{\pi/4} = 2$

29. $\int_{\pi/6}^{\pi/4} \csc x \cot x\,dx = [-\csc x]_{\pi/6}^{\pi/4} = -\sqrt{2} - (-2) = 2 - \sqrt{2}$

31. $\int_0^{2\pi} \sin x\,dx = [-\cos x]_0^{2\pi} = -1 - (-1) = 0$

33. $\int_0^{\pi/3} \left(\dfrac{2}{\pi}x - 2\sec^2 x\right) dx = \left[\dfrac{1}{\pi}x^2 - 2\tan x\right]_0^{\pi/3} = \dfrac{\pi}{9} - 2\sqrt{3}$

35. $\int_0^3 \left[\dfrac{d}{dx}\left(\sqrt{4 + x^2}\right)\right] dx = \left[\sqrt{4 + x^2}\right]_0^3 = \sqrt{13} - 2$

37. (a) $F(x) = \int_1^x (t + 2)^2\,dt \quad\Longrightarrow\quad F'(x) = (x + 2)^2$

   (b) $\int_1^x (t + 2)^2\,dt = \left[\dfrac{t^3}{3} + 2t^2 + 4t\right]_1^x = \dfrac{x^3}{3} + 2x^2 + 4x - 6\dfrac{1}{3}$

   $\Longrightarrow\quad F'(x) = x^2 + 4x + 4 = (x + 2)^2$

39. (a) $F(x) = \int_1^{2x+1} \dfrac{1}{2}\sec u \tan u\,du \quad\Rightarrow\quad F'(x) = \sec(2x + 1)\tan(2x + 1)$

   (b) $\int_1^{2x+1} \dfrac{1}{2}\sec u \tan u\,du = \left[\dfrac{1}{2}\sec u\right]_1^{2x+1} = \dfrac{1}{2}\sec(2x + 1) - \dfrac{1}{2}\sec 1$

   $\Longrightarrow\quad F'(x) = \sec(2x + 1)\tan(2x + 1)$

41. (a) $F(x) = \int_2^x \dfrac{dt}{t}$ (b) $F(x) = -3 + \int_2^x \dfrac{dt}{t}$

43. Area $= \int_0^4 (4x - x^2)\,dx = \left[2x^2 - \dfrac{x^3}{3}\right]_0^4 = \dfrac{32}{3}$

45. Area $= \int_{-\pi/2}^{\pi/4} 2\cos x\,dx = 2\,[\sin x]_{-\pi/2}^{\pi/4} = \sqrt{2} + 2$

47. (a) $\displaystyle\int_2^5 (x-3)\,dx = \left[\frac{x^2}{2} - 3x\right]_2^5 = \frac{3}{2}$

(b) $\displaystyle\int_2^5 |x-3|\,dx = \int_2^3 (3-x)\,dx + \int_3^5 (x-3)\,dx$

$$= \left[3x - \frac{x^2}{2}\right]_2^3 + \left[\frac{x^2}{2} - 3x\right]_3^5 = \frac{5}{2}$$

49. (a) $\displaystyle\int_{-2}^2 (x^2-1)\,dx = \left[\frac{x^3}{3} - x\right]_{-2}^2 = \frac{4}{3}$

(b)
$$\int_{-2}^2 |x^2-1|\,dx = \int_{-2}^{-1} (x^2-1)\,dx + \int_{-1}^1 (1-x^2)\,dx + \int_1^2 (x^2-1)\,dx$$

$$= \left[\frac{x^3}{3} - x\right]_{-2}^{-1} + \left[x - \frac{x^3}{3}\right]_{-1}^1 + \left[\frac{x^3}{3} - x\right]_1^2 = 4$$

51. (a) $\displaystyle x(t) = \int_0^t (10u - u^2)\,du = \left[5u^2 - \frac{u^3}{3}\right]_0^t = 5t^2 - \frac{t^3}{3}, \quad 0 \le t \le 10$

(b) $v'(t) = 10 - 2t$; v has an absolute maximum at $t = 5$. The object's position at $t = 5$ is

$$x(5) = \frac{250}{3}.$$

53. $\displaystyle\int_0^4 f(x)\,dx = \int_0^1 (2x+1)\,dx + \int_1^4 (4-x)\,dx = \left[x^2 + x\right]_0^1 + \left[4x - \frac{x^2}{2}\right]_1^4 = \frac{13}{2}$

55. $\displaystyle\int_{-\pi/2}^{\pi} f(x)\,dx = \int_{-\pi/2}^{\pi/3} \cos x\,dx + \int_{\pi/3}^{\pi} \left[\frac{3}{\pi}x + 1\right]\,dx = [\sin x]_{-\pi/2}^{\pi/3} + \left[\frac{3x^2}{2\pi} + x\right]_{\pi/3}^{\pi}$

$$= \frac{2 + \sqrt{3}}{2} + 2\pi$$

57. (a) f is continuous on $[-2, 2]$.

For $x \in [-2, 0]$, $\displaystyle g(x) = \int_{-2}^x (t+2)\,dt = \left[\frac{1}{2}t^2 + 2t\right]_{-2}^x = \frac{1}{2}x^2 + 2x + 2.$

For $x \in [0, 1]$, $\displaystyle g(x) = \int_{-2}^0 (t+2)\,dt + \int_0^x 2\,dt = 2 + [2t]_0^x = 2 + 2x.$

For $x \in [1, 2]$, $\displaystyle g(x) = \int_{-2}^0 (t+2)\,dt + \int_0^1 2\,dt + \int_1^x (4-2x)\,dx = 2 + 2 + \left[4t - t^2\right]_1^x = 1 + 4x - x^2.$

Thus $g(x) = \begin{cases} \frac{1}{2}x^2 + 2x + 2, & -2 \le x \le 0 \\ 2x + 2, & 0 \le x \le 1 \\ 1 + 4x - x^2, & 1 \le x \le 2 \end{cases}$

(b)

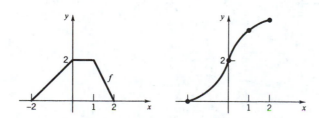

(c) f is continuous on $[-2, 2]$; f is differentiable on $(-2, 0)$, $(0, 1)$, and $(1, 2)$.

g is differentiable on $(-2, 2)$.

59. Follows from Theorem 5.3.2 since $f(x)$ is an antiderivative of $f'(x)$.

61. $\dfrac{d}{dx}\left[\displaystyle\int_a^x f(t)\,dt\right] = f(x);\quad \displaystyle\int_a^x \dfrac{d}{dt}[f(t)]\,dt = f(x) - f(a)$

SECTION 5.4

1. $A = \displaystyle\int_0^1 (2 + x^3)\,dx = \left[2x + \dfrac{x^4}{4}\right]_0^1 = \dfrac{9}{4}$

3. $A = \displaystyle\int_3^8 \sqrt{x+1}\,dx = \int_3^8 (x+1)^{1/2}\,dx = \left[\dfrac{2}{3}(x+1)^{3/2}\right]_3^8 = \dfrac{2}{3}[27 - 8] = \dfrac{38}{3}$

5. $A = \displaystyle\int_0^1 (2x^2 + 1)^2\,dx = \int_0^1 (4x^4 + 4x^2 + 1)\,dx = \left[\dfrac{4}{5}x^5 + \dfrac{4}{3}x^3 + x\right]_0^1 = \dfrac{47}{15}$

7. $A = \displaystyle\int_1^2 [0 - (x^2 - 4)]\,dx = \int_1^2 (4 - x^2)\,dx = \left[4x - \dfrac{x^3}{3}\right]_1^2 = \left[8 - \dfrac{8}{3}\right] - \left[4 - \dfrac{1}{3}\right] = \dfrac{5}{3}$

9. $A = \displaystyle\int_{\pi/3}^{\pi/2} \sin x\,dx = [-\cos x]_{\pi/3}^{\pi/2} = (0) - \left(-\dfrac{1}{2}\right) = \dfrac{1}{2}$

11.

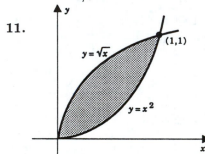

$A = \displaystyle\int_0^1 [x^{1/2} - x^2]\,dx$

$= \left[\tfrac{2}{3}x^{3/2} - \tfrac{1}{3}x^3\right]_0^1 = \tfrac{1}{3}$

13.

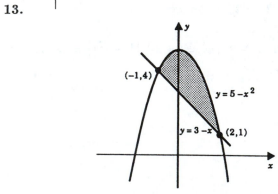

$A = \displaystyle\int_{-1}^2 [(5 - x^2) - (3 - x)]\,dx$

$= \displaystyle\int_{-1}^2 (2 + x - x^2)\,dx$

$= \left[2x + \dfrac{x^2}{2} - \dfrac{x^3}{3}\right]_{-1}^2$

$= \left[4 + 2 - \tfrac{8}{3}\right] - \left[-2 + \tfrac{1}{2} + \tfrac{1}{3}\right] = \tfrac{9}{2}$

15.

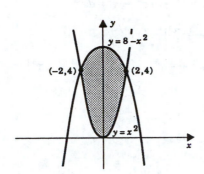

$$A = \int_{-2}^{2} [(8 - x^2) - (x^2)]\, dx$$

$$= \int_{-2}^{2} (8 - 2x^2)\, dx$$

$$= \left[8x - \tfrac{2}{3}x^3\right]_{-2}^{2}$$

$$= \left[16 - \tfrac{16}{3}\right] - \left[-16 + \tfrac{16}{3}\right] = \tfrac{64}{3}$$

17.

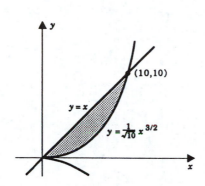

$$A = \int_{0}^{10} \left[x - \frac{1}{\sqrt{10}}\, x^{3/2}\right] dx$$

$$= \left[\frac{x^2}{2} - \frac{2\sqrt{10}}{50}\, x^{5/2}\right]_{0}^{10}$$

$$= 50 - \frac{2\sqrt{10}}{50}(10)^{5/2} = 10$$

19.

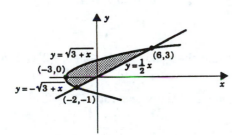

$$A = \int_{-3}^{-2} [(\sqrt{3+x}) - (-\sqrt{3+x})]\, dx + \int_{-2}^{6} \left[(\sqrt{3+x}) - \left(\frac{1}{2}x\right)\right] dx$$

$$= \int_{-3}^{-2} 2(3+x)^{1/2}\, dx + \int_{-2}^{6} \left[(3+x)^{1/2} - \frac{1}{2}x\right] dx$$

$$= \left[\frac{4}{3}(3+x)^{3/2}\right]_{-3}^{-2} + \left[\frac{2}{3}(3+x)^{3/2} - \frac{x^2}{4}\right]_{-2}^{6} = \left[\frac{4}{3} - 0\right] + \left[(18 - 9) - \left(\frac{2}{3} - 1\right)\right] = \frac{32}{3}$$

21.

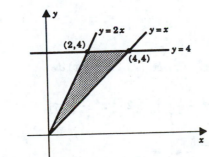

$$A = \int_{0}^{2} [2x - x]\, dx + \int_{2}^{4} [4 - x]\, dx$$

$$= \left[\tfrac{1}{2}x^2\right]_{0}^{2} + \left[4x - \tfrac{1}{2}x^2\right]_{2}^{4}$$

$$= 2 + [8 - 6] = 4$$

23.

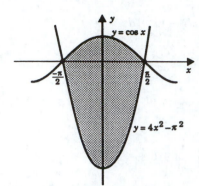

$$A = \int_{-\pi/2}^{\pi/2} [\cos x - (4x^2 - \pi^2)]\, dx$$

$$= \left[\sin x - \tfrac{4}{3}x^3 + \pi^2 x\right]_{-\pi/2}^{\pi/2}$$

$$= \left[1 - \tfrac{1}{6}\pi^3 + \tfrac{1}{2}\pi^3\right] - \left[-1 + \tfrac{1}{6}\pi^3 - \tfrac{1}{2}\pi^3\right]$$

$$= 2 + \tfrac{2}{3}\pi^3$$

25.

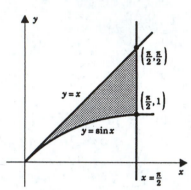

$$A = \int_0^{\pi/2} [x - \sin x]\, dx$$

$$= \left[\frac{x^2}{2} + \cos x\right]_0^{\pi/2}$$

$$= \frac{\pi^2}{8} - 1$$

27. (a) $\displaystyle \int_{-3}^{4} (x^2 - x - 6)\, dx = \left[\frac{1}{3}x^3 - \frac{1}{2}x^2 - 6x\right]_{-3}^{4} = -\frac{91}{6};$

the area of the region bounded by the graph of f and the x-axis for $x \in [-3, -2] \cup [3, 4]$

minus the area of the region bounded the graph of f and the x-axis for $x \in [-2, 3]$.

(b) $\displaystyle A = \int_{-3}^{-2} (x^2 - x - 6)\, dx + \int_{-2}^{3} (-x^2 + x + 6)\, dx + \int_{3}^{4} (x^2 - x - 6)\, dx$

$$= \left[\tfrac{1}{3}x^3 - \tfrac{1}{2}x^2 - 6x\right]_{-3}^{-2} + \left[-\tfrac{1}{3}x^3 + \tfrac{1}{2}x^2 + 6x\right]_{-2}^{3} + \left[\tfrac{1}{3}x^3 - \tfrac{1}{2}x^2 - 6x\right]_{3}^{4} = \frac{17}{6} + \frac{125}{6} + \frac{17}{6} = \frac{53}{2}$$

(c) $\displaystyle A = -\int_{-2}^{3} (x^2 - x - 6)\, dx = \frac{125}{6}$

29. (a) $\displaystyle \int_{-2}^{2} (x^3 - x)\, dx = \left[\frac{1}{4}x^4 - \frac{1}{2}x^2\right]_{-2}^{2} = 0$

(b)

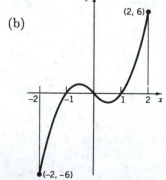

$$A = 2\left[-\int_0^1 (x^3 - x)\, dx + \int_1^2 (x^3 - x)\, dx\right]$$

$$= -2\left[\tfrac{1}{4}x^4 - \tfrac{1}{2}x^2\right]_0^1 + 2\left[\tfrac{1}{4}x^4 - \tfrac{1}{2}x^2\right]_1^2$$

$$= \frac{1}{2} + \frac{9}{2} = 5$$

31.   (a) $\int_{-2}^{3} (x^3 - 4x + 2)\, dx = \left[\dfrac{1}{4}\, x^4 - 2x^2 + 2x\right]_{-2}^{3} = \dfrac{65}{4}$

    (b)

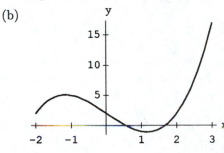

$A \cong \int_{-2}^{0.54} (x^3 - 4x + 2)\, dx - \int_{0.54}^{1.68} (x^3 - 4x + 2)\, dx + \int_{1.68}^{3} (x^3 - 4x + 2)\, dx$

$= \left[\tfrac{1}{4}\, x^4 - 2x^2 + 2x\right]_{-2}^{0.54} - \left[\tfrac{1}{4}\, x^4 - 2x^2 + 2x\right]_{0.54}^{1.68} + \left[\tfrac{1}{4}\, x^4 - 2x^2 + 2x\right]_{1.68}^{3}$

$= 8.52 + .81 + 8.54 = 17.87$

33.

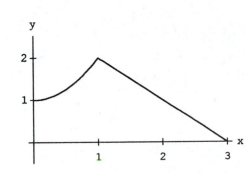

$A = \int_{0}^{1} (x^2 + 1)\, dx + \int_{1}^{3} (3 - x)\, dx$

$= \left[\tfrac{1}{3}\, x^3 + x\right]_{0}^{1} + \left[3x - \tfrac{1}{2}\, x^2\right]_{1}^{3}$

$= \dfrac{4}{3} + 2 = \dfrac{10}{3}$

35.

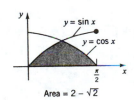

$y = \sin x$

$y = \cos x$

Area $= 2 - \sqrt{2}$

$A = \int_{0}^{\pi/4} \sin x\, dx + \int_{\pi/4}^{\pi/2} \cos x\, dx$

$= \left[-\cos x\right]_{0}^{\pi/4} + \left[\sin x\right]_{\pi/4}^{\pi/2}$

$= 2 - \sqrt{2}$

37.

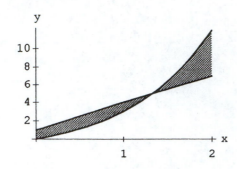

$$A \cong \int_0^{1.32} [3x + 1 - (x^3 + 2x)]\, dx + \int_{1.32}^2 [x^3 + 2x - (3x + 1)]\, dx$$

$$= \int_0^{1.32} (x + 1 - x^3)\, dx + \int_{1.32}^2 (x^3 - x - 1)\, dx$$

$$= \left[\tfrac{1}{2} x^2 + x - \tfrac{1}{4} x^4\right]_0^{1.32} + \left[\tfrac{1}{4} x^4 - \tfrac{1}{2} x^2 - x\right]_{1.32}^2 = 2.86$$

SECTION 5.5

1. $\displaystyle \int \frac{dx}{x^4} = \int x^{-4}\, dx = -\frac{1}{3} x^{-3} + C$

3. $\displaystyle \int (ax + b)\, dx = \frac{1}{2}\, ax^2 + bx + C$

5. $\displaystyle \int \frac{dx}{\sqrt{1+x}} = \int (1+x)^{-1/2}\, dx = 2(1+x)^{1/2} + C$

7. $\displaystyle \int \left(\frac{x^3 - 1}{x^2}\right) dx = \int (x - x^{-2})\, dx = \frac{1}{2}\, x^2 + x^{-1} + C$

9. $\displaystyle \int (t - a)(t - b)\, dt = \int [t^2 - (a + b)t + ab]\, dt = \frac{1}{3}\, t^3 - \frac{a+b}{2} t^2 + abt + C$

11. $\displaystyle \int \frac{(t^2 - a)(t^2 - b)}{\sqrt{t}}\, dt \quad = \int [t^{7/2} - (a + b)t^{3/2} + abt^{-1/2}]\, dt$

$$= \tfrac{2}{9} t^{9/2} - \tfrac{2}{5}(a + b)t^{5/2} + 2abt^{1/2} + C$$

13. $\displaystyle \int g(x)g'(x)\, dx = \frac{1}{2}[g(x)]^2 + C$

15. $\displaystyle \int \tan x \sec^2 x\, dx = \int \sec x\, \frac{d}{dx}\,[\sec x]\, dx = \frac{1}{2} \sec^2 x + C$

$$\int \tan x \sec^2 x\, dx = \int \tan x\, \frac{d}{dx}\,[\tan x]\, dx = \frac{1}{2} \tan^2 x + C$$

17. $\displaystyle \int \frac{4}{(4x + 1)^2}\, dx = \int 4(4x + 1)^{-2}\, dx = -(4x + 1)^{-1} + C$

19. $f(x) = \int f'(x)\,dx = \int (2x - 1)\,dx = x^2 - x + C.$

Since $f(3) = 4$, we get $4 = 9 - 3 + C$ so that $C = -2$ and

$$f(x) = x^2 - x - 2.$$

21. $f(x) = \int f'(x)\,dx = \int (ax + b)\,dx = \frac{1}{2}ax^2 + bx + C.$

Since $f(2) = 0$, we get $0 = 2a + 2b + C$ so that $C = -2a - 2b$ and

$$f(x) = \frac{1}{2}ax^2 + bx - 2a - 2b.$$

23. $f(x) = \int f'(x)\,dx = \int \sin x\,dx = -\cos x + C.$

Since $f(0) = 2$, we get $2 = -1 + C$ so that $C = 3$ and

$$f(x) = 3 - \cos x.$$

25. First,

$$f'(x) = \int f''(x)\,dx = \int (6x - 2)\,dx = 3x^2 - 2x + C.$$

Since $f'(0) = 1$, we get $1 = 0 + C$ so that $C = 1$ and

$$f'(x) = 3x^2 - 2x + 1.$$

Next,

$$f(x) = \int f'(x)\,dx = \int (3x^2 - 2x + 1)\,dx = x^3 - x^2 + x + K.$$

Since $f(0) = 2$, we get $2 = 0 + K$ so that $K = 2$ and

$$f(x) = x^3 - x^2 + x + 2.$$

27. First,

$$f'(x) = \int f''(x)\,dx = \int (x^2 - x)\,dx = \frac{1}{3}x^3 - \frac{1}{2}x^2 + C.$$

Since $f'(1) = 0$, we get $0 = \frac{1}{3} - \frac{1}{2} + C$ so that $C = \frac{1}{6}$ and

$$f'(x) = \frac{1}{3}x^3 - \frac{1}{2}x^2 + \frac{1}{6}.$$

Next,

$$f(x) = \int f'(x)\,dx = \int \left(\frac{1}{3}x^3 - \frac{1}{2}x^2 + \frac{1}{6} \right)\,dx = \frac{1}{12}x^4 - \frac{1}{6}x^3 + \frac{1}{6}x + K.$$

Since $f(1) = 2$, we get $2 = \frac{1}{12} - \frac{1}{6} + \frac{1}{6} + K$ so that $K = \frac{23}{12}$ and

$$f(x) = \frac{x^4}{12} - \frac{x^3}{6} + \frac{x}{6} + \frac{23}{12} = \frac{1}{12}(x^4 - 2x^3 + 2x + 23).$$

29. First,

$$f'(x) = \int f''(x)\,dx = \int \cos x\,dx = \sin x + C.$$

Since $f'(0) = 1$, we get $1 = 0 + C$ so that $C = 1$ and

$$f'(x) = \sin x + 1.$$

Next,

$$f(x) = \int f'(x)\, dx = \int (\sin x + 1)\, dx = -\cos x + x + K.$$

Since $f(0) = 2$, we get $2 = -1 + 0 + K$ so that $K = 3$ and

$$f(x) = -\cos x + x + 3.$$

31. First,

$$f'(x) = \int f''(x)\, dx = \int (2x - 3)\, dx = x^2 - 3x + C.$$

Then,

$$f(x) = \int f'(x)\, dx = \int (x^2 - 3x + C)\, dx = \frac{1}{3}x^3 - \frac{3}{2}x^2 + Cx + K.$$

Since $f(2) = -1$, we get

(1) $\qquad\qquad\qquad\qquad -1 = \frac{8}{3} - 6 + 2C + K;$

and, from $f(0) = 3$, we conclude that

(2) $\qquad\qquad\qquad\qquad 3 = 0 + K.$

Solving (1) and (2) simultaneously, we get $K = 3$ and $C = -\frac{1}{3}$ so that

$$f(x) = \tfrac{1}{3}x^3 - \tfrac{3}{2}x^2 - \tfrac{1}{3}x + 3.$$

33. $\dfrac{d}{dx}\left[\displaystyle\int f(x)\, dx\right] = f(x); \qquad \displaystyle\int \dfrac{d}{dx}[f(x)]\, dx = f(x) + C$

35. (a) $x(t) = \displaystyle\int v(t)\, dt = \int (6t^2 - 6)\, dt = 2t^3 - 6t + C.$

Since $x(0) = -2$, we get $-2 = 0 + C$ so that $C = -2$ and

$$x(t) = 2t^3 - 6t - 2. \quad \text{Therefore} \quad x(3) = 34.$$

Three seconds later the object is 34 units to the right of the origin .

(b) $s = \displaystyle\int_0^3 |v(t)|\, dt = \int_0^3 |6t^2 - 6|\, dt = \int_0^1 (6 - 6t^2)\, dt + \int_1^3 (6t^2 - 6)\, dt$

$$= [6t - 2t^3]_0^1 + [2t^3 - 6t]_1^3 = 4 + [36 - (-4)] = 44.$$

The object traveled 44 units.

37. (a) $v(t) = \displaystyle\int a(t)\, dt = \int (t+1)^{-1/2}\, dt = 2(t+1)^{1/2} + C.$

Since $v(0) = 1$, we get $1 = 2 + C$ so that $C = -1$ and

$$v(t) = 2(t+1)^{1/2} - 1.$$

(b) We know $v(t)$ by part (a). Therefore,

$$x(t) = \int v(t)\, dt = \int [2(t+1)^{1/2} - 1]\, dt = \frac{4}{3}(t+1)^{3/2} - t + C.$$

Since $x(0) = 0$, we get $0 = \frac{4}{3} - 0 + C$ so that

$$C = -\tfrac{4}{3} \quad \text{and} \quad x(t) = \tfrac{4}{3}(t+1)^{3/2} - t - \tfrac{4}{3}.$$

39. (a) $v_0 = 60$ mph $= 88$ feet per second. In general, $v(t) = at + v_0$. Here, in feet and seconds, $v(t) = -20t + 88$. Thus $v(t) = 0$ at $t = 4.4$ seconds.

(b) In general, $x(t) = \tfrac{1}{2}at^2 + v_0 t + x_0$. Here we take $x_0 = 0$. In feet and seconds
$$x(t) = -10(4.4)^2 + 88(4.4) = 10(4.4)^2 = 193.6 \text{ ft.}$$

41.
$$[v(t)]^2 = (at + v_0)^2 = a^2 t^2 + 2av_0 t + v_0{}^2$$
$$= v_0{}^2 + a(at^2 + 2v_0 t)$$
$$= v_0{}^2 + 2a(\tfrac{1}{2}at^2 + v_0 t)$$

$x(t) = \tfrac{1}{2}at^2 + v_0 t + x_0$
$$= v_0{}^2 + 2a\left[x(t) - x_0\right]$$

43. The car can accelerate to 60 mph (88 ft/sec) in 20 seconds thereby covering a distance of 880 ft. It can decelerate from 88 ft/sec to 0 ft/sec in 4 seconds thereby covering a distance of 176 ft. At full speed, 88 ft/sec, it must cover a distance of

$$\frac{5280}{2} - 880 - 176 = 1584 \text{ ft.}$$

This takes $\dfrac{1584}{88} = 18$ seconds. The run takes at least $20 + 18 + 4 = 42$ seconds.

45. $v(t) = \displaystyle\int a(t)\,dt = \int (2A + 6Bt)\,dt = 2At + 3Bt^2 + C.$

Since $v(0) = v_0$, we have $v_0 = 0 + C$ so that $v(t) = 2At + 3Bt^2 + v_0$.
$$x(t) = \int v(t)\,dt = \int (2At + 3Bt^2 + v_0)\,dt = At^2 + Bt^3 + v_0 t + K.$$

Since $x(0) = x_0$, we have $x_0 = 0 + K$ so that $K = x_0$ and
$$x(t) = x_0 + v_0 t + At^2 + Bt^3.$$

47.

$x'(t) = t^2 - 5,$	$y'(t) = 3t,$
$x(t) = \tfrac{1}{3}t^3 - 5t + C.$	$y(t) = \tfrac{3}{2}t^2 + K.$

When $t = 2$, the particle is at $(4, 2)$. Thus, $x(2) = 4$ and $y(2) = 2$.

$4 = \tfrac{8}{3} - 10 + C \implies C = \tfrac{34}{3}.$	$2 = 6 + K \implies K = -4.$
$x(t) = \tfrac{1}{3}t^3 - 5t + \tfrac{34}{3},$	$y(t) = \tfrac{3}{2}t^2 - 4.$

Four seconds later the particle is at $(x(6), y(6)) = (\tfrac{160}{3}, 50)$.

49. Since $v(0) = 2$, we have $2 = A \cdot 0 + B$ so that $B = 2$. Therefore

$$x(t) = \int v(t)\,dt = \int (At + 2)\,dt = \frac{1}{2}At^2 + 2t + C.$$

Since $x(2) = x(0) - 1$, we have

$$2A + 4 + C = C - 1 \quad \text{so that} \quad A = -\tfrac{5}{2}.$$

51. $$x(t) = \int v(t)\,dt = \int \sin t\,dt = -\cos t + C$$

Since $x(\pi/6) = 0$, we have $0 = -\dfrac{\sqrt{3}}{2} + C$ so that $C = \dfrac{\sqrt{3}}{2}$ and $x(t) = \dfrac{\sqrt{3}}{2} - \cos t.$

(a)   At $t = 11\pi/6$ sec.

(b)   We want to find the smallest $t_0 > \pi/6$ for which $x(t_0) = 0$ and $v(t_0) > 0$. We get

$$t_0 = 13\pi/6 \text{ seconds.}$$

53.   The mean-value theorem. With obvious notation

$$\frac{x(1/12) - x(0)}{1/12} = \frac{4}{1/12} = 48.$$

By the mean-value theorem there exists some time t_0 at which

$$x'(t_0) = \frac{x(1/12) - x(0)}{1/12}.$$

SECTION 5.6

1. $\left\{ \begin{array}{l} u = 2 - 3x \\ du = -3\,dx \end{array} \right\}$; $\displaystyle \int \frac{dx}{(2-3x)^2} = \int (2-3x)^{-2}\,dx = -\frac{1}{3}\int u^{-2}\,du = \frac{1}{3}u^{-1} + C$

$$= \tfrac{1}{3}(2-3x)^{-1} + C$$

3. $\left\{ \begin{array}{l} u = 2x + 1 \\ du = 2\,dx \end{array} \right\}$; $\displaystyle \int \sqrt{2x+1}\,dx = \int (2x+1)^{1/2}\,dx = \frac{1}{2}\int u^{1/2}\,du = \frac{1}{3}u^{3/2} + C$

$$= \frac{1}{3}(2x+1)^{3/2} + C$$

5. $\left\{ \begin{array}{l} u = ax + b \\ du = a\,dx \end{array} \right\}$; $\displaystyle \int (ax+b)^{3/4}\,dx = \frac{1}{a}\int u^{3/4}\,du = \frac{4}{7a}u^{7/4} + C$

$$= \frac{4}{7a}(ax+b)^{7/4} + C$$

7. $\left\{ \begin{array}{l} u = 4t^2 + 9 \\ du = 8t\,dt \end{array} \right\}$; $\displaystyle \int \frac{t}{(4t^2+9)^2}\,dt = \frac{1}{8}\int \frac{du}{u^2} = -\frac{1}{8}u^{-1} + C = -\frac{1}{8}\left(4t^2+9\right)^{-1} + C$

9. $\left\{ \begin{array}{l} u = 1 + x^3 \\ du = 3x^2\,dx \end{array} \right\}$; $\displaystyle \int x^2\left(1+x^3\right)^{1/4}\,dx = \frac{1}{3}\int u^{1/4}\,du = \frac{4}{15}u^{5/4} + C$

$$= \frac{4}{15}(1+x^3)^{5/4} + C$$

11. $\quad \left\{ \begin{array}{l} u = 1 + s^2 \\ du = 2s\,ds \end{array} \right\} ; \qquad \displaystyle\int \frac{s}{(1+s^2)^3}\,ds = \frac{1}{2}\int \frac{du}{u^3} = -\frac{1}{4}u^{-2} + C = -\frac{1}{4}(1+s^2)^{-2} + C$

13. $\quad \left\{ \begin{array}{l} u = x^2 + 1 \\ du = 2x\,dx \end{array} \right\} ; \qquad \displaystyle\int \frac{x}{\sqrt{x^2+1}}\,dx = \int (x^2+1)^{-1/2}\,x\,dx = \frac{1}{2}\int u^{-1/2}\,du$

$$= u^{1/2} + C = \sqrt{x^2+1} + C$$

15. $\quad \left\{ \begin{array}{l} u = x^2 + 1 \\ du = 2x \end{array} \right\} ; \qquad \displaystyle\int 5x\,(x^2+1)^{-3}\,dx = \frac{5}{2}\int u^{-3}\,du = -\frac{5}{4}u^{-2} + C$

$$= -\frac{5}{4}(x^2+1)^{-2} + C$$

17. $\quad \left\{ \begin{array}{l} u = x^{1/4} + 1 \\ du = \frac{1}{4}x^{-3/4}\,dx \end{array} \right\} ; \qquad \displaystyle\int x^{-3/4}\left(x^{1/4}+1\right)^{-2}\,dx \;= 4\int u^{-2}\,du = -4u^{-1} + C$

$$= -4(x^{1/4}+1)^{-1} + C$$

19. $\quad \left\{ \begin{array}{l} u = 1 - a^4 x^4 \\ du = -4a^4 x^3\,dx \end{array} \right\} ; \qquad \displaystyle\int \frac{b^3 x^3}{\sqrt{1-a^4 x^4}}\,dx = -\frac{b^3}{4a^4}\int u^{-1/2}\,du = -\frac{b^3}{2a^4}u^{1/2} + C$

$$= -\frac{b^3}{2a^4}\sqrt{1 - a^4 x^4} + C$$

21 $\quad \left\{ \begin{array}{l|l} u = x^2 + 1 & x = 0 \implies u = 1 \\ du = 2x\,dx & x = 1 \implies u = 2 \end{array} \right\} ; \qquad \displaystyle\int_0^1 x\,(x^2+1)^3\,dx = \frac{1}{2}\int_1^2 u^3\,du$

$$= \frac{1}{8}\left[u^4\right]_1^2 = \frac{15}{8}$$

23. $0;$ integrand is an odd function

25. $\quad \left\{ \begin{array}{l|l} u = a^2 - y^2 & y = 0 \implies u = a^2 \\ du = -2y\,dy & y = a \implies u = 0 \end{array} \right\} ; \qquad \displaystyle\int_0^a y\sqrt{a^2 - y^2}\,dy = -\frac{1}{2}\int_{a^2}^0 u^{1/2}\,du$

$$= -\frac{1}{3}\left[u^{3/2}\right]_{a^2}^0 = \frac{1}{3}\left(a^2\right)^{3/2} = \frac{1}{3}|a|^3$$

27.

$\quad \left\{ \begin{array}{l|l} u = 2x^2 + 1 & x = 0 \implies u = 1 \\ du = 4x\,dx & x = 2 \implies u = 9 \end{array} \right\} ; \qquad \displaystyle\int_0^2 x\sqrt{2x^2+1}\,dx = \frac{1}{4}\int_1^9 \sqrt{u}\,du$

$$= \left[\frac{1}{6}u^{\frac{3}{2}}\right]_1^9 = \frac{13}{3}$$

29.

$\quad \left\{ \begin{array}{l|l} u = 1 + x^{-2} & x = 1 \implies u = 2 \\ du = -2x^{-3}\,dx & x = 2 \implies u = \frac{5}{4} \end{array} \right\} ; \qquad \displaystyle\int_1^2 x^{-3}(1+x^{-2})^{-3}\,dx = -\frac{1}{2}\int_2^{\frac{5}{4}} u^{-3}\,du$

$$= \left[\frac{u^{-2}}{4}\right]_2^{\frac{5}{4}} = \frac{39}{400}$$

31. $\begin{cases} u = x+1 \\ du = dx \end{cases}$; $\displaystyle \int x\sqrt{x+1}\,dx = \int (u-1)\sqrt{u}\,du = \int \left(u^{3/2} - u^{1/2}\right)du$

$$= \tfrac{2}{5}u^{5/2} - \tfrac{2}{3}u^{3/2} + C = \tfrac{2}{5}(x+1)^{5/2} - \tfrac{2}{3}(x+1)^{3/2} + C$$

33. $\begin{cases} u = 2x-1 \\ du = dx \end{cases}$; $\displaystyle \int x\sqrt{2x-1}\,dx = \frac{1}{2}\int \frac{(u-1)}{2}\sqrt{u}\,du = \frac{1}{4}\int \left(u^{3/2} + u^{1/2}\right)du$

$$= \frac{1}{10}u^{5/2} + \frac{1}{6}u^{3/2} + C = \frac{1}{10}(2x-1)^{5/2} + \frac{1}{6}(2x-1)^{3/2} + C$$

35. $\begin{cases} u = x+1 \\ du = dx \end{cases} \Bigg| \begin{array}{l} x=0 \implies u=1 \\ x=1 \implies u=2 \end{array}$; $\displaystyle \int_0^1 \frac{x+3}{\sqrt{x+1}}\,dx = \int_1^2 \frac{u+2}{\sqrt{u}}\,du$

$$= \int_1^2 \left(u^{1/2} + 2u^{-1/2}\right)du$$

$$= \left[\tfrac{2}{3}u^{3/2} + 4u^{1/2}\right]_1^2 = \tfrac{2}{3}\sqrt{8} + 4\sqrt{2} - \tfrac{2}{3} - 4 = \tfrac{16}{3}\sqrt{2} - \tfrac{14}{3}$$

37. $\begin{cases} u = x^2+1 \\ du = 2xdx \end{cases}$; $\displaystyle \int x\sqrt{x^2+1}\,dx = \frac{1}{2}\int \sqrt{u}\,du = \frac{1}{3}u^{\frac{3}{2}} + C = \frac{1}{3}(x^2+1)^{\frac{3}{2}} + C.$

Also, $1 = \dfrac{1}{3}(0^2 + 1) + C \implies C = \dfrac{2}{3}.$ Thus $y = \dfrac{1}{3}(x^2+1)^{\frac{3}{2}} + \dfrac{2}{3}.$

39. $\displaystyle \int \cos(3x+1)\,dx = -frac13\sin(3x+1) + C$ **41.** $\displaystyle \int \csc^2 \pi x\,dx = -\frac{1}{\pi}\cot \pi x + C$

43. $\begin{cases} u = 3-2x \\ du = -2\,dx \end{cases}$; $\displaystyle \int \sin(3-2x)\,dx = \int -\frac{1}{2}\sin u\,du = \frac{1}{2}\cos u + C = \frac{1}{2}\cos(3-2x) + C$

45. $\begin{cases} u = \cos x \\ du = -\sin x\,dx \end{cases}$; $\displaystyle \int \cos^4 x \sin x\,dx = \int -u^4\,du = -\frac{1}{5}u^5 + C = -\frac{1}{5}\cos^5 x + C$

47. $\begin{cases} u = x^{1/2} \\ du = \frac{1}{2}x^{-1/2}\,dx \end{cases}$; $\displaystyle \int x^{-1/2}\sin x^{1/2}\,dx = \int 2\sin u\,du = -2\cos u + C$

$$= -2\cos x^{1/2} + C$$

49. $\begin{cases} u = 1+\sin x \\ du = \cos x\,dx \end{cases}$; $\displaystyle \int \sqrt{1+\sin x}\,\cos x\,dx = \int u^{1/2}\,du = \frac{2}{3}u^{3/2} + C$

$$= \tfrac{2}{3}(1+\sin x)^{3/2} + C$$

51. $\displaystyle \int \frac{1}{\cos^2 x}\,dx = \int \sec^2 x\,dx = \tan x + C$

53. $\left\{ \begin{array}{l} u = x^2 \\ du = 2x\,dx \end{array} \right\}$; $\displaystyle\int x\sin^3 x^2 \cos x^2\,dx = \int \frac{1}{2}\sin^3 u\cos u\,du = \frac{1}{8}\sin^4 u + C$

$$= \tfrac{1}{8}\sin^4 x^2 + C$$

55. $\left\{ \begin{array}{l} u = 1 + \tan x \\ du = \sec^2 x\,dx \end{array} \right\}$; $\displaystyle\int \frac{\sec^2 x}{\sqrt{1+\tan x}}\,dx = \int u^{-1/2}\,du = 2u^{1/2} + C$

$$= 2(1+\tan x)^{1/2} + C$$

57. 0;   the sine is an odd function   **59.** $\displaystyle\int_{1/4}^{1/3} \sec^2 \pi x\,dx = \frac{1}{\pi}\left[\tan\pi x\right]_{1/4}^{1/3} = \frac{1}{\pi}(\sqrt{3}-1)$

61. $\displaystyle\int_0^{\pi/2} \sin^3 x\cos x\,dx = \frac{1}{4}\left[\sin^4 x\right]_0^{\pi/2} = \frac{1}{4}$

63. $\displaystyle\int \sin^2 x\,dx = \int \frac{1-\cos 2x}{2}\,dx = \frac{1}{2}x - \frac{1}{4}\sin 2x + C$

65. $\displaystyle\int \cos^2 5x\,dx = \int \frac{1+\cos 10x}{2}\,dx = \frac{1}{2}x + \frac{1}{20}\sin 10x + C$

67. $\displaystyle\int_0^{\pi/2} \cos^2 2x\,dx = \int_0^{\pi/2} \frac{1+\cos 4x}{2}\,dx = \left[\frac{1}{2}x + \frac{1}{8}\sin 4x\right]_0^{\pi/2} = \frac{\pi}{4}$

69. $A = \displaystyle\int_0^{\frac{\pi}{2}} [\cos x - (-\sin x)]\,dx = [\sin x - \cos x]_0^{\frac{\pi}{2}} = 2$

71. $A = \displaystyle\int_0^{1/4} (\cos^2 \pi x - \sin^2 \pi x)\,dx = \int_0^{1/4} \cos 2\pi x\,dx = = \frac{1}{2\pi}\left[\sin 2\pi x\right]_0^{1/4} = \frac{1}{2\pi}$

73. $A = \displaystyle\int_{1/6}^{1/4} (\csc^2 \pi x - \sec^2 \pi x)\,dx = \frac{1}{\pi}\left[-\cot\pi x - \tan\pi x\right]_{1/6}^{1/4}$

$$= \frac{1}{\pi}\left(-2 + \cot\frac{\pi}{6} + \tan\frac{\pi}{6}\right)$$

$$= \frac{1}{\pi}\left(-2 + \sqrt{3} + \frac{1}{\sqrt{3}}\right) = \frac{1}{3\pi}(4\sqrt{3} - 6)$$

75. (a) $\left\{ \begin{array}{l} u = \sec x \\ du = \sec x\tan x\,dx \end{array} \right\}$; $\displaystyle\int \sec^2 x\tan x\,dx = \int u\,du = \frac{1}{2}u + C$

$$= \tfrac{1}{2}\sec^2 x + C$$

(b)
$$\left\{ \begin{array}{l} u = \tan x \\ du = \sec^2 x \, dx \end{array} \right\} ; \quad \int \sec^2 x \tan x \, dx = \int u \, du = \frac{1}{2} u^2 + C'$$
$$= \tfrac{1}{2} \tan^2 x + C'$$

(c) $C' = C + \frac{1}{2}$

77.
$$A = \frac{4b}{a} \int_0^a \sqrt{a^2 - x^2} \, dx = \frac{4b}{a} \left(\frac{\text{area of circle of radius } a}{4} \right)$$
$$= \frac{4b}{a} \left(\frac{\pi a^2}{4} \right) = \pi ab$$

SECTION 5.7

1.   Yes; $\displaystyle\int_a^b [f(x) - g(x)] \, dx = \int_a^b f(x) \, dx - \int_a^b g(x) \, dx > 0.$

3.   Yes;   otherwise we would have $f(x) \leq g(x)$ for all $x \in [a, b]$ and it would follow that
$$\int_a^b f(x) \, dx \leq \int_a^b g(x) \, dx.$$

5.   No;   take $f(x) = 0$, $g(x) = -1$ on $[0, 1]$.

7.   No;   take, for example, any odd function on an interval of the form $[-c, c]$.

9.   No; $\displaystyle\int_{-1}^1 x \, dx = 0$ but $\displaystyle\int_{-1}^1 |x| \, dx \neq 0.$

11.   Yes; $U_f(P) \geq \displaystyle\int_a^b f(x) \, dx = 0.$

13.   No; $L_f(P) \leq \displaystyle\int_a^b f(x) \, dx = 0.$

15.   Yes; $\displaystyle\int_a^b [f(x) + 1] \, dx = \int_a^b f(x) \, dx + \int_a^b 1 \, dx = 0 + b - a = b - a.$

17. $\dfrac{d}{dx} \left[\displaystyle\int_0^{1+x^2} \dfrac{dt}{\sqrt{2t + 5}} \right] = \dfrac{1}{\sqrt{2(1 + x^2) + 5}} \dfrac{d}{dx} (1 + x^2) = \dfrac{2x}{\sqrt{2x^2 + 7}}$

19. $\dfrac{d}{dx} \left[\displaystyle\int_x^a f(t) \, dt \right] = \dfrac{d}{dx} \left[-\int_a^x f(t) \, dt \right] = -f(x)$

21. $\dfrac{d}{dx} \left[\displaystyle\int_{x^2}^3 \dfrac{\sin t}{t} \, dt \right] = -\dfrac{d}{dx} \left[\displaystyle\int_3^{x^2} \dfrac{\sin t}{t} \, dt \right] = -\dfrac{\sin(x^2)}{x^2} (2x) = -\dfrac{2\sin(x^2)}{x}$

23. $\dfrac{d}{dx}\left[\displaystyle\int_1^{\sqrt{x}} \dfrac{t^2}{1+t^2}\, dt\right] = \dfrac{x}{1+x}\cdot\dfrac{1}{2\sqrt{x}} = \dfrac{\sqrt{x}}{2(1+x)}$

25. $\dfrac{d}{dx}\left[\displaystyle\int_x^{x^2} \dfrac{dt}{t}\right] = \dfrac{1}{x^2}\dfrac{d}{dx}\left(x^2\right) - \dfrac{1}{x}\dfrac{d}{dx}\left(x\right) = \dfrac{2x}{x^2} - \dfrac{1}{x} = \dfrac{1}{x}$

27. $\dfrac{d}{dx}\left[\displaystyle\int_{\tan x}^{2x} t\sqrt{1+t^2}\, dt\right] = 2x\sqrt{1+(2x)^2}\,(2) - \tan x\,\sqrt{1+\tan^2 x}\,(\sec^2 x)$

$$= 4x\sqrt{1+4x^2} - \tan x\,\sec^2 x\,|\sec x|$$

29. (a) With P a partition of $[a, b]$

$$L_f(P) \le \int_a^b f(x)\, dx.$$

If f is nonnegative on $[a, b]$, then $L_f(P)$ is nonnegative and, consequently, so is the integral. If f is positive on $[a, b]$, then $L_f(P)$ is positive and, consequently, so is the integral.

(b) Take F as an antiderivative of f on $[a, b]$. Observe that

$$F'(x) = f(x) \text{ on } (a, b) \quad \text{and} \quad \int_a^b f(x)\, dx = F(b) - F(a).$$

If $f(x) \ge 0$ on $[a, b]$, then F is nondecreasing on $[a, b]$ and $F(b) - F(a) \ge 0$.

If $f(x) > 0$ on $[a, b]$, then F is increasing on $[a, b]$ and $F(b) - F(a) > 0$.

31. Consider the trivial partition P of $[a, b]$ into the single interval $[a, b]$.

Then $\quad L_f(P) = m(b - a) \quad$ and $\quad U_f(P) = M(b - a)$.

Thus $m(b - a) \le \int_a^b f(x)\, dx \le M(b - a)$.

33.
$$H(x) = \int_{2x}^{x^3 - 4} \dfrac{x\, dt}{1 + \sqrt{t}} = x\int_{2x}^{x^3 - 4} \dfrac{dt}{1 + \sqrt{t}},$$

$$H'(x) = x\cdot\left[\dfrac{3x^2}{1 + \sqrt{x^3 - 4}} - \dfrac{2}{1 + \sqrt{2x}}\right] + 1\cdot\int_{2x}^{x^3 - 4} \dfrac{dt}{1 + \sqrt{t}},$$

$$H'(2) = 2\left[\dfrac{12}{3} - \dfrac{2}{3}\right] + \underbrace{\int_4^4 \dfrac{dt}{1 + \sqrt{t}}}_{=0} = \dfrac{20}{3}$$

35. (a) Let $u = -x$. Then $du = -dx$; and $u = 0$ when $x = 0$, $u = a$ when $x = -a$.

$$\int_{-a}^0 f(x)\, dx = -\int_a^0 f(-u)\, du = \int_0^a f(-u)\, du = \int_0^a f(-x)\, dx$$

37. $\displaystyle\int_{-\pi/4}^{\pi/4} (x + \sin 2x)\, dx = 0 \quad$ since $\quad f(x) = x + \sin 2x \quad$ is an odd function.

39. $\displaystyle\int_{-\pi/3}^{\pi/3} (1 + x^2 - \cos x)\, dx = 2\int_0^{\pi/3} (1 + x^2 - \cos x)\, dx \quad$ since $\quad f(x) = 1 + x^2 - \cos x \quad$ is an even function.

$$2\int_0^{\pi/3}(1+x^2-\cos x)\,dx = 2\left[x+\frac{1}{3}x^3-\sin x\right]_0^{\pi/3} = \frac{2}{3}\pi+\frac{2}{81}\pi^3-\sqrt{3}$$

PROJECT 5.7

1.   (a) Let $f(x)=x^4-3x^2+0.1x+1.$

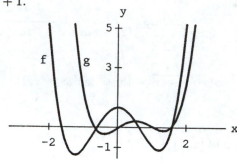

(b) dom $(f)=$ dom (g). A comparson of ranges is:

range $(f)=[-1.37,\infty)$

range $(g)=[-0.34,\infty)$

(c) Setting $g(x)=f(x)$ gives

$$\frac{f(x-1)+f(x)}{2}=f(x)$$

which implies $f(x-1)=f(x)$.   This occurs at

the points where $x\cong-0.6280,0.5200,1.6081.$

3. (a)

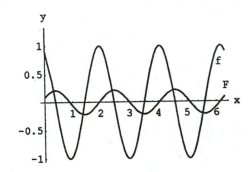

The moving average g smooths everything; the moving average F still oscillates, but it

oscillates "slower" and with less amplitude.

(b) In general, $F'(x)=\dfrac{1}{a}[f(x)-f(x-a)]$ and

$$G(x)=\frac{1}{a}\int_{x-a}^x f'(t)dt=\frac{1}{a}[f(x)-f(x-a)]=F'(x)$$

SECTION 5.8

1.   A.V. $= \dfrac{1}{c} \displaystyle\int_0^c (mx+b)\,dx = \dfrac{1}{c} \left[\dfrac{m}{2}x^2 + bx\right]_0^c = \dfrac{mc}{2} + b;$ at $x = c/2$

3.   A.V. $= \dfrac{1}{2} \displaystyle\int_{-1}^1 x^3\,dx = 0$ since the integrand is odd;   at $x = 0$

5.   A.V. $= \dfrac{1}{4} \displaystyle\int_{-2}^2 |x|\,dx = \dfrac{1}{2} \displaystyle\int_0^2 |x|\,dx = \dfrac{1}{2} \displaystyle\int_0^2 x\,dx = \dfrac{1}{2} \left[\dfrac{x^2}{2}\right]_0^2 = 1;$ at $x = \pm 1$

7.   A.V. $= \dfrac{1}{2} \displaystyle\int_0^2 (2x - x^2)\,dx = \dfrac{1}{2} \left[x^2 - \dfrac{x^3}{3}\right]_0^2 = \dfrac{2}{3};$ at $x = 1 \pm \dfrac{1}{3}\sqrt{3}$

9.   A.V. $= \dfrac{1}{9} \displaystyle\int_0^9 \sqrt{x}\,dx = \dfrac{1}{9} \left[\dfrac{2}{3}x^{3/2}\right]_0^9 = 2;$ at $x = 4$

11.   A.V. $= \dfrac{1}{2\pi} \displaystyle\int_0^{2\pi} \sin x\,dx = \dfrac{1}{2\pi} \left[-\cos x\right]_0^{2\pi} = 0;$ at $x = 0,\, \pi,\, 2\pi$

13.   (a) A.V. $= \dfrac{1}{8} \displaystyle\int_1^9 \sqrt{x}\,dx = \dfrac{1}{8} \left[\dfrac{2}{3}x^{3/2}\right]_1^9 = \dfrac{13}{6}$

   (b) $\sqrt{x} = \dfrac{13}{6} \implies x = 4.694$ (c)

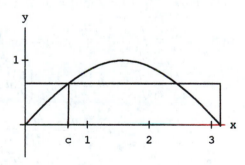

15.   (a) A.V. $= \dfrac{1}{\pi} \displaystyle\int_0^{\pi} \sin x\,dx = \dfrac{1}{\pi} \left[-\cos x\right]_0^{\pi} = \dfrac{2}{\pi}$

   (b) $\sin x = \dfrac{2}{\pi} \implies x = 0.691$ (c)

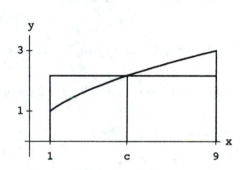

17.   A.V. $= \dfrac{1}{b-a} \displaystyle\int_a^b x^n\,dx = \dfrac{1}{b-a} \left[\dfrac{x^{n+1}}{n+1}\right]_a^b = \dfrac{b^{n+1} - a^{n+1}}{(n+1)(b-a)}.$

19.   Average of f' on $[a, b] = \dfrac{1}{b-a} \displaystyle\int_a^b f'(x)\,dx = \dfrac{1}{b-a} \left[f(x)\right]_a^b = \dfrac{f(b) - f(a)}{b-a}.$

21.   (a)   A.V. $= \dfrac{1}{\sqrt{3}} \displaystyle\int_0^{\sqrt{3}} y\,dx = \dfrac{1}{\sqrt{3}} \displaystyle\int_0^{\sqrt{3}} x^2\,dx = \dfrac{1}{\sqrt{3}} \left[\dfrac{x^3}{3}\right]_0^{\sqrt{3}} = 1$

(b)   A.V. $= \dfrac{1}{3} \displaystyle\int_0^3 x\, dy = \dfrac{1}{3}\int_0^3 \sqrt{y}\, dy = \dfrac{1}{3}\left[\dfrac{2}{3}y^{3/2}\right]_0^3 = \dfrac{2}{3}\sqrt{3}$

(c) $A.V. = \dfrac{1}{\sqrt{3}} \displaystyle\int_0^{\sqrt{3}} \sqrt{(x-0)^2 + (x^2-0)^2}\, dx = \dfrac{1}{\sqrt{3}}\int_0^{\sqrt{3}} x\sqrt{1+x^2}\, dx$

$\qquad = \dfrac{1}{\sqrt{3}}\left[\dfrac{1}{3}\left(1+x^2\right)^{3/2}\right]_0^{\sqrt{3}} = \dfrac{7}{9}\sqrt{3}$

23.   The distance the stone has fallen after t seconds is given by $s(t) = 16t^2$.

(a)   The terminal velocity after x seconds is $s'(x) = 32x$. The average velocity

is $\dfrac{s(x) - s(0)}{x - 0} = 16x$. Thus the terminal velocity is twice the average velocity.

(b)   For the first $\frac{1}{2}x$ seconds, aver. vel. $= \dfrac{s\left(\frac{1}{2}x\right) - s(0)}{\frac{1}{2}x - 0} = 8x$.

For the next $\frac{1}{2}x$ seconds, aver. vel. $= \dfrac{s(x) - s\left(\frac{1}{2}x\right)}{x - \frac{1}{2}x} = 24x$.

Thus, for the first $\frac{1}{2}x$ seconds the average velocity is one-third of the average velocity

during the next $\frac{1}{2}x$ seconds.

25.   Suppose $f(x) \neq 0$ for all x in (a, b). Then, since f is continuous, either

$$f(x) > 0 \text{ on } (a,b) \quad \text{or} \quad f(x) > 0 \text{ on } (a,b).$$

In either case, $\int_a^b f(x)\, dx \neq 0$.

27.   (a) $v(t) - v(0) = \displaystyle\int_0^t a\, du; \quad v(0) = 0$. Thus $v(t) = at$.

$x(t) - x(0) = \displaystyle\int_0^t v(u)\, du; \quad x(0) = x_0$. Thus $x(t) = \displaystyle\int_0^t au\, du + x_0 = \dfrac{1}{2}at^2 + x_0$.

(b) $v_{avg} = \dfrac{1}{t_2 - t_1} \displaystyle\int_{t_1}^{t_2} at\, dt = \dfrac{1}{t_2 - t_1}\left[\dfrac{1}{2}at^2\right]_{t_1}^{t_2}$

$\qquad = \dfrac{at_2^2 - at_1^2}{2(t_2 - t_1)} = \dfrac{v(t_1) + v(t_2)}{2}$

29.   (a) $M = \displaystyle\int_0^L k\sqrt{x}\, dx = k\left[\dfrac{2}{3}x^{3/2}\right]_0^L = \dfrac{2}{3}kL^{3/2}$

$x_M M = \displaystyle\int_0^L x\left(k\sqrt{x}\right) dx = \int_0^L kx^{3/2}\, dx = \left[\dfrac{2}{5}kx^{5/2}\right]_0^L = \dfrac{2}{5}kL^{5/2}$

$x_M = \left(\tfrac{2}{5}kL^{5/2}\right) / \left(\tfrac{2}{3}kL^{3/2}\right) = \tfrac{3}{5}L$

(b)
$$M = \int_0^L k\,(L-x)^2\,dx = \left[-\frac{1}{3}k\,(L-x)^3\right]_0^L = \frac{1}{3}kL^3$$

$$x_M M = \int_0^L x\left[k\,(L-x)^2\right]dx = \int_0^L k\,(L^2x - 2Lx^2 + x^3)\,dx$$

$$= k\left[\tfrac{1}{2}L^2x^2 - \tfrac{2}{3}Lx^3 + \tfrac{1}{4}x^4\right]_0^L = \tfrac{1}{12}kL^4$$

$$x_M = \left(\tfrac{1}{12}kL^4\right)\big/\left(\tfrac{1}{3}kL^3\right) = \tfrac{1}{4}L$$

31.
$$\tfrac{1}{4}LM = \tfrac{1}{8}LM_1 + x_{M_2}M_2$$

$$x_{M_2} = \frac{1}{M_2}\left(\frac{1}{4}LM - \frac{1}{8}LM_1\right) = \frac{L}{8M_2}(2M - M_1)$$

33. Let $M = \int_a^{a+L} kx\,dx$, where a is the point of the first cut.

Thus $M = \left[\dfrac{kx^2}{2}\right]_a^{a+L} = \dfrac{k}{2}(2aL + L^2)$.   Hence $a = \dfrac{2M - kL^2}{2kL}$,   and $a + L = \dfrac{2M + kL^2}{2kL}$.

35. If f is continuous on $[a, b]$, then, by Theorem 5.2.5, F satisfies the conditions of the mean-value theorem of differential calculus (Theorem 4.1.1). Therefore, by that theorem, there is at least one number c in (a, b) for which
$$F'(c) = \frac{F(b) - F(a)}{b - a}.$$
Then
$$\int_a^b f(x)\,dx = F(b) - F(a) = F'(c)(b - a) = f(c)(b - a).$$

37. If f and g take on the same average value on every interval $[a, x]$, then
$$\frac{1}{x - a}\int_a^x f(t)\,dt = \frac{1}{x - a}\int_a^x g(t)\,dt.$$

Multiplication by $(x - a)$ gives
$$\int_a^x f(t)\,dt = \int_a^x g(t)\,dt.$$

Differentiation with respect to x gives $f(x) = g(x)$. This shows that, if the averages are everywhere the same, then the functions are everywhere the same.

CHAPTER 6

SECTION 6.1

1.

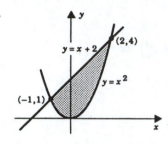

(a) $\int_{-1}^{2} [(x+2) - x^2]\, dx$

(b) $\int_{0}^{1} [(\sqrt{y}) - (-\sqrt{y})]\, dy + \int_{1}^{4} [(\sqrt{y}) - (y-2)]\, dy$

3.

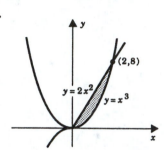

(a) $\int_{0}^{2} [(2x^2) - (x^3)]\, dx$

(b) $\int_{0}^{8} \left[(y^{1/3}) - \left(\frac{1}{2}y\right)^{1/2} \right]\, dy$

5.

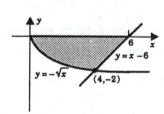

(a) $\int_{0}^{4} [(0) - (-\sqrt{x})]\, dx + \int_{4}^{6} [(0) - (x-6)]\, dx$

(b) $\int_{-2}^{0} [(y+6) - (y^2)]\, dy$

7.

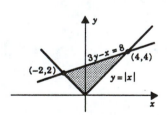

(a) $\int_{-2}^{0} \left[\left(\frac{8+x}{3}\right) - (-x) \right]\, dx + \int_{0}^{4} \left[\left(\frac{8+x}{3}\right) - (x) \right]\, dx$

(b) $\int_{0}^{2} [(y) - (-y)]\, dy + \int_{2}^{4} [(y) - (3y-8)]\, dy$

9.

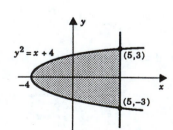

(a) $\int_{-4}^{5} [(\sqrt{4+x}) - (-\sqrt{4+x})]\, dx$

(b) $\int_{-3}^{3} [(5) - (y^2 - 4)]\, dy$

11.

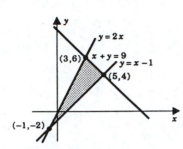

(a) $\displaystyle\int_{-1}^{3} [(2x) - (x-1)]\, dx + \int_{3}^{5} [(9-x) - (x-1)]\, dx$

(b) $\displaystyle\int_{-2}^{4} \left[(y+1) - \left(\frac{1}{2}y\right)\right] dy + \int_{4}^{6} \left[(9-y) - \left(\frac{1}{2}y\right)\right] dy$

13.

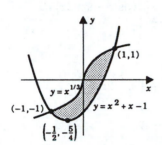

(a) $\displaystyle\int_{-1}^{1} \left[\left(x^{1/3}\right) - (x^2 + x - 1)\right] dx$

(b) $\displaystyle\int_{-5/4}^{-1} \left[\left(-\frac{1}{2} + \frac{1}{2}\sqrt{4y+5}\right) - \left(-\frac{1}{2} - \frac{1}{2}\sqrt{4y+5}\right)\right] dy$

$\displaystyle +\int_{-1}^{1} \left[\left(-\frac{1}{2} + \frac{1}{2}\sqrt{4y+5}\right) - (y^3)\right] dy$

15.

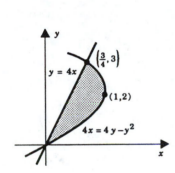

$\displaystyle A = \int_{0}^{3} \left[\left(\frac{4y - y^2}{4}\right) - \left(\frac{y}{4}\right)\right] dy$

$\displaystyle = \int_{0}^{3} \left(\frac{3}{4}y - \frac{1}{4}y^2\right) dy$

$\displaystyle = \left[\frac{3}{8}y^2 - \frac{1}{12}y^3\right]_{0}^{3} = \frac{9}{8}$

17.

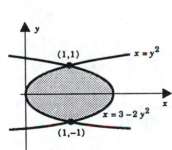

$\displaystyle A = 2\int_{0}^{1} [(3 - 2y^2) - (y^2)]\, dy$

$\displaystyle = 2\int_{0}^{1} (3 - 3y^2)\, dy$

$\displaystyle = 2\left[3y - y^3\right]_{0}^{1} = 2(2) = 4$

19.

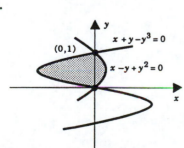

$\displaystyle A = \int_{0}^{1} [(y - y^2) - (y^3 - y)]\, dy$

$\displaystyle = \int_{0}^{1} (2y - y^2 - y^3)\, dy$

$\displaystyle = \left[y^2 - \frac{1}{3}y^3 - \frac{1}{4}y^4\right]_{0}^{1} = \frac{5}{12}$

21.

$$A = \int_{-\pi/4}^{\pi/4} \left[\sec^2 x - \cos x\right] dx$$

$$= 2 \int_0^{\pi/4} \left[\sec^2 x - \cos x\right] dx$$

$$= 2 \left[\tan x + \sin x\right]_0^{\pi/4} = 2 \left[1 + \sqrt{2}/2\right] = 2 + \sqrt{2}$$

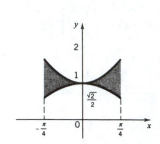

23.

$$A = \int_{-\pi}^{-\pi/2} \left[\sin 2x - 2\cos x\right] dx + \int_{-\pi/2}^{\pi/2} \left[2\cos x - \sin 2x\right] dx$$

$$+ \int_{\pi/2}^{\pi} \left[\sin 2x - 2\cos x\right] dx$$

$$= \left[-\tfrac{1}{2} \cos 2x - 2\sin x\right]_{\pi}^{\pi/2} + \left[2\sin x + \tfrac{1}{2} \cos 2x\right]_{-\pi/2}^{\pi/2}$$

$$+ \left[-\tfrac{1}{2} \cos 2x - 2\cos x\right]_{\pi/2}^{\pi} = 8$$

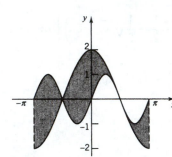

25.

$$A = \int_0^{\pi/2} \left(\sin^4 x \cos x\right) dx$$

$$= \int_0^1 u^4 \, du, \text{ where } u = \sin x$$

$$= \left[\frac{u^5}{5}\right]_0^1 = \frac{1}{5}$$

27.

$$A = \int_0^1 \left[3x - \frac{1}{3} x\right] dx + \int_1^3 \left[-x + 4 - \frac{1}{3} x\right] dx = \left[\frac{4}{3} x^2\right]_0^1 + \left[-\frac{2}{3} x^2 + 4x\right]_1^3 = 4$$

29.

$$A = \int_{-2}^1 \left[x - (-2)\right] dx + \int_1^5 \left[1 - (-2)\right] dx + \int_5^7 \left[-\frac{3}{2} x + \frac{17}{2} - (-2)\right] dx$$

$$= \left[\tfrac{1}{2} x^2 + 2x\right]_{-2}^1 + \left[3x\right]_1^5 \left[-\tfrac{3}{4} x^2 + \tfrac{21}{2} x\right]_5^7 = \frac{39}{2}$$

31.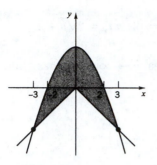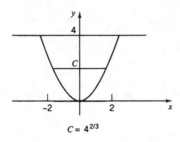

$$A = \int_{-3}^{0} [6 - x^2 - x] \, dx + \int_{0}^{3} [6 - x^2 - (-x)] \, dx$$

$$= \left[6x - \frac{1}{3} x^3 - \frac{1}{2} x^2 \right]_{-3}^{0} + \left[6x - \frac{1}{3} x^3 + \frac{1}{2} x^2 \right]_{0}^{3} = 27$$

33.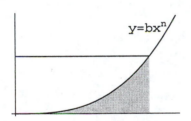

$C = 4^{2/3}$

$$\int_{0}^{\sqrt{c}} [c - x^2] \, dx = \frac{1}{2} \int_{0}^{2} [4 - x^2] \, dx$$

$$\left[cx - \tfrac{1}{3} x^3 \right]_{0}^{\sqrt{3}} = \tfrac{1}{2} \left[4x - \tfrac{1}{3} x^3 \right]_{0}^{2}$$

$$\tfrac{2}{3} c^{3/2} = \tfrac{8}{3} \quad \text{and} \quad c = 4^{2/3}$$

35. $\quad A = \int_{0}^{1} \left(\sqrt{4 - x^2} - \sqrt{3}x \right) dx$

37. $\quad A = \int_{0}^{2} \left[\sqrt{4 - x^2} - (2 - \sqrt{4x - x^2}) \right] dx$

39.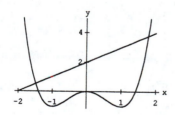

$y = bx^n$

The area under the curve is $A_c = \int_{0}^{a} bx^n \, dx = \dfrac{ba^{n+1}}{n+1}$.

For the rectangle, $A_r = ba^{n+1}$. Thus the ratio is $\dfrac{1}{n+1}$.

41.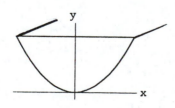

$$A \cong \int_{-1.49}^{1.79} \left[x + 2 - (x^4 - 2x^2) \right] dx$$

$$= \left[\tfrac{1}{2} x^2 - 2x - \tfrac{1}{5} x^5 + \tfrac{2}{3} x^3 \right]_{-1.49}^{1.78} \cong 7.93$$

43.

$$V = 8 \cdot 12 \int_{-3}^{3} \left[4 - \frac{4}{9} x^2 \right] dx$$

$$= 96 \cdot 2 \int_{0}^{3} \left[4 - \frac{4}{9} x^2 \right] dx$$

$$= 192 \left[4x - \tfrac{4}{27} x^3 \right]_{0}^{3} = 1536 \text{ cu. in.} \cong 0.89 \text{ cu. ft.}$$

45.

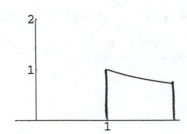

(a) $A = \int_1^b \frac{1}{\sqrt{x}}\,dx = \left[2\sqrt{x}\right]_1^b = 2\sqrt{b} - 2.$

(b) $2\sqrt{b} - 2 \to \infty$ as $b \to \infty$, so it has "infinite" area.

PROJECT 6.1

1. (a) $C = \dfrac{\int_0^1 [x - L(x)]\,dx}{\int_0^1 x\,dx} = \dfrac{\int_0^1 [x - L(x)]\,dx}{\left[\frac{x^2}{2}\right]_0^1} = 2\int_0^1 [x - L(x)]\,dx$

(b) The area between $y = x$ and the Lorenz curve must be between 0 and $\dfrac{1}{2}$. The area under

$y = x$ is equal to $\frac{1}{2}$. Thus their ratio must be between 0 and 1.

A coefficient close to 1 would mean the curve is close to the line $y = x$.

Thus the income distribution would be almost absolutely equal.

Similarly, a coefficient close to 0 would give a very unequal income distribution.

3. 1935, $L(x) = x^{2.4}$.

Thus $C = 2\int_0^1 (x - x^{2.4})\,dx = 2\left[\dfrac{x^2}{2} - \dfrac{x^{3.4}}{3.4}\right]_0^1 \cong 0.412$

1947, $L(x) = x^{1.6}$.

Thus $C = 2\int_0^1 (x - x^{1.6})\,dx = 2\left[\dfrac{x^2}{2} - \dfrac{x^{2.6}}{2.6}\right]_0^1 \cong 0.231$

SECTION 6.2

1.

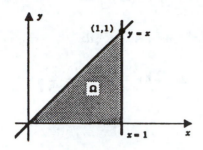

$$V = \int_0^1 \pi \left[(x)^2 - (0)^2 \right] dx = \pi \left[\frac{x^3}{3} \right]_0^1 = \frac{\pi}{3}$$

3.

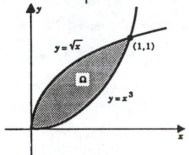

$$V = \int_{-3}^3 \pi \left[(9)^2 - \left(x^2 \right)^2 \right] dx = 2 \int_0^3 \pi \left(81 - x^4 \right) dx$$

$$= 2\pi \left[81x - \frac{x^5}{5} \right]_0^3 = \frac{1944\pi}{5}$$

5.

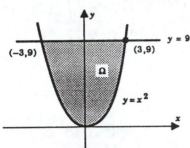

$$V = \int_0^1 \pi \left[\left(\sqrt{x} \right)^2 - \left(x^3 \right)^2 \right] dx$$

$$= \int_0^1 \pi \left(x - x^6 \right) dx$$

$$= \pi \left[\frac{1}{2}x^2 - \frac{1}{7}x^7 \right]_0^1 = \frac{5\pi}{14}$$

7.

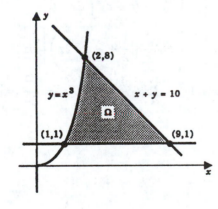

$$V = \int_1^2 \pi \left[\left(x^3 \right)^2 - (1)^2 \right] dx + \int_2^9 \pi \left[(10 - x)^2 - (1)^2 \right] dx$$

$$= \int_1^2 \pi \left(x^6 - 1 \right) dx + \int_2^9 \pi \left(99 - 20x + x^2 \right) dx$$

$$= \pi \left[\frac{1}{7}x^7 - x \right]_1^2 + \pi \left[99x - 10x^2 + \frac{1}{3}x^3 \right]_2^9 = \frac{3790\pi}{21}$$

9.

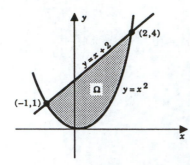

$$V = \int_{-1}^{2} \pi \left[(x+2)^2 - \left(x^2 \right)^2 \right] \, dx$$

$$= \int_{-1}^{2} \pi \left(x^2 + 4x + 4 - x^4 \right) \, dx$$

$$= \pi \left[\tfrac{1}{3} x^3 + 2x^2 + 4x - \tfrac{1}{5} x^5 \right]_{-1}^{2} = \frac{72}{5} \pi$$

11.

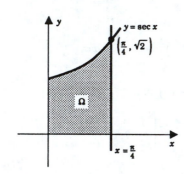

$$V = \int_{-2}^{2} \pi \left[\sqrt{4 - x^2} \right]^2 \, dx = 2 \int_{0}^{2} \pi \left(4 - x^2 \right) \, dx$$

$$= 2\pi \left[4x - \frac{x^3}{3} \right]_{0}^{2} = \frac{32}{3} \pi$$

13.

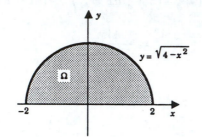

$$V = \int_{0}^{\pi/4} \pi \sec^2 x \, dx = \pi \left[\tan x \right]_{0}^{\pi/4} = \pi$$

15.

$$V = \int_{0}^{\pi/2} \pi \left[(x+1)^2 - (\cos x)^2 \right] \, dx$$

$$= \int_{0}^{\pi/2} \pi \left[(x+1)^2 - \left(\frac{1}{2} + \frac{1}{2} \cos 2x \right) \right] \, dx$$

$$= \pi \left[\frac{1}{3} (x+1)^3 - \frac{1}{2} x - \frac{1}{4} \sin 2x \right]_{0}^{\pi/2} = \frac{\pi^2}{24} \left(\pi^2 + 6\pi + 6 \right)$$

17.

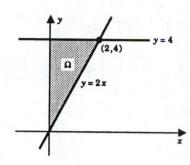

$$V = \int_0^4 \pi \left(\frac{y}{2}\right)^2 dy = \frac{\pi}{12} \left[y^3\right]_0^4 = \frac{16\pi}{3}$$

19.

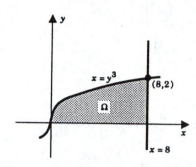

$$V = \int_0^2 \pi \left[(8)^2 - \left(y^3\right)^2\right] dy$$

$$= \int_0^2 \pi \left(64 - y^6\right) dy$$

$$= \pi \left[64y - \tfrac{1}{7}y^7\right]_0^2 = \frac{768}{7}\pi$$

21.

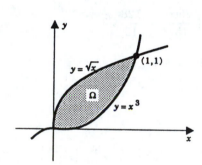

$$V = \int_0^1 \pi \left[\left(y^{1/3}\right)^2 - \left(y^2\right)^2\right] dy$$

$$= \int_0^1 \pi \left[y^{2/3} - y^4\right] dy$$

$$= \pi \left[\tfrac{3}{5}y^{5/3} - \tfrac{1}{5}y^5\right]_0^1 = \tfrac{2}{5}\pi$$

23.

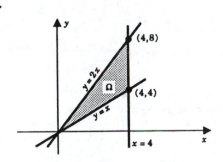

$$V = \int_0^4 \pi \left[y^2 - \left(\frac{y}{2}\right)^2\right] dy + \int_4^8 \pi \left[4^2 - \left(\frac{y}{2}\right)^2\right] dy$$

$$= \int_0^4 \pi \left[\frac{3}{4}y^2\right] dy + \int_4^8 \pi \left[16 - \frac{1}{4}y^2\right] dy$$

$$= \pi \left[\tfrac{1}{4}y^3\right]_0^4 + \pi \left[16y - \tfrac{1}{12}y^3\right]_4^8 = \frac{128}{3}\pi$$

25.

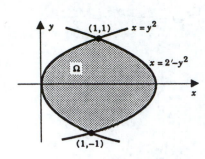

$$V = \int_{-1}^{1} \pi \left[(2 - y^2)^2 - (y^2)^2 \right] \, dy$$

$$= 2 \int_{0}^{1} \pi \left[4 - 4y^2 \right] \, dy = 2\pi \left[4y - \frac{4}{3} y^3 \right]_{0}^{1} = \frac{16}{3} \pi$$

27. (a) $V = \int_{-r}^{r} \left(2\sqrt{r^2 - x^2} \right)^2 \, dx = 8 \int_{0}^{r} (r^2 - x^2) \, dx = 8 \left[r^2 x - \frac{1}{3} x^3 \right]_{0}^{r} = \frac{16}{3} r^3$

  (b) $V = \int_{-r}^{r} \frac{\sqrt{3}}{4} \left(2\sqrt{r^2 - x^2} \right)^2 \, dx = 2\sqrt{3} \int_{0}^{r} (r^2 - x^2) \, dx = \frac{4\sqrt{3}}{3} r^3$

29. (a) $V = \int_{-2}^{2} (4 - x^2)^2 \, dx = 2 \int_{0}^{2} (16 - 8x^2 + x^4) \, dx = 2 \left[16x - \frac{8}{3} x^3 + \frac{1}{5} x^5 \right]_{0}^{2} = \frac{512}{15}$

  (b) $V = \int_{-2}^{2} \frac{\pi}{2} \left(\frac{4 - x^2}{2} \right)^2 \, dx = \frac{\pi}{4} \int_{0}^{2} (4 - x^2)^2 \, dx = \frac{\pi}{4} \left(\frac{256}{15} \right) = \frac{64}{15} \pi$

  (c) $V = \int_{-2}^{2} \frac{\sqrt{3}}{4} (4 - x^2)^2 \, dx = \frac{\sqrt{3}}{2} \int_{0}^{2} (4 - x^2)^2 \, dx = \frac{\sqrt{3}}{2} \left(\frac{256}{15} \right) = \frac{128}{15} \sqrt{3}$

31. (a) $V = \int_{0}^{4} [(\sqrt{y}) - (-\sqrt{y})]^2 \, dy = \int_{0}^{4} 4y \, dy = [2y^2]_{0}^{4} = 32$

  (b) $V = \int_{0}^{4} \frac{\pi}{2} (\sqrt{y})^2 \, dy = \frac{\pi}{2} \int_{0}^{4} y \, dy = \frac{\pi}{2} \left[\frac{1}{2} y^2 \right]_{0}^{4} = 4\pi$

  (c) $V = \int_{0}^{4} \frac{\sqrt{3}}{4} [(\sqrt{y}) - (-\sqrt{y})]^2 \, dy = \sqrt{3} \int_{0}^{4} y \, dy = 8\sqrt{3}$

33. (a) $V = \int_{0}^{4} (4 - x)^2 \, dx = \left[16x - 4x^2 + \frac{x^3}{3} \right]_{0}^{4} = \frac{64}{3}.$

  (b) $V = \frac{1}{4} \int_{0}^{4} (4 - x)^2 \, dx = \frac{1}{4} \left[16x - 4x^2 + \frac{x^3}{3} \right]_{0}^{4} = \frac{16}{3}.$

35. $V = \int_{-a}^{a} \pi \left(b\sqrt{1 - \frac{x^2}{a^2}} \right)^2 \, dx = \frac{2b^2}{a^2} \int_{0}^{a} \pi (a^2 - x^2)^2 \, dx = \frac{2b^2}{a^2} \pi \left[a^2 x - \frac{1}{3} x^3 \right]_{0}^{a}$

$$= \frac{2b^2}{a^2} \pi \left(\frac{2}{3} a^3 \right) = \frac{4}{3} \pi a b^2$$

37.

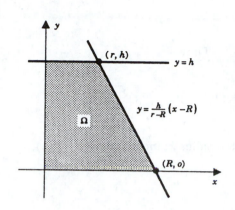

The specified frustum is generated by revolving the region Ω about the y-axis.

$$V = \int_0^h \pi \left[\frac{r-R}{h} y + R \right]^2 dy$$

$$= \pi \left[\frac{h}{3(r-R)} \left(\frac{r-R}{h} y + R \right)^3 \right]_0^h$$

$$= \frac{\pi h}{3(r-R)} \left(r^3 - R^3 \right) = \frac{\pi h}{3} \left(r^2 + rR + R^2 \right)$$

39. Capacity of basin $= \frac{1}{2} \left(\frac{4}{3} \pi r^3 \right) = \frac{2}{3} \pi r^3$.

(a)   Volume of water $= \int_{r/2}^r \pi \left[\sqrt{r^2 - x^2} \right]^2 dx$

$$= \pi \int_{r/2}^r \left(r^2 - x^2 \right) dx = \pi \left[r^2 x - \frac{1}{3} x^3 \right]_{r/2}^r = \frac{5}{24} \pi r^3.$$

The basin is $\left(\frac{5}{24} \pi r^3 \right) (100) / \left(\frac{2}{3} \pi r^3 \right) = 31\frac{1}{4}\%$ full.

(b) Volume of water $= \int_{2r/3}^r \pi \left[\sqrt{r^2 - x^2} \right]^2 dx = \pi \int_{2r/3}^r \left(r^2 - x^2 \right) dx = \frac{8}{81} \pi r^3.$

The basin is $\left(\frac{8}{81} \pi r^3 \right) (100) / \left(\frac{2}{3} \pi r^3 \right) = 14\frac{22}{27}\%$ full.

41. $V = \int_h^r \pi r^2 - y^2 \, dy = \pi \left[r^2 y - \frac{y^3}{3} \right]_h^r = \frac{\pi}{3} \left[2r^3 - 3r^2 h + h^3 \right].$

43. (a)

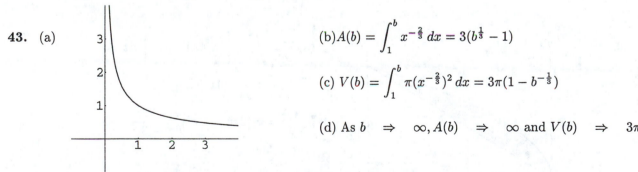

(b) $A(b) = \int_1^b x^{-\frac{2}{3}} dx = 3(b^{\frac{1}{3}} - 1)$

(c) $V(b) = \int_1^b \pi (x^{-\frac{2}{3}})^2 dx = 3\pi (1 - b^{-\frac{1}{3}})$

(d) As $b \Rightarrow \infty, A(b) \Rightarrow \infty$ and $V(b) \Rightarrow 3\pi$.

45. If the depth of the liquid in the container is h feet, then the volume of the liquid is:

$$V(h) = \int_0^h \pi \left(\sqrt{y+1} \right)^2 dy = \int_0^h [y+1] \, dy.$$

Differentiation with respect to t gives

$$\frac{dV}{dt} = \frac{dV}{dh} \cdot \frac{dh}{dt} = \pi(h+1)\frac{dh}{dt}.$$

Now, since $\dfrac{dV}{dt} = 2$, it follows that $\dfrac{dh}{dt} = \dfrac{2}{\pi(h+1)}$. Thus

$$\left.\frac{dh}{dt}\right|_{h=1} = \frac{2}{2\pi} = \frac{1}{\pi} \text{ ft/min} \quad \text{and} \quad \left.\frac{dh}{dt}\right|_{h=2} = \frac{2}{3\pi} \text{ ft/min}.$$

47. The cross section with coordinate x is a washer with outer radius k, inner radius $k - f(x)$,

and area

$$A(x) = \pi k^2 - \pi[k - f(x)]^2 = \pi\left(2kf(x) - [f(x)]^2\right)$$

Thus

$$V = \int_a^b \pi\left(2kf(x) - [f(x)]^2\right)\,dx$$

49. $V = \displaystyle\int_0^4 \pi\left[4\sqrt{x} - x\right]\,dx = \pi\left[-2\cos x\right]_0^\pi - \frac{\pi}{2}\int_0^\pi (1 - \cos 2x)\,dx$

51. $V = \displaystyle\int_0^\pi \pi\left[2\sin x - \sin^2 x\right]\,dx = \pi\left[\frac{8}{3}x^{3/2} - \frac{1}{2}x^2\right]_0^4 = \frac{40\pi}{3} = 4\pi - \frac{\pi}{2}\left[1 - \frac{1}{2}\sin 2x\right]_0^\pi = 4\pi - \frac{1}{2}\pi^2$

53.

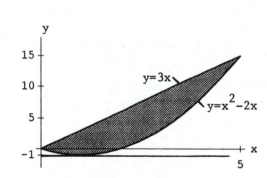

$$V = \int_0^5 \pi\left([(3x-)-1)]^2 - [x^2 - 2x - (-1)]^2\right)\,dx$$

$$= \pi\int_0^5 [3x+1]^2\,dx - \int_0^5 [x-1]^4\,dx$$

$$= \pi\left[\tfrac{1}{9}(3x+1)^3\right]_0^5 - \frac{1}{5}\left[x-1\right]_0^5$$

$$= 250\pi$$

55.

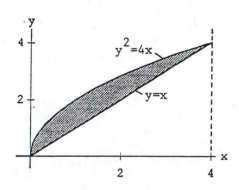

$$\text{(a) } V = \int_0^4 \pi\left[\left(\sqrt{4x}\right)^2 - x^2\right]\,dx$$

$$= \pi\int_0^4 [4x - x^2]\,dx$$

$$= \pi\left[2x^2 - \tfrac{1}{3}x^3\right]_0^4 = \frac{32\pi}{3}$$

(b) $V = \displaystyle\int_0^4 \pi \left[\left(\frac{1}{4} y^2 - 4 \right)^2 - (y-4)^2 \right] dy$

$= \pi \displaystyle\int_0^4 \left[\frac{1}{16} y^4 - 3y^2 + 8y \right] dy$

$= \pi \left[\frac{1}{80} y^5 - y^3 + 4y^2 \right]_0^4 = \dfrac{64\pi}{5}$

57. (a) $V = \displaystyle\int_0^4 \pi \left(x^{3/2} \right)^2 dx = \pi \int_0^4 x^3 \, dx = \pi \left[\frac{1}{4} x^4 \right]_0^4 = 64\pi$

(b) $V = \displaystyle\int_0^8 \pi \left(4 - y^{3/2} \right)^2 dy = \pi \int_0^8 \left(16 - 8y^{2/3} + y^{4/3} \right) dy$

$= \pi \left[16y - \frac{24}{5} y^{5/3} + \frac{3}{7} y^{7/3} \right]_0^8 = \frac{1024}{35} \pi$

(c) $V = \displaystyle\int_0^4 \pi \left[(8)^2 - \left(8 - x^{3/2} \right)^2 \right] dx = \pi \int_0^4 \left(16x^{3/2} - x^3 \right) dx$

$= \pi \left[\frac{32}{5} x^{5/2} - \frac{1}{4} x^4 \right]_0^4 = \frac{704}{5} \pi$

(d) $V = \displaystyle\int_0^8 \pi \left[(4)^2 - \left(y^{2/3} \right)^2 \right] dy = \pi \int_0^8 \left(16 - y^{4/3} \right) dy = \pi \left[16y - \frac{3}{7} y^{7/3} \right]_0^8 = \dfrac{512}{7} \pi$

SECTION 6.3

1.

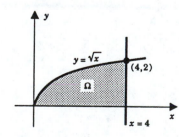

$V = \displaystyle\int_0^1 2\pi x \left[x - 0 \right] dx = 2\pi \int_0^1 x^2 \, dx$

$= 2\pi \left[\frac{1}{3} x^3 \right]_0^1 = \dfrac{2\pi}{3}$

3.

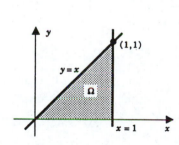

$V = \displaystyle\int_0^4 2\pi x \left[\sqrt{x} - 0 \right] dx = 2\pi \int_0^4 x^{3/2} \, dx$

$= 2\pi \left[\frac{2}{5} x^{5/2} \right]_0^4 = \dfrac{128}{5} \pi$

5.

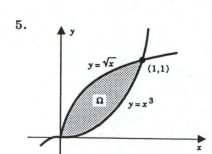

$$V = \int_0^1 2\pi x \left[\sqrt{x} - x^3 \right] dx$$

$$= 2\pi \int_0^1 \left(x^{3/2} - x^4 \right) dx$$

$$= 2\pi \left[\frac{2}{5} x^{5/2} - \frac{1}{5} x^5 \right]_0^1 = \frac{2\pi}{5}$$

7.

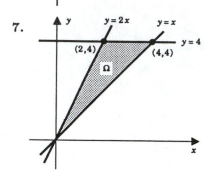

$$V = \int_0^2 2\pi x \left[2x - x \right] dx + \int_2^4 2\pi x \left[4 - x \right] dx$$

$$= 2\pi \int_0^2 x^2 \, dx + 2\pi \int_2^4 \left(4x - x^2 \right) dx$$

$$= 2\pi \left[\tfrac{1}{3} x^3 \right]_0^2 + 2\pi \left[2x^2 - \tfrac{1}{3} x^3 \right]_2^4 = 16\pi$$

9.

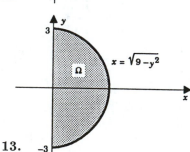

$$V = \int_0^1 2\pi x \left[(\sqrt{x}) - (-\sqrt{x}) \right] dx + \int_1^4 2\pi x \left[(\sqrt{x}) - (x - 2) \right] dx$$

$$= 4\pi \int_0^1 x^{3/2} \, dx + 2\pi \int_1^4 \left(x^{3/2} - x^2 + 2x \right) dx$$

$$= 4\pi \left[\tfrac{2}{5} x^{5/2} \right]_0^1 + 2\pi \left[\tfrac{2}{5} x^{5/2} - \tfrac{1}{3} x^3 + x^2 \right]_1^4 = \tfrac{72}{5} \pi$$

11.

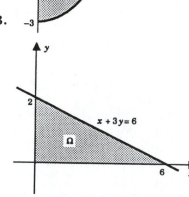

$$V = \int_0^3 2\pi x \left[\sqrt{9 - x^2} - \left(-\sqrt{9 - x^2} \right) \right] dx$$

$$= 4\pi \int_0^3 x \left(9 - x^2 \right)^{1/2} dx$$

$$= 4\pi \left[-\tfrac{1}{3} \left(9 - x^2 \right)^{3/2} \right]_0^3 = 36\pi$$

13.

$$V = \int_0^2 2\pi y \left[6 - 3y \right] dy$$

$$= 6\pi \int_0^2 \left(2y - y^2 \right) dy$$

$$= 6\pi \left[y^2 - \tfrac{1}{3} y^3 \right]_0^2 = 8\pi$$

15.

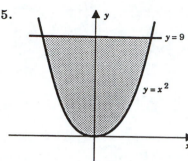

$$V = \int_0^9 2\pi y \left[(\sqrt{y}) - (-\sqrt{y}) \right] dy$$

$$= 4\pi \int_0^9 y^{3/2} dy$$

$$= 4\pi \left[\tfrac{2}{5} y^{5/2} \right]_0^9 = \tfrac{1944}{5} \pi$$

17.

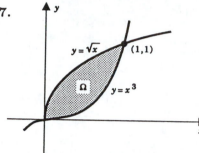

$$\int = \int_0^1 2\pi y \left[y^{1/3} - y^2 \right] dy$$

$$= 2\pi \int_0^1 \left(y^{4/3} - y^3 \right) dy$$

$$= 2\pi \left[\tfrac{3}{7} y^{7/3} - \tfrac{1}{4} y^4 \right]_0^1 = \tfrac{5}{14} \pi$$

19.

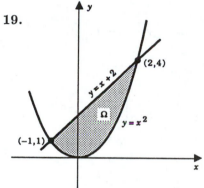

$$V = \int_0^1 2\pi y \left[(\sqrt{y}) - (-\sqrt{y}) \right] dy + \int_1^4 2\pi y \left[(\sqrt{y}) - (y-2) \right] dy$$

$$= 4\pi \int_0^1 y^{3/2} dy + 2\pi \int_1^4 \left(y^{3/2} - y^2 + 2y \right) dy$$

$$= 4\pi \left[\tfrac{2}{5} y^{5/2} \right]_0^1 + 2\pi \left[\tfrac{2}{5} y^{5/2} - \tfrac{1}{3} y^3 + y^2 \right]_1^4 = \tfrac{72}{5} \pi$$

21.

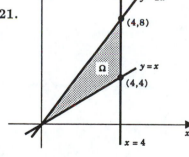

$$V = \int_0^4 2\pi y \left[y - \tfrac{y}{2} \right] dy + \int_4^8 2\pi y \left[4 - \tfrac{y}{2} \right] dy$$

$$= \pi \int_0^4 y^2 dy + \pi \int_4^8 \left(8y - y^2 \right) dy$$

$$= \pi \left[\tfrac{1}{3} y^3 \right]_0^4 + \pi \left[4y^2 - \tfrac{1}{3} y^3 \right]_4^8 = 64\pi$$

23.

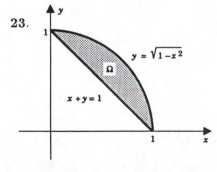

$$V = \int_0^1 2\pi y \left[\sqrt{1-y^2} - (1-y) \right] dy$$

$$= 2\pi \int_0^1 \left[y \left(1-y^2 \right)^{1/2} - y + y^2 \right] dy$$

$$= 2\pi \left[-\tfrac{1}{3} \left(1-y^2 \right)^{3/2} - \tfrac{1}{2} y^2 + \tfrac{1}{3} y^3 \right]_0^1 = \tfrac{\pi}{3}$$

25. (a) $V = \int_0^1 2\pi x \left[1 - \sqrt{x}\right] dx$

(b) $V = \int_0^1 \pi y^4 \, dy$

$= \pi \left[\frac{1}{5} y^5\right]_0^1 = \frac{1}{5}\pi$

27. (a) $V = \int_0^1 \pi \left(x - x^4\right) dx$

(b) $V = \int_0^1 2\pi y \left(\sqrt{y} - y^2\right) dy$

$= \pi \left[\frac{1}{2} x^2 - \frac{1}{5} x^5\right]_0^1 = \frac{3\pi}{10}$

29. (a) $V = \int_0^1 2\pi x \cdot x^2 \, dx$

(b) $V = \int_0^1 \pi(1 - y) \, dy$

$= 2\pi \left[\frac{1}{4}x^4\right]_0^1 = \frac{\pi}{2}$

31. $V = \int_0^a 2\pi x \left[2b\sqrt{1 - \frac{x^2}{a^2}}\right] dx = \frac{4\pi b}{a} \int_0^a x \left(a^2 - x^2\right)^{1/2} dx = \frac{4\pi b}{a} \left[-\frac{1}{3}\left(a^2 - x^2\right)^{3/2}\right]_0^a = \frac{4}{3}\pi a^2 b$

33.

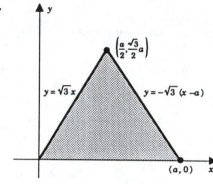

By the shell method

$V = \int_0^{a/2} 2\pi x \left(\sqrt{3}\,x\right) dx + \int_{a/2}^a 2\pi x \left[\sqrt{3}\,(a - x)\right] dx$

$= 2\pi\sqrt{3} \int_0^{a/2} x^2 \, dx + 2\pi\sqrt{3} \int_{a/2}^a \left(ax - x^2\right) dx$

$= 2\pi\sqrt{3} \left[\frac{1}{3}x^3\right]_0^{a/2} + 2\pi\sqrt{3} \left[\frac{a}{2}x^2 - \frac{1}{3}x^3\right]_{a/2}^a = \frac{\sqrt{3}}{4}a^3\pi$

35. (a) $V = \int_0^8 2\pi y \left[4 - y^{2/3}\right] dy = 2\pi \int_0^8 \left(4y - y^{5/3}\right) dy = 2\pi \left[2y^2 - \frac{3}{8}y^{8/3}\right]_0^8 = 64\pi$

(b) $V = \int_0^4 2\pi (4 - x) \left[x^{3/2}\right] dx = 2\pi \int_0^4 \left(4x^{3/2} - x^{5/2}\right) dx$

$= 2\pi \left[\frac{8}{5}x^{5/2} - \frac{2}{7}x^{7/2}\right]_0^4 = \frac{1024}{35}\pi$

(c) $V = \int_0^8 2\pi (8 - y) \left[4 - y^{2/3}\right] dy = 2\pi \int_0^8 \left(32 - 4y - 8y^{2/3} + y^{5/3}\right) dy$

$= 2\pi \left[32y - 2y^2 - \frac{24}{5}y^{5/3} + \frac{3}{8}y^{8/3}\right]_0^8 = \frac{704}{5}\pi$

(d) $V = \int_0^4 2\pi x \left[x^{3/2}\right] dx = 2\pi \int_0^4 x^{5/2} \, dx = 2\pi \left[\frac{2}{7}x^{7/2}\right]_0^4 = \frac{512}{7}\pi$

37. (a) $F'(x) = \sin x + x \cos x - \sin x = x \cos x = f(x)$.

(b) $V = \displaystyle\int_0^{\pi/2} 2\pi x \cdot \cos x \, dx = 2\pi \left[x \sin x + \cos x \right]_0^{\pi/2} = \pi^2 - 2\pi$

39. (a) $V = \displaystyle\int_0^1 2\sqrt{3}\pi x^2 \, dx + \int_1^2 2\pi x \sqrt{4 - x^2} \, dx$
    (b) $V = \displaystyle\int_0^{\sqrt{3}} \pi \left[4 - \frac{4}{3} y^2 \right] dy$

(c) $V = \displaystyle\int_0^{\sqrt{3}} \pi \left[4 - \frac{4}{3} y^2 \right] dy = \pi \left[4y - \frac{4}{9} y^3 \right]_0^{\sqrt{3}} = \frac{8\sqrt{3}\pi}{3}$

41. (a) $V = \displaystyle\int_0^4 2\sqrt{3}\,\pi x(2 - x) \, dx + \int_1^2 2\pi(2 - x)\sqrt{4 - x^2} \, dx$

(b) $V = \displaystyle\int_0^{\sqrt{3}} \pi \left[\left(2 - \frac{y}{\sqrt{3}} \right)^2 - \left(2 - \sqrt{4 - y^2} \right)^2 \right] dy$

43. (a) $V = \displaystyle\int_{b-a}^{b+a} 2\pi x \sqrt{a^2 - (x - b)^2} \, dx$

(b) $V = \displaystyle\int_a^a \pi \left[\left(b + \sqrt{a^2 - y^2} \right)^2 - \left(b - \sqrt{a^2 - y^2} \right)^2 \right] dy$

45. $V = \displaystyle\int_0^r 2\pi x \left(h - \frac{h}{r} x \right) dx = 2\pi h \left[\frac{x^2}{2} - \frac{x^3}{3r} \right]_0^r = \frac{\pi r^2 h}{3}.$

SECTION 6.4

1.

$A = \displaystyle\int_0^4 \sqrt{x} \, dx = \frac{16}{3}$

$\overline{x}A = \displaystyle\int_0^4 x\sqrt{x} \, dx = \frac{64}{5}, \quad \overline{x} = \frac{12}{5}$

$\overline{y}A = \displaystyle\int_0^4 \frac{1}{2} \left(\sqrt{x} \right)^2 dx = 4, \quad \overline{y} = \frac{3}{4}$

$V_x = 2\pi \overline{y} A = 8\pi, \quad V_y = 2\pi \overline{x} A = \frac{128}{5}\pi$

3.

$A = \displaystyle\int_0^1 \left(x^{1/3} - x^2 \right) dx = \frac{5}{12}$

$\overline{x}A = \displaystyle\int_0^1 x \left(x^{1/3} - x^2 \right) dx = \frac{5}{28}, \quad \overline{x} = \frac{3}{7}$

$\overline{y}A = \displaystyle\int_0^1 \frac{1}{2} \left[\left(x^{1/3} \right)^2 - \left(x^2 \right)^2 \right] dx = \frac{1}{5}, \quad \overline{y} = \frac{12}{25}$

$V_x = 2\pi \overline{y} A = \frac{2}{5}\pi, \quad V_y = 2\pi \overline{x} A = \frac{5}{14}\pi$

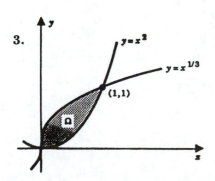

5.

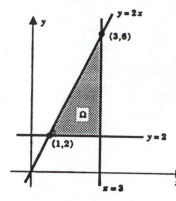

$$A = \int_1^3 (2x - 2)\, dx = 4$$

$$\overline{x}A = \int_1^3 x\,(2x - 2)\, dx = \frac{28}{3}, \quad \overline{x} = \frac{7}{3}$$

$$\overline{y}A = \int_1^3 \frac{1}{2}\left[(2x)^2 - (2)^2\right] dx = \frac{40}{3}, \quad \overline{y} = \frac{10}{3}$$

$$V_x = 2\pi\overline{y}A = \frac{80}{3}\pi, \quad V_y = 2\pi\overline{x}A = \frac{56}{3}\pi$$

7.

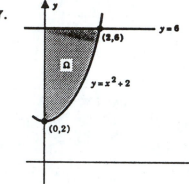

$$A = \int_0^2 \left[6 - (x^2 + 2x)\right] dx = \frac{16}{3}$$

$$\overline{x}A = \int_0^2 x\left[6 - (x^2 + 2)\right] dx = 4, \quad \overline{x} = \frac{3}{4}$$

$$\overline{y}A = \int_0^2 \frac{1}{2}\left[(6)^2 - (x^2 + 2)^2\right] dx = \frac{352}{15}, \quad \overline{y} = \frac{22}{5}$$

$$V_x = 2\pi\overline{y}A = \frac{704}{15}\pi, \quad V_y = 2\pi\overline{x}A = 8\pi$$

9.

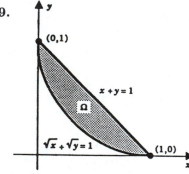

$$A = \int_0^1 \left[(1 - x) - (1 - \sqrt{x})^2\right] dx = \frac{1}{3}$$

$$\overline{x}A = \int_0^1 x\left[(1 - x) - (1 - \sqrt{x})^2\right] dx = \frac{2}{15}, \quad \overline{x} = \frac{2}{5}$$

$$\overline{y} = \frac{2}{5} \qquad \text{by symmetry}$$

$$V_x = 2\pi\overline{y}A = \frac{4}{15}\pi, \quad V_y = \frac{4}{15}\pi \qquad \text{by symmetry}$$

11.

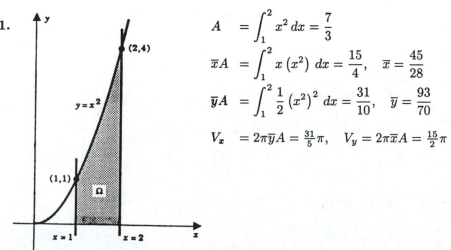

$$A = \int_1^2 x^2\, dx = \frac{7}{3}$$

$$\overline{x}A = \int_1^2 x\,(x^2)\, dx = \frac{15}{4}, \quad \overline{x} = \frac{45}{28}$$

$$\overline{y}A = \int_1^2 \frac{1}{2}\left(x^2\right)^2 dx = \frac{31}{10}, \quad \overline{y} = \frac{93}{70}$$

$$V_x = 2\pi\overline{y}A = \frac{31}{5}\pi, \quad V_y = 2\pi\overline{x}A = \frac{15}{2}\pi$$

13.

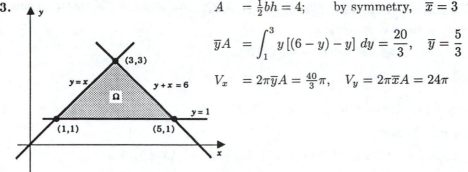

$$A = \tfrac{1}{2}bh = 4; \qquad \text{by symmetry,} \quad \overline{x} = 3$$

$$\overline{y}A = \int_1^3 y\,[(6-y)-y]\,dy = \frac{20}{3}, \quad \overline{y} = \frac{5}{3}$$

$$V_x = 2\pi\overline{y}A = \tfrac{40}{3}\pi, \quad V_y = 2\pi\overline{x}A = 24\pi$$

15. $\left(\tfrac{5}{2}, 5\right)$ **17.** $\left(1, \tfrac{8}{5}\right)$ **19.** $\left(\tfrac{10}{3}, \tfrac{40}{21}\right)$ **21.** $(2, 4)$ **23.** $\left(-\tfrac{3}{5}, 0\right)$

25. (a) $(0,0)$ by symmetry

(b) Ω_1 smaller quarter disc, Ω_2 the larger quarter disc

$$A_1 = \frac{1}{16}\pi, \quad A_2 = \pi; \quad \overline{x}_1 = \overline{y}_1 = \frac{2}{3\pi}, \quad \overline{x}_2 = \overline{y}_2 = \frac{8}{3\pi} \quad \text{(Problem 1)}$$

$$\overline{x}A = \left(\frac{8}{3\pi}\right)(\pi) - \frac{2}{3\pi}\left(\frac{1}{16}\pi\right)\frac{63}{24}, \quad A = \frac{15}{16}\pi$$

$$\overline{x} = \left(\frac{63}{24}\right) \bigg/ \left(\frac{15\pi}{16}\right) = \frac{14}{5\pi}, \quad \overline{y} = \overline{x} = \frac{14}{5\pi} \quad \text{(symmetry)}$$

(c) $\overline{x} = 0, \quad \overline{y} = \dfrac{14}{5\pi}$

27. Use theorem of Pappus. Centroid of rectangle is located

$$c + \sqrt{\left(\frac{a}{2}\right)^2 + \left(\frac{b}{2}\right)^2} \quad \text{units}$$

from line l. The area of the rectangle is ab. Thus,

$$\text{volume} = 2\pi \left[c + \sqrt{\left(\frac{a}{2}\right)^2 + \left(\frac{b}{2}\right)^2} \right] (ab) = \pi ab \left(2c + \sqrt{a^2 + b^2} \right).$$

29. (a) $\left(\tfrac{2}{3}a, \tfrac{1}{3}h\right)$ (b) $\left(\tfrac{2}{3}a + \tfrac{1}{3}b, \tfrac{1}{3}h\right)$ (c) $\left(\tfrac{1}{3}a + \tfrac{1}{3}b, \tfrac{1}{3}h\right)$

31. (a) $V = \tfrac{2}{3}\pi R^3 \sin^3\theta + \tfrac{1}{3}\pi R^3 \sin^2\theta\cos\theta = \tfrac{1}{3}\pi R^3 \sin^2\theta\,(2\sin\theta + \cos\theta)$

(b) $\overline{x} = \dfrac{V}{2\pi A} = \dfrac{\tfrac{1}{3}\pi R^3 \sin^2\theta\,(2\sin\theta + \cos\theta)}{2\pi\left(\tfrac{1}{2}R^2\sin\theta\cos\theta + \tfrac{1}{4}\pi R^2\sin^2\theta\right)} = \dfrac{2R\sin\theta\,(2\sin\theta + \cos\theta)}{3\left(\pi\sin\theta + 2\cos\theta\right)}$

33. (a) The mass contributed by $[x_{i-1}, x_i]$ is approximately $\lambda(x_i^*) \Delta x_i$ where x_i^* is the midpoint of $[x_{i-1}, x_i]$. The sum of these contributions,

$$\lambda(x_1^*) \Delta x_1 + \cdots + \lambda(x_n^*) \Delta x_n,$$

is a Riemann sum, which as $\|P\| \to 0$, tends to the given integral.

(b) Take M_i as the mass contributed by $[x_{i-1}, x_i]$. Then $x_{M_i} M_i \cong x_i^* \lambda(x_i^*) \Delta x_i$ where x_i^* is the midpoint of $[x_{i-1}, x_i]$. Therefore

$$x_M M = x_{M_1} M_1 + \cdots + x_{M_n} M_n \cong x_1^* \lambda(x_1^*) \Delta x_1 + \cdots + x_n^* \lambda(x_n^*) \Delta x_n.$$

As $\|P\| \to 0$, the sum on the right converges to the given integral.

PROJECT 6.4

1. Let $P = \{x_0, x_1, \ldots, x_n\}$ be a partition of $[a, b]$. P breaks up $[a, b]$ into n subintervals $[x_{i-1}, x_i]$.

Choose x_i^* as the midpoint of $[x_{i-1}, x_i]$. By revolving the ith midpoint rectangle about x-axis,

we obtain a solid cylinder of volume $V_i = \pi [f(x_i^*)]^2 \Delta x_i$ and centroid (center) on the x-axis at

$x = x_i^*$. The union of all these cylinders has centroid at $x = \overline{x}_P$ where

$$x_P V_p = \pi x_1^* [f(x_1^*)]^2 \Delta x_1 + \cdots + \pi x_n^* [f(x_n^*)]^2 \Delta x_n$$

(Here V_P represents the union of the n cylinders.) As $\|P\| \to 0$, the union of the cylinders

tends to the shape of S and the equation just derived tends to Formula 6.4.5.

3. (a)

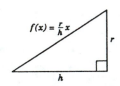

$f(x) = \frac{r}{h}x$

$$\overline{x}\left(\frac{1}{3}\pi r^2 h\right) = \int_0^h \pi x \left(\frac{r}{h}x\right)^2 dx = \frac{1}{4}\pi r^2 h^2$$

$$\overline{x} = \left(\frac{1}{4}\pi r^2 h^2\right) / \left(\frac{1}{3}\pi r^2 h\right) = \frac{3}{4}h.$$

The centroid of the cone lies on the axis of the cone at a distance $\frac{3}{4}h$ from the vertex.

(b) The hemisphere is obtained by rotating $f(x) = \sqrt{r^2 - x^2}, x \in [0, r]$, around the x-axis.

$$V_x = \frac{2}{3}\pi r^3; \quad \overline{x}V_x = \int_0^r \pi x(r^2 - x^2)\, dx = \pi \left[r^2\frac{x^2}{2} - \frac{x^4}{4}\right]_0^r = \pi\frac{r^4}{4}$$

$$\implies \overline{x} = \frac{3r}{8} \quad (\frac{3r}{8} \text{ units from the center of the base along the axis}).$$

(c) $V_x = \int_0^a \pi \frac{b^2}{a^2}(a^2 - x^2)\, dx = \frac{2}{3}\pi ab^2, \quad \overline{x}V_x = \int_0^a \pi x\frac{b^2}{a^2}(a^2 - x^2)\, dx = \frac{1}{4}\pi a^2 b^2$

$\overline{x} = \left(\frac{1}{4}\pi a^2 b^2\right) / \left(\frac{2}{3}\pi ab^2\right) = \frac{3}{8}a;$ centroid $\left(\frac{3}{8}a, 0\right)$

(d) (i) $V_x = \int_0^1 \pi \left(\sqrt{x}\right)^2 dx = \dfrac{1}{2}\pi, \quad \overline{x}V_x = \int_0^1 \pi x \left(\sqrt{x}\right)^2 dx = \dfrac{1}{3}\pi$

$\overline{x} = \left(\tfrac{1}{3}\pi\right)/\left(\tfrac{1}{2}\pi\right) = \tfrac{2}{3}; \quad$ centroid $\left(\tfrac{2}{3}, 0\right)$

(ii) $V_y = \int_0^1 2\pi x\sqrt{x}\, dx = \dfrac{4}{5}\pi, \quad \overline{y}V_y = \int_0^1 \pi x \left(\sqrt{x}\right)^2 dx = \dfrac{1}{3}\pi$

$\overline{y} = \left(\tfrac{1}{3}\pi\right)/\left(\tfrac{4}{5}\pi\right) = \tfrac{5}{12}; \quad$ centroid $\left(0, \tfrac{5}{12}\right)$

(e) (i) $V_x = \int_0^2 \pi(4 - x^2)^2\, dx = \dfrac{256}{15}\pi$

$\overline{x}V_x = \int_0^2 \pi x(4 - x^2)^2\, dx = \pi \left[-\dfrac{(4 - x^2)^3}{6}\right]_0^2 = \dfrac{32\pi}{3} \implies \overline{x} = \dfrac{5}{8}; \quad \overline{y} = 0$

(ii) $V_y = \int_0^2 2\pi x(4 - x^2)\, dx = 8\pi$

$\overline{y}V_y = \int_0^2 \pi x(4 - x^2)^2\, dx = \dfrac{32}{3}\pi \implies \overline{y} = \dfrac{4}{3}; \quad \overline{x} = 0$

SECTION 6.5

1. $W = \int_1^4 x\left(x^2 + 1\right)^2 dx = \dfrac{1}{6}\left[(x^2 + 1)^3\right]_1^4 = 817.5$ ft-lb

3. $W = \int_1^3 x\sqrt{x^2 + 7}\, dx = \dfrac{1}{3}\left[(x^2 + 7)^{\frac{3}{2}}\right]_0^3 = \dfrac{1}{3}(64 - 7^{\frac{3}{2}})$ft-lb

5. $W = \int_{\pi/6}^{pi} (x + \sin 2x)\, dx = \left[\dfrac{1}{2}x^2 - \dfrac{1}{2}\cos 2x\right]_{\pi/6}^{\pi} = \dfrac{35}{72}\pi^2 - \dfrac{1}{4}$ Newton-meters

7. By Hooke's law, we have $600 = -k(-1)$. Therefore $k = 600$.

The work required to compress the spring to 5 inches is given by

$$W = \int_{10}^5 600(x - 10)\, dx = 600\left[\dfrac{1}{2}x^2 - 10x\right]_{10}^5$$

$= 7500$ in-lb, or 625 ft-lb

9. To counteract the restoring force of the spring we must apply a force $F(x) = kx$.

Since $F(4) = 200$, we see that $k = 50$ and therefore $F(x) = 50x$.

(a) $W = \int_0^1 50x\, dx = 25$ ft-lb   (b) $W = \int_0^{3/2} 50x\, dx = \dfrac{225}{4}$ ft-lb

11.   Let L be the natural length of the spring.

$$\int_{2-L}^{2.1-L} kx\, dx = \frac{1}{2}\int_{2.1-L}^{2.2-L} kx\, dx$$

$$\left[\tfrac{1}{2}kx^2\right]_{2-L}^{2.1-L} = \tfrac{1}{2}\left[\tfrac{1}{2}kx^2\right]_{2.1-L}^{2.2-L}$$

$$(2.1-L)^2 - (2-L)^2 = \tfrac{1}{2}\left[(2.2-L)^2 - (2.1-L)^2\right].$$

Solve this equation for L and you will find that $L = 1.95$.

Answer: 1.95 ft

13.

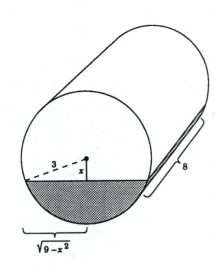

(a) $W = \displaystyle\int_0^3 (x+3)(60)(8)\left(2\sqrt{9-x^2}\right) dx$

$$= 960\int_0^3 x\left(9-x^2\right)^{1/2} dx$$

$$+2880\underbrace{\int_0^3 \sqrt{9-x^2}\, dx}_{\substack{\text{area of quarter}\\\text{circle of radius 3}}}$$

$$= 960\left[-\tfrac{1}{3}\left(9-x^2\right)^{3/2}\right]_0^3 + 2880\left[\tfrac{9}{4}\pi\right]$$

$$= (8640 + 6480\pi)\ \text{ft-lb}$$

(b) $W = \displaystyle\int_0^3 (x+7)(60)(8)\left(2\sqrt{9-x^2}\right) dx = 960\int_0^3 x\left(9-x^2\right)^{1/2} dx + 6720\int_0^3 \sqrt{9-x^2}\, dx$

$$= (8640 + 15120\pi)\ \text{ft-lb}$$

15.

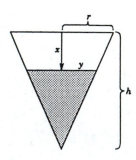

By similar triangles

$$\frac{h}{r} = \frac{h-x}{y} \quad \text{so that} \quad y = \frac{r}{h}(h-x).$$

Thus, the area of a cross section of the fluid at a depth of x feet is

$$\pi y^2 = \pi\frac{r^2}{h^2}(h-x)^2.$$

(a) $W = \displaystyle\int_0^{h/2} x\sigma\left[\pi\frac{r^2}{h^2}(h-x)^2\right] dx = \frac{\sigma\pi r^2}{h^2}\int_0^{h/2}\left(h^2x - 2hx^2 + x^3\right) dx = \frac{11}{192}\sigma\pi r^2 h^2\ \text{ft-lb}$

(b) $W = \displaystyle\int_0^{h/2} (x+k)\sigma \left[\pi \frac{r^2}{h^2} (h-x)^2 \right] dx = \frac{11}{192}\pi r^2 h^2 \sigma + \frac{7}{24}\pi r^2 hk\sigma$ ft-lb

17. $y = \dfrac{3}{4}x^2,\ 0 \le x \le 4$

(a) $W = \displaystyle\int_0^{12} \sigma(12-y)\pi x^2\, dy = \frac{4}{3}\pi\sigma \int_0^{12} (12y - y^2)\, dy$

$= \dfrac{4}{3}\pi\sigma \left[6y^2 - \dfrac{y^3}{3} \right]_0^{12} = \dfrac{4}{3}\pi\sigma(288) = 384\pi\sigma$ newton-meters.

(b) $W = \displaystyle\int_0^{12} \sigma(13-y)\pi x^2\, dy = \frac{4}{3}\pi\sigma \int_0^{12} (13y - y^2)\, dy$

$= \dfrac{4}{3}\pi\sigma \left[\dfrac{13y^2}{2} - \dfrac{y^3}{3} \right]_0^{12} = \dfrac{4}{3}\pi\sigma(360) = 480\pi\sigma$ newton-meters.

19. $W = \displaystyle\int_0^{80} (80-x)15\, dx = 15\left[80x - \frac{1}{2}x^2 \right]_0^{80} = 48,000$ ft-lb

21.   (a) $W = 200 \cdot 100 = 20,000$ ft-lb   (b) $W = \displaystyle\int_0^{100} \left[(100-x) + 200 \right] dx$

$= \displaystyle\int_0^{100} (400 - 2x)\, dx$

$= \left[400x - x^2 \right]_0^{100} = 30,000$ ft-lb

23.   The bag is raised 8 feet and loses a total of 3 pounds at a constant rate. Thus, the bag loses sand at the rate of 3/8 lb/ft. After the bag has been raised x feet it weighs $100 - \dfrac{3x}{8}$ pounds.

$$W = \int_0^8 \left(100 - \frac{3x}{8} \right) dx = \left[100 - \frac{3x^2}{16} \right]_0^8 = 788 \text{ ft-lb.}$$

25.   (a) $W = \displaystyle\int_0^l x\sigma\, dx = \frac{1}{2}\sigma l^2$ ft-lb   (b) $W = \displaystyle\int_0^l (x+l)\,\sigma\, dx = \frac{3}{2}\sigma l^2$ ft-lb

27.   Thirty feet of cable and the steel beam weighing a total of

$$800 + 30(6) = 980 \text{ lb}$$

are raised 20 feet. The work requires $(20)(980)$ ft-lb.

Next, the remaining 20 feet of cable is raised a varying distance and wound onto the steel drum. Thus the total work is given by

$$W = (20)(980) + \int_0^{20} 6x\, dx = 19,600 + 1,200 = 20,800 \text{ ft-lb.}$$

29. By the hint

$$W = \int_a^b F(x)\,dx = \int_a^b ma\,dx = \int_a^b mv\frac{dv}{dx}\,dx = \int_{v_a}^{v_b} mv\,dv = \left[\frac{1}{2}mv^2\right]_{v_a}^{v_b} = \frac{1}{2}v_b^2 - \frac{1}{2}v_a^2$$

31. (a) The work required to pump the water out of the tank is given by

$$W = \int_5^{10} (62.5)\pi\,5^2 x\,dx = 1562.5\pi \left[\frac{1}{2}x^2\right]_5^{10} \cong 184,078 \text{ ft-lb}$$

A $\frac{1}{2}$-horsepower pump can do 275 ft-lb of work per second. Therefore it will take

$$\frac{184,078}{275} \cong 669 \text{ seconds} \cong 11 \text{ min, 10 sec, to empty the tank.}$$

(b) The work required to pump the water to a point 5 feet above the top of the tank

is given by

$$W = \int_5^{10} (62.5)\pi\,5^2(x+5)\,dx = \int_5^{10} (62.5)\pi\,5^2 x\,dx + \int_5^{10} (62.5)\pi\,5^3\,dx \cong 306796 \text{ ft-lb}$$

It will take a $\frac{1}{2}$-horsepower pump approximately $1,116$ sec, or 18 min, 36 sec, in this case.

SECTION 6.6

1. $F = \int_0^6 (62.5)\cdot x \cdot 8\,dx = 250\left[x^2\right]_0^6 = 9000 \text{ lb}$

3. The width of the plate x meters below the surface is given by $w(x) = 60 + 2(20 - x) = 100 - 2x$

(see the figure). The force against the dam is

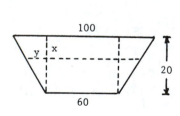

$$F = \int_0^{20} 9800x(100 - 2x)\,dx$$

$$= 9800\int_0^{20}(100x - 2x^2)\,dx$$

$$= 9800\left[50x^2 - \tfrac{2}{3}x^3\right]_0^{20}$$

$$\cong 1.437 \times 10^8 \text{ Newtons}$$

5. The width of the gate x meters below its top is given by $w(x) = 4 + \frac{2}{3}x$ (see the figure).

The force of the water against the gate is

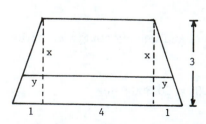

$$F = \int_0^3 9800(10 + x)\left(4 + \frac{2}{3}x\right)dx$$

$$= 9800\int_0^3 \left[\frac{2}{3}x^2 + \frac{32}{3} + 40\right]dx$$

$$= 9800\left[\tfrac{2}{9}x^3 + \tfrac{16}{3}x^2 + 40x\right]_0^3$$

$$\cong 1.7052 \times 10^6 \text{ Newtons}$$

7.

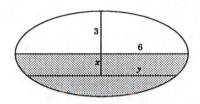

ellipse: $\dfrac{x^2}{3^2} + \dfrac{y^2}{6^2} = 1$

$$F = \int_0^3 (60)\, x \left[12\sqrt{1 - \frac{x^2}{9}} \right] dx$$

$$= 240 \int_0^3 x \left(9 - x^2\right)^{1/2} dx$$

$$= 240 \left[-\tfrac{1}{3} \left(9 - x^2\right)^{3/2} \right]_0^3 = 2160 \text{ lb}$$

9.

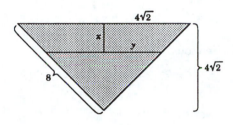

By similar triangles

$$\frac{4\sqrt{2}}{4\sqrt{2}} = \frac{y}{4\sqrt{2} - x} \quad \text{so} \quad y = 4\sqrt{2} - x.$$

$$F = \int_0^{4\sqrt{2}} (62.5)\, x \left[2\left(4\sqrt{2} - x\right) \right] dx$$

$$= 125 \int_0^{4\sqrt{2}} \left(4\sqrt{2}\, x - x^2\right) dx = \frac{8000}{3}\sqrt{2} \text{ lb}$$

11.

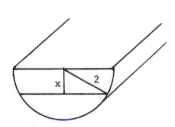

$$F = \int_0^2 (62.5) \cdot x \cdot 2\sqrt{4 - x^2}\, dx$$

$$= 125 \int_0^2 x\sqrt{4 - x^2}\, dx$$

$$= -\tfrac{125}{3} \left[\left(4 - x^2\right)^{3/2} \right]_0^2 = 333.33 \text{ lb}$$

13.

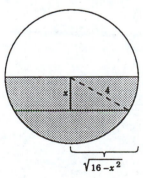

$$F = \int_0^4 60x \left(2\sqrt{16 - x^2}\right) dx$$

$$= 120 \int_0^4 x \left(16 - x^2\right)^{1/2} dx$$

$$= 120 \left[-\tfrac{1}{3}\left(16 - x^2\right)^{3/2} \right]_0^4 = 2560 \text{ lb}$$

15.   (a)   The width of the plate is 10 feet and the depth of the plate ranges from 8 feet to 14 feet. Thus

$$F = \int_8^{14} 62.5x\,(10)\, dx = 41,250 \text{ lb.}$$

(b)   The width of the plate is 6 feet and the depth of the plate ranges from 6 feet to 16 feet. Thus

$$F = \int_6^{16} 62.5x\,(6)\ dx = 41,250\ \text{lb.}$$

17.   (a)   Force on the sides:

$$F = \int_0^1 (9800)\,x\,14\,dx + \int_0^2 (9800)(1+x)\,7(2-x)\,dx$$

$$= 68,600\,\left[x^2\right]_0^1 + 68,600 \int_0^2 \left[2+x-x^2\right]\,dx$$

$$= 68,600 + 68,600\,\left[2x + \tfrac{1}{2}x^2 - \tfrac{1}{3}x^3\right]_0^2$$

$$\cong 297,267\ \text{Newtons}$$

(b)   Force at the shallow end:

$$F = \int_0^1 (9800)\cdot x\cdot 8\,dx = 39,200\,\left[x^2\right]_0^1 = 39,200\ \text{Newtons}$$

Force at the deep end:

$$F = \int_0^3 (9800)\cdot x\cdot 8\,dx = 39,200\,\left[x^2\right]_0^3 = 352,800\ \text{Newtons}$$

19. $F = \int_0^{14} (9800)\left(1 + \dfrac{1}{7}x\right)\cdot 8\dfrac{5\sqrt{2}}{7}\,dx = 392,000\dfrac{\sqrt{2}}{7}\left[x + \dfrac{1}{14}x^2\right]_0^1 4 \cong 2.217 \times 10^6\ \text{Newtons}$

21. $F = \int_a^b \sigma x w(x)\,dx = \sigma \int_a^b x w(x)\,dx = \sigma \overline{x} A$

where A is the area of the submerged surface and $\overline{x}$ is the depth of the centroid.

CHAPTER 7

SECTION 7.1

1. Suppose $f(x_1) = f(x_2)$ $x_1 \neq x_2$. Then

$5x_1 + 3 = 5x_2 + 3 \Rightarrow x_1 = x_2;$

f is one-to-one

$$f(t) = x$$

$$5t + 3 = x$$

$$5t = x - 3$$

$$t = \tfrac{1}{5}(x - 3)$$

$$f^{-1}(x) = \tfrac{1}{5}(x - 3)$$

3. Suppose $f(x_1) = f(x_2)$ $x_1 \neq x_2$. Then

$4x_1 - 7 = 4x_2 - 7 \Rightarrow x_1 = x_2;$

f is one-to-one

$$f(t) = x$$

$$4t - 7 = x$$

$$4t = x + 7$$

$$t = \tfrac{1}{4}(x + 7)$$

$$f^{-1}(x) = \tfrac{1}{4}(x + 7)$$

5. f is not one-to-one;   for instance, $f(1) = f(-1)$

7. $f'(x) = 5x^4 \geq 0$ on $(-\infty, \infty)$ and

$f'(x) = 0$ only at $x = 0$; f is increasing.

Therefore, f is one-to-one.

$$f(t) = x$$

$$t^5 + 1 = x$$

$$t^5 = x - 1$$

$$t = (x - 1)^{1/5}$$

$$f^{-1}(x) = (x - 1)^{1/5}$$

9. $f'(x) = 9x^2 \geq 0$ on $(-\infty, \infty)$ and

$f'(x) = 0$ only at $x = 0$; f is increasing.

Therefore, f is one-to-one.

$$f(t) = x$$

$$1 + 3t^3 = x$$

$$t^3 = \tfrac{1}{3}(x - 1)$$

$$t = \left[\tfrac{1}{3}(x - 1)\right]^{1/3}$$

$$f^{-1}(x) = \left[\tfrac{1}{3}(x - 1)\right]^{1/3}$$

11. $f'(x) = 3(1 - x)^2 \geq 0$ on $(-\infty, \infty)$ and

$f'(x) = 0$ only at $x = 1$; f is increasing.

Therefore, f is one-to-one.

$$f(t) = x$$

$$(1 - t)^3 = x$$

$$1 - t = x^{1/3}$$

$$t = 1 - x^{1/3}$$

$$f^{-1}(x) = 1 - x^{1/3}$$

13. $f'(x) = 3(x + 1)^2 \geq 0$ on $(-\infty, \infty)$ and

$f'(x) = 0$ only at $x = -1$; f is increasing.

Therefore, f is one-to-one.

$$f(t) = x$$

$$(t + 1)^3 + 2 = x$$

$$(t + 1)^3 = x - 2$$

$$t + 1 = (x - 2)^{1/3}$$

$$t = (x - 2)^{1/3} - 1$$

$$f^{-1}(x) = (x - 2)^{1/3} - 1$$

15. $f'(x) = \dfrac{3}{5x^{2/5}} > 0$ for all $x \neq 0$;

f is increasing on $(-\infty, \infty)$

$$f(t) = x$$
$$t^{3/5} = x$$
$$t = x^{5/3}$$
$$f^{-1}(x) = x^{5/3}$$

17. $f'(x) = 3(2 - 3x)^2 \geq 0$ for all x and

$f'(x) = 0$ only at $x = 2/3$; f is increasing

$$f(t) = x$$
$$(2 - 3t)^3 = x$$
$$2 - 3t = x^{1/3}$$
$$3t = 2 - x^{1/3}$$
$$t = \tfrac{1}{3}(2 - x^{1/3})$$
$$f^{-1}(x) = \tfrac{1}{3}(2 - x^{1/3})$$

19. $f'(x) = -\dfrac{1}{x^2} < 0$ for all $x \neq 0$;

f is decreasing on $(-\infty, 0) \cup (0, \infty)$

$$f(t) = x$$
$$\frac{1}{t} = x$$
$$t = \frac{1}{x}$$
$$f^{-1}(x) = \frac{1}{x}$$

21. f is not one-to-one;   for instance

$f\left(\tfrac{1}{2}\right) = f(2)$

23. $f'(x) = -\dfrac{3x^2}{(x^3 + 1)^2} \leq 0$ for all $x \neq -1$;

f is decreasing on $(-\infty, -1) \cup (-1, \infty)$

$$f(t) = x$$
$$\frac{1}{t^3 + 1} = x$$
$$t^3 + 1 = \frac{1}{x}$$
$$t^3 = \frac{1}{x} - 1$$
$$t = \left(\frac{1}{x} - 1\right)^{1/3}$$
$$f^{-1}(x) = \left(\frac{1}{x} - 1\right)^{1/3}$$

25. $f'(x) = \dfrac{1}{(x + 2)^2} > 0$ for all $x \neq -2$;

f is increasing on $(-\infty, -2) \cup (-2, \infty)$

$$f(t) = x$$
$$\frac{t + 2}{t + 1} = x$$
$$t + 2 = xt + x$$
$$t(1 - x) = x - 2$$
$$t = \frac{x - 2}{1 - x}$$
$$f^{-1}(x) = \frac{x - 2}{1 - x}$$

27.   they are equal

29.

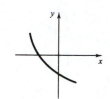

31.

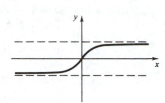

33. $f'(x) = 3x^2 \geq 0$ on $I = (-\infty, \infty)$ and $f'(x) = 0$ only at $x = 0$; f is increasing on I and so it has an inverse.

$$f(2) = 9 \text{ and } f'(2) = 12; \quad \left(f^{-1}\right)'(9) = \frac{1}{f'(2)} = \frac{1}{12}$$

35. $f'(x) = 1 + \frac{1}{\sqrt{x}} > 0$ on $I = (0, \infty)$; f is increasing on I and so it has an inverse.

$$f(4) = 8 \text{ and } f'(4) = 1 + \frac{1}{2} = \frac{3}{2}; \quad \left(f^{-1}\right)'(8) = \frac{1}{f'(4)} = \frac{1}{3/2} = \frac{2}{3}$$

37. $f'(x) = 2 - \sin x > 0$ on $I = (-\infty, \infty)$; f is increasing on I and so it has an inverse.

$$f(\pi/2) = \pi \text{ and } f'(\pi/2) = 1; \quad \left(f^{-1}\right)'(\pi) = \frac{1}{f'(\pi/2)} = 1$$

39. $f'(x) = \sec^2 x > 0$ on $I = (-\pi/2, \pi/2)$; f is increasing on I and so it has an inverse.

$$f(\pi/3) = \sqrt{3} \text{ and } f'(\pi/3) = 4; \quad \left(f^{-1}\right)'(\sqrt{3}) = \frac{1}{f'(\pi/3)} = \frac{1}{4}$$

41. $f'(x) = 3x^2 + \frac{3}{x^4} > 0$ on $I = (0, \infty)$; f is increasing on I and so it has an inverse.

$f(1) = 2$ and $f'(1) = 6$; $\quad f^{-1'}(2) = \frac{1}{f'(1)} = \frac{1}{6}.$

43. Let $x \in \text{dom}\left(f^{-1}\right)$ and let $f(z) = x$. Then

$$\left(f^{-1}\right)'(x) = \frac{1}{f'(z)} = \frac{1}{f(z)} = \frac{1}{x}$$

45. Let $x \in \text{dom}\left(f^{-1}\right)$ and let $f(z) = x$. Then

$$\left(f^{-1}\right)'(x) = \frac{1}{f'(z)} = \frac{1}{\sqrt{1 - [f(z)]^2}} = \frac{1}{\sqrt{1 - x^2}}$$

47. (a)

$$f'(x) = \frac{(cx + d)a - (ax + b)c}{(cx + d)^2} = \frac{ad - bc}{(cx + d)^2}, \quad x \neq -d/c$$

Thus, $f'(x) \neq 0$ iff $ad - bc \neq 0$.

(b)

$$\frac{at+b}{ct+d} = x$$

$$at + b = ctx + dx$$

$$(a - cx)t = dx - b$$

$$t = \frac{dx - b}{a - cx}; \quad f^{-1}(x) = \frac{dx - b}{a - cx}$$

49. (a) $f'(x) = \sqrt{1 + x^2} > 0$, so f is always increasing, hence one-to-one.

(b) $(f^{-1})'(0) = \dfrac{1}{f'(f^{-1}(0))} = \dfrac{1}{f'(2)} = \dfrac{1}{\sqrt{5}}.$

51. (a) $g(t) \geq 0$ or $g(t) \leq 0$ on $[a, b]$, with $g(t) = 0$ only at "isolated" points.

(b) $g(x) \neq 0$

(c) $(f^{-1})'(x) = \dfrac{1}{g(x)}$

53. (a)

$$g'(x) = \frac{1}{f'[g(x)]}$$

$$g''(x) = -\frac{1}{(f'[g(x)])^2} f''[g(x)]g'(x)$$

$$= -\frac{f''[g(x)]}{(f'[g(x)])^3}$$

(b) If f is increasing and its graph is concave up (down), then the graph of g is concave down (up). On the other hand, if f is decreasing then the graphs of f and g have the the same concavity.

55. Let $f(x) = \sin x$ and let $y = f^{-1}(x)$. Then

$$\sin y = x$$

$$\cos y \frac{dy}{dx} = 1$$

$$\frac{dy}{dx} = \frac{1}{\cos y} \quad (y \neq \pm \pi/2)$$

$$= \frac{1}{\sqrt{1 - \sin^2 y}} = \frac{1}{\sqrt{1 - x^2}} \quad (x \neq \pm 1)$$

57. $f'(x) = 3x^2 + 3 > 0$ for all x;

f is increasing on $(-\infty, \infty)$

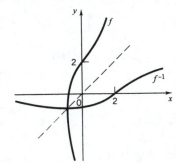

59. $f'(x) = 8 \cos 2x > 0$, $x \in (-\pi/4, \pi/4)$;

f is increasing on $[-\pi/4, \pi/4]$

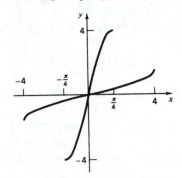

SECTION 7.2

1. $\ln 20 = \ln 2 + \ln 10 \cong 2.99$

3. $\ln 1.6 = \ln \frac{16}{10} = 2 \ln 4 - \ln 10 \cong 0.48$

5. $\ln 0.1 = \ln \frac{1}{10} = \ln 1 - \ln 10 \cong -2.30$

7. $\ln 7.2 = \ln \frac{72}{10} = \ln 8 + \ln 9 - \ln 10 \cong 1.98$

9. $\ln \sqrt{2} = \frac{1}{2} \ln 2 \cong 0.35$

11.

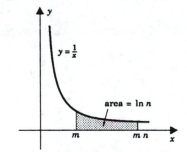

13. $\frac{1}{2}\left[L_f(P) + U_f(P)\right]$

$= \frac{1}{2}\left[\frac{763}{1980} + \frac{1691}{3960}\right]$

$\cong 0.406$

15. (a) $\ln 5.2 \cong \ln 5 + \frac{1}{5}(0.2) \cong 1.65$

(b) $\ln 4.8 \cong \ln 5 - \frac{1}{5}(0.2) \cong 1.57$

(c) $\ln 5.5 \cong \ln 5 + \frac{1}{5}(0.5) \cong 1.71$

17. $x = e^2$

19. $2 - \ln x = 0$ or $\ln x = 0$. Thus $x = e^2$ or $x = 1$.

21.

$$\ln\left[(2x + 1)(x + 2)\right] = 2 \ln (x + 2)$$

$$\ln\left[(2x + 1)(x + 2)\right] = \ln\left[(x + 2)^2\right]$$

$$(2x + 1)(x + 2) = (x + 2)^2$$

$$x^2 + x - 2 = 0$$

$$(x + 2)(x - 1) = 0$$

$$x = -2, 1$$

We disregard the solution $x = -2$ since it does not satisfy the initial equation.

Thus, the only solution is $x = 1$.

23. See Exercises 3.1, Definition (3.1.5).

$$\lim_{x\to 1}\frac{\ln x}{x-1} = \frac{d}{dx}(\ln x)\bigg|_{x=1} = \frac{1}{x}\bigg|_{x=1} = 1$$

25.   (a) Let $P = \{1, 2, \ldots, n\}$ be a regular partition of $[1, n]$. Then

$$L_f(P) = \frac{1}{2} + \frac{1}{3} + \cdots + \frac{1}{n} < \int_1^n \frac{1}{t}\, dt < 1 + \frac{1}{2} + \cdots + \frac{1}{n-1} = U_f(P)$$

(b) The sum of the shaded areas is give by

$$U_f(P) - \int_1^n \frac{1}{t}\, dt = 1 + \frac{1}{2} + \cdots + \frac{1}{n-1} - \ln n.$$

(c) Connect the points $(1, 1), \left(2, \frac{1}{2}\right), \ldots, \left(n, \frac{1}{n}\right)$ by straight line segments.

The sum of the areas of the triangles that are formed is:

$$\frac{1}{2} \cdot 1\left[\left(1 - \frac{1}{2}\right) + \left(\frac{1}{2} - \frac{1}{3}\right) + \cdots + \left(\frac{1}{n-1} - \frac{1}{n}\right)\right] = \frac{1}{2}\left(1 - \frac{1}{n}\right)$$

so

$$\frac{1}{2}\left(1 - \frac{1}{n}\right) < \gamma$$

The sum of the areas of the indicated rectangles is:

$$1\left[\left(1 - \frac{1}{2}\right) + \left(\frac{1}{2} - \frac{1}{3}\right) + \cdots + \left(\frac{1}{n-1} - \frac{1}{n}\right)\right] = 1 - \frac{1}{n}$$

so

$$\gamma < 1 - \frac{1}{n}$$

Letting $n \to \infty$ we have $\frac{1}{2} < \gamma < 1$.

27.   (a) Let $G(x) = \sin x - \ln x$. Then $G(3) = \sin 3 - \ln 3 \cong 0.96 > 0$ and $G(2) = \sin 2 - \ln 2 \cong -0.22 < 0$.

Thus, G has at least one zero on $[2, 3]$ which implies that there is at least one number $r \in [2, 3]$ such that $\sin r = \ln r$.

(b) $r \cong 2.2191$

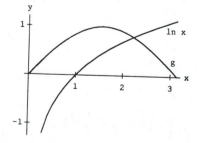

29. $L = 1$

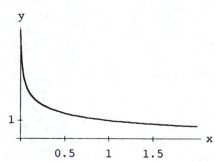

31. $L = 0$

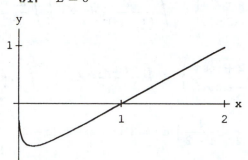

SECTION 7.3

1. $\mathrm{dom}\,(f) = (0, \infty)\,, \quad f'(x) = \dfrac{1}{4x}\,(4) = \dfrac{1}{x}$

3. $\mathrm{dom}\,(f) = (-1, \infty)\,, \quad f'(x) = \dfrac{1}{x^3 + 1}\dfrac{d}{dx}\left(x^3 + 1\right) = \dfrac{3x^2}{x^3 + 1}$

5. $\mathrm{dom}\,(f) = (-\infty, \infty)\,, \quad f(x) = \dfrac{1}{2}\ln\left(1 + x^2\right) \quad \text{so} \quad f'(x) = \dfrac{1}{2}\left[\dfrac{1}{1 + x^2}(2x)\right] = \dfrac{x}{1 + x^2}$

7. $\mathrm{dom}\,(f) = \{x \mid x \neq \pm 1\}\,, \quad f'(x) = \dfrac{1}{x^4 - 1}\dfrac{d}{dx}\left(x^4 - 1\right) = \dfrac{4x^3}{x^4 - 1}$

9. $\mathrm{dom}\,(f) = (0, \infty)\,, \quad f'(x) = x^2\dfrac{d}{dx}\,(\ln x) + 2x\,(\ln x) = x + 2x\,\ln x$

11. $\mathrm{dom}\,(f) = (0, 1) \cup (1, \infty)\,, \quad f(x) = (\ln x)^{-1} \quad \text{so} \quad f'(x) = -\,(\ln x)^{-2}\dfrac{d}{dx}\,(\ln x) = -\,\dfrac{1}{x\,(\ln x)^2}$

13. $\mathrm{dom}\,(f) = (0, \infty)\,, \quad f'(x) = \cos\,(\ln x)\left(\dfrac{1}{x}\right) = \dfrac{\cos\,(\ln x)}{x}$

15. $\displaystyle\int \dfrac{dx}{x + 1} = \ln|x + 1| + C$

17.

$$\left\{\begin{array}{l} u = 3 - x^2 \\ du = -2x\,dx \end{array}\right\}; \quad \int \dfrac{x}{3 - x^2}\,dx = -\dfrac{1}{2}\int \dfrac{du}{u} = -\dfrac{1}{2}\ln|u| + C = -\dfrac{1}{2}\ln|3 - x^2| + C$$

19.

$$\left\{\begin{array}{l} u = 3x \\ du = 3dx \end{array}\right\}; \quad \int \tan 3x\,dx = \dfrac{1}{3}\int \tan u\,du = \dfrac{1}{3}\ln|\sec u| + C = \dfrac{1}{3}\ln|\sec 3x| + C$$

21.

$$\left\{\begin{array}{l} u = x^2 \\ du = 2x\,dx \end{array}\right\}; \quad \int x\sec x^2\,dx = \dfrac{1}{2}\int \sec u\,du = \dfrac{1}{2}\ln|\sec u + \tan u| + C$$

$$= \tfrac{1}{2}\ln|\sec x^2 + \tan x^2| + C$$

23. $\left\{ \begin{array}{l} u = 3 - x^2 \\ du = -2x\,dx \end{array} \right\}$; $\displaystyle\int \frac{x}{(3-x^2)^2}\,dx = -\frac{1}{2}\int \frac{du}{u^2} = \frac{1}{2u} + C = \frac{1}{2(3-x^2)} + C$

25. $\left\{ \begin{array}{l} u = 2 + \cos x \\ du = -\sin x\,dx \end{array} \right\}$; $\displaystyle\int \frac{\sin x}{2+\cos x}\,dx = -\int \frac{1}{u}\,du = -\ln|u| + C = -\ln|2+\cos x| + C$

27. $\displaystyle\int \left(\frac{1}{x+2} - \frac{1}{x-2} \right)\,dx = \ln|x+2| - \ln|x-2| + C = \ln\left| \frac{x+2}{x-2} \right| + C$

29. $\left\{ u = \ln x,\ du = \dfrac{dx}{x} \right\}$; $\displaystyle\int \frac{dx}{x(\ln x)^2} = \int \frac{du}{u^2} = -\frac{1}{u} + C = -\frac{1}{\ln x} + C$

31. $\left\{ \begin{array}{l} u = \sin x + \cos x \\ du = (\cos x - \sin x)\,dx \end{array} \right\}$; $\displaystyle\int \frac{\sin x - \cos x}{\sin x + \cos x}\,dx = -\int \frac{1}{u}\,du = -\ln|u| + C$

$$= -\ln|\sin x + \cos x| + C$$

33. $\left\{ u = 1 + x\sqrt{x},\ du = \dfrac{3}{2}x^{1/2}\,dx \right\}$; $\displaystyle\int \frac{\sqrt{x}}{1+x\sqrt{x}}\,dx = \frac{2}{3}\int \frac{du}{u} = \frac{2}{3}\ln|u| + C$

$$= \frac{2}{3}\ln|1+x\sqrt{x}| + C$$

35.

$$\left\{ \begin{array}{l} u = \ln|\sin x| \\ \\ \\ du = \cot x\,dx \end{array} \right\} ; \quad \int \cot x \ln|\sin x|\,dx = \int u\,du = \frac{1}{2}u^2 + C$$

$$= \tfrac{1}{2}\left(\ln|\sin x|\right)^2 + C$$

37. $\displaystyle\int (1+\sec x)^2\,dx = \int \left(1 + 2\sec x + \sec^2 x\right)\,dx = x + 2\ln|\sec x + \tan x| + \tan x + C$

39.

$$\left\{ \begin{array}{l} u = \ln|\sec x + \tan x| \\ \\ \\ du = \sec x\,dx \end{array} \right\} ;$$

$$\int \frac{\sec x}{\sqrt{\ln|\sec x + \tan x|}}\,dx = \int \frac{du}{\sqrt{u}} = 2\sqrt{u} + C = 2\sqrt{\ln|\sec x + \tan x|} + C$$

41. $\displaystyle\int_1^e \frac{dx}{x} = [\ln x]_1^e = \ln e - \ln 1 = 1 - 0 = 1$

43. $\displaystyle\int_e^{e^2} \frac{dx}{x} = [\ln x]_e^{e^2} = \ln e^2 - \ln e = 2 - 1 = 1$

45. $\displaystyle\int_4^5 \frac{x}{x^2-1}\,dx = \left[\frac{1}{2}\ln|x^2-1|\right]_4^5 = \frac{1}{2}\left(\ln 24 - \ln 15\right) = \frac{1}{2}\ln\frac{8}{5}$

47. $\left\{\begin{array}{l} u = 1+\sin x \quad x = \pi/6 \quad\Longrightarrow\quad u = 3/2 \\[3mm] \\ du = \cos x\,dx \quad x = \pi/2 \quad\Longrightarrow\quad u = 2 \end{array}\right\}$; $\displaystyle\int_{\pi/6}^{\pi/2}\frac{\cos x}{1+\sin x}\,dx = \int_{3/2}^2 \frac{du}{u} = [\ln u]_{3/2}^2 = \ln\frac{4}{3}$

49. $\displaystyle\int_{\pi/4}^{\pi/2}\cot x\,dx = \left[\ln|\sin x|\right]_{\pi/4}^{\pi/2} = \ln 1 - \ln\frac{\sqrt 2}{2} = \ln\sqrt 2 = \frac{1}{2}\ln 2$

51. $\ln|g(x)| = 2\ln\left(x^2+1\right) + 5\ln|x-1| + 3\ln x$

$\displaystyle\frac{g'(x)}{g(x)} = 2\left(\frac{2x}{x^2+1}\right) + \frac{5}{x-1} + \frac{3}{x}$

$\displaystyle g'(x) = \left(x^2+1\right)^2 (x-1)^5\, x^3 \left(\frac{4x}{x^2+1} + \frac{5}{x-1} + \frac{3}{x}\right)$

53. $\ln|g(x)| = 4\ln|x| + \ln|x-1| - \ln|x+2| - \ln\left(x^2+1\right)$

$\displaystyle\frac{g'(x)}{g(x)} = \frac{4}{x} + \frac{1}{x-1} - \frac{1}{x+2} - \frac{2x}{x^2+1}$

$\displaystyle g'(x) = \frac{x^4(x-1)}{(x+2)\left(x^2+1\right)}\left(\frac{4}{x} + \frac{1}{x-1} - \frac{1}{x+2} - \frac{2x}{x^2+1}\right)$

55. $\ln|g(x)| = \frac{1}{2}\left(\ln|x-1| + \ln|x-2| - \ln|x-3| - \ln|x-4|\right)$

$\displaystyle\frac{g'(x)}{g(x)} = \frac{1}{2}\left(\frac{1}{x-1} + \frac{1}{x-2} - \frac{1}{x-3} - \frac{1}{x-4}\right)$

$\displaystyle g'(x) = \frac{1}{2}\sqrt{\frac{(x-1)(x-2)}{(x-3)(x-4)}}\left(\frac{1}{x-1} + \frac{1}{x-2} - \frac{1}{x-3} - \frac{1}{x-4}\right)$

57.

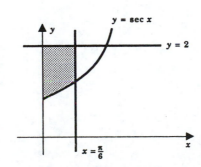

$\displaystyle A = \int_0^{\pi/6}(2 - \sec x)\,dx$

$\displaystyle = \left[2x - \ln|\sec x + \tan x|\right]_0^{\pi/6}$

$\displaystyle = \frac{\pi}{3} - \ln\left|\frac{2}{\sqrt 3} + \frac{1}{\sqrt 3}\right| = \frac{\pi}{3} - \frac{1}{2}\ln 3$

59.

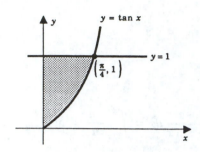

$$A = \int_0^{\pi/4} (1 - \tan x)\, dx$$

$$= [x - \ln|\sec x|]_0^{\pi/4}$$

$$= \frac{\pi}{4} - \ln\sqrt{2} = \frac{\pi}{4} - \frac{1}{2}\ln 2$$

61. $\quad A = \int_1^4 \left[\frac{5-x}{4} - \frac{1}{x}\right] dx = \left[\frac{5}{4}x - \frac{1}{8}x^2 - \ln x\right]_1^4 = \frac{15}{8} - \ln 4$

63. $\quad V = \int_0^8 \pi \left(\frac{1}{\sqrt{1+x}}\right)^2 dx = \pi \int_0^8 \frac{1}{1+x}\, dx = \pi\left[\ln|1+x|\right]_0^8 = \pi \ln 9$

65.

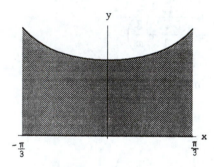

$$V = \int_{-\pi/3}^{\pi/3} \pi \left(\sqrt{\sec x}\right)^2 dx$$

$$= 2\pi \int_0^{\pi/3} \sec x\, dx$$

$$= 2\pi \left[\ln|\sec x + \tan x|\right]_0^{\pi/3} = 2\pi \ln\left(2 + \sqrt{3}\right)$$

67. $\quad v(t) = \int a(t)\, dt = \int -(t+1)^{-2}\, dt = \frac{1}{t+1} + C.$

Since $v(0) = 1$, we get $1 = 1 + C$ so that $C = 0$. Then

$$s = \int_0^4 |v(t)|\, dt = \int_0^4 \frac{dt}{t+1} = \left[\ln(t+1)\right]_0^4 = \ln 5.$$

The particle traveled $\ln 5$ ft.

69. $\quad \dfrac{d}{dx}(\ln x) = \dfrac{1}{x}$

$\dfrac{d^2}{dx^2}(\ln x) = -\dfrac{1}{x^2}$

$\dfrac{d^3}{dx^3}(\ln x) = \dfrac{2}{x^3}$

$\dfrac{d^4}{dx^4}(\ln x) = -\dfrac{2 \cdot 3}{x^4}$

$$\vdots$$

$$\frac{d^n}{dx^n}\left(\ln x\right) = (-1)^{n-1}\frac{(n-1)!}{x^n}$$

71. $\dfrac{d^n}{dx^n}\left(\ln 2x\right) = \dfrac{d^n}{dx^n}\left[\ln 2 + \ln x\right] = 0 + \dfrac{d^n}{dx^n}\left(\ln x\right) = (-1)^{n-1}\dfrac{(n-1)!}{x^n}$ (See Exercise 69)

73. $\displaystyle\int \csc x\,dx = \int \frac{\csc x(\csc x - \cot x)}{\csc x - \cot x}\,dx = \int \frac{\csc^2 - \csc x\,\cot x}{\csc x - \cot x}\,dx$

$$\left\{ \begin{aligned} u &= \csc x - \cot x \\[2mm] du &= (-\csc x\,\cot x + \csc^2 x)\,dx \end{aligned} \right\}; \quad \int \csc x\,dx = \int \frac{du}{u} = \ln|u| + C$$

$$= \ln|\csc x - \cot x| + C$$

75. $f(x) = \ln(4-x), \quad x < 4$ $f' :$ 

$$f'(x) = \frac{1}{x-4}$$

$f'' :$ 

$$f''(x) = \frac{-1}{(x-4)^2}$$

(i)   domain $(-\infty, 4)$

(ii)   decreases throughout

(iii)   no extreme values

(iv)   concave down throughout:

       no pts of inflection

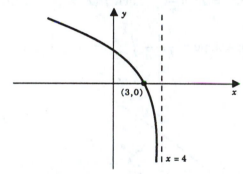

77. $f(x) = x^2 \ln x, \ x > 0$

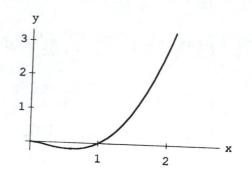

$f'(x) = 2x \ln x + x$

$f''(x) = 2 \ln x + 3$

(i)   domain $(0, \infty)$

(ii)   decreases on $(0, 1/\sqrt{e}]$,   increases on $[1/\sqrt{e}, \infty)$

(iii) $f(1/\sqrt{e}) = -1/2e$ local and absolute min

(iv)   concave down on $(0, 1/e^{3/2})$,

concave up on $(1/e^{3/2}, \infty)$;

pt of inflection at $(1/e^{3/2}, -3/2e^3)$

79.   Let $f(x) = x^2 \ln \dfrac{1}{x}$.   Then $f'(x) = -x + 2x \ln \dfrac{1}{x} = 0$ $\implies$ $ln \dfrac{1}{x} = \dfrac{1}{2}$ $\Rightarrow$ $x = \dfrac{1}{\sqrt{e}}$.

81.   Average slope $= \dfrac{1}{b-a} \displaystyle\int_a^b \dfrac{1}{x} \, dx = \dfrac{1}{(b-a)} \ln \dfrac{b}{a}$

83.

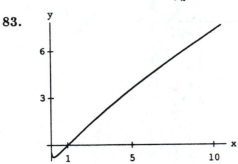

x-intercept: 1; abs min at $x = 1/e^2$;

abs max at $x = 10$

85.

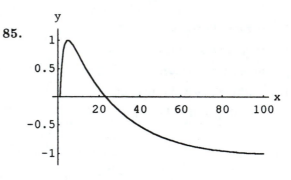

x-intercepts: $1, 23.1407$; abs min at $x = 100$;

abs max at $x = 4.8105$

87.   (a)

$$v(t) - v(0) = \int_0^t a(u)\,du, \quad 0 \le t \le 3$$

$$v(t) = \int_0^t \left[4 - 2(u+1) + \frac{3}{u+1} \right] du + 2$$

$$= \left[2u - u^2 + 3 \ln |u+1| \right]_0^t = 2 + 2t - t^2 + 3\ln(t+1)$$

(b)

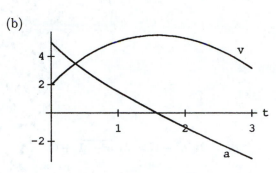

(c) max velocity at $t = 1.5811$; min velocity at $t = 0$

SECTION 7.4

1. $\dfrac{dy}{dx} = e^{-2x}\dfrac{d}{dx}(-2x) = -2e^{-2x}$

3. $\dfrac{dy}{dx} = e^{x^2-1}\dfrac{d}{dx}(x^2-1) = 2xe^{x^2-1}$

5. $\dfrac{dy}{dx} = e^x\dfrac{d}{dx}(\ln x) + \ln x\dfrac{d}{dx}(e^x) = e^x\left(\dfrac{1}{x} + \ln x\right)$

7. $\dfrac{dy}{dx} = x^{-1}\dfrac{d}{dx}(e^{-x}) + e^{-x}\dfrac{d}{dx}(x^{-1}) = -x^{-1}e^{-x} - e^{-x}x^{-2} = -\left(x^{-1} + x^{-2}\right)e^{-x}$

9. $\dfrac{dy}{dx} = \dfrac{1}{2}\left(e^x - e^{-x}\right)$

11. $\dfrac{dy}{dx} = e^{\sqrt{x}}\dfrac{d}{dx}\left(\ln\sqrt{x}\,\right) + \ln\sqrt{x}\,\dfrac{d}{dx}\left(e^{\sqrt{x}}\right) = e^{\sqrt{x}}\left(\dfrac{1}{\sqrt{x}}\cdot\dfrac{1}{2\sqrt{x}}\right) + \ln\sqrt{x}\dfrac{e^{\sqrt{x}}}{2\sqrt{x}} = \dfrac{1}{2}e^{\sqrt{x}}\left(\dfrac{1}{x} + \dfrac{\ln\sqrt{x}}{\sqrt{x}}\right)$

13. $\dfrac{dy}{dx} = 2\left(e^{x^2} + 1\right)\dfrac{d}{dx}\left(e^{x^2} + 1\right) = 2\left(e^{x^2} + 1\right)e^{x^2}\dfrac{d}{dx}(x^2) = 4xe^{x^2}\left(e^{x^2} + 1\right)$

15. $\dfrac{dy}{dx} = \left(x^2 - 2x + 2\right)\dfrac{d}{dx}(e^x) + e^x(2x - 2) = x^2e^x$

17. $\dfrac{dy}{dx} = \dfrac{(e^x + 1)e^x - (e^x - 1)e^x}{(e^x + 1)^2} = \dfrac{2e^x}{(e^x + 1)^2}$

19. $y = e^{4\ln x} = \left(e^{\ln x}\right)^4 = x^4$ so $\dfrac{dy}{dx} = 4x^3$.

21. $f'(x) = \cos\left(e^{2x}\right)e^{2x}2 = 2e^{2x}\cos\left(e^{2x}\right)$

23. $f'(x) = e^{-2x}(-\sin x) + e^{-2x}(-2)\cos x = -e^{-2x}(2\cos x + \sin x)$

25. $\displaystyle\int e^{2x}\,dx = \dfrac{1}{2}e^{2x} + C$

27. $\displaystyle\int e^{kx}\,dx = \dfrac{1}{k}e^{kx} + C$

29. $\left\{u = x^2,\quad du = 2x\,dx\right\};\quad \displaystyle\int xe^{x^2}\,dx = \dfrac{1}{2}\int e^u\,du = \dfrac{1}{2}e^u + C = \dfrac{1}{2}e^{x^2} + C$

31. $\left\{u = \dfrac{1}{x},\quad du = -\dfrac{1}{x^2}\,dx\right\};\quad \displaystyle\int\dfrac{e^{1/x}}{x^2}\,dx = -\int e^u\,du = -e^u + C = -e^{1/x} + C$

33. $\displaystyle\int \ln e^x\,dx = \int x\,dx = \frac{1}{2}x^2 + C$

35. $\displaystyle\int \frac{4}{\sqrt{e^x}}\,dx = \int 4e^{-x/2}\,dx = -8e^{-x/2} + C$

37. $\left\{\begin{array}{l} u = e^x + 1 \\ du = e^x\,dx \end{array}\right\};\quad \displaystyle\int \frac{e^x}{\sqrt{e^x+1}}\,dx = \int \frac{du}{\sqrt{u}} = \int u^{-1/2}\,du = 2u^{1/2} + C = 2\sqrt{e^x+1} + C$

39. $\left\{\begin{array}{l} u = 2e^{2x} + 3 \\ du = 4e^{2x}\,dx \end{array}\right\};\quad \displaystyle\int \frac{e^{2x}}{2e^{2x}+3}\,dx = \frac{1}{4}\int \frac{du}{u} = \frac{1}{4}\ln u + C = \frac{1}{4}\ln\left(2e^{2x}+3\right) + C$

41. $\{u = \sin x,\quad du = \cos x\,dx\};\quad \displaystyle\int \cos x\, e^{\sin x}\,dx = \int e^u\,du = e^u + C = e^{\sin x} + C$

43. $\{u = e^{-x},\quad du = -e^{-x}\,dx\};$

$\displaystyle\int e^{-x}\left[1 + \cos\left(e^{-x}\right)\right]\,dx = -\int(1 + \cos u)\,du = -u - \sin u + C = -e^{-x} - \sin\left(e^{-x}\right) + C$

45. $\displaystyle\int_0^1 e^x\,dx = [e^x]_0^1 = e - 1$

47. $\displaystyle\int_0^{\ln \pi} e^{-6x}\,dx = \left[-\frac{1}{6}e^{-6x}\right]_0^{\ln \pi} = -\frac{1}{6}e^{-6\ln \pi} + \frac{1}{6}e^0 = \frac{1}{6}\left(1 - \pi^{-6}\right)$

49. $\displaystyle\int_0^1 \frac{e^x + 1}{e^x}\,dx = \int_0^1 \left(1 + e^{-x}\right)\,dx = \left[x - e^{-x}\right]_0^1 = (1 - e^{-1}) - (0 - 1) = 2 - \frac{1}{e}$

51. $\displaystyle\int_0^{\ln 2} \frac{e^x}{e^x + 1}\,dx = [\ln(e^x + 1)]_0^{\ln 2} = \ln\left(e^{\ln 2} + 1\right) - \ln\left(e^0 + 1\right) = \ln 3 - \ln 2 = \ln\frac{3}{2}$

53. $\displaystyle\int_0^1 x\left(e^{x^2} + 2\right)\,dx = \int_0^1 \left(xe^{x^2} + 2x\right)\,dx = \left[\frac{1}{2}e^{x^2} + x^2\right]_0^1 = (\tfrac{1}{2}e + 1) - (\tfrac{1}{2} + 0) = \tfrac{1}{2}(e + 1)$

55. $y' = ake^{kx} = bke^{-kx} = 0 \quad\Longrightarrow\quad x = \frac{1}{2k}\ln\frac{b}{a}.$

At this point, $y = 2\sqrt{ab}$.

57. $A = 2xe^{-x^2}$

$A' = 2x(-2xe^{-x^2}) + 2e^{-x^2} = 2e^{-x^2}(1 - 2x^2) = 0 \quad\Longrightarrow\quad x = \pm\sqrt{\frac{1}{2}} \text{ and } y = \frac{1}{\sqrt{e}}.$

Put the vertices at $\left(\pm\dfrac{1}{\sqrt{2}}, \dfrac{1}{\sqrt{e}}\right)$.

59.
$$x(t) = Ae^{ct} + Be^{-ct}$$

$$x'(t) = Ace^{ct} - Bce^{-ct}$$

$$x''(t) = Ac^2e^{ct} + Bc^2e^{-ct}$$

$$= c^2\left(Ae^{ct} + Be^{-ct}\right)$$

$$= c^2 x(t)$$

61.

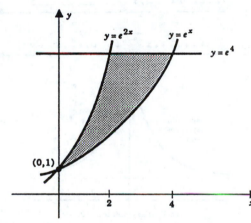

$$A = \int_0^2 \left(e^{2x} - e^x\right)\,dx + \int_2^4 \left(e^4 - e^x\right)\,dx$$

$$= \left[\tfrac{1}{2}e^{2x} - e^x\right]_0^2 + \left[e^4 x - e^x\right]_2^4$$

$$= \left(\tfrac{1}{2}e^4 - e^2 - \tfrac{1}{2} + 1\right) + \left(4e^4 - e^4 - 2e^4 + e^2\right)$$

$$= \tfrac{1}{2}\left(3e^4 + 1\right)$$

63.

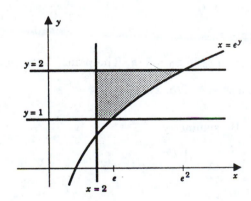

$$A = \int_1^2 \left(e^y - 2\right)\,dy$$

$$= \left[e^y - 2y\right]_1^2 = e^2 - e - 2$$

65. $f(x) = \frac{1}{2}(e^x + e^{-x})$

f' :

$f'(x) = \frac{1}{2}(e^x - e^{-x})$ f'' :

$f''(x) = \frac{1}{2}(e^x + e^{-x})$

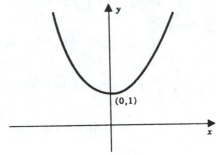

(i)   domain $(-\infty, \infty)$

(ii)   decreases on $(-\infty, 0]$,   increases on $[0, \infty)$

(iii) $f(0) = 1$ local and absolute min

(iv)   concave up everywhere

(0,1)

67. $f(x) = e^{(1/x)^2}$ f' :

$f'(x) = \frac{-2}{x^3}e^{(1/x)^2}$ f'' :

$f''(x) = \frac{6x^2 + 4}{x^6}e^{(1/x)^2}$

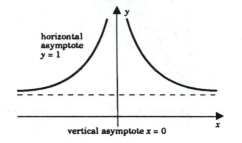

horizontal asymptote $y = 1$

(i)   domain $(-\infty, 0) \cup (0, \infty)$

(ii)   increases on $(-\infty, 0)$,   decreases on $(0, \infty)$

(iii) no extreme values

(iv)   concave up on $(-\infty, 0)$ and on $(0, \infty)$

vertical asymptote $x = 0$

69.   (a)   For $y = e^{ax}$ we have $dy/dx = ae^{ax}$. Therefore the line tangent to the curve $y = e^{ax}$

at an arbitrary point (x_0, e^{ax_0}) has equation

$$y - e^{ax_0} = ae^{ax_0}(x - x_0).$$

The line passes through the origin iff $e^{ax_0} = (ae^{ax_0})x_0$ iff $x_0 = 1/a$. The point of tangency

is $(1/a, e)$. This is point B. By symmetry, point A is $(-1/a, e)$.

(b)   The tangent line at B has equation $y = aex$.   By symmetry

$$A_I = 2\int_0^{1/a}(e^{ax} - aex)\, dx = 2\left[\frac{1}{a}e^{ax} - \frac{1}{2}aex^2\right]_0^{1/a} = \frac{1}{a}(e - 2).$$

(c)   The normal at B has equation

$$y - e = -\frac{1}{ae}\left(x - \frac{1}{a}\right).$$

This can be written
$$y = -\frac{1}{ae}x + \frac{a^2e^2 + 1}{a^2e}.$$

Therefore
$$A_{\text{II}} = 2\int_0^{1/a}\left(-\frac{1}{ae}x + \frac{a^2e^2 + 1}{a^2e} - e^{ax}\right)dx = \frac{1 + 2a^2e}{a^3e}.$$

71. For $x > (n+1)!$
$$e^x > 1 + x + \cdots + \frac{x^{n+1}}{(n+1)!} > \frac{x^{n+1}}{(n+1)!} = x^n\left[\frac{x}{(n+1)!}\right] > x^n.$$

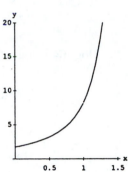

73. Numerically, $8.15 \le L \le 8.16$;
$$\lim_{x\to 1}\frac{e^{x^3} - e}{x - 1} = \lim_{x\to 1}\frac{e^{x^3} - e^1}{x - 1}$$

is the derivative of $f(x) = e^{x^3}$ at $x = 1$.

Note that $f'(x) = 3x^2e^{x^3}$; $f'(1) = 3e \cong 8.15485$.

75. (a) $\qquad\qquad\qquad\qquad\qquad\qquad$ (b) $x_1 \cong -1.9646$, $x_2 \cong 1.0580$

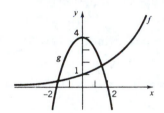

(c) $A \cong \displaystyle\int_{-1.9646}^{1.0580}\left[4 - x^2 - e^x\right]dx = \left[4x - \frac{1}{3}x^3 - e^x\right]_{-1.9646}^{1.0580} \cong 6.4240$

77. $\qquad\qquad\qquad\qquad\qquad\qquad\qquad$ **79.**

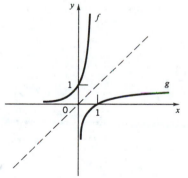 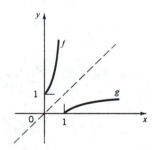

$f(g(x)) = e^{2\ln\sqrt{x}} = e^{2(1/2)\ln x} = e^{\ln x} = x$ $\qquad$ $f(g(x)) = e^{(\sqrt{\ln x})^2} = e^{\ln x} = x$

PROJECT 7.4

1. (a) $\ln a + \dfrac{1}{n} = \displaystyle\int_1^{1+\frac{1}{n}}\frac{dt}{t} \le \int_1^{1+\frac{1}{n}}1\,dt = 1$ $\quad$ (since $\dfrac{1}{t} \le 1$ throughout the interval of integration)

(b) $\ln a + \dfrac{1}{n} = \displaystyle\int_1^{1+\frac{1}{n}}\frac{dt}{t} \ge \int_1^{1+\frac{1}{n}}\frac{dt}{1+\frac{1}{n}}$

$$= \frac{1}{1+\frac{1}{n}} \frac{1}{n} \quad \text{(since } \frac{1}{t} \geq \frac{1}{1+\frac{1}{n}} \text{ throughout the interval of integration)} = \frac{1}{n+1}$$

3. For $n = 1000$, we get $e \cong 2.72$.

For $n = 10,000$, we get $e \cong 2.718$.

For $n = 100,000$, we get $e \cong 2.7183$.

For $n = 1,000,000$, we get $e \cong 2.71828$.

SECTION 7.5

1. $\log_2 64 = \log_2 \left(2^6\right) = 6$

3. $\log_{64}(1/2) = \dfrac{\ln(1/2)}{\ln 64} = \dfrac{-\ln 2}{6\ln 2} = -\dfrac{1}{6}$

5. $\log_5 1 = \log_5 \left(5^0\right) = 0$

7. $\log_5(125) = \log_5 \left(5^3\right) = 3$

9. $\log_p xy = \dfrac{\ln xy}{\ln p} = \dfrac{\ln x + \ln y}{\ln p} = \dfrac{\ln x}{\ln p} + \dfrac{\ln y}{\ln p} = \log_p x + \log_p y$

11. $\log_p x^y = \dfrac{\ln x^y}{\ln p} = y\dfrac{\ln x}{\ln p} = y\log_p x$

13. $10^x = e^x \implies \left(e^{\ln 10}\right) = e^x \implies e^{x\ln 10} = e^x \implies x\ln 10 = x \implies x(\ln 10 - 1) = 0$ Thus, $x = 0$.

15. $\log_x 10 = \log_4 100 \implies \dfrac{\ln 10}{\ln x} = \dfrac{\ln 100}{\ln 4} \implies \dfrac{\ln 10}{\ln x} = \dfrac{2\ln 10}{2\ln 2} \implies \ln x = \ln 2$ Thus, $x = 2$.

17. The logarithm function is increasing. Thus,

$$e^{t_1} < a < e^{t_2} \implies t_1 = \ln e^{t_1} < \ln a < \ln e^{t_2} = t_2.$$

19. $f'(x) = 3^{2x}(\ln 3)(2) = 2(\ln 3)3^{2x}$

21. $f'(x) = 2^{5x}(\ln 2)(5)3^{\ln x} + 2^{5x}3^{\ln x}(\ln 3)\dfrac{1}{x} = 2^{5x}3^{\ln x}\left(5\ln 2 + \dfrac{\ln 3}{x}\right)$

23. $g'(x) = \frac{1}{2}\left(\log_3 x\right)^{-1/2}\left(\dfrac{1}{\ln 3}\right)\dfrac{1}{x} = \dfrac{1}{2(\ln 3)x\sqrt{\log_3 x}}$

25. $f'(x) = \sec^2\left(\log_5 x\right)\left(1\ln 5\right)\dfrac{1}{x} = \dfrac{\sec^2\left(\log_5 x\right)}{x\ln 5}$

27. $F'(x) = -\sin\left(2^x + 2^{-x}\right)\left[2^x\ln 2 - 2^{-x}\ln 2\right] = \ln 2\left(2^{-x} - 2^x\right)\sin\left(2^x + 2^{-x}\right)$

29. $\displaystyle\int 3^x\, dx = \dfrac{3^x}{\ln 3} + C$

31. $\displaystyle\int \left(x^3 + 3^{-x}\right)dx = \dfrac{1}{4}x^4 - \dfrac{3^{-x}}{\ln 3} + C$

33. $\displaystyle\int \dfrac{dx}{x\ln 5} = \dfrac{1}{\ln 5}\int \dfrac{dx}{x} = \dfrac{\ln|x|}{\ln 5} + C = \log_5|x| + C$

35.
$$\int \frac{\log_2 x^3}{x} \, dx = \frac{1}{\ln 2} \int \frac{\ln x^3}{x} \, dx = \frac{3}{\ln 2} \int \frac{\ln x}{x} \, dx$$

$$= \frac{3}{\ln 2} \left[\frac{1}{2} (\ln x)^2 \right] + C = \frac{3}{\ln 4} (\ln x)^2 + C$$

37. $f'(x) = \dfrac{1}{x \ln 3}$ so $f'(e) = \dfrac{1}{e \ln 3}$

39. $f'(x) = \dfrac{1}{x \ln x}$ so $f'(e) = \dfrac{1}{e \ln e} = \dfrac{1}{e}$

41. $f(x) = p^x$

$$\ln f(x) = x \ln p$$

$$\frac{f'(x)}{f(x)} = \ln p$$

$$f'(x) = f(x) \ln p$$

$$f'(x) = p^x \ln p$$

43. $y = (x+1)^x$

$$\ln y = x \ln (x+1)$$

$$\frac{1}{y} \frac{dy}{dx} = \frac{x}{x+1} + \ln (x+1)$$

$$\frac{dy}{dx} = (x+1)^x \left[\frac{x}{x+1} + \ln (x+1) \right]$$

45. $y = (\ln x)^{\ln x}$

$$\ln y = \ln x \left[\ln (\ln x) \right]$$

$$\frac{1}{y} \frac{dy}{dx} = \ln x \left[\frac{1}{x \ln x} \right] + \frac{1}{x} \left[\ln (\ln x) \right]$$

$$\frac{dy}{dx} = (\ln x)^{\ln x} \left[\frac{1 + \ln (\ln x)}{x} \right]$$

47. $y = x^{\sin x}$

$$\ln y = (\sin x)(\ln x)$$

$$\frac{1}{y} \frac{dy}{dx} = (\cos x)(\ln x) + \sin x \left(\frac{1}{x} \right)$$

$$\frac{dy}{dx} = x^{\sin x} \left[(\cos x)(\ln x) + \frac{\sin x}{x} \right]$$

49. $y = (\sin x)^{\cos x}$

$$\ln y = (\cos x)(\ln[\sin x])$$

$$\frac{1}{y} \frac{dy}{dx} = (-\sin x)(\ln[\sin x]) + (\cos x) \left(\frac{1}{\sin x} \right) (\cos x)$$

$$\frac{dy}{dx} = (\sin x)^{\cos x} \left[\frac{\cos^2 x}{\sin x} - (\sin x)(\ln[\sin x]) \right]$$

51. $y = x^{2^x}$

$$\ln y = 2^x \ln x$$

$$\frac{1}{y} \frac{dy}{dx} = 2^x \ln 2 \ln x + 2^x \left(\frac{1}{x} \right)$$

$$\frac{dy}{dx} = x^{2^x} \left[2^x \ln 2 \ln x + \frac{2^x}{x} \right]$$

53.

55.

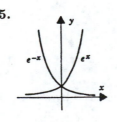

57.

59.

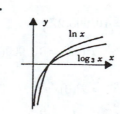

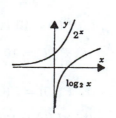

61. $\displaystyle\int_1^2 2^{-x} \, dx = \left[-\frac{2^{-x}}{\ln 2} \right]_1^2 = \frac{1}{4 \ln 2}$

62. $\displaystyle\int_0^1 4^x \, dx = \left[\frac{4^x}{\ln 4} \right]_0^1 = \frac{3}{\ln 4}$

63. $\displaystyle\int_1^4 \frac{dx}{x\ln 2} = [\log_2 x]_1^4 = \log_2 4 - 0 = 2$

65. $\displaystyle\int_0^1 x10^{1+x^2}\,dx = \left[\frac{1}{2\ln 10}10^{1+x^2}\right]_0^1 = \frac{1}{2\ln 10}(100-10) = \frac{45}{\ln 10}$

67. $\displaystyle\int_0^1 (2^x + x^2)\,dx = \left[\frac{2^x}{\ln 2} + \frac{x^3}{3}\right]_0^1 = \frac{1}{3} + \frac{1}{\ln 2}$

69. approx 16.99999; $5^{\ln 17/\ln 5} = \left(e^{\ln 5}\right)^{\ln 17/\ln 5} = e^{\ln 17} = 17$

SECTION 7.6

1. We begin with $A(t) = A_0 e^{rt}$

and take $A_0 = \$500$ and $t = 10$. The interest earned is given by

$$A(10) - A_0 = 500\left(e^{10r} - 1\right).$$

Thus, (a) $500\left(e^{0.6} - 1\right) \cong \$411.06$ (b) $500\left(e^{0.8} - 1\right) \cong \$612.77$

(c) $500\,(e - 1) \cong \$859.14.$

3. In general

$$A(t) = A_0 e^{rt}.$$

We set

$$3A_0 = A_0 e^{20r}$$

and solve for r:

$$3 = e^{20r}, \quad \ln 3 = 20r, \quad r = \frac{\ln 3}{20} \cong 5\tfrac{1}{2}\%.$$

5. Let $P = P(t)$ denote the bacteria population at time t. Then $P'(t) = kP(t)$ (k constant) $P(0) = 1000$, and $P(t) = 1000\,e^{kt}$. Since $P(1/2) = 2P(0) = P(0)e^{k/2}$, it follows that $e^{k/2} = 2$ or $k = 2\ln 2$. Now

$$P(t) = 1000e^{(2\ln 2)t} \quad \text{and} \quad P(2) = 1000\,e^{(2\ln 2)2} = 16{,}000$$

7. (a) $P(t) = 10{,}000e^{t\ln 2} = 10{,}000(2)^t$

(b) $P(26) = 10{,}000(2)^{26}, \quad P(52) = 10{,}000(2)^{52}$

9. (a) $P(10) = P(0)e^{0.035(10)t} = P(0)e^{0.35t}.$ Thus it increases by $e^{0.35}$.

(b) $2P(0) = P(0)e^{15k} \implies k = \dfrac{\ln 2}{15}.$

11. $P(40) = 206.94e^{40k} = 206.94(\dfrac{249}{227})^4 = 299.6$ million.

$$P(28) = 206.94e^{28k} = 206.94(\frac{249}{227})^{2.8} = 268.1 \text{ million.}$$

13. $4.5e^{0.0164t} = 30 \implies 0.0164t = \ln\frac{30}{4.5} \implies t \simeq 115.7 \text{ years.}$

Thus maximum population will be reached in 2095.

15.
$$V'(t) = ktV(t)$$

$$V'(t) - ktV(t) = 0$$

$$e^{-kt^2/2}V'(t) - kte^{-kt^2/2}V(t) = 0$$

$$\frac{d}{dt}\left[e^{-kt^2/2}V(t)\right] = 0$$

$$e^{-kt^2/2}V(t) = C$$

$$V(t) = Ce^{kt^2/2}.$$

Since $V(0) = C = 200$, $V(t) = 200e^{kt^2/2}$.

Since $V(5) = 160$, $200e^{k(25/2)} = 160$, $e^{k(25/2)} = \frac{4}{5}$, $e^k = \left(\frac{4}{5}\right)^{2/25}$

and therefore

$$V(t) = 200\left(\tfrac{4}{5}\right)^{t^2/25} \text{ liters.}$$

17. Take two years ago as time $t = 0$. In general

(*) $$A(t) = A_0 e^{kt}.$$

We are given that

$$A_0 = 5 \quad \text{and} \quad A(2) = 4.$$

Thus,

$$4 = 5e^{2k} \quad \text{so that} \quad \tfrac{4}{5} = e^{2k} \quad \text{or} \quad e^k = \left(\tfrac{4}{5}\right)^{1/2}.$$

We can write

$$A(t) = 5\left(\tfrac{4}{5}\right)^{t/2}$$

and compute $A(5)$ as follows:

$$A(5) = 5\left(\tfrac{4}{5}\right)^{5/2} = 5e^{\frac{5}{2}\ln(4/5)} \cong 5e^{-0.56} \cong 2.86.$$

About 2.86 gm will remain 3 years from now.

19. A fundamental property of radioactive decay is that the percentage of substance that decays during any year is constant:

$$100\left[\frac{A(t) - A(t+1)}{A(t)}\right] = 100\left[\frac{A_0 e^{kt} - A_0 e^{k(t+1)}}{A_0 e^{kt}}\right] = 100(1 - e^k)$$

If the half-life is n years, then

$$\tfrac{1}{2}A_0 = A_0 e^{kn} \quad \text{so that} \quad e^k = \left(\tfrac{1}{2}\right)^{1/n}.$$

Thus, $100\left[1 - \left(\tfrac{1}{2}\right)^{1/n}\right]$ % of the material decays during any one year.

21.   (a) $A(1620) = A_0 e^{1620k} = \dfrac{1}{2}A_0 \implies k = \dfrac{\ln\dfrac{1}{2}}{1620} \simeq -0.00043.$

Thus $A(500) = A_0 e^{500k} = 0.807 A_0.$ Hence 80.7% will remain.

(b) $0.25 A_0 = A_0 e^{kt} \implies t = 3240$ years.

23.   (a) $x_1(t) = 10^6 t, \quad x_2(t) = e^t - 1$

(b) $\dfrac{d}{dt}[x_1(t) - x_2(t)] = \dfrac{d}{dt}[10^6 t - (e^t - 1)] = 10^6 - e^t$

This derivative is zero at $t = 6\ln 10 \cong 13.8$. After that the derivative is negative.

(c) $x_2(15) < e^{15} = \left(e^3\right)^5 \cong 20^5 = 2^5\left(10^5\right) = 3.2\left(10^6\right) < 15\left(10^6\right) = x_1(15)$

$x_2(18) = e^{18} - 1 = \left(e^3\right)^6 - 1 \cong 20^6 - 1 = 64\left(10^6\right) - 1 > 18\left(10^6\right) = x_1(18)$

$x_2(18) - x_1(18) \cong 64\left(10^6\right) - 1 - 18\left(10^6\right) \cong 46\left(10^6\right)$

(d) If by time t_1 EXP has passed LIN, then $t_1 > 6\ln 10$. For all $t \geq t_1$ the speed of EXP is greater than the speed of LIN:

$$\text{for} \quad t \geq t_1 > 6\ln 10, \quad v_2(t) = e^t > 10^6 = v_1(t).$$

25.   Let $p(h)$ denote the pressure at altitude h. The equation $\dfrac{dp}{dh} = kp$ gives

$$(*) \qquad\qquad p(h) = p_0 e^{kh}$$

where p_0 is the pressure at altitude zero (sea level).

Since $p_0 = 15$ and $p(10000) = 10$,

$$10 = 15 e^{10000k}, \quad \tfrac{2}{3} = e^{10000k}, \quad \tfrac{1}{10000}\ln\tfrac{2}{3} = k.$$

Thus, $(*)$ can be written

$$p(h) = 15\left(\tfrac{2}{3}\right)^{h/10000}.$$

(a) $p(5000) = 15\left(\tfrac{2}{3}\right)^{1/2} \cong 12.25$ lb/in.^{2}.

(b) $p(15000) = 15\left(\tfrac{2}{3}\right)^{3/2} \cong 8.16$ lb/in.^{2}.

27.   From Exercise 26, we have $6000 = 10,000 e^{-8r}$. Thus

$$e^{-8r} = \dfrac{6000}{10,000} = \dfrac{3}{5} \implies -8r = \ln(3/5) \quad \text{and} \quad r \cong 0.064 \text{ or } r = 6.4\%$$

29.   The future value of $\$16,000$ at an interest rate r, t years from now is given by $Q(t) = 16,000\, e^{rt}$.

Thus

(a) For $r = 0.05$: $P(3) = 16,000\, e^{(0.05)3} \cong 18,589.35$ or $\$18,589.35$.

(b) For $r = 0.08$: $P(3) = 16,000\, e^{(0.08)3} \cong 20,339.99$ or $\$20,339.99$.

(c) For $r = 0.12$: $P(3) = 16,000\, e^{(0.12)3} \cong 22,933.27$ or $\$22,933.27$.

31. By Exercise 30

$(*)$ $v\,(t) = Ce^{-kt}, \quad t$ in seconds.

We use the initial conditions

$$v\,(0) = C = 4 \text{ mph } = \tfrac{1}{900} \text{ mi/sec} \quad \text{and} \quad v\,(60) = 2 = \tfrac{1}{1800} \text{ mi/sec}$$

to determine e^{-k}:

$$\tfrac{1}{1800} = \tfrac{1}{900}e^{-60k}, \qquad e^{60k} = 2, \qquad e^k = 2^{1/60}.$$

Thus, $(*)$ can be written

$$v\,(t) = \tfrac{1}{900}\, 2^{-t/60}.$$

The distance traveled by the boat is

$$s = \int_0^{60} \frac{1}{900} 2^{-t/60}\, dt = \frac{1}{900} \left[\frac{-60}{\ln 2} 2^{-t/60} \right]_0^{60} = \frac{1}{30 \ln 2} \text{ mi} = \frac{176}{\ln 2} \text{ ft} \quad \text{(about 254 ft).}$$

33. Let $A(t)$ denote the amount of ^{14}C remaining t years after the organism dies. Then $A(t) = A(0)e^{kt}$ for some constant k. Since the half-life of ^14C is 5700 years, we have

$$\frac{1}{2} = e^{5700k} \quad \Rightarrow k = -\frac{\ln 2}{5700} \cong 0.000122 \text{ and } A(t) = A(0)e^{-0.000122t}$$

If 25% of the original amount of ^{14}C remains after t years, then

$$0.25A(0) = A(0)e^{-0.000122t} \quad \Rightarrow \quad t = \frac{\ln 0.25}{-0.000122} \cong 11,400 \text{ (years)}$$

SECTION 7.7

1. 0   **3.** $\pi/3$ **5.** $2\pi/3$ **7.** $-\pi/4$ **9.** $-2/\sqrt{3}$

11. $1/2$ **13.** 1.1630   **15.** -0.4580 **17.** 1.2002

19. $\dfrac{dy}{dx} = \dfrac{1}{1 + (x+1)^2} = \dfrac{1}{x^2 + 2x + 2}$ **21.** $f'(x) = \dfrac{1}{|2x^2|\sqrt{(2x^2)^2 - 1}} \dfrac{d}{dx}\left(2x^2\right) = \dfrac{2x}{x\sqrt{4x^4 - 1}}$

23. $f'(x) = \sin^{-1} 2x + x\dfrac{1}{\sqrt{1 - (2x)^2}} \dfrac{d}{dx}(2x) = \sin^{-1} 2x + \dfrac{2x}{\sqrt{1 - 4x^2}}$

25. $\dfrac{du}{dx} = 2\left(\sin^{-1} x\right) \dfrac{d}{dx}\left(\sin^{-1} x\right) = \dfrac{2\sin^{-1} x}{\sqrt{1-x^2}}$

27. $\dfrac{dy}{dx} = \dfrac{x\left(\dfrac{1}{1+x^2}\right) - (1)\tan^{-1} x}{x^2} = \dfrac{x - \left(1+x^2\right)\tan^{-1} x}{x^2\left(1+x^2\right)}$

29. $f'(x) = \dfrac{1}{2}\left(\tan^{-1} 2x\right)^{-1/2}\dfrac{d}{dx}\left(\tan^{-1} 2x\right) = \dfrac{1}{2}\left(\tan^{-1} 2x\right)^{-1/2}\dfrac{2}{1+(2x)^2} = \dfrac{1}{\left(1+4x^2\right)\sqrt{\tan^{-1} 2x}}$

31. $\dfrac{dy}{dx} = \dfrac{1}{1+(\ln x)^2}\dfrac{d}{dx}(\ln x) = \dfrac{1}{x\left[1+(\ln x)^2\right]}$

33. $\dfrac{d\theta}{dr} = \dfrac{1}{\sqrt{1-\left(\sqrt{1-r^2}\,\right)^2}}\dfrac{d}{dr}\left(\sqrt{1-r^2}\,\right) = \dfrac{1}{\sqrt{r^2}}\cdot\dfrac{-r}{\sqrt{1-r^2}} = -\dfrac{r}{|r|\sqrt{1-r^2}}$

35. $g'(x) = 2x\sec^{-1}\left(\dfrac{1}{x}\right) + x^2\cdot\dfrac{1}{\left|\dfrac{1}{x}\right|\sqrt{\dfrac{1}{x^2}-1}}\cdot\left(-\dfrac{1}{x^2}\right) = 2x\sec^{-1}\left(\dfrac{1}{x}\right) - \dfrac{x^2}{\sqrt{1-x^2}}$

37. $\dfrac{dy}{dx} = \cos\left[\sec^{-1}(\ln x)\right]\cdot\dfrac{1}{|\ln x|\sqrt{(\ln x)^2 - 1}}\cdot\dfrac{1}{x} = \dfrac{\cos\left[\sec^{-1}(\ln x)\right]}{x\,|\ln x|\sqrt{(\ln x)^2 - 1}}$

39. $f'(x) = \dfrac{-x}{\sqrt{c^2-x^2}} + \dfrac{c}{\sqrt{1-(x/c)^2}}\cdot\left(\dfrac{1}{c}\right) = \dfrac{c-x}{\sqrt{c^2-x^2}} = \sqrt{\dfrac{c-x}{c+x}}$

41. $\dfrac{dy}{dx} = \dfrac{\sqrt{c^2-x^2}\,(1) - x\left(\dfrac{-x}{\sqrt{c^2-x^2}}\right)}{\left(\sqrt{c^2-x^2}\,\right)^2} - \dfrac{1}{\sqrt{1-(x/c)^2}}\left(\dfrac{1}{c}\right)$

$= \dfrac{c^2}{\left(c^2-x^2\right)^{3/2}} - \dfrac{1}{\left(c^2-x^2\right)^{1/2}} = \dfrac{x^2}{\left(c^2-x^2\right)^{3/2}}$

43. $\left\{\begin{array}{l} au = x+b \\ a\,du = dx \end{array}\right\}$; $\quad \displaystyle\int \dfrac{dx}{\sqrt{a^2-(x+b)^2}} = \int\dfrac{a\,du}{\sqrt{a^2-a^2u^2}} = \int\dfrac{du}{\sqrt{1-u^2}}$

$\qquad\qquad = \sin^{-1} u + C = \sin^{-1}\left(\dfrac{x+b}{a}\right) + C$

45. $\left\{\begin{array}{l} au = x+b \\ a\,du = dx \end{array}\right\}$; $\quad \displaystyle\int\dfrac{dx}{(x+b)\sqrt{(x+b)^2-a^2}} = \int\dfrac{a\,du}{au\sqrt{a^2u^2-a^2}} = \dfrac{1}{a}\int\dfrac{du}{u\sqrt{u^2-1}}$

$\qquad\qquad = \dfrac{1}{a}\sec^{-1}|u| + C = \dfrac{1}{a}\sec^{-1}\left(\dfrac{|x+b|}{a}\right) + C$

47. $\text{dom}\,(f) = (-\infty, \infty), \quad \text{range}\,(f) = (0, \pi)$

$$y = \cot^{-1} x$$

$$\cot y = x$$

$$-\csc^2 y \, \frac{dy}{dx} = 1$$

$$\frac{dy}{dx} = -\frac{1}{\csc^2 x} = -\frac{1}{1+x^2}$$

49. $\displaystyle\int_0^1 \frac{dx}{1+x^2} = \left[\tan^{-1} x\right]_0^1 = \frac{\pi}{4}$

51. $\displaystyle\int_0^{1/\sqrt{2}} \frac{dx}{\sqrt{1-x^2}} = \left[\sin^{-1} x\right]_0^{1/\sqrt{2}} = \frac{\pi}{4}$

53. $\displaystyle\int_0^5 \frac{dx}{25+x^2} = \left[\frac{1}{5}\tan^{-1}\frac{x}{5}\right]_0^5 = \frac{\pi}{20}$

$$(7.7.8)$$

55. $\left\{\begin{array}{ll} 3u = 2x & x = 0 \implies u = 0 \\ 3\,du = 2\,dx & x = 3/2 \implies u = 1 \end{array}\right\}$; $\displaystyle\int_0^{3/2} \frac{dx}{9+4x^2} = \frac{1}{6}\int_0^1 \frac{du}{1+u^2} = \frac{1}{6}\left[\tan^{-1} u\right]_0^1 = \frac{\pi}{24}$

57. $\left\{\begin{array}{ll} u = 4x & x = 3/4 \implies u = 3 \\ du = 4\,dx & x = 3 \implies u = 12 \end{array}\right\}$;

$$\int_{3/4}^3 \frac{dx}{x\sqrt{16x^2-9}} = \int_3^{12} \frac{du/4}{(u/4)\sqrt{u^2-9}} = \frac{1}{3}\left[\sec^{-1}\left(\frac{|u|}{3}\right)\right]_{3/4}^3 = \frac{1}{3}\sec^{-1} 4 \cong 0.4391$$

59. $\displaystyle\int_{-3}^{-2} \frac{dx}{\sqrt{4-(x+3)^2}} = \left[\sin^{-1}\left(\frac{x+3}{2}\right)\right]_{-3}^{-2} = \frac{\pi}{6}$

$$(7.7.13)$$

61. $\left\{\begin{array}{ll} u = e^x & x = 0 \implies u = 1 \\ du = e^x\,dx & x = \ln 2 \implies u = 2 \end{array}\right\}$;

$$\int_0^{\ln 2} \frac{e^x}{1+e^{2x}}\,dx = \int_1^2 \frac{du}{1+u^2} = \left[\tan^{-1} u\right]_1^2 = \tan^{-1} 2 - \frac{\pi}{4} \cong 0.322$$

63. $\left\{\begin{array}{l} u = x^2 \\ du = 2x\,dx \end{array}\right\}$; $\displaystyle\int \frac{x}{\sqrt{1-x^4}}\,dx = \frac{1}{2}\int \frac{du}{\sqrt{1-u^2}} = \frac{1}{2}\sin^{-1} u + C = \frac{1}{2}\sin^{-1} x^2 + C$

65. $\left\{\begin{array}{l} u = x^2 \\ du = 2x\,dx \end{array}\right\}$; $\displaystyle\int \frac{x}{1+x^4}\,dx = \frac{1}{2}\int \frac{du}{1+u^2} = \frac{1}{2}\tan^{-1} u + C = \frac{1}{2}\tan^{-1} x^2 + C$

67. $\left\{\begin{array}{l} u = \tan x \\ du = \sec^2 x\,dx \end{array}\right\}$; $\displaystyle\int \frac{\sec^2 x}{9+\tan^2 x}\,dx = \int \frac{du}{9+u^2} = \frac{1}{3}\tan^{-1}\left(\frac{u}{3}\right) + C = \frac{1}{3}\tan^{-1}\left(\frac{\tan x}{3}\right) + C$

69. $\left\{ \begin{array}{l} u = \sin 6 - 1x \\ du = \dfrac{1}{\sqrt{1 - x^2}}\, dx \end{array} \right\}$; $\displaystyle\int \dfrac{\sin^{-1} x}{\sqrt{1 - x^2}}\, dx = \int u\, du = \dfrac{1}{2} u^2 + C = \dfrac{1}{2}\left(\sin^{-1} x\right)^2 + C$

71. $\left\{ \begin{array}{l} u = \ln x \\ du = \dfrac{1}{x}\, dx \end{array} \right\}$; $\displaystyle\int \dfrac{dx}{x\sqrt{1 - (\ln x)^2}} = \int \dfrac{du}{\sqrt{1 - u^2}} = \sin^{-1} u + C = \sin^{-1}(\ln x) + C$

73. $A = \displaystyle\int_{-1}^{1} \dfrac{1}{\sqrt{4 - x^2}}\, dx = 2\int_{0}^{1} \dfrac{1}{\sqrt{4 - x^2}}\, dx$

$= 2\left[\sin^{-1}\left(\dfrac{x}{2}\right)\right]_{0}^{1} = \dfrac{\pi}{3}$

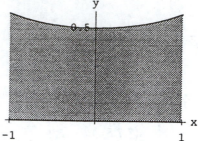

75. $\dfrac{8}{x^2 + 4} = \dfrac{1}{4} x^2 \;\Rightarrow\; x = \pm 2$

$A = \displaystyle\int_{-2}^{2} \left(\dfrac{8}{x^2 + 4} - \dfrac{1}{4} x^2\right) dx = 2\int_{0}^{2} \left(\dfrac{8}{x^2 + 4} - \dfrac{1}{4} x^2\right) dx$

$= 2\left[8 \cdot \dfrac{1}{2} \tan^{-1}\left(\dfrac{x}{2}\right) - \dfrac{1}{12} x^3\right]_{0}^{2} = 2\pi - \dfrac{4}{3}$

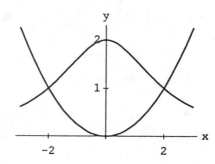

77. Let x be the distance between the motorist and the point on the road where the line determined by the sign intersects the road. Then, from the given figure,

$$\theta = \tan^{-1}\left(\dfrac{s + k}{x}\right) - \tan^{-1}\dfrac{s}{x}, \quad 0 < x < \infty$$

and

$$\dfrac{d\theta}{dx} = \dfrac{1}{1 + \dfrac{(s + k)^2}{x^2}}\left(-\dfrac{s + k}{x^2}\right) - \dfrac{1}{1 + \dfrac{s^2}{x^2}}\left(-\dfrac{s}{x^2}\right)$$

$$= \dfrac{-(s + k)}{x^2 + (s + k)^2} + \dfrac{s}{x^2 + s^2} = \dfrac{s^2 k + sk^2 - kx^2}{\left[x^2 + (s + k)^2\right]\left[x^2 + s^2\right]}$$

Setting $d\theta/dx = 0$ we get $x = \sqrt{s^2 + sk}$. Since θ is essentially 0 when x is close to 0 and when x is "large," we can conclude that θ is a maximum when $x = \sqrt{s^2 + sk}$.

79. (a) See graph on next page

(b) $\displaystyle\lim_{x \to (1/2)^+} f(x) = -\dfrac{\pi}{2}$; $\displaystyle\lim_{x \to (1/2)^-} f(x) = \dfrac{\pi}{2}$

(c) $f'(x) = \dfrac{1}{1 + \left(\dfrac{2 + x}{1 - 2x}\right)^2} \cdot \dfrac{(1 - 2x) - (2 + x)2}{(1 - 2x)^2}$

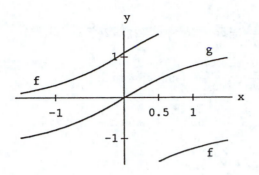

$$= \frac{(1-2x)^2}{(1-2x)^2 + (2+x)^2} \frac{5}{(1-2x)^2} = \frac{1}{1+x^2}$$

(d) This is clear from the graphs in part (a).

(e) Evaluating at $x = 0$ gives $C_1 = \tan^{-1} 2$;

   evaluating at $x = 1$ gives $C_2 = \tan^{-1} 3 - \pi/4$

81. (a)

(b) $\lim\limits_{x \to 0^+} f(x) = \frac{\pi}{2}$; $\lim\limits_{x \to 0^-} f(x) = -\frac{\pi}{2}$

(c) $f'(x) = \dfrac{1}{1 + \left(\dfrac{1}{x}\right)^2} \dfrac{-1}{x^2} = \dfrac{-1}{1+x^2}$

(d) This is clear from the graphs in part (a).

(e) Evaluating at $x = 1$ gives $C_1 = \pi/2$;

   evaluating at $x = -1$ gives $C_2 = -\pi/2$

83. $I = \displaystyle\int_0^{0.5} \frac{1}{\sqrt{1-x^2}}\, dx \cong \frac{1}{10}\left[f(0.05) + f(0.15) + f(0.25) + f(0.35) + f(0.45)\right] \cong 0.523;$

and $\sin(0.523) \cong 0.499.$ Explanation: $I = \sin^{-1}(0.5)$ and $\sin\left[\sin^{-1}(0.5)\right] = 0.5.$

PROJECT 7.7

1. (a) $n_1 \sin\theta_1 = n \sin\theta = n_2 \sin\theta_2$

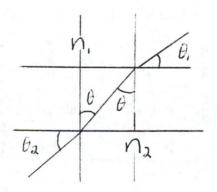

SECTION 7.8

1. $\dfrac{dy}{dx} = \cosh x^2 \dfrac{d}{dx}\left(x^2\right) = 2x \cosh x^2$

3. $\dfrac{dy}{dx} = \dfrac{1}{2}\left(\cosh ax\right)^{-1/2}\left(a \sinh ax\right) = \dfrac{a \sinh ax}{2\sqrt{\cosh ax}}$

5. $\dfrac{dy}{dx} = \dfrac{(\cosh x - 1)(\cosh x) - \sinh x\,(\sinh x)}{(\cosh x - 1)^2} = \dfrac{1}{1 - \cosh x}$

7. $\dfrac{dy}{dx} = ab \cosh bx - ab \sinh ax = ab\,(\cosh bx - \sinh ax)$

9. $\dfrac{dy}{dx} = \dfrac{1}{\sinh ax}(a \cosh ax) = \dfrac{a \cosh ax}{\sinh ax}$

11. $\dfrac{dy}{dx} = \cosh\left(e^{2x}\right)e^{2x}(2) = 2e^{2x}\cosh\left(e^{2x}\right)$

13. $\dfrac{dy}{dx} = -e^{-x}\cosh 2x + 2e^{-x}\sinh 2x$

15. $\dfrac{dy}{dx} = \dfrac{1}{\cosh x}(\sinh x) = \tanh x$

17. $\ln y = x \ln \sinh x;$ $\dfrac{1}{y}\dfrac{dy}{dx} = \ln \sinh x + x \dfrac{\cosh x}{\sinh x}$ and $\dfrac{dy}{dx} = (\sinh x)^x\left[\ln \sinh x + x \coth x\right]$

19.
$$\cosh^2 t - \sinh^2 t = \left(\frac{e^t + e^{-t}}{2}\right)^2 - \left(\frac{e^t + e^{-t}}{2}\right)^2$$
$$= \frac{1}{4}\left\{\left(e^{2t} + 2 + e^{-2t}\right) - \left(e^{2t} - 2 + e^{-2t}\right)\right\} = \frac{4}{4} = 1$$

21. $\cosh t \cosh s + \sinh t \sinh s$
$$= \left(\frac{e^t + e^{-t}}{2}\right)\left(\frac{e^s + e^{-s}}{2}\right) + \left(\frac{e^t - e^{-t}}{2}\right)\left(\frac{e^s - e^{-s}}{2}\right)$$
$$= \frac{1}{4}\left\{\left(e^{t+s} + e^{t-s} + e^{s-t} + e^{-t-s}\right) + \left(e^{t+s} - e^{t-s} - e^{s-t} + e^{-s-t}\right)\right\}$$
$$= \frac{1}{4}\left\{2e^{t+s} + 2e^{-(t+s)}\right\} = \frac{e^{t+s} + e^{-(t+s)}}{2} = \cosh(t+s)$$

23. Set $s = t$ in $\cosh(t+s) = \cosh t \cosh s + \sinh t \sinh s$ to get $\cosh(2t) = \cosh^2 t + \sinh^2 t$.

Then use Exercise 19 to obtain the other two identities.

25. $\sinh(-t) = \dfrac{e^{(-t)} - e^{-(-t)}}{2} = -\dfrac{e^t - e^{-t}}{2} = -\sinh t$

27.
$$y = -5\cosh x + 4\sinh x = -\frac{5}{2}\left(e^x + e^{-x}\right) + \frac{4}{2}\left(e^x - e^{-x}\right) = -\frac{1}{2}e^x - \frac{9}{2}e^{-x}$$

$$\frac{dy}{dx} = -\frac{1}{2}e^x + \frac{9}{2}e^{-x} = \frac{e^{-x}}{2}\left(9 - e^{2x}\right); \quad \frac{dy}{dx} = 0 \implies e^x = 3 \text{ or } x = \ln 3$$

$$\frac{d^2y}{dx^2} = -\frac{1}{2}e^x - \frac{9}{2}e^{-x} < 0 \quad \text{all} \quad x \quad \text{so abs max occurs when} \quad x = \ln 3.$$

The abs max is $y = -\frac{1}{2}e^{\ln 3} - \frac{9}{2}e^{-\ln 3} = -\frac{1}{2}(3) - \frac{9}{2}\left(\frac{1}{3}\right) = -3.$

29.
$$[\cosh x + \sinh x]^n = \left[\frac{e^x + e^{-x}}{2} + \frac{e^x - e^{-x}}{2}\right]^n$$
$$= [e^x]^n = e^{nx} = \frac{e^{nx} + e^{-nx}}{2} + \frac{e^{nx} - e^{-nx}}{2} = \cosh nx + \sinh nx$$

31.
$$y = A\cosh cx + B\sinh cx; \qquad\qquad y(0) = 2 \implies 2 = A.$$
$$y' = Ac\sinh cx + Bc\cosh cx; \qquad\quad y'(0) = 1 \implies 1 = Bc.$$
$$y'' = Ac^2\cosh cx + Bc^2\sinh cx = c^2 y; \quad y'' - 9y = 0 \implies (c^2 - 9)y = 0.$$

Thus, $c = 3$, $B = \frac{1}{3}$,   and $A = 2$.

33. $\frac{1}{a}\sinh ax + C$ **35.** $\frac{1}{3a}\sinh^3 ax + C$ **37.** $\frac{1}{a}\ln(\cosh ax) + C$ **39.** $-\dfrac{1}{a\cosh ax} + C$

41.   From the identity $\cosh 2t = 2\cosh^2 t - 1$ (Exercise 23), we get

$\cosh^2 t = \frac{1}{2}(1 + \cosh 2t)$. Thus,

$$\int \cosh^2 x \, dx = \frac{1}{2}\int (1 + \cosh 2x)\, dx$$
$$= \frac{1}{2}\left(x + \frac{1}{2}\sinh 2x\right) + C$$
$$= \frac{1}{2}(x + \sinh x \cosh x) + C$$

43. $\left\{\begin{array}{l} u = \sqrt{x} \\ du = dx/2\sqrt{x} \end{array}\right\}$; $\displaystyle\int \frac{\sinh \sqrt{x}}{\sqrt{x}}\, dx = 2\int \sinh u \, du = 2\cosh u + C = 2\cosh\sqrt{x} + C$

45. $f_{\text{avg}} = \dfrac{1}{1-(-1)}\displaystyle\int_{-1}^{1} \cosh x \, dx = \frac{1}{2}[\sinh x]_{-1}^{1} = \dfrac{e^2 - 1}{2e} \cong 1.175$

47. $V = \displaystyle\int_0^1 \pi\left(\cosh^2 x - \sinh^2 x\right) dx = \int_0^1 \pi \, dx = \pi$

SECTION 7.9

1. $\dfrac{dy}{dx} = 2\tanh x \,\text{sech}^2 x$ **3.** $\dfrac{dy}{dx} = \dfrac{1}{\tanh x}\,\text{sech}^2 x = \text{sech}\, x \,\text{csch}\, x$

5. $\dfrac{dy}{dx} = \cosh\left(\tan^{-1} e^{2x}\right)\dfrac{d}{dx}\left(\tan^{-1} e^{2x}\right) = \dfrac{2e^{2x}\cosh\left(\tan^{-1} e^{2x}\right)}{1 + e^{4x}}$

7. $\dfrac{dy}{dx} = -\text{csch}^2\left(\sqrt{x^2+1}\right)\dfrac{d}{dx}\left(\sqrt{x^2+1}\right) = -\dfrac{x}{\sqrt{x^2+1}}\text{csch}^2\left(\sqrt{x^2+1}\right)$

9. $\dfrac{dy}{dx} = \dfrac{(1+\cosh x)(-\text{sech}\, x \tanh x) - \text{sech}\, x\,(\sinh x)}{(1+\cosh x)^2}$

$= \dfrac{-\text{sech}\, x\,(\tanh x + \cosh x \tanh x + \sinh x)}{(1+\cosh x)^2} = \dfrac{-\text{sech}\, x\,(\tanh x + 2\sinh x)}{(1+\cosh x)^2}$

11. $\dfrac{d}{dx}(\coth x) = \dfrac{d}{dx}\left[\dfrac{\cosh x}{\sinh x}\right] = \dfrac{\sinh x\,(\sinh x) - \cosh x\,(\cosh x)}{\sinh^2 x}$

$= -\dfrac{\cosh^2 x - \sinh^2 x}{\sinh^2 x} = \dfrac{-1}{\sinh^2 x} = -\text{csch}^2 x$

13. $\dfrac{d}{dx}\left(\operatorname{csch} x\right) = \dfrac{d}{dx}\left[\dfrac{1}{\sinh x}\right] = -\dfrac{\cosh x}{\sinh^2 x} = -\operatorname{csch} x \coth x$

15. (a) By the hint $\operatorname{sech}^2 x_0 = \dfrac{9}{25}$. Take $\operatorname{sech} x_0 = \dfrac{3}{5}$ since $\operatorname{sech} x = \dfrac{1}{\cosh x} > 0$ for all x.

(b) $\cosh x_0 = \dfrac{1}{\operatorname{sech} x_0} = \dfrac{5}{3}$ (c) $\sinh x_0 = \cosh x_0 \tanh x_0 = \left(\dfrac{5}{3}\right)\left(\dfrac{4}{5}\right) = \dfrac{4}{3}$

(d) $\coth x_0 = \dfrac{\cosh x_0}{\sinh x_0} = \dfrac{5/3}{4/3} = \dfrac{5}{4}$ (e) $\operatorname{csch} x_0 = \dfrac{1}{\sinh x_0} = \dfrac{3}{4}$

17. If $x \le 0$, the result is obvious. Suppose then that $x > 0$. Since $x^2 \ge 1$, we have $x \ge 1$. Consequently

$$x - 1 = \sqrt{x-1}\,\sqrt{x-1} \le \sqrt{x-1}\,\sqrt{x+1} = \sqrt{x^2-1}$$

and therefore $\qquad x - \sqrt{x^2 - 1} \le 1.$

19. By Theorem 7.9.2,

$$\frac{d}{dx}\left(\sinh^{-1} x\right) = \frac{d}{dx}\left[\ln\left(x + \sqrt{x^2 + 1}\,\right)\right] = \frac{1}{x + \sqrt{x^2 + 1}}\left(1 + \frac{x}{\sqrt{x^2 + 1}}\right) = \frac{1}{\sqrt{x^2 + 1}}.$$

21. By Theorem 7.9.2

$$\frac{d}{dx}\left(\tan^{-1} x\right) = \frac{d}{dx}\left[\frac{1}{2}\ln\left(\frac{1+x}{1-x}\right)\right] = \frac{1}{2}\,\frac{1}{\left(\dfrac{1+x}{1-x}\right)}\left(\frac{(1-x)(1) - (1+x)(-1)}{(1-x)^2}\right)$$

$$= \frac{1}{\left(\dfrac{1+x}{1-x}\right)(1-x)^2} = \frac{1}{1 - x^2}.$$

23. Let $y = \operatorname{csch}^{-1} x$. Then $\operatorname{csch} y = x$ and $\sinh y = \dfrac{1}{x}$.

$$\sinh y = \frac{1}{x}$$

$$\cosh y\,\frac{dy}{dx} = -\frac{1}{x^2}$$

$$\frac{dy}{dx} = -\frac{1}{x^2 \cosh y} = -\frac{1}{x^2\sqrt{1 + (1/x)^2}} = -\frac{1}{|x|\sqrt{1 + x^2}}$$

25. (a) $\dfrac{dy}{dx} = -\operatorname{sech} x \tanh x = -\dfrac{\sinh x}{\cosh^2 x}$

$\dfrac{dy}{dx} = 0$ at $x = 0$; $\dfrac{dy}{dx} > 0$ if $x < 0$; $\dfrac{dy}{dx} < 0$ if $x > 0$

f is increasing on $(-\infty, 0]$ and decreasing on $[0, \infty)$; $f(0) = 1$ is the absolute maximum of f.

(b) $\dfrac{d^2y}{dx^2} = -\dfrac{\cosh^2 x - 2\sinh^2 x}{\cosh^3 x} = \dfrac{\sinh^2 x - 1}{\cosh^3 x}$

$\dfrac{d^2y}{dx^2} = 0 \;\Rightarrow\; \sinh x = \pm 1$

$$\sinh x = 1 \Rightarrow \frac{e^x - e^{-x}}{2} = 1 \Rightarrow e^{2x} - 2e^x - 1 = 0 \Rightarrow x = \ln\left(1 + \sqrt{2}\right) \cong 0.881;$$

$$\sinh x = -1 \Rightarrow \frac{e^x - e^{-x}}{2} = -1 \Rightarrow e^{2x} + 2e^x - 1 = 0 \Rightarrow x = -\ln\left(1 + \sqrt{2}\right) = -0.881$$

(c) The graph of f is concave up on $(-\infty, -0.881) \cup (0.881, \infty)$ and concave down on $(-0.881, 0.881)$; points of inflection at $x = \pm 0.881$

(d)

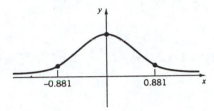

27. $y = \sinh x; \quad \dfrac{dy}{dx} = \cosh x; \quad \dfrac{d^2y}{dx^2} = \sinh x.$

$\dfrac{d^2y}{dx^2} = 0 \Rightarrow \sinh x = 0 \Rightarrow x = 0.$

$y = \sinh^{-1} x = \ln\left(x + \sqrt{x^2 + 1}\right); \quad \dfrac{dy}{dx} = \dfrac{1}{\sqrt{x^2 + 1}}; \quad \dfrac{d^2y}{dx^2} = -\dfrac{x}{(x^2 + 1)^{3/2}}.$

$\dfrac{d^2y}{dx^2} = 0 \Rightarrow -\dfrac{x}{(x^2 + 1)^{3/2}} = 0 \Rightarrow x = 0.$

It is easy to verify that $(0, 0)$ is a point of inflection for both graphs.

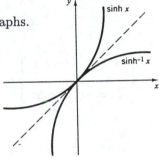

29.   (a) $\tan\phi = \sinh x$

$\phi = \tan^{-1}(\sinh x)$

$\dfrac{d\phi}{dx} = \dfrac{\cosh x}{1 + \sinh^2 x}$

$= \dfrac{\cosh x}{\cosh^2 x} = \dfrac{1}{\cosh x} = \operatorname{sech} x$

(b) $\sinh x = \tan\phi$

$x = \sinh^{-1}(\tan\phi)$

$= \ln\left(\tan\phi + \sqrt{\tan^2\phi + 1}\right)$

$= \ln(\tan\phi + \sec\phi)$

$= \ln(\sec\phi + \tan\phi)$

(c) $x = \ln(\sec\phi + \tan\phi)$

$\dfrac{dx}{d\phi} = \dfrac{\sec\phi\tan\phi + \sec^2\phi}{\tan\phi + \sec\phi} = \sec\phi$

31. $\displaystyle\int \tanh x \, dx = \int \dfrac{\sinh x}{\cosh x}\, dx$

$$\left\{ \begin{array}{l} u = \cosh x \\[2mm] du = \sinh x\, dx \end{array} \right\} \; ; \quad \int \frac{\sinh x}{\cosh x}\, dx = \int \frac{1}{u}\, du = \ln |u| + C = \ln \cosh x + C$$

33. $\displaystyle \int \operatorname{sech} x\, dx = \int \frac{1}{\cosh x}\, dx = \int \frac{2}{e^x + e^{-x}}\, dx = \int \frac{2e^x}{e^{2x}+1}\, dx$

$$\left\{ \begin{array}{l} u = e^x \\[2mm] du = e^x\, dx \end{array} \right\} \; ; \quad \int \frac{2e^x}{e^{2x}+1}\, dx = 2 \int \frac{1}{u^2+1}\, du = 2\tan^{-1} u + C = 2\tan^{-1}(e^x) + C$$

35. $\left\{ \begin{array}{l} u = \operatorname{sech} x \\[2mm] du = - \operatorname{sech} s \tanh x\, dx \end{array} \right\} \; ; \quad \int \operatorname{sech}^3 x \tanh x\, dx = - \int u^2\, du = -\frac{1}{3} u^3 + C$

$$= -\frac{1}{3} \operatorname{sech}^3 x + C$$

37. $\left\{ \begin{array}{l} u = \ln(\cosh x) \\[2mm] du = \tanh x\, dx \end{array} \right\} \; ; \quad \int \tanh x \ln(\cosh x)\, dx = \int u\, du = \frac{1}{2} u^2 + C$

$$= \frac{1}{2} \left[\ln(\cosh x) \right]^2 + C$$

39. $\left\{ \begin{array}{l} u = 1 + \tanh x \\[2mm] du = \operatorname{sech}^2 x\, dx \end{array} \right\} \; ; \quad \int \frac{\operatorname{sech}^2 x}{1 + \tanh x}\, dx = \int \frac{1}{u}\, du = \ln |u| + C$

$$= \ln |1 + \tanh x| + C$$

41. $\left\{ \begin{array}{l} x = a \sinh u \\[2mm] dx = a \cosh u\, du \end{array} \right\} \; ; \quad \int \frac{dx}{\sqrt{a^2 + x^2}}\, dx = \int \frac{a \cosh u}{\sqrt{a^2 + a^2 \sinh^2 u}}\, du$

$$= \int du = u + C = \sinh^{-1}\left(\frac{x}{a}\right) + C$$

43. Suppose $|x| < a$.

$$\left\{ \begin{array}{l} x = a \tanh u \\[2mm] dx = a \operatorname{sech}^2 u\, du \end{array} \right\} \; ; \quad \int \frac{dx}{a^2 - x^2}\, dx = \int \frac{a \operatorname{sech}^2 u}{a^2 - a^2 \tanh^2 u}\, du$$

$$= \frac{1}{a} \int du = \frac{u}{a} + C = \frac{1}{a} \tanh^{-1}\left(\frac{x}{a}\right) + C$$

The other case is done in the same way.

CHAPTER 8

SECTION 8.1

1. $\displaystyle\int e^{2-x}\,dx = -e^{2-x} + C$

3. $\displaystyle\int_0^1 \sin \pi x\,dx = \left[-\frac{1}{\pi}\cos \pi x\right]_0^1 = \frac{2}{\pi}$

5. $\displaystyle\int \sec^2(1-x)\,dx = -\tan(1-x) + C$

7. $\displaystyle\int_{\pi/6}^{\pi/3} \cot x\,dx = \left[\ln(\sin x)\right]_{\pi/6}^{\pi/3} = \ln\frac{\sqrt{3}}{2} - \ln\frac{1}{2} = \frac{1}{2}\ln 3$

9. $\displaystyle\left\{\begin{array}{l} u = 1 - x^2 \\ du = -2x\,dx \end{array}\right\}; \quad \int \frac{x\,dx}{\sqrt{1-x^2}} = -\frac{1}{2}\int u^{-1/2}\,du = -u^{1/2} + C = -\sqrt{1-x^2} + C$

11. $\displaystyle\int_{-\pi/4}^{\pi/4} \frac{\sin x}{\cos^2 x}\,dx = \int_{-\pi/4}^{\pi/4} \sec x \tan x\,dx = \left[\sec x\right]_{-\pi/4}^{\pi/4} = 0$

13. $\displaystyle\left\{\begin{array}{lll} u = 1/x \mid x = 1 & \Longrightarrow & u = 1 \\ du = -\dfrac{dx}{x^2} \mid x = 2 & \Longrightarrow & u = 1/2 \end{array}\right\};$

$$\int_1^2 \frac{e^{1/x}}{x^2}\,dx = \int_1^{1/2} -e^u\,du = \left[-e^u\right]_1^{1/2} = e - \sqrt{e}$$

15. $\displaystyle\int_0^c \frac{dx}{x^2 + c^2} = \left[\frac{1}{c}\tan^{-1}\left(\frac{x}{c}\right)\right]_0^c = \frac{\pi}{4c}$

17. $\displaystyle\left\{\begin{array}{l} u = 3\tan\theta + 1 \\ du = 3\sec^2\theta\,d\theta \end{array}\right\};$

$$\int \frac{\sec^2\theta}{\sqrt{3\tan\theta + 1}}\,d\theta = \frac{1}{3}\int u^{-1/2}\,du = \frac{2}{3}u^{1/2} + C = \frac{2}{3}\sqrt{3\tan\theta + 1} + C$$

19. $\displaystyle\int \frac{e^x}{ae^x - b}\,dx = \frac{1}{a}\ln|ae^x - b| + C$

21. $\displaystyle\left\{\begin{array}{l} u = x + 1 \\ du = dx \end{array}\right\};$

$$\int \frac{x}{(x+1)^2 + 4}\,dx = \int \frac{u-1}{u^2 + 4}\,du = \int \frac{u}{u^2 + 4}\,du - \int \frac{du}{u^2 + 4}$$

$$= \frac{1}{2}\ln|u^2 + 4| - \frac{1}{2}\tan^{-1}\frac{u}{2} + C$$

$$= \frac{1}{2}\ln|(x+1)^2 + 4| - \frac{1}{2}\tan^{-1}\left(\frac{x+1}{2}\right) + C$$

23. $\left\{ \begin{matrix} u = x^2 \\ du = 2x\,dx \end{matrix} \right\}$; $\displaystyle\int \frac{x}{\sqrt{1-x^4}}\,dx = \frac{1}{2}\int \frac{du}{\sqrt{1-u^2}} = \frac{1}{2}\sin^{-1}u + C$

$$= \tfrac{1}{2}\sin^{-1}(x^2) + C$$

25. $\left\{ \begin{matrix} u = x + 3 \\ du = dx \end{matrix} \right\}$; $\displaystyle\int \frac{dx}{x^2 + 6x + 10} = \int \frac{dx}{(x+3)^2 + 1} = \int \frac{du}{u^2 + 1}$

$$= \tan^{-1}u + C = \tan^{-1}(x + 3) + C$$

27. $\displaystyle\int x\sin x^2\,dx = -\frac{1}{2}\cos x^2 + C$

29. $\displaystyle\int \tan^2 x\,dx = \int (\sec^2 x - 1)\,dx = \tan x - x + C$

31. $\left\{ \begin{matrix} u = \ln x \mid x = 1 \;\Longrightarrow\; u = 0 \\ du = -\dfrac{dx}{x} \mid x = e \;\Longrightarrow\; u = 1 \end{matrix} \right\}$;

$$\int_1^e \frac{\ln x^3}{x}\,dx = \int_1^e \frac{3\ln x}{x}\,dx = 3\int_0^1 u\,du = 3\left[\frac{u^2}{2}\right]_0^1 = \frac{3}{2}$$

33. $\left\{ \begin{matrix} u = \sin^{-1}x \\ du = \dfrac{dx}{\sqrt{1-x^2}} \end{matrix} \right\}$; $\displaystyle\int \frac{\sin^{-1}x}{\sqrt{1-x^2}}\,dx = \int u\,du = \frac{1}{2}u^2 + C = \frac{1}{2}\left(\sin^{-1}x\right)^2 + C$

35. $\left\{ \begin{matrix} u = \ln x \\ du = \frac{1}{x}\,dx \end{matrix} \right\}$; $\displaystyle\int \frac{1}{x\ln x}\,dx = \int \frac{1}{u}\,du = \ln|u| + C = \ln|\ln x| + C$

37. $\left\{ \begin{matrix} u = \cos x \mid x = 0 \;\Longrightarrow\; u = 1 \\ du = -\sin x\,dx \mid x = \pi/4 \;\Longrightarrow\; u = \sqrt{2}/2 \end{matrix} \right\}$;

$$\int_0^{\pi/4} \frac{1 + \sin x}{\cos^2 x}\,dx = \int_0^{\pi/4} \sec^2 x\,dx + \int_0^{\pi/4} \frac{\sin x}{\cos^2 x}\,dx$$

$$= [\tan x]_0^{\pi/4} - \int_1^{\sqrt{2}/2} \frac{du}{u^2}$$

$$= 1 + \left[\frac{1}{u}\right]_1^{\sqrt{2}/2} = \sqrt{2}$$

39. (formula 99) $\displaystyle\int \sqrt{x^2 - 4}\,dx = \frac{x}{2}\sqrt{x^2 - 4} - 2\ln|x + \sqrt{x^2 - 4}| + C$

41. (formula 18) $\displaystyle\int \cos^2 2t\,dt = \frac{1}{2}[\sin 2t - \frac{1}{3}\sin^3 2t] + C$

43. (formula 108) $\displaystyle\int \frac{1}{x(2x + 3)}\,dx = \frac{1}{3}\ln\left|\frac{x}{2x + 3}\right| + C$

45. (formula 81) $\displaystyle\int \frac{\sqrt{x^2+9}}{x^2}\, dx = -\frac{\sqrt{x^2+9}}{x} + \ln\left|x + \sqrt{x^2+9}\right| + C$

47. (formula 11) $\displaystyle\int x^3 \ln x\, dx = x^4\left(\frac{\ln x}{4} - \frac{1}{16}\right) + C$

49.
$$\int_0^\pi \sqrt{1+\cos x}\, dx = \int_0^\pi \sqrt{2\cos^2\left(\frac{x}{2}\right)}\, dx$$
$$= \sqrt{2}\int_0^\pi \cos\left(\frac{x}{2}\right)\, dx \qquad \left[\cos\left(\frac{x}{2}\right) \geq 0 \ \text{ on } \ [0,\pi]\right]$$
$$= 2\sqrt{2}\left[\sin\left(\frac{x}{2}\right)\right]_0^\pi = 2\sqrt{2}$$

51. (a) $\displaystyle\int_0^\pi \sin^2 nx\, dx = \int_0^\pi \left[\frac{1}{2} - \frac{\cos 2nx}{2}\right] dx = \left[\frac{x}{2} - \frac{\sin 2nx}{4n}\right]_0^\pi = \frac{\pi}{2}$

(b) $\displaystyle\int_0^\pi \sin nx\, \cos nx\, dx = \frac{1}{2}\int_0^\pi \sin 2nx\, dx = -\left[\frac{\cos 2nx}{4n}\right]_0^\pi = 0$

(c) $\displaystyle\int_0^{\pi/n} \sin nx\, \cos nx\, dx = \frac{1}{2}\int_0^{\pi/n} \sin 2nx\, dx = -\left[\frac{\cos 2nx}{4n}\right]_0^{\pi/n} = 0$

53. (a) $\displaystyle\int \tan^3 x\, dx = \int \tan^2 x\, \tan x\, dx = \int (\sec^2 x - 1)\tan x\, dx$
$$= \int \sec^2 x\, \tan x\, dx - \int \tan x\, dx$$
$$= \int u\, du - \int \tan x\, dx \qquad (u = \tan x, \quad du = \sec^2 x\, dx)$$
$$= \frac{1}{2}u^2 - \ln|\sec x| + C = \frac{1}{2}\tan^2 x - \ln|\sec x| + C$$

(b) $\displaystyle\int \tan^5 x\, dx = \int \tan^3 x\, \tan^2 x\, dx = \int \tan^3 x\, (\sec^2 x - 1)\, dx$
$$= \int \tan^3 x\, \sec^2 x\, dx - \int \tan^3 x\, dx$$
$$= \int u^3\, du - \int \tan^3 x\, dx \qquad (u = \tan x \quad du = \sec^2 x\, dx)$$
$$= \frac{1}{4}u^4 - \frac{1}{2}\tan^2 x + \ln|\sec x| + C$$
$$= \frac{1}{4}\tan^4 x - \frac{1}{2}\tan^2 x + \ln|\sec x| + C$$

(c) $\displaystyle\int \tan^7 x\, dx = \int \tan^5 x\, \sec^2 x\, dx - \int \tan^5 x\, dx$
$$= \frac{1}{6}\tan^6 x - \frac{1}{4}\tan^4 x + \frac{1}{2}\tan^2 x - \ln|\sec x| + C$$

(d) $\displaystyle\int \tan^{2k+1} x\, dx = \int \tan^{2k-1} x\, \tan^2 x\, dx = \int \tan^{2k-1} x\, \sec^2 x\, dx - \int \tan^{2k-1} x\, dx$
$$= \frac{1}{2k}\tan^{2k} - \int \tan^{2k-1} x\, dx$$

55.   (a)

(b) $\sin x + \cos x = \sqrt{2}\left[\sin x \cos(\pi/4) + \cos x \sin(\pi/4)\right]$

$$= \sqrt{2}\sin\left(x + \frac{\pi}{4}\right);$$

$$A = \sqrt{2}, \quad B = \pi/4$$

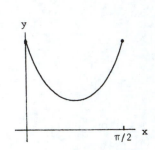

(c)   Area $= \displaystyle\int_0^{\pi/2} \frac{1}{\sin x + \cos x}\,dx = \frac{1}{\sqrt{2}}\int_0^{\pi/2}\frac{1}{\sin\left(x + \frac{\pi}{4}\right)}\,dx$

$$\left\{\begin{array}{l} u = x + \pi/4 \mid x = 0 \implies u = \pi/4 \\ du = dx \mid x = \pi/2 \implies u = 3\pi/4 \end{array}\right\};$$

$$\frac{1}{\sqrt{2}}\int_0^{\pi/2}\frac{1}{\sin\left(x + \frac{\pi}{4}\right)}\,dx = \frac{\sqrt{2}}{2}\int_{\pi/4}^{3\pi/4}\frac{1}{\sin u}\,du$$

$$= \frac{\sqrt{2}}{2}\int_{\pi/4}^{3\pi/4}\csc u\,du$$

$$= \frac{\sqrt{2}}{2}\left[\ln|\csc u - \cot u|\right]_{\pi/4}^{3\pi/4}$$

$$= \frac{\sqrt{2}}{2}\ln\left[\frac{\sqrt{2}+1}{\sqrt{2}-1}\right]$$

57.   (a)

(b) $x_1 \cong -0.80, \quad x_2 \cong 5.80$

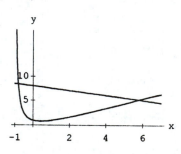

(c)   Area $\cong \displaystyle\int_{-0.80}^{5.80}\left[8 - \frac{1}{2}x - \frac{x^2+1}{x+1}\right]dx = \int_{-0.80}^{5.80}\left[8 - \frac{1}{2}x - x + 1 - \frac{2}{x+1}\right]dx$

$$= \int_{-0.80}^{5.80}\left[9 - \frac{3}{2}x - \frac{2}{x+1}\right]dx$$

$$= \left[9x - \frac{3}{4}x^2 - 2\ln|x+1|\right]_{-0.80}^{5.80} \cong 27.6$$

SECTION 8.2

1.

$$\boxed{\begin{array}{ll} u = x, & dv = e^{-x}\,dx \\ du = dx, & v = -e^{-x} \end{array}}$$

$$\int xe^{-x}\,dx = -xe^{-x} - \int -e^{-x}\,dx = -xe^{-x} - e^{-x} + C$$

3.

$$\left\{ \begin{array}{l} t = -x^3 \\ dt = -3x^2\,dx \end{array} \right\}; \quad \int x^2 e^{-x^3}\,dx = -\frac{1}{3}\int e^t\,dt = -\frac{1}{3}e^t + C = -\frac{1}{3}e^{-x^3} + C$$

5.

$$\int x^2 e^{-x}\,dx = -x^2 e^{-x} - \int -2xe^{-x}\,dx = -x^2 e^{-x} + 2\int xe^{-x}\,dx$$

$$\boxed{\begin{array}{ll} u = x^2, & dv = e^{-x}\,dx \\ du = 2x\,dx, & v = -e^{-x} \end{array}}$$

$$= -x^2 e^{-x} + 2\left[-xe^{-x} - \int -e^{-x}\,dx\right]$$

$$\boxed{\begin{array}{ll} u = x, & dv = e^{-x}\,dx \\ du = dx, & v = -e^{-x} \end{array}}$$

$$= -x^2 e^{-x} + 2\left(-xe^{-x} - e^{-x}\right) + C$$

$$= -e^{-x}\left(x^2 + 2x + 2\right) + C$$

$$\int_0^1 x^2 e^{-x}\,dx = \left[-e^{-x}\left(x^2 + 2x + 2\right)\right]_0^1 = 2 - 5e^{-1}$$

7.

$$\int x^2 (1-x)^{-1/2}\,dx = -2x^2 (1-x)^{1/2} + 4\int x (1-x)^{1/2}\,dx \cdot$$

$$\boxed{\begin{array}{ll} u = x^2, & dv = (1-x)^{-1/2}\,dx \\ du = 2x\,dx, & v = -2(1-x)^{1/2} \end{array}}$$

$$= -2x^2 (1-x)^{1/2} + 4\left[-\frac{2x}{3}(1-x)^{3/2} + \int \frac{2}{3}(1-x)^{3/2}\,dx\right]$$

$$\boxed{\begin{array}{ll} u = x, & dv = (1-x)^{1/2}\,dx \\ du = dx, & v = -\frac{2}{3}(1-x)^{3/2} \end{array}}$$

$$= -2x^2 (1-x)^{1/2} - \frac{8x}{3}(1-x)^{3/2} - \frac{16}{15}(1-x)^{5/2} + C$$

Or, use the substitution $t = 1 - x$ (no integration by parts needed) to obtain:

$$-2(1-x)^{1/2} + \tfrac{4}{3}(1-x)^{3/2} - \tfrac{2}{5}(1-x)^{5/2} + C.$$

9.

$$\int x \ln \sqrt{x}\,dx = \frac{1}{2}\int x \ln x\,dx = \frac{1}{2}\left[\frac{1}{2}x^2 \ln x - \frac{1}{2}\int x\,dx\right]$$

$$\boxed{\begin{array}{ll} u = \ln x, & dv = x\,dx \\ du = \dfrac{dx}{x}, & v = \dfrac{1}{2}x^2 \end{array}}$$

$$= \tfrac{1}{4}x^2 \ln x - \tfrac{1}{8}x^2 + C$$

$$\int_1^{e^2} x \ln \sqrt{x}\,dx = \left[\tfrac{1}{4}x^2 \ln x - \tfrac{1}{8}x^2\right]_1^{e^2} = \tfrac{3}{8}e^4 + \tfrac{1}{8}$$

11.

$$\int \frac{\ln(x+1)}{\sqrt{x+1}}\, dx = 2\sqrt{x+1}\ln(x+1) - \int \frac{2\, dx}{\sqrt{x+1}}$$

$$
\begin{array}{ll}
u = \ln(x+1), & dv = \dfrac{dx}{\sqrt{x+1}} \\[2mm]
du = \dfrac{dx}{x+1}, & v = 2\sqrt{x+1}
\end{array}
$$

$$= 2\sqrt{x+1}\ln(x+1) - 4\sqrt{x+1} + C$$

13.

$$\int (\ln x)^2\, dx = x(\ln x)^2 - 2\int \ln x\, dx$$

$$
\begin{array}{ll}
u = (\ln x)^2, & dv = dx \\[2mm]
du = \dfrac{2\ln x}{x}\, dx, & v = x
\end{array}
$$

$$= x(\ln x)^2 - 2\left[x\ln x - \int dx\right]$$

$$
\begin{array}{ll}
u = \ln x, & dv = dx \\[2mm]
du = \dfrac{dx}{x}, & v = x
\end{array}
$$

$$= x(\ln x)^2 - 2x\ln x + 2x + C$$

15.

$$\int x^3\, 3^x\, dx = \frac{x^3\, 3^x}{\ln 3} - \frac{3}{\ln 3}\int x^2\, 3^x\, dx$$

$$
\begin{array}{ll}
u = x^3, & dv = 3^x\, dx \\[2mm]
du = 3x^2\, dx, & v = \dfrac{3^x}{\ln 3}
\end{array}
$$

$$= \frac{x^3\, 3^x}{\ln 3} - \frac{3}{\ln 3}\left[\frac{x^2\, 3^x}{\ln 3} - \frac{2}{\ln 3}\int x\, 3^x\, dx\right]$$

$$
\begin{array}{ll}
u = x^2, & dv = 3^x\, dx \\[2mm]
du = 2x\, dx, & v = \dfrac{3^x}{\ln 3}
\end{array}
$$

$$= \frac{x^3\, 3^x}{\ln 3} - \frac{3x^2\, 3^x}{(\ln 3)^2} + \frac{6}{(\ln 3)^2}\int x\, 3^x\, dx$$

$$= \frac{x^3\, 3^x}{\ln 3} - \frac{3x^2\, 3^x}{(\ln 3)^2} + \frac{6}{(\ln 3)^2}\left[\frac{x3^x}{\ln 3} - \frac{1}{\ln 3}\int 3^x\, dx\right]$$

$$
\begin{array}{ll}
u = x, & dv = 3^x\, dx \\[2mm]
du = dx, & v = \dfrac{3^x}{\ln 3}
\end{array}
$$

$$= 3^x\left[\frac{x^3}{\ln 3} - \frac{3x^2}{(\ln 3)^2} + \frac{6x}{(\ln 3)^3} - \frac{6}{(\ln 3)^4}\right] + C$$

17.

$$\int x(x+5)^{14}\, dx = \frac{x}{15}(x+5)^{15} - \frac{1}{15}\int (x+5)^{15}\, dx$$

$$
\begin{array}{ll}
u = x, & dv = (x+5)^{14}\, dx \\[2mm]
du = dx, & v = \frac{1}{15}(x+5)^{15}
\end{array}
$$

$$= \tfrac{1}{15}x(x+5)^{15} - \tfrac{1}{240}(x+5)^{16} + C$$

Or, use the substitution $t = x + 5$ (integration by parts not needed) to obtain:

$$\frac{1}{16}(x+5)^{16} - \frac{1}{3}(x+5)^{15} + C.$$

19.
$$\int x \cos \pi x \, dx = \frac{1}{\pi} x \sin x - \frac{1}{\pi} \int \sin \pi x \, dx$$

$$\boxed{\begin{array}{ll} u = x, & dv = \cos \pi x \, dx \\ du = dx, & v = \dfrac{1}{\pi} \sin \pi x \end{array}}$$

$$= \frac{1}{\pi} x \sin \pi x + \frac{1}{\pi^2} \cos \pi x + C$$

$$\int_0^{1/2} x \cos \pi x \, dx = \left[\frac{1}{\pi} x \sin \pi x + \frac{1}{\pi^2} \cos \pi x \right]_0^{1/2} = \frac{1}{2\pi} - \frac{1}{\pi^2}$$

21.
$$\int x^2 (x+1)^9 \, dx = \frac{x^2}{10} (x+1)^{10} - \frac{1}{5} \int x (x+1)^{10} \, dx$$

$$\boxed{\begin{array}{ll} u = x^2, & dv = (x+1)^9 \, dx \\ du = 2x \, dx, & v = \frac{1}{10}(x+1)^{10} \end{array}}$$

$$= \frac{x^2}{10} (x+1)^{10} - \frac{1}{5} \left[\frac{x}{11}(x+1)^{11} - \frac{1}{11} \int (x+1)^{11} \, dx \right]$$

$$\boxed{\begin{array}{ll} u = x, & dv = (x+1)^{10} \, dx \\ du = dx, & v = \frac{1}{11}(x+1)^{11} \end{array}}$$

$$= \frac{x^2}{10} (x+1)^{10} - \frac{x}{55}(x+1)^{11} + \frac{1}{660}(x+1)^{12} + C$$

23.
$$\int e^x \sin x \, dx = -e^x \cos x + \int e^x \cos x \, dx$$

$$\boxed{\begin{array}{ll} u = e^x, & dv = \sin x \, dx \\ du = e^x \, dx, & v = -\cos x \end{array}}$$

$$= -e^x \cos x + e^x \sin x - \int e^x \sin x \, dx$$

$$\boxed{\begin{array}{ll} u = e^x, & dv = \cos x \, dx \\ du = e^x \, dx, & v = \sin x \end{array}}$$

Adding $\int e^x \sin x \, dx$ to both sides, we get

$$2 \int e^x \sin x \, dx = -e^x \cos x + e^x \sin x$$

so that

$$\int e^x \sin x \, dx = \frac{1}{2} e^x (\sin x - \cos x) + C.$$

25.
$$\int \ln\left(1+x^2\right) dx = x \ln\left(1+x^2\right) - 2\int \frac{x^2}{1+x^2}\, dx$$

$$\boxed{\begin{array}{ll} u = \ln\left(1+x^2\right), & dv = dx \\ du = \dfrac{2x}{1+x^2}\, dx, & v = x \end{array}}$$

$$= x \ln\left(1+x^2\right) - 2\int \frac{x^2+1-1}{1+x^2}\, dx$$

$$= x \ln\left(1+x^2\right) - 2\int \left(1 - \frac{1}{1+x^2}\right) dx$$

$$= x \ln\left(1+x^2\right) - 2x + 2\tan^{-1} x + C$$

$$\int_0^1 \ln(1+x^2)\, dx = \left[x\ln(1+x^2) - 2x + 2\tan^{-1} x\right]_0^1 = \ln 2 - 2 + \frac{\pi}{2}$$

27.
$$\int x^n \ln x\, dx = \frac{x^{n+1}\ln x}{n+1} - \frac{1}{n+1}\int x^n\, dx$$

$$\boxed{\begin{array}{ll} u = \ln x, & dv = x^n\, dx \\ du = \dfrac{dx}{x}, & v = \dfrac{x^{n+1}}{n+1} \end{array}}$$

$$= \frac{x^{n+1}\ln x}{n+1} - \frac{x^{n+1}}{(n+1)^2} + C$$

29.
$$\left\{t = x^2, \quad dt = 2x\, dx\right\}; \qquad \int x^3 \sin x^2\, dx = \frac{1}{2}\int t \sin t\, dt$$

$$\boxed{\begin{array}{ll} u = t, & dv = \sin t\, dt \\ du = dt, & v = -\cos t \end{array}}$$

$$= \frac{1}{2}\left[-t\cos t + \int \cos t\, dt\right]$$

$$= \tfrac{1}{2}\left(-t\cos t + \sin t\right) + C$$

$$= -\tfrac{1}{2}x^2\cos x^2 + \tfrac{1}{2}\sin x^2 + C$$

31.
$$\left\{\begin{array}{l|ll} u = 2x & x = 0 & \Longrightarrow \quad u = 0 \\ du = 2\, dx & x = 1/4 & \Longrightarrow \quad u = 1/2 \end{array}\right\};$$

$$\int_0^{1/4} \sin^{-1} 2x\, dx = \frac{1}{2}\int_0^{1/2} \sin^{-1} u\, du$$

$$= \frac{1}{2}\left[u\sin^{-1} u + \sqrt{1-u^2}\right]_0^{1/2} \qquad [\text{by}(8.2.5)]$$

$$= \frac{1}{2}\left[\frac{\pi}{12} + \frac{\sqrt{3}}{2} - 1\right] = \frac{\pi}{24} + \frac{\sqrt{3}-2}{4}$$

33.
$$\left\{\begin{array}{l|ll} u = x^2 & x = 0 & \Longrightarrow \quad u = 0 \\ du = 2x\, dx & x = 1 & \Longrightarrow \quad u = 1 \end{array}\right\};$$

$$\int_0^1 x\tan^{-1} x\, dx = \frac{1}{2}\int_0^1 \tan^{-1} u\, du$$

$$= \frac{1}{2}\left[u\tan^{-1} u - \tfrac{1}{2}(1+x^2)\right]_0^1 \qquad [\text{by } 8.2.6]$$

$$= \frac{\pi}{8} - \frac{1}{2}\ln 2$$

35.
$$\int x^2 \cosh 2x \, dx = \tfrac{1}{2}x^2 \sinh 2x - \int x \sinh 2x \, dx$$

$$\boxed{\begin{array}{ll} u = x^2, & dv = \cosh 2x \, dx \\ du = 2x \, dx, & v = \tfrac{1}{2}\sinh 2x \end{array}}$$

$$= \tfrac{1}{2}x^2 \sinh 2x - \tfrac{1}{2}x \cosh 2x + \tfrac{1}{2}\int \cosh 2x \, dx$$

$$\boxed{\begin{array}{ll} u = x, & dv = \sinh 2x \, dx \\ du = dx, & v = \tfrac{1}{2}\cosh 2x \end{array}}$$

$$= \tfrac{1}{2}x^2 \sinh 2x - \tfrac{1}{2}x \cosh 2x + \tfrac{1}{4}\sinh 2x + C$$

37. Let $u = \ln x$, $du = \frac{1}{x}\, dx$. Then

$$\int \frac{1}{x} \sin^{-1}(\ln x)\, dx = \int \sin^{-1} u \, du = u \sin^{-1} u + \sqrt{(1 - u^2)} + C$$

$$= (\ln x)\sin^{-1}(\ln x) + \sqrt{1 - (\ln x)^2} + C$$

39.
$$\int \sin(\ln x)\, dx = x \sin(\ln x) - \int \cos(\ln x)\, dx$$

$$\boxed{\begin{array}{ll} u = \sin(\ln x), & dv = dx \\ du = \cos(\ln x)\dfrac{1}{x}\, dx, & v = x \end{array}}$$

$$= x \sin(\ln x) - x \cos(\ln x) - \int \sin(\ln x)\, dx$$

$$\boxed{\begin{array}{ll} u = \cos(\ln x), & dv = dx \\ du = -\sin(\ln x)\dfrac{1}{x}\, dx, & v = x \end{array}}$$

Adding $\int \sin(\ln x)\, dx$ to both sides, we get

$$2\int \sin(\ln x)\, dx = x \sin(\ln x) - x \cos(\ln x)$$

so that

$$\int \sin(\ln x)\, dx = \tfrac{1}{2}\left[x \sin(\ln x) - x \cos(\ln x)\right] + C$$

41.
$$\boxed{\begin{array}{ll} u = \ln x & dv = dx \\ du = \dfrac{1}{x}\, dx & v = x \end{array}} \quad \int \ln x \, dx = x \ln x - \int dx = x \ln x - x + C$$

43. Cannot be solved

45.
$$\boxed{\begin{array}{ll} u = e^{ax} & dv = \sin bx \, dx \\ du = ae^{ax}\, dx & v = -\dfrac{1}{b}\cos bx \end{array}} \quad \int e^{ax} \sin bx \, dx = -\frac{1}{b}e^{ax}\cos bx + \frac{a}{b}\int e^{ax}\cos bx \, dx$$

$$\boxed{\begin{array}{ll} u = e^{ax} & dv = \cos bx \, dx \\ du = ae^{ax}\, dx & v = \dfrac{1}{b}\sin bx \end{array}} \quad = -\frac{1}{b}e^{ax}\cos bx + \frac{a}{b^2}e^{ax}\sin bx - \frac{a^2}{b^2}\int e^{ax}\sin bx \, dx.$$

$$\Longrightarrow \quad \int e^{ax} \sin bx \, dx = \frac{e^{ax}}{a^2 + b^2} (a \sin bx - b \cos bx) + C$$

47. The integrals cancel each other if you integrate by parts.

$$\int e^{ax} \cosh bx \, dx = \frac{1}{2a} e^{ax} - \frac{x}{2} + C$$

49. $A = \displaystyle\int_0^{1/2} \sin^{-1} x \, dx = \left[x \sin^{-1} x + \sqrt{1 - x^2} \right]_0^{1/2} = \frac{\pi}{12} + \frac{\sqrt{3} - 2}{2}$

51. (a) $A = \displaystyle\int_1^e \ln x \, dx = [x \ln x - x]_1^e = 1$

  (b) $\bar{x} A = \displaystyle\int_1^e x \ln x \, dx = \left[\frac{1}{2} x^2 \ln x - \frac{1}{4} x^2 \right]_1^e = \frac{1}{4} (e^2 + 1), \quad \bar{x} = \frac{1}{4} (e^2 + 1)$

  $\bar{y} A = \displaystyle\int_1^e \frac{1}{2} (\ln x)^2 \, dx = \frac{1}{2} \left[x (\ln x)^2 - 2x \ln x + 2x \right]_1^e = \frac{1}{2} e - 1, \quad \bar{y} = \frac{1}{2} e - 1$

  (c) $V_x = 2\pi \bar{y} A = \pi (e - 2), \quad V_y = 2\pi \bar{x} A = \frac{1}{2} \pi (e^2 + 1)$

53. $\bar{x} = \dfrac{1}{e - 1}, \qquad \bar{y} = \dfrac{1}{4} (e + 1)$ **55.** $\bar{x} = \frac{1}{2}\pi, \qquad \bar{y} = \frac{1}{8}\pi$

57. (a) $M = \displaystyle\int_0^1 e^{kx} \, dx = \frac{1}{k} (e^k - 1)$

  (b) $x_M M = \displaystyle\int_0^1 x e^{kx} \, dx = \frac{(k - 1) e^k + 1}{k^2}, \qquad x_M = \frac{(k - 1) e^k + 1}{k (e^k - 1)}$

59. $V_y = \displaystyle\int_0^1 2\pi x \cos \frac{1}{2} \pi x \, dx = \left[4x \sin \frac{1}{2}\pi x + \frac{8}{\pi} \cos \frac{1}{2}\pi x \right]_0^1 = 4 - \frac{8}{\pi}$

61. $V_y = \displaystyle\int_0^1 2\pi x^2 e^x \, dx = 2\pi (e - 2)$ (see Example 6)

63.

$$V_x = \int_0^1 \pi e^{2x} \, dx = \pi \left[\frac{1}{2} e^{2x} \right]_0^1 = \frac{1}{2} \pi (e^2 - 1)$$

$$\bar{x} V_x = \int_0^1 \pi x e^{2x} \, dx = \pi \left[\frac{1}{2} x e^{2x} - \frac{1}{4} e^{2x} \right]_0^1 = \frac{1}{4} \pi (e^2 + 1)$$

$$\bar{x} = \frac{e^2 + 1}{2 (e^2 - 1)}$$

65.

$$A = \int_0^1 \cosh x \, dx = [\sinh x]_0^1 = \sinh 1 = \frac{e - e^{-1}}{2} = \frac{e^2 - 1}{2e}$$

$$\bar{x}A = \int_0^1 x \cosh x \, dx = [x \sinh x - \cosh x]_0^1 = \sinh 1 - \cosh 1 + 1 = \frac{2(e-1)}{2e}$$

$$\bar{y}A = \int_0^1 \frac{1}{2} \cosh^2 x \, dx = \frac{1}{4} [\sinh x \cosh x + x]_0^1 = \frac{1}{4}(\sinh 1 \cosh 1 + 1) = \frac{e^4 + 4e^2 - 1}{16e^2}$$

Therefore $\bar{x} = \dfrac{2}{e+1}$ and $\bar{y} = \dfrac{e^4 + 4e^2 - 1}{8e(e^2 - 1)}$.

67.

$$\boxed{\begin{array}{ll} u = x^2, & dv = e^{-x} \, dx \\ du = 2x \, dx, & v = -e^{-x} \end{array}} \qquad \int x^n e^{ax} \, dx = \frac{x^n e^{ax}}{a} - \frac{n}{a} \int x^{n-1} e^{ax} \, dx$$

69.

$$\int x^3 e^{2x} \, dx = \frac{1}{2} x^3 e^{2x} - \frac{3}{2} \int x^2 e^{2x} \, dx = \frac{1}{2} x^3 e^{2x} - \frac{3}{2} \left[\frac{1}{2} x^2 e^{2x} - \int x e^{2x} \, dx \right]$$

$$= \frac{1}{2} x^3 e^{2x} - \frac{3}{4} x^2 e^{2x} - \frac{3}{4} x e^{2x} - \frac{3}{4} \int e^{2x} \, dx$$

$$= \frac{1}{2} x^3 e^{2x} - \frac{3}{4} x^2 e^{2x} - \frac{3}{4} x e^{2x} - \frac{3}{8} e^{2x} + C$$

71.

$$\int (\ln x)^3 \, dx = x(\ln x)^3 - 3 \int (\ln x)^2 \, dx = x(\ln x)^3 - 3x(\ln x)^2 + 6 \int \ln x \, dx$$

$$= x(\ln x)^3 - 3x(\ln x)^2 + 6x \ln x - 6 \int dx$$

$$= x(\ln x)^3 - 3x(\ln x)^2 + 6x \ln x - 6x + C$$

73. (a) Differentiating, $x^3 e^x = Ax^3 e^x + 3Ax^2 e^x + 2Bxe^x + Bx^2 e^x + Ce^x + Cxe^x + De^x$

    $\implies$ $A = 1$, $B = -3$, $C = 6$, and $D = -6$

(b)

$$\int x^3 e^x \, dx = x^3 e^x - 3 \int x^2 e^x \, dx$$

$$= x^3 e^x - 3x^2 e^x + 6 \int x e^x \, dx$$

$$= x^3 e^x - 3x^2 e^x + 6x e^x - 6 \int e^x \, dx$$

$$= x^3 e^x - 3x^2 e^x + 6x e^x - 6e^x + C$$

75. Let $u = f(x)$, $dv = g''(x) \, dx$. Then $du = f'(x) \, dx$, $v = g'(x)$, and

$$\int_a^b f(x) g''(x) \, dx = [f(x) g'(x)]_a^b - \int_a^b f'(x) g'(x) \, dx = - \int_a^b f'(x) g'(x) \, dx \quad \text{since } f(a) = f(b) = 0$$

Now let $u = f'(x)$, $dv = g'(x)\,dx$. Then $du = f''(x)\,dx$, $v = g(x)$, and

$$-\int_a^b f'(x)g'(x)\,dx = [-f'(x)g(x)]_a^b + \int f''(x)g(x)\,dx = \int g(x)f''(x)\,dx \quad \text{since } g(a) = g(b) = 0$$

Therefore, if f and g have continuous second derivatives, and if $f(a) = g(a) = f(b) = g(b) = 0$, then

$$\int_a^b f(x)g''(x)\,dx = \int_a^b g(x)f''(x)\,dx$$

PROJECT 8.2

1. Let P be a regular partition of the interval $[0, n]$. The present value of R dollars continuously compounded at the rate r on the interval $[t_{i-1}, t_i]$ is approximately $Re^{-rt}\,\Delta t$. Therefore, it follows that the present value of the revenue stream over the time interval $[0, n]$ is given by the definite integral

$$P.V. = \int_0^n Re^{-rt}\,dt$$

3. (a) $P.V. = \displaystyle\int_0^2 (1000 + 60t)\,e^{-0.05t}\,dt$

    $$= 1000 \int_0^2 e^{-t/20}\,dt + 60 \int_0^2 te^{-t/20}\,dt$$

    $$= 1000 \left[-20e^{-t/20}\right]_0^2 + 60 \left[-20te^{-t/20} - 400e^{-t/20}\right]_0^2$$

    (by parts)

    $$= \left[-(44000 + 1200t)\,e^{-t/20}\right]_0^2 = 44000 - 46400\,e^{-0.10} \cong \$2016$$

    (b) $P.V. = \displaystyle\int_0^2 (1000 + 60t)\,e^{-0.1t}\,dt$

    $$= 1000 \left[-10e^{-t/10}\right]_0^2 + 60 \left[-10te^{-t/10} - 100e^{-t/10}\right]_0^2$$

    (by parts)

    $$= \left[-(16000 + 600t)\,e^{-t/10}\right]_0^2 = 16000 - 17200e^{-0.20} \cong \$1918$$

5. $r(t) = a\sin\omega t$, with a, ω positive constants.

    $$P.V. = \int_0^n a\sin\omega t e^{-rt}\,dt = a \int_0^n e^{-rt}\sin\omega t\,dt = \frac{ae^{-rt}}{r^2 + \omega^2}\left[-r\sin\omega t - \omega\cos\omega t\right]_0^n$$

    $$= \frac{a\omega}{r^2 + \omega^2} - \frac{ae^{-rn}}{r^2 + \omega^2}(r\sin\omega n - \omega\cos\omega n).$$

SECTION 8.3

1. $\displaystyle\int \sin^3 x\, dx = \int (1 - \cos^2 x)\sin x\, dx = \frac{1}{3}\cos^3 x - \cos x + C$

3. $\displaystyle\int_0^{\pi/6} \sin^2 3x\, dx = \int_0^{\pi/6} \frac{1 - \cos 6x}{2}\, dx = \left[\frac{1}{2}x - \frac{1}{12}\sin 6x\right]_0^{\pi/6} = \frac{\pi}{12}$

5.
$$\int \cos^4 x \sin^3 x\, dx = \int \cos^4 x\,(1 - \cos^2 x)\sin x\, dx$$
$$= \int (\cos^4 x - \cos^6 x)\sin x\, dx$$
$$= -\tfrac{1}{5}\cos^5 x + \tfrac{1}{7}\cos^7 x + C$$

7.
$$\int \sin^3 x \cos^3 x\, dx = \int \sin^3 x\,(1 - \sin^2 x)\cos x\, dx$$
$$= \int (\sin^3 x - \sin^5 x)\cos x\, dx$$
$$= \tfrac{1}{4}\sin^4 x - \tfrac{1}{6}\sin^6 x + C$$

9. $\displaystyle\int \sec^2 \pi x\, dx = \frac{1}{\pi}\tan \pi x + C$

11.
$$\int \tan^3 x\, dx = \int (\sec^2 x - 1)\tan x\, dx$$
$$= \int \tan x \sec^2 x\, dx - \int \tan x\, dx$$
$$= \tfrac{1}{2}\tan^2 x + \ln|\cos x| + C$$

13.
$$\int \sin^4 x\, dx = \int \left(\frac{1 - \cos 2x}{2}\right)^2 dx$$
$$= \frac{1}{4}\int \left(1 - 2\cos 2x + \cos^2 2x\right)\, dx$$
$$= \frac{1}{4}\int \left(1 - 2\cos 2x + \frac{1 + \cos 4x}{2}\right) dx$$
$$= \int \left(\frac{3}{8} - \frac{1}{2}\cos 2x + \frac{1}{8}\cos 4x\right) dx$$
$$= \tfrac{3}{8}x - \tfrac{1}{4}\sin 2x + \tfrac{1}{32}\sin 4x + C$$

$$\int_0^\pi \sin^4 x \, dx = \left[\tfrac{3}{8}x - \tfrac{1}{4}\sin 2x + \tfrac{1}{32}\sin 4x\right]_0^\pi = \tfrac{3}{8}\pi$$

15.

$$\int \sin 2x \cos 3x \, dx = \int \frac{1}{2}\left[\sin(-x) + \sin 5x\right] dx$$

$$= \int \frac{1}{2}\left(-\sin x + \sin 5x\right) dx$$

$$= \tfrac{1}{2}\cos x - \tfrac{1}{10}\cos 5x + C$$

17. $\displaystyle\int \tan^2 x \sec^2 x \, dx = \frac{1}{3}\tan^3 x + C$

19.

$$\int \csc^3 x \, dx = -\csc x \cot x - \int \csc x \cot^2 x \, dx$$

$u = \csc x,$	$dv = \csc^2 x \, dx$
$du = -\csc x \cot x \, dx,$	$v = -\cot x$

$$= -\csc x \cot x - \int \csc x \left(\csc^2 x - 1\right) dx$$

Thus

$$2\int \csc^3 x \, dx = -\csc x \cot x + \int \csc x \, dx$$

so that

$$\int \csc^3 x \, dx = -\frac{1}{2}\csc x \cot x + \frac{1}{2}\ln|\csc x - \cot x| + C.$$

21.

$$\int \sin^2 x \sin 2x \, dx = \int \sin^2 x \left(2\sin x \cos x\right) dx$$

$$= 2\int \sin^3 x \cos x \, dx$$

$$= \tfrac{1}{2}\sin^4 x + C$$

23.

$$\int \sin^6 x \, dx = \int \left(\frac{1 - \cos 2x}{2}\right)^3 dx$$

$$= \frac{1}{8}\int \left(1 - 3\cos 2x + 3\cos^2 2x - \cos^3 2x\right) dx$$

$$= \frac{1}{8}\int \left[1 - 3\cos 2x + 3\left(\frac{1 + \cos 4x}{2}\right) - \cos 2x \left(1 - \sin^2 2x\right)\right] dx$$

$$= \frac{1}{8}\int \left(\frac{5}{2} - 4\cos 2x + \frac{3}{2}\cos 4x + \sin^2 2x \cos 2x\right) dx$$

$$= \tfrac{5}{16}x - \tfrac{1}{4}\sin 2x + \tfrac{3}{64}\sin 4x + \tfrac{1}{48}\sin^3 2x + C$$

25. $\displaystyle\int_{\pi/6}^{\pi/2} \cot^2 x\, dx = \int_{\pi/6}^{\pi/2} (\csc^2 x - 1)\, dx = [-\cot x - x]_{\pi/6}^{\pi/2} = \sqrt{3} - \dfrac{\pi}{3}$

27.
$$\int \cot^3 x \csc^3 x\, dx = \int (\csc^2 x - 1)\csc^3 x \cot x\, dx$$
$$= \int (\csc^4 x - \csc^2 x)\csc x \cot x\, dx$$
$$= -\frac{1}{5}\csc^5 x + \frac{1}{3}\csc^3 x + C$$

29.
$$\int \sin 5x \sin 2x\, dx = \int \frac{1}{2}(\cos 3x - \cos 7x)\, dx$$
$$= \tfrac{1}{6}\sin 3x - \tfrac{1}{14}\sin 7x + C$$

31. $\displaystyle\int \csc^4 2x\, dx = \int (1 + \cot^2 2x)\csc^2 2x\, dx = -\frac{1}{2}\cot 2x - \frac{1}{6}\cot^3 2x + C$

33.
$$\int \tan^5 3x\, dx = \int \tan^3 3x\,(\sec^2 3x - 1)\, dx$$
$$= \int \tan^3 3x \sec^2 3x\, dx - \int \tan^3 3x\, dx$$
$$= \int \tan^3 3x \sec^2 3x\, dx - \int (\tan 3x \sec^2 3x - \tan 3x)\, dx$$
$$= \tfrac{1}{12}\tan^4 3x - \tfrac{1}{6}\tan^2 3x + \tfrac{1}{3}\ln|\sec 3x| + C$$

35.
$$\int_{-1/6}^{1/3} \sin^4 3\pi x \cos^3 3\pi x\, dx = \int_{-1/6}^{1/3} \sin^4 3\pi x \cos^2 3\pi x \cos 3\pi x\, dx$$
$$= \int_{-1/6}^{1/3} \sin^4 3\pi x\,(1 - \sin^2 3\pi x \cos 3\pi x\, dx$$
$$= \int_{-1}^{0} u^4(1 - u^2)\frac{1}{3\pi}\, du \quad [u = \sin 3\pi x, \quad du = 3\pi \cos 3\pi x\, dx]$$
$$= \frac{1}{3\pi}\left[\tfrac{1}{5}u^5 - \tfrac{1}{7}u^7\right]_{-1}^{0} = \frac{2}{105\pi}$$

37.
$$\int_{0}^{\pi/4} \cos 4x \sin 2x\, dx = \int_{0}^{\pi/4} \frac{1}{2}(\sin 6x - \sin 2x)\, dx$$
$$= \left[-\tfrac{1}{12}\cos 6x + \tfrac{1}{4}\cos 2x\right]_{0}^{\pi/4} = -\tfrac{1}{6}$$

39.
$$\int \sec^5 x \, dx = \tan x \sec^3 x - 3 \int \sec^3 x \tan^2 x \, dx$$

$$\boxed{\begin{array}{ll} u = \sec^3 x, & dv = \sec^2 x \, dx \\ du = 3\sec^3 x \tan x \, dx, & v = \tan x \end{array}} = \tan x \sec^3 x - 3 \int \sec^5 x \, dx + 3 \int \sec^3 x \, dx.$$

Rearranging terms, we get

$$4 \int \sec^5 x \, dx = \tan x \sec^3 x + 3 \int \sec^3 x \, dx.$$

We have already seen that

$$\int \sec^3 x \, dx = \frac{1}{2} \sec x \tan x + \frac{1}{2} \ln |\sec x + \tan x| + C.$$

Therefore

$$\int \sec^5 x \, dx = \frac{1}{4} \tan x \sec^3 x + \frac{3}{8} \sec x \tan x + \frac{3}{8} \ln |\sec x + \tan x| + C.$$

41.
$$\int \tan^4 x \sec^4 x \, dx = \int \tan^4 x \left(\tan^2 x + 1\right) \sec^2 x \, dx$$

$$= \int \left(\tan^6 x + \tan^4 x\right) \sec^2 x \, dx$$

$$= \frac{1}{7} \tan^7 x + \frac{1}{5} \tan^5 x + C$$

43.
$$\int \sin(x/2) \cos 2x \, dx = \int \tfrac{1}{2}\left(\sin\left(\tfrac{5}{2}x\right) - \sin\left(\tfrac{3}{2}x\right)\right) \, dx$$

$$= \tfrac{1}{3} \cos\left(\tfrac{3}{2}x\right) - \tfrac{1}{5} \cos\left(\tfrac{5}{2}x\right) + C$$

45.
$$\left\{\begin{array}{lll} u = \tan x | \, x = 0 & \Longrightarrow & u = 0 \\ du = \sec^2 x \, dx | \, x = \pi/4 & \Longrightarrow & u = 1 \end{array}\right\};$$

$$\int_0^{\pi/4} \tan^3 x \sec^2 x \, dx = \int_0^1 u^3 \, du = \left[\tfrac{1}{4} u^4\right]_0^1 = \tfrac{1}{4}$$

47.
$$\int_0^{\pi/6} \tan^2 2x \, dx = \int_0^{\pi/6} \left(\sec^2 2x - 1\right) \, dx = \left[\tfrac{1}{2} \tan 2x - x\right]_0^{\pi/6} = \frac{\sqrt{3}}{2} - \frac{\pi}{6}$$

49.
$$A = \int_0^{\pi} \sin^2 x \, dx = \int_0^{\pi} \tfrac{1}{2}(1 - \cos 2x) \, dx = \tfrac{1}{2} \left[x - \tfrac{1}{2} \sin 2x\right]_0^{\pi} = \frac{\pi}{2}$$

51.
$$V = \int_0^{\pi} \pi \left(\sin^2 x\right)^2 \, dx = \pi \int_0^{\pi} \sin^4 x \, dx = \pi \int_0^{\pi} \left(\tfrac{1}{2}(1 - \cos 2x)\right)^2 \, dx$$

$$= \frac{\pi}{4} \int_0^{\pi} \left(1 - 2\cos 2x + \cos^2 2x\right) \, dx$$

$$= \frac{\pi}{4} \left[x - \sin 2x\right]_0^{\pi} + \frac{\pi}{8} \int_0^{\pi} (1 + \cos 4x) \, dx$$

$$= \frac{\pi^2}{4} + \frac{\pi}{8} \left[x + \tfrac{1}{4} \sin 4x\right]_0^{\pi} = \frac{3\pi^2}{8}$$

53. $V = \int_0^{\pi/4} \pi \left[1^2 - \tan^2 x\right] dx = \pi \int_0^{\pi/4} \left[2 - \sec^2 x\right] dx = \pi \left[2x - \tan x\right]_0^{\pi/4} = \dfrac{\pi^2}{2} - \pi$

55. $V = \int_0^{\pi/4} \pi \left[(\tan x + 1)^2 - 1^2\right] dx = \pi \int_0^{\pi/4} \left[\tan^2 x + 2 \tan x\right] dx$

$$= \pi \int_0^{\pi/4} \left(\sec^2 x + 2 \tan x - 1\right) dx$$

$$= \pi \left[\tan x + 2 \ln |\sec x| - x\right]_0^{\pi/4} = \pi \left[\ln 2 + 1 - \dfrac{\pi}{4}\right]$$

57. Suppose $m \neq n$:

$$\int \sin mx \sin nx \, dx = \int \tfrac{1}{2}[\cos(m-n)x - \cos(m+n)x] \, dx$$

$$= \dfrac{\sin(m-n)x}{2(m-n)} - \dfrac{\sin(m+n)x}{2(m+n)} + C$$

Now suppose that $m = n$:

$$\int \sin mx \sin nx \, dx = \int \sin^2 mx \, dx = \int \tfrac{1}{2}(1 - \cos 2mx) \, dx$$

$$= \dfrac{1}{2}\left(x - \dfrac{1}{2m} \sin 2mx\right) + C = \dfrac{x}{2} - \dfrac{\sin 2mx}{4m} + C$$

59. Suppose $m \neq n$:

$$\int_{-\pi}^{\pi} \sin mx \sin nx \, dx = \left[\dfrac{\sin(m-n)x}{2(m-n)} - \dfrac{\sin(m+n)x}{2(m+n)}\right]_{-\pi}^{\pi} = 0$$

Suppose $m = n$:

$$\int_{-\pi}^{\pi} \sin^2 mx \, dx = \left[\dfrac{x}{2} - \dfrac{\sin 2mx}{4m}\right]_{-\pi}^{\pi} = \pi$$

61. Let $u = \cos^{n-1} x$, $dv = \cos x \, dx$. Then $du = (n-1)\cos^{n-2} x(-\sin x)\, dx$, $v = \sin x$ and

$$\int \cos^n x \, dx = \cos^{n-1} x \sin x + (n-1) \int \cos^{n-2} x \sin^2 x \, dx$$

$$= \cos^{n-1} x \sin x + (n-1) \int \cos^{n-2} x (1 - \cos^2 x) \, dx$$

$$= \cos^{n-1} x \sin x + (n-1) \int \cos^{n-2} x \, dx - (n-1) \int \cos^n x \, dx$$

Adding $(n-1) \int \cos^n x \, dx$ to both sides, we get

$$n \int \cos^n x \, dx = \cos^{n-1} x \sin x + (n-1) \int \cos^{n-2} x \, dx$$

so that

$$\int \cos^n x \, dx = \dfrac{1}{n} \cos^{n-1} x \sin x + \dfrac{(n-1)}{n} \int \cos^{n-2} x \, dx$$

63. $\displaystyle\int_0^{\pi/2} \sin^7 x \, dx = \dfrac{6 \cdot 4 \cdot 2}{7 \cdot 5 \cdot 3} = \dfrac{16}{35}$

65.

$$\int \cot^n x \, dx = \int \cot^{n-2} x \cot^2 x \, dx$$

$$= \int \cot^{n-2} x \left(\csc^2 x - 1 \right) \, dx$$

$$= -\frac{\cot^{n-1} x}{n-1} - \int \cot^{n-2} x \, dx$$

67. $\displaystyle \int \csc^n x \, dx = \int \csc^{n-2} x \csc^2 x \, dx$

Let $u = \csc^{n-2} x$, $dv = \csc^2 x \, dx$. Then $du = -(n-2)\csc^{n-2} x \cot x \, dx$, $v = -\cot x$ and

$$\int \csc^n x \, dx = -(n-2)\csc^{n-2} x \cot x - (n-2)\int \csc^{n-2} x \cot^2 x \, dx$$

$$= -\csc^{n-2} x \cot x - (n-2)\int \csc^{n-2} x \left(\csc^2 x - 1 \right) \, dx$$

$$= -\csc^{n-2} x \cot x - (n-2)\int \left(\csc^n x - \csc^{n-2} x \right) \, dx$$

Adding $(n-2)\displaystyle\int \csc^n x \, dx$ to both sides, we get

$$(n-1)\int \csc^n x \, dx = -\csc^{n-2} x \cot x + (n-2)\int \csc^{n-2} x \, dx.$$

Thus,

$$\int \csc^n x \, dx = \frac{-\csc^{n-2} x \cot x}{n-1} + \frac{n-2}{n-1}\int \csc^{n-2} x \, dx.$$

SECTION 8.4

1. $\left\{ \begin{array}{l} x = a \sin u \\ dx = a \cos u \, du \end{array} \right\};$ $\displaystyle \int \frac{dx}{\sqrt{a^2 - x^2}} = \int \frac{a \cos u \, du}{a \cos u}$

$$= \int du = u + C = \sin^{-1}\left(\frac{x}{a}\right) + C$$

3. $\left\{ \begin{array}{l} x = \sqrt{5} \sin u \\ dx = \sqrt{5} \cos u \, du \end{array} \right\};$ $\displaystyle \int \frac{dx}{(5 - x^2)^{3/2}} = \int \frac{\sqrt{5} \cos u \, du}{(5 \cos^2 u)^{3/2}}$

$$= \frac{1}{5}\int \sec^2 u \, du$$

$$= \frac{1}{5}\tan u + C = \frac{x}{5\sqrt{5 - x^2}} + C$$

$$\int_0^1 \frac{dx}{(5 - x^2)^{3/2}} = \left[\frac{x}{5\sqrt{5 - x^2}} \right]_0^1 = \frac{1}{10}$$

5. $\left\{ \begin{array}{l} x = \sec u \\ dx = \sec u \tan u \, du \end{array} \right\};$ $\displaystyle \int \sqrt{x^2 - 1} \, dx = \int \tan^2 u \sec u \, du$

$$= \int (\sec^3 u - \sec u) \, du$$

$$= \frac{1}{2}\sec u \tan u - \frac{1}{2}\ln|\sec u + \tan u| + C$$

$$= \frac{1}{2}x\sqrt{x^2 - 1} - \frac{1}{2}\ln|x + \sqrt{x^2 - 1}| + C$$

7. $\left\{ \begin{array}{l} x = 2\sin u \\ dx = 2\cos u\, du \end{array} \right\};$ $\quad \displaystyle\int \frac{x^2}{\sqrt{4 - x^2}}\, dx = \int \frac{4\sin^2 u}{2\cos u} 2\cos u\, du$

$$= 2\int (1 - \cos 2u)\, du$$

$$= 2u - \sin 2u + C$$

$$= 2u - 2\sin u \cos u + C$$

$$= 2\sin^{-1}\left(\frac{x}{2}\right) - \frac{1}{2}x\sqrt{4 - x^2} + C$$

9. $\left\{ \begin{array}{l} u = 1 - x^2 \\ du = -2x\, dx \end{array} \right\};$ $\quad \displaystyle\int \frac{x}{(1 - x^2)^{3/2}}\, dx = -\frac{1}{2}\int \frac{du}{u^{3/2}} = u^{-1/2} + C = \frac{1}{\sqrt{1 - x^2}} + C$

11. $\left\{ \begin{array}{l} x = \sin u \\ dx = \cos u\, du \end{array} \right\};$ $\quad \displaystyle\int \frac{x^2}{(1 - x^2)^{3/2}}\, dx = \int \frac{\sin^2 u}{\cos^3 u}\cos u\, du = \int \tan^2 u\, du$

$$= \int (\sec^2 u - 1)\, du = \tan u - u + C$$

$$= \frac{x}{\sqrt{1 - x^2}} - \sin^{-1} x + C$$

$$\int_0^{1/2} \frac{x^2}{(1 - x^2)^{3/2}}\, dx = \left[\frac{x}{\sqrt{1 - x^2}} - \sin^{-1} x \right]_0^{1/2} = \frac{2\sqrt{3} - \pi}{6}$$

13. $\left\{ \begin{array}{l} u = 4 - x^2 \\ du = -2x\, dx \end{array} \right\};$ $\quad \displaystyle\int x\sqrt{4 - x^2}\, dx = -\frac{1}{2}\int u^{1/2}\, du = -\frac{1}{3}u^{3/2} + C$

$$= -\frac{1}{3}(4 - x^2)^{3/2} + C$$

15. $\left\{ \begin{array}{l} x = 5\sin u| \; x = 0 \implies u = 0 \\ dx = 5\cos u\, du| \; x = 5 \implies u = \pi/2 \end{array} \right\};$

$$\int_0^5 x^2\sqrt{25 - x^2}\, dx = \int_0^{\pi/2} (5\sin u)^2 (5\cos u)^2\, du$$

$$= 625 \int_0^{\pi/2} (\sin^2 u - \sin^4 u)\, du$$

$$= 625 \left[\frac{1}{2}\cdot\frac{\pi}{2} - \frac{3\cdot 1}{4\cdot 2}\cdot\frac{\pi}{2} \right] = \frac{625\pi}{16} \qquad \text{[see Exercise 62, Section 8.3]}$$

17. $\left\{ \begin{array}{l} x = \sqrt{8}\tan u \\ dx = \sqrt{8}\sec^2 u\, du \end{array} \right\};$ $\quad \displaystyle\int \frac{x^2}{(x^2 + 8)^{3/2}}\, dx = \int \frac{8\tan^2 u}{(8\sec^2 u)^{3/2}}\sqrt{8}\sec^2 u\, du$

$$= \int \frac{\tan^2 u}{\sec u}\, du = \int \frac{\sec^2 u - 1}{\sec u}\, du$$

$$= \int (\sec u - \cos u)\, du$$

$$= \ln|\sec u + \tan u| - \sin u + C$$

$$= \ln \left(\frac{\sqrt{x^2 + 8} + x}{\sqrt{8}} \right) - \frac{x}{\sqrt{x^2 + 8}} + C$$

(absorb $- \ln \sqrt{8}$ in C)

$$= \ln \left(\sqrt{x^2 + 8} + x \right) - \frac{x}{\sqrt{x^2 + 8}} + C$$

19. $\left\{ \begin{array}{l} x = a \sin u \\ dx = a \cos u \, du \end{array} \right\}$; $\displaystyle\int \frac{dx}{x\sqrt{a^2 - x^2}} = \int \frac{a \cos u \, du}{a \sin u \, (a \cos u)} = \frac{1}{a} \int \csc u \, du$

$$= \frac{1}{a} \ln |\csc u - \cot u| + C$$

$$= \frac{1}{a} \ln \left| \frac{a - \sqrt{a^2 - x^2}}{x} \right| + C$$

21. $\left\{ \begin{array}{l} x = a \sec u \\ dx = a \sec u \tan u \, du \end{array} \right\}$; $\displaystyle\int \frac{dx}{\sqrt{x^2 - a^2}} = \int \frac{a \sec u \tan u \, du}{a \tan u} = \int \sec u \, du$

$$= \ln |\sec u + \tan u| + C$$

$$= \ln \left| \frac{x + \sqrt{x^2 - a^2}}{a} \right| + C$$

(absorb $- \ln a$ in C)

$$= \ln \left| x + \sqrt{x^2 - a^2} \right| + C$$

23. $\left\{ \begin{array}{l} x = 3 \tan u \\ dx = 3 \sec^2 u \, du \end{array} \right\}$; $\displaystyle\int \frac{x^3}{\sqrt{9 + x^2}} \, dx = \int \frac{27 \tan^3 u}{3 \sec u} \cdot 3 \sec^2 u \, du$

$$= 27 \int \tan^3 u \sec u \, du$$

$$= 27 \int \left(\sec^2 u - 1 \right) \sec u \tan u \, du$$

$$= 27 \left[\tfrac{1}{3} \sec^3 u - \sec u \right] + C$$

$$= \tfrac{1}{3} \left(9 + x^2 \right)^{3/2} - 9 \left(9 + x^2 \right)^{1/2} + C$$

$$\int_0^3 \frac{x^3}{\sqrt{9 + x^2}} \, dx = \left[\tfrac{1}{3} \left(9 + x^2 \right)^{3/2} - 9 \left(9 + x^2 \right)^{1/2} \right]_0^3 = 18 - 9\sqrt{2}$$

25. $\left\{ \begin{array}{l} x = a \tan u \\ dx = a \sec^2 u \, du \end{array} \right\}$; $\displaystyle\int \frac{dx}{x^2 \sqrt{a^2 + x^2}} = \int \frac{a \sec^2 u \, du}{a^2 \tan^2 u \, (a \sec u)}$

$$= \frac{1}{a^2} \int \frac{\sec u}{\tan^2 u} \, du$$

$$= \frac{1}{a^2} \int \cot u \csc u \, du$$

$$= -\frac{1}{a^2}\cos u + C = -\frac{1}{a^2 x}\sqrt{a^2 + x^2} + C$$

27. $\left\{\begin{array}{l} x = a\sec u \\ dx = a\sec u\tan u\,du \end{array}\right\};\ \displaystyle\int \frac{dx}{x^2\sqrt{x^2 - a^2}} = \int \frac{a\sec u\tan u\,du}{a^2\sec^2 u\,(a\tan u)}$

$$= \frac{1}{a^2}\int \cos u\,du$$

$$= \frac{1}{a^2}\sin u + C$$

$$= \frac{1}{a^2 x}\sqrt{x^2 - a^2} + C$$

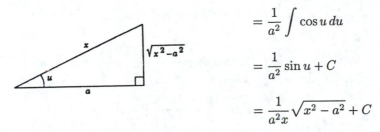

29. $\left\{\begin{array}{l} e^x = 3\sec u \\ e^x\,dx = 3\sec u\tan u\,du \end{array}\right\};\ \displaystyle\int \frac{dx}{e^x\sqrt{e^{2x} - 9}} = \int \frac{\tan u\,du}{3\sec u\,(3\tan u)}$

$$= \frac{1}{9}\int \cos u\,du$$

$$= \tfrac{1}{9}\sin u + C$$

$$= \tfrac{1}{9}e^{-x}\sqrt{e^{2x} - 9} + C$$

31. $(x^2 - 4x + 4)^{\frac{3}{2}} = \left\{\begin{array}{ll} (x - 2)^3, & x > 2 \\ (2 - x)^3, & x < 2 \end{array}\right.$

$$\int \frac{dx}{(x^2 - 4x + 4)^{3/2}} = \left\{\begin{array}{ll} -\dfrac{1}{2(x - 2)^2} + C, & x > 2 \\[2mm] \dfrac{1}{2(2 - x)^2} + C, & x < 2 \end{array}\right.$$

33. $\left\{\begin{array}{l} x - 3 = \sin u \\ dx = \cos u\,du \end{array}\right\};\ \displaystyle\int x\sqrt{6x - x^2 - 8}\,dx = \int x\sqrt{1 - (x - 3)^2}\,dx$

$$= \int (3 + \sin u)(\cos u)\cos u\,du$$

$$= \int (3\cos^2 u + \cos^2 u\sin u)\,du$$

$$= \int \left[3\left(\frac{1 + \cos 2u}{2}\right) + \cos^2 u\sin u\right] du$$

$$= \frac{3u}{2} + \frac{3}{4}\sin 2u - \frac{1}{3}\cos^3 u + C$$

$$= \frac{3}{2}\sin^{-1}(x - 3) + \frac{3}{2}(x - 3)\sqrt{6x - x^2 - 8} - \frac{1}{3}(6x - x^2 - 8)^{3/2} + C$$

35. $\left\{\begin{array}{l} x + 1 = 2\tan u \\ dx = 2\sec^2 u\,du \end{array}\right\};\ \displaystyle\int \frac{x}{(x^2 + 2x + 5)^2}\,dx = \int \frac{x}{[(x + 1)^2 + 4]^2}\,dx$

$$= \int \frac{2\tan u - 1}{(4\sec^2 u)^2} 2\sec^2 u\, du$$

$$= \frac{1}{8} \int \frac{2\tan u - 1}{\sec^2 u}\, du$$

$$= \frac{1}{8} \int (2\sin u \cos u - \cos^2 u)\, du$$

$$= \frac{1}{8} \int \left(2\sin u \cos u - \frac{1 + \cos 2u}{2} \right) du$$

$$= \frac{1}{8} \left(\sin^2 u - \frac{u}{2} - \frac{\sin 2u}{4} \right) + C$$

$$= \frac{1}{8} \left[\left(\frac{x+1}{\sqrt{x^2+2x+5}} \right)^2 - \frac{1}{2} \tan^{-1}\left(\frac{x+1}{2} \right) - \frac{1}{4} \overbrace{(2)\left(\frac{x+1}{\sqrt{x^2+2x+5}} \right)\left(\frac{2}{\sqrt{x^2+2x+5}} \right)}^{\sin 2u = 2\sin u \cos u} \right] + C$$

$$= \frac{x^2 + x}{8(x^2 + 2x + 5)} - \frac{1}{16}\tan^{-1}\left(\frac{x+1}{2} \right) + C$$

37. $\left\{ \begin{array}{l} x - 3 = \sin u \\ dx = \cos u\, du \end{array} \right\};\quad \int \sqrt{6x - x^2 - 8}\, dx = \int \sqrt{1 - (x-3)^2}\, dx$

$$= \int \cos^2 u\, du$$

$$= \int \frac{1 + \cos 2u}{2}\, du$$

$$= \frac{u}{2} + \frac{\sin 2u}{4} + C$$

$$= \tfrac{1}{2} \left[\sin^{-1}(x - 3) + (x - 3)\sqrt{6x - x^2 - 8} \right] + C$$

39. Let $x = a\tan u$. Then $dx = a\sec^2 u\, du,\ \sqrt{x^2 + a^2} = a\sec u,$ and

$$\int \frac{dx}{(x^2 + a^2)^n} = \int \frac{a\sec^2 u}{a^{2n}\sec^{2n} u}\, du = \frac{1}{a^{2n-1}} \int \cos^{2n-2} u\, du$$

41. $\displaystyle \int \frac{1}{(x^2 + 1)^3}\, dx = \int \cos^4 u\, du$

$$= \frac{1}{4}\cos^3 u \sin u + \frac{3}{8}\cos u \sin u + \frac{3}{8}u + C \quad \text{[by the reduction formula (8.3.2)]}$$

$$= \frac{3}{8}\tan^{-1} x + \frac{3x}{8(x^2 + 1)} + \frac{x}{4(x^2 + 1)^2} + C$$

43.

$$\boxed{\begin{array}{ll} u = \sin^{-1} x & dv = x\,dx \\[2mm] du = \dfrac{1}{\sqrt{1-x^2}}\,dx & v = \dfrac{x^2}{2} \end{array}}$$

$$\int x\sin^{-1} x\,dx = \frac{x^2}{2}\sin^{-1} x - \frac{1}{2}\int \frac{x^2}{\sqrt{1-x^2}}\,dx$$

$$\boxed{x = \sin u \quad dx = \cos u\,du}$$

$$= \frac{x^2}{2}\sin^{-1} x - \frac{1}{2}\int \sin^2 u\,du$$

$$= \frac{x^2}{2}\sin^{-1} x - \frac{1}{2}\left[\frac{1}{2}u - \frac{1}{4}\sin 2u\right] + C$$

$$= \frac{x^2}{2}\sin^{-1} x - \frac{1}{4}\sin^{-1} x + \frac{1}{8}(2x\sqrt{1-x^2}) + C$$

$$= \sin^{-1} x\left(\frac{x^2}{2} - \frac{1}{4}\right) + \frac{1}{4}x\sqrt{1-x^2} + C$$

45.

$$V = \int_0^1 \pi\left(\frac{1}{1+x^2}\right)^2 dx = \pi\int_0^1 \frac{1}{(1+x^2)^2}\,dx$$

$$= \pi\int_0^{\pi/4} \cos^2 u\,du \qquad [x = \tan u, \;\; \text{see Ex.45}]$$

$$= \frac{\pi}{2}\int_0^{\pi/4} (1 + \cos 2u)\,du$$

$$= \frac{\pi}{2}\left[u + \tfrac{1}{2}\sin 2u\right]_0^{\pi/4} = \frac{\pi^2}{8} + \frac{\pi}{4}$$

47.   We need only consider angles θ between 0 and π. Assume first that $0 \le \theta \le \frac{\pi}{2}$.

The area of the triangle is: $\frac{1}{2}r^2\sin\theta\,\cos\theta$.

The area of the other region is given by:

$$\int_{r\cos\theta}^{r}\sqrt{r^2 - x^2}\,dx = \left[\frac{x}{2}\sqrt{r^2 - x^2} + \frac{r^2}{2}\sin^{-1}\frac{x}{r}\right]_{r\cos\theta}^{r}$$

$$= \frac{\pi r^2}{4} - \frac{r^2}{2}\sin\theta\,\cos\theta - \frac{r^2}{2}\sin^{-1}(\cos\theta)$$

$$= \frac{r^2\theta}{2} - \frac{r^2}{2}\sin\theta\,\cos\theta$$

Thus, the area of the sector is $A = \frac{1}{2}r^2\theta$.

Now suppose that $\dfrac{\pi}{2} < \theta \le \pi$. Then $A = \frac{1}{2}\pi r^2 - \frac{1}{2}r^2(\pi - \theta) = \frac{1}{2}r^2\theta$.

49. $A = 2\int_3^5 4\sqrt{\dfrac{x^2}{9} - 1}\,dx = 24\int_1^{\frac{5}{3}} \sqrt{u^2 - 1}\,du$ (where $u = \frac{x}{3}$)

$\qquad = 24\left[\dfrac{u}{2}\sqrt{u^2 - 1} - \dfrac{1}{2}\ln|u + \sqrt{u^2 - 1}|\right]_1^{\frac{5}{3}} = \dfrac{80}{3} - 12\ln 3.$

51. $\qquad M = \int_0^a \dfrac{dx}{\sqrt{x^2 + a^2}} = \left[\ln\left(x + \sqrt{x^2 + a^2}\right)\right]_0^a = \ln\left(1 + \sqrt{2}\right)$

$\qquad x_M M = \int_0^a \dfrac{x}{\sqrt{x^2 + a^2}}\,dx = \left[\sqrt{x^2 + a^2}\right]_0^a = (\sqrt{2} - 1)a \qquad x_M = \dfrac{(\sqrt{2} - 1)a}{\ln\left(1 + \sqrt{2}\right)}$

53.

$A = \int_a^{\sqrt{2}a} \sqrt{x^2 - a^2}\,dx = \left[\dfrac{1}{2}x\sqrt{x^2 - a^2} - \dfrac{1}{2}a^2\ln\left|x + \sqrt{x^2 - a^2}\right|\right]_a^{\sqrt{2}a}$

$\qquad\qquad = \frac{1}{2}a^2[\sqrt{2} - \ln(\sqrt{2} + 1)]$

$\bar{x}A = \int_a^{\sqrt{2}a} x\sqrt{x^2 - a^2}\,dx = \dfrac{1}{3}a^3, \qquad \bar{y}A = \int_a^{\sqrt{2}a}\left[\dfrac{1}{2}(x^2 - a^2)\right]dx = \dfrac{1}{6}a^3(2 - \sqrt{2})$

$\qquad \bar{x} = \dfrac{2a}{3[\sqrt{2} - \ln(\sqrt{2} + 1)]}, \qquad \bar{y} = \dfrac{(2 - \sqrt{2})a}{3[\sqrt{2} - \ln(\sqrt{2} + 1)]}$

55. $V_y = 2\pi\overline{R}A = \dfrac{2}{3}\pi a^3 \qquad \bar{y}V_y = \int_a^{\sqrt{2}a} \pi x(x^2 - a^2)\,dx = \dfrac{1}{4}\pi a^4, \quad \bar{y} = \dfrac{3}{8}a$

57. (a)

(b) $A = \int_3^6 \dfrac{\sqrt{x^2 - 9}}{x^2}\,dx$

$\qquad = \int_0^{\pi/3} \dfrac{3\tan u}{9\sec^2 u} \cdot 3\sec u \tan u\,du \quad [x = 3\sec u]$

$\qquad = \int_0^{\pi/3} (\sec u - \cos u)\,du$

$\qquad = [\,\ln|\sec u + \tan u| - \sin u\,]_0^{\pi/3}$

$\qquad = \ln(2 + \sqrt{3}) - \dfrac{\sqrt{3}}{2}$

(c) $\bar{x}A = \int_3^6 x \cdot \dfrac{\sqrt{x^2 - 9}}{x^2}\,dx = \int_0^{\pi/3} (3\sec^2 u - 3)\,du = [3\tan u - 3u]_0^{\pi/3} = 3\sqrt{3} - \pi$

Thus, $\bar{x} = \dfrac{2(3\sqrt{3} - \pi)}{2\ln(2 + \sqrt{3}) - \sqrt{3}}.$

$\bar{y}A = \int \dfrac{1}{2}\left(\dfrac{\sqrt{x^2 - 9}}{x^2}\right)^2 dx = \dfrac{1}{2}\int_3^6\left(\dfrac{1}{x^2} - \dfrac{9}{x^4}\right)dx = \dfrac{1}{2}\left[-\dfrac{1}{x} + \dfrac{3}{x^3}\right]_3^6 = \dfrac{5}{144}$

Thus, $\bar{y} = \dfrac{5}{72\left[2\ln\left(2+\sqrt{3}\right) - \sqrt{3}\right]}$.

SECTION 8.5

1. $\dfrac{1}{x^2 + 7x + 6} = \dfrac{1}{(x+1)(x+6)} = \dfrac{A}{x+1} + \dfrac{B}{x+6}$

$1 = A(x+6) + B(x+1)$

$x = -6: \quad 1 = -5B \quad \Longrightarrow \quad B = -1/5$

$x = -1: \quad 1 = 5A \quad \Longrightarrow \quad A = 1/5$

$\dfrac{1}{x^2 + 7x + 6} = \dfrac{1/5}{x+1} - \dfrac{1/5}{x+6}$

3. $\dfrac{x}{x^4 - 1} = \dfrac{x}{(x^2+1)(x+1)(x-1)} = \dfrac{Ax+B}{x^2+1} + \dfrac{C}{x+1} + \dfrac{D}{x-1}$

$x = (Ax+B)(x^2-1) + C(x-1)(x^2+1) + D(x+1)(x^2+1)$

$x = 1: \quad 1 = 4D \quad \Longrightarrow \quad D = 1/4$

$x = -1: \quad -1 = -4C \quad \Longrightarrow \quad C = 1/4$

$x = 0: \quad -B - C + D = 0 \Longrightarrow B = 0$

$x = 2: \quad 6A + 5C + 15D = 2 \Longrightarrow A = -1/2$

$\dfrac{x}{x^4 - 1} = \dfrac{1/4}{x-1} + \dfrac{1/4}{x+1} - \dfrac{x/2}{x^2+1}$

5. $\dfrac{x^2 - 3x - 1}{x^3 + x^2 - 2x} = \dfrac{x^2 - 3x - 1}{x(x+2)(x+1)} = \dfrac{A}{x} + \dfrac{B}{x+2} + \dfrac{C}{x-1}$

$x^2 - 3x - 1 = A(x+2)(x-1) + Bx(x-1) + Cx(x+2)$

$x = 0: \quad -1 = -2A \quad \Longrightarrow \quad A = 1/2$

$x = -2: \quad 9 = 6B \quad \Longrightarrow \quad B = 3/2$

$x = 1: \quad -3 = 3C \Longrightarrow C = -1$

$\dfrac{x^2 - 3x - 1}{x^3 + x^2 - 2x} = \dfrac{1/2}{x} + \dfrac{3/2}{x+2} - \dfrac{1}{x-1}$

7. $\dfrac{2x^2 + 1}{x^3 - 6x^2 + 11x - 6} = \dfrac{2x^2 + 1}{(x-1)(x-2)(x-3)} = \dfrac{A}{x-1} + \dfrac{B}{x-2} + \dfrac{C}{x-3}$

$2x^2 + 1 = A(x-2)(x-3) + B(x-1)(x-3) + C(x-1)(x-2)$

$x = 1: \quad 3 = 2A \quad \Longrightarrow \quad A = 3/2$

$x = 2: \quad 9 = -B \quad \Longrightarrow \quad B = -9$

$x = 3: \quad 19 = 2C \Longrightarrow C = 19/2$

$$\frac{2x^2 + 1}{x^3 - 6x^2 + 11x - 6} = \frac{3/2}{x - 1} - \frac{9}{x - 2} + \frac{19/2}{x - 3}$$

9.
$$\frac{7}{(x - 2)(x + 5)} = \frac{A}{x - 2} + \frac{B}{x + 5}$$

$$7 = A(x + 5) + B(x - 2)$$

$x = -5: \quad 7 = -7B \quad \Longrightarrow \quad B = -1$

$x = 2: \quad 7 = 7A \quad \Longrightarrow \quad A = 1$

$$\int \frac{7}{(x - 2)(x + 5)}\, dx = \int \left(\frac{1}{x - 2} - \frac{1}{x + 5} \right) = \ln|x - 2| - \ln|x + 5| + C = \ln\left| \frac{x - 2}{x + 5} \right| + C$$

11.
$$\frac{2x^2 + 3}{x^2(x - 1)} = \frac{A}{x} + \frac{B}{x^2} + \frac{C}{x - 1}$$

$$2x^2 + 3 = Ax(x - 1) + B(x - 1) + Cx^2$$

$x = 0: \quad 3 = -B \quad \Longrightarrow \quad B = -3$

$x = 1: \quad 5 = C \quad \Longrightarrow \quad C = 5$

$x = -1: \quad 5 = 2A - 2B + C \quad \Longrightarrow \quad A = -3$

$$\int \frac{2x^2 + 3}{x^2(x - 1)}\, dx = \int \left(-\frac{3}{x} - \frac{3}{x^2} + \frac{5}{x - 1} \right) dx = -3\ln|x| + \frac{3}{x} + 5\ln|x - 1| + C$$

13. We carry out the division until the numerator has degree smaller than the denominator:

$$\frac{x^5}{(x - 2)^2} = \frac{x^5}{x^2 - 4x + 4} = x^3 + 4x^2 + 12x + 32 + \frac{80x - 128}{(x - 2)^2}.$$

Then,

$$\frac{80x - 128}{(x - 2)^2} = \frac{80x - 160 + 32}{(x - 2)^2} = \frac{80}{x - 2} + \frac{32}{(x - 2)^2}$$

$$\int \frac{x^5}{(x - 2)^2}\, dx = \int \left(x^3 + 4x^2 + 12x + 32 + \frac{80}{x - 2} + \frac{32}{(x - 2)^2} \right) dx$$

$$= \frac{1}{4}x^4 + \frac{4}{3}x^3 + 6x^2 + 32x + 80\ln|x - 2| - \frac{32}{x - 2} + C.$$

15.
$$\frac{x + 3}{x^2 - 3x + 2} = \frac{A}{x - 1} + \frac{B}{x - 2}$$

$$x + 3 = A(x - 2) + B(x - 1)$$

$x = 1: \quad 4 = -A \quad \Longrightarrow \quad A = -4$

$x = 2: \quad 5 = B \quad \Longrightarrow \quad B = 5$

$$\int \frac{x + 3}{x^2 - 3x + 2}\, dx = \int \left(\frac{-4}{x - 1} + \frac{5}{x - 2} \right) dx = -4 \ln |x - 1| + 5 \ln |x - 2| + C$$

17. $\displaystyle \int \frac{dx}{(x - 1)^3} = \int (x - 1)^{-3}\, dx = -\frac{1}{2}(x - 1)^{-2} + C = -\frac{1}{2(x - 1)^2} + C$

19.
$$\frac{x^2}{(x - 1)^2(x + 1)} = \frac{A}{x - 1} + \frac{B}{(x - 1)^2} + \frac{C}{x + 1}$$

$$x^2 = A(x - 1)(x + 1) + B(x + 1) + C(x - 1)^2$$

$x = 1: \quad 1 = 2B \quad \Longrightarrow \quad B = 1/2$

$x = -1: \quad 1 = 4C \quad \Longrightarrow \quad C = 1/4$

$x = 0: \quad 0 = -A + B + C \quad \Longrightarrow \quad A = 3/4$

$$\int \frac{x^2}{(x - 1)^2(x + 1)}\, dx = \int \left(\frac{3/4}{x - 1} + \frac{1/2}{(x - 1)^2} + \frac{1/4}{x + 1} \right) dx$$

$$= \frac{3}{4} \ln |x - 1| - \frac{1}{2(x - 1)} + \frac{1}{4} \ln |x + 1| + C$$

21.
$$x^4 - 16 = (x^2 - 4)(x^2 + 4) = (x - 2)(x + 2)(x^2 + 4)$$

$$\frac{1}{x^4 - 16} = \frac{A}{x - 2} + \frac{B}{x + 2} + \frac{Cx + D}{x^2 + 4}$$

$$1 = A(x + 2)(x^2 + 4) + B(x - 2)(x^2 + 4) + (Cx + D)(x^2 - 4)$$

$x = 2: \quad 1 = 32A \quad \Longrightarrow \quad A = 1/32$

$x = -2: \quad 1 = -32B \quad \Longrightarrow \quad B = -1/32$

$x = 0: \quad 1 = 8A - 8B - 4D \quad \Longrightarrow \quad D = -1/8$

$x = 1: \quad 1 = 15A - 5B - 3C - 3D \quad \Longrightarrow \quad C = 0$

$$\int \frac{dx}{x^4 - 16} = \int \left(\frac{1/32}{x - 2} - \frac{1/32}{x + 2} - \frac{1/8}{x^2 + 4} \right) dx$$

$$= \frac{1}{32} \ln |x - 2| - \frac{1}{32} \ln |x + 2| - \frac{1}{8} \left(\frac{1}{2} \tan^{-1} \frac{x}{2} \right) + C$$

$$= \frac{1}{32} \ln \left| \frac{x - 2}{x + 2} \right| - \frac{1}{16} \tan^{-1} \frac{x}{2} + C$$

23.
$$\frac{x^3 + 4x^2 - 4x - 1}{(x^2 + 1)^2} = \frac{Ax + B}{x^2 + 1} + \frac{Cx + D}{(x^2 + 1)^2}$$

$$x^3 + 4x^2 - 4x - 1 = (Ax + B)(x^2 + 1) + (Cx + D)$$

$x = 0$: $-1 = B + D$ $\implies D = -B - 1$

$x = 1$: $0 = 2A + 2B + C + D$ $\implies 6 = 4B + 2D$ $\implies B = 4, \ D = -5$

$x = -1$: $6 = -2A + 2B - C + D$ $6 = -2A + 8 - C - 5$

$x = 2$: $15 = 10A + 5B + 2C + D$ $15 = 10A + 20 + 2C - 5$ $\implies A = 1, \ C = -5$

$$\int \frac{x^3 + 4x^2 - 4x - 1}{(x^2 + 1)^2} \, dx = \int \left(\frac{x}{x^2 + 1} + \frac{4}{x^2 + 1} - \frac{5x}{(x^2 + 1)^2} - \frac{5}{(x^2 + 1)^2} \right) dx$$

(∗)
$$= \frac{1}{2} \ln(x^2 + 1) + 4 \tan^{-1} x + \frac{5}{2(x^2 + 1)} - 5 \int \frac{dx}{(x^2 + 1)^2}$$

For this last integral we set

$$\left\{ \begin{array}{l} x = \tan u \\ dx = \sec^2 u \, du \end{array} \right\}; \quad \int \frac{dx}{(x^2 + 1)^2} = \int \frac{\sec^2 u \, du}{(1 + \tan^2 u)^2} = \int \cos^2 u \, du$$

$$= \frac{1}{2} \int (1 + \cos 2u) \, du$$

$$= \frac{1}{2} \left(u + \frac{1}{2} \sin 2u \right) + C = \frac{1}{2} (u + \sin u \cos u) + C$$

$$= \frac{1}{2} \left(\tan^{-1} x + \frac{x}{1 + x^2} \right) + C.$$

Substituting this result in (∗) and rearranging the terms, we get

$$\int \frac{x^3 + 4x^2 - 4x + 1}{(x^2 + 1)^2} \, dx = \frac{1}{2} \ln(x^2 + 1) + \frac{3}{2} \tan^{-1} x + \frac{5(1 - x)}{2(1 + x^2)} + C.$$

25.
$$\frac{1}{x^4 + 4} = \frac{Ax + B}{x^2 + 2x + 2} + \frac{Cx + D}{x^2 - 2x + 2} \qquad \text{(using the hint)}$$

$$1 = (Ax + B)(x^2 - 2x + 2) + (Cx + D)(x^2 + 2x + 2)$$

$x = \ 0$: $1 = 2B + 2D$ $A = \ 1/8$
$x = \ 1$: $1 = A + B + 5C + 5D$ $\implies$ $B = \ 1/4$
$x = -1$: $1 = -5A + 5B - C + D$ $C = -1/8$
$x = \ 2$: $1 = 4A + 2B + 20C + 10D$ $D = \ 1/4$

$$\int \frac{dx}{x^4 + 4} = \frac{1}{8} \int \frac{x + 2}{x^2 + 2x + 2} \, dx - \frac{1}{8} \int \frac{x - 2}{x^2 - 2x + 2} \, dx$$

$$= \frac{1}{8} \int \frac{x + 1}{x^2 + 2x + 2} \, dx + \frac{1}{8} \int \frac{dx}{(x + 1)^2 + 1} - \frac{1}{8} \int \frac{x - 1}{x^2 - 2x + 2} \, dx + \frac{1}{8} \int \frac{dx}{(x - 1)^2 + 1}$$

$$= \frac{1}{16} \ln \left(x^2 + 2x + 2 \right) + \frac{1}{8} \tan^{-1} \left(x + 1 \right) - \frac{1}{16} \ln \left(x^2 - 2x + 2 \right) + \frac{1}{8} \tan^{-1} \left(x - 1 \right) + C$$

$$= \frac{1}{16} \ln \left(\frac{x^2 + 2x + 2}{x^2 - 2x + 2} \right) + \frac{1}{8} \tan^{-1} \left(x + 1 \right) + \frac{1}{8} \tan^{-1} \left(x - 1 \right) + C$$

27.
$$\frac{x - 3}{x^3 + x^2} = \frac{x - 3}{x^2 (x + 1)} = \frac{A}{x} + \frac{B}{x^2} + \frac{C}{x + 1}$$

$$x - 3 = Ax(x + 1) + B(x + 1) + Cx^2$$

$x = 0:$ $-3 = B$

$x = -1:$ $-4 = C$

$x = 1:$ $-2 = 2A + 2B + C \implies A = 4$

$$\int \frac{x - 3}{x^3 + x^2} \, dx = 4 \int \frac{1}{x} \, dx - 3 \int \frac{1}{x^2} \, dx - 4 \int \frac{1}{x + 1} \, dx$$

$$= 4 \ln |x| + \frac{3}{x} - 4 \ln |x + 1| + C = \frac{3}{x} + 4 \ln \left| \frac{x}{x + 1} \right| + C$$

29.
$$\frac{x + 1}{x^3 + x^2 - 6x} = \frac{x + 1}{x(x - 2)(x + 3)} = \frac{A}{x} + \frac{B}{x - 2} + \frac{C}{x + 3}$$

$$x + 1 = A(x - 2)(x + 3) + Bx(x + 3) + Cx(x - 2)$$

$x = 0:$ $1 = -6A \implies A = -1/6$

$x = 2:$ $3 = 10B \implies B = 3/10$

$x = -3:$ $-2 = 15C \implies C = -2/15$

$$\int \frac{x + 1}{x^3 + x^2 - 6x} \, dx = -\frac{1}{6} \int \frac{1}{x} \, dx + \frac{3}{10} \int \frac{1}{x - 2} \, dx - \frac{2}{15} \int \frac{1}{x + 3} \, dx$$

$$= -\frac{1}{6} \ln |x| + \frac{3}{10} \ln |x - 2| - \frac{2}{15} \ln |x + 3| + C$$

31.
$$\int_0^2 \frac{x}{x^2 + 5x + 6} \, dx = \int_0^2 \frac{x}{(x + 2)(x + 3)} \, dx = \int_0^2 \left(\frac{3}{x + 3} - \frac{2}{x + 2} \right) dx$$

$$= [3 \ln |x + 3| - 2 \ln |x + 2|]_0^2 = \ln \left(\frac{125}{108} \right)$$

33.

$$\int_1^3 \frac{x^2 - 4x + 3}{x^3 + 2x^2 + x}\, dx = \int_1^3 \frac{x^2 - 4x + 3}{x(x+1)^2}\, dx = \int_1^3 \left(\frac{3}{x} - \frac{2}{x+1} - \frac{8}{(x+1)^2} \right) dx$$

$$= \left[3\ln|x| - 2\ln|x+1| + \frac{8}{x+1} \right]_1^3 = \ln\left(\frac{27}{4} \right) - 2$$

35.

$$\int \frac{\cos\theta}{\sin^2\theta - 2\sin\theta - 8}\, d\theta = \frac{1}{6}\int \frac{\cos\theta}{\sin\theta - 4}\, d\theta - \frac{1}{6}\int \frac{\cos\theta}{\sin\theta + 2}\, d\theta$$

$$= \frac{1}{6}\ln|\sin\theta - 4| - \frac{1}{6}\ln|\sin\theta + 2| + C$$

$$= \frac{1}{6}\ln\left| \frac{\sin\theta - 4}{\sin\theta + 2} \right| + C$$

37.

$$\int \frac{1}{t[(\ln t)^2 - 4]}\, dt = \frac{1}{4}\int \frac{1}{t(\ln t - 2)}\, dt - \frac{1}{4}\int \frac{1}{t(\ln t + 2)}\, dt$$

$$= \frac{1}{4}\ln|\ln t - 2| - \frac{1}{4}\ln|\ln t + 2| + C$$

$$= \frac{1}{4}\ln\left| \frac{\ln t - 2}{\ln t + 2} \right| + C$$

39.

$$\int \frac{u}{a+bu}\, du = \frac{1}{b}\int \left[1 - \frac{a}{a+bu} \right] du$$

$$= \frac{1}{b}\left[u - \frac{a}{b}\ln|a+bu| \right] + C$$

$$= \frac{1}{b^2}\left[a + bu - a\ln|a+bu| \right] + C$$

41.

$$\int \frac{1}{u^2(a+bu)}\, du = -\frac{b}{a^2}\int \frac{1}{u}\, du + \frac{1}{a}\int \frac{1}{u^2}\, du + \frac{b}{a^2}\int \frac{b}{a+bu}\, du$$

$$= \frac{b}{a^2}(\ln|a+bu| - \ln|u|) - \frac{1}{au} + C$$

$$= \frac{b}{a^2}\ln\left| \frac{u}{a+bu} \right| - \frac{1}{au} + C$$

43.

$$\int \frac{1}{(a+bu)(c+du)}\, du = -\frac{1}{ad-bc}\int \frac{b}{a+bu}\, du + \frac{1}{ad-bc}\int \frac{d}{c+du}\, du$$

$$= \frac{1}{ad-bc}\left(\ln\left| \frac{c+du}{a+bu} \right| \right) + C$$

45.

$$\int \frac{u}{a^2 - u^2}\, du = -\frac{1}{2} \int \frac{1}{v}\, dv, \quad \text{where} \quad v = a^2 - u^2$$

$$= -\frac{1}{2} \ln v + C$$

$$= -\frac{1}{2} \ln |a^2 - u^2| + C$$

47. Note that

$$y = \frac{1}{x^2 - 1} = \frac{1}{2}\left[\frac{1}{x-1} - \frac{1}{x+1}\right]$$

and thus

$$\frac{d^0 y}{dx^0} = \left(\frac{1}{2}\right)(-1)^0\, 0!\left[\frac{1}{(x-1)^{0+1}} - \frac{1}{(x+1)^{0+1}}\right].$$

The rest is a routine induction.

49.

$$A = \int_0^1 \frac{dx}{x^2 + 1} = \left[\tan^{-1} x\right]_0^1 = \frac{1}{4}\pi$$

$$\bar{x} A = \int_0^1 \frac{x}{x^2 + 1}\, dx = \frac{1}{2}\left[\ln(x^2 + 1)\right]_0^1 = \frac{1}{2}\ln 2$$

$$\bar{y} A = \int_0^1 \frac{dx}{2(x^2 + 1)^2} = \frac{1}{2}\left[\tan^{-1} x + \frac{x}{x^2 + 1}\right]_0^1 = \frac{1}{8}(\pi + 2)$$

$$\bar{x} = \frac{2\ln 2}{\pi}; \quad \bar{y} = \frac{\pi + 2}{2\pi}$$

51.   (a)

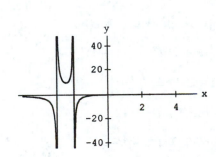

(b) $A = \displaystyle\int_0^4 \frac{x}{x^2 + 5x + 6}\, dx$

$$= \int_0^4 \frac{x}{(x+2)(x+3)}\, dx$$

$$= \int_0^4 \left(\frac{3}{x+3} - \frac{2}{x+2}\right) dx$$

$$= \left[3\ln|x+3| - 2\ln|x+2|\right]_0^4 = 3\ln 7 - 5\ln 3$$

53.   (a)

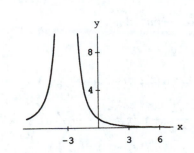

(b) $A = \displaystyle\int_{-2}^9 \frac{9 - x}{(x+3)^2}\, dx$

$$= \int_{-2}^9 \left(\frac{12}{(x+3)^2} - \frac{1}{x+3}\right) dx$$

$$= \left[-\ln|x+3| - \frac{12}{x+3}\right]_{-2}^9 = 11 - \ln 12$$

1. $\left\{\begin{array}{l} x = u^2 \\ dx = 2u\,du \end{array}\right\};$ $\displaystyle\int \frac{dx}{1-\sqrt{x}} = \int \frac{2u\,du}{1-u} = 2\int\left(\frac{1}{1-u}-1\right)du$

$$= -2(\ln|1-u|+u) + C = -2(\sqrt{x}+\ln|1-\sqrt{x}|) + C$$

3. $\left\{\begin{array}{l} u^2 = 1+e^x \\ 2u\,du = e^x\,dx \end{array}\right\};$ $\displaystyle\int \sqrt{1+e^x}\,dx = \int u\cdot\frac{2u\,du}{u^2-1} = 2\int\left(1+\frac{1}{u^2-1}\right)du$

$$= 2\int\left(1+\frac{1}{2}\left[\frac{1}{u-1}-\frac{1}{u+1}\right]\right)du$$

$$= 2u + \ln|u-1| - \ln|u+1| + C$$

$$= 2\sqrt{1+e^x} + \ln\left[\frac{\sqrt{1+e^x}-1}{\sqrt{1+e^x}+1}\right] + C$$

$$= 2\sqrt{1+e^x} + \ln\left[\frac{\left(\sqrt{1+e^x}-1\right)^2}{e^x}\right] + C$$

$$= 2\sqrt{1+e^x} + 2\ln\left(\sqrt{1+e^x}-1\right) - x + C$$

5. (a) $\left\{\begin{array}{l} u^2 = 1+x \\ 2u\,du = dx \end{array}\right\};$ $\displaystyle\int x\sqrt{1+x}\,dx = \int (u^2-1)(u)2u\,du$

$$= \int (2u^4 - 2u^2)\,du$$

$$= \tfrac{2}{5}u^5 - \tfrac{2}{3}u^3 + C$$

$$= \tfrac{2}{5}(x+1)^{5/2} - \tfrac{2}{3}(x+1)^{3/2} + C$$

(b) $\left\{\begin{array}{l} u = 1+x \\ du = dx \end{array}\right\};$ $\displaystyle\int x\sqrt{1+x}\,dx = \int (u-1)\sqrt{u}\,du = \int\left(u^{3/2}-u^{1/2}\right)du$

$$= \tfrac{2}{5}u^{5/2} - \tfrac{2}{3}u^{3/2} + C$$

$$= \tfrac{2}{5}(1+x)^{5/2} - \tfrac{2}{3}(1+x)^{3/2} + C$$

7. $\left\{\begin{array}{l} u^2 = x-1 \\ 2u\,du = dx \end{array}\right\};$ $\displaystyle\int (x+2)\sqrt{x-1}\,dx = \int (u+3)(u)2u\,du$

$$= \int (2u^4 + 6u^2)\,du$$

$$= \tfrac{2}{5}u^5 + 2u^3 + C$$

$$= \tfrac{2}{5}(x-1)^{5/2} + 2(x-1)^{3/2} + C$$

9. $\left\{\begin{array}{l} u^2 = 1+x^2 \\ 2u\,du = 2x\,dx \end{array}\right\};$ $\displaystyle\int \frac{x^3}{(1+x^2)^3}\,dx = \int \frac{x^2}{(1+x^2)^3}x\,dx = \int \frac{u^2-1}{u^6}u\,du$

$$= \int (u^{-3}-u^{-5})\,du = \frac{1}{2}u^{-2} + \frac{1}{4}u^{-4} + C$$

$$= \frac{1}{4(1+x^2)^2} - \frac{1}{2(1+x^2)} + C$$

$$= -\frac{1+2x^2}{4(1+x^2)^2} + C$$

11. $\left\{ \begin{matrix} u^2 = x \\ 2u\,du = dx \end{matrix} \right\}$; $\displaystyle\int \frac{\sqrt{x}}{\sqrt{x}-1}\,dx = \int \left(\frac{u}{u-1}\right) 2u\,du = 2\int \left(u+1+\frac{1}{u-1}\right) du$

$$= u^2 + 2u + 2\ln|u-1| + C$$

$$= x + 2\sqrt{x} + 2\ln|\sqrt{x}-1| + C$$

13. $\left\{ \begin{matrix} u^2 = x-1 \\ 2u\,du = dx \end{matrix} \right\}$; $\displaystyle\int \frac{\sqrt{x-1}+1}{\sqrt{x-1}-1}\,dx = \int \frac{u+1}{u-1} 2u\,du = \int \left(2u+4+\frac{4}{u-1}\right) du$

$$= u^2 + 4u + 4\ln|u-1| + C$$

$$= x - 1 + 4\sqrt{x-1} + 4\ln|\sqrt{x-1}-1| + C$$

(absorb -1 in C)

$$= x + 4\sqrt{x-1} + 4\ln|\sqrt{x-1}-1| + C$$

15. $\left\{ \begin{matrix} u^2 = 1+e^x \\ 2u\,du = e^x\,dx \end{matrix} \right\}$; $\displaystyle\int \frac{dx}{\sqrt{1+e^x}} = \int \left(\frac{1}{u}\right) \frac{2u\,du}{u^2-1} = \int \left[\frac{1}{u-1} - \frac{1}{u+1}\right] du$

$$= \ln|u-1| - \ln|u+1| + C$$

$$= \ln\left[\frac{\sqrt{1+e^x}-1}{\sqrt{1+e^x}+1}\right] + C$$

$$= \ln\left[\frac{(\sqrt{1+e^x}-1)^2}{e^x}\right] + C$$

$$= 2\ln\left(\sqrt{1+e^x}-1\right) - x + C$$

17. $\left\{ \begin{matrix} u^2 = x+4 \\ 2u\,du = dx \end{matrix} \right\}$; $\displaystyle\int \frac{x}{\sqrt{x+4}}\,dx = \int \frac{u^2-4}{u} 2u\,du = \int (2u^2-8)\,du$

$$= \tfrac{2}{3}u^3 - 8u + C$$

$$= \tfrac{2}{3}(x+4)^{3/2} - 8(x+4)^{1/2} + C$$

$$= \tfrac{2}{3}(x-8)\sqrt{x+4} + C$$

19. $\left\{ \begin{matrix} u^2 = 4x+1 \\ 2u\,du = 4\,dx \end{matrix} \right\}$; $\displaystyle\int 2x^2(4x+1)^{-5/2}\,dx = \int 2\left(\frac{u^2-1}{4}\right)^2 (u^{-5})\frac{u}{2}\,du$

$$= \frac{1}{16}\int (1 - 2u^{-2} + u^{-4})\,du = \frac{1}{16}u + \frac{1}{8}u^{-1} - \frac{1}{48}u^{-3} + C$$

$$= \tfrac{1}{16}(4x+1)^{1/2} + \tfrac{1}{8}(4x+1)^{-1/2} - \tfrac{1}{48}(4x+1)^{-3/2} + C$$

21. $\left\{ \begin{array}{l} u^2 = ax + b \\ 2u\,du = a\,dx \end{array} \right\}$; $\displaystyle\int \frac{x}{(ax+b)^{3/2}}\,dx = \int \dfrac{\dfrac{u^2-b}{a}}{u^3}\,\dfrac{2u}{a}\,du$

$$= \frac{2}{a^2}\int (1 - bu^{-2})\,du$$

$$= \frac{2}{a^2}(u + bu^{-1}) + C = \frac{2u^2 + 2b}{a^2 u} + C$$

$$= \frac{4b + 2ax}{a^2\sqrt{ax+b}} + C$$

23. $\left\{ \begin{array}{l} u = \tan(x/2), \quad dx = \dfrac{2}{1+u^2}\,du \\[2mm] \sin x = \dfrac{2u}{1+u^2}, \quad \cos x = \dfrac{1-u^2}{1+u^2} \end{array} \right\}$;

$$\int \frac{1}{1 + \cos x - \sin x}\,dx = \int \frac{1}{1 + \dfrac{1-u^2}{1+u^2} - \dfrac{2u}{1+u^2}}\cdot \frac{2}{1+u^2}\,du$$

$$= \int \frac{1}{1-u}\,du$$

$$= -\ln|1-u| + C = -\ln\left|1 - \tan\left(\frac{x}{2}\right)\right| + C$$

25. $\left\{ \begin{array}{l} u = \tan(x/2), \quad dx = \dfrac{2}{1+u^2}\,du \\[2mm] \sin x = \dfrac{2u}{1+u^2} \end{array} \right\}$;

$$\int \frac{1}{2 + \sin x}\,dx = \int \frac{1}{2 + \dfrac{2u}{1+u^2}}\cdot \frac{2}{1+u^2}\,du$$

$$= \int \frac{1}{u^2 + u + 1}\,du = \int \frac{1}{\left(u + \frac{1}{2}\right)^2 + \left(\frac{\sqrt{3}}{2}\right)^2}\,du$$

$$= \frac{2}{\sqrt{3}}\tan^{-1}\left(\frac{u + \frac{1}{2}}{\frac{\sqrt{3}}{2}}\right) + C$$

$$= \frac{2}{\sqrt{3}}\tan^{-1}\left[\frac{1}{\sqrt{3}}\left(2\tan(x/2) + 1\right)\right] + C$$

27. $\left\{ \begin{array}{l} u = \tan(x/2), \quad dx = \dfrac{2}{1+u^2}\,du \\[2mm] \sin x = \dfrac{2u}{1+u^2}, \quad \tan x = \dfrac{2u}{1-u^2} \end{array} \right\}$;

$$\int \frac{1}{\sin x + \tan x}\,dx = \int \frac{1}{\dfrac{2u}{1+u^2} + \dfrac{2u}{1-u^2}}\cdot \frac{2}{1+u^2}\,du$$

$$= \int \frac{1-u^2}{2u}\,du = \frac{1}{2}\int \left(\frac{1}{u} - u\right)du$$

$$= \frac{1}{2}\left(\ln|u| - \frac{1}{2}u^2\right) + C = \frac{1}{2}\ln\|\tan(x/2)\| - \frac{1}{4}[\tan(x/2)]^2 + C$$

29.
$$\left\{ \begin{array}{cc} u = \tan(x/2), & dx = \dfrac{2}{1+u^2}\,du \\[2mm] \sin x = \dfrac{2u}{1+u^2}, & \cos x = \dfrac{1-u^2}{1+u^2} \end{array} \right\};$$

$$\int \frac{1-\cos x}{1+\sin x}\,dx = \int \frac{1-\dfrac{1-u^2}{1+u^2}}{1+\dfrac{2u}{1+u^2}} \cdot \frac{2}{1+u^2}\,du$$

$$= \int \frac{4u^2}{(1+u^2)(u+1)^2}\,du$$

$$= \int \left[\frac{2u}{1+u^2} - \frac{2}{u+1} + \frac{2}{(u+1)^2} \right] du$$

$$= \ln(u^2+1) - 2\ln|u+1| - \frac{2}{u+1} + C$$

$$= \ln\left[\frac{u^2+1}{(u+1)^2} - \frac{2}{u+1} \right] + C$$

$$= \ln\left[\frac{\tan^2(x/2)+1}{(\tan(x/2)+1)^2} \right] - \frac{2}{\tan(x/2)+1} + C = \ln\left| \frac{1}{1+\sin x} \right| - \frac{2}{\tan(x/2)+1} + C$$

31.
$$\left\{ \begin{array}{c} u^2 = x \\[1mm] 2u\,du = dx \end{array} \right\}; \qquad \int \frac{x^{3/2}}{x+1}\,dx = \int \frac{u^3}{u^2+1}\,2u\,du$$

$$= \int \frac{2u^4}{u^2+1}\,du = \int \left[2u^2 - 2 + \frac{2}{u^2+1} \right] du$$

$$= \frac{2}{3}u^3 - 2u + 2\tan^{-1}u + C = \frac{2}{3}x^{3/2} - 2x^{1/2} + 2\tan^{-1}x^{1/2} + C$$

$$\int_0^4 \frac{x^{3/2}}{x+1}\,dx = \left[\frac{2}{3}x^{3/2} - 2x^{1/2} + 2\tan^{-1}x^{1/2} \right]_0^4 = \frac{4}{3} + 2\tan^{-1}2$$

33.
$$\int_0^{\pi/2} \frac{\sin 2x}{2+\cos x}\,dx = \int_0^{\pi/2} \frac{2\sin x \cos x}{2+\cos x}\,dx$$

$$= \int_0^1 \frac{2u}{2+u}\,du \qquad [u = \cos x, \quad du = -\sin x\,dx]$$

$$= \int_0^1 \left(2 - \frac{4}{2+u} \right) du$$

$$= [2u - 4\ln|2+u|]_0^1 = 2 + 4\ln\left(\frac{2}{3}\right)$$

35.
$$\left\{ \begin{array}{cc} u = \tan(x/2), & dx = \dfrac{2}{1+u^2}\,du \\[2mm] \sin x = \dfrac{2u}{1+u^2}, & \cos x = \dfrac{1-u^2}{1+u^2} \end{array} \right\};$$

$$\int \frac{1}{\sin x - \cos x - 1}\,dx = \int \frac{1}{\dfrac{2u}{1+u^2} - \dfrac{1-u^2}{1+u^2} - 1} \cdot \frac{2}{1+u^2}\,du$$

$$= \int \frac{1}{u-1}\,du$$

$$= \ln|u-1| + C = \ln|\tan(x/2) - 1| + C$$

$$\int_0^{\pi/3} \frac{1}{\sin x - \cos x - 1}\, dx = \left[\ln|\tan(x/2) - 1|\right]_0^{\pi/3} = \ln\left(\frac{\sqrt{3}-1}{\sqrt{3}}\right)$$

37.
$$\left\{ \begin{array}{ll} u = \tan(x2), & dx = \dfrac{2}{1+u^2}\, du \\[3mm] \sin x = \dfrac{2u}{1+u^2}, & \cos x = \dfrac{1-u^2}{1+u^2} \end{array} \right\};$$

$$\int \sec x\, dx = \int \frac{1}{\cos x}\, dx = \int \frac{1}{\dfrac{1-u^2}{1+u^2}} \cdot \frac{2}{1+u^2}\, du$$

$$= 2\int \frac{1}{1-u^2}\, du = 2\int \left[\frac{1/2}{1-u} + \frac{1/2}{1+u}\right] du$$

$$= \int \left[\frac{1}{1-u} + \frac{1}{1+u}\right] du = -\ln|1-u| + \ln|1+u| + C$$

$$= \ln\left|\frac{1+\tan(x/2)}{1-\tan(x/2)}\right| + C$$

39.

$$\int \csc x\, dx = \int \frac{\sin x}{\sin^2 x}\, dx = \int \frac{\sin x}{1-\cos^2 x}\, dx$$

$$= -\int \frac{1}{1-u^2}\, du \qquad [u = \cos x, \quad du = -\sin x\, dx]$$

$$= \frac{1}{2}\int \left[\frac{1}{u-1} - \frac{1}{u+1}\right] du$$

$$= \frac{1}{2}\left[\ln|u-1| - \ln|u+1|\right] + C = \ln\sqrt{\frac{1-\cos x}{1+\cos x}} + C$$

41.
$$\left\{ \begin{array}{ll} u = \tanh(x/2), & dx = \dfrac{2}{1-u^2}\, du \\[3mm] \cosh x = \dfrac{1+u^2}{1-u^2}, & \operatorname{sech} x = \dfrac{1-u^2}{1+u^2} \end{array} \right\};$$

$$\int \operatorname{sech} x\, dx = \int \frac{1-u^2}{1+u^2} \cdot \frac{2}{1-u^2}\, du$$

$$= \int \frac{2}{1+u^2}\, du = 2\tan^{-1} u + C = 2\tan^{-1}(\tanh(x/2)) + C$$

43.
$$\left\{ \begin{array}{ll} u = \tanh(x/2), & dx = \dfrac{2}{1-u^2}\, du \\[3mm] \sinh x = \dfrac{2u}{1-u^2}, & \cosh x = \dfrac{1+u^2}{1-u^2} \end{array} \right\};$$

$$\int \frac{1}{\sinh x + \cosh x}\, dx = \int \frac{1}{\dfrac{2u}{1-u^2} + \dfrac{1+u^2}{1-u^2}} \cdot \frac{2}{1-u^2}\, du$$

$$= \int \frac{2}{(1+u)^2}\, du = \frac{-2}{u+1} + C = \frac{-2}{\tanh(x/2)+1} + C$$

SECTION 8.7

1. (a) $L_{12} = \frac{12}{12}[0 + 1 + 4 + 9 + 16 + 25 + 36 + 49 + 64 + 81 + 100 + 121] = 506$

   (b) $R_{12} = \frac{12}{12}[1 + 4 + 9 + 16 + 25 + 36 + 49 + 64 + 81 + 100 + 121 + 144] = 650$

   (c) $M_6 = \frac{12}{6}[1 + 9 + 25 + 49 + 81 + 121] = 572$

   (d) $T_{12} = \frac{12}{24}[0 + 2(1 + 4 + 9 + 16 + 25 + 36 + 49 + 64 + 81 + 100 + 121) + 144] = 578$

   (e) $S_6 = \frac{12}{36}[0 + 144 + 2(4 + 16 + 36 + 64 + 100) + 4(1 + 9 + 25 + 49 + 81 + 121)] = 576$

   $$\int_0^{12} x^2\, dx = \left[\frac{1}{3}x^3\right]_0^{12} = 576$$

3. (a) $L_6 = \frac{3}{6}\left[\frac{1}{1+0} + \frac{1}{1+1/8} + \frac{1}{1+1} + \frac{1}{1+27/8} + \frac{1}{1+8} + \frac{1}{1+125/8}\right]$

   $= \frac{1}{2}\left[1 + \frac{8}{9} + \frac{1}{2} + \frac{8}{35} + \frac{1}{9} + \frac{8}{133}\right] \cong 1.394$

   (b) $R_6 = \frac{3}{6}\left[\frac{1}{1+1/8} + \frac{1}{1+1} + \frac{1}{1+27/8} + \frac{1}{1+8} + \frac{1}{1+125/8} + \frac{1}{1+27}\right]$

   $= \frac{1}{2}\left[\frac{8}{9} + \frac{1}{2} + \frac{8}{35} + \frac{1}{9} + \frac{8}{133} + \frac{1}{28}\right] \cong 0.9122$

   (c) $M_3 = \frac{3}{3}\left[\frac{1}{1+1/8} + \frac{1}{1+27/8} + \frac{1}{1+125/8}\right] = \frac{8}{9} + \frac{8}{35} + \frac{8}{133} \cong 1.1852$

   (d) $T_6 = \frac{3}{12}\left[1 + 2\left(\frac{8}{9} + \frac{1}{2} + \frac{8}{35} + \frac{1}{9} + \frac{8}{133}\right) + \frac{1}{28}\right] \cong 1.1533$

   (e) $S_3 = \frac{3}{18}\left\{1 + \frac{1}{28} + 2\left[\frac{1}{2} + \frac{1}{9}\right] + 4\left[\frac{8}{9} + \frac{8}{35} + \frac{8}{133}\right]\right\} \cong 1.1614$

5. (a) $\frac{1}{4}\pi \cong T_4 = \frac{1}{8}\left[1 + 2\left(\frac{1}{1+1/16} + \frac{1}{1+1/4} + \frac{1}{1+9/16}\right) + \frac{1}{1+1}\right]$

   $= \frac{1}{8}\left[1 + 2\left(\frac{16}{17} + \frac{4}{5} + \frac{16}{25}\right) + \frac{1}{2}\right] \cong 0.7828$

   (b) $\frac{1}{4}\pi \cong S_4 = \frac{1}{24}\left[1 + \frac{1}{2} + 2\left(\frac{16}{17} + \frac{4}{5} + \frac{16}{25}\right) + 4\left(\frac{64}{65} + \frac{64}{73} + \frac{64}{89} + \frac{64}{113}\right)\right] \cong 0.7854$

7. (a) $M_4 = \frac{2}{4}\left[\cos\left(\frac{-3}{4}\right)^2 + \cos\left(\frac{-1}{4}\right)^2 + \cos\left(\frac{1}{4}\right)^2 + \cos\left(\frac{3}{4}\right)^2\right] \cong 1.8440$

   (b) $T_8 = \frac{2}{16}\left[\cos(-1)^2 + 2\cos\left(\frac{-3}{4}\right)^2 + 2\cos\left(\frac{-1}{2}\right)^2 + 2\cos\left(\frac{-1}{4}\right)^2 + \right.$

   $\left. 2\cos(0)^2 + 2\cos\left(\frac{1}{4}\right)^2 + 2\cos\left(\frac{1}{2}\right)^2 + 2\cos\left(\frac{3}{4}\right)^2 + \cos(1)^2\right] \cong 1.7915$

(c) $S_4 = \frac{2}{24}\left\{\cos(-1)^2 + \cos(1)^2 + 2\left[\cos\left(\frac{-1}{2}\right)^2 + \cos(0)^2 + \cos\left(\frac{1}{2}\right)^2\right] + \right.$

$$\left. 4\left[\cos\left(\frac{-3}{4}\right)^2 + \cos\left(\frac{-1}{4}\right)^2 + \cos\left(\frac{1}{4}\right)^2 + \cos\left(\frac{3}{4}\right)^2\right]\right\} \cong 1.8090$$

9. (a) $T_{10} = \frac{2}{20}\left[e^{-0^2} + 2e^{-(1/5)^2} + 2e^{-(2/5)^2} + 2e^{-(3/5)^2} + 2e^{-(4/5)^2} + 2e^{-1^2} + 2e^{-(6/5)^2} + \right.$

$$\left. 2e^{-(7/5)^2} + 2e^{-(8/5)^2} + 2e-(9/5)^2 + e-2^2\right] \cong 0.8818$$

(b) $S_5 = \frac{2}{30}\left\{e^{-0^2} + e^{-2^2} + 2\left[e^{-(2/5)^2} + e^{-(4/5)^2} + e^{-(6/5)^2} + e^{-(8/5)^2}\right] + \right.$

$$\left. 4\left[e^{-(1/5)^2} + e^{-(3/5)^2} + e^{-1^2} + e^{-(7/5)^2} + e^{-(9/5)^2}\right]\right\} \cong 0.8821$$

11. Such a curve passes through the three points

$$(a_1, b_1), \quad (a_2, b_2), \quad (a_3, b_3)$$

iff

$$b_1 = a_1{}^2 A + a_1 B + C, \quad b_2 = a_2{}^2 A + a_2 B + C, \quad b_3 = a_3{}^2 A + a_3 B + C,$$

which happens iff

$$A = \frac{b_1(a_2 - a_3) - b_2(a_1 - a_3) + b_3(a_1 - a_2)}{(a_1 - a_3)(a_1 - a_2)(a_2 - a_3)},$$

$$B = -\frac{b_1(a_2{}^2 - a_3{}^2) - b_2(a_1{}^2 - a_3{}^2) + b_3(a_1{}^2 - a_2{}^2)}{(a_1 - a_3)(a_1 - a_2)(a_2 - a_3)},$$

$$C = \frac{a_1{}^2(a_2 b_3 - a_3 b_2) - a_2{}^2(a_1 b_3 - a_3 b_1) + a_3{}^2(a_1 b_2 - a_2 b_1)}{(a_1 - a_3)(a_1 - a_2)(a_2 - a_3)}.$$

13. (a) $\left|\dfrac{(b-a)^3}{12n^2}f''(c)\right| = \dfrac{27}{12n^2}\dfrac{1}{4c^{3/2}} \leq \dfrac{9}{16n^2} < 0.01 \implies n^2 > \left(\dfrac{15}{2}\right)^2 \implies n \geq 8$

(b) $\left|\dfrac{(b-a)^5}{2880n^4}f^{(4)}(c)\right| = \dfrac{243}{2880n^4}\dfrac{15}{16c^{7/2}} \leq \dfrac{81}{1024n^4} < 0.01 \implies n \geq 2$

15. (a) $\left|\dfrac{(b-a)^3}{12n^2}f''(c)\right| = \dfrac{27}{12n^2}\dfrac{1}{4c^{3/2}} \leq \dfrac{9}{16n^2} < 0.00001 \implies n > 75\sqrt{10} \implies n \geq 238$

(b) $\left|\dfrac{(b-a)^5}{2880n^4}f^{(4)}(c)\right| = \dfrac{243}{2880n^4}\dfrac{15}{16c^{7/2}} \leq \dfrac{81}{1024n^4} < 0.00001 \implies n \geq 10$

17. (a) $\left|\dfrac{(b-a)^3}{12n^2}f''(c)\right| = \dfrac{\pi^3}{12n^2}\sin c \leq \dfrac{\pi^3}{12n^2} < 0.001 \implies n > 5\pi\sqrt{\dfrac{10\pi}{3}} \implies n \geq 51$

(b) $\left|\dfrac{(b-a)^5}{2880n^4}f^{(4)}(c)\right| = \dfrac{\pi^5}{2880n^4}\sin c \leq \dfrac{\pi^5}{2880n^4} < 0.001 \implies n \geq 4$

19. (a) $\left|\dfrac{(b-a)^3}{12n^2}f''(c)\right| = \dfrac{8}{12n^2}e^c \le \dfrac{8}{12n^2}e^3 < 0.01 \implies n > 10e\sqrt{\dfrac{2e}{3}} \implies n \ge 37$

(b) $\left|\dfrac{(b-a)^5}{2880n^4}f^{(4)}(c)\right| = \dfrac{32}{2880n^4}e^c \le \dfrac{e^3}{90n^4} < 0.01 \implies n \ge 3$

21. (a) $\left|\dfrac{(b-a)^3}{12n^2}f''(c)\right| = \left|\dfrac{8}{12n^2}2e^{-c^2}(2c^2-1)\right| \le \dfrac{8}{3n^2}e^{-3/2} < 0.0001$

$\implies n > 100\sqrt{\tfrac{8}{3}e^{-3/2}} \implies n \ge 78$

(b) $\left|\dfrac{(b-a)^5}{2880n^4}f^{(4)}(c)\right| = \left|\dfrac{32}{2880n^4}4e^{-c^2}\left(4c^4-12c^2+3\right)\right| \le \dfrac{32}{2880n^4}12 < 0.0001$

$\implies n > 10\left[\dfrac{32\cdot 12}{2880}\right]^{1/4} \implies n \ge 7$

23. $f^{(4)}(x) = 0$ for all x; therefore by (8.7.3) the theoretical error is zero

25. (a) $\left|T_2 - \displaystyle\int_0^1 x^2\,dx\right| = \dfrac{3}{8} - \dfrac{1}{3} = \dfrac{1}{24} = E_2^T$

(b) $\left|S_1 - \displaystyle\int_0^1 x^4\,dx\right| = \dfrac{5}{24} - \dfrac{1}{5} = \dfrac{1}{120} = E_1^S$

27. Let f be twice differentiable on $[a,b]$ with $f(x) > 0$ and $f''(x) > 0$, and let $P = \{x_0, x_1, x_2, \ldots, x_n\}$ be a regular partition of $[a,b]$. Figure A shows a typical subinterval with the approximating trapezoid ABCD. Since the area under the curve is less than the area of the trapezoid, we can conclude that

$$\int_a^b f(x)\,dx \le T_n.$$

Figure A   **Figure B**

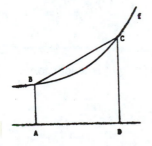 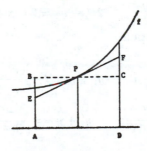

Now consider Figure B. Since the triangles EBP and PFC are congruent, the area of the rectangle ABCD equals the area of the trapezoid AEFD, and since the area under the curve is greater than the area of AEFD it follows that

$$M_n \le \int_a^b f(x)\,dx.$$

PROJECT 8.7

1. (a) Since $f(t) = e^{-t^2}$ is an even function,

$$\int_{-x}^{0} e^{-t^2}\, dt = \int_0^x e^{-t^2}\, dt \quad \text{for all } x.$$

Now, $B(-x) = \int_0^{-x} e^{-t^2}\, dt = -\int_{-x}^0 e^{-t^2}\, dt = -\int_0^x e^{-t^2}\, dt = -B(x)$.

Thus B is an odd function.

(b) Since $f(t) = e^{-t^2}$ is continuous on $(-\infty, \infty)$, $B(x) = \int_0^x e^{-t^2}\, dt$ is differentiable

(Theorem 5.2.5)

(c) By Theorem 5.2.5, $B'(x) = e^{-x^2}$. Since $B'(x) > 0$, B is increasing on $(-\infty, \infty)$.

(d) $B''(x) = -2xe^{-x^2}$; the graph of B has a point of inflection at $(0,1)$.

3. (a)

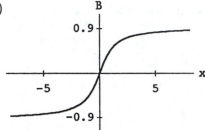

(b)

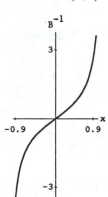

Since B is increasing on $(-\infty, \infty)$, B^{-1} exists

SECTION 8.8

1. $y_1'(x) = \frac{1}{2} e^{x/2}$; $2y_1' - y_1 = 2\left(\frac{1}{2}\right) e^{x/2} - e^{x/2} = 0$; y_1 is a solution.

$y_2'(x) = 2x + e^{x/2}$; $2y_2' - y_2 = 2\left(2x + e^{x/2}\right) - \left(x^2 + 2e^{x/2}\right) = 4x - x^2 \neq 0$;

y_2 is not a solution.

3. $y_1'(x) = \frac{-e^x}{(e^x + 1)^2}$; $y_1' + y_1 = \frac{-e^x}{(e^x + 1)^2} + \frac{1}{e^x + 1} = \frac{1}{(e^x + 1)^2} = y_1^2$; y_1 is a solution.

$y_2'(x) = \frac{-Ce^x}{(Ce^x + 1)^2}$; $y_2' + y_2 = \frac{-Ce^x}{(Ce^x + 1)^2} + \frac{1}{Ce^x + 1} = \frac{1}{(Ce^x + 1)^2} = y_2^2$;

y_2 is a solution.

5. $y_1'(x) = 2e^{2x}$, $y_1'' = 4e^{2x}$; $y_1'' - 4y_1 = 4e^{2x} - 4e^{2x} = 0$; y_1 is a solution.

$y_2'(x) = 2C \cosh 2x$, $y_2'' = 4C \sinh 2x$; $y_2'' - 4y_2 = 4C \sinh 2x - 4C \sinh 2x = 0$;

y_2 is a solution.

7. $y' - 2y = 1;$ $H(x) = \int (-2)\, dx = -2x,$ integrating factor: e^{-2x}

$$e^{-2x}y' - 2e^{-2x}y = e^{-2x}$$

$$\frac{d}{dx}\left[e^{-2x}y\right] = e^{-2x}$$

$$e^{-2x}y = -\frac{1}{2}e^{-2x} + C$$

$$y = -\frac{1}{2} + Ce^{2x}$$

9. $y' + \frac{5}{2}y = 1;$ $H(x) = \int \left(\frac{5}{2}\right) dx = \frac{5}{2}x,$ integrating factor: $e^{5x/2}$

$$e^{5x/2}y' + \frac{5}{2}e^{5x/2}y = e^{5x/2}$$

$$\frac{d}{dx}\left[e^{5x/2}y\right] = e^{5x/2}$$

$$e^{5x/2}y = \frac{2}{5}e^{5x/2} + C$$

$$y = \frac{2}{5} + Ce^{-5x/2}$$

11. $y' - 2y = 1 - 2x;$ $H(x) = \int (-2)\, dx = -2x,$ integrating factor: e^{-2x}

$$e{-2x}y' - 2e^{-2x}y = e^{-2x} - 2xe^{-2x}$$

$$\frac{d}{dx}\left[e^{-2x}y\right] = e^{-2x} - 2xe^{-2x}$$

$$e^{-2x}y = -\frac{1}{2}e^{-2x} + \frac{1}{2}xe^{-2x} + \frac{1}{2}e^{-2x} + C = xe^{-2x} + C$$

$$y = x + Ce^{2x}$$

13. $y' - \frac{4}{x}y = -2n;$ $H(x) = \int \left(-\frac{4}{x}\right) dx = -4\ln x = \ln x^{-4},$ integrating factor: $e^{\ln x^{-4}} = x^{-4}$

$$x^{-4}y' - \frac{4}{x}x^{-4}y = -2nx^{-4}$$

$$\frac{d}{dx}\left[x^{-4}y\right] = -2nx^{-4}$$

$$x^{-4}y = \frac{2}{3}nx^{-3} + C$$

$$y = \frac{2}{3}nx + Cx^4$$

15. $y' - e^x y = 0;$ $H(x) = \int -e^x\, dx = -e^x,$ integrating factor: e^{-e^x}

$$e^{-e^x}y' - e^x e^{-e^x}y = 0$$

$$\frac{d}{dx}\left[e^{-e^x}y\right] = 0$$

$$e^{-e^x}y = C$$

$$y = Ce^{e^x}$$

17. $y' + \dfrac{1}{1+e^x}\, y = \dfrac{1}{1+e^x};$ $H(x) = \displaystyle\int \dfrac{1}{1+e^x}\, dx = \ln \dfrac{e^x}{1+e^x},$

integrating factor: $e^{H(x)} = \dfrac{e^x}{1+e^x}$

$$\dfrac{e^x}{1+e^x}\, y' + \dfrac{1}{1+e^x} \cdot \dfrac{e^x}{1+e^x}\, y = \dfrac{1}{1+e^x} \cdot \dfrac{e^x}{1+e^x}$$

$$\dfrac{d}{dx}\left[\dfrac{e^x}{1+e^x}\, y\right] = \dfrac{e^x}{(1+e^x)^2}$$

$$\dfrac{e^x}{1+e^x}\, y = -\dfrac{1}{1+e^x} + C$$

$$y = -e^{-x} + C\left(1+e^{-x}\right)$$

This solution can also be written: $y = 1 + K\left(e^{-x}+1\right),$ where K is an arbitrary constant.

19. $y' + 2xy = xe^{-x^2};$ $H(x) = \displaystyle\int 2x\, dx = x^2,$ integrating factor: e^{x^2}

$$e^{x^2}\, y' + 2xe^{x^2}\, y = x$$

$$\dfrac{d}{dx}\left[e^{x^2}\, y\right] = x$$

$$e^{x^2}\, y = \dfrac{1}{2}\, x^2 + C$$

$$y = e^{-x^2}\left(\tfrac{1}{2}\, x^2 + C\right)$$

21. $y' + \dfrac{2}{x+1}\, y = 0;$ $H(x) = \displaystyle\int \dfrac{2}{x+1}\, dx = 2\ln(x+1) = \ln(x+1)^2,$

integrating factor: $e^{\ln(x+1)^2} = (x+1)^2$

$$(x+1)^2\, y' + 2(x+1)\, y = 0$$

$$\dfrac{d}{dx}\left[(x+1)^2\, y\right] = 0$$

$$(x+1)^2\, y = C$$

$$y = \dfrac{C}{(x+1)^2}$$

23. $y' + y = x;$ $H(x) = \displaystyle\int 1\, dx = x,$ integrating factor : e^x

$$e^x\, y' + e^x\, y = xe^x$$

$$\dfrac{d}{dx}\left[e^x\, y\right] = xe^x$$

$$e^x\, y = xe^x - e^x + C$$

$$y = (x-1) + Ce^{-x}$$

$y(0) = -1 + C = 1$ $\Longrightarrow$ $C = 2.$ Therefore, $y = 2e^{-x} + x - 1$ is the solution which satisfies the side condition.

25. $y' + y = \dfrac{1}{1+e^x};$ $\quad H(x) = \displaystyle\int 1\,dx = x,$ integrating factor : $\quad e^x$

$$e^x\,y' + e^x\,y = \frac{e^x}{1+e^x}$$

$$\frac{d}{dx}\,[e^x\,y] = \frac{e^x}{1+e^x}$$

$$e^x\,y = \ln\,(1+e^x) + C$$

$$y = e^{-x}\,[\ln\,(1+e^x) + C]$$

$y(0) = \ln 2 + C = e \implies C = e - \ln 2.$ Therefore, $\quad y = e^{-x}\,[\ln\,(1+e^x) + e - \ln 2]$ is the solution which satisfies the side condition.

27. $y' - \dfrac{2}{x}\,y = x^2 e^x;$ $\quad H(x) = \displaystyle\int\left(-\frac{2}{x}\right)\,dx = -2\ln x = \ln x^{-2},$

integrating factor: $\quad e^{\ln x^{-2}} = x^{-2}$

$$x^{-2}\,y' - 2x^{-3}\,y = e^x$$

$$\frac{d}{dx}\,[x^{-2}\,y] = e^x$$

$$x^{-2}\,y = e^x + C$$

$$y = x^2\,(e^x + C)$$

$y(1) = e + C = 0 \implies C = -e.$ Therefore, $\quad y = x^2\,(e^x - e)$ is the solution which satisfies the side condition.

29. (a) You can determine that

$$v(t) = \frac{32}{K}\left(1 - e^{-Kt}\right).$$

(b) At each time $t,$ $\quad 1 - e^{-Kt} < 1.$ With $K > 0,$

$$v(t) = \frac{32}{K}\left(1 - e^{-Kt}\right) < \frac{32}{K}\quad\text{and}\quad \lim_{t\to\infty} v(t) = \frac{32}{K}$$

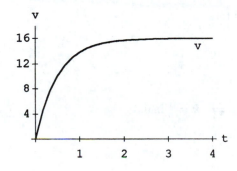

31.   **(a)**

$$\frac{di}{dt} + \frac{R}{L}i = \frac{E}{L}; \quad H(t) = \int \frac{R}{L}\,dt = \frac{R}{L}, \quad \text{integrating factor :} \quad e^{\frac{R}{L}t}$$

$$e^{\frac{R}{L}t}\frac{di}{dt} + \frac{R}{L}e^{\frac{R}{L}t}i = \frac{E}{L}e^{\frac{R}{L}t}$$

$$\frac{d}{dt}\left[e^{\frac{R}{L}t}i\right] = \frac{E}{L}e^{\frac{R}{L}t}$$

$$e^{\frac{R}{L}t}i = \frac{E}{R}e^{\frac{R}{L}t} + C$$

$$i(t) = \frac{E}{R} + Ce^{-\frac{R}{L}t}$$

$$i(0) = 0 \implies C = -\frac{E}{R}, \quad \text{so} \quad i(t) = \frac{E}{R}\left[1 - e^{-(R/L)\,t}\right].$$

(b) $\displaystyle\lim_{t\to\infty} i(t) = \lim_{t\to\infty} \frac{E}{R}\left(1 - e^{-(R/L)t}\right) = \frac{E}{R}$ amps

(c) $i(t) = 0.9\frac{E}{R} \implies e^{-(R/L)t} = \frac{1}{10} \implies -\frac{R}{L}t = -\ln 10 \implies t = \frac{L}{R}\ln 10$ seconds.

33.   **(a)**

$$V'(t) = ktV(t)$$

$$V'(t) - ktV(t) = 0$$

$$e^{-kt^2/2}V'(t) - kte^{-kt^2/2}V(t) = 0$$

$$\frac{d}{dt}\left[e^{-kt^2/2}V(t)\right] = 0$$

$$e^{-kt^2/2}V(t) = C$$

$$V(t) = Ce^{kt^2/2}.$$

Since $V(0) = C = 200$,

$$V(t) = 200e^{kt^2/2}.$$

Since $V(5) = 160$,

$$200e^{k(25/2)} = 160, \quad e^{k(25/2)} = \frac{4}{5}, \quad e^k = \left(\frac{4}{5}\right)^{2/25}$$

and therefore

$$V(t) = 200\left(\frac{4}{5}\right)^{t^2/25} \text{ liters.}$$

(b) $V'(t) = kV(t) \implies V(t) = V_0 e^{kt}$

Loses 20% in 5 minutes, so $V(5) = V_0 e^{5k} = 0.8V_0 \implies k = \frac{1}{5}\ln 0.8$

$\implies V(t) = V_0 e^{\frac{1}{5}(\ln 0.8)t} = V_0\left(e^{\ln 0.8}\right)^{t/5} = V_0(0.8)^{t/5} = V_0\left(\frac{4}{5}\right)^{t/5}.$

Since $V_0 = 200$ liters, we get $V(t) = 200\left(\frac{4}{5}\right)^{t/5}$

35.   **(a)** $\dfrac{dP}{dt} = k(M - P)$

(b) $\dfrac{dP}{dt} + kP = kM; \qquad H(t) = \displaystyle\int k\,dt = kt, \quad$ integrating factor : $\quad e^{kt}$

$$e^{kt}\frac{dP}{dt} + ke^{kt}P = kM\,e^{kt}$$

$$\frac{d}{dt}\left[e^{kt}P\right] = kM\,e^{kt}$$

$$e^{kt}P = M\,e^{kt} + C$$

$$P = M + Ce^{-kt}$$

$P(0) = M + C = 0 \implies C = -M \quad$ and $\quad P(t) = M\left(1 - e^{-kt}\right)$

$P(10) = M\left(1 - e^{-10k}\right) = 0.3M \implies k \cong 0.0357 \quad$ and $\quad P(t) = M\left(1 - e^{-0.0357t}\right)$

(c) $P(t) = M\left(1 - e^{-0.0357t}\right) = 0.9M \implies e^{-0.0357t} = 0.1 \implies t \cong 65$

Therefore, it will take approximately 65 days for 90 % of the population to be aware of the product.

37. (a)

$\dfrac{dP}{dt} - 2\cos 2\pi t\,P = 0 \implies P = Ce^{-\frac{1}{\pi}\sin 2\pi t}.$

$P(0) = C = 1000 \implies P = 1000e^{-\frac{1}{\pi}\sin 2\pi t}.$

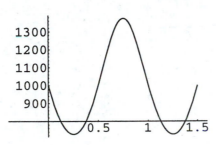

(b)

$\dfrac{dP}{dt} - 2\cos 2\pi t\,P = 2000\cos 2\pi t \implies P = Ce^{-\frac{1}{\pi}\sin 2\pi t} - 1000.$

$P(0) = 1000 \implies C = 2000 \implies P = 2000e^{-\frac{1}{\pi}\sin 2\pi t} - 1000.$

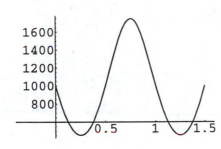

SECTION 8.9

1.

$$y' = y \sin(2x + 3)$$

$$\frac{1}{y} dy = \sin(2x + 3) \, dx$$

$$\int \frac{1}{y} dy = \int \sin(2x + 3) \, dx$$

$$\ln |y| = -\frac{1}{2} \cos(2x + 3) + C$$

This solution can also be written: $y = Ce^{-(1/2)\cos(2x+3)}$.

3.

$$y' = (xy)^3$$

$$\frac{1}{y^3} dy = x^3 \, dx, \qquad y \neq 0$$

$$\int \frac{1}{y^3} dy = \int x^3 \, dx$$

$$-\frac{1}{2} y^{-2} = \frac{1}{4} x^4 + C$$

This solution can also be written: $x^4 + \dfrac{2}{y^2} = C,$ or $y^2 = \dfrac{2}{C - x^4};$

$y = 0$ is a singular solution.

5.

$$y' = -\frac{\sin(1/x)}{x^2 y \cos y}$$

$$y \cos y \, dy = -\frac{1}{x^2} \sin(1/x) \, dx$$

$$\int y \cos y \, dy = \int -\frac{1}{x^2} \sin(1/x) \, dx$$

$$y \sin y + \cos y = -\cos(1/x) + C$$

7.

$$y' = x \, e^{y-x}$$

$$e^{-y} \, dy = xe^{-x} \, dx$$

$$\int e^{-y} \, dy = \int xe^{-x} \, dx$$

$$e^{-y} = xe^{-x} + e^{-x} + C$$

9.

$$(y \ln x)y' = \frac{(y + 1)^2}{x}$$

$$\frac{y}{(y + 1)^2} dy = \frac{1}{x \ln x} \, dx$$

$$\int \frac{y}{(y + 1)^2} dy = \int \frac{1}{x \ln x} \, dx$$

$$\ln |y + 1| + \frac{1}{y + 1} = \ln |\ln x| + C$$

11.
$$y' = x\sqrt{\frac{1-y^2}{1-x^2}}, \qquad y(0) = 0$$

$$\frac{1}{\sqrt{1-y^2}}\, dy = \frac{x}{\sqrt{1-x^2}}\, dx$$

$$\int \frac{1}{\sqrt{1-y^2}}\, dy = \int \frac{x}{\sqrt{1-x^2}}\, dx$$

$$\sin^{-1} y = -\sqrt{1-x^2} + C$$

$$y(0) = 0 \quad\Longrightarrow\quad \sin^{-1} 0 = -1 + C \quad\Longrightarrow\quad C = 1$$

Thus, $\quad \sin^{-1} y = 1 - \sqrt{1-x^2}$.

13.
$$y' = \frac{x^2 y - y}{y+1}, \qquad y(3) = 1$$

$$\frac{y+1}{y}\, dy = (x^2 - 1)\, dx, \qquad y \neq 0$$

$$\int \frac{y+1}{y}\, dy = \int (x^2 - 1)\, dx$$

$$y + \ln|y| = \frac{1}{3}x^3 - x + C$$

$$y(3) = 1 \quad\Longrightarrow\quad 1 + \ln 1 = \frac{1}{3}(3)^3 - 3 + C \quad\Longrightarrow\quad C = -5.$$

Thus, $\quad y + \ln|y| = \frac{1}{3}x^3 - x - 5.$

15. $\quad (xy^2 + y^2 + x + 1)\, dx + (y-1)\, dy = 0, \qquad y(2) = 0$

$$(x+1)(y^2+1)\, dx + (y-1)\, dy = 0$$

$$(x+1)\, dx + \frac{y-1}{y^2+1}\, dy = 0$$

$$\int (x+1)\, dx + \int \frac{y-1}{y^2+1}\, dy = C$$

$$\frac{x^2}{2} + x + \frac{1}{2}\ln(y^2+1) - \tan^{-1} y = C$$

$y(2) = 0 \quad\Longrightarrow\quad C = 4.$ Thus, $\frac{1}{2}x^2 + x + \frac{1}{2}\ln(y^2+1) - \tan^{-1} y = 4$

17. $\quad 2x + 3y = C \quad\Longrightarrow\quad 2 + 3y' = 0 \quad\Longrightarrow\quad y' = -\frac{2}{3}$

The orthogonal trajectories are the solutions of:
$$y' = \frac{3}{2}.$$

$y' = \frac{3}{2} \quad\Longrightarrow\quad y = \frac{3}{2}x + C$

19. $\quad xy = C \quad\Longrightarrow\quad y + xy' = 0 \quad\Longrightarrow\quad y' = -\frac{y}{x}$

The orthogonal trajectories are the solutions of:

$$y' = \frac{x}{y}.$$

$$y' = \frac{x}{y}$$

$$\int x\,dx = \int y\,dy$$

$$\tfrac{1}{2}x^2 = \tfrac{1}{2}y^2 + C$$

or $\quad x^2 - y^2 = C$

21. $\quad y = Ce^x \quad \Longrightarrow \quad y' = Ce^x = y$

The orthogonal trajectories are the solutions of:

$$y' = -\frac{1}{y}.$$

$$y' = -\frac{1}{y}$$

$$\int y\,dy = -\int dx$$

$$\tfrac{1}{2}y^2 = -x + K$$

or $\quad y^2 = -2x + C$

23. A differential equation for the given family is:

$$y^2 = 2xyy' + y^2(y')^2$$

A differential equation for the family of orthogonal trajectories is found by replacing y' by $-1/y'$. The result is:

$$y^2 = -\frac{2xy}{y'} + \frac{y^2}{(y')^2} \qquad \text{which simplifies to} \qquad y^2 = 2xyy' + y^2(y')^2$$

Thus, the given family is self-orthogonal.

25. We assume that $C = 0$ at time $t = 0$. $\qquad$ (a) Let $A_0 = B_0$. Then

$$\frac{dC}{dt} = k(A_0 - C)^2 \quad \text{and} \quad \frac{dC}{(A_0 - C)^2} = k\,dt \qquad \text{see Section 7.6.}$$

Integrating, we get

$$\int \frac{1}{(A_0 - C)^2}\,dC = \int k\,dt$$

$$\frac{1}{A_0 - C} = kt + M \qquad M \text{ an arbitrary constant.}$$

Since $C(0) = 0$, $\ M = \dfrac{1}{A_0}$ and

$$\frac{1}{A_0 - C} = kt + \frac{1}{A_0}.$$

Solving this equation for C gives

$$C(t) = \frac{kA_0^2 t}{1 + kA_0 t}.$$

(b) Suppose that $A_0 \neq B_0$. Then

$$\frac{dC}{dt} = k(A_0 - C)(B_0 - C) \quad \text{and} \quad \frac{dC}{(A_0 - C)(B_0 - C)} = k\,dt.$$

Integrating, we get

$$\int \frac{1}{(A_0 - C)(B_0 - C)}\,dC = \int k\,dt$$

$$\frac{1}{B_0 - A_0} \int \left(\frac{1}{A_0 - C} - \frac{1}{B_0 - C} \right) dC = \int k\,dt$$

$$\frac{1}{B_0 - A_0} \left[-\ln(A_0 - C) + \ln(B_0 - C) \right] = kt + M$$

$$\frac{1}{B_0 - A_0} \ln\left(\frac{B_0 - C}{A_0 - C} \right) = kt + M \qquad M \text{ an arbitrary constant}$$

Since $C(0) = 0$, $M = \frac{1}{B_0 - A_0} \ln\left(\frac{B_0}{A_0} \right)$ and

$$\frac{1}{B_0 - A_0} \ln\left(\frac{B_0 - C}{A_0 - C} \right) = kt + \frac{1}{B_0 - A_0}.$$

Solving this equation for C, gives

$$C(t) = \frac{A_0 B_0 \left(e^{kA_0 t} - e^{kB_0 t} \right)}{A_0 e^{kA_0 t} - B_0 e^{kB_0 t}}.$$

27.   (a)

$$m\frac{dv}{dt} = -\alpha v - \beta v^2$$

$$\frac{dv}{v(\alpha + \beta v)} = -\frac{1}{m}\,dt$$

$$\int \frac{1}{v(\alpha + \beta v)}\,dv = -\int \frac{1}{m}\,dt$$

$$\frac{1}{\alpha} \int \frac{1}{v}\,dv - \frac{\beta}{\alpha} \int \frac{1}{\alpha + \beta v}\,dv = -\int \frac{1}{m}\,dt$$

$$\frac{1}{\alpha} \ln v - \frac{1}{\alpha} \ln(\alpha + \beta v) = -\frac{1}{m} t + M, \quad M \text{ an arbitrary constant}$$

$$\ln\left(\frac{v}{\alpha + \beta v} \right) = -\frac{\alpha}{m} t + M$$

$$\frac{v}{\alpha + \beta v} = K e^{-\alpha t/m} \quad \left[K = e^M \right]$$

Solving this equation for v we get $\quad v(t) = \dfrac{\alpha K}{e^{\alpha t/m} - \beta K} = \dfrac{\alpha}{C e^{\alpha t/m} - \beta} \quad [C = 1/K].$

(b) Setting $v(0) = v_0$, we get   (c) $\displaystyle \lim_{t \to \infty} v(t) = 0$

$$C = \frac{\alpha + \beta v_0}{v_0} \quad \text{and}$$

$$v(t) = \frac{\alpha v_0}{(\alpha + \beta v_0) e^{\alpha t/m} - \beta v_0}$$

29.   (a) Let $P = P(t)$ denote the number of people who have the disease at time t. Then

$$\frac{dP}{dt} = kP(25,000 - P) \quad k > 0 \text{ constant}$$

$$\frac{dP}{P(25,000 - P)} = k\,dt$$

$$\int \frac{1}{P(25,000 - P)}\,dP = \int k\,dt$$

$$\frac{1}{25,000} \ln\left| \frac{P}{25,000 - P} \right| = kt + M$$

Solving for P, we get

$$P(t) = \frac{25,000}{1 + Ce^{25,000\,kt}}$$

Now, $P(0) = \dfrac{25,000}{1 + C} = 100 \implies C = 249.$

Also, $P(10) = \dfrac{25,000}{1 + 249e^{25,000\,(10k)}} = 400 \implies 25,000k \cong -0.1382.$

Therefore, $P(t) = \dfrac{25,000}{1 + 249e^{-0.1382\,t}}.$

$$P(20) = \frac{25,000}{1 + 249e^{-0.1382\,(20)}} \cong 1498; \quad 1498 \text{ people will have the disease after 20 days.}$$

(b) $\dfrac{25,000}{1 + 249e^{-0.1382\,t}} = 12,500 \implies t \cong 40;$ (c)

It will take 40 days for half the
population to have the disease.

31.   Assume that the package is dropped from rest.

(a) Let $v = v(t)$ be the velocity at time t. Then

$$100\frac{dv}{dt} = 100g - 2v \quad \text{or} \quad \frac{dv}{dt} + \frac{1}{50}v = g \quad (g = 9.8\text{m/sec}^2)$$

This is a linear differential equation; $e^{t/50}$ is an integrating factor.

$$e^{t/50}\frac{dv}{dt} + \frac{1}{50}e^{t/50}v = g\,e^{t/50}g$$

$$\frac{d}{dt}\left[e^{t/50}v\right] = g\,e^{t/50}$$

$$e^{t/50}v = 50g\,e^{t/50} + C$$

$$v = 50g + Ce^{-t/50}$$

Now, $v(0) = 0 \implies C = -50g$ and $v(t) = 50g\left(1 - e^{-t/50}\right).$

At the instant the parachute opens, $v(10) = 50g\left(1 - e^{-1/5}\right) \cong 50g(0.1813) \cong 88.82$ m/sec.

(b) Now let $v = v(t)$ denote the velocity of the package t seconds after the parachute opens. Then

$$100\frac{dv}{dt} = 100g - 4v^2 \quad \text{or} \quad \frac{dv}{dt} = g - \frac{1}{25}v^2$$

This is a separable differential equation:

$$\frac{dv}{dt} = g - \frac{1}{25}v^2 \quad \text{set } u = v/5, \ du = (1/5)dv$$

$$\frac{du}{g - u^2} = \frac{1}{5}dt$$

$$\frac{1}{2\sqrt{g}}\ln\left|\frac{u + \sqrt{g}}{u - \sqrt{g}}\right| = \frac{t}{5} + K$$

$$\ln\left|\frac{u + \sqrt{g}}{u - \sqrt{g}}\right| = \frac{2\sqrt{g}}{5}t + M$$

$$\frac{u + \sqrt{g}}{u - \sqrt{g}} = Ce^{2\sqrt{g}t/5} \cong Ce^{1.25\,t}$$

$$u = \sqrt{g}\,\frac{Ce^{1.25\,t} + 1}{Ce^{1.25t} - 1}$$

$$v = 5\sqrt{g}\,\frac{Ce^{1.25\,t} + 1}{Ce^{1.25t} - 1}$$

Now, $v(0) = 88.82 \implies 5\sqrt{g}\,\dfrac{C + 1}{C - 1} = 88.82 \implies C \cong 1.43$.

Therefore, $v(t) = 5\sqrt{g}\,\dfrac{1.43e^{1.25\,t} + 1}{1.43e^{1.25t} - 1} = \dfrac{15.65\left(1 + 0.70e^{-1.25\,t}\right)}{1 - 0.70e^{-1.25\,t}}$

(c) From part (b), $\displaystyle\lim_{t\to\inf} v(t) = 15.65\text{m/sec}$.

CHAPTER 9

SECTION 9.1

1. $8x^6 + 64 = 8(x^2 + 2)(x^4 - 2x^2 + 4)$ **3.** $4x^2 + 12x + 9 = (2x + 3)^2$

5. $x^2 - x - 2 = (x - 2)(x + 1) = 0;\quad x = 2, -1$

7. Adjust the sign of A and B so that the equation reads $Ax + By = |C|$.   Then we have

$$x\frac{A}{\sqrt{A^2 + B^2}} + y\frac{B}{\sqrt{A^2 + B^2}} = \frac{|C|}{\sqrt{A^2 + B^2}}.$$

Now set $\dfrac{A}{\sqrt{A^2 + B^2}} = \cos\alpha,\quad \dfrac{B}{\sqrt{A^2 + B^2}} = \sin\alpha,\quad \dfrac{|C|}{\sqrt{A^2 + B^2}} = p$

p is the length of $\overline{OQ}$, the distance
between the line and the origin; α is
the angle from the positive x-axis to
the line segment $\overline{OQ}$.

9. $y^2 = 8x$

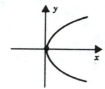

11. $(x + 1)^2 = -12(y - 3)$

13. $(x - 1)^2 = 4y$

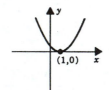

15. $(y - 1)^2 = -2(x - \frac{3}{2})$

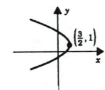

17. $y^2 = 2x$

vertex $(0,0)$

focus $(\frac{1}{2}, 0)$

axis $y = 0$

directrix $x = -\frac{1}{2}$

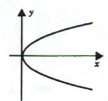

19. $x^2 = \frac{1}{2}(y + \frac{1}{2})$

vertex $(0, -\frac{1}{2})$

focus $(0, -\frac{3}{8})$

axis $x = 0$

directrix $y = -\frac{5}{8}$

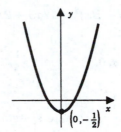

21. $(x + 2)^2 = -8(y - \frac{3}{2})$

vertex $(-2, \frac{3}{2})$

focus $(-2, -\frac{1}{2})$

axis $x = -2$

directrix $y = \frac{7}{2}$

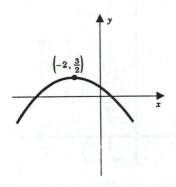

23. $(y + \frac{1}{2})^2 = x - \frac{3}{4}$

vertex $(\frac{3}{4}, -\frac{1}{2})$

focus $(1, -\frac{1}{2})$

axis $y = -\frac{1}{2}$

directrix $x = \frac{1}{2}$

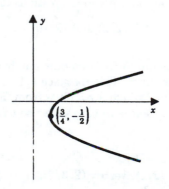

25. $\sqrt{(x - 1)^2 + (y - 2)^2} = \dfrac{|x + y + 1|}{\sqrt{2}}$ simplifies to $(x - y)^2 = 6x + 10y - 9$

27. Directrix has equation $x - y - 6 = 0$ since it has slope 1 and passes through the point $(4, -2)$.

$\sqrt{x^2 + (y - 2)^2} = \dfrac{|x - y - 6|}{\sqrt{2}}$ simplifies to $(x + y)^2 = -12x + 20y + 28$.

29. $P(x,y)$ is on the parabola with directrix $l: Ax + By + C = 0$ and focus $F(a,b)$ iff $d(P,l) = d(P,F)$

which happens iff $\quad \dfrac{|Ax + By + C|}{\sqrt{A^2 + B^2}} = \sqrt{(x-a)^2 + (y-b)^2}.$

Squaring both sides of this equation and simplifying, we obtain
$$(Ay - Bx)^2 = (2aS + 2AC)x + (2bS + 2BC)y + c^2 - (a^2 + b^2)S$$

with $S = A^2 + B^2 \neq 0.$

31. We can choose the coordinate system so that the parabola has an equation of the form $y = \alpha x^2, \alpha > 0$. One of the points of intersection is then the origin and the other is of the form $(c, \alpha c^2)$. We will assume that $c > 0.$

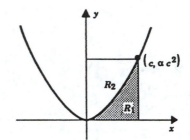

$$\text{area of } R_1 = \int_0^c \alpha x^2 \, dx = \frac{1}{3}\alpha c^3 = \frac{1}{3}A,$$
$$\text{area of } R_2 = A - \tfrac{1}{3}A = \tfrac{2}{3}A.$$

33. There are two possible positions for the focus:
$$(2,2) \text{ and } (2,10).$$

[The point $(5,6)$ is equidistant from the focus and the directrix. This distance is 5. The points on the line $x = 2$ which are 5 units from $(5,6)$ are $(2,2)$ and $(2,10)$.] These in turn give rise to two parabolas.

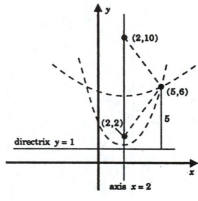

Focus $(2,2)$, vertex $(2,3/2)$:
$$(x-2)^2 = 4(\tfrac{1}{2})(y - \tfrac{3}{2}), \quad \text{which simplifies to} \quad x^2 - 4x + 7 = 2y.$$

Focus $(2,10)$, vertex $(2,11/2)$:
$$(x-2)^2 = 4(\tfrac{9}{2})(y - \tfrac{11}{2}), \quad \text{which simplifies to} \quad x^2 - 4x + 103 = 18y.$$

35. In this case the length of the latus rectum is the width of the parabola at height $y = c$. With $y = c$, $4c^2 = x^2$, and $x = \pm 2c$. The length of the latus rectum is thus $4c$.

37.
$$A = \int_{-2c}^{2c} \left(c - \frac{x^2}{4c} \right) dx = 2 \int_0^{2c} \left(c - \frac{x^2}{4c} \right) dx = 2 \left[cx - \frac{x^3}{12c} \right]_0^{2c} = \frac{8}{3}c^2$$

$\overline{x} = 0$ by symmetry

$$\overline{y}A = \int_{-2c}^{2c} \frac{1}{2} \left(c^2 - \frac{x^4}{16c^2} \right) dx = \int_0^{2c} \left(c^2 - \frac{x^4}{16c^2} \right) dx = \left[c^2 x - \frac{x^5}{80c^2} \right]_0^{2c} = \frac{8}{5}c^3$$

$$\overline{y} = (\frac{8}{5}c^3)/(\frac{8}{3}c^2) = \frac{3}{5}c$$

39. $\dfrac{kx}{p(0)} = \tan\theta = \dfrac{dy}{dx}, \quad y = \dfrac{k}{2\,p(0)}x^2 + C$

In our figure $C = y(0) = 0$. Thus the equation of the cable is $y = kx^2/2p(0)$, the equation of a parabola.

41.   Start with any two parabolas γ_1, γ_2. By moving them we can see to it that they have equations of the following form:

$$\gamma_1: x^2 = 4c_1 y, \quad c_1 > 0; \qquad \gamma_2: x^2 = 4c_2 y, \quad c_2 > 0.$$

Now we change the scale for γ_2 so that the equation for γ_2 will look exactly like the equation for γ_1. Set $X = (c_1/c_2)\,x, \quad Y = (c_1/c_2)\,y$. Then

$$x^2 = 4c_2 y \quad \Longrightarrow \quad (c_2/c_1)^2\,X^2 = 4c_2\,(c_2/c_1)\,Y \quad \Longrightarrow \quad X^2 = 4c_1 Y.$$

Now γ_2 has exactly the same equation as γ_1; only the scale, the units by which we measure distance, has changed.

SECTION 9.2

1. $\dfrac{x^2}{9} + \dfrac{y^2}{4} = 1$

center $(0,0)$

foci $(\pm\sqrt{5}, 0)$

length of major axis 6

length of minor axis 4

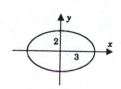

3. $\dfrac{x^2}{4} + \dfrac{y^2}{6} = 1$

center $(0,0)$

foci $(0, \pm\sqrt{2})$

length of major axis $2\sqrt{6}$

length of minor axis 4

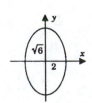

5. $\dfrac{x^2}{9} + \dfrac{(y-1)^2}{4} = 1$

center $(0, 1)$

foci $(\pm\sqrt{5},\, 1)$

length of major axis 6

length of minor axis 4

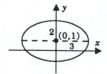

7. $\dfrac{(x-1)^2}{16} + \dfrac{y^2}{64} = 1$

center $(1, 0)$

foci $(1, \pm 4\sqrt{3})$

length of major axis 16

length of minor axis 8

9. Foci $(-1, 0), (1, 0)$ $\implies$ center $(0, 0)$, $c = 1$, and major axis parallel to x-axis.

Major axis 6 $\implies$ $a = 3$. Thus, $b = \sqrt{8}$.

Equation: $\dfrac{x^2}{9} + \dfrac{y^2}{8} = 1$.

11. Foci at $(1, 3)$ and $(1, 9)$ $\implies$ center $(1, 6)$, $c = 3$, and major axis parallel to y-axis.

Minor axis 8 $\implies$ $b = 4$. Thus, $a = 5$.

Equation: $\dfrac{(x-1)^2}{16} + \dfrac{(y-6)^2}{25} = 1$.

13. Focus $(1, 1)$ and center $(1, 3)$ $\implies$ $c = 2$ and major axis parallel to y-axis.

Major axis 10 $\implies$ $a = 5$. Thus, $b = \sqrt{21}$.

Equation: $\dfrac{(x-1)^2}{21} + \dfrac{(y-3)^2}{25} = 1$.

15. Major axis 10 $\implies$ $a = 5$. Vertices at $(3, 2)$ and $(3, -4)$ are then on minor axis parallel to y-axis. Then, $b = 3$ and center is $(3, -1)$.

Equation: $\dfrac{(x-3)^2}{25} + \dfrac{(y+1)^2}{9} = 1$.

17. Foci $(-5, 0)$ and $(5, 0)$ $\implies$ $c = 5$ and center $(0, 0)$.

Transverse axis 6 $\implies$ $a = 3$. Thus, $b = 4$.

Equation: $\dfrac{x^2}{9} - \dfrac{y^2}{16} = 1$.

19. Foci $(0, -13)$ and $(0, 13)$ $\implies$ $c = 13$ and center $(0, 0)$.

Transverse axis 10 $\implies$ $a = 5$. Thus, $b = 12$.

Equation: $\dfrac{y^2}{25} - \dfrac{x^2}{144} = 1$.

21. Foci $(-5, 1)$ and $(5, 1)$ $\implies$ $c = 5$ and center $(0, 1)$.

Transverse axis 6 $\implies$ $a = 3$. Thus, $b = 4$.

Equation: $\dfrac{x^2}{9} - \dfrac{(y-1)^2}{16} = 1$.

23. Foci $(-1, -1)$ and $(-1, 1)$ $\implies$ $c = 1$ and center $(-1, 0)$.

Transverse axis $\frac{1}{2}$ $\implies$ $a = \frac{1}{4}$. Thus, $b = \frac{1}{4}\sqrt{15}$.

Equation: $\dfrac{y^2}{1/16} - \dfrac{(x+1)^2}{15/16} = 1$.

25. $x^2 - y^2 = 1$

center $(0, 0)$
transverse axis 2
vertices $(\pm 1, 0)$
foci $(\pm\sqrt{2}, 0)$
asymptotes $y = \pm x$

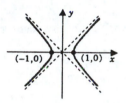

27. $\dfrac{x^2}{9} - \dfrac{y^2}{16} = 1$

center $(0, 0)$
transverse axis 6
vertices $(\pm 3, 0)$
foci $(\pm 5, 0)$
asymptotes $y = \pm\frac{4}{3}x$

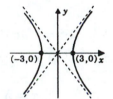

29. $\dfrac{y^2}{16} - \dfrac{x^2}{9} = 1$

center $(0, 0)$
transverse axis 8
vertices $(0, \pm 4)$
foci $(0, \pm 5)$
asymptotes $y = \pm\frac{4}{3}x$

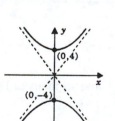

31. $\dfrac{(x-1)^2}{9} - \dfrac{(y-3)^2}{16} = 1$

center $(1, 3)$
transverse axis 6
vertices $(4, 3)$ and $(-2, 3)$
foci $(6, 3)$ and $(-4, 3)$
asymptotes $y - 3 = \pm\frac{4}{3}(x - 1)$

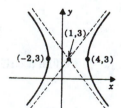

33. $\dfrac{(y-3)^2}{4} - \dfrac{(x-1)^2}{1} = 1$

center $(1, 3)$

transverse axis 4

vertices $(1, 5)$ and $(1, 1)$

foci $(1, 3 \pm \sqrt{5})$

asymptotes $y - 3 = \pm 2(x - 1)$

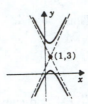

35. The length of the string is $d(F_1, F_2) + k = 2(c + a)$.

37. $2\sqrt{\pi^2 a^4 - A^2} / \pi a$

39. The equation of the ellipse is of the form

$$\frac{(x-5)^2}{25} + \frac{y^2}{25 - c^2} = 1.$$

Substitute $x = 3$ and $y = 4$ in that equation and you find that $c = \pm\frac{5}{21}\sqrt{5}$.
The foci are at $\left(5 \pm \frac{5}{21}\sqrt{5}, 0\right)$.

41. By the hint, $xy = X^2 - Y^2 = 1$. In the XY-system $a = 1, \quad b = 1, \quad c = \sqrt{2}$. We have center $(0, 0)$, vertices $(\pm 1, 0)$, foci $(\pm\sqrt{2}, 0)$ and asymptotes $Y = \pm X$. Using

$$x = X + Y \quad \text{and} \quad y = X - Y$$

to convert to the xy-system, we find center $(0, 0)$, vertices $(1, 1)$ and $(-1, -1)$, foci $(\sqrt{2}, \sqrt{2})$ and $(-\sqrt{2}, -\sqrt{2})$, asymptotes $y = 0$ and $x = 0$, transverse axis $2\sqrt{2}$.

43.
$$A = \frac{2b}{a} \int_a^{2a} \sqrt{x^2 - a^2}\, dx = \frac{2b}{a}\left[\frac{x}{2}\sqrt{x^2 - a^2} - \frac{a^2}{2}\ln\left(x + \sqrt{x^2 - a^2}\right)\right]_a^{2a}$$
$$= [2\sqrt{3} - \ln(2 + \sqrt{3})]ab$$

45. $e = \dfrac{\sqrt{25 - 16}}{\sqrt{25}} = \dfrac{3}{5}$ **47.** $e = \dfrac{\sqrt{25 - 9}}{\sqrt{25}} = \dfrac{4}{5}$

49. E_1 is fatter than E_2, more like a circle.   **51.** The ellipse tends to a line segment of length $2a$.

53. $x^2/9 + y^2 = 1$

55. $e = \frac{5}{3}$ **57.** $e = \sqrt{2}$

59. The branches of H_1 open up less quickly than the branches of H_2.

61. The hyperbola tends to a pair of parallel lines separated by the transverse axis.

63. Measure distances in miles and time in seconds. Place the origin at A and let $P(x, y)$ be the site of the crash. Then

$$d(P, B) - d(P, A) = (4)(0.20) = 0.80.$$

This places P on the right branch of the hyperbola

$$\frac{(x+1)^2}{(0.4)^2} - \frac{y^2}{1 - (0.4)^2} = 1.$$

Also

$$d(P, C) - d(P, A) = 6(0.20) = 1.20.$$

This places P on the left branch of the hyperbola

$$\frac{(x-1)^2}{(0.6)^2} - \frac{y^2}{1 - (0.6)^2} = 1.$$

Solve the two equations simultaneously keeping in mind the conditions of the problem and you will find that $x \cong -0.248$ and $y \cong 1.459$. The impact takes place about a quarter of a mile west of A and one and a half miles north.

SECTION 9.3

1–7.

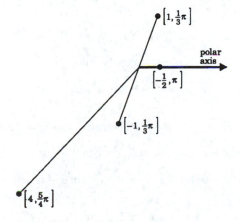

9. $x = 3\cos\frac{1}{2}\pi = 0$

$y = 3\sin\frac{1}{2}\pi = 3$

11. $x = -\cos(-\pi) = 1$

$y = -\sin(-\pi) = 0$

13. $x = -3\cos\left(-\frac{1}{3}\pi\right) = -\frac{3}{2}$

$y = -3\sin\left(-\frac{1}{3}\pi\right) = \frac{3}{2}\sqrt{3}$

15. $x = 3\cos\left(-\frac{1}{2}\pi\right) = 0$

$y = 3\sin\left(-\frac{1}{2}\pi\right) = -3$

17. $\left.\begin{array}{l} r^2 = 0^2 + 1^2, \quad r = \pm 1 \\ r = 1: \quad \cos\theta = 0 \text{ and } \sin\theta = 1 \\ \theta = \frac{1}{2}\pi \end{array}\right\}$ $\left[1, \frac{1}{2}\pi + 2n\pi\right], \quad \left[-1, \frac{3}{2}\pi + 2n\pi\right]$

19. $\left.\begin{array}{l} r^2 = (-3)^2 + 0^2 = 9, \quad r = \pm 3 \\ r = 3: \quad \cos\theta = -1 \text{ and } \sin\theta = 0 \\ \theta = \pi \end{array}\right\}$ $\left[3, \pi + 2n\pi\right], \quad \left[3, 2n\pi\right]$

21. $r^2 = 2^2 + (-2)^2 = 8, \quad r = \pm 2\sqrt{2}$
$\left. \begin{array}{l} r + 2\sqrt{2}: \quad \cos\theta = \frac{1}{2}\sqrt{2}, \quad \sin\theta = -\frac{1}{2}\sqrt{2} \\ \theta = \frac{7}{4}\pi \end{array} \right\}$ $\left[2\sqrt{2}, \frac{7}{4}\pi + 2n\pi \right], \quad \left[-2\sqrt{2}, \frac{3}{4}\pi + 2n\pi \right]$

23. $r^2 = \left(4\sqrt{3}\right)^2 + 4^2 = 64, \quad r \pm 8$
$\left. \begin{array}{l} r = 8: \quad \cos\theta = \frac{1}{2}\sqrt{3}, \quad \sin\theta = \frac{1}{2} \\ r = \frac{1}{6}\pi \end{array} \right\}$ $\left[8, \frac{1}{6}\pi + 2n\pi \right], \quad \left[-8, \frac{7}{6}\pi + 2n\pi \right]$

25. $d^2 = (x_1 - x_2)^2 + (y_1 - y_2)^2 = (r_1 \cos\theta_1 - r_2 \cos\theta_2)^2 + (r_1 \sin\theta_1 - r_2 \sin\theta_2)^2$

$$= r_1{}^2 \cos^2\theta_1 - 2r_1 r_2 \cos\theta_1 \cos\theta_2 + r_2{}^2 \cos^2\theta_2$$

$$+ r_1{}^2 \sin^2\theta_1 - 2r_1 r_2 \sin\theta_1 \sin\theta_2 + r_2{}^2 \sin^2\theta_2$$

$$= r_1{}^2 + r_2{}^2 - 2r_1 r_2 \left(\cos\theta_1 \cos\theta_2 + \sin\theta_1 \sin\theta_2\right)$$

$$= r_1{}^2 + r_2{}^2 - 2r_1 r_2 \cos\left(\theta_1 - \theta_2\right)$$

$$d = \sqrt{r_1{}^2 + r_2{}^2 - 2r_1 r_2 \cos\left(\theta_1 - \theta_2\right)}$$

27. (a) $\left[\frac{1}{2}, \frac{11}{6}\pi \right]$ (b) $\left[\frac{1}{2}, \frac{5}{6}\pi \right]$ (c) $\left[\frac{1}{2}, \frac{7}{6}\pi \right]$

29. (a) $\left[2, \frac{2}{3}\pi \right]$ (b) $\left[2, \frac{5}{3}\pi \right]$ (c) $\left[2, \frac{1}{3}\pi \right]$

31. about the x-axis?: $r = 2 + \cos\left(-\theta\right) \implies r = 2 + \cos\theta, \quad$ yes.

about the y-axis?: $r = 2 + \cos\left(\pi - \theta\right) \implies r = 2 - \cos\theta, \quad$ no.

about the origin?: $r = 2 + \cos\left(\pi + \theta\right) \implies r = 2 - \cos\theta, \quad$ no.

33. about the x-axis?: $r\left(\sin\left(-\theta\right) + \cos\left(-\theta\right)\right) = 1 \implies r\left(-\sin\theta + \cos\theta\right) = 1, \quad$ no.

about the y-axis?: $r\left(\sin\left(\pi - \theta\right) + \cos\left(\pi - \theta\right)\right) = 1 \implies r\left(\sin\theta - \cos\theta\right) = 1, \quad$ no.

about the origin?: $r\left(\sin\left(\pi + \theta\right) + \cos\left(\pi + \theta\right)\right) = 1 \implies r\left(-\sin\theta - \cos\theta\right) = 1, \quad$ no.

35. about the x-axis?: $r^2 \sin\left(-2\theta\right) = 1 \implies -r^2 \sin 2\theta = 1, \quad$ no.

about the y-axis?: $r^2 \sin\left(2\left(\pi - \theta\right)\right) = 1 \implies -r^2 \sin 2\theta = 1, \quad$ no.

about the origin?: $r^2 \sin\left(2\left(\pi + \theta\right)\right) = 1 \implies r^2 \sin 2\theta = 1, \quad$ yes.

37. $x = 2$ **39.** $2xy = 1$

$r \cos\theta = 2$ $2\left(r \cos\theta\right)\left(r \sin\theta\right) = 1$

$r^2 \sin 2\theta = 1$

41. $x^2 + (y - 2)^2 = 4$

$x^2 + y^2 - 4y = 0$

$r^2 - 4r \sin \theta = 0$

$r = 4 \sin \theta$

[note: division by r okay
since $[0, 0,]$ is on the curve]

43. $y = x$

$r \sin \theta = r \cos \theta$

$\tan \theta = 1$

$\theta = \pi/4$

45. $x^2 + y^2 + x = \sqrt{x^2 + y^2}$

$r^2 + r \cos \theta = r$

$r = 1 - \cos \theta$

47. $(x^2 + y^2)^2 = 2xy$

$r^4 = 2(r \cos \theta)(r \sin \theta)$

$r^2 = \sin 2\theta$

49.   The horizontal line $y = 4$

51.   the line $y = \sqrt{3}x$

53. $r = 2(1 - \cos \theta)^{-1}$

$r - r \cos \theta = 2$

$\sqrt{x^2 + y^2} - x = 2$

$x^2 + y^2 = (x + 2)^2$

$y^2 = 4(x + 1)$

a parabola

55. $r = 3 \cos \theta$

$r^2 = 3r \cos \theta$

$x^2 + y^2 = 3x$

a circle

57.   the line $y = 2x$

59. $r = \dfrac{4}{2 - \cos \theta}$

$2r - r \cos \theta = 4$

$2\sqrt{x^2 + y^2} - x = 4$

$4(x^2 + y^2) = (x + 4)^2$

$3x^2 + 4y^2 - 8x = 16$

an ellipse

61. $r = \dfrac{4}{1 - \cos \theta}$

$r - r \cos \theta = 4$

$\sqrt{x^2 + y^2} - x = 4$

$x^2 + y^2 = (x + 4)^2$

$y^2 = 8x + 16$

a parabola

63.

$r = a \sin \theta + b \cos \theta$

$r^2 = ar \sin \theta + br \cos \theta$

$x^2 + y^2 = ay + bz$

$\left(x - \dfrac{b}{2}\right)^2 + \left(y - \dfrac{a}{2}\right)^2 = \dfrac{a^2 + b^2}{4}$

center: $(b/2, a/2)$;   radius: $\frac{1}{2}\sqrt{a^2 + b^2}$

65. $\frac{1}{2}(r\cos\theta + d) = r$

  $r = \dfrac{d}{2 - \cos\theta}$

SECTION 9.4

1.

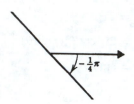

3.

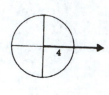

5.

7.

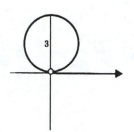

9.

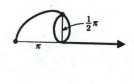

11.

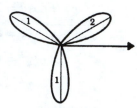

13.

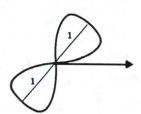

15.

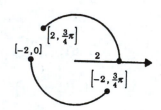

17.

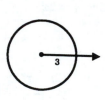

19.

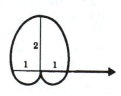

21.

23.

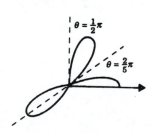

25.

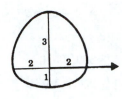

27.

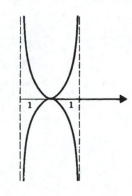

29.

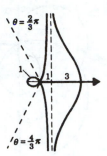

31.

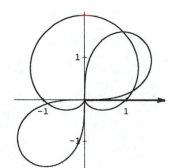

33.   yes; $[1, \pi] = [-1, 0]$ and the pair $r = -1, \theta = 0$ satisfies the equation

35.   yes;   the pair $r = \frac{1}{2}, \theta = \frac{1}{2}\pi$ satisfies the equation

37. $[2, \pi] = [-2, 0]$. The coordinates $[-2, 0]$ satisfy the equation $r^2 = 4\cos\theta$, and the coordinates $[2, \pi]$ satisfy the equation $r = 3 + \cos\theta$.

39. $(0,0)$, $\left(-\frac{1}{2}, \frac{1}{2}\right)$

41. $(1, 0)$, $(-1, 0)$

43. $(0, 0)$, $\left(\frac{1}{4}, \frac{1}{4}\sqrt{3}\right)$, $\left(\frac{1}{4}, -\frac{1}{4}\sqrt{3}\right)$

45. $(0, 0)$, $\left(\pm\frac{\sqrt{3}}{4}, \frac{3}{4}\right)$

47.   (a)

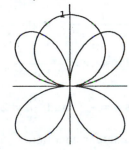

(b)   The curves intersect at the pole and at:

$[1.172, 0.173]$, $[1.86, 1.036]$, $[0.90, 3.245]$

49.   (a)

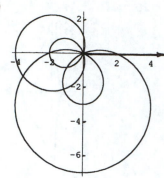

(b)   The curves intersect at the pole and at:

$r = 1 - 3\sin\theta$ $\qquad$ $r = 2 - 5\sin\theta$

$[-2, 0]$ $\qquad\qquad$ $[2, \pi]$

$[3.800, 3.510]$ $\qquad$ $[3.800, 3.510]$

$[2.412, 4.223]$ $\qquad$ $[-2.412, 1.081]$

$[-1.267, 0.713]$ $\qquad$ $[-1.267, 0.713]$

51.   "Butterfly" curves. The graph for the case $k = 2$ is:

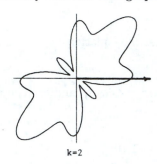

k=2

PROJECT 9.4

1. $e = \dfrac{r}{d - e\cos\theta}$ $\implies$ $r = ed - er\cos\theta$ $\implies$ $r(1 + e\cos\theta) = ed$ $\implies$ $r = \dfrac{ed}{1 + e\cos\theta}$

3.   (a) ellipse: $r = \dfrac{8}{4 + 3\cos\theta} = \dfrac{2}{1 + \frac{3}{4}\cos\theta}$.

Thus $e = \dfrac{3}{4}$ and $\dfrac{3}{4}d = 2$ $\implies$ $d = \dfrac{8}{3}$.

Rectangular equation:

$a = \dfrac{32}{7}$, $\quad c = \dfrac{24}{7}$, $\quad$ so $\dfrac{\left(x + \frac{24}{7}\right)^2}{\left(\frac{32}{7}\right)^2} + \dfrac{y^2}{\left(\frac{24}{7}\right)^2} = 1$

(b) hyperbola: $r = \dfrac{6}{1 + 2\cos\theta}$

Thus $e = 2$ and $2d = 6$ $\implies$ $d = 3$.

Rectangular equation:

$a = 2$, $\quad c = 4$, $\quad$ so $\dfrac{(x - 4)^2}{4} - \dfrac{y^2}{12} = 1$

(c) parabola: $r = \dfrac{6}{2 + 2\cos\theta} = \dfrac{3}{1 + \cos\theta}$

Thus $e = 1$ and $d = 3$.

Rectangular equation:

$$y^2 = -4\left(\frac{3}{2}\right)\left(x - \frac{3}{2}\right) + -6\left(x - \frac{3}{2}\right)$$

SECTION 9.5

1.

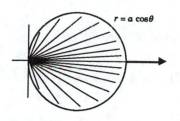

$$A = \int_{-\pi/2}^{\pi/2} \frac{1}{2}\left[a\cos\theta\right]^2 d\theta$$

$$= a^2 \int_0^{\pi/2} \frac{1 + \cos 2\theta}{2}\, d\theta$$

$$= a^2 \left[\frac{\theta}{2} + \frac{\sin 2\theta}{4}\right]_0^{\pi/2} = \frac{1}{4}\pi a^2$$

3.

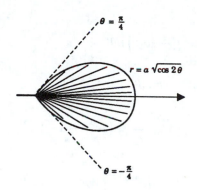

$$A = \int_{-\pi/4}^{\pi/4} \frac{1}{2}\left[a\sqrt{\cos 2\theta}\right]^2 d\theta$$

$$= a^2 \int_0^{\pi/4} \cos 2\theta\, d\theta$$

$$= a^2 \left[\frac{\sin 2\theta}{2}\right]_0^{\pi/4} = \frac{1}{2}a^2$$

5.

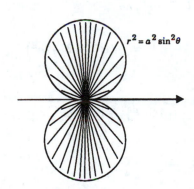

$$A = 2\int_0^{\pi} \frac{1}{2}\left(a^2\sin^2\theta\right) d\theta$$

$$= a^2 \int_0^{\pi} \frac{1 - \cos 2\theta}{2}\, d\theta$$

$$= a^2 \left[\frac{\theta}{2} - \frac{\sin 2\theta}{4}\right]_0^{\pi} = \frac{1}{2}\pi a^2$$

7.

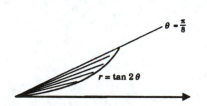

$$A = \int_0^{\pi/8} \frac{1}{2}\left[\tan 2\theta\right]^2 d\theta$$

$$= \frac{1}{2}\int_0^{\pi/8} \left(\sec^2 2\theta - 1\right) d\theta$$

$$= \frac{1}{2}\left[\frac{1}{2}\tan 2\theta - \theta\right]_0^{\pi/8} = \frac{1}{4} - \frac{\pi}{16}$$

9.

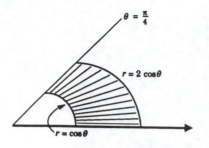

$$A = \int_0^{\pi/4} \frac{1}{2} \left([2\cos\theta]^2 - [\cos\theta]^2 \right) d\theta$$

$$= \frac{3}{2} \int_0^{\pi/4} \frac{1 + \cos 2\theta}{2} d\theta$$

$$= \frac{3}{2} \left[\frac{\theta}{2} + \frac{\sin 2\theta}{4} \right]_0^{\pi/4} = \frac{3}{16}\pi + \frac{3}{8}$$

11.

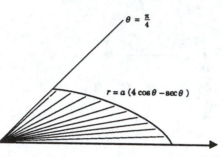

$$A = \int_0^{\pi/4} \frac{1}{2} \left[a\left(4\cos\theta - \sec\theta\right) \right]^2 d\theta$$

$$= \frac{a^2}{2} \int_0^{\pi/4} \left[16\cos^2\theta - 8 + \sec^2\theta \right] d\theta$$

$$= \frac{a^2}{2} \int_0^{\pi/4} \left[8\left(1 + \cos 2\theta\right) - 8 + \sec^2\theta \right] d\theta$$

$$= \frac{a^2}{2} \left[4\sin 2\theta + \tan\theta \right]_0^{\pi/4} = \frac{5}{2}a^2$$

13.

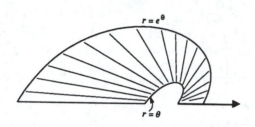

$$A = \int_0^\pi \frac{1}{2} \left([e^\theta]^2 - [\theta]^2 \right) d\theta$$

$$= \frac{1}{2} \int_0^\pi \left(e^{2\theta} - \theta^2 \right) d\theta$$

$$= \tfrac{1}{2} \left[\tfrac{1}{2}e^{2\theta} - \tfrac{1}{3}\theta^3 \right]_0^\pi = \tfrac{1}{12}\left(3e^{2\pi} - 3 - 2\pi^3\right)$$

15.

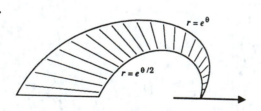

$$A = \int_0^\pi \frac{1}{2} \left([e^\theta]^2 - \left[e^{\theta/2}\right]^2 \right) d\theta$$

$$= \frac{1}{2} \int_0^\pi \left(e^{2\theta} - e^\theta \right) d\theta$$

$$= \tfrac{1}{2} \left[\tfrac{1}{2}e^{2\theta} - e^\theta \right]_0^\pi = \tfrac{1}{4}\left(e^{2\pi} + 1 - 2e^\pi\right)$$

17.

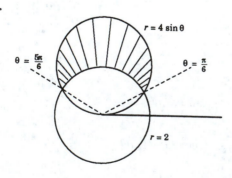

$$A = \int_{\pi/6}^{5\pi/6} \frac{1}{2} \left([4\sin\theta]^2 - [2]^2 \right) d\theta$$

19.

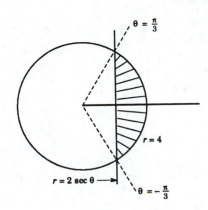

$$A = \int_{-\pi/3}^{\pi/3} \frac{1}{2} \left([\,4\,]^2 - [\,2\sec\theta\,]^2 \right)\, d\theta$$

21.

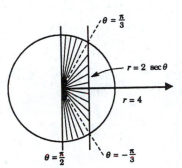

$$A = 2\left\{ \int_0^{\pi/3} \frac{1}{2}(2\sec\theta)^2\, d\theta + \int_{\pi/3}^{\pi/2} \frac{1}{2}(4)^2\, d\theta \right\}$$

23.

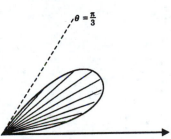

$$A = \int_0^{\pi/3} \frac{1}{2}(2\sin 3\theta)^2\, d\theta$$

25.

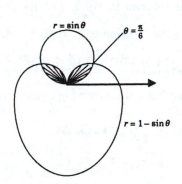

$$A = 2\left\{ \int_0^{\pi/6} \frac{1}{2}(\sin\theta)^2\, d\theta + \int_{\pi/6}^{\pi/2} \frac{1}{2}(1-\sin\theta)^2\, d\theta \right\}$$

27.

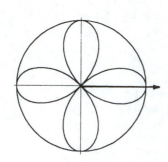

$$A = \pi - 8 \int_0^{\pi/4} \tfrac{1}{2} (\cos 2\theta)^2 \, d\theta$$

29. The area of one petal of the curve $r = a \cos 2n\theta$ is given by:

$$2 \int_0^{\pi/4n} \tfrac{1}{2}(a \cos 2n\theta)^2 \, d\theta = a^2 \int_0^{\pi/4n} \cos^2 2n\theta \, d\theta$$

$$= a^2 \int_0^{\pi/4n} \left(\frac{1}{2} + \frac{\cos 4n\theta}{2} \right) d\theta$$

$$= a^2 \left[\frac{1}{2} \theta + \frac{\sin 4n\theta}{8n} \right]_0^{\pi/4n} = \frac{\pi a^2}{8n}$$

The total area enclosed by $r = a \cos 2n\theta$ is $\dfrac{\pi a^2}{2}$.

The area of one petal of the curve $r = a \sin 2n\theta$ is given by:

$$A = 2 \int_0^{\pi/4n} \tfrac{1}{2}(a \sin 2n\theta)^2 \, d\theta = a^2 \int_0^{\pi/4n} \left(\frac{1}{2} + \frac{\cos 4n\theta}{2} \right) d\theta = \frac{\pi a^2}{8n}$$

and the total area enclosed by the curve is $\dfrac{\pi a^2}{2}$.

31. Let $P = \{\alpha = \theta_0, \theta_1, \theta_2, \ldots, \theta_n = \beta\}$ be a partition of the interval $[\alpha, \beta]$. Let θ_i^* be the midpoint of $[\theta_{i-1}, \theta_i]$ and let $r_i^* = f(\theta_i^*)$. The area of the ith "triangular" region is $\tfrac{1}{2}(r_i^*)\Delta\theta_i$, where $\Delta\theta_i = \theta_i - \theta_{i-1}$, and the rectangular coordinates of its centroid are(approximately) $\left(\tfrac{2}{3} r_i^* \cos \theta_i^*, \tfrac{2}{3} r_i^* \sin \theta_i^* \right)$.

The centroid $(\overline{x}_p, \overline{y}_p)$ of the union of the triangular regions satisfies the following equations

$$\overline{x}_p A_p = \frac{1}{3}(r_1^*)^3 \cos \theta_1 \Delta\theta_1 + \frac{1}{3}(r_2^*)^3 \cos \theta_2 \Delta\theta_2 + \cdots + \frac{1}{3}(r_n^*)^3 \cos \theta_n \Delta\theta_n$$

$$\overline{y}_p A_p = \frac{1}{3}(r_1^*)^3 \sin \theta_1 \Delta\theta_1 + \frac{1}{3}(r_2^*)^3 \sin \theta_2 \Delta\theta_2 + \cdots + \frac{1}{3}(r_n^*)^3 \sin \theta_n \Delta\theta_n$$

As $\|P\| \to 0$, the union of the triangular regions tends to the region Ω and the equations above tend to

$$\overline{x} A = \int_\alpha^\beta \frac{1}{3} r^3 \cos \theta \, d\theta$$

$$\bar{x}\, A = \int_{\alpha}^{\beta} \frac{1}{3} r^3 \cos\theta \, d\theta$$

The result follows from the fact that $A = \int_{\alpha}^{\beta} \frac{1}{2} r^2 \cos\theta \, d\theta$.

33. Since the region enclosed by the cardioid $r = 1 + \cos\theta$ is symmetric with respect to the x-axis, $\bar{y} = 0$.

To find $\bar{x}$:

$$A = \int_{0}^{2\pi} r^2 \, d\theta = \int_{0}^{2\pi} (1 + \cos\theta)^2 \, d\theta$$

$$= \int_{0}^{2\pi} (1 + 2\cos\theta + \cos^2\theta) \, d\theta$$

$$= \int_{0}^{2\pi} \left(\frac{3}{2} + 2\cos\theta \frac{1}{2}\cos 2\theta \right) d\theta$$

$$= \left[\frac{3}{2} + 2\sin\theta + \frac{1}{4}\sin 2\theta \right]_{0}^{2\pi} = 3\pi$$

and

$$\frac{2}{3} \int_{0}^{2\pi} r^3 \cos\theta \, d\theta = \frac{2}{3} \int_{0}^{2\pi} (1 + \cos\theta)^3 \cos\theta \, d\theta$$

$$= \frac{2}{3} \int_{0}^{2\pi} (\cos\theta + 3\cos^2\theta + 3\cos^3\theta + \cos^4\theta) \, d\theta$$

$$= \frac{2}{3} \int_{0}^{2\pi} \left(\frac{15}{8} + 4\cos\theta + 2\cos 2\theta + \frac{1}{8}\cos 4\theta - 3\sin^2\theta \cos\theta \right) d\theta$$

$$= \frac{2}{3} \left[\frac{15}{8}\theta + 4\sin\theta + \sin 2\theta + \frac{1}{32}\sin 4\theta - \sin^3\theta \right]_{0}^{2\pi} = \frac{5}{2}\pi$$

Thus $\quad \bar{x} = \dfrac{5\pi/2}{3\pi} = \dfrac{5}{6}$.

35. (a) $\quad y^2 = x^2 \left(\dfrac{a - x}{a + x} \right)$ $\qquad\qquad$ (b) Let $a = 2$

$$r^2 \sin^2\theta = r^2 \cos^2\theta \left(\frac{a - r \cos\theta}{a + r \cos\theta} \right)$$

$$\sin^2\theta (a + r \cos\theta) = \cos^2\theta (a - r \cos\theta)$$

$$r \cos\theta = a \cos 2\theta$$

$$r = a \cos 2\theta \sec\theta$$

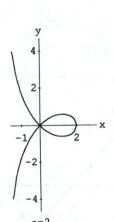

a=2

(c) $A = \int_{3\pi/4}^{5\pi/4} \frac{1}{2} a^2 \cos^2 2\theta \sec^2 \theta \, d\theta$

$\qquad = 2 \int_{3\pi/4}^{5\pi/4} \cos^2 2\theta \sec^2 \theta \, d\theta \qquad (a=2)$

$\qquad = 2 \int_{3\pi/4}^{5\pi/4} \frac{\left(2 \cos^2 \theta - 1\right)^2}{\cos^2 \theta} \, d\theta$

$\qquad = 2 \int_{3\pi/4}^{5\pi/4} \left(4 \cos^2 \theta - 4 + \sec^2 \theta\right) \, d\theta$

$\qquad = 2 \int_{3\pi/4}^{5\pi/4} \left(-2 + 2 \cos 2\theta + \sec^2 \theta\right) \, d\theta$

$\qquad = 2 \left[-2\theta + \sin 2\theta + \tan \theta\right]_{3\pi/4}^{5\pi/4} = 8 - 2\pi$

SECTION 9.6

1. $4x = (y-1)^2$

3. $y = 4x^2 + 1, \quad x \geq 0$

5. $9x^2 + 4y^2 = 36$

7. $1 + x^2 = y^2$

9. $y = 2 - x^2, \quad -1 \leq x \leq 1$

11. $2y - 6 = x, \quad -4 \leq x \leq 4$

13. $y = x - 1$

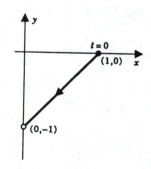

15. $xy = 1$

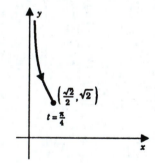

17. $2x + y = 11$

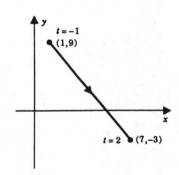

19. $x = \sin \frac{1}{2}\pi y$

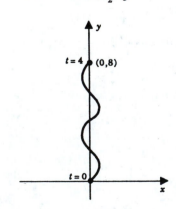

21. $1 + x^2 = y^2$

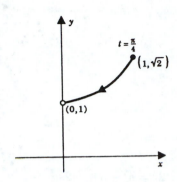

23.   (a) $x(t) = -\sin 2\pi t, \quad y(t) = \cos 2\pi t$ (b) $x(t) = \sin 4\pi t, \quad y(t) = \cos 4\pi t$

(c) $x(t) = \cos \frac{1}{2}\pi t, \quad y(t) = \sin \frac{1}{2}\pi t$ (d) $x(t) = \cos \frac{3}{2}\pi t, \quad y(t) = -\sin \frac{3}{2}\pi t$

25. $x(t) = \tan \frac{1}{2}\pi t, \quad y(t) = 2$ **27.** $x(t) = 3 + 5t, \quad y(t) = 7 - 2t$

29. $x(t) = \sin^2 \pi t, \quad y(t) = -\cos \pi t$ **31.** $x(t) = (2 - t)^2, \quad y(t) = (2 - t)^3$

33. $x(t) = t(b - a) + a, \quad y(t) = f(t(b - a) + a)$

35.

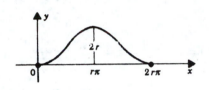

$$A = \int_0^{2\pi} x(t)\, y'(t)\, dt$$

$$= r^2 \int_0^{2\pi} (1 - \cos t)\, dt$$

$$= r^2 \left[t - \sin t \right]_0^{2\pi} = 2\pi r^2$$

37.   (a) $V_x = 2\pi \overline{y} A = 2\pi \left(\frac{3}{4} r \right) \left(2\pi r^2 \right) = 3\pi^2 r^3$

(b) $V_y = 2\pi \overline{x} A = 2\pi (\pi r)\, 2\pi r^2 = 4\pi^3 r^3$

39. $x(t) = -a \cos t, \quad y(t) = b \sin t \qquad t \in [0, \pi]$

41.   (a)   Equation for the ray: $y + 2x = 17, \quad x \geq 6$.

Equation for the circle: $(x - 3)^2 + (y - 1)^2 = 25$.

Simultaneous solution of these equations gives the points of intersection: $(6, 5)$ and $(8, 1)$.

(b)   The particle on the ray is at $(6, 5)$ when $t = 0$.   However, when $t = 0$ the particle on the circle is at the point $(-2, 1)$. Thus, the intersection point $(6, 5)$ is not a collision point.

The particle on the ray is at $(8, 1)$ when $t = 1$. Since the particle on the circle is also at $(8, 1)$ when $t = 1$, the intersection point $(8, 1)$ is a collision point.

43.   If $x(r) = x(s)$ and $r \neq s$, then

$$r^2 - 2r = s^2 - 2s$$

$$r^2 - s^2 = 2r - 2s$$

(1) $$r + s = 2.$$

If $y(r) = y(s)$ and $r \neq s$, then

$$r^3 - 3r^2 + 2r = s^3 - 3s^2 + 2s$$

$$\left(r^3 - s^3\right) - 3\left(r^2 - s^2\right) + 2\left(r - s\right) = 0$$

(2) $$\left(r^2 + rs + s^2\right) - 3\left(r + s\right) + 2 = 0.$$

Simultaneous solution of (1) and (2) gives $r = 0$ and $r = 2$. Since $(x(0), y(0)) = (0,0) = (x(2), y(2))$, the curve intersects itself at the origin.

45.   Suppose that $r, s \in [0, 4]$ and $r \neq s$.

$$x(r) = x(s) \quad \Longrightarrow \quad \sin 2\pi r = \sin 2\pi s.$$

$$y(r) = y(s) \quad \Longrightarrow \quad 2r - r^2 = 2s - s^2 \quad \Longrightarrow \quad 2(r - s) = r^2 - s^2 \quad \Longrightarrow \quad 2 = r + s.$$

Now we solve the equations simultaneously:

$$\sin 2\pi r = \sin\left[2\pi\left(2 - r\right)\right] = -\sin 2\pi r$$

$$2 \sin 2\pi r = 0$$

$$\sin 2\pi r = 0.$$

Since $r \in [0, 4]$, $r = 0, \frac{1}{2}, 1, \frac{3}{2}, 2, \frac{5}{2}, 3, \frac{7}{2}, 4$.

Since $s \in [0, 4]$ and $r \neq s$ and $r + s = 2$,   we are left with $r = 0, \frac{1}{2}, \frac{3}{2}, 2$. Note that

$$(x(0), y(0)) = (0, 0) = (x(2), y(2)) \quad \text{and} \quad \left(x\left(\tfrac{1}{2}\right), y\left(\tfrac{1}{2}\right)\right) = \left(0, \tfrac{3}{4}\right) = \left(x\left(\tfrac{3}{2}\right), y\left(\tfrac{3}{2}\right)\right).$$

The curve intersects itself at $(0, 0)$ and $\left(0, \frac{3}{4}\right)$.

47.   (a)   The coefficient a affects the amplitude and the period.

(b) $$\frac{dy}{dx} = \frac{dy/d\theta}{dx/d\theta}$$

$$= \frac{a \sin \theta}{a(1 - \cos \theta)} = \frac{\sin \theta}{1 - \cos \theta}$$

You can verify that $\dfrac{dy}{dx} \to -\infty$ as $\theta \to 2\pi^-$; $\dfrac{dy}{dx} \to \infty$ as $\theta \to 2\pi^+$.

(c)   The curve has a vertical cusp at $\theta = 2\pi$.

49.   See the answer section in the text.

PROJECT 9.6

1.   Since $x''(t) = 0$, we have $x'(t) = C$.

Since $x'(0) = v_0 \cos\theta$, we have $x'(t) = v_0 \cos\theta$.

Integrating again, $x(t) = (v_0 \cos\theta)t + x_0$ (since $x(0) = x_0$)

Similarly, since $y''(t) = -g$, we have $y'(t) = -gt + C$.

Since $y'(0) = v_0 \sin\theta$, we have $y'(t) = -gt + v_0 \sin\theta$.

Integrating again, $y(t) = -\frac{1}{2}gt^2 + (v_0 \sin\theta)t + y_0$ (since $y(0) = y_0$)

3.   (a) Using $x_0 = 0$, $y_0 = 0$, $g = 32$: $y = -\dfrac{16}{v_0{}^2}(\sec^2\theta)x^2 + (\tan\theta)x$.

(b) At maximum height, $y'(t) = 0$. $y(t) = -16t^2 + (v_0 \sin\theta)\,t$; $y'(t) = -32t + v_0 \sin\theta$.

$y'(t) = 0 \implies t = \dfrac{v_0 \sin\theta}{32}$;   max height: $y\left(\dfrac{v_0 \sin\theta}{32}\right) = \dfrac{v_0{}^2}{64}\sin^2\theta$.

(c) $y = 0$ (and $x \neq 0$) $\implies$ $\dfrac{v_0^2}{16}\cos\theta \sin\theta$

(d) $y(t) = 0$ (and $t \neq 0$) when $t = \dfrac{v_0}{16}\sin\theta$

(e) The range $\frac{1}{16}v_0{}^2 \sin\theta \cos\theta = \frac{1}{32}v_0{}^2 \sin 2\theta$ is clearly maximal when $\theta = \frac{1}{4}\pi$ for then $\sin 2\theta = 1$.

(f) We want $\dfrac{1}{32}v_0{}^2 \sin 2\theta = b$, $\theta = \dfrac{1}{2}\sin^{-1}\left(\dfrac{32b}{v_0{}^2}\right)$.

SECTION 9.7

1. $x'(1) = 1$, $y'(1) = 3$,   slope 3,   point $(1,0)$;   tangent $y = 3(x-1)$

3. $x'(0) = 2$, $y'(0) = 0$,   slope 0,   point $(0,1)$;   tangent $y = 1$

5. $x'(1/2) = 1$, $y'(1/2) = -3$,   slope -3,   point $\left(\frac{1}{4}, \frac{9}{4}\right)$;   tangent $y - \frac{9}{4} = -3\left(x - \frac{1}{4}\right)$

7. $x'\left(\dfrac{\pi}{4}\right) = -\dfrac{3}{4}\sqrt{2}$, $y'\left(\dfrac{\pi}{4}\right) = \dfrac{3}{4}\sqrt{2}$,   slope -1,   point $\left(\dfrac{1}{4}\sqrt{2}, \dfrac{1}{4}\sqrt{2}\right)$;

tangent $y - \frac{1}{4}\sqrt{2} = -\left(x - \frac{1}{4}\sqrt{2}\right)$

9. $x(\theta) = \cos\theta\,(4 - 2\sin\theta)$, $y(\theta) = \sin\theta\,(4 - 2\sin\theta)$,   point $(4,0)$

$x'(\theta) = -4\sin\theta - 2\left(\cos^2\theta - \sin^2\theta\right), \quad y'(\theta) = 4\cos\theta - 4\sin\theta\cos\theta$

$x'(0) = -2, \quad y'(0) = 4, \quad \text{slope} -2, \quad \text{tangent } y = -2\left(x - 4\right)$

11. $x(\theta) = \dfrac{4\cos\theta}{5 - \cos\theta}, \quad y(\theta) = \dfrac{4\sin\theta}{5 - \cos\theta}, \quad \text{point } \left(0, \dfrac{4}{5}\right)$

$x'(\theta) = \dfrac{-20\sin\theta}{(5 - \cos\theta)^2}, \quad y'(\theta) = \dfrac{4\left(5\cos\theta - 1\right)}{(5 - \cos\theta)^2}$

$x'\left(\dfrac{\pi}{2}\right) = -\dfrac{4}{5}, \quad y'\left(\dfrac{\pi}{2}\right) = -\dfrac{4}{25}, \quad \text{slope } \dfrac{1}{5}, \quad \text{tangent } y - \dfrac{4}{5} = \dfrac{1}{5}x$

13. $x(\theta) = \dfrac{\cos\theta\left(\sin\theta - \cos\theta\right)}{\sin\theta + \cos\theta}, \quad y(\theta) = \dfrac{\sin\theta\left(\sin\theta - \cos\theta\right)}{\sin\theta + \cos\theta}, \quad \text{point } (-1, 0)$

$x'(\theta) = \dfrac{\sin\theta\,\cos 2\theta + 2\cos\theta}{(\sin\theta + \cos\theta)^2}, \quad y'(\theta) = \dfrac{2\sin\theta - \cos\theta\,\cos 2\theta}{(\sin\theta + \cos\theta)^2}$

$x'(0) = 2, \quad y'(0) = -1, \quad \text{slope } -\tfrac{1}{2}, \quad \text{tangent } y = -\tfrac{1}{2}(x + 1)$

15. $x(t) = t, \quad y(t) = t^3$

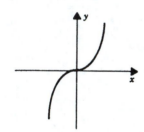

$x'(0) = 1, \quad y'(0) = 0, \quad \text{slope } 0$

tangent $y = 0$

17. $x(t) = t^{5/3}, \quad y(t) = t$

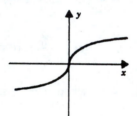

$x'(0) = 0, \quad y'(0) = 1, \quad \text{slope undefined}$

tangent $x = 0$

19. $x'(t) = 3 - 3t^2, \quad y'(t) = 1$

$x'(t) = 0 \quad \Longrightarrow \quad t = \pm 1; \quad y'(t) \neq 0$

(a)   none

(b)   at $(2, 2)$ and $(-2, 0)$

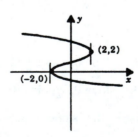

21. curve traced once completely with $t \in [0, 2\pi)$

$$x'(t) = -4\cos t, \quad y'(t) = -3\sin t$$

$$x'(t) = 0 \implies t = \frac{\pi}{2}, \frac{3\pi}{2};$$

$$y'(t) = 0 \implies t = 0, \pi$$

(a) at (3, 7) and (3, 1)

(b) at (−1, 4) and (7, 4)

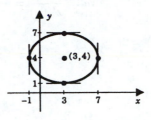

23. $x'(t) = 2t - 2, \quad y'(t) = 3t^2 - 6t + 2$

$$x'(t) = 0 \implies t = 1$$

$$y'(t) = 0 \implies t = 1 \pm \tfrac{1}{3}\sqrt{3}$$

(a) at $\left(-\tfrac{2}{3}, \pm\tfrac{2}{9}\sqrt{3}\right)$

(b) at (−1, 0)

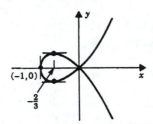

25. curve traced completely with $t \in [0, 2\pi)$

$$x'(t) = -\sin t, \quad y'(t) = 2\cos 2t$$

$$x'(t) = 0 \implies t = 0, \pi$$

$$y'(t) = 0 \implies t = \frac{\pi}{4}, \frac{3\pi}{4}, \frac{5\pi}{4}, \frac{7\pi}{4}$$

(a) at $\left(\pm\tfrac{1}{2}\sqrt{2}, \pm 1\right)$

(b) at $(\pm 1, 0)$

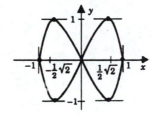

27. First, we find the values of t when the curve passes through $(2, 0)$.

$$y(t) = 0 \implies t^4 - 4t^2 = 0 \implies t = 0, \pm 2.$$

$$x(-2) = 2, \quad x(0) = 2, \quad x(2) = -2.$$

The curve passes through $(2, 0)$ at $t = -2$ and $t = 0$.

$$x'(t) = -1 - \frac{\pi}{2}\sin\frac{\pi t}{4}, \quad y'(t) = 4t^3 - 8t.$$

At $t = -2$, $x'(-2) = \dfrac{\pi}{2} - 1$, $y'(t) = -16$, tangent $y = \dfrac{32}{2 - \pi}(x - 2)$.

At $t = 0$, $x'(0) = -1$, $y'(0) = 0$, tangent $y = 0$.

29. The slope of $\overline{OP}$ is $\tan\theta_1$. The curve $r = f(\theta)$ can be parametrized by setting

$$x(\theta) = f(\theta)\cos\theta, \quad y(\theta) = f(\theta)\sin\theta.$$

Differentiation gives

$$x'(\theta) = -f(\theta)\sin\theta + f'(\theta)\cos\theta, \quad y'(\theta) = f(\theta)\cos\theta + f'(\theta)\sin\theta.$$

If $f'(\theta_1) = 0$, then

$$x'(\theta_1) = -f(\theta_1)\sin\theta_1, \quad y'(\theta_1) = f(\theta_1)\cos\theta_1.$$

Since $f(\theta_1) \neq 0$, we have

$$m = \frac{y'(\theta_1)}{x'(\theta_1)} = -\cot\theta_1 = -\frac{1}{\text{slope of } \overline{OP}}.$$

31. $x'(t) = 3t^2, \quad y'(t) = 2t$

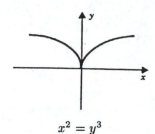

$$x^2 = y^3$$

33. $x'(t) = 5t^4, \quad y'(t) = 3t^2$

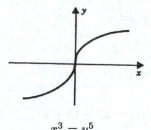

$$x^3 = y^5$$

35. $x'(t) = 2t, \quad y'(t) = 2t$

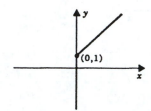

ray: $y = x + 1, \quad x \geq 0$

37. By (9.7.5), $\dfrac{d^2y}{dx^2} = \dfrac{(-\sin t)(-\sin t) - (\cos t)(-\cos t)}{(-\sin t)^3} = \dfrac{-1}{\sin^3 t}.$ At $t = \dfrac{\pi}{6}$, $\dfrac{d^2y}{dx^2} = -8.$

39. By (9.7.5), $\dfrac{d^2y}{dx^2} = \dfrac{(e^t)(e^{-t}) - (-e^{-t})(e^t)}{(e^t)^3} = 2e^{-3t}.$ At $t = 0$, $\dfrac{d^2y}{dx^2} = 2.$

SECTION 9.8

1. $L = \displaystyle\int_0^1 \sqrt{1 + 2^2}\, dx = \sqrt{5}$

3. $L = \displaystyle\int_1^4 \sqrt{1 + \left[\frac{3}{2}\left(x - \frac{4}{9}\right)^{1/2}\right]^2}\, dx = \int_1^4 \frac{3}{2}\sqrt{x}\, dx = \left[x^{3/2}\right]_1^4 = 7$

5. $L = \displaystyle\int_0^3 \sqrt{1 + \left(\frac{1}{2}\sqrt{x} - \frac{1}{2\sqrt{x}}\right)^2}\, dx = \int_0^3 \left(\frac{1}{2}\sqrt{x} + \frac{1}{2\sqrt{x}}\right) dx = \left[\frac{1}{3}x^{3/2} + x^{1/2}\right]_0^3 = 2\sqrt{3}$

7. $L = \displaystyle\int_0^1 \sqrt{1 + \left[x\,(x^2 + 2)^{1/2}\right]^2}\, dx = \int_0^1 (x^2 + 1)\, dx = \left[\frac{1}{3}x^3 + x\right]_0^1 = \frac{4}{3}$

9. $L = \displaystyle\int_1^5 \sqrt{1 + \left[\frac{1}{2}\left(x - \frac{1}{x}\right)\right]^2}\, dx = \int_1^5 \frac{1}{2}\left(x + \frac{1}{x}\right) dx = \left[\frac{1}{2}\left(\frac{1}{2}x^2 + \ln x\right)\right]_1^5 = 6 + \frac{1}{2}\ln 5 \cong 6.80$

11. $L = \int_1^8 \sqrt{1 + \left[\frac{1}{2}\left(x^{1/3} - x^{-1/3}\right)\right]^2} \, dx = \int_1^8 \frac{1}{2}\left(x^{1/3} + x^{-1/3}\right) dx = \frac{1}{2}\left[\frac{3}{4}x^{4/3} + \frac{3}{2}x^{2/3}\right]_1^8 = \frac{63}{8}$

13. $L = \int_0^{\pi/4} \sqrt{1 + \tan^2 x} \, dx = \int_0^{\pi/4} \sec x \, dx = [\ln|\sec x + \tan x|]_0^{\pi/4} = \ln\left(1 + \sqrt{2}\right) \cong 0.88$

15. $L = \int_1^2 \sqrt{1 + \left(\sqrt{x^2 - 1}\right)^2} \, dx = \int_1^2 x \, dx = \left[\frac{1}{2}x^2\right]_1^2 = \frac{3}{2}$

17. $L = \int_0^1 \sqrt{1 + \left[\sqrt{3 - x^2}\right]^2} \, dx = \int_0^1 \sqrt{4 - x^2} \, dx = \int_0^{\pi/6} 4\cos^2 u \, du$

$$(x = 2\sin u)$$

$$= 2\int_0^{\pi/6} [1 + \cos 2u] \, du = [2u + \sin 2u]_0^{\pi/6} = \frac{\pi}{3} + \frac{\sqrt{3}}{2}$$

19. $v(t) = \sqrt{(2t)^2 + 2^2} = 2\sqrt{t^2 + 1}$

initial speed $= v(0) = 2$, terminal speed $= v\left(\sqrt{3}\right) = 4$

$$s = \int_0^{\sqrt{3}} 2\sqrt{t^2 + 1} \, dt = 2\int_0^{\pi/3} \sec^3 u \, du = 2\left[\frac{1}{2}\sec u \tan u + \frac{1}{2}\ln|\sec u + \tan u|\right]_0^{\pi/3}$$

$$(t = \tan u) \qquad\qquad\qquad \text{(by parts)}$$

$$= 2\sqrt{3} + \ln\left(2 + \sqrt{3}\right) \cong 4.78$$

21. $v(t) = \sqrt{(2t)^2 + (3t^2)^2} \, dt = t\left(4 + 9t^2\right)^{1/2}$

initial speed $= v(0) = 0$, terminal speed $= v(1) = \sqrt{13}$

$$s = \int_0^1 t\left(4 + 9t^2\right)^{1/2} \, dt = \left[\frac{1}{27}\left(4 + 9t^2\right)^{3/2}\right]_0^1 = \frac{1}{27}\left(13\sqrt{13} - 8\right)$$

· **23.** $v(t) = \sqrt{\left[e^t \cos t + e^t \sin t\right]^2 + \left[e^t \cos t - e^t \sin t\right]^2} = \sqrt{2}\, e^t$

initial speed $= v(0) = \sqrt{2}$, terminal speed $= \sqrt{2}\, e^\pi$

$$s = \int_0^\pi \sqrt{2}\, e^t \, dt = \left[\sqrt{2}\, e^t\right]_0^\pi = \sqrt{2}\, (e^\pi - 1)$$

25. $L = \int_0^{2\pi} \sqrt{[x'(\theta)]^2 + [y'(\theta)]^2} \, d\theta = \int_0^{2\pi} \sqrt{a^2(1 - \cos\theta)^2 + a^2\sin^2\theta} \, d\theta$

$$= a\int_0^{2\pi} \sqrt{2(1 - \cos\theta)} \, d\theta = 2a\int_0^{2\pi} \sin\frac{\theta}{2} \, d\theta = -4a\left[\cos\frac{\theta}{2}\right]_0^{2\pi} = 8a$$

27. (a) $L = \int_0^{2\pi} \sqrt{(-3a\sin\theta - 3a\sin 3\theta)^2 + (3a\cos\theta - 3a\cos 3\theta)^2} \; d\theta$

$= 3a \int_0^{2\pi} \sqrt{\sin^2\theta + 2\sin\theta\sin 3\theta + \sin^2 3\theta + \cos^2\theta - 2\cos\theta\cos 3\theta + \cos^2 3\theta} \; d\theta$

$= 3a \int_0^{2\pi} \sqrt{2(1-\cos 4\theta)} \; d\theta = 6a \int_0^{2\pi} |\sin 2\theta| \, d\theta$

$= 24a \int_0^{\pi/2} \sin 2\theta \, d\theta = -12a \, [\cos 2\theta]_0^{\pi/2} = 24a$

(b) The result follows from the identities: $\cos 3\theta = 4\cos^3\theta - 3\cos\theta;$ $\sin 3\theta = 3\sin\theta - 4\sin^3\theta$

29. $L =$ circumference of circle of radius $1 = 2\pi$

31. $L = \int_0^{4\pi} \sqrt{[e^\theta]^2 + [e^\theta]^2} \; d\theta = \int_0^{4\pi} \sqrt{2} \, e^\theta d\theta = \left[\sqrt{2}\, e^\theta\right]_0^{4\pi} = \sqrt{2}\left(e^{4\pi} - 1\right)$

33. $L = \int_0^{2\pi} \sqrt{[e^{2\theta}]^2 + [2e^{2\theta}]^2} \; d\theta = \int_0^{2\pi} \sqrt{5} \, e^{2\theta} d\theta = \left[\frac{1}{2}\sqrt{5}\, e^{2\theta}\right]_0^{2\pi} = \frac{1}{2}\sqrt{5}\left(e^{4\pi} - 1\right)$

35. $L = \int_0^{\pi/2} \sqrt{(1-\cos\theta)^2 + \sin^2\theta} \; d\theta = \int_0^{\pi/2} \sqrt{2 - 2\cos\theta} \; d\theta$

$= \int_0^{\pi/2} \left(2\sin\frac{1}{2}\theta\right) d\theta = \left[-4\cos\frac{1}{2}\theta\right]_0^{\pi/2} = 4 - 2\sqrt{2}$

37. $s = \int_0^1 \sqrt{\left[\frac{1}{1+t^2}\right]^2 + \left[\frac{-t}{1+t^2}\right]^2} \; dt = \int_0^1 \frac{dt}{\sqrt{1+t^2}}$

$= \int_0^{\pi/4} \sec u \, du = [\ln|\sec u + \tan u|]_0^{\pi/4} = \ln\left(1 + \sqrt{2}\right)$

$\quad (t = \tan u)$

initial speed $= v(0) = 1,$ terminal speed $= v(1) = \frac{1}{2}\sqrt{2}$

39. $c = 1;$ the curve $y = e^x$ is the curve $y = \ln x$ reflected in the line $y = x$

41. $L = \int_a^b \sqrt{1 + \sinh^2 x} \; dx = \int_a^b \sqrt{\cosh^2 x} \; dx = \int_a^b \cosh x \, dx = A$

43. $\sqrt{1 + [f(x)]^2} = \sqrt{1 + \tan^2[\alpha(x)]} = |\sec[\alpha(x)]|$

45.   **(a)**

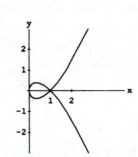

     (b) $L = \displaystyle\int_{-1}^{1} \sqrt{9t^4 - 2t^2 + 1}\, dt \cong 2.7156$

47.   **(a)** $L = \displaystyle\int_{0}^{2\pi} \sqrt{a^2 \sin^2 t + b^2 \cos^2 t}\, dt = 4 \int_{0}^{\pi/2} \sqrt{a^2(1 - \cos^2 t) + b^2 \cos^2 t}\, dt$

$$= 4a \int_{0}^{\pi/2} \sqrt{1 - e^2 \cos^2 t}\, dt, \quad \text{where} \quad e = \frac{\sqrt{a^2 - b^2}}{a}$$

     (b) $L = 4 \displaystyle\int_{0}^{\pi/2} \sqrt{25 - 9 \cos^2 t}\, dt \cong 28.3617$

SECTION 9.9

1. $L = $ length of the line segment $= 1$

$(\bar{x}, \bar{y}) = \left(\frac{1}{2}, 4\right)$ (the midpoint of the line segment)

$A_x = $ lateral surface area of cylinder of radius 4 and side $1 = 16\pi$.

3. $L = \displaystyle\int_{0}^{3} \sqrt{1 + \left(\frac{4}{3}\right)^2}\, dx = \left(\frac{5}{3}\right)^3 = 5$

$\bar{x}L = \displaystyle\int_{0}^{3} x\sqrt{1 + \left(\frac{4}{3}\right)^2}\, dx = \frac{5}{3}\left[\frac{1}{2}x^2\right]_{0}^{3} = \frac{15}{2}, \quad \bar{x} = \frac{3}{2}$

$\bar{y}L = \displaystyle\int_{0}^{3} \frac{4}{3}x\sqrt{1 + \left(\frac{4}{3}\right)^2}\, dx = \left(\frac{4}{3}\right)\left(\frac{15}{2}\right) = 10, \quad \bar{y} = 2$

$A_x = 2\pi\bar{y}L = 2\pi(2)(5) = 20\pi$

5. $L = \displaystyle\int_{0}^{2} \sqrt{(3)^2 + (4)^2}\, dt = (2)(5) = 10$

$\bar{x}L = \displaystyle\int_{0}^{2} 3t\sqrt{(3)^2 + (4)^2}\, dt = 15\left[\frac{1}{2}t^2\right]_{0}^{2} = 30, \quad \bar{x} = 3$

$\bar{y}L = \displaystyle\int_{0}^{2} 4t\sqrt{(3)^2 + (4)^2}\, dt = 20\left[\frac{1}{2}t^2\right]_{0}^{2} = 40, \quad \bar{y} = 4$

$A_x = 2\pi\bar{y}L = 2\pi(4)(10) = 80\pi$

7. $$L = \int_0^{\pi/6} \sqrt{4\sin^2 t + 4\cos^2 t}\, dt = 2\left(\frac{\pi}{6}\right) = \frac{1}{3}\pi$$

$$\bar{x}L = \int_0^{\pi/6} 2\cos t\sqrt{4\sin^2 t + 4\cos^2 t}\, dt = 4\left[\sin t\right]_0^{\pi/6} = 2, \quad \bar{x} = \frac{6}{\pi}$$

$$\bar{y}L = \int_0^{\pi/6} 2\sin t\sqrt{4\sin^2 t + 4\cos^2 t}\, dt = 4\left[-\cos t\right]_0^{\pi/6} = 4 - 2\sqrt{3}, \quad \bar{y} = 6\left(2 - \sqrt{3}\right)/\pi$$

$$A_x = 2\pi\bar{y}L = 2\pi(6(2-\sqrt{3})/\pi)\tfrac{1}{3}\pi = 4\pi(2-\sqrt{3})$$

9. $$x(t) = a\cos t, \quad y = a\sin t; \quad t \in [\tfrac{1}{3}\pi, \tfrac{2}{3}\pi]$$

$$L = \int_{\pi/3}^{2\pi/3} \sqrt{a^2\sin^2 t + a^2\cos^2 t}\, dt = \frac{1}{3}\pi a$$

by symmetry $\bar{x} = 0$

$$\bar{y}L = \int_{\pi/3}^{2\pi/3} a\sin t\sqrt{a^2\sin^2 t + a^2\cos^2 t}\, dt = a^2\int_{\pi/3}^{2\pi/3} \sin t\, dt$$

$$= a^2\left[-\cos t\right]_{\pi/3}^{2\pi/3} = a^2, \quad \bar{y} = 3a/\pi$$

$$A_x = 2\pi\bar{y}L = 2\pi a^2$$

11. $$A_x = \int_0^2 \frac{2}{3}\pi x^3\sqrt{1+x^4}\, dx = \frac{1}{9}\pi\left[(1+x^4)^{3/2}\right]_0^2 = \frac{1}{9}(17\sqrt{17}-1) \cong 7.68$$

13. $$A_x = \int_0^1 \frac{1}{2}\pi x^3\sqrt{1+\frac{9}{16}x^4}\, dx = \frac{4}{27}\pi\left[\left(1+\frac{9}{16}x^4\right)\right]_0^1 = \frac{61}{432}\pi$$

15. $$A_x = \int_0^{\pi/2} 2\pi\cos x\sqrt{1+\sin^2 x}\, dx = \int_0^1 2\pi\sqrt{1+u^2}\, du$$

$$u = \sin x$$

$$= 2\pi\left[\tfrac{1}{2}u\sqrt{1+u^2} + \tfrac{1}{2}\ln\left(u+\sqrt{1+u^2}\right)\right]_0^1 = \pi\left[\sqrt{2}+\ln\left(1+\sqrt{2}\right)\right]$$

(8.5.1)

17.
$$A_x = \int_0^{\pi/2} 2\pi(e^\theta \sin\theta)\sqrt{[e^\theta \cos\theta - e^\theta \sin\theta]^2 + [e^\theta \sin\theta + e^\theta \cos\theta]^2}\, d\theta$$

$$= 2\pi\sqrt{2} \int_0^{\pi/2} e^{2\theta} \sin\theta\, d\theta$$

$$= 2\pi\sqrt{2} \left[\tfrac{1}{5}\left(2e^{2\theta}\sin\theta - e^{2\theta}\cos\theta\right)\right]_0^{\pi/2} = \tfrac{2}{5}\sqrt{2}\,\pi\,(2e^\pi + 1)$$

(by parts twice)

19. (a) $A = \int_0^{2\pi} y(\theta)x'(\theta)\, d\theta$ [see (9.6.4)]

$$= \int_0^{2\pi} a^2(1 - \cos\theta)^2\, d\theta$$

$$= a^2 \int_0^{2\pi} (1 - 2\cos\theta + \cos^2\theta)\, d\theta$$

$$= a^2 \int_0^{2\pi} \left(\frac{3}{2} - 2\cos\theta + \frac{1}{2}\cos 2\theta\right) d\theta$$

$$= a^2 \left[\frac{3}{2}\theta - 2\sin\theta + \frac{1}{4}\sin 2\theta\right]_0^{2\pi}$$

$$= 3\pi\, a^2$$

(b) $A = \int_0^{2\pi} 2\pi\, y(\theta)\sqrt{[x'(\theta)]^2 + [y'(\theta)]^2}\, d\theta$ (9.9.2)

$$= \int_0^{2\pi} 2\pi\, a(1 - \cos\theta)\sqrt{a^2(1 - \cos\theta)^2 + a^2\sin^2\theta}\, d\theta$$

$$= 2\pi\, a^2 \int_0^{2\pi} (1 - \cos\theta)\sqrt{2 - 2\cos\theta}\, d\theta$$

$$= 4\pi\, a^2 \int_0^{2\pi} (1 - \cos\theta)\sin\frac{\theta}{2}\, d\theta$$

$$= 4\pi\, a^2 \int_0^{2\pi} \left(2\sin\frac{\theta}{2} - 2\cos^2\frac{\theta}{2}\sin\frac{\theta}{2}\right) d\theta$$

$$= 4\pi\, a^2 \left[-4\cos\frac{\theta}{2}\right]_0^{2\pi} + \frac{16\pi\, a^2}{3}\left[\cos^3(\theta/2)\right]_0^{2\pi}$$

$$= \frac{64\pi\, a^2}{3}$$

21.
$$A = \tfrac{1}{2}\theta s_2{}^2 - \tfrac{1}{2}\theta s_1{}^2$$

$$= \tfrac{1}{2}(\theta s_2 + \theta s_1)(s_2 - s_1)$$

$$= \tfrac{1}{2}(2\pi R + 2\pi r)s = \pi(R + r)s$$

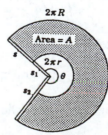

23. (a) The centroids of the 3, 4, 5 sides are the midpoints $\left(\frac{3}{2}, 0\right)$, $(3, 2)$, $\left(\frac{3}{2}, 2\right)$.

(b) $\bar{x}(3 + 4 + 5) = \frac{3}{2}(3) + 3(4) + \frac{3}{2}(5),\quad 12\bar{x} = 24,\quad \bar{x} = 2$

$\bar{y}(3 + 4 + 5) = 0(3) + 2(4) + 2(5),\quad 12\bar{y} = 18,\quad \bar{y} = \frac{3}{2}$

(c) $A = \frac{1}{3}(3)(4) = 6$

$$\bar{x}A = \int_0^3 x\left(\frac{4}{3}x\right)\,dx = \int_0^3 \frac{4}{3}x^2\,dx = \frac{4}{9}\left[x^3\right]_0^3 = 12,\quad \bar{x} = 2$$

$$\bar{y}A = \int_0^3 \frac{1}{2}\left(\frac{4}{3}x\right)^2\,dx = \int_0^3 \frac{8}{9}x^2\,dx = \frac{8}{27}\left[x^3\right]_0^3 = 8,\quad \bar{x} = \frac{4}{3}$$

(d) $\bar{x}(4 + 5) = 3(4) + \frac{3}{2}(5),\quad 9\bar{x} = \frac{39}{2},\quad \bar{x} = \frac{13}{6}$

$\bar{y}(4 + 5) = 2(4) + 2(5),\quad 9\bar{y} = 18,\quad \bar{y} = 2$

(e) $A_x = 2\pi(2)(5) = 20\pi$ (f) $A_x = 2\pi(2)(4 + 5) = 36\pi$

25. $A_x = 2\pi\bar{y}L = 2\pi(b)(2\pi a) = 4\pi^2 ab$

27. The band can be obtained by revolving about the x-axis the graph of the function

$$f(x) = \sqrt{r^2 - x^2},\qquad x \in [a, b].$$

A straightforward calculation shows that the surface area of the band is $2\pi r(b - a)$.

29. (a) Parametrize the upper half of the ellipse by

$$x(t) = a\cos t,\quad y(t) = b\sin t;\qquad t \in [0, \pi].$$

Here

$$\sqrt{[x'(t)]^2 + [y'(t)]^2} = \sqrt{a^2 \sin^2 t + b^2 \cos^2 t} = \sqrt{a^2 - (a^2 - b^2)\cos^2 t},$$

which, with $c = \sqrt{a^2 - b^2}$, can be written $\sqrt{a^2 - c^2 \cos^2 t}$.   Therefore,

$$A = \int_0^\pi 2\pi b \sin t \sqrt{a^2 - c^2 \cos^2 t}\,dt = 4\pi b \int_0^{\pi/2} \sin t \sqrt{a^2 - c^2 \cos^2 t}\,dt.$$

Setting $u = c\cos t$, we have $du = -c\sin t$ and

$$A = -\frac{4\pi b}{c}\int_c^0 \sqrt{a^2 - u^2}\,du = \frac{4\pi b}{c}\left[\frac{u}{2}\sqrt{a^2 - u^2} + \frac{a^2}{2}\sin^{-1}\left(\frac{u}{a}\right)\right]_0^c$$

$$= 2\pi b^2 + \frac{2\pi a^2 b}{c}\sin^{-1}\left(\frac{c}{a}\right) = 2\pi b^2 + \frac{2\pi ab}{e}\sin^{-1} e$$

where e is the eccentricity of ellipse: $e = c/a$.

(b) Parametrize the right half of the ellipse by

$$x(t) = a\cos t, \quad y(t) = b\sin t; \qquad t \in [-\tfrac{1}{2}\pi, \tfrac{1}{2}\pi].$$

Again $\sqrt{[x'(t)]^2 + [y'(t)]^2} = \sqrt{a^2 - c^2 \cos^2 t}$ where $c = \sqrt{a^2 - b^2}$.

Therefore

$$A = \int_{-\pi/2}^{\pi/2} 2\pi a \cos t \sqrt{a^2 - c^2 \cos^2 t}\, dt.$$

Set $u = c\sin t$. Then $du = c\cos t\, dt$ and

$$A = \frac{2\pi a}{c} \int_{-c}^{c} \sqrt{b^2 + u^2}\, du = \frac{2\pi a}{c} \left[\frac{u}{2}\sqrt{b^2 + u^2} + \frac{b^2}{2} \ln\left|u + \sqrt{b^2 + u^2}\right| \right]_{-c}^{c}$$

Routine calculation gives

$$A = 2\pi a^2 + \frac{\pi b^2}{e} \ln\left|\frac{1+e}{1-e}\right|.$$

31. Such a hemisphere can be obtained by revolving about the x-axis the curve

$$x(t) = r\cos t, \quad y(t) = r\sin t; \quad t \in [0, \tfrac{1}{2}\pi].$$

Therefore,
$$\overline{x}A = \int_0^{\pi/2} 2\pi(r\cos t)(r\sin t)\sqrt{r^2 \sin^2 t + r^2 \cos^2 t}\, dt$$

$$= \int_0^{\pi/2} 2\pi r^3 \sin t \cos t\, dt = \pi r^3 \left[\sin^2 t\right]_0^{\pi/2} = \pi r^3.$$

$$A = 2\pi r^2; \quad \overline{x} = \overline{x}A/A = \tfrac{1}{2}r.$$

The centroid lies on the midpoint of the axis of the hemisphere.

33. Such a surface can be obtained by revolving about the x-axis the graph of the function

$$f(x) = \left(\frac{R-r}{h}\right)x + r, \quad x \in [0, h].$$

Formula (9.9.8) gives

$$\overline{x}A = \int_0^h 2\pi x f(x)\sqrt{1 + [f'(x)]^2}\, dx$$

$$= \frac{2\pi}{h}\sqrt{h^2 + (R-h)^2} \int_0^h \left[\left(\frac{R-r}{h}\right)x^2 + rx\right] dx$$

$$= \frac{\pi}{3}\sqrt{h^2 + (R-r)^2}\,(2R+r)h$$

$$A = \pi(R+r)s = \pi(R+r)\sqrt{h^2 + (R-r)^2} \quad \text{and} \quad \overline{x} = \frac{\overline{x}A}{A} = \left(\frac{2R+r}{R+r}\right)\frac{h}{3}.$$

The centroid of the surface lies on the axis of the cone $\left(\dfrac{2R+r}{R+r}\right)\dfrac{h}{3}$ units from the base of radius r.

PROJECT 9.9

1.   Referring to the figure we have

$$x(\theta) = \overline{OB} - \overline{AB} = R\theta - R\sin\theta = R\,(\theta - \sin\theta)$$

$$y(\theta) = \overline{BQ} - \overline{QC} = R - R\cos\theta = R\,(1 - \cos\theta).$$

3.

(a)

$\bar{x} = \pi R$ by symmetry

$$\bar{y}A = \int_0^{2\pi} \frac{1}{2}[y(\theta)]^2 x'(\theta)\,d\theta$$

$$= \int_0^{2\pi} \frac{1}{2}R^2(1 - \cos\theta)^2\,[R(1 - \cos\theta)]\,d\theta$$

$$= \frac{1}{2}R^3 \int_0^{2\pi} (1 - 3\cos\theta + 3\cos^2\theta - \cos^3\theta)\,d\theta = \frac{5}{2}\pi R^3$$

$A = 3\pi R^2$ (by Exercise 2b) $\bar{y} = \left(\frac{5}{2}\pi R^3\right)/\left(3\pi R^2\right) = \frac{5}{6}R$

(b) $V_x = 2\pi\bar{y}A = 2\pi\left(\frac{5}{6}R\right)(3\pi R^2) = 5\pi^2 R^3$

(c) $V_y = 2\pi\bar{x}A = 2\pi(\pi R)(3\pi R^2) = 6\pi^3 R^3$

5.   (a) $\dfrac{dy}{dx} = \dfrac{y'(\phi)}{x'(\phi)} = \dfrac{R\sin\phi}{R\,(1 + \cos\phi)} = \dfrac{2\sin\frac{1}{2}\phi\cos\frac{1}{2}\phi}{2\cos^2\frac{1}{2}\phi} = \tan\frac{1}{2}\phi; \quad \alpha = \frac{1}{2}\phi$

(b) $$s = \int_0^\phi \sqrt{[x'(t)]^2 + [y'(t)]^2}\,dt = \int_0^\phi \sqrt{[R(1 + \cos t)]^2 + [R\sin t]^2}\,dt$$

$$= R\int_0^\phi \sqrt{2 + 2\cos t}\,dt = R\int_0^\phi 2\cos\frac{1}{2}t\,dt = 4R\left[\sin\frac{1}{2}t\right]_0^\phi$$

$$= 4R\sin\frac{1}{2}\phi = 4R\sin\alpha$$

Now note that the tangent at $(x(\phi), y(\phi))$ has slope

$$m = \frac{y'(\phi)}{x'(\phi)} = \frac{\sin\phi}{1 + \cos\phi} = \frac{\sin\phi(1 - \cos\phi)}{1 - \cos^2\phi} = \frac{1 - \cos\phi}{\sin\phi} = \frac{2\sin^2\frac{\phi}{2}}{2\sin\frac{\phi}{2}\cos\frac{\phi}{2}} = \tan\frac{\phi}{2}$$

So the inclination of the tangent is $\dfrac{\phi}{2} = \alpha$

7. $\dfrac{d^2 s}{dt^2} = -g\sin\alpha = -g\cdot\dfrac{2R}{\sqrt{\pi^2 R^2 + 4R^2}} = -\dfrac{2g}{\sqrt{\pi^2 + 4}}$

Since $\dfrac{ds}{dt} = 0$ and $s = R\sqrt{\pi^2 + 4}$ when $t = 0$, integrating twice gives

$$s = -\frac{g}{\sqrt{\pi^2 + 4}}\, t^2 + R\sqrt{\pi^2 + 4}$$

$s = 0$ at $t^2 = \dfrac{1}{g} R(\pi^2 + 4)$, so $t = \sqrt{R(\pi^2 + 4)/g}$

CHAPTER 10

SECTION 10.1

1. lub $= 2$; glb $= 0$ **3.** no lub; glb $= 0$

5. lub $= 2$; glb $= -2$ **7.** no lub; glb $= 2$

9. lub $= 2\frac{1}{2}$; glb $= 2$ **11.** lub $= 1$; glb $= 0.9$

13. lub $= e$; glb $= 0$ **15.** lub $= \frac{1}{2}(-1 + \sqrt{5})$; glb $= \frac{1}{2}(-1 - \sqrt{5})$

17. no lub; no glb   **19.** no lub; no glb

21. glb $S = 0$, $0 \leq \left(\frac{1}{11}\right)^3 < 0 + 0.001$

23. glb $S = 0$, $0 \leq \left(\dfrac{1}{10^{2n-1}}\right) < 0 + \left(\dfrac{1}{10^k}\right)$ $\left(n > \dfrac{k+1}{2}\right)$

25. Let $\epsilon > 0$. The condition $m \leq s$ is satisfied by all numbers s in S. All we have to show therefore is that there is some number s in S such that

$$s < m + \epsilon.$$

Suppose on the contrary that there is no such number in S. We then have

$$m + \epsilon \leq x \quad \text{for all} \quad x \in S.$$

This makes $m + \epsilon$ a lower bound for S. But this cannot be, for then $m + \epsilon$ is a lower bound for S that is *greater* than m, and by assumption, m is the *greatest* lower bound.

27. Let $c = \text{lub } S$. Since $b \in S$, $b \leq c$. Since b is an upper bound for S, $c \leq b$. Thus, $b = c$.

29. (a) Suppose that K is an upper bound for S and k is a lower bound. Let t be any element of T. Then $t \in S$ which implies that $k \leq t \leq K$. Thus K is an upper bound for T and k is a lower bound, and T is bounded.

(b) Let $a = \text{glb } S$. Then $a \leq t$ for all $t \in T$. Therefore, $a \leq \text{glb } T$. Similarly, if $b = \text{lub } S$, then $t \leq b$ for all $t \in T$, so lub $T \leq b$. It now follows that glb $S \leq$ glb $T \leq$ lub $T \leq$ lub S.

31. Let c be a positive number and let $S = \{c, 2c, 3c, \ldots\}$. Choose any positive number M and consider the positive number M/c. Since the set of positive integers is not bounded above, there exists a positive integer k such that $k \geq M/c$. This implies that $kc \geq M$. Since $kc \in S$, it follows that S is not bounded above.

33. (a)

a_1	a_2	a_3	a_4	a_5
1.4142	1.6818	1.8340	1.9152	1.9571

a_6	a_7	a_8	a_9	a_{10}
1.9785	1.9892	1.9946	1.9973	1.9986

(b)   Let S be the set of positive integers for which $u_n < 2$. Then $1 \in S$ since

$$a_1 = \sqrt{2} \cong 1.4142 < 2.$$

Assume that $k \in S$. Since $a_{k+1}^2 = 2a_k < 4$, it follows that $a_{k+1} < 2$. Thus $k+1 \in S$ and S is the set of positive integers.

(c)   Yes, $a_n \to 2$ as $n \to \infty$.

(d)   Let c be a positive number. Then c is the least upper bound of the set

$$S = \left\{ \sqrt{c}, \; \sqrt{c\sqrt{c}}, \; \sqrt{c\sqrt{c\sqrt{c}}}, \dots \right\}.$$

SECTION 10.2

1. $a_n = 2 + 3(n-1), \quad n = 1, 2, 3, \dots$

3. $a_n = \dfrac{(-1)^{n-1}}{2n-1}, \quad n = 1, 2, 3, \dots$

5. $a_n = \dfrac{n^2 + 1}{n}, \quad n = 1, 2, 3, \dots$

7. $a_n = \begin{cases} n, & \text{if } n = 2k-1 \\ 1/n, & \text{if } n = 2k, \end{cases}$ where $k = 1, 2, 3, \dots$

9.   decreasing;   bounded below by 0 and above by 2

11. $\dfrac{n + (-1)^n}{n} = 1 + (-1)^n \dfrac{1}{n}$:   not monotonic;   bounded below by 0 and above by $\dfrac{3}{2}$

13.   decreasing; bounded below by 0 and above by 0.9

15. $\dfrac{n^2}{n+1} = n - 1 + \dfrac{1}{n+1}$:   increasing;   bounded below by $\dfrac{1}{2}$ but not bounded above

17. $\dfrac{4n}{\sqrt{4n^2 + 1}} = \dfrac{2}{\sqrt{1 + 1/4n^2}}$ and $\dfrac{1}{4n^2}$ decreases to 0:   increasing;

bounded below by $\frac{4}{5}\sqrt{5}$ and above by 2

19.   increasing;   bounded below by $\frac{2}{51}$ but not bounded above

21. $\dfrac{2n}{n+1} = 2 - \dfrac{2}{n+1}$ increases toward 2:   increasing;   bounded below by 0 and above by $\ln 2$

23.   decreasing;   bounded below by 1 and above by 4

25.   increasing;   bounded below by $\sqrt{3}$ and above by 2

27. $(-1)^{2n+1}\sqrt{n} = -\sqrt{n}$:   decreasing;   bounded above by -1 but not bounded below

29. $\dfrac{2^n - 1}{2^n} = 1 - \dfrac{1}{2^n}$:   increasing;   bounded below by $\dfrac{1}{2}$ and above by 1

31.   consider $\sin x$ as $x \to 0^+$:   decreasing;   bounded below by 0 and above by 1

33.   decreasing;   bounded below by 0 and above by $\frac{5}{6}$

35. $\dfrac{1}{n} - \dfrac{1}{n+1} = \dfrac{1}{n(n+1)}$: decreasing;   bounded below by 0 and above by $\dfrac{1}{2}$

37.   Set $f(x) = \dfrac{\ln x}{x}$.   Then, $f'(x) = \dfrac{1 - \ln x}{x^2} < 0$ for $x > e$: decreasing; bounded below by 0 and above by $\frac{1}{3}\ln 3$.

39.   Set $a_n = \dfrac{3^n}{(n+1)^2}$.   Then, $\dfrac{a_{n+1}}{a_n} = 3\left(\dfrac{n+1}{n+2}\right)^2 > 1$: increasing; bounded below by $\frac{3}{4}$ but not bounded above.

41.   For $n \geq 5$

$$\frac{a_{n+1}}{a_n} = \frac{5^{n+1}}{(n+1)!} \cdot \frac{n!}{5^n} = \frac{5}{n+1} < 1 \quad \text{and thus} \quad a_{n+1} < a_n.$$

Sequence is not nonincreasing: $a_1 = 5 < \frac{25}{2} = a_2$.

43.   boundedness: $0 < (c^n + d^n)^{1/n} < (2d^n)^{1/n} = 2^{1/n}d \leq 2d$

monotonicity : $a_{n+1}^{n+1} = c^{n+1} + d^{n+1} = cc^n + dd^n$

$$< (c^n + d^n)^{1/n}c^n + (c^n + d^n)^{1/n}d^n$$

$$= (c^n + d^n)^{1+1/n}$$

$$= (c^n + d^n)^{(n+1)/n}$$

$$= a_n^{n+1}$$

Taking the $(n+1)$st root of each side we have $a_{n+1} < a_n$.   The sequence is monotonic decreasing.

45. $a_1 = 1$, $a_2 = \frac{1}{2}$, $a_3 = \frac{1}{6}$, $a_4 = \frac{1}{24}$, $a_5 = \frac{1}{120}$, $a_6 = \frac{1}{720}$; $a_n = 1/n!$

47. $a_1 = a_2 = a_3 = a_4 = a_5 = a_6 = 1$; $a_n = 1$

49. $a_1 = 1$, $a_2 = 3$, $a_3 = 5$, $a_4 = 7$, $a_5 = 9$, $a_6 = 11$; $a_n = 2n - 1$

51. $a_1 = 1$, $a_2 = 4$, $a_3 = 9$, $a_4 = 16$, $a_5 = 25$, $a_6 = 36$; $a_n = n^2$

53. $a_1 = 1$, $a_2 = 3$, $a_3 = 4$, $a_4 = 8$, $a_5 = 16$, $a_6 = 32$; $a_n = 2^{n-1}$ $(n \geq 3)$

55. $a_1 = 1$, $a_2 = 3$, $a_3 = 5$, $a_4 = 7$, $a_5 = 9$, $a_6 = 11$; $a_n = 2n - 1$

57.   First $a_1 = 2^1 - 1 = 1$.   Next suppose $a_k = 2^k - 1$ for some $k \geq 1$.   Then

$$a_{k+1} = 2a_k + 1 = 2\left(2^k - 1\right) + 1 = 2^{k+1} - 1.$$

59.   First $a_1 = \dfrac{1}{2^0} = 1$.   Next suppose $a_k = \dfrac{k}{2^{k-1}}$ for some $k \geq 1$.   Then

$$a_{k+1} = \frac{k+1}{2k}a_k = \frac{k+1}{2k}\frac{k}{2^{k-1}} = \frac{k+1}{2^k}.$$

61.   (a)   If $r = 1$ then $S_n = n$ for $n = 1, 2, 3, \ldots$

(b)

$$S_n = 1 + r + r^2 + \cdots + r^{n-1}$$
$$rS_n = r + r^2 + \cdots + r^n$$
$$S_n - rS_n = 1 - r^n$$
$$S_n = \frac{1 - r^n}{1 - r}, \qquad r \neq 1.$$

63.   (a)   Let S_n denote the distance traveled between the nth and $(n+1)$st bounce. Then

$$S_1 = 75 + 75 = 150, \qquad S_2 = \tfrac{3}{4}(75) + \tfrac{3}{4}(75) = 150\left(\frac{3}{4}\right), \ldots, \qquad S_n = 150\left(\frac{3}{4}\right)^{n-1}$$

(b)   An object dropped from rest from a height h above the ground will hit the ground in $\frac{1}{4}\sqrt{h}$ seconds. Therefore it follows that the ball will be in the air

$$T_n = 2\left(\tfrac{1}{4}\right)\sqrt{\frac{S_n}{2}} = \frac{5\sqrt{3}}{2}\left(\frac{3}{4}\right)^{(n-1)/2} \qquad \text{seconds.}$$

65.   (a)   Let S be the set of positive integers for which $a_{n+1} > a_n$. Since $a_2 = 1 + \sqrt{a_1} = 2 > 1$, $\ 1 \in S$. Assume that $a_k = 1 + \sqrt{a_{k-1}} > a_{k-1}$. Then

$$a_{k+1} = 1 + \sqrt{a_k} > 1 + \sqrt{a_{k-1}} = a_k.$$

Thus, $k \in S$ implies $k + 1 \in S$. It now follows that $\{a_n\}$ is an increasing sequence.

(b)   Since $\{a_n\}$ is an increasing sequence,

$$a_n = 1 + \sqrt{a_{n-1}} < 1 + \sqrt{a_n}, \quad \text{or} \quad a_n - \sqrt{a_n} - 1 < 0.$$

Rewriting the second inequality as

$$\left(\sqrt{a_n}\right)^2 - \sqrt{a_n} - 1 < 0$$

and solving for $\sqrt{a_n}$ it follows that $\sqrt{a_n} < \tfrac{1}{2}(1 + \sqrt{5})$. Hence, $a_n < \tfrac{1}{2}(3 + \sqrt{5})$ for all n.

(c) $a_2 = 2, \quad a_3 \cong 2.4142, \quad a_4 \cong 2.5538, \quad a_5 \cong 2.6118, \ldots, \quad a_9 \cong 2.6179, \quad \ldots, \quad a_{15} \cong 2.6180;$
lub $\{a_n\} = \tfrac{1}{2}(3 + \sqrt{5}) \cong 2.6180$

SECTION 10.3

1.   diverges   **3.**   converges to 0

5.   converges to 1: $\dfrac{n-1}{n} = 1 - \dfrac{1}{n} \to 1$

7.   converges to 0: $\dfrac{n+1}{n^2} = \dfrac{1}{n} + \dfrac{1}{n^2} \to 0$

9.   converges to 0: $0 < \dfrac{2^n}{4^n + 1} < \dfrac{2^n}{4^n} = \dfrac{1}{2^n} \to 0$

11.   diverges   **13.**   converges to 0

15.   converges to 1: $\dfrac{n\pi}{4n+1} \to \dfrac{\pi}{4}$ so $\tan \dfrac{n\pi}{4n+1} \to \tan \dfrac{\pi}{4} = 1$

17.   converges to $\dfrac{4}{9}$: $\dfrac{(2n+1)^2}{(3n-1)^2} = \dfrac{4 + 4/n + 1/n^2}{9 - 6/n + 1/n^2} \to \dfrac{4}{9}$

19.   converges to $\dfrac{1}{2}\sqrt{2}$: $\dfrac{n^2}{\sqrt{2n^4 + 1}} = \dfrac{1}{\sqrt{2 + 1/n^4}} \to \dfrac{1}{\sqrt{2}}$

21.   diverges: $\cos n\pi = (-1)^n$

23.   converges to 1: $\dfrac{1}{\sqrt{n}} \to 0$ so $e^{1/\sqrt{n}} \to e^0 = 1$

25.   diverges

27.   converges to 0 : $\ln n - \ln (n+1) = \ln \left(\dfrac{n}{n+1} \right) \to \ln 1 = 0$

29.   converges to $\dfrac{1}{2}$: $\dfrac{\sqrt{n+1}}{2\sqrt{n}} = \dfrac{1}{2}\sqrt{1 + \dfrac{1}{n}} \to \dfrac{1}{2}$

31.   converges to e^2: $\left(1 + \dfrac{1}{n} \right)^{2n} = \left[\left(1 + \dfrac{1}{n} \right)^n \right]^2 \to e^2$

33.   diverges;   since $2^n > n^3$ for $n \geq 10$, $\dfrac{2^n}{n^2} > \dfrac{n^3}{n^2} = n$

35.   converges to 0: $\left| \dfrac{\sqrt{n}\sin(e^n\pi)}{n+1} \right| = \dfrac{|\sin(e^n\pi)|}{\sqrt{n} + \dfrac{1}{\sqrt{n}}} \leq \dfrac{1}{\sqrt{n} + \dfrac{1}{\sqrt{n}}} \to 0$

37.   Set $\epsilon > 0$.   Since $a_n \to L$, there exists N_1 such that

$$\text{if} \quad n \leq N_1, \quad \text{then} \quad |a_n - L| < \epsilon/2.$$

Since $b_n \to M$, there exists N_2 such that

if $n \geq N_2$, then $|b_n - M| < \epsilon/2$.

Now set $N = \max\{N_1, N_2\}$. Then, for $n \geq N$,

$$|(a_n + b_n) - (L + M)| \leq |a_n - L| + |b_n - M| < \frac{\epsilon}{2} + \frac{\epsilon}{2} = \epsilon.$$

39. Since $\left(1 + \dfrac{1}{n}\right) \to 1$ and $\left(1 + \dfrac{1}{n}\right)^n \to e$,

$$\left(1 + \frac{1}{n}\right)^{n+1} = \left(1 + \frac{1}{n}\right)^n \left(1 + \frac{1}{n}\right) \to (e)(1) = e.$$

41. Suppose that $\{a_n\}$ is bounded and non-increasing. If L is the greatest lower bound of the range of this sequence, then $a_n \geq L$ for all n. Set $\epsilon > 0$. By Theorem 10.1.5 there exists a_k such that $a_k < L + \epsilon$. Since the sequence is non-increasing, $a_n \leq a_k$ for all $n \geq k$. Thus,

$$L \leq a_n < L + \epsilon \quad \text{or} \quad |a_n - L| < \epsilon \quad \text{for all} \quad n \geq k$$

and $a_n \to L$.

43. Let $\epsilon > 0$. Choose k so that, for $n \geq k$,

$$L - \epsilon < a_n < L + \epsilon, \quad L - \epsilon < c_n < L + \epsilon \quad \text{and} \quad a_n \leq b_n \leq c_n.$$

For such n,

$$L - \epsilon < b_n < L + \epsilon.$$

45. Let $\epsilon > 0$. Since $a_n \to L$, there exists a positive integer N such that $L - \epsilon < a_n < L + \epsilon$ for all $n \geq N$. Now $a_n \leq M$ for all n, so $L - \epsilon < M$, or $L < M + \epsilon$. Since ϵ is arbitrary, $L \leq M$.

47. By the continuity of f, $f(L) = f\left(\lim_{n\to\infty} a_n\right) = \lim_{n\to\infty} f(a_n) = \lim_{n\to\infty} a_{n+1} = L$.

49. Set $f(x) = x^{1/p}$. Since $\dfrac{1}{n} \to 0$ and f is continuous at 0, it follows by Theorem 10.3.12 that

$$\left(\frac{1}{n}\right)^{1/p} \to 0.$$

51. $a_n = e^{1-n} \to 0$

53. $a_n = \dfrac{1}{n!} \to 0$

55. $a_n = \dfrac{1}{2}[1 - (-1)^n]$ diverges

57. $a_n = \dfrac{2^n - 1}{2^{n-1}} \to 2$

59. $L = 0$, $n = 32$

61. $L = 0$, $n = 4$

63. $L = 0$, $n = 7$

65. $L = 0$, $n = 65$

67. (a) $a_n = 1 + \sqrt{a_{n-1}}$ Suppose that $a_n \to L$ as $n \to \infty$. Then $a_{n-1} \to L$ as $n \to \infty$. Therefore
$L = 1 + \sqrt{L}$ which implies that $L = \frac{1}{2}(3 + \sqrt{5})$.

(b) $a_n = \sqrt{3a_{n-1}}$ Suppose that $a_n \to L$ as $n \to \infty$. Then $a_{n-1} \to L$ as $n \to \infty$. Therefore

$L = \sqrt{3L}$ which implies that $L = 3$.

69.   (a)

a_2	a_3	a_4	a_5	a_6
0.540302	0.857553	0.654290	0.793480	0.701369

a_7	a_8	a_9	a_{10}
0.763960	0.722102	0.750418	0.73140

(b) L is a fixed point of $f(x) = \cos x$,   that is, $\cos L = L$; $L \cong 0.739085$.

71.   (a)

a_2	a_3	a_4	a_5	a_6	a_7	a_8
2.000000	1.750000	1.732143	1.732051	1.732051	1.732051	1.732051

(b) $L = \dfrac{1}{2}\left(L + \dfrac{3}{L}\right)$ which implies $L^2 = 3$ or $L = \sqrt{3}$.

(c)   Newton's method applied to the function $f(x) = x^2 - R$ gives

$$a_n = a_{n-1} - \frac{f(a_{n-1})}{f'(a_{n-1})} = a_{n-1} - \frac{a_{n-1}^2 - R}{2a_{n-1}}$$

$$= \frac{1}{2}a_{n-1} + \frac{1}{2}\frac{R}{a_{n-1}} = \frac{1}{2}\left(a_{n-1} + \frac{R}{a_{n-1}}\right), \quad n = 2, 3, \ldots .$$

PROJECT 10.3

1.   (a)

a_2	a_3	a_4	a_5	a_6	a_7	a_8
2.000000	1.750000	1.732143	1.732051	1.732051	1.732051	1.732051

(b) $L = \dfrac{1}{2}\left(L + \dfrac{3}{L}\right)$ which implies $L^2 = 3$ or $L = \sqrt{3}$.

3.   (a) $f(x) = x^3 - 8$,   so $x_n \to 2$

(b) $f(x) = \sin x - \frac{1}{2}$,   so $x_n \to \frac{\pi}{6}$

(c) $f(x) = \ln x - 1$,   so $x_n \to e$

SECTION 10.4

1.   converges to 1: $2^{2/n} = (2^{1/n})^2 \to 1^2 = 1$

3.   converges to 0:   for $n > 3$, $0 < \left(\dfrac{2}{n}\right)^n < \left(\dfrac{2}{3}\right)^n \to 0$

5.   converges to 0: $\dfrac{\ln(n+1)}{n} = \left[\dfrac{\ln(n+1)}{n+1}\right]\left(\dfrac{n+1}{n}\right) \to (0)(1) = 0$

7.   converges to 0: $\dfrac{x^{100n}}{n!} = \dfrac{(x^{100})^n}{n!} \to 0$

9.   converges to 1: $n^{\alpha/n} = (n^{1/n})^\alpha \to 1^\alpha = 1$

11.   converges to 0: $\dfrac{3^{n+1}}{4^{n-1}} = 12\left(\dfrac{3^n}{4^n}\right) = 12\left(\dfrac{3}{4}\right)^n \to 12(0) = 0$

13.   converges to 1: $(n+2)^{1/n} = e^{\frac{1}{n}\ln(n+2)}$ and, since

$$\frac{1}{n}\ln(n+2) = \left[\frac{\ln(n+2)}{n+2}\right]\left(\frac{n+2}{n}\right) \to (0)(1) = 0,$$

it follows that

$$(n+2)^{1/n} \to e^0 = 1.$$

15.   converges to 1: $\displaystyle\int_0^n e^{-x}\,dx = 1 - \frac{1}{e^n} \to 1$

17.   converges to π: integral $= 2\displaystyle\int_0^n \frac{dx}{1+x^2} = 2\tan^{-1}n \to 2\left(\dfrac{\pi}{2}\right) = \pi$

19.   converges to 1:   recall (10.4.6)

21.   converges to 0: $\dfrac{\ln(n^2)}{n} = 2\dfrac{\ln\,n}{n} \to 2(0) = 0$

23.   diverges: since $\displaystyle\lim_{x\to 0}\frac{\sin x}{x} = 1,$

$$\frac{n}{\pi}\sin\frac{\pi}{n} = \frac{\sin(\pi/n)}{\pi/n} \to 1$$

and, for n sufficiently large,

$$n^2\sin\frac{\pi}{n} = n\pi\left(\frac{n}{\pi}\sin\frac{\pi}{n}\right) > n\pi\left(\frac{1}{2}\right) = \frac{n\pi}{2}$$

25.   converges to 0: $\dfrac{5^{n+1}}{4^{2n-1}} = 20\left(\dfrac{5}{16}\right)^n \to 0$

27.   converges to e^{-1}: $\left(\dfrac{n+1}{n+2}\right)^n = \left(1 - \dfrac{1}{n+2}\right)^n = \dfrac{\left(1 + \dfrac{(-1)}{n+2}\right)^{n+2}}{\left(1 + \dfrac{(-1)}{n+2}\right)^2} \to \dfrac{e^{-1}}{1} = e^{-1}$

29.   converges to 0: $0 < \displaystyle\int_n^{n+1} e^{-x^2}\,dx \le e^{-n^2}[(n+1) - n] = e^{-n^2} \to 0$

31.   converges to 0: $\dfrac{n^n}{2^{n^2}} = \left(\dfrac{n}{2^n}\right)^n \to 0$ since $\dfrac{n}{2^n} \to 0$

33.   converges to e^x: use (10.4.7)

35. converges to 0: $\left|\int_{-1/n}^{1/n}\sin x^2\,dx\right| \le \int_{-1/n}^{1/n}|\sin x^2|\,dx \le \int_{-1/n}^{1/n}1\,dx = \dfrac{2}{n} \to 0$

37. $\sqrt{n+1}-\sqrt{n} = \dfrac{\sqrt{n+1}-\sqrt{n}}{\sqrt{n+1}+\sqrt{n}}\left(\sqrt{n+1}+\sqrt{n}\right) = \dfrac{1}{\sqrt{n+1}+\sqrt{n}} \to 0$

39. (a) The length of each side of the polygon is $2r\sin(\pi/n)$. Therefore the perimeter, p_n, of the polygon is given by: $p_n = 2rn\sin(\pi/n)$.

(b) $2rn\sin(\pi/n) \to 2\pi r$ as $n \to \infty$: The number $2rn\sin(\pi/n)$ is the perimeter of a regular polygon of n sides inscribed in a circle of radius r. As n tends to ∞, the perimeter of the polygon tends to the circumference of the circle.

41. By the hint, $\displaystyle\lim_{n\to\infty}\dfrac{1+2+\cdots+n}{n^2} = \lim_{n\to\infty}\dfrac{n(n+1)}{2n^2} = \lim_{n\to\infty}\dfrac{1+1/n}{2} = \dfrac{1}{2}.$

43. By the hint, $\displaystyle\lim_{n\to\infty}\dfrac{1^3+2^3+\cdots+n^3}{2n^4+n-1} = \lim_{n\to\infty}\dfrac{n^2(n+1)^2}{4(2n^4+n-1)} = \lim_{n\to\infty}\dfrac{1+2/n+1/n^2}{8+4/n^3-4/n^4} = \dfrac{1}{8}.$

45. (a)
$$m_{n+1}-m_n = \frac{1}{n+1}(a_1+\cdots+a_n+a_{n+1}) - \frac{1}{n}(a_1+\cdots+a_n)$$
$$= \frac{1}{n(n+1)}\left[na_{n+1} - (\overbrace{a_1+\cdots+a_n}^{n})\right]$$
$$> 0 \quad \text{since } \{a_n\} \text{ is increasing.}$$

(b)   We begin with the hint
$$m_n < \frac{|a_1+\cdots+a_j|}{n} + \frac{\epsilon}{2}\left(\frac{n-j}{n}\right).$$
Since j is fixed,
$$\frac{|a_1+\cdots+a_j|}{n} \to 0$$
and therefore for n sufficiently large
$$\frac{|a_1+\cdots+a_j|}{n} < \frac{\epsilon}{2}.$$
Since
$$\frac{\epsilon}{2}\left(\frac{n-j}{n}\right) < \frac{\epsilon}{2},$$
we see that, for n sufficiently large, $|m_n| < \epsilon$. This shows that $m_n \to 0$.

47. (a) Let S be the set of positive integers n ($n \ge 2$) for which the inequalities hold. Since
$$\left(\sqrt{b}\right)^2 - 2\sqrt{ab} + \left(\sqrt{a}\right)^2 > 0 = \left(\sqrt{b}-\sqrt{a}\right)^2 > 0,$$
it follows that $\dfrac{a+b}{2} > \sqrt{ab}$ and so $a_1 > b_1$. Now,
$$a_2 = \frac{a_1+b_1}{2} < a_1 \quad \text{and} \quad b_2 = \sqrt{a_1 b_1} > b_1.$$

Also, by the argument above,

$$a_2 = \frac{a_1 + b_1}{2} > \sqrt{a_1 b_1} = b_2,$$

and so $a_1 > a_2 > b_2 > b_1$. Thus $2 \in S$. Assume that $k \in S$. Then

$$a_{k+1} = \frac{a_k + b_k}{2} < \frac{a_k + a_k}{2} = a_k, \quad b_{k+1} = \sqrt{a_k b_k} > \sqrt{b_k^2} = b_k,$$

and

$$a_{k+1} = \frac{a_k + b_k}{2} > \sqrt{a_k b_k} = b_{k+1}.$$

Thus $k + 1 \in S$. Therefore, the inequalities hold for all $n \geq 2$.

(b) $\{a_n\}$ is a decreasing sequence which is bounded below.

$\{b_n\}$ is an increasing sequence which is bounded above.

Let $L_a = \lim\limits_{n \to \infty} a_n$, $L_b = \lim\limits_{n \to \infty} b_n$. Then

$$a_n = \frac{a_{n-1} + b_{n-1}}{2} \quad \text{implies} \quad L_a = \frac{L_a + L_b}{2} \quad \text{and} \quad L_a = L_b.$$

49.   The numerical work suggests $L \cong 1$. Justification: Set $f(x) = \sin x - x^2$. Note that $f(0) = 0$ and $f'(x) = \cos x - 2x > 0$ for x close to 0. Therefore $\sin x - x^2 > 0$ for x close to 0 and $\sin 1/n - 1/n^2 > 0$ for n large. Thus, for n large,

$$\frac{1}{n^2} < \sin \frac{1}{n} < \frac{1}{n}$$

$$|\sin x| \leq |x| \quad \text{for all } x$$

$$\left(\frac{1}{n^2} \right)^{1/n} < \left(\sin \frac{1}{n} \right)^{1/n} < \left(\frac{1}{n} \right)^{1/n}$$

$$\left(\frac{1}{n^{1/n}} \right)^2 < \left(\sin \frac{1}{n} \right)^{1/n} < \frac{1}{n^{1/n}}.$$

As $n \to \infty$ both bounds tend to 1 and therefore the middle term also tends to 1.

51. (a)

a_3	a_4	a_5	a_6	a_7	a_8	a_9	a_{10}
2	3	5	8	13	21	34	55

(b)

r_1	r_2	r_3	r_4	r_5	r_6
1	2	1.2	1.667	1.600	1.625

(c) Following the hint,

$$1 + \frac{1}{r_{n-1}} = 1 + \frac{1}{\dfrac{a_n}{a_{n-1}}} = 1 + \frac{a_n - 1}{a_n} = \frac{a_n + a_{n-1}}{a_n} = \frac{a_{n+1}}{a_n} = r_n.$$

Now, if $r_n \to L$, then $r_{n-1} \to L$ and

$$1 + \frac{1}{L} = L \quad \text{which implies} \quad L = \frac{1 + \sqrt{5}}{2} \cong 1.618034.$$

SECTION 10.5

(We'll use $\star$ to indicate differentiation of numerator and denominator.)

1. $\displaystyle\lim_{x \to 0^+} \frac{\sin x}{\sqrt{x}} \overset{\star}{=} \lim_{x \to 0^+} 2\sqrt{x}\cos x = 0$

3. $\displaystyle\lim_{x \to 0} \frac{e^x - 1}{\ln(1 + x)} \overset{\star}{=} \lim_{x \to 0}(1 + x)e^x = 1$

5. $\displaystyle\lim_{x \to \pi/2} \frac{\cos x}{\sin 2x} \overset{\star}{=} \lim_{x \to \pi/2} \frac{-\sin x}{2\cos 2x} = \frac{1}{2}$

7. $\displaystyle\lim_{x \to 0} \frac{2^x - 1}{x} \overset{\star}{=} \lim_{x \to 0} 2^x \ln 2 = \ln 2$

9. $\displaystyle\lim_{x \to 1} \frac{x^{1/2} - x^{1/4}}{x - 1} \overset{\star}{=} \lim_{x \to 1}\left(\frac{1}{2}x^{-1/2} - \frac{1}{4}x^{-3/4}\right) = \frac{1}{4}$

11. $\displaystyle\lim_{x \to 0} \frac{e^x - e^{-x}}{\sin x} \overset{\star}{=} \lim_{x \to 0} \frac{e^x + e^{-x}}{\cos x} = 2$

13. $\displaystyle\lim_{x \to 0} \frac{x + \sin \pi x}{x - \sin \pi x} \overset{\star}{=} \lim_{x \to 0} \frac{1 + \pi \cos \pi x}{1 - \pi \cos \pi x} = \frac{1 + \pi}{1 - \pi}$

15. $\displaystyle\lim_{x \to 0} \frac{e^x + e^{-x} - 2}{1 - \cos 2x} \overset{\star}{=} \lim_{x \to 0} \frac{e^x - e^{-x}}{2\sin 2x} \overset{\star}{=} \lim_{x \to 0} \frac{e^x + e^{-x}}{4\cos 2x} = \frac{1}{2}$

17. $\displaystyle\lim_{x \to 0} \frac{\tan \pi x}{e^x - 1} \overset{\star}{=} \lim_{x \to 0} \frac{\pi \sec^2 \pi x}{e^x} = \pi$

19. $\displaystyle\lim_{x \to 0} \frac{1 + x - e^x}{x(e^x - 1)} \overset{\star}{=} \lim_{x \to 0} \frac{1 - e^x}{xe^x + e^x - 1} \overset{\star}{=} \lim_{x \to 0} \frac{-e^x}{xe^x + 2e^x} = -\frac{1}{2}$

21. $\displaystyle\lim_{x \to 0} \frac{x - \tan x}{x - \sin x} \overset{\star}{=} \lim_{x \to 0} \frac{1 - \sec^2 x}{1 - \cos x} \overset{\star}{=} \lim_{x \to 0} \frac{-2\sec^2 x \tan x}{\sin x} = \lim_{x \to 0} \frac{-2\sec^2 x}{\cos x} = -2$

23. $\displaystyle\lim_{x \to 1^-} \frac{\sqrt{1 - x^2}}{\sqrt{1 - x^3}} = \lim_{x \to 1^-} \sqrt{\frac{1 - x^2}{1 - x^3}} = \sqrt{\frac{2}{3}} = \frac{1}{3}\sqrt{6} \quad \text{since} \quad \lim_{x \to 1^-} \frac{1 - x^2}{1 - x^3} \overset{\star}{=} \lim_{x \to 1^-} \frac{2x}{3x^2} = \frac{2}{3}$

25. $\displaystyle\lim_{x \to \pi/2} \frac{\ln(\sin x)}{(\pi - 2x)^2} \overset{\star}{=} \lim_{x \to \pi/2} \frac{-\cot x}{4(\pi - 2x)} \overset{\star}{=} \lim_{x \to \pi/2} \frac{\csc^2 x}{-8} = -\frac{1}{8}$

27. $\displaystyle\lim_{x \to 0} \frac{\cos x - \cos 3x}{\sin(x^2)} \overset{\star}{=} \lim_{x \to 0} \frac{-\sin x + 3\sin 3x}{2x\cos(x^2)} \overset{\star}{=} \lim_{x \to 0} \frac{-\cos x + 9\cos 3x}{2\cos(x^2) - 4x^2\sin(x^2)} = 4$

29. $\displaystyle\lim_{x \to \pi/4} \frac{\sec^2 x - 2\tan x}{1 + \cos 4x} \overset{\star}{=} \lim_{x \to \pi/4} \frac{2\sec^2 x \,\tan x - 2\sec^2 x}{-4\sin 4x}$

$$\overset{\star}{=} \lim_{x \to \pi/4} \frac{2\sec^4 x + 4\sec^2 x \tan^2 x - 4\sec^2 x \tan x}{-16\cos 4x} = \frac{1}{2}$$

31. $\displaystyle\lim_{x\to 0}\frac{\tan^{-1}x}{\tan^{-1}2x}\overset{*}{=}\lim_{x\to 0}\frac{1}{1+x^2}\frac{1+4x^2}{2}=\frac{1}{2}$

33. $1;\quad \displaystyle\lim_{x\to\infty}\frac{\pi/2-\tan^{-1}x}{1/x}\overset{*}{=}\lim_{x\to\infty}\frac{x^2}{1+x^2}=1$

35. $1;\quad \displaystyle\lim_{x\to\infty}\frac{1}{x[\ln(x+1)-\ln x]}=\lim_{x\to\infty}\frac{1/x}{\ln(1+1/x)}=\lim_{t\to 0^+}\frac{t}{\ln(1+t)}\overset{*}{=}\lim_{t\to 0^+}(1+t)=1$

37. $\displaystyle\lim_{x\to 0}(2+x+\sin x)\neq 0,\quad \lim_{x\to 0}(x^3+x-\cos x)\neq 0$

39. The limit does not exist if $b\neq 1$. Therefore, $b=1$.

$$\lim_{x\to 0}\frac{\cos ax-1}{2x^2}\overset{*}{=}\lim_{x\to 0}\frac{-a\sin ax}{4x}\overset{*}{=}\lim_{x\to 0}\frac{-a^2\cos ax}{4}=-\frac{a^2}{4}$$

Now, $-\dfrac{a^2}{4}=-4$ implies $a=\pm 4$.

41. $\displaystyle\lim_{x\to 0}\frac{1}{x}\int_0^x f(t)\,dt\overset{*}{=}\lim_{x\to 0}\frac{f(x)}{1}=f(0)$

43. $A(b)=2\displaystyle\int_0^{\sqrt{b}}(b-x^2)\,dx=2\left[bx-x^2\right]_0^{\sqrt{b}}=\frac{4}{3}b\sqrt{b}$ and $T(b)=\frac{1}{2}\left(2\sqrt{b}\right)b=b\sqrt{b}.$

Thus, $\displaystyle\lim_{b\to 0}\frac{T(b)}{A(b)}=\frac{b\sqrt{b}}{\frac{4}{3}b\sqrt{b}}=\frac{3}{4}.$

45. (a) $\qquad\qquad\qquad\qquad\qquad\qquad f(x)\to\infty$ as $x\to\pm\infty$

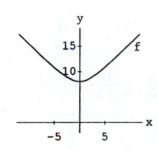

(b) $f(x)\to 10$ as $x\to 4$

Confirmation: $\displaystyle\lim_{x\to 4}\frac{x^2-16}{\sqrt{x^2+9}-5}\overset{*}{=}\lim_{x\to 4}\frac{2x}{x\left(x^2+9\right)^{-1/2}}=\lim_{x\to 4}2\sqrt{x^2+9}=10$

47. (a) $f(x) \to 0.7$ as $x \to 0$

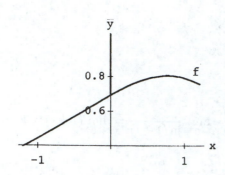

(b) Confirmation: $\displaystyle\lim_{x \to 0} \frac{2^{\sin x} - 1}{x} \overset{\star}{=} \lim_{x \to 0} \frac{\ln(2)\, 2^{\sin x}\, \cos x}{1} = \ln 2 \cong 0.6931$

SECTION 10.6

(We'll use $\star$ to indicate differentiation of numerator and denominator.)

1. $\displaystyle\lim_{x \to -\infty} \frac{x^2 + 1}{1 - x} \overset{\star}{=} \lim_{x \to -\infty} \frac{2x}{-1} = \infty$

3. $\displaystyle\lim_{x \to \infty} \frac{x^3}{1 - x^3} = \lim_{x \to \infty} \frac{1}{1/x^3 - 1} = -1$

5. $\displaystyle\lim_{x \to \infty} x^2 \sin \frac{1}{x} = \lim_{h \to 0^+} \left[\left(\frac{1}{h}\right)\left(\frac{\sin h}{h}\right) \right] = \infty$

7. $\displaystyle\lim_{x \to \frac{\pi}{2}^-} \frac{\tan 5x}{\tan x} = \lim_{x \to \frac{\pi}{2}^-} \left[\left(\frac{\sin 5x}{\sin x}\right)\left(\frac{\cos x}{\cos 5x}\right) \right] = \frac{1}{5}$ since

$$\lim_{x \to \frac{\pi}{2}^-} \frac{\sin 5x}{\sin x} = 1 \quad \text{and} \quad \lim_{x \to \frac{\pi}{2}^-} \frac{\cos x}{\cos 5x} \overset{\star}{=} \lim_{x \to \frac{\pi}{2}^-} \frac{\sin x}{5 \sin 5x} = \frac{1}{5}$$

9. $\displaystyle\lim_{x \to 0^+} x^{2x} = \lim_{x \to 0^+} (x^x)^2 = 1^2 = 1$ [see (10.6.4)]

11. $\displaystyle\lim_{x \to 0} x(\ln |x|)^2 = \lim_{x \to 0} \frac{(\ln |x|)^2}{1/x} \overset{\star}{=} \lim_{x \to 0} \frac{2 \ln |x|}{-1/x} \overset{\star}{=} \lim_{x \to 0} \frac{2}{1/x} = 0$

13. $\displaystyle\lim_{x \to \infty} \frac{1}{x} \int_0^x e^{t^2}\, dt \overset{\star}{=} \lim_{x \to \infty} \frac{e^{x^2}}{1} = \infty$

15.
$$\lim_{x\to 0}\left[\frac{1}{\sin^2 x}-\frac{1}{x^2}\right]=\lim_{x\to 0}\frac{x^2-\sin^2 x}{x^2\sin^2 x}\stackrel{*}{=}\lim_{x\to 0}\frac{2x-2\sin x\cos x}{2x^2\sin x\cos x+2x\sin^2 x}$$

$$=\lim_{x\to 0}\frac{2x-\sin 2x}{x^2\sin 2x+2x\sin^2 x}$$

$$\stackrel{*}{=}\lim_{x\to 0}\frac{2-2\cos 2x}{2x^2\cos 2x+4x\sin 2x+2\sin^2 x}$$

$$\stackrel{*}{=}\lim_{x\to 0}\frac{4\sin 2x}{-4x^2\sin 2x+12x\cos 2x+6\sin 2x}$$

$$\stackrel{*}{=}\lim_{x\to 0}\frac{8\cos 2x}{-8x^2\cos 2x-32x\sin 2x+24\cos 2x}=\frac{1}{3}$$

17. $\displaystyle\lim_{x\to 1}x^{1/(x-1)}=e$ since $\displaystyle\lim_{x\to 1}\ln\left[x^{1/(x-1)}\right]=\lim_{x\to 1}\frac{\ln x}{x-1}\stackrel{*}{=}\lim_{x\to 1}\frac{1}{x}=1$

19. $\displaystyle\lim_{x\to\infty}\left(\cos\frac{1}{x}\right)^x=1$ since $\displaystyle\lim_{x\to\infty}\ln\left[\left(\cos\frac{1}{x}\right)^x\right]=\lim_{x\to\infty}\frac{\ln\left(\cos\dfrac{1}{x}\right)}{(1/x)}$

$$\stackrel{*}{=}\lim_{x\to\infty}\left(-\frac{\sin(1/x)}{\cos(1/x)}\right)=0$$

21.
$$\lim_{x\to 0}\left[\frac{1}{\ln(1+x)}-\frac{1}{x}\right]=\lim_{x\to 0}\frac{x-\ln(1+x)}{x\ln(1+x)}\stackrel{*}{=}\lim_{x\to 0}\frac{x}{x+(1+x)\ln(1+x)}$$

$$\stackrel{*}{=}\lim_{x\to 0}\frac{1}{1+1+\ln(1+x)}=\frac{1}{2}$$

23.
$$\lim_{x\to 0}\left[\frac{1}{x}-\cot x\right]=\lim_{x\to 0}\frac{\sin x-x\cos x}{x\sin x}\stackrel{*}{=}\lim_{x\to 0}\frac{x\sin x}{\sin x+x\cos x}$$

$$\stackrel{*}{=}\lim_{x\to 0}\frac{\sin x+x\cos x}{2\cos x-x\sin x}=0$$

25.
$$\lim_{x\to\infty}\left(\sqrt{x^2+2x}-x\right)=\lim_{x\to\infty}\left[\left(\sqrt{x^2+2x}-x\right)\left(\frac{\sqrt{x^2+2x}+x}{\sqrt{x^2+2x}+x}\right)\right]$$

$$=\lim_{x\to\infty}\frac{2x}{\sqrt{x^2+2x}+x}=\lim_{x\to\infty}\frac{2}{\sqrt{1+2/x}+1}=1$$

27. $\displaystyle\lim_{x\to\infty}\left(x^3+1\right)^{1/\ln x}=e^3$ since

$$\lim_{x\to\infty} \ln\left[(x^3+1)^{1/\ln x}\right] = \lim_{x\to\infty} \frac{\ln\left(x^3+1\right)}{\ln x} \overset{\star}{=} \lim_{x\to\infty} \frac{\left(\dfrac{3x^2}{x^3+1}\right)}{1/x} = \lim_{x\to\infty} \frac{3}{1+1/x^3} = 3.$$

29. $\displaystyle\lim_{x\to\infty} (\cosh x)^{1/x} = e$ since

$$\lim_{x\to\infty} \ln\left[(\cosh x)^{1/x}\right] = \lim_{x\to\infty} \frac{\ln\left(\cosh x\right)}{x} \overset{\star}{=} \lim_{x\to\infty} \frac{\sinh x}{\cosh x} = 1.$$

31.

$$\lim_{x\to 0}\left(\frac{1}{\sin x} - \frac{1}{x}\right) = \lim_{x\to 0} \frac{x-\sin x}{x\sin x} \overset{\star}{=} \lim_{x\to 0} \frac{1-\cos x}{\sin x + x\cos x}$$

$$\overset{\star}{=} \lim_{x\to 0} \frac{\sin x}{2\cos x - x\sin x} = 0$$

33.

$$\lim_{x\to 1}\left(\frac{1}{\ln x} - \frac{x}{x-1}\right) = \lim_{x\to 1} \frac{x-1\ln x}{(x-1)\ln x} \overset{\star}{=} \lim_{x\to 1} \frac{-\ln x}{(x-1)(1/x)+\ln x}$$

$$= \lim_{x\to 1} \frac{-x\ln x}{x-1+x\ln x} \overset{\star}{=} \lim_{x\to 1} \frac{-\ln x - 1}{2+\ln x} = -\frac{1}{2}$$

35. $0;\quad \dfrac{1}{n}\ln\dfrac{1}{n} = -\dfrac{\ln n}{n} \to 0$ **37.** $1;\quad \ln\left[(\ln n)^{1/n}\right] = \dfrac{1}{n}\ln\left(\ln n\right) \to 0$

39. $1;\quad \ln\left[\left(n^2+n\right)^{1/n}\right] = \dfrac{1}{n}\ln\left[n(n+1)\right] = \dfrac{\ln n}{n} + \dfrac{\ln\left(n+1\right)}{n} \to 0$

41. $0;\quad 0 \le \dfrac{n^2\ln n}{e^n} < \dfrac{n^3}{e^n},\quad \displaystyle\lim_{x\to\infty} \frac{x^3}{e^x} = 0$

43.   **45.**

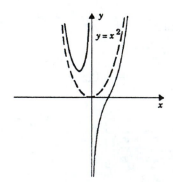

vertical asymptote y-axis

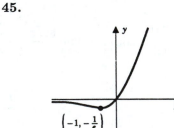

horizontal asymptote x-axis

47.

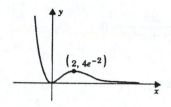

horizontal asymptote x-axis

49. $\dfrac{b}{a}\sqrt{x^2-a^2}-\dfrac{b}{a}x = \dfrac{\sqrt{x^2-a^2}+x}{\sqrt{x^2-a^2}+x}\left(\dfrac{b}{a}\right)\left(\sqrt{x^2-a^2}-x\right) = \dfrac{-ab}{\sqrt{x^2-a^2}+x} \to 0$ as $x \to \infty$

51. for instance, $f(x) = x^2 + \dfrac{(x-1)(x-2)}{x^3}$

53. $\displaystyle\lim_{x\to 0^-} -\dfrac{2x}{\cos x} \neq \lim_{x\to 0^-} \dfrac{2}{-\sin x}$.   L'Hospital's rule does not apply here since $\displaystyle\lim_{x\to 0^-} \cos x = 1$.

55.   (a) Let S be the set of positive integers for which the statement is true. Since $\displaystyle\lim_{x\to\infty} \dfrac{\ln x}{x} = 0$, $1 \in S$. Assume that $k \in S$. By L'Hospital's rule,

$$\lim_{x\to\infty} \dfrac{(\ln x)^{k+1}}{x} \stackrel{\star}{=} \lim_{x\to\infty} \dfrac{(k+1)(\ln x)^k}{x} = 0 \quad (\text{since} \ \ k \in S).$$

Thus $k+1 \in S$, and S is the set of positive integers.

(b) Choose any positive number α. Let $k-1$ and k be positive integers such that $k-1 \leq \alpha \leq k$. Then, for $x > e$,

$$\dfrac{(\ln x)^{k-1}}{x} \leq \dfrac{(\ln x)^{\alpha}}{x} \leq \dfrac{(\ln x)^k}{x}$$

and the result follows by the pinching theorem.

57.   (a) $\displaystyle\lim_{x\to 0^+} \left(1+x^2\right)^{1/x} = 1$.

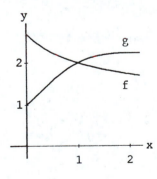

(b) $\displaystyle\lim_{x\to 0^+} \left(1+x^2\right)^{1/x} = 1.$ since

$$\lim_{x\to 0^+} \ln\left[\left(1+x^2\right)^{1/x}\right] = \lim_{x\to 0^+} \frac{\ln\left(1+x^2\right)}{x} \overset{*}{=} \lim_{x\to 0^+} \frac{2x}{(1+x^2)} = 0$$

59.   (a) $\displaystyle\lim_{x\to\infty} g(x) \cong -1.7.$

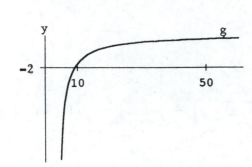

(b) $\displaystyle\lim_{x\to\infty} g(x) = \lim_{x\to\infty}\left[\sqrt[3]{x^3 - 5x^2 + 2x + 1} - x\right]$

$$= \lim_{x\to\infty} \frac{-5x^2 + 2x + 1}{\left(\sqrt[3]{x^3 - 5x^2 + 2x + 1}\right)^2 + x\sqrt[3]{x^3 - 5x^2 + 2x + 1} + x^2}$$

$$= -\frac{5}{3} \cong -1.667$$

PROJECT 10.6

1.   (a) $\displaystyle\lim_{x\to\infty} \frac{2x}{\sqrt{x^2+1}} \overset{*}{=} \lim_{x\to\infty} \frac{2}{\dfrac{x}{\sqrt{x^2+1}}} = \lim_{x\to\infty} \frac{2\sqrt{x^2+1}}{x} \overset{*}{=} \lim_{x\to\infty} \frac{\dfrac{2x}{\sqrt{x^2+1}}}{1}$ and so on.

(b)

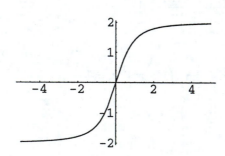

(c) $\displaystyle\lim_{x\to\infty} \frac{2x}{\sqrt{x^2+1}} = \lim_{x\to\infty} \frac{2x}{x\sqrt{1+(1/x^2)}} = \lim_{x\to\infty} \frac{2}{\sqrt{1+(1/x^2)}} = 2$

3. $\displaystyle\lim_{x\to 0} \frac{f(x) - f(0)}{x} = \lim_{x\to 0} \frac{e^{-1/x^2}}{x} = 0.$ Therefore, $f'(0) = 0.$

SECTION 10.7

1. 1; $\displaystyle\int_1^\infty \frac{dx}{x^2} = \lim_{b\to\infty}\int_1^b \frac{dx}{x^2} = \lim_{b\to\infty}\left[-\frac{1}{x}\right]_1^b = \lim_{b\to\infty}\left[1-\frac{1}{b}\right] = 1$

3. $\dfrac{\pi}{4}$; $\displaystyle\int_0^\infty \frac{dx}{4+x^2} = \lim_{b\to\infty}\int_0^b \frac{dx}{4+x^2} = \lim_{b\to\infty}\left[\frac{1}{2}\tan^{-1}\frac{x}{2}\right]_0^b = \lim_{b\to\infty}\frac{1}{2}\tan^{-1}\left(\frac{b}{2}\right) = \frac{\pi}{4}$

5. diverges; $\displaystyle\int_0^\infty e^{px}\,dx = \lim_{b\to\infty}\int_0^b e^{px}\,dx = \lim_{b\to\infty}\left[\frac{1}{p}e^{px}\right]_0^b = \lim_{b\to\infty}\frac{1}{p}\left(e^{pb}-1\right) = \infty$

7. 6; $\displaystyle\int_0^8 \frac{dx}{x^{2/3}} = \lim_{a\to 0^+}\int_a^8 x^{-2/3}\,dx = \lim_{a\to 0^+}\left[3x^{1/3}\right]_a^8 = \lim_{a\to 0^+}\left[6-3a^{1/3}\right] = 6$

9. $\dfrac{\pi}{2}$; $\displaystyle\int_0^1 \frac{dx}{\sqrt{1-x^2}} = \lim_{b\to 1^-}\int_0^b \frac{dx}{\sqrt{1-x^2}} = \lim_{b\to 1^-}\sin^{-1}b = \frac{\pi}{2}$

11. 2; $\displaystyle\int_0^2 \frac{x}{\sqrt{4-x^2}}\,dx = \lim_{b\to 2^-}\int_0^b x\left(4-x^2\right)^{-1/2}\,dx = \lim_{b\to 2^-}\left[-\left(4-x^2\right)^{1/2}\right]_0^b$

$$= \lim_{b\to 2^-}\left(2-\sqrt{4-b^2}\right) = 2$$

13. diverges; $\displaystyle\int_e^\infty \frac{\ln x}{x}\,dx = \lim_{b\to\infty}\int_e^b \frac{\ln x}{x}\,dx = \lim_{b\to\infty}\left[\frac{1}{2}\left(\ln x\right)^2\right]_e^b$

$$= \lim_{b\to\infty}\left[\frac{1}{2}\left(\ln b\right)^2 - \frac{1}{2}\right] = \infty$$

15. $-\dfrac{1}{4}$;

$$\int_0^1 x\ln x\,dx = \lim_{a\to 0^+}\int_a^1 x\ln x\,dx = \lim_{a\to 0^+}\left[\frac{1}{2}x^2\ln x - \frac{1}{4}x^2\right]_a^1$$

(by parts)

$$= \lim_{a\to 0^+}\left[\frac{1}{4}a^2 - \frac{1}{2}a^2\ln a - \frac{1}{4}\right] = -\frac{1}{4}$$

Note: $\displaystyle\lim_{t\to 0^+} t^2\ln t = \lim_{t\to 0^+}\frac{\ln t}{1/t^2} \overset{*}{=} \lim_{t\to 0^+}\frac{1/t}{-2/t^3} = -\frac{1}{2}\lim_{t\to 0^+}t^2 = 0.$

17. π; $\displaystyle\int_{-\infty}^\infty \frac{dx}{1+x^2} = \lim_{a\to -\infty}\int_a^0 \frac{dx}{1+x^2} + \lim_{b\to\infty}\int_0^b \frac{dx}{1+x^2}$

$$= \lim_{a\to -\infty}\left[\tan^{-1}x\right]_a^0 + \lim_{b\to\infty}\left[\tan^{-1}x\right]_0^b = -\left(-\frac{\pi}{2}\right) + \frac{\pi}{2} = \pi$$

19. diverges; $\displaystyle\int_{-\infty}^{\infty} \frac{dx}{x^2} = \lim_{a\to-\infty} \int_a^{-1} \frac{dx}{x^2} + \lim_{b\to0^-} \int_{-1}^b \frac{dx}{x^2} + \lim_{c\to0^+} \int_c^1 \frac{dx}{x^2} + \lim_{d\to\infty} \int_1^d \frac{dx}{x^2}$;

and, $\displaystyle\lim_{c\to0^+} \int_c^1 \frac{dx}{x^2} = \lim_{c\to0^+} \left[-\frac{1}{x}\right]_c^1 = \lim_{c\to0^+} \left[\frac{1}{c} - 1\right] = \infty$

21. $\ln 2$; $\displaystyle\int_1^{\infty} \frac{dx}{x(x+1)} = \lim_{b\to\infty} \int_1^b \left[\frac{1}{x} - \frac{1}{x+1}\right] dx$

$$= \lim_{b\to\infty} \left[\ln\left(\frac{x}{x+1}\right)\right]_1^b = \lim_{b\to\infty} \left[\ln\left(\frac{b}{b+1}\right) - \ln\left(\frac{1}{2}\right)\right]$$

$$= 0 - \ln\tfrac{1}{2} = \ln 2$$

23. 4; $\displaystyle\int_3^5 \frac{x}{\sqrt{x^2-9}} dx = \lim_{a\to3^-} \int_a^5 x\left(x^2-9\right)^{-1/2} dx$

$$= \lim_{a\to3^-} \left[\left(x^2-9\right)^{1/2}\right]_a^5 = \lim_{a\to3^-} \left[4 - \left(a^2-9\right)^{1/2}\right] = 4$$

25. $\displaystyle\int_{-3}^3 \frac{dx}{x(x+1)}$ diverges since $\displaystyle\int_0^3 \frac{dx}{x(x+1)}$ diverges:

$$\int_0^3 \frac{dx}{x(x+1)} = \int_0^3 \left(\frac{1}{x} - \frac{1}{x+1}\right) dx = \lim_{a\to0^+} \left[\ln|x| - \ln|x+1|\right]_a^3$$

$$= \lim_{a\to0^+} \left[\ln 3 - \ln 4 - \ln a + \ln(a+1)\right] = \infty.$$

27. $\displaystyle\int_{-3}^1 \frac{dx}{x^2-4}$ diverges since $\displaystyle\int_{-2}^1 \frac{dx}{x^2-4}$ diverges:

$$\int_{-2}^1 \frac{dx}{x^2-4} = \int_{-2}^1 \frac{1}{4}\left[\frac{1}{x-2} - \frac{1}{x+2}\right] dx$$

$$= \lim_{a\to-2^+} \left[\frac{1}{4}\left(\ln|x-2| - \ln|x+2|\right)\right]_a^1$$

$$= \lim_{a\to-2^+} \frac{1}{4}\left[-\ln 3 - \ln|a-2| + \ln|a+2|\right] = -\infty.$$

29. diverges: $\displaystyle\int_0^{\infty} \cosh x\, dx = \lim_{b\to\infty} \int_0^b \cosh x\, dx = \lim_{b\to\infty} \left[\sinh x\right]_0^b = \infty$

31. $\frac{1}{2}$; $\displaystyle\int_0^\infty e^{-x}\sin x\,dx = \lim_{b\to\infty}\int_0^b e^{-x}\sin x\,dx - \lim_{b\to\infty} -\frac{1}{2}\left[e^{-x}\cos x + e^{-x}\sin x\right]_0^b$

(by parts)

$$= \lim_{b\to\infty}\frac{1}{2}\left[1 - e^{-b}\cos b - e^{-b}\sin b\right] = \frac{1}{2}$$

33. $2e - 2$; $\displaystyle\int_0^1 \frac{e^{\sqrt{x}}}{\sqrt{x}}\,dx = \lim_{a\to 0^+}\int_a^1 \frac{e^{\sqrt{x}}}{\sqrt{x}}\,dx = \lim_{a\to 0^+}\left[2\,e^{\sqrt{x}}\right]_a^1 = 2(e-1)$

35. $\displaystyle\int_0^1 \sin^{-1} x\,dx = \left[x\sin^{-1} x\right]_0^1 - \int_0^1 \frac{x}{\sqrt{1-x^2}}\,dx = \frac{\pi}{2} - \lim_{a\to 1^-}\int_0^a \frac{x}{\sqrt{1-x^2}}\,dx$

(by parts)

Now, $\displaystyle\int_0^a \frac{x}{\sqrt{1-x^2}}\,dx = -\frac{1}{2}\int_1^{1-a^2}\frac{1}{\sqrt{u}}\,du = \left[-\sqrt{u}\right]_1^{1-a^2} = 1 - \sqrt{1-a^2}$

$u = 1 - x^2$

Thus, $\displaystyle\int_0^1 \sin^{-1} x\,dx = \frac{\pi}{2} - \lim_{a\to 1^-}\left(1 - \sqrt{1-a^2}\right) = \frac{\pi}{2} - 1.$

37. $\displaystyle\int_0^\infty \frac{1}{\sqrt{x}\,(1+x)}\,dx = \int_0^1 \frac{1}{\sqrt{x}\,(1+x)}\,dx + \int_1^\infty \frac{1}{\sqrt{x}\,(1+x)}\,dx$

$$= \lim_{a\to 0^+}\int_a^1 \frac{1}{\sqrt{x}\,(1+x)}\,dx + \lim_{b\to\infty}\int_1^b \frac{1}{\sqrt{x}\,(1+x)}\,dx$$

Now, $\displaystyle\int \frac{1}{\sqrt{x}\,(1+x)}\,dx = \int \frac{2}{1+u^2}\,du = 2\tan^{-1} u + C = 2\tan^{-1}\sqrt{x} + C.$

$u = \sqrt{x}$

Therefore, $\displaystyle\lim_{a\to 0^+}\int_a^1 \frac{1}{\sqrt{x}\,(1+x)}\,dx = \lim_{a\to 0^+}\left[2\tan^{-1}\sqrt{x}\right]_a^1 = \lim_{a\to 0^+} 2\left[\pi/4 - \tan^{-1} a\right)\right] = \frac{\pi}{2}$

and $\displaystyle\lim_{b\to\infty}\int_1^b \frac{1}{\sqrt{x}\,(1+x)}\,dx = \lim_{b\to\infty}\left[2\tan^{-1}\sqrt{x}\right]_1^b = \lim_{b\to\infty} 2\left[\tan^{-1} b - \pi/4\right] = \frac{\pi}{2}.$

Thus, $\displaystyle\int_0^\infty \frac{1}{\sqrt{x}\,(1+x)}\,dx = \pi.$

39. (a)

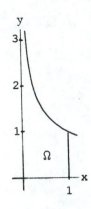

(b) $A = \int_0^1 \frac{1}{\sqrt{x}}\,dx = \lim_{a \to 0+} \int_a^1 \frac{1}{\sqrt{x}}\,dx$

$= \lim_{a \to 0+} \left[2\sqrt{x}\right]_a^1 = 2$

(c) $V = \int_0^1 \pi \left(\frac{1}{\sqrt{x}}\right)^2 dx = \pi \int_0^1 \frac{1}{x}\,dx = \pi \lim_{a \to 0+} \int_a^1 \frac{1}{x}\,dx = \pi \lim_{a \to 0+} \left[\ln x\right]_a^1$ diverges

41. (a)

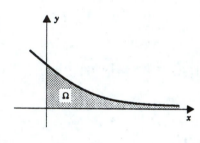

(b) $A = \int_0^\infty e^{-x}\,dx = 1$

(c) $V_x = \int_0^\infty \pi e^{-2x}\,dx = \pi/2$

(d) $V_y = \int_0^\infty 2\pi x e^{-x}\,dx = \lim_{b \to \infty} \int_0^b 2\pi x e^{-x}\,dx = \lim_{b \to \infty} \left[2\pi(-x-1)e^{-x}\right]_0^b$

(by parts)

$= 2\pi \left(1 - \lim_{b \to \infty} \frac{b+1}{e^b}\right) = 2\pi(1-0) = 2\pi$

(e) $A = \int_0^\infty 2\pi e^{-x}\sqrt{1+e^{-2x}}\,dx = \lim_{b \to \infty} \int_0^b 2\pi e^{-x}\sqrt{1+e^{-2x}}\,dx$

$\int_0^b 2\pi e^{-x}\sqrt{1+e^{-2x}}\,dx = -2\pi \int_1^{e^{-b}} \sqrt{1+u^2}\,du$

$u = e^{-x}$

$= -\pi \left[u\sqrt{1+u^2} + \ln\left(1+\sqrt{1+u^2}\right)\right]_1^{e^{-b}}$

$= \pi \left[\sqrt{2} + \ln\left(1+\sqrt{2}\right) - e^{-b}\sqrt{1+e^{-2b}} - \ln\left(1+\sqrt{1+e^{-2b}}\right)\right]$

Taking the limit of this last expression as $b \to \infty$, we have

$$A = \pi \left[\sqrt{2} + \ln\left(1+\sqrt{2}\right)\right].$$

43. (a) The interval $[0,1]$ causes no problem. For $x \ge 1$, $e^{-x^2} \le e^{-x}$ and $\int_1^\infty e^{-x}\,dx$ is finite.

(b) $V_y = \int_0^\infty 2\pi x e^{-x^2}\,dx = \lim_{b\to\infty}\int_0^b 2\pi x e^{-x^2}\,dx = \lim_{b\to\infty}\pi\left[-e^{-x^2}\right]_0^b = \lim_{b\to\infty}\pi\left(1 - e^{-b^2}\right) = \pi$

45. (a)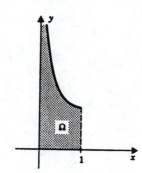

(b) $A = \lim_{a\to 0^+}\int_a^1 x^{-1/4}\,dx = \lim_{a\to 0^+}\left[\frac{4}{3}x^{3/4}\right]_a^1 = \frac{4}{3}$

(c) $V_x = \lim_{a\to 0^+}\int_a^1 \pi x^{-1/2}\,dx = \lim_{a\to 0^+}\left[2\pi x^{1/2}\right]_a^1 = 2\pi$

(d) $V_y = \lim_{a\to 0^+}\int_a^1 2\pi x^{3/4}\,dx = \lim_{a\to 0^+}\left[\frac{8\pi}{7}x^{7/4}\right]_a^1 = \frac{8}{7}\pi$

47. converges by comparison with $\displaystyle\int_1^\infty \frac{dx}{x^{3/2}}$

49. diverges since for x large the integrand is greater than $\dfrac{1}{x}$ and $\displaystyle\int_1^\infty \frac{dx}{x}$ diverges

51. converges by comparison with $\displaystyle\int_1^\infty \frac{dx}{x^{3/2}}$

53. $r(\theta) = ae^{c\theta}, \quad r'(\theta) = ace^{c\theta}$

$$L = \int_{-\infty}^{\theta_1}\sqrt{a^2 e^{2c\theta} + a^2 c^2 e^{2c\theta}}\,d\theta$$

(9.9.3)

$$= \left(a\sqrt{1 + c^2}\right)\left(\lim_{b\to-\infty}\int_b^{\theta_1} e^{c\theta}\,d\theta\right)$$

$$= \left(a\sqrt{1 + c^2}\right)\left(\lim_{b\to-\infty}\left[\frac{e^{c\theta}}{c}\right]_b^{\theta_1}\right)$$

$$= \left(\frac{a\sqrt{1+c^2}}{c}\right)\left(\lim_{b\to-\infty}\left[e^{c\theta_1} - e^{cb}\right]\right) = \left(\frac{a\sqrt{1+c^2}}{c}\right)e^{c\theta_1}$$

55. $F(s) = \displaystyle\int_0^\infty e^{-sx}\cdot 1\,dx = \lim_{b\to\infty}\int_0^b e^{-sx}\,dx = \lim_{b\to\infty}\left[-\frac{1}{s}e^{-sx}\right]_0^b = \frac{1}{s}$ provided $s > 0$.

Thus, $F(s) = \dfrac{1}{s}; \qquad \mathrm{dom}(F) = (0,\infty)$.

57. $F(s) = \displaystyle\int_0^\infty e^{-sx}\cos 2x\,dx = \lim_{b\to\infty}\int_0^b e^{-sx}\cos 2x\,dx$

Using integration by parts $\displaystyle\int e^{-sx}\cos 2x\,dx = \frac{4}{s^2 + 4}\left[\frac{1}{2}e^{-sx}\sin 2x - \frac{s}{4}e^{-sx}\cos 2x\right] + C.$

Therefore,

$$F(s) = \lim_{b \to \infty} \frac{4}{s^2 + 4} \left[\frac{1}{2} e^{-sx} \sin 2x - \frac{s}{4} e^{-sx} \cos 2x \right]_0^b$$

$$= \frac{4}{s^2 + 4} \lim_{b \to \infty} \left[\frac{1}{2} e^{-sb} \sin 2b - \frac{s}{4} e^{-sb} \cos 2b + \frac{s}{4} \right] = \frac{4}{s^2 + 4} \cdot \frac{s}{4} = \frac{s}{s^2 + 4} \qquad \text{provided} \quad s > 0.$$

Thus, $\quad F(s) = \dfrac{s}{s^2 + 4}; \qquad \text{dom}(F) = (0, \infty).$

59. The function f is nonnegative on $(-\infty, \infty)$ and

$$\int_{-\infty}^{\infty} f(x)\, dx = \int_{-\infty}^{0} 0\, dx + \int_0^{\infty} \frac{6x}{(1 + 3x^2)^2}\, dx = \int_0^{\infty} \frac{6x}{(1 + 3x^2)^2}\, dx$$

Now, $\quad \displaystyle\int \frac{6x}{(1 + 3x^2)^2}\, dx = -\frac{1}{1 + 3x^2} + C.$

Therefore,

$$\int_{-\infty}^{\infty} f(x)\, dx = \lim_{b \to \infty} \left[-\frac{1}{1 + 3x^2} \right]_0^b = \lim_{b \to \infty} \left(1 - \frac{1}{1 + 3b^2} \right) = 1.$$

61. $\quad \mu = \displaystyle\int_{-\infty}^{\infty} x f(x)\, dx = \int_{-\infty}^{0} 0\, dx + \int_0^{\infty} kxe^{-kx}\, dx = \lim_{b \to \infty} \int_0^b kxe^{-kx}\, dx$

Using integration by parts, $\quad \displaystyle\int kxe^{-kx}\, dx = -xe^{-kx} - \frac{1}{k} e^{-kx} + C.$

Therefore,

$$\mu = \int_{-\infty}^{\infty} x f(x)\, dx = \lim_{b \to \infty} \left[-xe^{-kx} - \frac{1}{k} e^{-kx} \right]_0^b = \lim_{b \to \infty} \left[-be^{-kb} - \frac{1}{k} e^{-kb} + \frac{1}{k} \right] = \frac{1}{k}$$

63. Observe that

$$F(t) = \int_1^t f(x)\, dx$$

is continuous and increasing, that

$$a_n = \int_1^n f(x)\, dx$$

is increasing, and that

$(*) \qquad a_n \le \displaystyle\int_1^t f(x)\, dx \le a_{n+1} \quad \text{for} \quad t \in [n, n+1].$

If

$$\int_1^\infty f(x)\,dx$$

converges, then F, being continuous, is bounded and, by $(*)$, $\{a_n\}$ is bounded and therefore convergent.

If $\{a_n\}$ converges, then $\{a_n\}$ is bounded and, by $(*)$, F is bounded. Being increasing, F is also convergent; i.e., $\int_1^\infty f(x)\,dx$ converges.

CHAPTER 11

SECTION 11.1

1. $1 + 4 + 7 = 12$

3. $1 + 2 + 4 + 8 = 15$

5. $-\frac{1}{6} + \frac{1}{24} - \frac{1}{120} = -\frac{2}{15}$

7. $\displaystyle\sum_{n=1}^{11}(2n - 1)$

9. $\displaystyle\sum_{k=1}^{35} k(k + 1)$

11. $\displaystyle\sum_{k=1}^{n} M_k \Delta x_k$

13. $\displaystyle\sum_{k=3}^{10} \frac{1}{2^k}, \quad \sum_{i=0}^{7} \frac{1}{2^{i+3}}$

15. $\displaystyle\sum_{k=3}^{10}(-1)^{k+1}\frac{k}{k+1}, \quad \sum_{i=0}^{7}(-1)^i\frac{i+3}{i+4}$

17. Set $k = n + 3$. Then $n = -1$ when $k = 2$ and $n = 7$ when $k = 10$.

$$\sum_{k=2}^{10} \frac{k}{k^2 + 1} = \sum_{n=-1}^{7} \frac{n + 3}{(n + 3)^2 + 1} = \sum_{n=-1}^{7} \frac{n + 3}{n^2 + 6n + 10}$$

19. Set $k = n - 3$. Then $n = 7$ when $k = 4$ and $n = 28$ when $k = 25$.

$$\sum_{k=4}^{25} \frac{1}{k^2 - 9} = \sum_{n=7}^{28} \frac{1}{(n - 3)^2 - 9} = \sum_{n=7}^{28} \frac{1}{n^2 - 6n}$$

21. $\displaystyle\sum_{k=1}^{10}(2k + 3) = 2\sum_{k=1}^{10} k + \sum_{k=1}^{10} 3 = 2 \cdot \frac{(10)(11)}{2} + 3 \cdot 10 = 140$

23. $\displaystyle\sum_{k=1}^{8}(2k - 1)^2 = \sum_{k=1}^{8}(4k^2 - 4k + 1) = 4\sum_{k=1}^{8} k^2 - 4\sum_{k=1}^{8} k + \sum_{k=1}^{8} 1$

$$= 4 \cdot \frac{(8)(9)(17)}{6} - 4 \cdot \frac{(8)(9)}{2} + 8 = 680$$

25. $\dfrac{1}{2}; \qquad s_n = \dfrac{1}{2}\left[\dfrac{1}{1 \cdot 2} + \dfrac{1}{2 \cdot 3} + \cdots + \dfrac{1}{(n)(n + 1)}\right]$

$$= \frac{1}{2}\left[\left(1 - \frac{1}{2}\right) + \left(\frac{1}{2} - \frac{1}{3}\right) + \cdots + \left(\frac{1}{n} - \frac{1}{n + 1}\right)\right] = \frac{1}{2}\left[1 - \frac{1}{n + 1}\right] \to \frac{1}{2}$$

27. $\dfrac{11}{18}; \qquad s_n = \dfrac{1}{1 \cdot 4} + \dfrac{1}{2 \cdot 5} + \cdots + \dfrac{1}{n(n + 3)}$

$$= \frac{1}{3}\left[\left(1 - \frac{1}{4}\right) + \left(\frac{1}{2} - \frac{1}{5}\right) + \cdots + \left(\frac{1}{n} - \frac{1}{n + 3}\right)\right]$$

$$= \frac{1}{3}\left[1 + \frac{1}{2} + \frac{1}{3} - \frac{1}{n + 1} - \frac{1}{n + 2} - \frac{1}{n + 3}\right] \to \frac{1}{3}\left(1 + \frac{1}{2} + \frac{1}{3}\right) = \frac{11}{18}$$

29. $\dfrac{10}{3}; \qquad \displaystyle\sum_{k=0}^{\infty} \frac{3}{10^k} = 3\sum_{k=0}^{\infty}\left(\frac{1}{10}\right)^k = 3\left(\frac{1}{1 - 1/10}\right) = \frac{30}{9} = \frac{10}{3}$

31. $-\dfrac{3}{2}; \qquad \displaystyle\sum_{k=0}^{\infty} \frac{1 - 2^k}{3^k} = \sum_{k=0}^{\infty}\left(\frac{1}{3}\right)^k - \sum_{k=0}^{\infty}\left(\frac{2}{3}\right)^k = \frac{1}{1 - 1/3} - \frac{1}{1 - 2/3} = \frac{3}{2} - 3 = -\frac{3}{2}$

33. $\frac{1}{2}$; geometric series with $a = \frac{1}{4}$ and $r = \frac{1}{2}$, sum $= \frac{a}{1-r} = \frac{1}{2}$

35. 24; geometric series with $a = 8$ and $r = \frac{2}{3}$, sum $= \frac{a}{1-r} = 24$

37. $\displaystyle\sum_{k=1}^{\infty} \frac{7}{10^k} = \frac{7/10}{1-1/10} = \frac{7}{9}$ **39.** $\displaystyle\sum_{k=1}^{\infty} \frac{24}{100^k} = \frac{24/100}{1-1/100} = \frac{8}{33}$

41. $\dfrac{62}{100} + \dfrac{1}{100}\displaystyle\sum_{k=1}^{\infty}\dfrac{45}{100^k} = \dfrac{62}{100} + \dfrac{1}{100}\left(\dfrac{45/100}{1-1/100}\right) = \dfrac{687}{1100}$

43. Let $x = 0.\overset{\frown}{a_1 a_2 \cdots a_n}\,\overset{\frown}{a_1 a_2 \cdots a_n}\cdots$. Then

$$x = \sum_{k=1}^{\infty}\frac{a_1 a_2 \cdots a_n}{(10^n)^k} = a_1 a_2 \cdots a_n \sum_{k=1}^{\infty}\left(\frac{1}{10^n}\right)^k$$

$$= a_1 a_2 \cdots a_n \left[\frac{1}{1-1/10^n} - 1\right] = \frac{a_1 a_2 \cdots a_n}{10^n - 1}.$$

45. $\dfrac{1}{1+x} = \dfrac{1}{1-(-x)} = \displaystyle\sum_{k=0}^{\infty}(-x)^k = \sum_{k=0}^{\infty}(-1)^k x^k$

47. $\dfrac{x}{1-x} = x\left(\dfrac{1}{1-x}\right) = x\displaystyle\sum_{k=0}^{\infty}(x^k) = \sum_{k=0}^{\infty}x^{k+1}$

49. $\dfrac{x}{1+x^2} = x\left[\dfrac{1}{1-(-x^2)}\right] = x\displaystyle\sum_{k=0}^{\infty}(-x^2)^k = \sum_{k=0}^{\infty}(-1)^k x^{2k+1}$

51. $1 + \dfrac{3}{2} + \dfrac{9}{4} + \dfrac{27}{8} + \dfrac{81}{16} + \cdots = \displaystyle\sum_{k=0}^{\infty}\left(\dfrac{3}{2}\right)^k$

This is a geometric series with $x = \frac{3}{2} > 1$. Therefore the series diverges.

53. $\displaystyle\lim_{k\to\infty}\left(\frac{k+1}{k}\right)^k = e \neq 0$

55. Rebounds to half its previous height:

$$s = 6 + 3 + 3 + \frac{3}{2} + \frac{3}{2} + \frac{3}{4} + \frac{3}{4} + \cdots = 6 + 6 \sum_{k=0}^{\infty} \frac{1}{2^k} = 6 + \frac{6}{1 - \frac{1}{2}} = 18 \text{ ft.}$$

57. A principal x deposited now at $r\%$ interest compounded annually will grow in k years to

$$x \left(1 + \frac{r}{100}\right)^k.$$

This means that in order to be able to withdraw n_k dollars after k years one must place

$$n_k \left(1 + \frac{r}{100}\right)^{-k}$$

dollars on deposit today. To extend this process in perpetuity as described in the text, the total deposit must be

$$\sum_{k=1}^{\infty} n_k \left(1 + \frac{r}{100}\right)^{-k}.$$

59. $\displaystyle\sum_{n=1}^{\infty} \left(\frac{9}{10}\right)^n = \frac{\frac{9}{10}}{1 - \frac{9}{10}} = 9$ or 9

61. $$A = 4^2 + (2\sqrt{2})^2 + 2^2 + (\sqrt{2})^2 + 1^2 + \cdots + \left[4\left(\frac{1}{\sqrt{2}}\right)^n\right]^2 + \cdots$$

$$= \sum_{n=0}^{\infty} \left[4\left(\frac{1}{\sqrt{2}}\right)^n\right]^2 = 16 \sum_{n=0}^{\infty} \left(\frac{1}{2}\right)^n = 16 \cdot \frac{1}{1 - \frac{1}{2}} = 32$$

63. Let $L = \displaystyle\sum_{k=0}^{\infty} a_k$. Then

$$L = \sum_{k=0}^{\infty} a_k = \sum_{k=0}^{n} a_k + \sum_{k=n+1}^{\infty} a_k = s_n + R_n.$$

Therefore, $R_n = L - s_n$ and since $s_n \to L$ as $n \to \infty$, it follows that $R_n \to 0$ as $n \to \infty$.

65. $s_0 = 1, \ s_1 = 0, \ s_2 = 1, \ \ldots,; \quad s_n = \dfrac{1 + (-1)^n}{2}, \ n = 0, 1, 2, \ldots$.

67. $s_n = \displaystyle\sum_{k=1}^{n} \ln\left(\frac{k+1}{k}\right) = [\ln(n+1) - \ln(n)] + [\ln n - \ln(n-1)] + \cdots + [\ln 2 - \ln 1] = \ln(n+1) \to \infty$

69. (a) $s_n = \displaystyle\sum_{k=1}^{n} (d_k - d_{k+1}) = d_1 - d_{n+1} \to d_1$

(b)   We use part (a).

(i) $\displaystyle\sum_{k=1}^{\infty} \frac{\sqrt{k+1}-\sqrt{k}}{\sqrt{k(k+1)}} = \sum_{k=1}^{\infty}\left[\frac{1}{\sqrt{k}} - \frac{1}{\sqrt{k+1}}\right] = 1$

(ii) $\displaystyle\sum_{k=1}^{\infty} \frac{2k+1}{2k^2(k+1)^2} = \sum_{k=1}^{\infty} \frac{1}{2}\left[\frac{1}{k^2} - \frac{1}{(k+1)^2}\right] = \frac{1}{2}$

71. $\displaystyle R_n = \sum_{k=n+1}^{\infty} \frac{1}{4^k} = \frac{\left(\frac{1}{4}\right)^{n+1}}{1 - \frac{1}{4}} = \frac{1}{3 \cdot 4^n};$

$\displaystyle \frac{1}{3\cdot 4^n} < 0.0001 \implies 4^n > 3333.33 \implies n > \frac{\ln 3333.33}{\ln 4} \cong 5.85$

Take $N = 6$.

73. $\displaystyle R_n = \sum_{k=n+1}^{\infty} \frac{1}{k(k+2)} = \frac{1}{2}\sum_{k=n+1}^{\infty}\left(\frac{1}{k} - \frac{1}{k+2}\right) = \frac{1}{2}\left(\frac{1}{n+1} + \frac{1}{n+2}\right);$

$\displaystyle \frac{1}{2}\left(\frac{1}{n+1} + \frac{1}{n+2}\right) < 0.0001 \implies n \geq 9999.$ Take $N = 9999$.

75. $\displaystyle |R_n| = \left| \sum_{k=n+1}^{\infty} x^k \right| = \left|\frac{x^{n+1}}{1-x}\right| = \frac{|x|^{n+1}}{1-x};$

$$\frac{|x|^{n+1}}{1-x} < \epsilon$$

$$|x|^{n+1} < \epsilon(1-x)$$

$$(n+1)\ln|x| < \ln\epsilon(1-x)$$

$$n+1 > \frac{\ln\epsilon(1-x)}{\ln|x|} \qquad [\text{recall} \ \ \ln|x| < 0]$$

$$n > \frac{\ln\epsilon(1-x)}{\ln|x|} - 1$$

Take N to be smallest integer which is greater than $\dfrac{\ln\epsilon(1-x)}{\ln|x|}$.

SECTION 11.2

1.   converges;   basic comparison with $\displaystyle\sum \frac{1}{k^2}$ **3.**   converges;   basic comparison with $\displaystyle\sum \frac{1}{k^2}$

5.   diverges;   basic comparison with $\displaystyle\sum \frac{1}{k+1}$ **7.**   diverges;   limit comparison with $\displaystyle\sum \frac{1}{k}$

9. converges; integral test, $\displaystyle\int_1^\infty \frac{\tan^{-1} x}{1+x^2}\, dx = \lim_{b\to\infty}\left[\frac{1}{2}(\tan^{-1} x)^2\right]_1^b = \frac{3\pi^2}{32}$

11. diverges; p-series with $p = \frac{2}{3} \le 1$ **13.** diverges; divergence test, $(\frac{3}{4})^{-k} \not\to 0$

15. diverges; basic comparison with $\displaystyle\sum \frac{1}{k}$

17. diverges; divergence test, $\displaystyle\frac{1}{2+3^{-k}} \to \frac{1}{2} \ne 0$

19. converges; limit comparison with $\displaystyle\sum \frac{1}{k^2}$

21. diverges; integral test, $\displaystyle\int_2^\infty \frac{dx}{x \ln x} = \lim_{b\to\infty}\left[\ln(\ln x)\right]_2^b = \infty$

23. converges; limit comparison with $\displaystyle\sum \frac{1}{k^2}$ **25.** diverges; limit comparison with $\displaystyle\sum \frac{1}{k}$

27. converges; limit comparison with $\displaystyle\sum \frac{1}{k^{3/2}}$

29. converges; integral test, $\displaystyle\int_1^\infty xe^{-x^2}\, dx = \lim_{b\to\infty}\left[-\frac{1}{2}e^{-x^2}\right]_1^b = \frac{1}{2e}$

31. Converges; basic comparison with $\displaystyle\sum \frac{3}{k^2}$, $2 + \sin k \le 3$ for all k.

33. Recall that $\displaystyle 1 + 2 + 3 + \cdots + k = \frac{k(k+1)}{2}$. Therefore

$$\sum \frac{1}{1+2+3+\cdots+k} = \sum \frac{2}{k(k+1)}.$$ This series converges; direct comparison with $\displaystyle\sum \frac{2}{k^2}$

35. Use the integral test:

Let $u = \ln x$, $du = \frac{1}{x}\, dx$: $\displaystyle\int \frac{1}{x(\ln x)^p}\, dx = \int u^{-p}\, du = \frac{u^{1-p}}{1-p} + C.$

$$\int_1^\infty \frac{1}{x(\ln x)^p}\, dx = \lim_{b\to\infty} \int_1^b \frac{1}{x(\ln x)^p}\, dx = \lim_{b\to\infty} \frac{1}{1-p}(\ln a)^{1-p}$$

The series converges for $p > 1$.

37. (a) Use the integral test: $\displaystyle\int_0^\infty e^{-\alpha x}\, dx = \lim_{b\to\infty}\left[-\frac{1}{\alpha}e^{-\alpha x}\right]_0^b = \frac{1}{\alpha}$ converges.

(b) Use the integral test: $\displaystyle\int_0^\infty xe^{-\alpha x}\, dx = \lim_{b\to\infty}\left[-\frac{1}{\alpha x e^{-\alpha x}} - \frac{1}{\alpha^2}e^{-\alpha x}\right]_0^b = \frac{1}{\alpha^2}$ converges.

(c) The proof follows by induction using parts (a) and (b) and the reduction formula

$$\int x^n e^{\alpha x}\, dx = \frac{x^n e^{\alpha x}}{a} - \frac{n}{a}\int x^{n-1} e^{\alpha x}\, dx \quad \text{[see Exercise 67, Section 8.2]}$$

39. (a) $\displaystyle\sum_{k=1}^{4}\frac{1}{k^3} \cong 1.1777$ (b) $\dfrac{1}{2\cdot 5^2} < R_4 < \dfrac{1}{2\cdot 4^2}$

$$0.02 < R_4 < 0.0313$$

(c) $1.1777 + 0.02 = 1.1977 < \displaystyle\sum_{k=1}^{\infty}\frac{1}{k^3} < 1.1777 + 0.0313 = 1.2090$

41. (a) Put $p=2$ and $n=100$ in the estimates in Exercise 38. The result is: $\dfrac{1}{101} < R_{100} < \dfrac{1}{100}$.

(b) $R_n < \dfrac{1}{(2-1)n^{2-1}} < 0.0001 \implies n > 10,000$ Take $n = 10,001$.

43. (a) $R_n < \dfrac{1}{(4-1)n^{4-1}} < 0.0001 \implies n^3 > 3333 \implies n > 14.94 :$ Take $n = 15$.

(b) $R_n < \dfrac{1}{(4-1)n^{4-1}} < 0.0005 \implies n^3 > 666.67 \implies n > 8.74 :$ Take $n = 9$.

$$\sum_{k=1}^{\infty}\frac{1}{k^4} \cong \sum_{k=1}^{9}\frac{1}{k^4} \cong 1.082$$

45. (a) If $a_k/b_k \to 0$, then $a_k/b_k < 1$ for all $k \geq K$ for some K. But then $a_k < b_k$ for all $k \geq K$ and, since $\sum b_k$ converges, $\sum a_k$ converges. [The Basic Comparison Theorem 11.2.5.]

(b) Similar to (a) except that this time we appeal to part (ii) of Theorem 11.2.5.

(c) $\displaystyle\sum a_k = \sum\frac{1}{k^2}$ converges, $\displaystyle\sum b_k = \sum\frac{1}{k^{3/2}}$ converges, $\dfrac{1/k^2}{1/k^{3/2}} = \dfrac{1}{\sqrt{k}} \to 0$

$\displaystyle\sum a_k = \sum\frac{1}{k^2}$ converges, $\displaystyle\sum b_k = \sum\frac{1}{\sqrt{k}}$ diverges, $\dfrac{1/k^2}{1/\sqrt{k}} = \dfrac{1}{k^{3/2}} \to 0$

(d) $\displaystyle\sum b_k = \sum\frac{1}{\sqrt{k}}$ diverges, $\displaystyle\sum a_k = \sum\frac{1}{k^2}$ converges, $\dfrac{1/k^2}{1/\sqrt{k}} = 1/k^{3/2} \to 0$

$\displaystyle\sum b_k = \sum\frac{1}{\sqrt{k}}$ diverges, $\displaystyle\sum a_k = \sum\frac{1}{k}$ diverges, $\dfrac{1/k}{1/\sqrt{k}} = \dfrac{1}{\sqrt{k}} \to 0$

47. (a) Since $\sum a_k$ converges, $a_k \to 0$. Therefore there exists a positive integer N such that $0 < a_k < 1$ for $k \geq N$. Thus, for $k \geq N$, $a_k^2 < a_k$ and so $\sum a_k^2$ converges by the comparison test.

(b) $\sum a_k$ may either converge or diverge: $\sum 1/k^4$ and $\sum 1/k^2$ both converge; $\sum 1/k^2$ converges and $\sum 1/k$ diverges.

49. $0 < L - \displaystyle\sum_{k=1}^{n} f(k) = L - s_n = \sum_{k=n+1}^{\infty} f(k) < \int_{n}^{\infty} f(x)\,dx$ [see the proof of the integral test]

51. $L - s_n < \displaystyle\int_{n}^{\infty} xe^{-x^2}\,dx = \lim_{b\to\infty}\int_{n}^{b} xe^{-x^2}\,dx$

$$= \lim_{b\to\infty}\left[-\frac{1}{2}e^{-x^2}\right]_{n}^{b} = \frac{1}{2}e^{-n^2}$$

$$\frac{1}{2}e^{-n^2} < 0.001 \quad \Longrightarrow \quad e^{n^2} > 500 \quad \Longrightarrow \quad n > 2.49; \text{ take } n=3.$$

53.   (a)   Set $f(x) = x^{1/4} - \ln x$. Then

$$f'(x) = \frac{1}{4}x^{-3/4} - \frac{1}{x} = \frac{1}{4x}(x^{1/4} - 4).$$

Since $f(e^{12}) = e^3 - 12 > 0$ and $f'(x) > 0$ for $x > e^{12}$, we have that

$$n^{1/4} > \ln n \quad \text{and therefore} \quad \frac{1}{n^{5/4}} > \frac{\ln n}{n^{3/2}}$$

for sufficiently large n. Since $\sum \frac{1}{n^{5/4}}$ is a convergent p-series, $\sum \frac{\ln n}{n^{3/2}}$ converges

by the basic comparison test.

(b)   By L'Hospital's rule

$$\lim_{x \to \infty} \frac{(\ln x)/x^{3/2}}{1/x^{5/4}} = \lim_{x \to \infty} \frac{\ln x}{x^{1/4}} \overset{\star}{=} \lim_{x \to \infty} \frac{1/x}{\frac{1}{4}x^{-3/4}} = \lim_{x \to \infty} \frac{4}{x^{1/4}} = 0.$$

Thus, the limit comparison test does not apply.

SECTION 11.3

1.   converges;   ratio test: $\dfrac{a_{k+1}}{a_k} = \dfrac{10}{k+1} \to 0$ **3.**   converges;   root test: $(a_k)^{1/k} = \dfrac{1}{k} \to 0$

5.   diverges;   divergence test: $\dfrac{k!}{100^k} \to \infty$ **7.**   diverges;   limit comparison with $\sum \dfrac{1}{k}$

9.   converges;   root test: $(a_k)^{1/k} = \dfrac{2}{3}k^{1/k} \to \dfrac{2}{3}$ **11.**   diverges;   limit comparison with $\sum \dfrac{1}{\sqrt{k}}$

13.   diverges;   ratio test: $\dfrac{a_{k+1}}{a_k} = \dfrac{k+1}{10^4} \to \infty$ **15.**   converges;   basic comparison with $\sum \dfrac{1}{k^{3/2}}$

17.   converges;   basic comparison with $\sum \dfrac{1}{k^2}$

19.   diverges;   integral test: $\displaystyle\int_2^\infty \frac{1}{x}(\ln x)^{-1/2}dx = \lim_{b \to \infty}\left[2(\ln x)^{1/2}\right]_2^b = \infty$

21.   diverges;   divergence test: $\left(\dfrac{k}{k+100}\right)^k = \left(1 + \dfrac{100}{k}\right)^{-k} \to e^{-100} \neq 0$

23.   diverges;   limit comparison with $\sum \dfrac{1}{k}$ **25.**   converges;   ratio test: $\dfrac{a_{k+1}}{a_k} = \dfrac{\ln(k+1)}{e \ln k} \to \dfrac{1}{e}$

27.   converges;   basic comparison with $\sum \dfrac{1}{k^{3/2}}$

29.   converges;   ratio test: $\dfrac{a_{k+1}}{a_k} = \dfrac{2(k+1)}{(2k+1)(2k+2)} \to 0$

31.   converges;   ratio test: $\dfrac{a_{k+1}}{a_k} = \dfrac{(k+1)(2k+1)(2k+2)}{(3k+1)(3k+2)(3k+3)} \to \dfrac{4}{27}$

33.   converges;   ratio test: $\dfrac{a_{k+1}}{a_k} = \dfrac{1}{(k+1)^{1/2}} \left(\dfrac{k+1}{k} \right)^{k/2} \to 0 \cdot \sqrt{e} = 0$

35.   converges;   root test: $(a_k)^{1/k} = \dfrac{k}{3^k} \to 0$

37. $\dfrac{1}{2} + \dfrac{2}{3^2} + \dfrac{4}{4^3} + \dfrac{8}{5^4} + \cdots = \displaystyle\sum_{k=0}^{\infty} \dfrac{2^k}{(k+2)^{k+1}}$

converges;   root test: $(a_k)^{1/k} = \dfrac{2}{(k+2)^{1+1/k}} \to 0$

39. $\dfrac{1}{4} + \dfrac{1 \cdot 3}{4 \cdot 7} + \dfrac{1 \cot 3 \cdot 5}{4 \cdot 7 \cdot 10} + \cdots = \displaystyle\sum_{k=0}^{\infty} \dfrac{1 \cdot 3 \cdots (1+2k)}{4 \cdot 7 \cdots (4+3k)}$

converges;   ratio test: $\dfrac{a_{k+1}}{a_k} = \dfrac{3+2k}{7+3k} \to \dfrac{2}{3}$

41.   By the hint

$$\sum_{k=1}^{\infty} k \left(\dfrac{1}{10} \right)^k = \dfrac{1}{10} \sum_{k=1}^{\infty} k \left(\dfrac{1}{10} \right)^{k-1} = \dfrac{1}{10} \left[\dfrac{1}{1-1/10} \right]^2 = \dfrac{10}{81}.$$

43.   The series $\displaystyle\sum_{k=0}^{\infty} \dfrac{k!}{k^k}$ converges (see Exercise 26). Therefore, $\displaystyle\lim_{k \to \infty} \dfrac{k!}{k^k} = 0$ by Theorem 11.1.6.

45.   Use the ratio test:

$$\dfrac{a_{k+1}}{a_k} = \dfrac{\dfrac{[(k+1)!]^2}{[p(k+1)]!}}{\dfrac{(k!)^2}{(pk)!}} = (k+1)^2 \dfrac{(pk)!}{(pk)!(pk+1)\cdots(pk+p)} = \dfrac{(k+1)^2}{(pk+1)\cdots(pk+p)}$$

Thus

$$\dfrac{a_{k+1}}{a_k} \to \begin{cases} \dfrac{1}{4}, & \text{if } p = 2 \\ 0, & \text{if } p > 2 \end{cases}$$

The series converges for all $p \geq 2$.

47.   Set $b_k = a_k r^k$.   If $(a_k)^{1/k} \to \rho$ and $\rho < \dfrac{1}{r}$,   then

$$(b_k)^{1/k} = (a_k r^k)^{1/k} = (a_k)^{1/k} r \to \rho r < 1$$

and thus, by the root test, $\Sigma b_k = \Sigma a_k r^k$ converges.

SECTION 11.4

1. diverges; $a_k \not\to 0$ **3.** diverges; $\dfrac{k}{k+1} \to 1 \neq 0$

5. (a) does not converge absolutely; integral test,

$$\int_1^\infty \frac{\ln x}{x}\,dx = \lim_{b \to \infty} \left[\frac{1}{2}(\ln x)^2\right]_1^b = \infty$$

     (b) converges conditionally; Theorem 11.4.4

7. diverges; limit comparison with $\displaystyle\sum \frac{1}{k}$

     another approach: $\displaystyle\sum \left(\frac{1}{k} - \frac{1}{k!}\right) = \sum \frac{1}{k} - \sum \frac{1}{k!}$ diverges since $\displaystyle\sum \frac{1}{k}$ diverges and

     $\displaystyle\sum \frac{1}{k!}$ converges

9. (a) does not converge absolutely; limit comparison with $\displaystyle\sum \frac{1}{k}$

     (b) converges conditionally; Theorem 11.4.4

11. diverges; $a_k \not\to 0$

13. (a) does not converge absolutely;

$$(\sqrt{k+1} - \sqrt{k}) \cdot \frac{(\sqrt{k+1} + \sqrt{k})}{(\sqrt{k+1} + \sqrt{k})} = \frac{1}{\sqrt{k+1} + \sqrt{k}}$$

     and

$$\sum \frac{1}{\sqrt{k} + \sqrt{k+1}} > \sum \frac{1}{2\sqrt{k+1}} = 2\sum \frac{1}{\sqrt{k+1}} \quad \text{(a p-series with $p < 1$)}$$

     (b) converges conditionally; Theorem 11.4.4

15. converges absolutely (terms already positive); basic comparison,

$$\sum \sin\left(\frac{\pi}{4k^2}\right) \le \sum \frac{\pi}{4k^2} = \frac{\pi}{4} \sum \frac{1}{k^2} \quad (|\sin x| \le |x|)$$

17. converges absolutely; ratio test, $\dfrac{a_{k+1}}{a_k} = \dfrac{k+1}{2k} \to \dfrac{1}{2}$

19. (a) does not converge absolutely; limit comparison with $\displaystyle\sum \frac{1}{k}$

     (b) converges conditionally; Theorem 11.4.4

21. diverges; $a_k = \dfrac{4^{k-2}}{e^k} = \dfrac{1}{16}\left(\dfrac{4}{e}\right)^k \not\to 0$

23. diverges; $a_k = k\sin(1/k) = \dfrac{\sin(1/k)}{1/k} \to 1 \neq 0$

25. converges absolutely;   ratio test, $\dfrac{a_{k+1}}{a_k} = \dfrac{(k+1)e^{-(k+1)}}{ke^{-k}} = \dfrac{k+1}{k}\dfrac{1}{e} \to \dfrac{1}{e}$

27. diverges; $\sum(-1)^k\dfrac{\cos\pi k}{k} = \sum(-1)^k\dfrac{(-1)^k}{k} = \sum\dfrac{1}{k}$

29. converges absolutely;   basic comparison

$$\sum\left|\dfrac{\sin(\pi k/4)}{k^2}\right| \le \sum\dfrac{1}{k^2}$$

31. diverges; $a_k \not\to 0$

33. Use (11.4.5); $|s - s_{80}| < a_{81} = \dfrac{1}{\sqrt{82}} \cong 0.1104$

35. Use (11.4.5); $|s - s_9| < a_{10} = \dfrac{1}{10^3} = 0.001$

37. $\dfrac{10}{11}$;   geometric series with $a = 1$ and $r = -\dfrac{1}{10}$,   sum $= \dfrac{a}{1-r} = \dfrac{10}{11}$

39. Use (11.4.5); $|s - s_n| < a_{n+1} = \dfrac{1}{\sqrt{n+2}} < 0.005 \implies n \ge 39{,}998$

41. Use (11.4.5).

   (a) $n = 4$; $\dfrac{1}{(n+1)!} < 0.01 \implies 100 < (n+1)!$

   (b) $n = 6$; $\dfrac{1}{(n+1)!} < 0.001 \implies 1000 < (n+1)!$

43. No.   For instance, set $a_{2k} = 2/k$ and $a_{2k+1} = 1/k$.

45. (a) Since $\sum|a_k|$ converges, $\sum|a_k|^2 = \sum a_k^2$ converges (Exercise 47, Section 11.2).

   (b) $\sum\dfrac{1}{k^2}$ converges, $\sum(-1)^k\dfrac{1}{k}$ is not absolutely convergent.

47. See the proof of Theorem 11.7.2.

49. (a) $\displaystyle\sum_{k=1}^{\infty}\dfrac{(-1)^{k-1}(a+b) + (a-b)}{2k} = \sum_{k=1}^{\infty}\dfrac{(-1)^{k-1}(a+b)}{2k} + \sum_{k=1}^{\infty}\dfrac{a-b}{2k}$

   (b) The series is absolutely convergent if $a = b = 0$;   conditionally convergent if $a = b \neq 0$;

     divergent if $a \neq b$.

PROJECT 11.4

n	a_n	s_n	err s_n	t_n	err t_n	b_n	u_n
9	0.112329	−0.053					
10	0.100899	0.394036	0.047898	0.346242	0.000103	0.525290	0.343568
19	0.052770	0.320450	−0.025690				
20	0.050119	0.370569	0.024431	0.346152	1.4e − 05	0.512544	0.345510
29	0.034522	0.329175	−0.016960				
30	0.033369	0.362544	0.016416	0.346142	4.2e − 06	0.508350	0.345860
39	0.025657	0.333474	−0.012660				
40	0.025015	0.358489	0.012351	0.346140	1.7e − 06	0.506260	0.345982
49	0.020416	0.336034	−0.0101				
50	0.020008	0.356042	0.009904	0.346139	8.3e − 07	0.505009	0.346038
59	0.016954	0.337733	−0.00841				
60	0.016671	0.354404	0.008266	0.346139	4.3e − 07	0.504176	0.346069
69	0.014496	0.338943	−0.0072				
70	0.014289	0.353231	0.007903	0.346138	2.3e − 07	0.503581	0.346087
79	0.012660	0.338848	−0.00629				
80	0.012502	0.352350	0.006212	0.346138	1.1e − 07	0.503135	0.346099
89	0.011237	0.340551	−0.00559				
90	0.011112	0.351663	0.005525	0.346138	4.5e − 08	0.502789	0.346107
99	0.010102	0.341113	−0.00503				
100	0.010001	0.351114	0.004975	0.346138	0	0.502513	0.346113

1. Column s_n gives these estimates.

3. $t_n = \dfrac{s_n + s_{n-1}}{2} = \dfrac{1}{2}(s_n + s_{n-1})$

$$\lim_{n\to\infty} t_n = \frac{1}{2}\left(\lim_{n\to\infty} s_n + \lim_{n\to\infty} s_{n-1}\right) = \frac{1}{2}(2s) = s.$$

$u_n = \dfrac{a_n s_n + a_{n-1} s_{n-1}}{a_n + a_{n-1}}$

Let $\epsilon > 0$. Since $\lim_{n\to\infty} = s$, there exists a positive integer N such that

$$|s_n - s| < \epsilon \text{ whenever } n - 1 > N.$$

Choose any integer $k > N + 1$. Then

$$
\begin{aligned}
|u_k - s| &= \left| \frac{a_k s_k + a_{k-1} s_{k-1}}{a_k + a_{k-1}} - s \right| \\
&= \left| \frac{a_k(s_k - s) + a_{k-1}(s_{k-1} - s)}{a_k + a_{k-1}} \right| \\
&\leq \frac{a_k|s_k - s| + a_{k-1}|s_{k-1} - s|}{a_k + a_{k-1}} \\
&< \frac{\epsilon(a_k + a_{k-1})}{a_k + a_{k-1}} = \epsilon
\end{aligned}
$$

Therefore $\lim_{n\to\infty} u_n = s$.

SECTION 11.5

1. $-1 + x + \frac{1}{2}x^2 - \frac{1}{24}x^4$

3. $-\frac{1}{2}x^2 - \frac{1}{12}x^4$

5. $1 - x + x^2 - x^3 + x^4 - x^5$

7. $x + \frac{1}{3}x^3 + \frac{2}{15}x^5$

9. $P_0(x) = 1, \quad P_1(x) = 1 - x, \quad P_2(x) = 1 - x + 3x^2, \quad P_3(x) = 1 - x + 3x^2 + 5x^3$

11. $\sum_{k=0}^{n}(-1)^k \dfrac{x^k}{k!}$

13. $\sum_{k=0}^{m} \dfrac{x^{2k}}{(2k)!}$ where $m = \dfrac{n}{2}$ and n is even

15. $f^{(k)}(x) = r^k e^{rx}$ and $f^{(k)}(0) = r^k$, $k = 0, 1, 2, \ldots$. Thus, $P_n(x) = \sum_{k=0}^{n} \dfrac{r^k}{k!}x^k$

17. The Taylor polynomial

$$P_n(0.5) = 1 + (0.5) + \frac{(0.5)^2}{2!} + \cdots + \frac{(0.5)^n}{n!}$$

estimates $e^{0.5}$ within

$$|R_{n+1}(0.5)| \le e^{0.5}\frac{|0.5|^{n+1}}{(n+1)!} < 2\frac{(0.5)^{n+1}}{(n+1)!}.$$

Since

$$2\frac{(0.5)^4}{4!} = \frac{1}{8(24)} < 0.01,$$

we can take $n = 3$ and be sure that

$$P_3(0.5) = 1 + (0.5) + \frac{(0.5)^2}{2} + \frac{(0.5)^3}{6} = \frac{79}{48}$$

differs from $\sqrt{e}$ by less than 0.01. Our calculator gives

$$\tfrac{79}{48} \cong 1.645833 \quad \text{and} \quad \sqrt{e} \cong 1.6487213.$$

19. At $x = 1$, the sine series gives

$$\sin 1 = 1 - \frac{1}{3!} + \frac{1}{5!} - \frac{1}{7!} + \cdots .$$

This is a convergent alternating series with decreasing terms. The first term of magnitude less than 0.01 is $1/5! = 1/120$. Thus

$$1 - \frac{1}{3!} = 1 - \frac{1}{6} = \frac{5}{6}$$

differs from $\sin 1$ by less than 0.01. Our calculator gives

$$\tfrac{5}{6} \cong 0.8333333 \quad \text{and} \quad \sin 1 \cong 0.84114709.$$

The estimate

$$1 - \frac{1}{3!} + \frac{1}{5!} = \frac{101}{120}$$

is much more accurate:

$$\tfrac{101}{120} \cong 0.8416666.$$

21. At $x = 1$, the cosine series gives

$$\cos 1 = 1 - \frac{1}{2!} + \frac{1}{4!} - \frac{1}{6!} + \frac{1}{8!} + \cdots .$$

This is a convergent alternating series with decreasing terms. The first term of magnitude less than 0.01 is $1/6! = 1/720$. Thus

$$1 - \frac{1}{2!} + \frac{1}{4!} = 1 - \frac{1}{2} + \frac{1}{24} = \frac{13}{24}$$

differs from $\cos 1$ by less than 0.01. Our calculator gives

$$\tfrac{13}{24} \cong 0.5416666 \quad \text{and} \quad \cos 1 \cong 0.5403023.$$

23. First convert $10°$ to radians: $10° = \dfrac{10}{180}\pi \cong 0.1745$ radians

At $x = 0.1745$, the sine series gives

$$\sin 0.1745 = 0.1745 - \frac{(0.1745)^3}{3!} + \frac{(0.1745)^5}{5!} - \cdots .$$

This is a convergent alternating series with decreasing terms. The first term of magnitude less than 0.01 is $(0.1745)^3/3! \cong 0.00089$. Thus 0.1745 differs from $\sin 10°$ by less than 0.01. Our calculator gives $\sin 10° \cong 0.1736$

25. $f(x) = e^{2x}; \quad f^{(5)}(x) = 2^5 e^{2x}; \quad R_5(x) = \dfrac{2^5 e^{2c}}{5!} x^5 = \dfrac{4}{15} e^{2c} x^5,$ where c is between 0 and x.

27. $f(x) = \cos 2x; \quad f^{(5)}(x) = -2^5 \sin 2x$

$$R_5(x) = \frac{-2^5 \sin 2c}{5!} x^5 = -\frac{4}{15} \sin(2c) x^5,$$

where c is between 0 and x.

29. $f(x) = \tan x; \quad f'''(x) = 6\sec^4 x - 4\sec^2 x$

$$R_3(x) = \frac{6\sec^4 c - 4\sec^2 c}{3!} x^3 = \frac{3\sec^4 c - 2\sec^2 c}{3} x^3,$$

where c is between 0 and x.

31. $f(x) = \tan^{-1} x; \quad f'''(x) = \dfrac{6x^2 - 2}{(1 + x^2)^3}$

$$R_3(x) = \frac{6c^2 - 2}{3! (1 + c^2)^3} x^3 = \frac{3c^2 - 1}{3 (1 + c^2)^3} x^3,$$

where c is between 0 and x.

33. $f(x) = e^{-x}; \quad f^{(k)}(x) = (-1)^k e^{-x}, \; k = 0, 1, 2, \ldots$

$$R_{n+1}(x) = \frac{(-1)^{n+1} e^{-c}}{(n + 1)!} x^{n+1},$$

where c is between 0 and x.

35. $f(x) = \dfrac{1}{1 - x}; \quad f^{(k)}(x) = \dfrac{k!}{(1 - x)^{k+1}}, \; k = 0, 1, 2, \ldots$

$$R_{n+1}(x) = \frac{(n + 1)!}{(1 - c)^{n+2}(n + 1)!} x^{n+1} = \frac{1}{(1 - c)^{n+2}} x^{n+1},$$

where c is between 0 and x.

37. By (11.5.8)

$$P_n(x) = x - \frac{x^2}{2} + \frac{x^3}{3} - \frac{x^4}{4} + \cdots + (-1)^{n+1} \frac{x^n}{n}.$$

For $0 \leq x \leq 1$ we know from (11.4.5) that

$$|P_n(x) - \ln(1+x)| < \frac{x^{n+1}}{n+1}.$$

(a) $n = 4; \dfrac{(0.5)^{n+1}}{n+1} \leq 0.01 \implies 100 \leq (n+1)2^{n+1} \implies n \geq 4$

(b) $n = 2; \dfrac{(0.3)^{n+1}}{n+1} \leq 0.01 \implies 100 \leq (n+1)\left(\dfrac{10}{3}\right)^{n+1} \implies n \geq 2$

(c) $n = 999; \dfrac{(1)^{n+1}}{n+1} \leq 0.001 \implies 1000 \leq n+1 \implies n \geq 999$

39. $f(x) = e^x; \quad f^{(n)}(x) = e^x; \quad R_{n+1}(x) = \dfrac{e^c}{(n+1)!}\,x^{n+1}, \quad |c| < |x|$

(a) We want $|R_{n+1}(1/2)| < .00005$: for $0 < c < \frac{1}{2}$, we have

$$|R_{n+1}(1/2)| = \frac{e^c}{(n+1)!}\left(\frac{1}{2}\right)^{n+1} < \frac{e^{1/2}}{(n+1)!}\left(\frac{1}{2}\right)^{n+1} < \frac{2}{2^{n+1}(n+1)!} < 0.00005$$

You can verify that this inequality is satisfied if $n \geq 5$.

$$P_5(x) = 1 + x + \frac{x^2}{2!} + \frac{x^3}{3!} + \frac{x^4}{4!} + \frac{x^5}{5!}$$

$$P_5(1/2) = 1 + \frac{1}{2} + \frac{1}{8} + \frac{1}{48} + \frac{1}{320} + \frac{1}{3840} \cong 1.6492$$

(b) We want $|R_{n+1}(-1)| < .0005$: for $-1 < c < 0$, we have

$$|R_{n+1}(-1)| = \frac{e^c}{(n+1)!}\,|(-1)^{n+1}| < \frac{1}{(n+1)!} < 0.0005$$

You can verify that this inequality is satisfied if $n \geq 7$.

$$P_7(x) = \sum_{k=0}^{7} \frac{x^k}{k!}; \quad P_7(-1) = \sum_{k=0}^{7} \frac{(-1)^k}{k!} \cong 0.368$$

41.   The result follows from the fact that $P^{(k)}(0) = \begin{cases} k!a_k, & 0 \leq k \leq n \\ 0, & n < k \end{cases}$.

43.

$$\frac{d^k}{dx^k}(\sinh x) = \begin{cases} \sinh x, & \text{if } k \text{ is odd} \\ \cosh x, & \text{if } k \text{ is even} \end{cases}$$

Thus

$$\frac{d^k}{dx^k}(\sinh x)\Big|_{x=0} = \begin{cases} 0, & \text{if } k \text{ is odd} \\ 1, & \text{if } k \text{ is even} \end{cases}$$

and

$$\sinh x = x + \frac{x^3}{3!} + \frac{x^5}{5!} + \cdots = \sum_{k=0}^{\infty} \frac{x^{(2k+1)}}{(2k+1)!}$$

45. Set $t = ax$.   Then, $e^{ax} = e^t = \sum_{k=0}^{\infty} \dfrac{t^k}{k!} = \sum_{k=0}^{\infty} a^k \dfrac{x^k}{k!}$, $(-\infty, \infty)$.

47. Set $t = ax$.   Then, $\cos ax = \cos t = \sum_{k=0}^{\infty} \dfrac{(-1)^k}{(2k)!} t^{2k} = \sum_{k=0}^{\infty} \dfrac{(-1)^k a^{2k}}{(2k)!} x^{2k}$, $(-\infty, \infty)$.

49. By the hint

$$\ln(a + x) = \ln\left[a\left(1 + \dfrac{x}{a}\right)\right] = \ln a + \ln\left(1 + \dfrac{x}{a}\right) = \ln a + \sum_{k=1}^{\infty} \dfrac{(-1)^{k+1}}{ka^k} x^k.$$

By (11.5.8) the series converges for $-1 < \dfrac{x}{a} \leq 1$;   that is, $-a < x \leq a$.

51. $\ln 2 = \ln\left(\dfrac{1 + 1/3}{1 - 1/3}\right) \cong 2\left[\dfrac{1}{3} + \dfrac{1}{3}\left(\dfrac{1}{3}\right)^3 + \dfrac{1}{5}\left(\dfrac{1}{3}\right)^5\right] = \dfrac{842}{1215}$.

Our calculator gives $\dfrac{842}{1215} \cong 0.6930041$ and $\ln 2 \cong 0.6931471$.

53. Set $u = (x - t)^k$, $dv = f^{(k+1)}(t)\, dt$

$\quad\quad du = -k(x - t)^{k-1}\, dt$, $v = f^{(k)}(t)$.

Then, $-\dfrac{1}{k!} \displaystyle\int_0^x f^{(k+1)}(t)(x - t)^k\, dt$

$$= -\dfrac{1}{k!}\left[(x - t)^k f^{(k)}(t)\right]_0^x - \dfrac{1}{k!}\int_0^x k(x - t)^{k-1} f^{(k)}(t)\, dt$$

$$= \dfrac{f^{(k)}(0)}{k!} x^k - \dfrac{1}{(k-1)!}\int_0^x f^{(k)}(t)(x - t)^{k-1}\, dt.$$

The given identity follows.

55. (a)

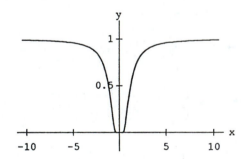

(b) Let $g(x) = \dfrac{x^{-n}}{e^{1/x^2}}$. Then $\lim_{x \to 0} g(x)$ has the form ∞/∞. Successive applications of L'Hospital's rule will finally produce a quotient of the form $\dfrac{cx^k}{e^{1/x^2}}$, where k is a nonnegative integer and c is a constant. It follows that $\lim_{x \to 0} g(x) = 0$.

(c) $f'(0) = \lim_{x \to 0} \dfrac{e^{-1/x^2} - 0}{x} = 0$ by part (b). Assume that $f^{(k)}(0) = 0$. Then

$$f^{(k+1)}(0) = \lim_{x \to 0} \dfrac{f^{(k)}(x) - 0}{x} = \lim_{x \to 0} \dfrac{f^{(k)}(x)}{x}.$$

Now, $f^{(k)}(x)/x$ is a sum of terms of the form $ce^{-1/x^2}/x^n$, where n is a positive integer and c is a constant.

Again by part (b), $f^{(k+1)}(0) = 0$. Therefore, $f^{(n)}(0) = 0$ for all n.

(d) 0   (e) x=0

57.

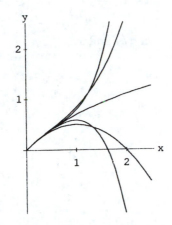

SECTION 11.6

1. $f(x) = \sqrt{x} = x^{1/2};$ $f(4) = 2$

$f'(x) = \dfrac{1}{2}x^{-1/2};$ $f'(4) = \dfrac{1}{4}$

$f''(x) = -\dfrac{1}{4}x^{-3/2};$ $f''(4) = -\dfrac{1}{32}$

$f'''(x) = \dfrac{3}{8}x^{-5/2};$ $f'''(4) = \dfrac{3}{256}$

$f^{(4)}(x) = -\dfrac{15}{16}x^{-7/2}$

$P_3(x) = 2 + \dfrac{1}{4}(x-4) - \dfrac{1/32}{2!}(x-4)^2 + \dfrac{3/256}{3!}(x-4)^3$

$\qquad = 2 + \dfrac{1}{4}(x-4) - \dfrac{1}{64}(x-4)^2 + \dfrac{1}{512}(x-4)^3$

$R_4(x) = \dfrac{f^{(4)}(c)}{4!}(x-4)^4 = -\dfrac{15}{16}\cdot\dfrac{1}{4!}c^{-7/2}(x-4)^4 = -\dfrac{5}{128c^{7/2}}(x-4)^4,$ where c is between 4 and x.

3. $f(x) = \sin x;$ $f(\pi/4) = \dfrac{\sqrt{2}}{2}$

$f'(x) = \cos x;$ $f'(\pi/4) = \dfrac{\sqrt{2}}{2}$

$f''(x) = -\sin x;$ $f''(\pi/4) = -\dfrac{\sqrt{2}}{2}$

$f'''(x) = -\cos x;$ $f'''(\pi/4) = -\dfrac{\sqrt{2}}{2}$

$f^{(4)}(x) = \sin x;$ $f^{(4)}(\pi/4) = \dfrac{\sqrt{2}}{2}$

$$f^{(5)}(x) = \cos x$$

$$P_4(x) = \frac{\sqrt{2}}{2} + \frac{\sqrt{2}}{2}\left(x - \frac{\pi}{4}\right) - \frac{\sqrt{2}/2}{2!}\left(x - \frac{\pi}{4}\right)^2 - \frac{\sqrt{2}/2}{3!}\left(x - \frac{\pi}{4}\right)^3 + \frac{\sqrt{2}/2}{4!}\left(x - \frac{\pi}{4}\right)^4$$

$$= \frac{\sqrt{2}}{2} + \frac{\sqrt{2}}{2}\left(x - \frac{\pi}{4}\right) - \frac{\sqrt{2}}{4}\left(x - \frac{\pi}{4}\right)^2 - \frac{\sqrt{2}}{12}\left(x - \frac{\pi}{4}\right)^3 + \frac{\sqrt{2}}{48}\left(x - \frac{\pi}{4}\right)^4$$

$$R_5(x) = \frac{f^{(5)}(c)}{5!}\left(x - \frac{\pi}{4}\right)^5 = \frac{\cos c}{120}\left(x - \frac{\pi}{4}\right)^5, \quad \text{where } c \text{ is between } \pi/4 \text{ and } x.$$

5. $f(x) = \tan^{-1}(x)$ $\qquad\qquad\qquad\qquad f(1) = \dfrac{\pi}{4}$

$f'(x) = \dfrac{1}{1 + x^2}$ $\qquad\qquad\qquad\qquad f'(1) = \frac{1}{2}$

$f''(x) = \dfrac{-2x}{(1 + x^2)^2}$ $\qquad\qquad\qquad f''(1) = -\frac{1}{2}$

$f'''(x) = \dfrac{6x^2 - 2}{(1 + x^2)^2}$ $\qquad\qquad\qquad f'''(1) = \frac{1}{2}$

$f^{(4)}(x) = \dfrac{24(x - x^3)}{(1 + x^2)^3}$

$$P_3(x) = \frac{\pi}{4} + \frac{1}{2}(x - 1) - \frac{1/2}{2!}(x - 1)^2 + \frac{1/2}{3!}(x - 1)^3 = \frac{\pi}{4} + \frac{1}{2}(x - 1) - \frac{1}{4}(x - 1)^2 + \frac{1}{12}(x - 1)^3$$

$$R_4(x) = \frac{f^{(4)}(c)}{4!}(x - 1)^4 = \frac{24(c - c^3)}{(1 + c^2)^3} \cdot \frac{1}{4!}(x - 1)^4 = \frac{c - c^3}{(1 + c^2)^3}(x - 1)^4, \quad \text{where } c \text{ is between } 1 \text{ and } x.$$

7. $g(x) = 6 + 9(x - 1) + 7(x - 1)^2 + 3(x - 1)^3, \quad (-\infty, \infty)$

9. $g(x) = -3 + 5(x + 1) - 19(x + 1)^2 + 20(x + 1)^3 - 10(x + 1)^4 + 2(x + 1)^5, \quad (-\infty, \infty)$

11. $g(x) = \dfrac{1}{1 + x} = \dfrac{1}{2 + (x - 1)} = \dfrac{1}{2}\left[\dfrac{1}{1 + \left(\dfrac{x - 1}{2}\right)}\right] = \dfrac{1}{2}\displaystyle\sum_{k=0}^{\infty}(-1)^k\left(\dfrac{x - 1}{2}\right)^k$

(geometric series)

$$= \sum_{k=0}^{\infty}(-1)^k\frac{(x - 1)^k}{2^{k+1}} \quad \text{for} \quad \left|\frac{x - 1}{2}\right| < 1 \quad \text{and thus for} \quad -1 < x < 3$$

13. $g(x) = \dfrac{1}{1 - 2x} = \dfrac{1}{5 - 2(x + 2)} = \dfrac{1}{5}\left[\dfrac{1}{1 - \frac{2}{5}(x + 2)}\right] = \dfrac{1}{5}\displaystyle\sum_{k=0}^{\infty}\left[\dfrac{2}{5}(x + 2)\right]^k$

(geometric series)

$$= \sum_{k=0}^{\infty}\frac{2^k}{5^{k+1}}(x + 2)^k \quad \text{for} \quad \left|\frac{2}{5}(x + 2)\right| < 1 \quad \text{and thus for} \quad -\frac{9}{2} < x < \frac{1}{2}$$

15. $g(x) = \sin x = \sin\left[(x - \pi) + \pi\right] = \sin(x - \pi)\cos\pi + \cos(x - \pi)\sin\pi$

$$= -\sin(x - \pi) = -\sum_{k=0}^{\infty}(-1)^k \frac{(x - \pi)^{2k+1}}{(2k + 1)!}$$

(11.5.6)

$$= \sum_{k=0}^{\infty}(-1)^{k+1}\frac{(x - \pi)^{2k+1}}{(2k + 1)!}, \quad (-\infty, \infty)$$

17. $g(x) = \cos x = \cos\left[(x - \pi) + \pi\right] = \cos(x - \pi)\cos\pi - \sin(x - \pi)\sin\pi$

$$= -\cos(x - \pi) = -\sum_{k=0}^{\infty}(-1)^k \frac{(x - \pi)^{2k}}{(2k)!} = \sum_{k=0}^{\infty}(-1)^{k+1}\frac{(x - \pi)^{2k}}{(2k)!}, \quad (-\infty, \infty)$$

(11.5.7)

19. $g(x) = \sin\frac{1}{2}\pi x = \sin\left[\frac{\pi}{2}(x - 1) + \frac{\pi}{2}\right]$

$$= \sin\left[\frac{\pi}{2}(x - 1)\right]\cos\frac{\pi}{2} + \cos\left[\frac{\pi}{2}(x - 1)\right]\sin\frac{\pi}{2}$$

$$= \cos\left[\frac{\pi}{2}(x - 1)\right] = \sum_{k=0}^{\infty}(-1)^k \left(\frac{\pi}{2}\right)^{2k}\frac{(x - 1)^{2k}}{(2k)!}, \quad (-\infty, \infty)$$

(11.5.7)

21. $g(x) = \ln(1 + 2x) = \ln\left[3 + 2(x - 1)\right] = \ln\left[3\left(1 + \frac{2}{3}(x - 1)\right)\right]$

$$= \ln 3 + \ln\left[1 + \frac{2}{3}(x - 1)\right] = \ln 3 + \sum_{k=1}^{\infty}\frac{(-1)^{k+1}}{k}\left[\frac{2}{3}(x - 1)\right]^k.$$

(11.5.8)

$$= \ln 3 + \sum_{k=1}^{\infty}\frac{(-1)^{k+1}}{k}\left(\frac{2}{3}\right)^k(x - 1)^k.$$

This result holds if $-1 < \frac{2}{3}(x - 1) \le 1$,   which is to say, if $-\frac{1}{2} < x \le \frac{5}{2}$.

23.

$$\begin{aligned}
g(x) &= x\ln x \\
g'(x) &= 1 + \ln x \\
g''(x) &= x^{-1} \\
g'''(x) &= -x^{-2} \\
g^{(iv)}(x) &= 2x^{-3} \\
&\;\;\vdots \\
g^{(k)}(x) &= (-1)^k(k - 2)!\,x^{1-k}, \quad k \ge 2.
\end{aligned}$$

Then, $g(2) = 2\ln 2$, $g'(2) = 1 + \ln 2$,   and $g^{(k)}(2) = \dfrac{(-1)^k(k - 2)!}{2^{k-1}}, \quad k \ge 2.$

Thus, $g(x) = 2 \ln 2 + (1 + \ln 2)(x - 2) + \displaystyle\sum_{k=2}^{\infty} \frac{(-1)^k}{k(k-1)2^{k-1}} (x-2)^k.$

25. $g(x) = x \sin x = x \displaystyle\sum_{k=0}^{\infty} (-1)^k \frac{x^{2k+1}}{(2k+1)!} = \displaystyle\sum_{k=0}^{\infty} (-1)^k \frac{x^{2k+2}}{(2k+1)!}$

27.

$$\begin{aligned}
g(x) &= (1 - 2x)^{-3} \\
g'(x) &= -2(-3)(1-2x)^{-4} \\
g''(x) &= (-2)^2(4 \cdot 3)(1-2x)^{-5} \\
g'''(x) &= (-2)^3(-5 \cdot 4 \cdot 3)(1-2x)^{-6} \\
&\vdots \\
g^{(k)}(x) &= (-2)^k \left[(-1)^k \frac{(k+2)!}{2} \right] (1-2x)^{-k-3}, \quad k \geq 0.
\end{aligned}$$

Thus, $g^{(k)}(-2) = (-2)^k \left[(-1)^k \dfrac{(k+2)!}{2} \right] 5^{-k-3} = \dfrac{2^{k-1}}{5^{k+3}}(k+2)!$

and $g(x) = \displaystyle\sum_{k=0}^{\infty} (k+2)(k+1) \frac{2^{k-1}}{5^{k+3}} (x-2)^k.$

29. $g(x) = \cos^2 x = \dfrac{1 + \cos 2x}{2} = \dfrac{1}{2} + \dfrac{1}{2} \cos\left[2(x - \pi) + 2\pi\right]$

$$= \frac{1}{2} + \frac{1}{2} \cos\left[2(x-\pi)\right] = \frac{1}{2} + \frac{1}{2} \sum_{k=0}^{\infty} (-1)^k \frac{[2(x-\pi)]^{2k}}{(2k)!}$$

$$= 1 + \sum_{k=1}^{\infty} \frac{(-1)^k 2^{2k-1}}{(2k)!} (x - \pi)^{2k}$$

$\left(k = 0 \ \text{ term is } \ \frac{1}{2} \right)$

31.

$$\begin{aligned}
g(x) &= x^n \\
g'(x) &= nx^{n-1} \\
g''(x) &= n(n-1)x^{n-2} \\
g'''(x) &= n(n-1)(n-2)x^{n-3} \\
&\vdots \\
g^{(k)}(x) &= n(n-1)\cdots(n-k+1)x^{n-k}, \qquad 0 \leq k \leq n \\
g^{(k)}(x) &= 0, \qquad k > n.
\end{aligned}$$

Thus,

$$g^{(k)}(1) = \left\{ \begin{array}{ll} \dfrac{n!}{(n-k)!}, & 0 \leq k \leq n \\ 0, & k > n \end{array} \right]   \text{and}   g(x) = \sum_{k=0}^{n} \frac{n!}{(n-k)!k!} (x-1)^k.$$

33.   (a) $\dfrac{e^x}{e^a} = e^{x-a} = \displaystyle\sum_{k=0}^{\infty} \frac{(x-a)^k}{k!}, \quad e^x = e^a \displaystyle\sum_{k=0}^{\infty} \frac{(x-a)^k}{k!}$

(b) $e^{a+(x-a)} = e^x = e^a \displaystyle\sum_{k=0}^{\infty} \frac{(x-a)^k}{k!}, \quad e^{x_1+x_2} = e^{x_1} \displaystyle\sum_{k=0}^{\infty} \frac{x_2^k}{k!} = e^{x_1}e^{x_2}$

(c) $\quad e^{-a} \displaystyle\sum_{k=0}^{\infty} (-1)^k \dfrac{(x-a)^k}{k!}$

35.   (a) Let $g(x) = \sin x$ and $a = \pi/6$. Then

$$\left| R_{n+1}(x) \right| = \frac{\left| g^{(n+1)}(c) \right|}{(n+1)!} \left| \left(x - \frac{\pi}{6} \right)^{n+1} \right|$$

$$\leq \frac{\left| \left(x - \dfrac{\pi}{6} \right) \right|^{n+1}}{(n+1)!} \quad (g^{(n+1)}(c) = \pm \sin c \text{ or } \pm \cos c)$$

Now, $35° = \dfrac{35\pi}{180}$ radians. We want to find the smallest positive integer n such that

$\left| R_{n+1}(35\pi/180) \right| < 0.00005.$

$$\left| R_{n+1}(35\pi/180) \right| \leq \frac{\left(\dfrac{35\pi}{180} - \dfrac{\pi}{6} \right)^{n+1}}{(n+1)!} \cong \frac{(0.087266)^{n+1}}{(n+1)!} < 0.00005 \quad \Longrightarrow \quad n \geq 3$$

$g(x) = \sin x;$ $g(\pi/6) = \dfrac{1}{2}$

$g'(x) = \cos x;$ $g'(\pi/6) = \dfrac{\sqrt{3}}{2}$

$g''(x) = -\sin x;$ $g(\pi/6) = -\dfrac{1}{2}$

$g'''(x) = -\cos x;$ $g(\pi/6) = -\dfrac{\sqrt{3}}{2}$

$$P_3(x) = \frac{1}{2} + \frac{\sqrt{3}}{2}\left(x - \frac{\pi}{6}\right) - \frac{1/2}{2!}\left(x - \frac{\pi}{6}\right)^2 - \frac{\sqrt{3}/2}{3!}\left(x - \frac{\pi}{6}\right)^3$$

$$= \frac{1}{2} + \frac{\sqrt{3}}{2}\left(x - \frac{\pi}{6}\right) - \frac{1}{4}\left(x - \frac{\pi}{6}\right)^2 - \frac{\sqrt{3}}{12}\left(x - \frac{\pi}{6}\right)^3$$

(b) $P_3(35\pi/180) \cong 0.5736$

37.   Let $g(x) = \sqrt{x} = x^{1/2}$ and $a = 36$.

(a) $g(x) = x^{1/2}$ $g(36) = 6$

$\quad g'(x) = \dfrac{1}{2}x^{-1/2}$ $g'(36) = \dfrac{1}{12}$

$\quad g''(x) = -\dfrac{1}{4}x^{-3/2}$ $g''(36) = -\dfrac{1}{864}$

$\quad g'''(x) = \dfrac{3}{8}x^{-5/2}$ $g'''(36) = \dfrac{1}{20,736}$

We want to find the smallest positive integer n such that $\left| R_{n+1}(38) \right| < 0.0005:$

$$n = 1: \quad |R_2(38)| = \frac{c^{-3/2}/4}{2!}(38 - 36)^2 = \frac{4}{8c^{3/2}} = \frac{1}{2c^{3/2}}, \quad \text{where } 36 \le c \le 38,$$

and

$$|R_2(38)| \le \frac{1}{2(36)^{3/2}} = \frac{1}{432} \cong 0.0023.$$

$$n = 2: \quad |R_3(38)| = \frac{3c^{-5/2}/8}{3!}(38 - 36)^3 = \frac{8}{16c^{5/2}} = \frac{1}{2c^{5/2}}, \quad \text{where } 36 \le c \le 38,$$

and

$$|R_3(38)| \le \frac{1}{2(36)^{5/2}} = \frac{1}{15,552} \cong 0.000064.$$

Thus, we take $n = 2$:

$$P_2(x) = 6 + \frac{1}{12}(x - 36) - \frac{1}{1728}(x - 36)^2 \quad \text{and} \quad P_2(38) \cong 6.164$$

SECTION 11.7

1. (a) converges

   (b) absolutely converges

   (c) ?

   (d) diverges

3. $(-1, 1)$;   ratio test: $\dfrac{b_{k+1}}{b_k} = \dfrac{k+1}{k}|x| \to |x|$, series converges for $|x| < 1$.

   At the endpoints $x = 1$ and $x = -1$ the series diverges since at those points $b_k \not\to 0$.

5. $(-\infty, \infty)$;   ratio test: $\dfrac{b_{k+1}}{b_k} = \dfrac{|x|}{(2k+1)(2k+2)} \to 0$, series converges all x.

7. Converges only at 0;   divergence test: $(-k)^{2k}x^{2k} \to 0$ only if $x = 0$, and series clearly converges at $x = 0$.

9. $[-2, 2)$;   root test: $(b_k)^{1/k} = \dfrac{|x|}{2k^{1/k}} \to \dfrac{|x|}{2}$, series converges for $|x| < 2$.

   At $x = 2$ series becomes $\sum \dfrac{1}{k}$, the divergent harmonic series.

At $x = -2$ series becomes $\sum (-1)^k \dfrac{1}{k}$, a convergent alternating series.

11. Converges only at 0; divergence test: $\left(\dfrac{k}{100}\right)^k x^k \to 0$ only if $x = 0$, and series clearly converges at $x = 0$.

13. $\left[-\dfrac{1}{2}, \dfrac{1}{2}\right)$;   root test: $(b_k)^{1/k} = \dfrac{2|x|}{\sqrt{k^{1/k}}} \to 2|x|$, series converges for $|x| < \dfrac{1}{2}$.

At $x = \dfrac{1}{2}$ series becomes $\sum \dfrac{1}{\sqrt{k}}$, a divergent p-series.

At $x = -\dfrac{1}{2}$ series becomes $\sum (-1)^k \dfrac{1}{\sqrt{k}}$, a convergent alternating series.

15. $(-1, 1)$;   ratio test: $\dfrac{b_{k+1}}{b_k} = \dfrac{k^2}{(k+1)(k-1)}|x| \to |x|$, series converges for $|x| < 1$.

At the endpoints $x = 1$ and $x = -1$ the series diverges since there $b_k \not\to 0$.

17. $(-10, 10)$;   root test: $(b_k)^{1/k} = \dfrac{k^{1/k}}{10}|x| \to \dfrac{|x|}{10}$, series converges for $|x| < 10$.

At the endpoints $x = 10$ and $x = -10$ the series diverges since there $b_k \not\to 0$.

19. $(-\infty, \infty)$;   root test: $(b_k)^{1/k} = \dfrac{|x|}{k} \to 0$, series converges all x.

21. $(-\infty, \infty)$;   root test: $(b_k)^{1/k} = \dfrac{|x - 2|}{k} \to 0$, series converges all x.

23. $\left(-\dfrac{3}{2}, \dfrac{3}{2}\right)$;   ratio test: $\dfrac{b_{k+1}}{b_k} = \dfrac{\dfrac{2^{k+1}}{3^{k+2}}|x|}{\dfrac{2^k}{3^{k+1}}} = \dfrac{2}{3}|x|$, series converges for $|x| < \dfrac{3}{2}$.

At the endpoints $x = 3/2$ and $x = -3/2$, the series diverges since there $b_k \not\to 0$.

25. Converges only at $x = 1$;   ratio test: $\dfrac{b_{k+1}}{b_k} = \dfrac{k^3}{(k+1)^2}|x - 1| \to \infty$ if $x \neq 1$

The series clearly converges at $x = 1$; otherwise it diverges.

27. $(-4, 0)$;   ratio test: $\dfrac{b_{k+1}}{b_k} = \dfrac{k^2 - 1}{2k^2}|x + 2| \to \dfrac{|x + 2|}{2}$, series converges for $|x + 2| < 2$.

At the endpoints $x = 0$ and $x = -4$, the series diverges since there $b_k \not\to 0$.

29. $(-\infty, \infty)$;   ratio test: $\dfrac{b_{k+1}}{b_k} = \dfrac{(k+1)^2}{k^2(k+2)}|x+3| \to 0$, series converges for all x.

31. $(-1, 1)$;   root test: $(b_k)^{1/k} = \left(1 + \dfrac{1}{k}\right)|x| \to |x|$, series converges for $|x| < 1$.

At the endpoints $x = 1$ and $x = -1$, the series diverges since there $b_k \not\to 0$

$[\text{recall } \left(1 + \dfrac{1}{k}\right)^k \to e]$

33. $(0, 4)$;   ratio test: $\dfrac{b_{k+1}}{b_k} = \dfrac{\ln(k+1)}{\ln k} \dfrac{|x-2|}{2} \to \dfrac{|x-2|}{2}$,   series converges for $|x-2| < 2$.

At the endpoints $x = 0$ and $x = 4$ the series diverges since there $b_k \not\to 0$.

35. $\left(-\dfrac{5}{2}, \dfrac{1}{2}\right)$;   root test: $(b_k)^{1/k} = \dfrac{2}{3}|x+1| \to \dfrac{2}{3}|x+1|$,   series converges for $|x+1| < \dfrac{3}{2}$.

At the endpoints $x = -\dfrac{5}{2}$ and $x = \dfrac{1}{2}$ the series diverges since there $b_k \not\to 0$.

37. $1 - \dfrac{x}{2} + \dfrac{2x^2}{4} - \dfrac{3x^3}{8} + \dfrac{4x^4}{16} - \cdots = 1 + \displaystyle\sum_{k=1}^{\infty} (-1)^k \dfrac{kx^k}{2^k}$

$(-2, 2)$;   ratio test: $\dfrac{b_{k+1}}{b_k} = \dfrac{k+1}{2k}|x| \to \dfrac{|x|}{2}$, series converges for $|x| < 2$.

At the endpoints $x = 2$ and $x = -2$ the series diverges since there $b_k \not\to 0$.

39. $\dfrac{3x^2}{4} + \dfrac{9x^4}{9} + \dfrac{27x^6}{16} + \dfrac{81x^8}{25} + \cdots = \displaystyle\sum_{k=1}^{\infty} \dfrac{3^k}{(k+1)^2} x^{2k}$

$\left[-\dfrac{1}{\sqrt{3}}, \dfrac{1}{\sqrt{3}}\right]$;   ratio test: $\dfrac{b_{k+1}}{b_k} = \dfrac{3(k+1)^2}{(k+2)^2}x^2 \to 3x^2$, series converges for $x^2 < \dfrac{1}{3}$

or $|x| < \dfrac{1}{\sqrt{3}}$.

At $x = \pm\dfrac{1}{\sqrt{3}}$, the series becomes $\displaystyle\sum \dfrac{1}{(k+1)^2} \cong \sum \dfrac{1}{n^2}$, a convergent series p-series.

41. (a)   absolutely converges

(b)   absolutely converges

(c)   ?

43. Examine the convergence of $\sum |a_k x^k|$; for (a) use the root test and for (b) use the ratio rest.

45. $\displaystyle\sum |a_k r^k| = \sum |a_k(-r)^k|$

SECTION 11.8

1. Use the fact that $\displaystyle\frac{d}{dx}\left(\frac{1}{1-x}\right) = \frac{1}{(1-x)^2}$:

$$\frac{1}{(1-x)^2} = \frac{d}{dx}(1 + x + x^2 + x^3 + \cdots + x^n + \cdots) = 1 + 2x + 3x^2 + \cdots + nx^{n-1} + \cdots.$$

3. Use the fact that $\displaystyle\frac{d^{(k-1)}}{dx^{(k-1)}}\left[\frac{1}{1-x}\right] = \frac{(k-1)!}{(1-x)^k}$:

$$\frac{1}{(1-x)^k} = \frac{1}{(k-1)!}\frac{d^{(k-1)}}{dx^{(k-1)}}\left[1 + x + \cdots + x^{k-1} + x^k + x^{k+1} + \cdots + x^{n+k-1} + \cdots\right]$$

$$= \frac{1}{(k-1)!}\frac{d^{(k-1)}}{dx^{(k-1)}}\left[x^{k-1} + x^k + x^{k+1} + \cdots + x^{n+k-1} + \cdots\right]$$

$$= 1 + kx + \frac{(k+1)k}{2}x^2 + \cdots + \frac{(n+k-1)(n+k-2)\cdots(n+1)}{(k-1)!}x^n + \cdots$$

$$= 1 + kx + \frac{(k+1)k}{2!}x^2 + \cdots + \frac{(n+k-1)!}{n!(k-1)!}x^n + \cdots.$$

5. Use the fact that $\displaystyle\frac{d}{dx}[\ln(1-x^2)] = \frac{-2x}{1-x^2}$:

$$\frac{1}{1-x^2} = 1 + x^2 + x^4 + \cdots + x^{2n} + \cdots$$

$$\frac{-2x}{1-x^2} = -2x - 2x^3 - 2x^5 - \cdots - 2x^{2n+1} - \cdots.$$

By integration

$$\ln(1-x^2) = \left(-x^2 - \frac{1}{2}x^4 - \frac{1}{3}x^6 - \cdots - \frac{x^{2n+2}}{n+1} - \cdots\right) + C.$$

At $x = 0$, both $\ln(1-x^2)$ and the series are 0. Thus, $C = 0$ and

$$\ln(1-x^2) = -x^2 - \frac{1}{2}x^4 - \frac{1}{3}x^6 - \cdots - \frac{1}{n+1}x^{2n+2} - \cdots.$$

7. $\displaystyle\sec^2 x = \frac{d}{dx}(\tan x) = \frac{d}{dx}\left(x + \frac{1}{3}x^3 + \frac{2}{15}x^5 + \frac{17}{315}x^7 + \cdots\right) = 1 + x^2 + \frac{2}{3}x^4 + \frac{17}{45}x^6 + \cdots$

9. On its interval of convergence a power series is the Taylor series of its sum. Thus,

$$f(x) = x^2 \sin^2 x = x^2 \left(x - \frac{x^3}{3!} + \frac{x^5}{5!} - \frac{x^7}{7!} + \cdots \right)$$

$$= x^3 - \frac{x^5}{3!} + \frac{x^7}{5!} - \frac{x^9}{7!} + \cdots = \sum_{n=0}^{\infty} f^{(n)}(0) \frac{x^n}{n!}$$

implies $f^{(9)}(0) = -9!/7! = -72.$

11. $\sin x^2 = \sum_{k=0}^{\infty} (-1)^k \frac{(x^2)^{2k+1}}{(2k+1)!} = \sum_{k=0}^{\infty} (-1)^k \frac{x^{4k+2}}{(2k+1)!}$

13. $e^{3x^3} = \sum_{k=0}^{\infty} \frac{(3x^3)^k}{k!} = \sum_{k=0}^{\infty} \frac{3^k}{k!} x^{3k}$

15. $\frac{2x}{1-x^2} = 2x \left(\frac{1}{1-x^2} \right) = 2x \sum_{k=0}^{\infty} (x^2)^k = \sum_{k=0}^{\infty} 2x^{2k+1}$

17. $\frac{1}{1-x} + e^x = \sum_{k=0}^{\infty} x^k + \sum_{k=0}^{\infty} \frac{x^k}{k!} = \sum_{k=0}^{\infty} \frac{(k!+1)}{k!} x^k$

19. $x \ln(1+x^3) = x \sum_{k=1}^{\infty} \frac{(-1)^{k+1}}{k} (x^3)^k = \sum_{k=1}^{\infty} \frac{(-1)^{k+1}}{k} x^{3k+1}$

    (11.5.8)

21. $x^3 e^{-x^3} = x^3 \sum_{k=0}^{\infty} \frac{(-x^3)^k}{k!} = \sum_{k=0}^{\infty} \frac{(-1)^k}{k!} x^{3k+3}$

23.   (a) $\lim_{x \to 0} \frac{1 - \cos x}{x^2} \overset{\star}{=} \lim_{x \to 0} \frac{\sin x}{2x} = \frac{1}{2}$ ($\star$ indicates differentiation of **numerator and denominator**).

     (b) $\lim_{x \to 0} \frac{1 - \cos x}{x^2} = \lim_{x \to 0} \frac{\frac{x^2}{2!} - \frac{x^4}{4!} + \frac{x^6}{6!} - \cdots}{x^2} = \lim_{x \to 0} \left(\frac{1}{2} - \frac{x^2}{4!} + \frac{x^4}{6!} - \cdots \right) = \frac{1}{2}$

25.   (a) $\lim_{x \to 0} \frac{\cos x - 1}{x \sin x} \overset{\star}{=} \lim_{x \to 0} \frac{-\sin x}{\sin x + x \cos x} \overset{\star}{=} \lim_{x \to 0} \frac{-\cos x}{2 \cos x - x \sin x} = -\frac{1}{2}$

     (b)

$$\lim_{x \to 0} \frac{\cos x - 1}{x \sin x} = \frac{-\frac{x^2}{2!} + \frac{x^4}{4!} - \frac{x^6}{6!} + \cdots}{x^2 - \frac{x^4}{3!} + \frac{x^6}{5!} \cdots}$$

$$= \frac{-\frac{1}{2} + \frac{x^2}{4!} - \frac{x^4}{6!} + \cdots}{1 - \frac{x^2}{3!} + \frac{x^4}{5!} \cdots} = -\frac{1}{2}$$

27.
$$\int_0^x \frac{\ln(1+t)}{t}\,dt = \int_0^x \frac{1}{t}\left(\sum_{k=1}^{\infty}\frac{(-1)^{k-1}}{k}t^k\right)dt = \int_0^x \left(\sum_{k=1}^{\infty}\frac{(-1)^{k-1}}{k}t^{k-1}\right)dt$$

$$= \sum_{k=1}^{\infty}\frac{(-1)^{k-1}}{k}\int_0^x t^{k-1}\,dt$$

$$= \sum_{k=1}^{\infty}\frac{(-1)^{k-1}}{k^2}x^k, \quad -1 \le x \le 1$$

29.
$$\int_0^x \frac{\tan^{-1}t}{t}\,dt = \int_0^x \frac{1}{t}\left(\sum_{k=0}^{\infty}\frac{(-1)^k}{2k+1}t^{2k+1}\right)dt = \int_0^x \left(\sum_{k=0}^{\infty}\frac{(-1)^k}{2k+1}t^{2k}\right)dt$$

$$= \sum_{k=0}^{\infty}\frac{(-1)^k}{2k+1}\int_0^x t^{2k}\,dt$$

$$= \sum_{k=0}^{\infty}\frac{(-1)^k}{(2k+1)^2}x^{2k+1}, \quad -1 \le x \le 1$$

31. $0.804 \le I \le 0.808;$ $I = \displaystyle\int_0^1 \left(1 - x^3 + \frac{x^6}{2!} - \frac{x^9}{3!} + \cdots\right)dx$

$$= \left[x - \frac{x^4}{4} + \frac{x^7}{14} - \frac{x^{10}}{60} + \frac{x^{13}}{(13)(24)} - \cdots\right]_0^1$$

$$= 1 - \tfrac{1}{4} + \tfrac{1}{14} - \tfrac{1}{60} + \tfrac{1}{312} - \cdots.$$

Since $\tfrac{1}{312} < 0.01,$ we can stop there:

$$1 - \frac{1}{4} + \frac{1}{14} - \frac{1}{60} \le I \le 1 - \frac{1}{4} + \frac{1}{14} - \frac{1}{60} + \frac{1}{312} \quad \text{gives} \quad 0.804 \le I \le 0.808.$$

33. $0.600 \le I \le 0.603;$ $I = \displaystyle\int_0^1 \left(x^{1/2} - \frac{x^{3/2}}{3!} + \frac{x^{5/2}}{5!} - \cdots\right)dx$

$$= \left[\tfrac{2}{3}x^{3/2} - \tfrac{1}{15}x^{5/2} + \tfrac{1}{420}x^{7/2} - \cdots\right]_0^1$$

$$= \tfrac{2}{3} - \tfrac{1}{15} + \tfrac{1}{420} - \cdots.$$

Since $\tfrac{1}{420} < 0.01,$ we can stop there:

$$\tfrac{2}{3} - \tfrac{1}{15} \le I \le \tfrac{2}{3} - \tfrac{1}{15} + \tfrac{1}{420} \quad \text{gives} \quad 0.600 \le I \le 0.603.$$

35. $0.294 \le I \le 0.304;$ $I = \displaystyle\int_0^1 \left(x^2 - \frac{x^6}{3} + \frac{x^{10}}{5} - \frac{x^{14}}{7} + \cdots\right)dx$

               (11.8.7)

$$= \left[\tfrac{1}{3}x^3 - \tfrac{1}{21}x^7 + \tfrac{1}{55}x^{11} - \tfrac{1}{105}x^{15} + \cdots\right]_0^1$$

$$= \tfrac{1}{3} - \tfrac{1}{21} + \tfrac{1}{55} - \tfrac{1}{105} + \cdots.$$

Since $\frac{1}{105} < 0.01,$ we can stop there:

$$\tfrac{1}{3} - \tfrac{1}{21} + \tfrac{1}{55} - \tfrac{1}{105} \le I \le \tfrac{1}{3} - \tfrac{1}{21} + \tfrac{1}{55} \quad \text{gives} \quad 0.294 \le I \le 0.304.$$

37. $I \cong 0.9461;$

$$I = \int_0^1 \left(1 - \frac{x^2}{3!} + \frac{x^4}{5!} - \cdots\right) dx$$

$$= \left[x - \frac{x^3}{3 \cdot 3!} + \frac{x^5}{5 \cdot 5!} - \cdots\right]_0^1$$

$$= 1 - \frac{1}{3 \cdot 3!} + \frac{1}{5 \cdot 5!} - \frac{1}{7 \cdot 7!} \cdots.$$

Since $\dfrac{1}{7 \cdot 7!} = \dfrac{1}{35,280} \cong 0.000028 < 0.0001,$ we can stop there:

$$1 - \frac{1}{3 \cdot 3!} + \frac{1}{5 \cdot 5!} - \frac{1}{7 \cdot 7!} < I < 1 - \frac{1}{3 \cdot 3!} + \frac{1}{5 \cdot 5!}; \quad I \cong 0.9461$$

39. $I \cong 0.4485;$

$$I = \int_0^{0.5} \left(1 - \frac{x}{2} + \frac{x^2}{3} - \frac{x^3}{4} + \cdots\right) dx$$

$$= \left[x - \frac{x^2}{2^2} + \frac{x^3}{3^2} - \frac{x^4}{4^2} + \cdots\right]_0^{1/2}$$

$$= \frac{1}{2} - \frac{1}{2^2 \cdot 2^2} + \frac{1}{3^2 \cdot 2^3} - \frac{1}{4^2 \cdot 2^4} + \cdots = \sum_{k=1}^{\infty} \frac{(-1)^{k-1}}{k^2 \cdot 2^k}$$

Now, $\dfrac{1}{8^2 \cdot 2^8} = \dfrac{1}{16,384} \cong 0.000061$ is the first term which is less than 0.0001. Thus

$$\sum_{k=1}^{7} \frac{(-1)^{k-1}}{k^2 \cdot 2^k} < I < \sum_{k=1}^{8} \frac{(-1)^{k-1}}{k^2 \cdot 2^k}; \quad I \cong 0.4485$$

41. $e^{x^3};$ by (11.5.5)

43. $3x^2 e^{x^3} = \dfrac{d}{dx}(e^{x^3})$

45. (a) $f(x) = xe^x = x\displaystyle\sum_{k=0}^{\infty} \frac{x^k}{k!} = \sum_{k=0}^{\infty} \frac{x^{k+1}}{k!}$

(b) Using integration by parts: $\displaystyle\int_0^1 xe^x \, dx = [xe^x - e^x]_0^1 = e - e + 1 = 1.$

Using the power series representation:

$$\int_0^1 xe^x \, dx = \int_0^1 \left(\sum_{k=0}^{\infty} \frac{x^{k+1}}{k!}\right) dx = \sum_{k=0}^{\infty} \int_0^1 \left(\frac{x^{k+1}}{k!}\right) dx$$

$$= \sum_{k=0}^{\infty} \frac{1}{k!}\left[\frac{x^{k+2}}{k+2}\right]_0^1$$

$$= \sum_{k=0}^{\infty} \frac{1}{k!(k+2)}$$

$$= \frac{1}{2} + \sum_{k=1}^{\infty} \frac{1}{k!(k+2)}$$

Thus, $1 = \dfrac{1}{2} + \displaystyle\sum_{k=1}^{\infty} \frac{1}{k!(k+2)}$ and $\displaystyle\sum_{k=1}^{\infty} \frac{1}{k!(k+2)} = \frac{1}{2}.$

47.   Let $f(x)$ be the sum of these series; a_k and b_k are both $\dfrac{f^{(k)}(0)}{k!}$.

49.   (a) If f is even, then the odd ordered derivatives $f^{(2k-1)}$, $k = 1, 2, \ldots$ are odd. This implies
that $f^{(2k-1)}(0) = 0$ for all k and so $a_{2k-1} = f^{(2k-1)}(0)/(2k-1)! = 0$ for all k.

(b) If f is odd, then all the even ordered derivatives $f^{(2k)}$, $k = 1, 2, \ldots$ are odd. This implies
that $f^{(2k)}(0) = 0$ for all k and so $a_{2k} = f^{(2k)}(0)/(2k)! = 0$ for all k.

51. $0.0352 \le I \le 0.0359$; $I = \displaystyle\int_0^{1/2} \left(x^2 - \dfrac{x^3}{2} + \dfrac{x^4}{3} - \dfrac{x^5}{4} + \cdots \right) dx$

$$= \left[\dfrac{x^3}{3} - \dfrac{x^4}{8} + \dfrac{x^5}{15} - \dfrac{x^6}{24} + \cdots \right]_0^{1/2}$$

$$= \dfrac{1}{3(2^3)} - \dfrac{1}{8(2^4)} + \dfrac{1}{15(2^5)} - \dfrac{1}{24(2^6)} + \cdots .$$

Since $\dfrac{1}{24(2^6)} = \dfrac{1}{1536} < 0.001$, we can stop there:

$$\dfrac{1}{3(2^3)} - \dfrac{1}{8(2^4)} + \dfrac{1}{15(2^5)} - \dfrac{1}{24(2^6)} \le I \le \dfrac{1}{3(2^3)} - \dfrac{1}{8(2^4)} + \dfrac{1}{15(2^5)}$$

gives $0.0352 \le I \le 0.0359$.   Direct integration gives

$$I = \int_0^{1/2} x \ln(1+x)\, dx = \left[\dfrac{1}{2}(x^2 - 1)\ln(1+x) - \dfrac{1}{4}x^2 + \dfrac{1}{2}x \right]_0^{1/2} = \dfrac{3}{16} - \dfrac{3}{8}\ln 1.5 \cong 0.0354505.$$

53. $0.2640 \le I \le 0.2643$;

$$I = \int_0^1 \left(x - x^2 + \dfrac{x^3}{2!} - \dfrac{x^4}{3!} + \dfrac{x^5}{4!} - \dfrac{x^6}{5!} + \dfrac{x^7}{6!} - \cdots \right) dx$$

$$= \left[\dfrac{x^2}{2} - \dfrac{x^3}{3} + \dfrac{x^4}{4(2!)} - \dfrac{x^5}{5(3!)} + \dfrac{x^6}{6(4!)} - \dfrac{x^7}{7(5!)} + \dfrac{x^8}{8(6!)} - \cdots \right]_0^1$$

$$= \dfrac{1}{2} - \dfrac{1}{3} + \dfrac{1}{4(2!)} - \dfrac{1}{5(3!)} + \dfrac{1}{6(4!)} - \dfrac{1}{7(5!)} + \dfrac{1}{8(6!)} - \cdots .$$

Note that $\dfrac{1}{8(6!)} = \dfrac{1}{5760} < 0.001$.   The integral lies between

$$\dfrac{1}{2} - \dfrac{1}{3} + \dfrac{1}{4(2!)} - \dfrac{1}{5(3!)} + \dfrac{1}{6(4!)} - \dfrac{1}{7(5!)}$$

and

$$\dfrac{1}{2} - \dfrac{1}{3} + \dfrac{1}{4(2!)} - \dfrac{1}{5(3!)} + \dfrac{1}{6(4!)} - \dfrac{1}{7(5!)} + \dfrac{1}{8(6!)}.$$

The first sum is greater than 0.2640 and the second sum is less than 0.2643.

Direct integration gives

$$\int_0^1 x e^{-x}\, dx = \left[-x e^{-x} - e^{-x} \right]_0^1 = 1 - 2/e \cong 0.2642411.$$

PROJECT 11.8

1. $f(x) = \dfrac{4}{1 + x^2}$

(a) $T_8 = \dfrac{1-0}{2 \cdot 8} \left[f(0) + 2 \sum\limits_{I=1}^{7} f\left(\tfrac{i}{8}\right) + f(1) \right] \simeq 3.13899$

(b) $S_4 = \dfrac{1-0}{6 \cdot 4} \left[f(0) + 2 \sum\limits_{I=1}^{3} f\left(\tfrac{i}{4}\right) + 4 \sum\limits_{I=1}^{4} f\left(\tfrac{2i-1}{8}\right) + f(1) \right] \simeq 3.141592$

3.   (a) $|E_n^T| \le \dfrac{(1-0)^3}{12n^2} (8) < 0.000005$

$12n^2 > \dfrac{8}{0.00005} = 1,600,000 \implies n^2 > 133,333.33 \implies n > 365.15$

Thus n must be at least 366.

(b) $|E_n^S| \le \dfrac{(1-0)^5}{2880n^4} (96) < 5 \cdot 10^{-11}$

$n^4 > \dfrac{96(10)^{11}}{5(2880)} \implies n > 160.68.$

Thus n must be at least 161.

5.   (a) $4\tan^{-1}\left(\tfrac{1}{5}\right) - \tan^{-1}\left(\tfrac{1}{239}\right) < 4[\tfrac{1}{5} - \tfrac{1}{3}(\tfrac{1}{5})^3 + \tfrac{1}{5}(\tfrac{1}{5})^5] - [\tfrac{1}{239} - \tfrac{1}{3}(\tfrac{1}{239})^3]$
$4\tan^{-1}\left(\tfrac{1}{5}\right) - \tan^{-1}\left(\tfrac{1}{239}\right) > 4[\tfrac{1}{5} - \tfrac{1}{3}(\tfrac{1}{5})^3] - \tfrac{1}{239}$
These inequalities imply $3.1406 < \pi < 3.1416.$

(b) $4\tan^{-1}\left(\tfrac{1}{5}\right) - \tan^{-1}\left(\tfrac{1}{239}\right) < 4 \sum\limits_{k=1}^{5} \dfrac{(-1)^{k-1}}{2k-1}(\tfrac{1}{5})^{2k-1} - [\tfrac{1}{239} - \tfrac{1}{3}(\tfrac{1}{239})^3]$

$\qquad = 0.789582246 - 0.004184076 = 0.78539817.$

$4\tan^{-1}\tfrac{1}{5} - \tan^{-1}\tfrac{1}{239} > 4 \sum\limits_{k=1}^{5} \dfrac{(-1)^{k-1}}{2k-1}(\tfrac{1}{5})^{2k-1} - \tfrac{1}{239}$

$\qquad = 0.789582238 - 0.0041841 = 0.785398138.$

These inequalities imply $3.14159255 < \pi < 3.14159268.$

SECTION 11.9

1.   Take $\alpha = 1/2$ in (11.9.2) to obtain $1 + \tfrac{1}{2}x - \tfrac{1}{8}x^2 + \tfrac{1}{16}x^3 - \tfrac{5}{128}x^4$.

3.   In (11.9.2), replace x by x^2 and take $\alpha = 1/2$ to obtain $1 + \tfrac{1}{2}x^2 - \tfrac{1}{8}x^4$.

5.   Take $\alpha = -1/2$ in (11.9.2) to obtain $1 - \tfrac{1}{2}x + \tfrac{3}{8}x^2 - \tfrac{5}{16}x^3 + \tfrac{35}{128}x^4$.

7.   In (11.9.2), replace x by $-x$ and take $\alpha = 1/4$ to obtain $1 - \tfrac{1}{4}x - \tfrac{3}{32}x^2 - \tfrac{7}{128}x^3 - \tfrac{77}{2048}x^4$.

9. $f(x) = (4+x)^{3/2} = 8\left(1 + \dfrac{x}{4}\right)^{3/2}$

In 11.9.2, replace x by $x/4$ and take $\alpha = 3/2$ to obtain

$$8 \left[1 + \frac{3}{8}\left(\frac{x}{4}\right) + \frac{1}{2!}\left(\frac{3}{2}\right)\left(\frac{1}{2}\right)\left(\frac{x}{4}\right)^2 + \frac{1}{3!}\left(\frac{3}{2}\right)\left(\frac{1}{2}\right)\left(-\frac{1}{2}\right)\left(\frac{x}{4}\right)^3 \right]$$

$$\left[+ \frac{1}{4!}\left(\frac{3}{2}\right)\left(\frac{1}{2}\right)\left(-\frac{1}{2}\right)\left(-\frac{3}{2}\right)\left(\frac{x}{4}\right)^4 \right]$$

$$= 8 + 3x + \frac{3}{16}x^2 - \frac{1}{128}x^3 + \frac{3}{4096}x^4$$

11. (a) $f(x) = \dfrac{1}{\sqrt{1 - x^2}} = (1 - x^2)^{-1/2}$

In 11.9.2, replace x by x^2 and take $\alpha = -1/2$ to obtain

$$\frac{1}{\sqrt{1 - x^2}} = \sum_{k=0}^{\infty} \binom{-1/2}{k}(-1)^k x^{2k}$$

By Problem 2, this series has radius of convergence $R = 1$.

(b)
$$\sin^{-1} x = \int_0^x \frac{1}{\sqrt{1 - x^2}}\, dt = \int_0^x \sum_{k=0}^{\infty} \binom{-1/2}{k}(-1)^k t^{2k}\, dt$$

$$= \sum_{k=0}^{\infty} \binom{-1/2}{k}(-1)^k \int_0^x t^{2k}\, dt$$

$$= \sum_{k=0}^{\infty} \binom{-1/2}{k}\frac{(-1)^k}{2k + 1} x^{2k+1}$$

By Theorem 11.8.4, the radius of convergence of this series is $R = 1$.

13. 9.8995; $\sqrt{98} = (100 - 2)^{1/2} = 10\left(1 - \frac{1}{50}\right)^{1/2} \cong 10\left[1 - \frac{1}{100} - \frac{1}{20000}\right] = 9.8995$

15. 2.0799; $\sqrt[3]{9} = (8 + 1)^{1/3} = 2\left(1 + \frac{1}{8}\right)^{1/3} \cong 2\left[1 + \frac{1}{24} - \frac{1}{576}\right] \cong 2.0799$

17. 0.4925; $17^{-1/4} = (16 + 1)^{-1/4} = \frac{1}{2}\left(1 + \frac{1}{16}\right)^{-1/4} \cong \frac{1}{2}\left[1 - \frac{1}{64} + \frac{5}{8192}\right] \cong 0.4925$

19.
$$I = \int_0^{1/3} \sqrt{1 + x^3}\, dx = \int_0^{1/3} \sum_{k=0}^{\infty} \binom{1/2}{k} x^{3k}\, dx$$

$$= \sum_{k=0}^{\infty} \binom{1/2}{k} \int_0^{1/3} x^{3k}\, dx$$

$$= \sum_{k=0}^{\infty} \binom{1/2}{k} \left[\frac{x^{3k+1}}{3k + 1}\right]_0^{1/3}$$

$$= \sum_{k=0}^{\infty} \binom{1/2}{k} \frac{1}{3k + 1} \left(\frac{1}{3}\right)^{3k+1}$$

$$= \frac{1}{3} + \left(\frac{1}{2}\right)\left(\frac{1}{4}\right)\left(\frac{1}{3}\right)^4 + \frac{1}{2!}\left(\frac{1}{2}\right)\left(-\frac{1}{2}\right)\left(\frac{1}{7}\right)\left(\frac{1}{3}\right)^7 + \cdots$$

Now, $I - \dfrac{1}{3} = \left(\dfrac{1}{2}\right)\left(\dfrac{1}{4}\right)\left(\dfrac{1}{3}\right)^4 + \dfrac{1}{2!}\left(\dfrac{1}{2}\right)\left(-\dfrac{1}{2}\right)\left(\dfrac{1}{7}\right)\left(\dfrac{1}{3}\right)^7 + \cdots$

is an alternating series and $\dfrac{1}{2!}\left(\dfrac{1}{2}\right)\left(-\dfrac{1}{2}\right)\left(\dfrac{1}{7}\right)\left(\dfrac{1}{3}\right)^7 \cong 8.2 \times 10^{-6} < 0.001$

Therefore, $\displaystyle\int_0^{1/3} \sqrt{1 + x^3}\, dx \cong \dfrac{1}{3} + \left(\dfrac{1}{2}\right)\left(\dfrac{1}{4}\right)\left(\dfrac{1}{3}\right)^4 \cong 0.3349$

21.

$$\int_0^{1/2} \frac{1}{\sqrt{1+x^2}}\,dx = \int_0^{1/2} (1+x^2)^{-1/2}\,dx = \int_0^{1/2} \sum_{k=0}^{\infty} \binom{-1/2}{k} x^{2k}\,dx$$

$$= \sum_{k=0}^{\infty} \binom{-1/2}{k} \int_0^{1/2} x^{2k}\,dx$$

$$= \sum_{k=0}^{\infty} \binom{-1/2}{k} \left[\frac{1}{2k+1} x^{2k+1} \right]_0^{1/2}$$

$$= \sum_{k=0}^{\infty} \binom{-1/2}{k} \frac{1}{2k+1} \left(\frac{1}{2}\right)^{2k+1}$$

Now

$$\sum_{k=0}^{\infty} \binom{-1/2}{k} \frac{1}{2k+1} \left(\frac{1}{2}\right)^{2k+1} = \frac{1}{2} + \left(-\frac{1}{2}\right)\left(\frac{1}{3}\right)\left(\frac{1}{2}\right)^3 + \frac{1}{2!}\left(-\frac{1}{2}\right)\left(-\frac{3}{2}\right)\left(\frac{1}{5}\right)\left(\frac{1}{2}\right)^5$$

$$+ \frac{1}{3!}\left(-\frac{1}{2}\right)\left(-\frac{3}{2}\right)\left(-\frac{5}{2}\right)\left(\frac{1}{7}\right)\left(\frac{1}{2}\right)^7 + \cdots$$

is an alternating series and

$$\frac{1}{3!}\left(-\frac{1}{2}\right)\left(-\frac{3}{2}\right)\left(-\frac{5}{2}\right)\left(\frac{1}{7}\right)\left(\frac{1}{2}\right)^7 \cong 3.5 \times 10^{-4} < 0.001$$

Therefore,

$$\int_0^{1/2} \frac{1}{\sqrt{1+x^2}}\,dx \cong \frac{1}{2} + \left(-\frac{1}{2}\right)\left(\frac{1}{3}\right)\left(\frac{1}{2}\right)^3 + \frac{1}{2!}\left(-\frac{1}{2}\right)\left(-\frac{3}{2}\right)\left(\frac{1}{5}\right)\left(\frac{1}{2}\right)^5 \cong 0.4815$$

CHAPTER 12

SECTION 12.1

1.

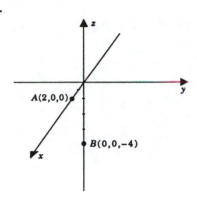

3.

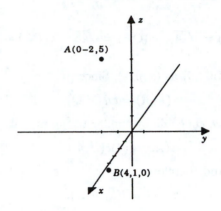

length $\overline{AB}$: $2\sqrt{5}$

midpoint: $(1, 0, -2)$

length $\overline{AB}$: $5\sqrt{2}$

midpoint: $\left(2, -\frac{1}{2}, \frac{5}{2}\right)$

5. $z = -2$ **7.** $y = 1$ **9.** $x = 3$

11. $x^2 + (y-2)^2 + (z+1)^2 = 9$ **13.** $(x-2)^2 + (y-4)^2 + (z+4)^2 = 36$

15. $(x-3)^2 + (y-2)^2 + (z-2)^2 = 13$ **17.** $(x-2)^2 + (y-3)^2 + (z+4)^2 = 25$

19.
$$x^2 + y^2 + z^2 + 4x - 8y - 2z + 5 = 0$$

$$x^2 + 4x + 4 + y^2 - 8y + 16 + z^2 - 2z + 1 = -5 + 4 + 16 + 1$$

$$(x+2)^2 + (y-4)^2 + (z-1)^2 = 16$$

center: $(-2, 4, 1)$,   radius: 4

21. $(2, 3, -5)$ **23.** $(-2, 3, 5)$ **25.** $(-2, 3, -5)$

27. $(-2, -3, -5)$ **29.** $(2, -5, 5)$ **31.** $(-2, 1, -3)$

33. Each such sphere has an equation of the form
$$(x-a)^2 + (y-a)^2 + (z-a)^2 = a^2.$$

Substituting $x = 5$, $y = 1$, $z = 4$ we get
$$(5-a)^2 + (1-a)^2 + (4-a)^2 = a^2.$$

This reduces to $a^2 - 10a + 21 = 0$ and gives $a = 3$ or $a = 7$. The equations are:
$$(x-3)^2 + (y-3)^2 + (z-3)^2 = 9; \quad (x-7)^2 + (y-7)^2 + (z-7)^2 = 49$$

35. Not a sphere; this equation is equivalent to:

$$(x - 2)^2 + (y + 2)^2 + (z + 3)^2 = -3$$

which has no (real) solutions.

37. $d(PR) = \sqrt{14}, \quad d(QR) = \sqrt{45}, \quad d(PQ) = \sqrt{59}; \qquad [d(PR)]^2 + [d(QR)]^2 = [d(PQ)]^2$

39. (a) Take R as (x, y, z). Since

$$d(P, R) = t \, d(P, Q)$$

we conclude by similar triangles that

$$d(AR) = t \, d(B, Q)$$

and therefore

$$z - a_3 = t(b_3 - a_3).$$

Thus

$$z = a_3 + t(b_3 - a_3).$$

In similar fashion

$$x = a_1 + t(b_1 - a_1) \quad \text{and} \quad y = a_2 + t(b_2 - a_2).$$

(b) The midpoint of PQ, $\left(\dfrac{a_1 + b_1}{2}, \dfrac{a_2 + b_2}{2}, \dfrac{a_3 + b_3}{2} \right)$, occurs at $t = \dfrac{1}{2}$.

SECTION 12.3

1. $\overrightarrow{PQ} = (3, 4, -2); \qquad \|\overrightarrow{PQ}\| = \sqrt{29}$ **3.** $\overrightarrow{PQ} = (-2, 1); \qquad \|\overrightarrow{PQ}\| = \sqrt{5}$

5. $2\mathbf{a} - \mathbf{b} = (2 \cdot 1 - 3, 2 \cdot [-2] - 0, 2 \cdot 3 + 1) = (-1, -4, 7)$

7. $-2\mathbf{a} + \mathbf{b} - \mathbf{c} = [-2(\mathbf{a} - \mathbf{b})] - \mathbf{c} = (1 + 4, 4 - 2, -7 - 1) = (5, 2, -8)$

9. $3\mathbf{i} - 4\mathbf{j} + 6\mathbf{k}$ **11.** $-3\mathbf{i} - \mathbf{j} + 8\mathbf{k}$

13. 5 **15.** 3 **17.** $\sqrt{6}$

19. (a) $\mathbf{a}, \mathbf{c}, \text{ and } \mathbf{d}$ since $\mathbf{a} = \frac{1}{3}\mathbf{c} = -\frac{1}{2}\mathbf{d}$

(b) $\mathbf{a}$ and $\mathbf{c}$ since $\mathbf{a} = \frac{1}{3}\mathbf{c}$

(c) $\mathbf{a}$ and $\mathbf{c}$ both have direction opposite to $\mathbf{d}$

21. $\|\mathbf{a}\| = 5; \qquad \dfrac{\mathbf{a}}{\|\mathbf{a}\|} = \left(\dfrac{3}{5}, -\dfrac{4}{5} \right)$ **23.** $\|\mathbf{a}\| = 3; \qquad \dfrac{\mathbf{a}}{\|\mathbf{a}\|} = \dfrac{1}{3}\mathbf{i} - \dfrac{2}{3}\mathbf{j} + \dfrac{2}{3}\mathbf{k}$

25. $\|\mathbf{a}\| = \sqrt{14}; \quad -\dfrac{\mathbf{a}}{\|\mathbf{a}\|} = \dfrac{1}{\sqrt{14}}\mathbf{i} - \dfrac{3}{\sqrt{14}}\mathbf{j} - \dfrac{2}{\sqrt{14}}\mathbf{k}$

27. (i) $\mathbf{a} + \mathbf{b}$ (ii) $-(\mathbf{a} + \mathbf{b})$ (iii) $\mathbf{a} - \mathbf{b}$ (iv) $\mathbf{b} - \mathbf{a}$

29. (a) $\mathbf{a} - 3\mathbf{b} + 2\mathbf{c} + 4\mathbf{d} = (2\mathbf{i} - \mathbf{k}) - 3(\mathbf{i} + 3\mathbf{j} + 5\mathbf{k}) + 2(-\mathbf{i} + \mathbf{j} + \mathbf{k}) + 4(\mathbf{i} + \mathbf{j} + 6\mathbf{k})$

$$= \mathbf{i} - 3\mathbf{j} + 10\mathbf{k}$$

(b) The vector equation

$$(1, 1, 6) = A(2, 0, -1) + B(1, 3, 5) + C(-1, 1, 1)$$

implies

$$\begin{aligned} 1 &= 2A + B - C, \\ 1 &= 3B + C, \\ 6 &= -A + 5B + C. \end{aligned}$$

Simultaneous solution gives $A = -2, \quad B = \frac{3}{2}, \quad C = -\frac{7}{2}.$

31. $\|3\mathbf{i} + \mathbf{j}\| = \|\alpha\mathbf{j} - \mathbf{k}\| \implies 10 = \alpha^2 + 1$ so $\alpha = \pm 3$

33. $\|\alpha\mathbf{i} + (\alpha - 1)\mathbf{j} + (\alpha + 1)\mathbf{k}\| = 2 \implies \alpha^2 + (\alpha - 1)^2 + (\alpha + 1)^2 = 4$

$$\implies 3\alpha^2 = 2 \text{ so } \alpha = \pm\tfrac{1}{3}\sqrt{6}$$

35. $\pm\frac{2}{13}\sqrt{13}\,(3\mathbf{j} + 2\mathbf{k})$ since $\|\alpha(3\mathbf{j} + 2\mathbf{k})\| = 2 \implies \alpha = \pm\frac{2}{13}\sqrt{13}$

37. $\mathbf{v} = (2\cos 30°)\,\mathbf{i} + (2\sin 30°)\,\mathbf{j} = \sqrt{3}\,\mathbf{i} + \mathbf{j}$

39. Since the $\mathbf{i}$ component is twice the $\mathbf{j}$ component, $\mathbf{v} = 2y\,\mathbf{i} + y\,\mathbf{j}$. Now, $\|\mathbf{v}\| = \sqrt{4y^2 + y^2} = 3$ which implies that $y = \dfrac{3}{\sqrt{5}}$. Thus, $\mathbf{v} = \dfrac{6}{\sqrt{5}}\mathbf{i} + \dfrac{3}{\sqrt{5}}\mathbf{j}$ or $\mathbf{v} = -\dfrac{6}{\sqrt{5}}\mathbf{i} - \dfrac{3}{\sqrt{5}}\mathbf{j}$.

41. If $\mathbf{a}$ and $\mathbf{b}$ are the sides of a triangle, then $\mathbf{b} - \mathbf{a}$ is the third side. Now $\|\mathbf{a}\| = \sqrt{2^2 + (-1)^2} = \sqrt{5}$, $\|\mathbf{b}\| = \sqrt{1^2 + 2^2} = \sqrt{5}$, and $\|\mathbf{b} - \mathbf{a}\| = \sqrt{(1-2)^2 + (2+1)^2} = \sqrt{10}$. The triangle is a right triangle since $\|\mathbf{a}\|^2 + \|\mathbf{b}\|^2 = \|\mathbf{b} - \mathbf{a}\|^2$.

43. (a) Since $\|\mathbf{a} - \mathbf{b}\|$ and $\|\mathbf{a} + \mathbf{b}\|$ are the lengths of the diagonals of the parallelogram, the parallelogram must be a rectangle.

(b) Simplify

$$\sqrt{(a_1 - b_1)^2 + (a_2 - b_2)^2 + (a_3 - b_3)^2} = \sqrt{(a_1 + b_1)^2 + (a_2 + b_2)^2 + (a_3 + b_3)^2}.$$

45.   (a)

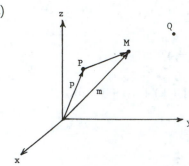

(b) Let $P = (x_1, y_1, z_1)$, $Q = (x_2, y_2, z_2)$, and

$M = (x_m, y_m, z_m)$. Then

$$(x_m, y_m, z_m) = (x_1, y_1, z_1) + \tfrac{1}{2}(x_2 - x_1, y_2 - y_1, z_2 - z_1)$$

$$= \left(\frac{x_1 + x_2}{2}, \frac{y_1 + y_2}{2}, \frac{z_1 + z_2}{2} \right)$$

47. $\|\mathbf{F}_1\| \sin 40° + \|\mathbf{F}_2\| \sin 25° = 200$ and $\|\mathbf{F}_1\| \cos 40° = \|\mathbf{F}_2\| \cos 25°$

$\implies$ $\|\mathbf{F}_1\| = 200.02$ and $\|\mathbf{F}_2\| = 169.05$

$\mathbf{F}_1 = -\|\mathbf{F}_1\| \cos 40° \, \mathbf{i} + \|\mathbf{F}_1\| \sin 40° \, \mathbf{j} = -153.21 \, \mathbf{i} + 128.56 \, \mathbf{j}$

$\mathbf{F}_2 = \|\mathbf{F}_2\| \cos 25° \, \mathbf{i} + \|\mathbf{F}_2\| \sin 25° \, \mathbf{j} = 153.21 \, \mathbf{i} + 71.44 \, \mathbf{j}$

49. $\mathbf{V}_1 = 600 \sin 30° \, \mathbf{i} + 600 \cos 30° \, \mathbf{j} = 300 \, \mathbf{i} + 300\sqrt{3} \, \mathbf{j}$ and

$\mathbf{V}_2 = 50 \sin 45° \, \mathbf{i} - 50 \cos 45° \, \mathbf{j} = 25\sqrt{2} \, \mathbf{i} - 25\sqrt{2} \, \mathbf{j}$

$\mathbf{V} = \mathbf{V}_1 + \mathbf{V}_2 = (300 + 25\sqrt{2}) \, \mathbf{i} + (300\sqrt{3} - 25\sqrt{2}) \, \mathbf{j} \cong 335.36 \, \mathbf{i} + 484.26 \, \mathbf{j}$

true course: $\theta = \tan^{-1} \dfrac{335.36}{484.26} = 34.70°$;   or $N \, 34.70° \, E.$

ground speed: $\|\mathbf{V}\| = \sqrt{(335.56)^2 + (484.26)^2} \cong 589.05$ mi/hr

51.   (a) $\|\mathbf{r} - \mathbf{a}\| = 3$ where $\mathbf{a} = a_1 \mathbf{i} + a_2 \mathbf{j} + a_3 \mathbf{k}$

(b) $\|\mathbf{r}\| \le 2$ (c) $\|\mathbf{r} - \mathbf{a}\| \le 1$ where $\mathbf{a} = a_1 \mathbf{i} + a_2 \mathbf{j} + a_3 \mathbf{k}$

(d) $\|\mathbf{r} - \mathbf{a}\| = \|\mathbf{r} - \mathbf{b}\|$ (e) $\|\mathbf{r} - \mathbf{a}\| + \|\mathbf{r} - \mathbf{b}\| = k$

SECTION 12.4

1. $\mathbf{a} \cdot \mathbf{b} = (2)(-2) + (-3)(0) + (1)(3) = -1$ **3.** $\mathbf{a} \cdot \mathbf{b} = (2)(1) + (-4)(1/2) = 0$

5. $\mathbf{a} \cdot \mathbf{b} = (2)(1) + (1)(1) - (2)(2) = -1$ **7.** $\mathbf{a} \cdot \mathbf{b}$

9. $(\mathbf{a} - \mathbf{b}) \cdot \mathbf{c} + \mathbf{b} \cdot (\mathbf{c} + \mathbf{a}) = \mathbf{a} \cdot \mathbf{c} - \mathbf{b} \cdot \mathbf{c} + \mathbf{b} \cdot \mathbf{c} + \mathbf{b} \cdot \mathbf{a} = \mathbf{a} \cdot (\mathbf{b} + \mathbf{c})$

11.   (a) $\mathbf{a} \cdot \mathbf{b} = (2)(3) + (1)(-1) + (0)(2) = 5$

$\mathbf{a} \cdot \mathbf{c} = (2)(4) + (1)(0) + (0)(3) = 8$

$\mathbf{b} \cdot \mathbf{c} = (3)(4) + (-1)(0) + (2)(3) = 18$

(b) $\|\mathbf{a}\| = \sqrt{5}$, $\|\mathbf{b}\| = \sqrt{14}$, $\|\mathbf{c}\| = 5.$ Then,

$$\cos \sphericalangle(\mathbf{a}, \mathbf{b}) = \frac{\mathbf{a} \cdot \mathbf{b}}{\|\mathbf{a}\| \, \|\mathbf{b}\|} = \frac{5}{(\sqrt{5})(\sqrt{14})} = \frac{1}{14}\sqrt{70},$$

$$\cos \sphericalangle(\mathbf{a}, \mathbf{c}) = \frac{8}{(\sqrt{5})(5)} = \frac{8}{25}\sqrt{5},$$

$$\cos \sphericalangle(\mathbf{b}, \mathbf{c}) = \frac{18}{(\sqrt{14})(5)} = \frac{9}{35}\sqrt{14}.$$

(c) $\mathbf{u_b} = \dfrac{1}{\sqrt{14}}(3\mathbf{i} - \mathbf{j} + 2\mathbf{k}), \quad \text{comp}_\mathbf{b}\,\mathbf{a} = \mathbf{a} \cdot \mathbf{u_b} = \dfrac{1}{\sqrt{14}}(6 - 1) = \dfrac{5}{14}\sqrt{14},$

$\mathbf{u_c} = \frac{1}{5}(4\mathbf{i} + 3\mathbf{k}), \quad \text{comp}_\mathbf{c}\,\mathbf{a} = \mathbf{a} \cdot \mathbf{u_c} = \frac{8}{5}$

(d) $\textbf{proj}_\mathbf{b}\,\mathbf{a} = (\text{comp}_\mathbf{b}\,\mathbf{a})\,\mathbf{u_b} = \frac{5}{14}(3\mathbf{i} - \mathbf{j} + 2\mathbf{k})$

$\textbf{proj}_\mathbf{c}\,\mathbf{a} = (\text{comp}_\mathbf{c}\,\mathbf{a})\,\mathbf{u_c} = \frac{8}{25}(4\mathbf{i} + 3\mathbf{k})$

13. $\mathbf{u} = \cos\dfrac{\pi}{3}\mathbf{i} + \cos\dfrac{\pi}{4}\mathbf{j} + \cos\dfrac{2\pi}{3}\mathbf{k} = \dfrac{1}{2}\mathbf{i} + \dfrac{1}{2}\sqrt{2}\,\mathbf{j} - \dfrac{1}{2}\mathbf{k}$

15. $\cos\theta = \dfrac{(3\mathbf{i} - \mathbf{j} - 2\mathbf{k}) \cdot (\mathbf{i} + 2\mathbf{j} - 3\mathbf{k})}{\|3\mathbf{i} - \mathbf{j} - 2\mathbf{k}\| \, \|\mathbf{i} + 2\mathbf{j} - 3\mathbf{k}\|} = \dfrac{7}{\sqrt{14}\,\sqrt{14}} = \dfrac{1}{2}, \quad \theta = \dfrac{\pi}{3}$

17. Since $\|\mathbf{i} - \mathbf{j} + \sqrt{2}\,\mathbf{k}\| = 2,$ we have $\cos\alpha = \frac{1}{2}, \;\; \cos\beta = -\frac{1}{2}, \;\; \cos\gamma = \frac{1}{2}\sqrt{2}.$
The direction angles are $\frac{1}{3}\pi, \quad \frac{2}{3}\pi, \quad \frac{1}{4}\pi.$

19. $\theta = \cos^{-1}\dfrac{\mathbf{a} \cdot \mathbf{b}}{\|\mathbf{a}\|\|\mathbf{b}\|} = \cos^{-1}\left(\dfrac{-1}{\sqrt{231}}\right) \cong 2.2 \text{ radians} \quad\text{or}\quad 126.3°$

21. $\theta = \cos^{-1}\dfrac{\mathbf{a} \cdot \mathbf{b}}{\|\mathbf{a}\|\|\mathbf{b}\|} = \cos^{-1}\left(\dfrac{-13}{5\sqrt{10}}\right) \cong 2.5 \text{ radians} \quad\text{or}\quad 145.3°$

23. $\|\mathbf{a}\| = \sqrt{1^2 + 2^2 + 2^2} = 3; \qquad\qquad \cos\alpha = \dfrac{1}{3}, \;\; \cos\beta = \dfrac{2}{3}, \;\; \cos\gamma = \dfrac{2}{3}$

$\qquad\qquad\qquad\qquad\qquad\qquad\qquad\qquad\quad \alpha \cong 70.5° \quad \beta \cong 48.2°, \quad \gamma \cong 48.2°$

25. $\|\mathbf{a}\| = \sqrt{3^2 + (12)^2 + 4^2} = 13; \qquad \cos\alpha = \dfrac{3}{13}, \;\; \cos\beta = \dfrac{12}{13} \;\; \cos\gamma = \dfrac{4}{13}$

$\qquad\qquad\qquad\qquad\qquad\qquad\qquad\qquad\quad \alpha \cong 76.7° \quad \beta \cong 22.6°, \quad \gamma \cong 72.1°$

27. (a) $\textbf{proj}_\mathbf{b}\,\alpha\mathbf{a} = (\alpha\mathbf{a} \cdot \mathbf{u_b})\mathbf{u_b} = \alpha(\mathbf{a} \cdot \mathbf{u_b})\mathbf{u_b} = \alpha\,\textbf{proj}_\mathbf{b}\,\mathbf{a}$

(b) $\qquad\quad \textbf{proj}_\mathbf{b}\,(\mathbf{a} + \mathbf{c}) = [(\mathbf{a} + \mathbf{c}) \cdot \mathbf{u_b}]\,\mathbf{u_b}$

$\qquad\qquad\qquad\qquad\quad = (\mathbf{a} \cdot \mathbf{u_b} + \mathbf{c} \cdot \mathbf{u_b})\mathbf{u_b}$

$\qquad\qquad\qquad\qquad\quad = (\mathbf{a} \cdot \mathbf{u_b})\mathbf{u_b} + (\mathbf{c} \cdot \mathbf{u_b})\mathbf{u_b} = \textbf{proj}_\mathbf{b}\,\mathbf{a} + \textbf{proj}_\mathbf{b}\,\mathbf{c}$

29.   (a)   For $\mathbf{a} \neq 0$ the following statements are equivalent:

$$\mathbf{a} \cdot \mathbf{b} = \mathbf{a} \cdot \mathbf{c}, \quad \mathbf{b} \cdot \mathbf{a} = \mathbf{c} \cdot \mathbf{a},$$

$$\mathbf{b} \cdot \frac{\mathbf{a}}{\|\mathbf{a}\|} = \mathbf{c} \cdot \frac{\mathbf{a}}{\|\mathbf{a}\|}, \quad \mathbf{b} \cdot \mathbf{u_a} = \mathbf{c} \cdot \mathbf{u_a}$$

$$(\mathbf{b} \cdot \mathbf{u_a})\mathbf{u_a} = (\mathbf{c} \cdot \mathbf{u_a})\mathbf{u_a},$$

$$\mathrm{proj_a}\,\mathbf{b} = \mathrm{proj_a}\,\mathbf{c}.$$

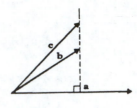

$$\mathbf{a} \cdot \mathbf{b} = \mathbf{a} \cdot \mathbf{c} \quad \text{but} \quad \mathbf{b} \neq \mathbf{c}$$

(b) $\mathbf{b} = (\mathbf{b} \cdot \mathbf{i})\mathbf{i} + (\mathbf{b} \cdot \mathbf{j})\mathbf{j} + (\mathbf{b} \cdot \mathbf{k})\mathbf{k} = (\mathbf{c} \cdot \mathbf{i})\mathbf{i} + (\mathbf{c} \cdot \mathbf{j})\mathbf{j} + (\mathbf{c} \cdot \mathbf{k})\mathbf{k} = \mathbf{c}$

(12.4.13)   (12.4.13)

31.   (a) $\|\mathbf{a}+\mathbf{b}\|^2 - \|\mathbf{a}-\mathbf{b}\|^2 = (\mathbf{a}+\mathbf{b}) \cdot (\mathbf{a}+\mathbf{b}) - (\mathbf{a}-\mathbf{b}) \cdot (\mathbf{a}-\mathbf{b})$

$$= [(\mathbf{a} \cdot \mathbf{a}) + 2(\mathbf{a} \cdot \mathbf{b}) + (\mathbf{b} \cdot \mathbf{b})] - [(\mathbf{a} \cdot \mathbf{a}) - 2(\mathbf{a} \cdot \mathbf{b}) + (\mathbf{b} \cdot \mathbf{b})] = 4(\mathbf{a} \cdot \mathbf{b})$$

(b)   The following statements are equivalent:

$$\mathbf{a} \perp \mathbf{b}, \quad \mathbf{a} \cdot \mathbf{b} = 0, \quad \|\mathbf{a}+\mathbf{b}\|^2 - \|\mathbf{a}-\mathbf{b}\|^2 = 0, \quad \|\mathbf{a}+\mathbf{b}\| = \|\mathbf{a}-\mathbf{b}\|.$$

(c)   By (b), the relation $\|\mathbf{a}+\mathbf{b}\| = \|\mathbf{a}-\mathbf{b}\|$ gives $\mathbf{a} \perp \mathbf{b}$. The relation $\mathbf{a}+\mathbf{b} \perp \mathbf{a}-\mathbf{b}$ gives

$$0 = (\mathbf{a}+\mathbf{b}) \cdot (\mathbf{a}-\mathbf{b}) = \|\mathbf{a}\|^2 - \|\mathbf{b}\|^2 \quad \text{and thus} \quad \|\mathbf{a}\| = \|\mathbf{b}\|.$$

The parallelogram is a square since it has two adjacent sides of equal length and these meet at right angles.

33. $\|\mathbf{a}+\mathbf{b}\|^2 = (\mathbf{a}+\mathbf{b}) \cdot (\mathbf{a}+\mathbf{b}) = \mathbf{a} \cdot \mathbf{a} + 2\mathbf{a} \cdot \mathbf{b} + \mathbf{b} \cdot \mathbf{b} = \|\mathbf{a}\|^2 + 2\mathbf{a} \cdot \mathbf{b} + \|\mathbf{b}\|^2$

$\|\mathbf{a}-\mathbf{b}\|^2 = (\mathbf{a}-\mathbf{b}) \cdot (\mathbf{a}-\mathbf{b}) = \mathbf{a} \cdot \mathbf{a} - 2\mathbf{a} \cdot \mathbf{b} + \mathbf{b} \cdot \mathbf{b} = \|\mathbf{a}\|^2 - 2\mathbf{a} \cdot \mathbf{b} + \|\mathbf{b}\|^2$

Add the two equations and the result follows.

35.   Let $\theta_1, \theta_2, \theta_3$ be the direction angles of $-\mathbf{a}$. Then

$$\theta_1 = \cos^{-1}\left[\frac{(-\mathbf{a} \cdot \mathbf{i})}{\|-\mathbf{a}\|}\right] = \cos^{-1}\left[-\frac{(\mathbf{a} \cdot \mathbf{i})}{\|\mathbf{a}\|}\right] = \cos^{-1}(-\cos\alpha) = \cos^{-1}(\pi - \alpha) = \pi - \alpha.$$

Similarly $\theta_2 = \pi - \beta$ and $\theta_3 = \pi - \gamma$.

37.   If $\mathbf{a} \perp \mathbf{b}$ and $\mathbf{a} \perp \mathbf{c}$,   then

$$\mathbf{a} \cdot \mathbf{b} = 0, \quad \mathbf{a} \cdot \mathbf{c} = 0$$

$$\mathbf{a} \cdot (\alpha\mathbf{b} + \beta\mathbf{c}) = \alpha(\mathbf{a} \cdot \mathbf{b}) + \beta(\mathbf{a} \cdot \mathbf{c}) = 0$$

$$\mathbf{a} \perp (\alpha\mathbf{b} + \beta\mathbf{c}).$$

39.   Existence of decomposition:

$$\mathbf{a} = (\mathbf{a} \cdot \mathbf{u_b})\mathbf{u_b} + [\mathbf{a} - (\mathbf{a} \cdot \mathbf{u_b})\mathbf{u_b}].$$

Uniqueness of decomposition: suppose that

$$\mathbf{a} = \mathbf{a}_\| + \mathbf{a}_\perp = \mathbf{A}_\| + \mathbf{A}_\perp.$$

Then the vector $\mathbf{a}_\parallel - \mathbf{A}_\parallel = \mathbf{A}_\perp - \mathbf{a}_\perp$ is both parallel to $\mathbf{b}$ and perpendicular to $\mathbf{b}$. (Exercises 37 and 38.) Therefore it is zero. Consequently $\mathbf{A}_\parallel = \mathbf{a}_\parallel$ and $\mathbf{A}_\perp = \mathbf{a}_\perp$.

41. $\cos \dfrac{\pi}{3} = \dfrac{\mathbf{c} \cdot \mathbf{d}}{\|\mathbf{c}\| \, \|\mathbf{d}\|}, \quad \dfrac{1}{2} = \dfrac{2x+1}{x^2+2}, \quad x^2 = 4x; \quad x = 0, 4$

43.

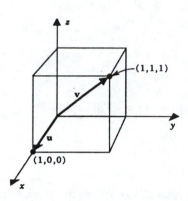

We take $\mathbf{u} = \mathbf{i}$ as an edge and $\mathbf{v} = \mathbf{i} + \mathbf{j} + \mathbf{k}$ as a diagonal of a cube. Then,

$$\cos \theta = \frac{\mathbf{u} \cdot \mathbf{v}}{\|\mathbf{u}\| \, \|\mathbf{v}\|} = \frac{1}{3}\sqrt{3},$$

$$\theta = \cos^{-1}\left(\tfrac{1}{3}\sqrt{3}\right) \cong 0.96 \text{ radians.}$$

45. (a) The direction angles of a vector always satisfy

$$\cos^2 \alpha + \cos^2 \beta + \cos^2 \gamma = 1$$

and, as you can check,

$$\cos^2 \tfrac{1}{4}\pi + \cos^2 \tfrac{1}{6}\pi + \cos^2 \tfrac{2}{3}\pi \neq 1.$$

(b) The relation

$$\cos^2 \alpha + \cos^2 \tfrac{1}{4}\pi + \cos^2 \tfrac{1}{4}\pi = 1$$

gives

$$\cos^2 \alpha + \tfrac{1}{2} + \tfrac{1}{2} = 1, \quad \cos \alpha = 0, \quad a_1 = \|\mathbf{a}\| \cos \alpha = 0.$$

47. Set $\mathbf{u} = a\mathbf{i} + b\mathbf{j} + c\mathbf{k}$. The relations

$$(a\mathbf{i} + b\mathbf{j} + c\mathbf{k}) \cdot (\mathbf{i} + 2\mathbf{j} + \mathbf{k}) = 0 \quad \text{and} \quad (a\mathbf{i} + b\mathbf{j} + c\mathbf{k}) \cdot (3\mathbf{i} - 4\mathbf{j} + 2\mathbf{k}) = 0$$

give

$$a + 2b + c = 0 \qquad 3a - 4b + 2c = 0$$

so that $b = \tfrac{1}{8}a$ and $c = -\tfrac{5}{4}a$.

Then, since $\mathbf{u}$ is a unit vector,

$$a^2 + b^2 + c^2 = 1, \quad a^2 + \left(\frac{a}{8}\right)^2 + \left(\frac{-5a}{4}\right)^2 = 1, \quad \frac{165}{64}a^2 = 1.$$

Thus, $\quad a = \pm\dfrac{8}{165}\sqrt{165} \quad$ and $\quad \mathbf{u} = \pm\dfrac{\sqrt{165}}{165}\,(8\mathbf{i} + \mathbf{j} - 10\mathbf{k}).$

49. Place center of sphere at the origin.

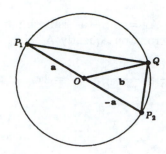

$$\overrightarrow{P_1Q} \cdot \overrightarrow{P_2Q} = (-\mathbf{a} + \mathbf{b}) \cdot (\mathbf{a} + \mathbf{b})$$

$$= -\|\mathbf{a}\|^2 + \|\mathbf{b}\|^2$$

$$= 0.$$

PROJECT 12.4

1. (a) $W = \mathbf{F} \cdot \mathbf{r}$ (b) 0 (c) $\|\mathbf{F}\| \mathbf{i} \cdot (b - a)\mathbf{i} = \|\mathbf{F}\|(b - a)$

3. $F \cos 40° = 50 \implies F \cong 65.3$ pounds.

5. Since the object returns to its starting point, the total displacement is zero, so the work done is zero.

SECTION 12.5

1. $(\mathbf{i} + \mathbf{j}) \times (\mathbf{i} - \mathbf{j}) = [\mathbf{i} \times (\mathbf{i} - \mathbf{j})] + [\mathbf{j} \times (\mathbf{i} - \mathbf{j})] = (0 - \mathbf{k}) + (-\mathbf{k} - 0) = -2\mathbf{k}$

3. $(\mathbf{i} - \mathbf{j}) \times (\mathbf{j} - \mathbf{k}) = [\mathbf{i} \times (\mathbf{j} - \mathbf{k})] - [\mathbf{j} \times (\mathbf{j} - \mathbf{k})] = (\mathbf{j} + \mathbf{k}) - (0 - \mathbf{i}) = \mathbf{i} + \mathbf{j} + \mathbf{k}$

5. $(2\mathbf{j} - \mathbf{k}) \times (\mathbf{i} - 3\mathbf{j}) = [2\mathbf{j} \times (\mathbf{i} - 3\mathbf{j})] - [\mathbf{k} \times (\mathbf{i} - 3\mathbf{j})] = (-2\mathbf{k}) - (\mathbf{j} + 3\mathbf{i}) = -3\mathbf{i} - \mathbf{j} - 2\mathbf{k}$

or

$$(2\mathbf{j} - \mathbf{k}) \times (\mathbf{i} - 3\mathbf{j}) = \begin{vmatrix} \mathbf{i} & \mathbf{j} & \mathbf{k} \\ 0 & 2 & -1 \\ 1 & -3 & 0 \end{vmatrix} = \mathbf{i} \begin{vmatrix} 2 & -1 \\ -3 & 0 \end{vmatrix} - \mathbf{j} \begin{vmatrix} 0 & -1 \\ 1 & -3 \end{vmatrix} + \mathbf{k} \begin{vmatrix} 0 & 2 \\ 1 & -3 \end{vmatrix} = -3\mathbf{i} - \mathbf{j} - 2\mathbf{k}$$

7. $\mathbf{j} \cdot (\mathbf{i} \times \mathbf{k}) = \mathbf{j} \cdot (-\mathbf{j}) = -1$ **9.** $(\mathbf{i} \times \mathbf{j}) \times \mathbf{k} = \mathbf{k} \times \mathbf{k} = 0$ **11.** $\mathbf{j} \cdot (\mathbf{k} \times \mathbf{i}) = \mathbf{j} \cdot (\mathbf{j}) = 1$

13. $(\mathbf{i} + 3\mathbf{j} - \mathbf{k}) \times (\mathbf{i} + \mathbf{k}) = \begin{vmatrix} \mathbf{i} & \mathbf{j} & \mathbf{k} \\ 1 & 3 & -1 \\ 2 & 0 & 1 \end{vmatrix} = [(3)(1) - (-1)(0)]\mathbf{i} - [(1)(1) - (-1)(1)]\mathbf{j} + [(1)0 - (3)(1)]\mathbf{k}$

$$= 3\mathbf{i} - 2\mathbf{j} - 3\mathbf{k}$$

15. $(\mathbf{i} + \mathbf{j} + \mathbf{k}) \times (2\mathbf{i} + \mathbf{k}) = \begin{vmatrix} \mathbf{i} & \mathbf{j} & \mathbf{k} \\ 1 & 1 & 1 \\ 2 & 0 & 1 \end{vmatrix} = [(1)(1) - (1)(0)]\mathbf{i} - [(1)(1) - (1)(2)]\mathbf{j} + [(1)(0) - (1)(2)]\mathbf{k}$

$$= \mathbf{i} + \mathbf{j} - 2\mathbf{k}$$

17. $[2\mathbf{i} + \mathbf{j}] \cdot [(\mathbf{i} - 3\mathbf{j} + \mathbf{k}) \times (4\mathbf{i} + \mathbf{k})] = \begin{vmatrix} 1 & -3 & 1 \\ 4 & 0 & 1 \\ 2 & 1 & 0 \end{vmatrix} =$

$$[(0)(0) - (1)(1)] - (-3)[(4)(0) - (1)(2)] + [(4)(1) - (0)(2)] = -3$$

19.
$$[(\mathbf{i} - \mathbf{j}) \times (\mathbf{j} - \mathbf{k})] \times [\mathbf{i} + 5\mathbf{k}] = \{[\mathbf{i} \times (\mathbf{j} - \mathbf{k})] - [\mathbf{j} \times (\mathbf{j} - \mathbf{k})]\} \times [\mathbf{i} + 5\mathbf{k}]$$

$$= [(\mathbf{k} + \mathbf{j}) - (-\mathbf{i})] \times [\mathbf{i} + 5\mathbf{k}]$$

$$= (\mathbf{i} + \mathbf{j} + \mathbf{k}) \times (\mathbf{i} + 5\mathbf{k})$$

$$= [(\mathbf{i} + \mathbf{j} + \mathbf{k}) \times \mathbf{i}] + [(\mathbf{i} + \mathbf{j} + \mathbf{k}) \times 5\mathbf{k}]$$

$$= (-\mathbf{k} + \mathbf{j}) + (-5\mathbf{j} + 5\mathbf{i})$$

$$= 5\mathbf{i} - 4\mathbf{j} - \mathbf{k}$$

21. $\mathbf{a} \times \mathbf{b} = \begin{vmatrix} 1 & -3 & 1 \\ 4 & 0 & 1 \\ 2 & 1 & 0 \end{vmatrix} = 3\mathbf{i} - 3\mathbf{j} - 6\mathbf{k}$

$\dfrac{\mathbf{a} \times \mathbf{b}}{\|\mathbf{a} \times \mathbf{b}\|} = \dfrac{1}{\sqrt{6}}\mathbf{i} - \dfrac{1}{\sqrt{6}}\mathbf{j} - \dfrac{2}{\sqrt{6}}\mathbf{k}; \quad \dfrac{\mathbf{b} \times \mathbf{a}}{\|\mathbf{b} \times \mathbf{a}\|} = -\dfrac{1}{\sqrt{6}}\mathbf{i} + \dfrac{1}{\sqrt{6}}\mathbf{j} + \dfrac{2}{\sqrt{6}}\mathbf{k}$

23. Set $\mathbf{a} = \overrightarrow{PQ} = -\mathbf{i} + 2\mathbf{k}$ and $\mathbf{b} = \overrightarrow{PR} = 2\mathbf{i} - \mathbf{k}$. Then

$$\mathbf{N} = \overrightarrow{PQ} \times \overrightarrow{PR} = \begin{vmatrix} \mathbf{i} & \mathbf{j} & \mathbf{k} \\ -1 & 0 & 2 \\ 2 & 0 & -1 \end{vmatrix} = 3\mathbf{j}$$

and $A = \frac{1}{2}\|\mathbf{a} \times \mathbf{b}\| = \frac{1}{2}\|3\mathbf{j}\| = \frac{3}{2}$.

25. Set $\mathbf{a} = \overrightarrow{PQ} = \mathbf{i} + \mathbf{j} - 3\mathbf{k}$ and $\mathbf{b} = \overrightarrow{PR} = -\mathbf{i} + 3\mathbf{j} - \mathbf{k}$. Then

$$\mathbf{N} = \overrightarrow{PQ} \times \overrightarrow{PR} = \begin{vmatrix} \mathbf{i} & \mathbf{j} & \mathbf{k} \\ 1 & 1 & -3 \\ -1 & 3 & -1 \end{vmatrix} = 8\mathbf{j} + 4\mathbf{j} + 4\mathbf{k}$$

and $A = \frac{1}{2}\|\mathbf{a} \times \mathbf{b}\| = \frac{1}{2}\|8\mathbf{i} + 4\mathbf{j} + 4\mathbf{k}\| = \frac{1}{2}\sqrt{8^2 + 4^2 + 4^2} = 2\sqrt{6}$.

27. $V = \left| [(\mathbf{i} + \mathbf{j}) \times (2\mathbf{i} - \mathbf{k})] \cdot (3\mathbf{j} + \mathbf{k}) \right| = \left| (-\mathbf{i} + \mathbf{j} - 2\mathbf{k}) \cdot (3\mathbf{j} + \mathbf{k}) \right| = 1$

29. $V = \overrightarrow{OP} \cdot \left(\overrightarrow{OQ} \times \overrightarrow{OR} \right) = \begin{vmatrix} 1 & 2 & 3 \\ 1 & 1 & 2 \\ 2 & 1 & 1 \end{vmatrix} = 2$

31.
$$(\mathbf{a} + \mathbf{b}) \times (\mathbf{a} - \mathbf{b}) = [\mathbf{a} \times (\mathbf{a} - \mathbf{b})] + [\mathbf{b} \times (\mathbf{a} - \mathbf{b})]$$

$$= [\mathbf{a} \times (-\mathbf{b})] + [\mathbf{b} \times \mathbf{a}]$$

$$= -(\mathbf{a} \times \mathbf{b}) - (\mathbf{a} \times \mathbf{b}) = -2(\mathbf{a} \times \mathbf{b})$$

33. $\mathbf{a} \times \mathbf{i} = 0, \quad \mathbf{a} \times \mathbf{j} = 0 \implies \mathbf{a} \| \mathbf{i}$ and $\mathbf{a} \| \mathbf{j} \implies \mathbf{a} = 0$

35.
$$(\alpha\mathbf{a} + \beta\mathbf{b}) \times (\gamma\mathbf{a} + \delta\mathbf{b}) = (\alpha\mathbf{a} \times \delta\mathbf{b}) + (\beta\mathbf{b} \times \gamma\mathbf{a})$$

$$= \alpha\delta(\mathbf{a} \times \mathbf{b}) - \beta\gamma(\mathbf{a} \times \mathbf{b})$$

$$= (\alpha\delta - \beta\gamma)(\mathbf{a} \times \mathbf{b}) = \begin{vmatrix} \alpha & \beta \\ \gamma & \delta \end{vmatrix}(\mathbf{a} \times \mathbf{b})$$

37. $\mathbf{a} \cdot (\mathbf{b} \times \mathbf{c}) = (\mathbf{a} \times \mathbf{b}) \cdot \mathbf{c} = (\mathbf{c} \times \mathbf{a}) \cdot \mathbf{b} = (\mathbf{b} \times \mathbf{c}) \cdot \mathbf{a} = (\mathbf{a} \times -\mathbf{c}) \cdot \mathbf{b}$

   $\mathbf{a} \cdot (\mathbf{c} \times \mathbf{b}) = \mathbf{c} \cdot (\mathbf{b} \times \mathbf{a}) = (-\mathbf{a} \times \mathbf{b}) \cdot \mathbf{c}$

39. $\mathbf{a} \times \mathbf{b}$ is perpendicular to the plane determined by $\mathbf{a}$ and $\mathbf{b}$;

   $\mathbf{c}$ is in this plane iff $\mathbf{a} \times \mathbf{b} \cdot \mathbf{c} = 0$.

41. $\mathbf{a} \cdot \mathbf{b} = \mathbf{a} \cdot \mathbf{c} \implies \mathbf{a} \cdot (\mathbf{b} - \mathbf{c}) = 0;$ $\mathbf{a}$ is perpendicular to $\mathbf{b} - \mathbf{c}$.

   $\mathbf{a} \times \mathbf{b} = \mathbf{a} \times \mathbf{c} \implies \mathbf{a} \times (\mathbf{b} - \mathbf{c}) = \mathbf{0};$ $\mathbf{a}$ is parallel to $\mathbf{b} - \mathbf{c}$.

   Since $\mathbf{a} \neq \mathbf{0}$ it follows that $\mathbf{b} - \mathbf{c} = \mathbf{0}$ or $\mathbf{b} = \mathbf{c}$.

43. $\mathbf{c} \times \mathbf{a} = (\mathbf{a} \times \mathbf{b}) \times \mathbf{a} = (\mathbf{a} \cdot \mathbf{a})\mathbf{b} - (\mathbf{a} \cdot \mathbf{b})\mathbf{a} = (\mathbf{a} \cdot \mathbf{a})\mathbf{b} = \|\mathbf{a}\|^2\mathbf{b}$

   Exercise 42(a) $\mathbf{a} \cdot \mathbf{b} = 0$

45.   Expanding the determinant by the bottom row gives
$$\begin{vmatrix} a_1 & a_2 & a_3 \\ b_1 & b_2 & b_3 \\ c_1 & c_2 & c_3 \end{vmatrix} = c_1\begin{vmatrix} a_2 & a_3 \\ b_2 & b_3 \end{vmatrix} - c_2\begin{vmatrix} a_1 & a_3 \\ b_1 & b_3 \end{vmatrix} + c_3\begin{vmatrix} a_1 & a_2 \\ b_1 & b_2 \end{vmatrix}$$

PROJECT 12.5

1. $\mathbf{r} = \overrightarrow{OP} = \mathbf{i} + \mathbf{j} + \mathbf{k}$

   $\|\tau\| = \|\mathbf{r} \times \mathbf{F}\| = \|(\mathbf{i} + \mathbf{j} + \mathbf{k}) \times (\mathbf{i} + 2\mathbf{j} + \mathbf{k})\| = \|-\mathbf{i} + \mathbf{k}\| = \sqrt{2}$

3. $\|\tau\| = \|\mathbf{r}\|\|\mathbf{F}\| \sin\theta = (10)(20) \sin 130° \cong 152.2$ inch-lb $= 12.68$ ft-lb;

   the bolt will move into the plane of the paper.

SECTION 12.6

1. P (when $t = 0$) and Q (when $t = -1$)

3.   Take $\mathbf{r}_0 = \overrightarrow{OP} = 3\mathbf{i} + \mathbf{j}$ and $\mathbf{d} = \mathbf{k}.$ Then, $\mathbf{r}(t) = (3\mathbf{i} + \mathbf{j}) + t\mathbf{k}.$

5. Take $\mathbf{r}_0 = 0$ and $\mathbf{d} = \overrightarrow{OQ}$. Then, $\mathbf{r}(t) = t(x_1\mathbf{i} + y_1\mathbf{j} + z_1\mathbf{k})$.

7. $\overrightarrow{PQ} = \mathbf{i} - \mathbf{j} + \mathbf{k}$ so direction numbers are $1, -1, 1$. Using P as a point on the line, we have

$$x(t) = 1 + t, \quad y(t) = -t, \quad z(t) = 3 + t.$$

9. The line is parallel to the y-axis so we can take $0, 1, 0$ as direction numbers. Therefore

$$x(t) = 2, \quad y(t) = -2 + t, \quad z(t) = 3.$$

11. Since the line $2(x + 1) = 4(y - 3) = z$ can be written

$$\frac{x + 1}{2} = \frac{y - 3}{1} = \frac{z}{4},$$

it has direction numbers $2, 1, 4$. The line through $P(-1, 2, -3)$ with direction vector

$2\mathbf{i} + \mathbf{j} + 4\mathbf{k}$ can be parametrized

$$\mathbf{r}(t) = (-\mathbf{i} + 2\mathbf{j} - 3\mathbf{k}) + t(2\mathbf{i} + \mathbf{j} + 4\mathbf{k}).$$

13. We set $\mathbf{r}_1(t) = \mathbf{r}_2(u)$ and solve for t and u:

$$\mathbf{i} + t\mathbf{j} = \mathbf{j} + u(\mathbf{i} + \mathbf{j}),$$

$$(1 - u)\mathbf{i} + (-1 - u + t)\mathbf{j} = 0.$$

Thus,

$$1 - u = 0 \quad \text{and} \quad -1 - u + t = 0.$$

The equation gives $u = 1, \ t = 2$. The point of intersection is $P(1, 2, 0)$.

As direction vectors for the lines we can take $\mathbf{u} = \mathbf{j}$ and $\mathbf{v} = \mathbf{i} + \mathbf{j}$. Thus

$$\cos\theta = \frac{\mathbf{u} \cdot \mathbf{v}}{\|\mathbf{u}\|\,\|\mathbf{v}\|} = \frac{1}{(1)(\sqrt{2})} = \frac{1}{2}\sqrt{2}.$$

The angle of intersection is $\frac{1}{4}\pi$ radians.

15. We solve the system

$$3 + t = 1, \quad 1 - t = 4 + u, \quad 5 + 2t = 2 + u$$

for t and u to find that $t = -2, \ u = -1$. The point of intersection is $(1, 3, 1)$.

Since $\mathbf{i} - \mathbf{j} + 2\mathbf{k}$ is a direction vector for l_1 and $\mathbf{j} + \mathbf{k}$ is a direction vector for l_2,

$$\cos\theta = \frac{(\mathbf{i} - \mathbf{j} + 2\mathbf{k}) \cdot (\mathbf{j} + \mathbf{k})}{\sqrt{6}\sqrt{2}} = \frac{1}{2\sqrt{3}} = \frac{1}{6}\sqrt{3} \quad \text{and} \quad \theta \cong 1.28 \text{ radians}.$$

17. $\left(x_0 - \dfrac{d_1}{d_3}z_0, \ \ y_0 - \dfrac{d_2}{d_3}z_0, \ \ 0 \right)$ **19.** The lines are parallel.

21. $\mathbf{r}(t) = (2\mathbf{i} + 7\mathbf{j} - \mathbf{k}) + t(2\mathbf{i} - 5\mathbf{j} + 4\mathbf{k}), \quad 0 \leq t \leq 1$

23. Set
$$\mathbf{u} = \frac{\overrightarrow{PQ}}{\|\overrightarrow{PQ}\|} = \frac{-4\mathbf{i} + 2\mathbf{j} + 4\mathbf{k}}{\| - 4\mathbf{i} + 2\mathbf{j} + 4\mathbf{k}\|} = -\frac{2}{3}\mathbf{i} + \frac{1}{3}\mathbf{j} + \frac{2}{3}\mathbf{k}.$$

Then $\mathbf{r}(t) = (6\mathbf{i} - 5\mathbf{j} + \mathbf{k}) + t\mathbf{u}$ is $\overrightarrow{OP}$ at $t = 9$ and it is $\overrightarrow{OQ}$ at $t = 15$. (Check this.)

Answer: $\mathbf{u} = -\frac{2}{3}\mathbf{i} + \frac{1}{3}\mathbf{j} + \frac{2}{3}\mathbf{k}, \quad 9 \leq t \leq 15.$

25. The given line, call it l, has direction vector $2\mathbf{i} - 4\mathbf{j} + 6\mathbf{k}$.

If $a\mathbf{i} + b\mathbf{j} + c\mathbf{k}$ is a direction vector for a line perpendicular to l, then

$$(2\mathbf{i} - 4\mathbf{j} + 6\mathbf{k}) \cdot (a\mathbf{i} + b\mathbf{j} + c\mathbf{k}) = 2a - 4b + 6c = 0.$$

The lines through $P(3, -1, 8)$ perpendicular to l can be parametrized

$$X(u) = 3 + au, \quad Y(u) = -1 + bu, \quad Z(u) = 8 + cu$$

with $2a - 4b + 6c = 0$.

27. $d(P, l) = \dfrac{\|(\mathbf{i} + 2\mathbf{k}) \times (2\mathbf{i} - \mathbf{j} + 2\mathbf{k})\|}{\|2\mathbf{i} - \mathbf{j} + 2\mathbf{k}\|} = 1$

29. The line contains the point $P_0(1, 0, 2)$. Therefore

$$d(P, l) = \frac{\|(2\mathbf{j} + \mathbf{k}) \times (\mathbf{i} - 2\mathbf{j} + 3\mathbf{k})\|}{\|\mathbf{i} - 2\mathbf{j} + 3\mathbf{k}\|} = \sqrt{\frac{69}{14}} \cong 2.22$$

31. The line contains the point $P_0(2, -1, 0)$. Therefore

$$d(P, l) = \frac{\|(\mathbf{i} - \mathbf{j} - \mathbf{k}) \times (\mathbf{i} + \mathbf{j})\|}{\|\mathbf{i} + \mathbf{j}\|} = \sqrt{3} \cong 1.73.$$

33. (a) The line passes through $P(1, 1, 1)$ with direction vector $\mathbf{i} + \mathbf{j}$. Therefore

$$d(0, l) = \frac{\|(\mathbf{i} + \mathbf{j} + \mathbf{k}) \times (\mathbf{i} + \mathbf{j})\|}{\|\mathbf{i} + \mathbf{j}\|} = 1.$$

(b) The distance from the origin to the line segment is $\sqrt{3}$.

Solution. The line segment can be parametrized

$$\mathbf{r}(t) = \mathbf{i} + \mathbf{j} + \mathbf{k} + t(\mathbf{i} + \mathbf{j}), \quad t \in [0, 1].$$

This is the set of all points $P(1 + t, 1 + t, 1)$ with $t \in [0, 1]$.

The distance from the origin to such a point is

$$f(t) = \sqrt{2\left(1 + t^2\right) + 1}\,.$$

The minimum value of this function is $f(0) = \sqrt{3}$.

Explanation. The point on the line through P and Q closest to the origin is not on the line

segment $\overline{PQ}$.

35. We begin with $\mathbf{r}(t) = \mathbf{j} - 2\mathbf{k} + t(\mathbf{i} - \mathbf{j} + 3\mathbf{k})$. The scalar t_0 for which $\mathbf{r}(t_0) \perp l$ can be found by solving the equation

$$[\mathbf{j} - 2\mathbf{k} + t_0(\mathbf{i} - \mathbf{j} + 3\mathbf{k})] \cdot [\mathbf{i} - \mathbf{j} + 3\mathbf{k}] = 0.$$

This equation gives $-7 + 11t_0 = 0$ and thus $t_0 = 7/11$. Therefore

$$\mathbf{r}(t_0) = \mathbf{j} - 2\mathbf{k} + \tfrac{7}{11}(\mathbf{i} - \mathbf{j} + 3\mathbf{k}) = \tfrac{7}{11}\mathbf{i} + \tfrac{4}{11}\mathbf{j} - \tfrac{1}{11}\mathbf{k}.$$

The vectors of norm 1 parallel to $\mathbf{i} - \mathbf{j} + 3\mathbf{k}$ are

$$\pm \frac{1}{\sqrt{11}}(\mathbf{i} - \mathbf{j} + 3\mathbf{k}).$$

The standard parametrizations are

$$\mathbf{R}(t) = \frac{7}{11}\mathbf{i} + \frac{4}{11}\mathbf{j} - \frac{1}{11}\mathbf{k} \pm \frac{t}{\sqrt{11}}(\mathbf{i} - \mathbf{j} + 3\mathbf{k})$$

$$= \frac{1}{11}(7\mathbf{i} + 4\mathbf{j} - \mathbf{k}) \pm t\left[\frac{\sqrt{11}}{11}(\mathbf{i} - \mathbf{j} + 3\mathbf{k})\right].$$

37. $0 < t < s$

By similar triangles, if $0 < s < 1$, the tip of $\overrightarrow{OA} + s\overrightarrow{AB} + s\overrightarrow{BC}$ falls on $\overline{AC}$. If $0 < t < s$, then the tip of $\overrightarrow{OA} + s\overrightarrow{AB} + t\overrightarrow{BC}$ falls short of $\overline{AC}$ and stays within the triangle. Clearly all points in the interior of the triangle can be reached in this manner.

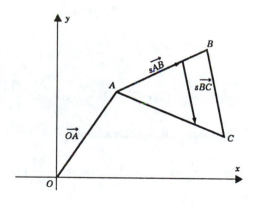

SECTION 12.7

1. Q

3. Since $\mathbf{i} - 4\mathbf{j} + 3\mathbf{k}$ is normal to the plane, we have

$$(x - 2) - 4(y - 3) + 3(z - 4) = 0 \quad \text{and thus} \quad x - 4y + 3z - 2 = 0.$$

5. The vector $3\mathbf{i} - 2\mathbf{j} + 5\mathbf{k}$ is normal to the given plane and thus to every parallel plane: the equation we want can be written

$$3(x - 2) - 2(y - 1) + 5(z - 1) = 0, \quad 3x - 2y + 5z - 9 = 0.$$

7. The point $Q(0, 0, -2)$ lies on the line l; and $\mathbf{d} = \mathbf{i} + \mathbf{j} + \mathbf{k}$ is a direction vector for l.

We want an equation for the plane which has the vector

$$\mathbf{N} = \overrightarrow{PQ} \times \mathbf{d} = (\mathbf{i} + 3\mathbf{j} + 3\mathbf{k}) \times (\mathbf{i} + \mathbf{j} + \mathbf{k})$$

as a normal vector:

$$\mathbf{N} = \begin{vmatrix} \mathbf{i} & \mathbf{j} & \mathbf{k} \\ 1 & 3 & 3 \\ 1 & 1 & 1 \end{vmatrix} = 2\mathbf{j} - 2\mathbf{k}$$

An equation for the plane is: $2(y - 3) - 2(z - 1) = 0$ or $y - z - 2 = 0$

9. $\overrightarrow{OP_0} = x_0\mathbf{i} + y_0\mathbf{j} + z_0\mathbf{k}$ An equation for the plane is:

$$x_0(x - x_0) + y_0(y - y_0) + z_0(z - z_0)$$

11. The vector $\mathbf{N} = 2\mathbf{i} - \mathbf{j} + 5\mathbf{k}$ is normal to the plane $2x - y + 5z - 10 = 0$. The unit normals are:

$$\frac{\mathbf{N}}{\|\mathbf{N}\|} = \frac{1}{\sqrt{30}}(2\mathbf{i} - \mathbf{j} + 5\mathbf{k}) \quad \text{and} \quad -\frac{\mathbf{N}}{\|\mathbf{N}\|} = -\frac{1}{\sqrt{30}}(2\mathbf{i} - \mathbf{j} + 5\mathbf{k})$$

13. Intercept form: $\dfrac{x}{15} + \dfrac{y}{12} - \dfrac{z}{10} = 1$ x-intercept: $(15, 0, 0)$

y-intercept: $(0, 12, 0)$

z-intercept: $(0, 0, 10)$

15. $\mathbf{u_{N_1}} = \dfrac{\sqrt{38}}{38}(5\mathbf{i} - 3\mathbf{j} + 2\mathbf{k}), \quad \mathbf{u_{N_2}} = \dfrac{\sqrt{14}}{14}(\mathbf{i} + 3\mathbf{j} + 2\mathbf{k}), \quad \cos\theta = |\mathbf{u_{N_1}} \cdot \mathbf{u_{N_2}}| = 0.$

Therefore $\theta = \pi/2$ radians.

17. $\mathbf{u_{N_1}} = \dfrac{\sqrt{3}}{3}(\mathbf{i} - \mathbf{j} + \mathbf{k}), \quad \mathbf{u_{N_2}} = \dfrac{\sqrt{14}}{14}(2\mathbf{i} + \mathbf{j} + 3\mathbf{k}), \quad \cos\theta = |\mathbf{u_{N_1}} \cdot \mathbf{u_{N_2}}| = \dfrac{2}{21}\sqrt{42} \cong 0.617.$

Therefore $\theta \cong 0.91$ radians.

19. coplanar since $0(4\mathbf{j} - \mathbf{k}) + 0(3\mathbf{i} + \mathbf{j} + 2\mathbf{k}) + 1(0) = 0$

21.   We need to determine whether there exist scalars s, t, u not all zero such that

$$s(\mathbf{i} + \mathbf{j} + \mathbf{k}) + t(2\mathbf{i} - \mathbf{j}) + u(3\mathbf{i} - \mathbf{j} - \mathbf{k}) = 0$$

$$(s + 2t + 3u)\,\mathbf{i} + (s - t - u)\,\mathbf{j} + (s - u)\,\mathbf{k} = 0.$$

The only solution of the system

$$s + 2t + 3u = 0, \quad s - t - u = 0, \quad s - u = 0$$

is $s = t = u = 0.$ Thus, the vectors are not coplanar.

23.   By (12.7.7), $d(P, p) = \dfrac{|2(2) + 4(-1) - (3) + 1|}{\sqrt{4 + 16 + 1}} = \dfrac{2}{\sqrt{21}} = \dfrac{2}{21}\sqrt{21}.$

25.   By (12.7.7), $d(P, p) = \dfrac{|(-3)(1) + 0(-3) + 4(5) + 5|}{\sqrt{9 + 16}} = \dfrac{22}{5}.$

27. $\overrightarrow{P_1 P} = (x - 1)\,\mathbf{i} + y\mathbf{j} + (z - 1)\,\mathbf{k}, \quad \overrightarrow{P_1 P_2} = \mathbf{i} + \mathbf{j} - \mathbf{k}, \quad \overrightarrow{P_1 P_3} = \mathbf{j}.$

Therefore

$$(\overrightarrow{P_1 P_2} \times \overrightarrow{P_1 P_3}) = (\mathbf{i} + \mathbf{j} - \mathbf{k}) \times \mathbf{j} = \mathbf{i} + \mathbf{k}$$

and

$$\overrightarrow{P_1 P} \cdot (\overrightarrow{P_1 P_2} \times \overrightarrow{P_1 P_3}) = [(x - 1)\,\mathbf{i} + y\mathbf{j} + (z - 1)\,\mathbf{k}] \cdot [\mathbf{i} + \mathbf{k}] = x - 1 + z - 1.$$

An equation for the plane can be written $x + z = 2.$

29. $\overrightarrow{P_1 P} = (x - 3)\,\mathbf{i} + (y + 4)\,\mathbf{j} + (z - 1)\,\mathbf{k}, \quad \overrightarrow{P_1 P_2} = 6\mathbf{j}, \quad \overrightarrow{P_1 P_3} = -4\mathbf{i} + 5\mathbf{j} - 3\mathbf{k}.$

Therefore

$$(\overrightarrow{P_1 P_2} \times \overrightarrow{P_1 P_3}) = 6\mathbf{j} \times (-4\mathbf{i} + 5\mathbf{j} - 3\mathbf{k}) = -18\mathbf{i} + 24\mathbf{k}$$

and

$$\overrightarrow{P_1 P} \cdot (\overrightarrow{P_1 P_2} \times \overrightarrow{P_1 P_3}) = [(x - 3)\,\mathbf{i} + (y + 4)\,\mathbf{j} + (z - 1)\,\mathbf{k}] \cdot [-18\mathbf{i} + 24\mathbf{k}]$$

$$= -18(x - 3) + 24(z - 1)$$

An equation for the plane can be written $-18(x - 3) + 24(z - 1) = 0$ or $3x - 4z - 5 = 0.$

31.   The line passes through the point $P_0\,(x_0, y_0, z_0)$ with direction numbers: A, B, C.

Equations for the line written in symmetric form are:

$$\frac{x - x_0}{A} = \frac{y - y_0}{B} = \frac{z - z_0}{C}, \quad \text{provided}\quad A \neq 0,\ B \neq 0,\ C \neq 0.$$

33. $\dfrac{x-x_0}{d_1} = \dfrac{y-y_0}{d_2}, \quad \dfrac{y-y_0}{d_2} = \dfrac{z-z_0}{d_3}$

35. Following the hint we take $x = 0$ and find that $P_0(0,0,0)$ lies on the line of intersection. As normals to the plane we use

$$\mathbf{N}_1 = \mathbf{i} + 2\mathbf{j} + 3\mathbf{k} \quad \text{and} \quad \mathbf{N}_2 = -3\mathbf{i} + 4\mathbf{j} + \mathbf{k}.$$

Note that

$$\mathbf{N}_1 \times \mathbf{N}_2 = (\mathbf{i} + 2\mathbf{j} + 3\mathbf{k}) \times (-3\mathbf{i} + 4\mathbf{j} + \mathbf{k}) = -10\mathbf{i} - 10\mathbf{j} + 10\mathbf{k}.$$

We take $-\frac{1}{10}(\mathbf{N}_1 \times \mathbf{N}_2) = \mathbf{i} + \mathbf{j} - \mathbf{k}$ as a direction vector for the line through $P_0(0,0,0)$. Then

$$x(t) = t, \quad y(t) = t, \quad z(t) = -t.$$

37. Straightforward computations give us

$$l : x(t) = 1 - 3t, \quad y(t) = -1 + 4t, \quad z(t) = 2 - t$$

and

$$p : x + 4y - z = 6.$$

Substitution of the scalar parametric equations for l in the equation for p gives

$$(1 - 3t) + 4(-1 + 4t) - (2 - t) = 6 \quad \text{and thus} \quad t = 11/14.$$

Using $t = 11/14$, we get $\quad x = -19/14, \quad y = 15/7, \quad z = 17/14.$

39. Let $\quad \mathbf{N} = A\mathbf{i} + B\mathbf{j} + C\mathbf{k} \quad$ be normal to the plane. Then

$$\mathbf{N} \cdot \mathbf{d} = (\mathbf{i} + B\mathbf{j} + C\mathbf{k}) \cdot (\mathbf{i} + 2\mathbf{j} + 4\mathbf{k}) = 1 + 2B + 4C = 0$$

and

$$\mathbf{N} \cdot \mathbf{D} = (\mathbf{i} + B\mathbf{j} + C\mathbf{k}) \cdot (-\mathbf{i} - \mathbf{j} + 3\mathbf{k}) = -1 - B + 3C = 0.$$

This gives $\quad B = -7/10 \quad$ and $\quad C = 1/10$. The equation for the plane can be written

$$1(x - 0) - \tfrac{7}{10}(y - 0) + \tfrac{1}{10}(z - 0) = 0, \quad \text{which simplifies to} \quad 10x - 7y + z = 0.$$

41. $\mathbf{N} + \overrightarrow{PQ}$ and $\mathbf{N} - \overrightarrow{PQ}$ are the diagonals of a rectangle with sides $\mathbf{N}$ and $\overrightarrow{PQ}$. Since the diagonals are perpendicular, the rectangle is a square; that is $\|\mathbf{N}\| = \|\overrightarrow{PQ}\|$. Thus, the points Q form a circle centered at P with radius $\|\mathbf{N}\|$.

43. If $\alpha > 0$, then P_1 lies on the same side of the plane as the tip of $\mathbf{N}$; if $\alpha < 0$, then P_1 and the tip of $\mathbf{N}$ lie on opposite sides of the plane.

To see this, suppose that the tip of $\mathbf{N}$ is at $P_0(x_0, y_0, z_0)$. Then

$$\mathbf{N} \cdot \overrightarrow{P_0 P_1} = A(x_1 - x_0) + B(y_1 - y_0) + C(z_1 - z_0) = Ax_1 + By_1 + Cz_1 + D = \alpha.$$

If $\alpha > 0$, $0 \leq \sphericalangle \left(\mathbf{N}, \overrightarrow{P_0 P_1} \right) < \pi/2$; if $\alpha < 0$, then $\pi/2 < \sphericalangle \left(\mathbf{N}, \overrightarrow{P_0 P_1} \right) < \pi$. Since $\mathbf{N}$ is perpendicular to the plane, the result follows.

45. $\mathbf{a} \cdot (\mathbf{b} \times \mathbf{c}) = 0$

47.

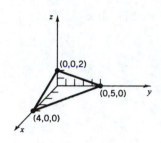

49.

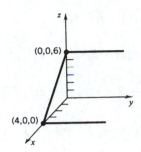

51. $\dfrac{x}{2} + \dfrac{y}{5} + \dfrac{z}{4} = 1$

$10x + 4y + 5z = 20$

53. $\dfrac{x}{3} + \dfrac{y}{5} = 1$

$5x + 3y = 15$

SECTION 13.1

1. $\mathbf{f}'(t) = 2\mathbf{i} - \mathbf{j} + 3\mathbf{k}$

3. $\mathbf{f}'(t) = -\dfrac{1}{2\sqrt{1-t}}\,\mathbf{i} + \dfrac{1}{2\sqrt{1+t}}\,\mathbf{j} + \dfrac{1}{(1-t)^2}\,\mathbf{k}$

5. $\mathbf{f}'(t) = \cos t\,\mathbf{i} - \sin t\,\mathbf{j} + \sec^2 t\,\mathbf{k}$

7. $\mathbf{f}'(t) = \dfrac{-1}{1-t}\,\mathbf{i} - \sin t\,\mathbf{j} + 2t\,\mathbf{k}$

9. $\mathbf{f}'(t) = 4\,\mathbf{i} + 6t^2\mathbf{j} + (2t+2)\mathbf{k}$; $\mathbf{f}''(t) = 12t\,\mathbf{j} + 2\,\mathbf{k}$

11. $\mathbf{f}'(t) = -2\sin 2t\,\mathbf{i} + 2\cos 2t\,\mathbf{j} + 4t\,\mathbf{k}$; $\mathbf{f}''(t) = -4\cos 2t\,\mathbf{i} - 4\sin 2t\,\mathbf{j}$

13. $\displaystyle\int_1^2 (\mathbf{i} + 2t\,\mathbf{j})\,dt = \left[t\,\mathbf{i} + t^2\,\mathbf{j}\right]_1^2 = \mathbf{i} + 3\mathbf{j}$

15. $\displaystyle\int_0^1 \left(e^t\mathbf{i} + e^{-t}\mathbf{k}\right) dt = \left[e^t\mathbf{i} - e^{-t}\mathbf{k}\right]_0^1 = (e-1)\,\mathbf{i} + \left(1 - \dfrac{1}{e}\right)\mathbf{k}$

17. $\displaystyle\int_0^1 \left(\dfrac{1}{1+t^2}\,\mathbf{i} + \sec^2 t\,\mathbf{j}\right) dt = \left[\tan^{-1} t\,\mathbf{i} + \tan t\,\mathbf{j}\right]_0^1 = \dfrac{\pi}{4}\,\mathbf{i} + \tan(1)\,\mathbf{j}$

19. $\displaystyle\lim_{t\to 0}\mathbf{f}(t) = \left(\lim_{t\to 0}\dfrac{\sin t}{2t}\right)\mathbf{i} + \left(\lim_{t\to 0} e^{2t}\right)\mathbf{j} + \left(\lim_{t\to 0}\dfrac{t^2}{e^t}\right)\mathbf{k} = \dfrac{1}{2}\mathbf{i} + \mathbf{j}$

21. $\displaystyle\lim_{t\to 0}\mathbf{f}(t) = \left(\lim_{t\to 0} t^2\right)\mathbf{i} + \left(\lim_{t\to 0}\dfrac{1-\cos t}{3t}\right)\mathbf{j} + \left(\lim_{t\to 0}\dfrac{t}{t+1}\right)\mathbf{k} = 0\,\mathbf{i} + \dfrac{1}{3}\left(\lim_{t\to 0}\dfrac{1-\cos t}{t}\right)\mathbf{j} + 0\,\mathbf{k} = \mathbf{0}$

23.

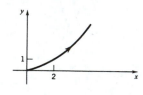

25.

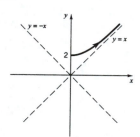

27.

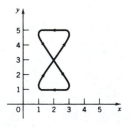

29.

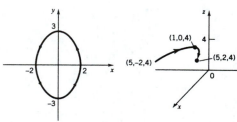

31.

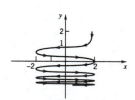

33.

35. (a) $\mathbf{f}(t) = 3\cos t\,\mathbf{i} + 2\sin t\,\mathbf{j}$ 　　　　　 (b) $\mathbf{f}(t) = 3\cos t\,\mathbf{i} - 2\sin t\,\mathbf{j}$

37. (a) $\mathbf{f}(t) = t\mathbf{i} + t^2\mathbf{j}$ 　　　　　 (b) $\mathbf{f}(t) = -t\mathbf{i} + t^2\mathbf{j}$

39. $\mathbf{f}(t) = (1 + 2t)\,\mathbf{i} + (4 + 5t)\,\mathbf{j} + (-2 + 8t)\,\mathbf{k}, \quad 0 \le t \le 1$

41. $\mathbf{f}'(t_0) = \mathbf{i} + m\mathbf{j},$

$$\int_a^b \mathbf{f}(t)\,dt = \left[\tfrac{1}{2}t^2\mathbf{i}\right]_a^b + \left[\int_a^b f(t)\,dt\right]\mathbf{j} = \tfrac{1}{2}\,(b^2 - a^2)\,\mathbf{i} + A\mathbf{j},$$

$$\int_a^b \mathbf{f}'(t)\,dt = [t\mathbf{i} + f(t)\,\mathbf{j}]_a^b = (b - a)\,\mathbf{i} + (d - c)\,\mathbf{j}$$

43.
$$\mathbf{f}'(t) = \mathbf{i} + t^2\mathbf{j}$$
$$\mathbf{f}(t) = (t + C_1)\,\mathbf{i} + \left(\tfrac{1}{3}t^3 + C_2\right)\mathbf{j} + C_3\mathbf{k}$$
$$\mathbf{f}(0) = \mathbf{j} - \mathbf{k} \implies C_1 = 0, \quad C_2 = 1, \quad C_3 = -1$$
$$\mathbf{f}(t) = t\mathbf{i} + \left(\tfrac{1}{3}t^3 + 1\right)\mathbf{j} - \mathbf{k}$$

45. $\mathbf{f}'(t) = \alpha\,\mathbf{f}(t) \implies \mathbf{f}(t) = e^{\alpha t}\,\mathbf{f}(0) = e^{\alpha t}\mathbf{c}$

47. (a) If $\mathbf{f}'(t) = \mathbf{0}$ on an interval, then the derivative of each component is 0 on that interval, each component is constant on that interval, and therefore $\mathbf{f}$ itself is constant on that interval.

(b) Set $\mathbf{h}(t) = \mathbf{f}(t) - \mathbf{g}(t)$ and apply part (a).

49. If $\mathbf{f}$ is differentiable at t, then each component is differentiable at t, each component is continuous at t, and therefore $\mathbf{f}$ is continuous at t.

51. no; as a counter-example, set $\mathbf{f}(t) = \mathbf{i} = \mathbf{g}(t)$.

53. Suppose $\mathbf{f}(t) = f_1(t)\mathbf{i} + f_2(t)\mathbf{j} + f_3(t)\mathbf{k}$. Then $\|\mathbf{f}(t)\| = \sqrt{f_1^2(t) + f_2^2(t) + f_3^2(t)}$ and

$$\frac{d}{dt}\left(\|\mathbf{f}\|\right) = \frac{1}{2}\left[f_1^2 + f_2^2 + f_3^2\right]^{-1/2}\left(2f_1 \cdot f_1' + 2f_2 \cdot f_2' + 2f_3 \cdot f_3'\right) = \frac{\mathbf{f}(t) \cdot \mathbf{f}'(t)}{\|\mathbf{f}(t)\|}$$

SECTION 13.2

1. $\mathbf{f}'(t) = \mathbf{b}, \quad \mathbf{f}''(t) = \mathbf{0}$

3. $\mathbf{f}'(t) = 2e^{2t}\,\mathbf{i} - \cos t\,\mathbf{j}, \quad \mathbf{f}''(t) = 4e^{2t}\,\mathbf{i} + \sin t\,\mathbf{j}$

5. $\mathbf{f}'(t) = [(t^2\mathbf{i} - 2t\mathbf{j}) \cdot (\mathbf{i} + 3t^2\mathbf{j}) + (2t\,\mathbf{i} - 2\mathbf{j}) \cdot (t\,\mathbf{i} + t^3\mathbf{j}]\mathbf{j} = [3t^2 - 8t^3]\mathbf{j}$

$\mathbf{f}''(t) = (6t - 24t^2)\,\mathbf{j}$

7.
$$\mathbf{f}'(t) = \left[(e^t\mathbf{i} + t\mathbf{k}) \times \frac{d}{dt}\,(t\mathbf{j} + e^{-t}\mathbf{k})\right] + \left[\frac{d}{dt}\,(e^t\mathbf{i} + t\mathbf{k}) \times (t\mathbf{j} + e^{-t}\mathbf{k})\right]$$

$$= \left[(e^t\mathbf{i} + t\mathbf{k}) \times (\mathbf{j} - e^{-t}\mathbf{k})\right] + \left[(e^t\mathbf{i} + \mathbf{k}) \times (t\mathbf{j} + e^{-t}\mathbf{k})\right]$$

$$= (-t\mathbf{i} + \mathbf{j} + e^t\mathbf{k}) + (-t\mathbf{i} - \mathbf{j} + te^t\mathbf{k})$$

$$= -2t\mathbf{i} + e^t(t+1)\mathbf{k}$$

$$\mathbf{f}''(t) = -2\mathbf{i} + e^t(t+2)\mathbf{k}$$

9. $\mathbf{f}'(t) = (\cos t\,\mathbf{i} + \sin t\,\mathbf{j} + \mathbf{k}) \times (2\sin 2t\,\mathbf{i} - 2\cos 2t\,\mathbf{j} + \mathbf{k}) + (-\sin t\,\mathbf{i} + \cos t\,\mathbf{j}) \times (\sin 2t\,\mathbf{i} + \cos 2t\,\mathbf{j} + t\,\mathbf{k})$

$\quad = (\sin t + t\cos t + 2\sin 2t)\mathbf{i} + (2\cos 2t - \cos t + t\sin t)\mathbf{j} - 3\sin 3t\,\mathbf{k}$

$\mathbf{f}''(t) = (2\cos t - t\sin t + 4\cos t)\mathbf{i} + (-4\sin 2t + 2\sin t + t\cos t)\mathbf{j} - 9\cos 3t\,\mathbf{k}$

11. $\mathbf{f}'(t) = \dfrac{1}{2}\sqrt{t}\,\mathbf{g}'\left(\sqrt{t}\right) + \mathbf{g}\left(\sqrt{t}\right), \quad \mathbf{f}''(t) = \dfrac{1}{4}\mathbf{g}''\left(\sqrt{t}\right) + \dfrac{3}{4\sqrt{t}}\mathbf{g}'\left(\sqrt{t}\right)$

13. $-\sin t\,e^{\cos t}\,\mathbf{i} + \cos t\,e^{\sin t}\,\mathbf{j}$

15. $(e^t\mathbf{i} + e^{-t}\mathbf{j}) \cdot (e^t\mathbf{i} - e^{-t}\mathbf{j}) = e^{2t} - e^{-2t}; \quad$ therefore

$$\frac{d^2}{dt^2}\left[(e^t\mathbf{i} + e^{-t}\mathbf{j}) \cdot (e^t\mathbf{i} - e^{-t}\mathbf{j})\right] = \frac{d^2}{dt^2}\left[e^{2t} - e^{-2t}\right] = \frac{d}{dt}\left[2e^{2t} + 2e^{-2t}\right] = 4e^{2t} - 4e^{-2t}$$

17. $\dfrac{d}{dt}\left[(\mathbf{a} + t\mathbf{b}) \times (\mathbf{c} + t\mathbf{d})\right] = \left[(\mathbf{a} + t\mathbf{b}) \times \mathbf{d}\right] + \left[\mathbf{b} \times (\mathbf{c} + t\mathbf{d})\right] = (\mathbf{a} \times \mathbf{d}) + (\mathbf{b} \times \mathbf{c}) + 2t\,(\mathbf{b} \times \mathbf{d})$

19. $\dfrac{d}{dt}\left[(\mathbf{a} + t\mathbf{b}) \cdot (\mathbf{c} + t\mathbf{d})\right] = \left[(\mathbf{a} + t\mathbf{b}) \cdot \mathbf{d}\right] + \left[\mathbf{b} \cdot (\mathbf{c} + t\mathbf{d})\right] = (\mathbf{a} \cdot \mathbf{d}) + (\mathbf{b} \cdot \mathbf{c}) + 2t\,(\mathbf{b} \cdot \mathbf{d})$

21. $\mathbf{r}(t) = \mathbf{a} + t\mathbf{b}$ **23.** $\mathbf{r}(t) = \frac{1}{2}t^2\mathbf{a} + \frac{1}{6}t^3\mathbf{b} + t\mathbf{c} + \mathbf{d}$

25. $\mathbf{r}(t) = \sin t\,\mathbf{i} + \cos t\,\mathbf{j}, \quad \mathbf{r}'(t) = \cos t\,\mathbf{i} - \sin t\,\mathbf{j}, \quad \mathbf{r}''(t) = -\sin t\,\mathbf{i} - \cos t\,\mathbf{j} = -\mathbf{r}(t).$

Thus $\mathbf{r}(t)$ and $\mathbf{r}''(t)$ are parallel, and they always point in opposite directions.

27.
$$\mathbf{r}(t) \cdot \mathbf{r}'(t) = (\cos t\,\mathbf{i} + \sin t\,\mathbf{j}) \cdot (-\sin t\,\mathbf{i} + \cos t\,\mathbf{j}) = 0$$

$$\mathbf{r}(t) \times \mathbf{r}'(t) = (\cos t\,\mathbf{i} + \sin t\,\mathbf{j}) \times (-\sin t\,\mathbf{i} + \cos t\,\mathbf{j})$$

$$= \cos^2 t\,\mathbf{k} + \sin^2 t\,\mathbf{k} = \left(\cos^2 t + \sin^2 t\right)\mathbf{k} = \mathbf{k}$$

29. $\dfrac{d}{dt}\left[\mathbf{f}(t) \times \mathbf{f}'(t)\right] = \left[\mathbf{f}(t) \times \mathbf{f}''(t)\right] + \underbrace{\left[\mathbf{f}'(t) \times \mathbf{f}'(t)\right]}_{\mathbf{0}} = \mathbf{f}(t) \times \mathbf{f}''(t).$

31. $[\mathbf{f} \cdot \mathbf{g} \times \mathbf{h}]' = \mathbf{f}' \cdot (\mathbf{g} \times \mathbf{h}) + \mathbf{f} \cdot (\mathbf{g} \times \mathbf{h})' = \mathbf{f}' \cdot (\mathbf{g} \times \mathbf{h}) + \mathbf{f} \cdot [\mathbf{g}' \times \mathbf{h} + \mathbf{g} \times \mathbf{h}']$

and the result follows.

33. $\|\mathbf{r}(t)\|$ is constant $\iff$ $\|\mathbf{r}(t)\|^2 = \mathbf{r}(t) \cdot \mathbf{r}(t)$ is constant

$$\iff \frac{d}{dt}[\mathbf{r}(t) \cdot \mathbf{r}(t)] = 2[\mathbf{r}(t) \cdot \mathbf{r}'(t)] = 0 \text{ identically}$$

$$\iff \mathbf{r}(t) \cdot \mathbf{r}'(t) = 0 \text{ identically}$$

35.   Write

$$\frac{[\mathbf{f}(t+h) \times \mathbf{g}(t+h)] - [\mathbf{f}(t) \times \mathbf{g}(t)]}{h}$$

as

$$\left(\mathbf{f}(t+h) \times \left[\frac{\mathbf{g}(t+h) - \mathbf{g}(t)}{h}\right]\right) + \left(\left[\frac{\mathbf{f}(t+h) - \mathbf{f}(t)}{h}\right] \times \mathbf{g}(t)\right)$$

and take the limit as $h \to 0$. (Appeal to Theorem 13.1.3.)

SECTION 13.3

1. $\mathbf{r}'(t) = -\pi \sin \pi t\, \mathbf{i} + \pi \cos \pi t\, \mathbf{j} + \mathbf{k}$, $\mathbf{r}'(2) = \pi \mathbf{j} + \mathbf{k}$

$\mathbf{R}(u) = (\mathbf{i} + 2\mathbf{k}) + u(\pi \mathbf{j} + \mathbf{k})$

3. $\mathbf{r}'(t) = \mathbf{b} + 2t\mathbf{c}$, $\mathbf{r}'(-1) = \mathbf{b} - 2\mathbf{c}$, $\mathbf{R}(u) = (\mathbf{a} - \mathbf{b} + \mathbf{c}) + u(\mathbf{b} - 2\mathbf{c})$

5. $\mathbf{r}'(t) = 4t\mathbf{i} - \mathbf{j} + 4t\mathbf{k}$, P is tip of $\mathbf{r}(1)$, $\mathbf{r}'(1) = 4\mathbf{i} - \mathbf{j} + 4\mathbf{k}$

$\mathbf{R}(u) = (2\mathbf{i} + 5\mathbf{k}) + u(4\mathbf{i} - \mathbf{j} + 4\mathbf{k})$

7. $\mathbf{r}'(t) = -2 \sin t\, \mathbf{i} + 3 \cos t\, \mathbf{j} + \mathbf{k}$, $\mathbf{r}'(\pi/4) = -\sqrt{2}\,\mathbf{i} + \frac{3}{2}\sqrt{2}\,\mathbf{j} + \mathbf{k}$

$\mathbf{R}(u) = \left(\sqrt{2}\,\mathbf{i} + \frac{3}{2}\sqrt{2}\,\mathbf{j} + \frac{\pi}{4}\mathbf{k}\right) + u\left(-\sqrt{2}\,\mathbf{i} + \frac{3}{2}\sqrt{2}\,\mathbf{j} + \mathbf{k}\right)$

9.   The scalar components $x(t) = at$ and $y(t) = bt^2$ satisfy the equation

$$a^2 y(t) = a^2(bt^2) = b(a^2 t^2) = b[x(t)]^2$$

and generate the parabola $a^2 y = bx^2$.

11. $\mathbf{r}(t) = t\mathbf{i} + (1 + t^2)\mathbf{j}$, $\mathbf{r}'(t) = \mathbf{i} + 2t\mathbf{j}$

(a) $\mathbf{r}(t) \perp \mathbf{r}'(t) \implies \mathbf{r}(t) \cdot \mathbf{r}'(t) = [t\mathbf{i} + (1+t^2)\mathbf{j}] \cdot (\mathbf{i} + 2t\mathbf{j})$

$$= t(2t^2 + 3) = 0 \implies t = 0$$

$\mathbf{r}(t)$ and $\mathbf{r}'(t)$ are perpendicular at $(0, 1)$.

(b) and (c) $\mathbf{r}(t) = \alpha \mathbf{r}'(t)$ with $\alpha \neq 0 \implies t = \alpha$ and $1 + t^2 = 2t\alpha \implies t = \pm 1$.

If $\alpha > 0$, then $t = 1$. $\mathbf{r}(t)$ and $\mathbf{r}'(t)$ have the same direction at $(1, 2)$.

If $\alpha < 0$, then $t = -1$. $\mathbf{r}(t)$ and $\mathbf{r}'(t)$ have opposite directions at $(-1, 2)$.

13. The tangent line at $t = t_0$ has the form $\mathbf{R}(u) = \mathbf{r}(t_0) + u\mathbf{r}'(t_0)$. If $\mathbf{r}'(t_0) = \alpha \mathbf{r}(t_0)$, then

$$\mathbf{R}(u) = \mathbf{r}(t_0) + u\,\alpha \mathbf{r}(t_0) = (1 + u\alpha)\,\mathbf{r}(t_0).$$

The tangent line passes through the origin at $u = -1/\alpha$.

15. $\mathbf{r}_1(t)$ passes through $P(0, 0, 0)$ at $t = 0$; $\mathbf{r}_2(u)$ passes through $P(0, 0, 0)$ at $u = -1$.

$$\mathbf{r}_1'(t) = e^t\,\mathbf{i} + 2\,\cos t\,\mathbf{j} + \frac{1}{t+1}\,\mathbf{k}; \quad \mathbf{r}_1'(0) = \mathbf{i} + 2\mathbf{j} + \mathbf{k}$$

$$\mathbf{r}_2'(u) = \mathbf{i} + 2u\,\mathbf{j} + 3u^2\,\mathbf{k}; \quad \mathbf{r}_2'(-1) = \mathbf{i} - 2\mathbf{j} + 3\mathbf{k}$$

$$\cos\theta = \frac{\mathbf{r}_1'(0) \cdot \mathbf{r}_2'(1)}{\|\mathbf{r}_1'(0)\|\,\|\mathbf{r}_2'(1)\|} = 0; \quad \theta = \frac{\pi}{2} \cong 1.57, \text{ or } 90°.$$

17. $\mathbf{r}_1(t) = \mathbf{r}_2(u)$ implies

$$\left.\begin{array}{c} e^t = u \\ 2\sin\left(t + \tfrac{1}{2}\pi\right) = 2 \\ t^2 - 2 = u^2 - 3 \end{array}\right\} \quad \text{so that} \quad t = 0, \quad u = 1.$$

The point of intersection is $(1, 2, -2)$.

$$\mathbf{r}_1'(t) = e^t\mathbf{i} + 2\cos\left(t + \frac{\pi}{2}\right)\mathbf{j} + 2t\mathbf{k}, \quad \mathbf{r}_1'(0) = \mathbf{i}$$

$$\mathbf{r}_2'(u) = \mathbf{i} + 2u\mathbf{k}, \quad \mathbf{r}_2'(1) = \mathbf{i} + 2\mathbf{k}$$

$$\cos\theta = \frac{\mathbf{r}_1'(0) \cdot \mathbf{r}_2'(1)}{\|\mathbf{r}_1'(0)\|\,\|\mathbf{r}_2'(1)\|} = \frac{1}{5}\sqrt{5} \cong 0.447, \quad \theta \cong 1.11 \text{ radians}$$

19. (a) $\mathbf{r}(t) = a\cos t\,\mathbf{i} + b\sin t\,\mathbf{j}$ (b) $\mathbf{r}(t) = a\cos t\,\mathbf{i} - b\sin t\,\mathbf{j}$

   (c) $\mathbf{r}(t) = a\cos 2t\,\mathbf{i} + b\sin 2t\,\mathbf{j}$ (d) $\mathbf{r}(t) = a\cos 3t\,\mathbf{i} - b\sin 3t\,\mathbf{j}$

21. $\mathbf{r}'(t) = t^3\,\mathbf{i} + 2t\,\mathbf{j}$ **23.** $\mathbf{r}'(t) = 2e^{2t}\,\mathbf{i} - 4e^{-4t}\,\mathbf{j}$ **25.** $\mathbf{r}'(t) = -2\,\sin t\,\mathbf{i} + 3\,\cos t\,\mathbf{j}$

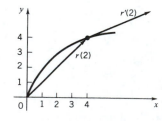 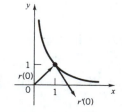 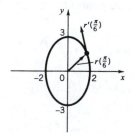

27. $\mathbf{r}(t) = (t^2 + 1)\,\mathbf{i} + t\mathbf{j}, \quad t \geq 1;$ or, $\mathbf{r}(t) = \sec^2 t\,\mathbf{i} + \tan t\,\mathbf{j}, \quad t \in \left[\tfrac{1}{4}\pi, \tfrac{1}{2}\pi\right)$

29. $\mathbf{r}(t) = \cos t \sin 3t\,\mathbf{i} = \sin t \sin 3t\,\mathbf{j}, \quad t \in [0, \pi]$

31. $y^3 = x^2$

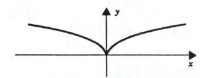

There is no tangent vector at the origin.

33.   We substitute $x = t$, $y = t^2$, $z = t^3$ in the plane equation to obtain

$$4t + 2t^2 + t^3 = 24, \quad (t - 2)\left(t^2 + 4t + 12\right) = 0, \quad t = 2.$$

The twisted cubic intersects the plane at the tip of $\mathbf{r}(2)$, the point $(2, 4, 8)$.

The angle between the curve and the normal line at the point of intersection is the angle between the tangent vector $\mathbf{r}'(2) = \mathbf{i} + 4\mathbf{j} + 12\mathbf{k}$ and the normal $\mathbf{N} = 4\mathbf{i} + 2\mathbf{j} + \mathbf{k}$:

$$\cos\theta = \frac{(\mathbf{i} + 4\mathbf{j} + 12\mathbf{k}) \cdot (4\mathbf{i} + 2\mathbf{j} + \mathbf{k})}{\|\mathbf{i} + 4\mathbf{j} + 12\mathbf{k}\|\,\|4\mathbf{i} + 2\mathbf{j} + \mathbf{k}\|} = \frac{24}{\sqrt{161}\,\sqrt{21}} \cong 0.412, \quad \theta \cong 1.15 \text{ radians.}$$

35. $\mathbf{r}'(t) = 2\mathbf{j} + 2t\,\mathbf{k}, \quad \|\mathbf{r}'(t)\| = 2\sqrt{1 + t^2}$

$$\mathbf{T}(t) = \frac{\mathbf{r}'(t)}{\|\mathbf{r}'(t)\|} = \frac{1}{\sqrt{1 + t^2}}\,(\mathbf{j} + t\,\mathbf{k}),$$

$$\mathbf{T}'(t) = \frac{1}{(1 + t^2)^{3/2}}\,[-t\,\mathbf{j} + \mathbf{k}]$$

at $t = 1$:   tip of $\mathbf{r} = (1, 2, 1)$, $\quad \mathbf{T} = \mathbf{T}(1) = \dfrac{1}{\sqrt{2}}\mathbf{j} + \dfrac{1}{\sqrt{2}}\mathbf{k};$

$$\mathbf{T}'(1) = -\frac{1}{2\sqrt{2}}\mathbf{j} + \frac{1}{2\sqrt{2}}\mathbf{k}; \quad \|\mathbf{T}'(1)\| = \frac{1}{2}; \quad \mathbf{N} = \mathbf{N}(1) = \frac{\mathbf{T}'(1)}{\|\mathbf{T}'(1)\|} = -\frac{1}{\sqrt{2}}\mathbf{j} + \frac{1}{2\sqrt{2}}\mathbf{k}$$

normal for osculating plane:

$$\mathbf{T} \times \mathbf{N} = \left(\frac{1}{\sqrt{2}}\mathbf{j} + \frac{1}{\sqrt{2}}\mathbf{k}\right) \times \left(-\frac{1}{\sqrt{2}}\mathbf{j} + \frac{1}{2\sqrt{2}}\mathbf{k}\right) = \frac{1}{2}\mathbf{i}$$

equation for osculating plane:

$$\frac{1}{2}(x - 1) + 0(y - 2) + 0(z - 1) = 0, \quad \text{which gives} \quad x - 1 = 0$$

37.

$$\mathbf{r}'(t) = -2\sin 2t\,\mathbf{i} + 2\cos 2t\,\mathbf{j} + \mathbf{k}, \quad \|\mathbf{r}'(t)\| = \sqrt{5}$$

$$\mathbf{T}(t) = \frac{\mathbf{r}'(t)}{\|\mathbf{r}'(t)\|} = \frac{1}{5}\sqrt{5}\,(-2\sin 2t\,\mathbf{i} + 2\cos 2t\,\mathbf{j} + \mathbf{k})$$

$$\mathbf{T}'(t) = -\tfrac{4}{5}\sqrt{5}\,(\cos 2t\,\mathbf{i} + \sin 2t\,\mathbf{j}), \quad \|\mathbf{T}'(t)\| = \tfrac{4}{5}\sqrt{5}$$

$$\mathbf{N}(t) = \frac{\mathbf{T}'(t)}{\|\mathbf{T}'(t)\|} = -(\cos 2t\,\mathbf{i} + \sin 2t\,\mathbf{j})$$

at $t = \pi/4$:   tip of $\mathbf{r} = (0, 1, \pi/4)$, $\mathbf{T} = \tfrac{1}{5}\sqrt{5}\,(-2\mathbf{i} + \mathbf{k})$, $\mathbf{N} = -\mathbf{j}$

normal for osculating plane:

$$\mathbf{T} \times \mathbf{N} = \tfrac{1}{5}\sqrt{5}\,(-2\mathbf{i} + \mathbf{k}) \times (-\mathbf{j}) = \tfrac{1}{5}\sqrt{5}\,\mathbf{i} + \tfrac{2}{5}\sqrt{5}\,\mathbf{k}$$

equation for osculating plane:

$$\frac{1}{5}\sqrt{5}\,(x - 0) + \frac{2}{5}\sqrt{5}\left(z - \frac{\pi}{4}\right) = 0, \quad \text{which gives} \quad x + 2z = \frac{\pi}{2}$$

39.

$$\mathbf{r}'(t) = \mathbf{i} + 2t\mathbf{j} + 3t^2\mathbf{k}, \quad \|\mathbf{r}'(t)\| = \sqrt{1 + 4t^2 + 9t^4}$$

$$\mathbf{T}(t) = \frac{\mathbf{r}'(t)}{\|\mathbf{r}'(t)\|} = \frac{1}{\sqrt{1 + 4t^2 + 9t^4}}\,(\mathbf{i} + 2t\mathbf{j} + 3t^2\mathbf{k}),$$

$$\mathbf{T}'(t) = \frac{1}{(1 + 4t^2 + 9t^4)^{3/2}}\left[(-4t - 18t^3)\,\mathbf{i} + (2 - 18t^4)\,\mathbf{j} + (6t + 12t^3)\,\mathbf{k}\right]$$

at $t = 1$:   tip of $\mathbf{r} = (1, 1, 1)$, $\mathbf{T} = \dfrac{1}{\sqrt{14}}\,(\mathbf{i} + 2\mathbf{j} + 3\mathbf{k})$,

$$\mathbf{T}' = \frac{1}{7\sqrt{14}}\,(-11\mathbf{i} - 8\mathbf{j} + 9\mathbf{k}), \quad \|\mathbf{T}'\| = \frac{\sqrt{266}}{7\sqrt{14}}, \quad \mathbf{N} = \frac{1}{\sqrt{266}}\,(-11\mathbf{i} - 8\mathbf{j} + 9\mathbf{k})$$

normal for osculating plane:

$$\mathbf{T} \times \mathbf{N} = \frac{1}{\sqrt{14}}\,(\mathbf{i} + 2\mathbf{j} + 3\mathbf{k}) \times \frac{1}{\sqrt{266}}\,(-11\mathbf{i} - 8\mathbf{j} + 9\mathbf{k}) = \frac{\sqrt{19}}{19}\,(3\mathbf{i} - 3\mathbf{j} + \mathbf{k})$$

equation for osculating plane:

$$3(x - 1) - 3(y - 1) + (z - 1) = 0, \quad \text{which gives} \quad 3x - 3y + z = 1$$

41.

$$\mathbf{r}'(t) = e^t\left[(\sin t + \cos t)\,\mathbf{i} + (\cos t - \sin t)\,\mathbf{j} + \mathbf{k}\right], \quad \|\mathbf{r}'(t)\| = e^t\sqrt{3}$$

$$\mathbf{T}(t) = \frac{\mathbf{r}'(t)}{\|\mathbf{r}'(t)\|} = \frac{1}{\sqrt{3}}\left[(\sin t + \cos t)\,\mathbf{i} + (\cos t - \sin t)\,\mathbf{j} + \mathbf{k}\right],$$

$$\mathbf{T}'(t) = \frac{1}{\sqrt{3}}\left[(\cos t - \sin t)\,\mathbf{i} - (\sin t + \cos t)\,\mathbf{j}\right]$$

at $t = 0$:   tip of $\mathbf{r} = (0, 1, 1)$, $\mathbf{T} = \mathbf{T}(0) = \dfrac{1}{\sqrt{3}}(\mathbf{i} + \mathbf{j} + \mathbf{k})$;

$\mathbf{T}'(0) = \dfrac{1}{\sqrt{3}}(\mathbf{i} - \mathbf{j})$; $\|\mathbf{T}'(0)\| = \dfrac{\sqrt{2}}{\sqrt{3}}$; $\mathbf{N} = \mathbf{N}(0) = \dfrac{\mathbf{T}'(0)}{\|\mathbf{T}'(0)\|} = \dfrac{1}{\sqrt{2}}(\mathbf{i} - \mathbf{j})$

normal for osculating plane:

$$\mathbf{T} \times \mathbf{N} = \frac{1}{\sqrt{3}}(\mathbf{i} + \mathbf{j} + \mathbf{k}) \times \frac{1}{\sqrt{2}}(\mathbf{i} - \mathbf{j}) = \frac{1}{\sqrt{6}}(\mathbf{i} + \mathbf{j} - 2\mathbf{k})$$

equation for osculating plane:

$$\frac{1}{\sqrt{6}}(x - 0) + \frac{1}{\sqrt{6}}(y - 1) - \frac{2}{\sqrt{6}}(z - 1) = 0, \quad \text{which gives} \quad x + y - 2z + 1 = 0$$

43. $\mathbf{T}_1 = \dfrac{\mathbf{R}'(u)}{\|\mathbf{R}'(u)\|} = -\dfrac{\mathbf{r}'(a + b - u)}{\|\mathbf{r}'(a + b - u)\|} = -\mathbf{T}.$

Therefore $\mathbf{T}_1'(u) = \mathbf{T}'(a + b - u)$ and $\mathbf{N}_1 = \mathbf{N}.$

45.   Let $\mathbf{T}$ be the unit tangent at the tip of $\mathbf{R}(u) = \mathbf{r}(\phi(u))$ as calculated from the parametrization $\mathbf{r}$ and let $\mathbf{T}_1$ be the unit tangent at the same point as calculated from the parametrization $\mathbf{R}$.   Then

$$\mathbf{T}_1 = \frac{\mathbf{R}'(u)}{\|\mathbf{R}'(u)\|} = \frac{\mathbf{r}'(\phi(u))\,\phi'(u)}{\|\mathbf{r}'(\phi(u))\,\phi'(u)\|} = \frac{\mathbf{r}'(\phi(u))}{\|\mathbf{r}'(\phi(u))\|} = \mathbf{T}.$$

$$\phi'(u) > 0$$

This shows the invariance of the unit tangent.

The invariance of the principal normal and the osculating plane follows directly from the invariance of the unit tangent.

47.   (a)   Let $t = \Psi(v) = 2\pi - v^2$. When t increases from 0 to 2π, v decreases from $\sqrt{2\pi}$ to 0.

(b) $\mathbf{r}(t) = 2\cos t\,\mathbf{i} + 2\sin t\,\mathbf{j} + 4t\,\mathbf{k}$, $\mathbf{r}'(t) = -2\sin t\,\mathbf{i} + 2\cos t\,\mathbf{j} + 4\,\mathbf{k}$, $\|\mathbf{r}'(t)\| = 2\sqrt{5}$

$\mathbf{T_r}(t) = -\dfrac{1}{\sqrt{5}}\sin t\,\mathbf{i} + \dfrac{1}{\sqrt{5}}\cos t\,\mathbf{j} + \dfrac{2}{\sqrt{5}}\,\mathbf{k}$, $\mathbf{T_r}'(t) = -\dfrac{1}{\sqrt{5}}\cos t\,\mathbf{i} - \dfrac{1}{\sqrt{5}}\sin t\,\mathbf{j}$,

$\|\mathbf{T_r}'(t)\| = 1/\sqrt{5}$

$\mathbf{N_r}(t) = -\cos t\,\mathbf{i} - \sin t\,\mathbf{j}$,

$\mathbf{R}(v) = 2\cos(2\pi - v^2)\,\mathbf{i} + 2\sin(2\pi - v^2)\,\mathbf{j} + 4(2\pi - v^2)\,\mathbf{k}$

$\mathbf{R}'(t) = 4v\sin(2\pi - v^2)\,\mathbf{i} - 4v\cos(2\pi - v^2)\,\mathbf{j} + -8v\,\mathbf{k}$, $\|\mathbf{R}'(t)\| = 4v\sqrt{5}$

$\mathbf{T_R}(t) = \dfrac{1}{\sqrt{5}}\sin(2\pi - v^2)\,\mathbf{i} - \dfrac{1}{\sqrt{5}}\cos(2\pi - v^2)\,\mathbf{j} - \dfrac{2}{\sqrt{5}}\,\mathbf{k}$

$\mathbf{T_R}'(t) = -\dfrac{2v}{\sqrt{5}}\cos(2\pi - v^2)\,\mathbf{i} - \dfrac{2v}{\sqrt{5}}\sin(2\pi - v^2)\,\mathbf{j}$, $\|\mathbf{T_R}'(t)\| = 2v/\sqrt{5}$

$$\mathbf{N_R}(t) = -\cos(2\pi - v^2)\,\mathbf{i} - \sin(2\pi - v^2)\,\mathbf{j}$$

$$\mathbf{T_r}(\pi/4) = -\frac{1}{\sqrt{5}}\sin(\pi/4)\,\mathbf{i} + \frac{1}{\sqrt{5}}\cos(\pi/4)\,\mathbf{j} + \frac{2}{\sqrt{5}}\,\mathbf{k} = -\frac{1}{\sqrt{10}}\,\mathbf{i} + \frac{1}{\sqrt{10}}\,\mathbf{j} + \frac{2}{\sqrt{5}}\,\mathbf{k}$$

$$\mathbf{T_R}(\sqrt{7\pi}/2) = \frac{1}{\sqrt{5}}\sin(2\pi - 7\pi/4)\,\mathbf{i} - \frac{1}{\sqrt{5}}\cos(2\pi - 7\pi/4)\,\mathbf{j} - \frac{2}{\sqrt{5}}\,\mathbf{k} = \frac{1}{\sqrt{10}}\,\mathbf{i} - \frac{1}{\sqrt{10}}\,\mathbf{j} - \frac{2}{\sqrt{5}}\,\mathbf{k} = -\mathbf{T_r}$$

Similarly, $\quad \mathbf{N_r}(\pi/4) = -\dfrac{1}{\sqrt{2}}\,\mathbf{i} - \dfrac{1}{\sqrt{2}}\,\mathbf{j} = \mathbf{N_R}(\sqrt{7\pi}/2)$

SECTION 13.4

1. $\quad \mathbf{r}'(t) = \mathbf{i} + t^{1/2}\,\mathbf{j}, \quad \|\mathbf{r}'(t)\| = \sqrt{1+t}$

$$L = \int_0^8 \sqrt{1+t}\,dt = \left[\frac{2}{3}(1+t)^{3/2}\right]_0^8 = \frac{52}{3}$$

3. $\quad \mathbf{r}'(t) = -a\sin t\,\mathbf{i} + a\cos t\,\mathbf{j} + b\mathbf{k}, \quad \|\mathbf{r}'(t)\| = \sqrt{a^2 + b^2}$

$$L = \int_0^{2\pi} \sqrt{a^2 + b^2}\,dt = 2\pi\sqrt{a^2 + b^2}$$

5. $\quad \mathbf{r}'(t) = \mathbf{i} + \tan t\,\mathbf{j}, \quad \|\mathbf{r}'(t)\| = \sqrt{1 + \tan^2 t} = |\sec t|$

$$L = \int_0^{\pi/4} |\sec t|\,dt = \int_0^{\pi/4} \sec t\,dt = [\ln|\sec t + \tan t|]_0^{\pi/4} = \ln(1 + \sqrt{2})$$

7. $\quad \mathbf{r}'(t) = 3t^2\mathbf{i} + 2t\mathbf{j}, \quad \|\mathbf{r}'(t)\| = \sqrt{9t^4 + 4t^2} = |t|\sqrt{4 + 9t^2}$

$$L = \int_0^1 \left|t\sqrt{4 + 9t^2}\right|\,dt = \int_0^1 t\sqrt{4 + 9t^2}\,dt = \left[\frac{1}{27}(4 + 9t^2)^{3/2}\right]_0^1 = \frac{1}{27}\left(13\sqrt{13} - 8\right)$$

9. $\quad \mathbf{r}'(t) = (\cos t - \sin t)e^t\mathbf{i} + (\sin t + \cos t)e^t\mathbf{j}, \quad \|\mathbf{r}'(t)\| = \sqrt{2}\,e^t$

$$L = \int_0^{\pi} \sqrt{2}\,e^t\,dt = \sqrt{2}\,(e^{\pi} - 1)$$

11. $\quad \mathbf{r}'(t) = 2\mathbf{i} + 2t\mathbf{j} - 2t\mathbf{k}, \quad \|\mathbf{r}'(t)\| = 2\sqrt{1 + 2t^2}$

$$L = \int_0^2 2\sqrt{1 + 2t^2}\,dt = \sqrt{2}\int_0^{\tan^{-1}(2\sqrt{2})} \sec^3 u\,du$$

$$(t\sqrt{2} = \tan u)$$

$$= \tfrac{1}{2}\sqrt{2}\,[\sec u\tan u + \ln|\sec u + \tan u|]_0^{\tan^{-1}(2\sqrt{2})} = 6 + \tfrac{1}{2}\sqrt{2}\,\ln(3 + 2\sqrt{2})$$

13. $\mathbf{r}'(t) = \dfrac{1}{t}\mathbf{i} + 2\mathbf{j} + 2t\,\mathbf{k}, \quad \|\mathbf{r}'(t)\| = \sqrt{\dfrac{1}{t^2} + 4 + 4t^2}$

$$L = \int_1^e \sqrt{\dfrac{1}{t^2} + 4 + 4t^2}\, dt = \int_1^e \left(\dfrac{1}{t} + 2t\right) dt = \left[\ln|t| + t^2\right]_1^e = e^2$$

15.
$$s = s(t) = \int_a^t \|\mathbf{r}'(u)\|\, du$$

$$s'(t) = \|\mathbf{r}'(t)\| = \|x'(t)\,\mathbf{i} + y'(t)\,\mathbf{j} + z'(t)\,\mathbf{k}\|$$

$$= \sqrt{[x'(t)]^2 + [y'(t)]^2 + [z'(t)]^2}.$$

In the Leibniz notation this translates to

$$\dfrac{ds}{dt} = \sqrt{\left(\dfrac{dx}{dt}\right)^2 + \left(\dfrac{dy}{dt}\right)^2 + \left(\dfrac{dz}{dt}\right)^2}.$$

17.
$$s = s(x) = \int_a^x \sqrt{1 + [f'(t)]^2}\, dt$$

$$s'(x) = \sqrt{1 + [f'(x)]^2}.$$

In the Leibniz notation this translates to

$$\dfrac{ds}{dx} = \sqrt{1 + \left(\dfrac{dy}{dx}\right)^2}.$$

19. Let L be the length as computed from $\mathbf{r}$ and L^* the length as computed from $\mathbf{R}$. Then

$$L^* = \int_c^d \|\mathbf{R}'(u)\|\, du = \int_c^d \|\mathbf{r}'(\phi(u))\|\, \phi'(u)\, du = \int_a^b \|\mathbf{r}'(t)\|\, dt = L.$$

$$t = \phi(u)$$

21. (a) $s = \displaystyle\int_0^t \sqrt{(-3\sin t)^2 + (3\cos t)^2 + 4^2}\, dt = \int_0^t 5\, dt = 5t$

(b) $t = \dfrac{s}{5}; \qquad R(s) = 3\cos\left(\dfrac{s}{5}\right)\mathbf{i} + 3\sin\left(\dfrac{s}{5}\right)\mathbf{j} + \dfrac{4s}{5}\mathbf{k}$

(c) $\mathbf{r}(t) = 3\cos t\,\mathbf{i} + 3\sin t\,\mathbf{j} + 4t\,\mathbf{k} = 3\,\mathbf{i} + 0\,\mathbf{j} + 0\,\mathbf{k} \implies t = 0$

From part (a), the arc length $s = 5t$ and $5t = 5\pi \implies t = \pi; \quad \mathbf{r}(\pi) = -3\,\mathbf{i} + 0\,wj + 4\pi\,\mathbf{k}$

$\implies Q(-3, 0, 4\pi).$

(d) $\mathbf{R}'(s) = -\frac{3}{5} \sin\left(\frac{s}{5}\right) \mathbf{i} + \frac{3}{5} \cos\left(\frac{s}{5}\right) \mathbf{j} + \frac{4}{5} \mathbf{k};$

$$\|\mathbf{R}'(s)\| = \sqrt{\left[-\frac{3}{5} \sin\left(\frac{s}{5}\right)\right]^2 + \left[\frac{3}{5} \cos\left(\frac{s}{5}\right)\right]^2 + \left[\frac{4}{5}\right]^2} = 1$$

23. $\mathbf{r}'(t) = t^{3/2} \mathbf{j} + \mathbf{k}, \quad \|\mathbf{r}'(t)\| = \sqrt{\left(t^{3/2}\right)^2 + 1} = \sqrt{t^3 + 1}$

$$s = \int_0^{1/2} \sqrt{t^3 + 1} \, dt \cong 0.5077$$

25. $\mathbf{r}'(t) = -3\sin t \, \mathbf{i} + 4\cos t \, \mathbf{j}, \quad \|\mathbf{r}'(t)\| = \sqrt{9\sin^2 t + 16\cos^2 t} \, dt$

$$s = \int_0^{2\pi} \sqrt{9\sin^2 t + 16\cos^2 t} \, dt \cong 22.0939$$

SECTION 13.5

1. $\mathbf{r}(t) = r[\cos\theta(t)\,\mathbf{i} + \sin\theta(t)\,\mathbf{j}]$

$\mathbf{r}'(t) = r[-\sin\theta(t)\,\mathbf{i} + \cos\theta(t)\,\mathbf{j})]\theta'(t)$

$\|\mathbf{r}'(t)\| = v \implies r|\theta'(t)| = v \implies |\theta'(t)| = v/r$

$\mathbf{r}''(t) = r[-\cos\theta(t)\,\mathbf{i} - \sin\theta(t)\,\mathbf{j}]\,[\theta'(t)]^2$

$\|\mathbf{r}''(t)\| = r[\theta'(t)]^2 = v^2/r$

3. $\mathbf{r}(t) = at\mathbf{i} + b\sin at\,\mathbf{j}$

$\mathbf{r}'(t) = a\mathbf{i} + ab\cos at\,\mathbf{j}$

$\mathbf{r}''(t) = -a^2 b\sin at\,\mathbf{j}$

$\|\mathbf{r}''(t)\| = a^2|b\sin at|$

$= a^2|y(t)|$

5. $y = \cos\pi x, \quad 0 \le x \le 2$

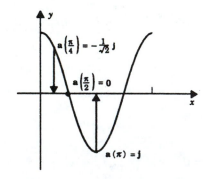

7. $x = \sqrt{1 + y^2}, \quad y \ge -1$

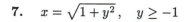
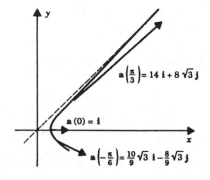

9. (a) initial position is tip of $\mathbf{r}(0) = x_0\,\mathbf{i} + y_0\,\mathbf{j} + z_0\,\mathbf{k}$

    (b) $\mathbf{r}'(t) = (\alpha\cos\theta)\,\mathbf{j} + (\alpha\sin\theta - 32t)\,\mathbf{k}, \quad \mathbf{r}'(0) = (\alpha\cos\theta)\,\mathbf{j} + (\alpha\sin\theta)\,\mathbf{k}$

(c) $|\mathbf{r}'(0)| = |\alpha|$ (d) $\mathbf{r}''(t) = -32\mathbf{k}$

(e) a parabolic arc from the parabola

$$z = z_0 + (\tan\theta)(y - y_0) - 16\frac{(y - y_0)^2}{\alpha^2 \cos^2\theta}$$

in the plane $x = x_0$

11. (a) $\mathbf{r}'(t) = \dfrac{a\omega}{2}\left(e^{\omega t} - e^{-\omega t}\right)\mathbf{i} + \dfrac{b\omega}{2}\left(e^{\omega t} + e^{-\omega t}\right)\mathbf{j}, \quad \mathbf{r}'(0) = b\omega\mathbf{j}$

(b) $\mathbf{r}''(t) = \dfrac{a\omega^2}{2}\left(e^{\omega t} + e^{-\omega t}\right)\mathbf{i} + \dfrac{b\omega^2}{2}\left(e^{\omega t} - e^{-\omega t}\right)\mathbf{j} = \omega^2\mathbf{r}(t)$

(c) The torque $\boldsymbol{\tau}$ is $\mathbf{0}$: $\quad \boldsymbol{\tau}(t) = \mathbf{r}(t) \times m\mathbf{a}(t) = \mathbf{r}(t) \times m\omega^2\mathbf{r}(t) = \mathbf{0}.$

The angular momentum $\mathbf{L}(t)$ is constant since $\mathbf{L}'(t) = \boldsymbol{\tau}(t) = \mathbf{0}.$

13. We begin with the force equation $\mathbf{F}(t) = \alpha\mathbf{k}$. In general, $\mathbf{F}(t) = m\,\mathbf{a}(t)$, so that here

$$\mathbf{a}(t) = \frac{\alpha}{m}\,\mathbf{k}.$$

Integration gives

$$\mathbf{v}(t) = C_1\mathbf{i} + C_2\mathbf{j} + \left(\frac{\alpha}{m}\,t + C_3\right)\mathbf{k}.$$

Since $\mathbf{v}(0) = 2\mathbf{j}$, we can conclude that $C_1 = 0$, $C_2 = 2$, $C_3 = 0$. Thus

$$\mathbf{v}(t) = 2\mathbf{j} + \frac{\alpha}{m}\,t\mathbf{k}.$$

Another integration gives

$$\mathbf{r}(t) = D_1\mathbf{i} + (2t + D_2)\mathbf{j} + \left(\frac{\alpha}{2m}\,t^2 + D_3\right)\mathbf{k}.$$

Since $\mathbf{r}(0) = y_0\mathbf{j} + z_0\mathbf{k}$, we have $D_1 = 0$, $D_2 = y_0$, $D_3 = z_0$, and therefore

$$\mathbf{r}(t) = (2t + y_0)\mathbf{j} + \left(\frac{\alpha}{2m}\,t^2 + z_0\right)\mathbf{k}.$$

The conditions of the problem require that t be restricted to nonnegative values.
To obtain an equation for the path in Cartesian coordinates, we write out the components

$$x(t) = 0, \quad y(t) = 2t + y_0, \quad z(t) = \frac{\alpha}{2m}\,t^2 + z_0. \quad (t \geq 0)$$

From the second equation we have

$$t = \tfrac{1}{2}\,[y(t) - y_0]. \quad (y(t) \geq y_0)$$

Substituting this into the third equation, we get

$$z(t) = \frac{\alpha}{8m}\,[y(t) - y_0]^2 + z_0. \quad (y(t) \geq y_0)$$

Eliminating t altogether, we have

$$z = \frac{\alpha}{8m}\,(y - y_0)^2 + z_0. \quad (y \geq y_0)$$

Since $x = 0$, the path of the object is a parabolic arc in the yz-plane.

Answers to (a) through (d):

(a)   velocity: $\mathbf{v}(t) = 2\mathbf{j} + \dfrac{\alpha}{m}\, t\mathbf{k}.$ (b)   speed: $v(t) = \dfrac{1}{m}\sqrt{4m^2 + \alpha^2 t^2}.$

(c)   momentum: $\mathbf{p}(t) = 2m\mathbf{j} + \alpha\, t\mathbf{k}.$

(d)   path in vector form: $\mathbf{r}(t) = (2t + y_0)\mathbf{j} + \left(\dfrac{\alpha}{2m}\, t^2 + z_0\right)\mathbf{k}, \quad t \geq 0.$

   path in Cartesian coordinates: $z = \dfrac{\alpha}{8m}(y - y_0)^2 + z_0, \quad y \geq y_0, \quad x = 0.$

15. $\mathbf{F}(t) = m\,\mathbf{a}(t) = m\,\mathbf{r}''(t) = 2m\mathbf{k}$

17.   From $\mathbf{F}(t) = m\,\mathbf{a}(t)$ we obtain

$$\mathbf{a}(t) = \pi^2[a\cos\pi t\,\mathbf{i} + b\sin\pi t\,\mathbf{j}].$$

By direct calculation using $\mathbf{v}(0) = -\pi b\mathbf{j} + \mathbf{k}$ and $\mathbf{r}(0) = b\mathbf{j}$ we obtain

$$\mathbf{v}(t) = a\pi\sin\pi t\,\mathbf{i} - b\pi\cos\pi t\,\mathbf{j} + \mathbf{k}$$

$$\mathbf{r}(t) = a(1 - \cos\pi t)\,\mathbf{i} + b(1 - \sin\pi t)\,\mathbf{j} + t\mathbf{k}.$$

(a) $\mathbf{v}(1) = b\pi\mathbf{j} + \mathbf{k}$ (b) $\|\mathbf{v}(1)\| = \sqrt{\pi^2 b^2 + 1}$

(c) $\mathbf{a}(1) = -\pi^2 a\mathbf{i}$ (d) $m\mathbf{v}(1) = m(\pi b\mathbf{j} + \mathbf{k})$

(e) $\mathbf{L}(1) = \mathbf{r}(1) \times m\mathbf{v}(1) = [2a\mathbf{i} + b\mathbf{j} + \mathbf{k}] \times [m(b\pi\mathbf{j} + \mathbf{k})]$

   $= m[b(1 - \pi)\mathbf{i} - 2a\mathbf{j} + 2ab\pi\mathbf{k}]$

(f) $\boldsymbol{\tau}(1) = \mathbf{r}(1) \times \mathbf{F}(1) = [2a\mathbf{i} + b\mathbf{j} + \mathbf{k}] \times [-m\pi^2 a\mathbf{i}] = -m\pi^2 a[\mathbf{j} - b\mathbf{k}]$

19.   We have $m\mathbf{v} = m\mathbf{v}_1 + m\mathbf{v}_2$ and $\tfrac{1}{2}mv^2 = \tfrac{1}{2}mv_1^2 + \tfrac{1}{2}mv_2^2.$

   Therefore $\mathbf{v} = \mathbf{v}_1 + \mathbf{v}_2$ and $v^2 = v_1^2 + v_2^2.$

   Since $v^2 = \mathbf{v} \cdot \mathbf{v} = (\mathbf{v}_1 + \mathbf{v}_2) \cdot (\mathbf{v}_1 + \mathbf{v}_2) = v_1^2 + v_2^2 + 2(\mathbf{v}_1 \cdot \mathbf{v}_2),$

   we have $\mathbf{v}_1 \cdot \mathbf{v}_2 = 0$ and $\mathbf{v}_1 \perp \mathbf{v}_2.$

21. $\mathbf{r}''(t) = \mathbf{a}, \quad \mathbf{r}'(t) = \mathbf{v}(0) + t\mathbf{a}, \quad \mathbf{r}(t) = \mathbf{r}(0) + t\,\mathbf{v}(0) + \tfrac{1}{2}t^2\,\mathbf{a}.$

If neither $\mathbf{v}(0)$ nor $\mathbf{a}$ is zero, the displacement $\mathbf{r}(t) - \mathbf{r}(0)$ is a linear combination of $\mathbf{v}(0)$ and $\mathbf{a}$ and thus remains on the plane determined by these vectors. The equation of this plane can be written

$$[\mathbf{a} \times \mathbf{v}(0)] \cdot [\mathbf{r} - \mathbf{r}(0)] = 0.$$

(If either $\mathbf{v}(0)$ or $\mathbf{a}$ is zero, the motion is restricted to a straight line; if both of these vectors are zero, the particle remains at its initial position $\mathbf{r}(0)$.)

23. $\mathbf{r}(t) = \mathbf{i} + t\mathbf{j} + \left(\dfrac{qE_0}{2m}\right)t^2\mathbf{k}$

25. $\mathbf{r}(t) = \left(1 + \dfrac{t^3}{6m}\right)\mathbf{i} + \dfrac{t^4}{12m}\mathbf{j} + t\mathbf{k}$

27.   (13.2.3)

$$\frac{d}{dt}\left(\frac{1}{2}mv^2\right) = mv\frac{dv}{dt} = m\left(\mathbf{v}\cdot\frac{d\mathbf{v}}{dt}\right) = m\frac{d\mathbf{v}}{dt}\cdot\mathbf{v} = \mathbf{F}\cdot\frac{d\mathbf{r}}{dt}$$

$$= 4r^2\left(\mathbf{r}\cdot\frac{d\mathbf{r}}{dt}\right) = 4r^2\left(r\frac{dr}{dt}\right) = 4r^3\frac{dr}{dt} = \frac{d}{dt}\left(r^4\right).$$

Therefore $d/dt\left(\frac{1}{2}mv^2 - r^4\right) = 0$ and $\frac{1}{2}mv^2 - r^4$ is a constant E.   Evaluating E from $t = 0$,   we find that $E = 2m$.

Thus $\frac{1}{2}mv^2 - r^4 = 2m$ and $v = \sqrt{4 + (2/m)\,r^4}$.

SECTION 13.6

1.   On Earth: year of length T,   average distance from sun d.

On Venus: year of length αT,   average distance from sun $0.72d$.

Therefore

$$\frac{(\alpha T)^2}{T^2} = \frac{(0.72d)^3}{d^3}.$$

This gives $\alpha^2 = (0.72)^3 \cong 0.372$ and $\alpha \cong 0.615$.   Answer: about 61.5% of an Earth year.

3. $\left(\dfrac{dx}{dt}\right)^2 + \left(\dfrac{dy}{dt}\right)^2 = \left[\dfrac{d}{dt}(r\cos\theta)\right]^2 + \left[\dfrac{d}{dt}(r\sin\theta)\right]^2$

$$= \left[r\,(-\sin\theta)\frac{d\theta}{dt} + \frac{dr}{dt}\cos\theta\right]^2 + \left[r\cos\theta\,\frac{d\theta}{dt} + \frac{dr}{dt}\sin\theta\right]^2$$

$$= r^2\sin^2\theta\left(\frac{d\theta}{dt}\right)^2 + \left(\frac{dr}{dt}\right)^2\cos^2\theta + r^2\cos^2\theta\left(\frac{d\theta}{dt}\right)^2 + \left(\frac{dr}{dt}\right)^2\sin^2\theta$$

$$= \left(\frac{dr}{dt}\right)^2 + r^2\left(\frac{d\theta}{dt}\right)^2$$

5.   Substitute

$$r = \frac{a}{1 + e\cos\theta}, \quad \left(\frac{dr}{d\theta}\right)^2 = \left[\frac{-a}{(1 + e\cos\theta)^2}\cdot(-e\sin\theta)\right]^2 = \frac{(ae\sin\theta)^2}{(1 + e\cos\theta)^4}$$

into the right side of the equation and you will see that, with a and e^2 as given, the expression reduces to E.

SECTION 13.7

1. $k = \dfrac{e^{-x}}{(1 + e^{-2x})^{3/2}}$

3. $y' = \dfrac{1}{2x^{1/2}}; \quad y'' = \dfrac{-1}{4x^{3/2}}$

$$k = \dfrac{\left| -1/4x^{3/2} \right|}{\left[1 + (1/2x^{1/2})^2 \right]^{3/2}} = \dfrac{2}{(1 + 4x)^{3/2}}$$

5. $k = \dfrac{\sec^2 x}{(1 + \tan^2 x)^{3/2}} = |\cos x|$

7. $k = \dfrac{|\sin x|}{(1 + \cos^2 x)^{3/2}}$

9. $k = \dfrac{|x|}{(1 + x^4/4)^{3/2}}; \quad$ at $\left(2, \dfrac{4}{3} \right), \quad \rho = \dfrac{5}{2}\sqrt{5}$

11. $k = \dfrac{\left| -1/y^3 \right|}{(1 + 1/y^2)^{3/2}} = \dfrac{1}{(1 + y^2)^{3/2}}; \quad$ at $(2, 2), \quad \rho = 5\sqrt{5}$

13. $y'(x) = \dfrac{1}{x + 1}, \quad y'(2) = \dfrac{1}{3}; \quad y''(x) = \dfrac{-1}{(x+1)^2}, \quad y''(2) = -\dfrac{1}{9}.$

At $x = 2, \quad k = \dfrac{\left| -\dfrac{1}{9} \right|}{\left[1 + \left(\dfrac{1}{3} \right)^2 \right]^{3/2}} = \dfrac{3}{10\sqrt{10}}; \quad \rho = \dfrac{10\sqrt{10}}{3}$

15. $k(x) = \dfrac{\left| -1/x^2 \right|}{(1 + 1/x^2)^{3/2}} = \dfrac{x}{(x^2 + 1)^{3/2}}, \quad x > 0$

$k'(x) = \dfrac{(1 - 2x^2)}{(x^2 + 1)^{5/2}}, \quad k'(x) = 0 \implies x = \dfrac{1}{2}\sqrt{2}$

Since k increases on $\left(0, \dfrac{1}{2}\sqrt{2} \right]$ and decreases on $\left[\dfrac{1}{2}\sqrt{2}, \infty \right)$, k is maximal at $\left(\dfrac{1}{2}\sqrt{2}, \dfrac{1}{2}\ln\dfrac{1}{2} \right)$.

17. $x(t) = t, \quad x'(t) = 1, \quad x''(t) = 0; \qquad y(t) = \dfrac{1}{2}t^2, \quad y'(t) = t, \quad y''(t) = 1$

$k = \dfrac{1}{(1 + t^2)^{3/2}}$

19. $x(t) = 2t, \quad x'(t) = 2, \quad x''(t) = 0; \qquad y(t) = t^3, \quad y'(t) = 3t^2, \quad y''(t) = 6t$

$k = \dfrac{12|t|}{(4 + 9t^4)^{3/2}}$

21. $x(t) = e^t \cos t, \quad x'(t) = e^t(\cos t - \sin t), \quad x''(t) = -2e^t \sin t$

$y(t) = e^t \sin t, \quad y'(t) = e^t(\sin t + \cos t), \quad y''(t) = 2e^t \cos t$

$$k = \frac{\left|2e^{2t}\cos t\,(\cos t - \sin t) + 2e^{2t}\sin t\,(\cos t + \sin t)\right|}{\left[\,e^{2t}(\cos t - \sin t)^2 + e^{2t}(\cos t + \sin t)^2\,\right]^{3/2}} = \frac{2e^{2t}}{(2e^{2t})^{3/2}} = \frac{1}{2}\sqrt{2}\,e^{-t}$$

23. $x(t) = t\cos t, \quad x'(t) = \cos t - t\sin t, \quad x''(t) = -2\sin t - t\cos t$

$y(t) = t\sin t, \quad y'(t) = \sin t + t\cos t, \quad y''(t) = 2\cos t - t\sin t$

$$k = \frac{\left|(\cos t - t\sin t)(2\cos t - t\sin t) - (\sin t + t\cos t)(-2\sin t - t\cos t)\right|}{\left[\,(\cos t - t\sin t)^2 + (\sin t + t\cos t)^2\,\right]^{3/2}} = \frac{2 + t^2}{[1 + t^2]^{3/2}}$$

25. $k = \dfrac{\left|2/x^3\right|}{[1 + 1/x^4]^{3/2}} = \dfrac{2\left|x^3\right|}{(x^4 + 1)^{3/2}}; \quad \text{at } x = \pm 1,\ \rho = \dfrac{2^{3/2}}{2} = \sqrt{2}$

27.   We use (13.7.3) and the hint to obtain

$$k = \frac{\left|ab\sinh^2 t - ab\cosh^2 t\right|}{\left[a^2\sinh^2 t + b^2\cosh^2 t\right]^{3/2}} = \frac{\left|\dfrac{a}{b}y^2 - \dfrac{b}{a}x^2\right|}{\left[\left(\dfrac{ay}{b}\right)^2 + \left(\dfrac{bx}{a}\right)^2\right]^{3/2}}$$

$$= \frac{a^3 b^3\left|\dfrac{a}{b}y^2 - \dfrac{b}{a}x^2\right|}{[a^4 y^2 + b^4 x^2]^{3/2}} = \frac{a^4 b^4}{[a^4 y^2 + b^4 x^2]^{3/2}}.$$

29.   By the hint and the fact that $\|\mathbf{T} \times \mathbf{N}\| = 1$,

$$\frac{\|\mathbf{v} \times \mathbf{a}\|}{(ds/dt)^3} = \frac{\left\|\left(\dfrac{ds}{dt}\mathbf{T}\right) \times \left(\dfrac{d^2 s}{dt^2}\mathbf{T} + k\left(\dfrac{ds}{dt}\right)^2 \mathbf{N}\right)\right\|}{(ds/dt)^3}$$

$$\mathbf{T} \times \mathbf{T} = 0 \qquad = \frac{\|\,k\,(ds/dt)^3\,(\mathbf{T} \times \mathbf{N})\,\|}{(ds/dt)^3} = k.$$

31. $\mathbf{r}'(t) = e^t(\cos t - \sin t)\,\mathbf{i} + e^t(\sin t + \cos t)\,\mathbf{j} + e^t\mathbf{k}$

$\dfrac{ds}{dt} = \|\mathbf{r}'(t)\| = \sqrt{3}\,e^t, \quad \dfrac{d^2 s}{dt^2} = \sqrt{3}\,e^t$

$\mathbf{T}(t) = \dfrac{\mathbf{r}'(t)}{\|\mathbf{r}'(t)\|} = \dfrac{1}{\sqrt{3}}\left[(\cos t - \sin t)\,\mathbf{i} + (\sin t + \cos t)\,\mathbf{j} + \mathbf{k}\right]$

$\mathbf{T}'(t) = \dfrac{1}{\sqrt{3}}\left[(-\sin t - \cos t)\,\mathbf{i} + (\cos t - \sin t)\,\mathbf{j}\right]$

Then,
$$k = \frac{\|\mathbf{T}'(t)\|}{ds/dt} = \frac{\sqrt{2/3}}{\sqrt{3}\,e^t} = \frac{1}{3}\sqrt{2}\,e^{-t},$$

$$\mathbf{a_T} = \frac{d^2 s}{dt^2} = \sqrt{3}\,e^t, \quad \mathbf{a_N} = k\left(\frac{ds}{dt}\right)^2 = \sqrt{2}\,e^t.$$

33. $\mathbf{r}'(t) = -2\sin 2t\,\mathbf{i} + 2\cos 2t\,\mathbf{j}; \quad \dfrac{ds}{dt} = \|\mathbf{r}'(t)\| = 2, \quad \dfrac{d^2 s}{dt^2} = 0$

$\mathbf{T}(t) = \dfrac{\mathbf{r}'(t)}{\|\mathbf{r}'(t)\|} = -\sin 2t\,\mathbf{i} + \cos 2t\,\mathbf{j}$

$$\mathbf{T}'(t) = -2\left(\cos 2t\,\mathbf{i} + \sin 2t\,\mathbf{j}\right)$$

Then,

$$k = \frac{\|\mathbf{T}'(t)\|}{ds/dt} = \frac{2}{2} = 1,$$

$$\mathbf{a_T} = \frac{d^2 s}{dt^2} = 0, \quad \mathbf{a_N} = k\left(\frac{ds}{dt}\right)^2 = 1\cdot 4 = 4.$$

35. $\mathbf{r}'(t) = \mathbf{i} + t\mathbf{j} + t^2\mathbf{k}, \quad \dfrac{ds}{dt} = \|\mathbf{r}'(t)\| = \sqrt{t^4 + t^2 + 1}, \quad \dfrac{d^2 s}{dt^2} = \dfrac{2t^3 + t}{\sqrt{t^4 + t^2 + 1}}$

$$\mathbf{T}(t) = \frac{\mathbf{r}'(t)}{\|\mathbf{r}'(t)\|} = \frac{1}{\sqrt{t^4 + t^2 + 1}}\left(\mathbf{i} + t\mathbf{j} + t^2\mathbf{k}\right),$$

$$\mathbf{T}'(t) = \frac{1}{(t^4 + t^2 + 1)^{3/2}}\left[-t\left(2t^2 + 1\right)\mathbf{i} + \left(1 - t^4\right)\mathbf{j} + t\left(t^2 + 2\right)\mathbf{k}\right].$$

Then,

$$k = \frac{\|\mathbf{T}'(t)\|}{ds/dt} = \frac{\sqrt{t^2\left(2t^2 + 1\right)^2 + \left(1 + t^4\right)^2 + t^2\left(t^2 + 2\right)^2}}{\left(t^4 + t^2 + 1\right)^2}$$

$$= \frac{\sqrt{\left(t^4 + 4t^2 + 1\right)\left(t^4 + t^2 + 1\right)}}{\left(t^4 + t^2 + 1\right)^2} = \frac{\sqrt{t^4 + 4t^2 + 1}}{\left(t^4 + t^2 + 1\right)^{3/2}},$$

$$\mathbf{a_T} = \frac{d^2 s}{dt^2} = \frac{2t^3 + t}{\sqrt{t^4 + t^2 + 1}}, \quad \mathbf{a_N} = k\left(\frac{ds}{dt}\right)^2 = \frac{\sqrt{t^4 + 4t^2 + 1}}{\sqrt{t^4 + t^2 + 1}}.$$

37. By Exercise 36

$$k = \frac{\left|\left(e^{a\theta}\right)^2 + 2\left(ae^{a\theta}\right)^2 - \left(e^{a\theta}\right)\left(a^2 e^{a\theta}\right)\right|}{\left[\left(e^{a\theta}\right)^2 + \left(ae^{a\theta}\right)^2\right]^{3/2}} = \frac{e^{-a\theta}}{\sqrt{1 + a^2}}.$$

39. By Exercise 36

$$k = \frac{\left|a^2(1 - \cos\theta)^2 + 2a^2\sin^2\theta - a^2(1 - \cos\theta)(\cos\theta)\right|}{\left[a^2(1 - \cos\theta)^2 + a^2\sin^2\theta\right]^{3/2}}$$

$$= \frac{3a^2(1 - \cos\theta)}{[2a^2(1 - \cos\theta)]^{3/2}} = \frac{3ar}{[2ar]^{3/2}} = \frac{3}{2\sqrt{2ar}}.$$

41. (a) For $0 \le \theta \le \pi$,

$$s(\theta) = \int_\theta^\pi \sqrt{[x'(t)]^2 + [y'(t)]^2}\,dt = \int_\theta^\pi \sqrt{R^2(1 - \cos t)^2 + R^2\sin^2 t}\,dt$$

$$= \int_\theta^\pi R\sqrt{2(1 - \cos t)}\,dt = \int_\theta^\pi 2R\sin\frac{1}{2}t\,dt = 4R\cos\frac{1}{2}\theta = 4R\left|\cos\frac{1}{2}\theta\right|.$$

For $\pi \le \theta \le 2\pi$,

$$s(\theta) = \int_\pi^\theta 2R\sin\frac{1}{2}t\,dt = -4R\cos\frac{1}{2}\theta = 4R\left|\cos\frac{1}{2}\theta\right|.$$

(b) $k(\theta) = \dfrac{|x'(\theta)y''(\theta) - y'(\theta)x''(\theta)|}{\{[x'(\theta)]^2 + [y'(\theta)]\}^{3/2}} = \dfrac{|R(1 - \cos\theta)R\cos\theta - R\sin\theta\,(R\sin\theta)|}{8R^3\sin^3\frac{1}{2}\theta}.$

This reduces to $k(\theta) = 1/(4R\sin\frac{1}{2}\theta)$ and gives $\rho(\theta) = 4R\sin\frac{1}{2}\theta$.

(c) $\rho^2 + s^2 = 16R^2$

43.   Straightforward calculation gives

$$s(\theta) = 4a\left|\cos\tfrac{1}{2}\theta\right| \quad\text{and}\quad \rho(\theta) = \tfrac{4}{3}a\sin\tfrac{1}{2}\theta.$$

Therefore

$$9\rho^2 + s^2 = 16a^2.$$

PROJECT 13.7

1.   The system of equations generated by the specified conditions is:

$a + b + c + d = 3$ $\qquad\qquad\qquad$ $27a + 9b + 3c + d = 7$

$6a + 2b = 0$ $\qquad\qquad\qquad\qquad\quad$ $27\alpha + 9\beta + 3\gamma + \delta = 7$

$729\alpha + 81\beta + 9\gamma + \delta = -2$ $\qquad\quad$ $54\alpha + 2\beta = 0$

$27a + 6b + c = 27\alpha + 6\beta + \gamma$ $\qquad$ $18a + 2b = 18\alpha + 2\beta$

$a \cong -0.1094 \qquad b \cong 0.3281 \qquad\qquad c \cong 2.1094 \qquad\qquad d \cong 0.6719$

$\alpha \cong 0.0365 \qquad \beta \cong -0.9844 \qquad\quad \gamma \cong 6.0469 \qquad\qquad \delta \cong -3.2656$

3. $k = \dfrac{|y''|}{[1+(y')^2]^{3/2}} = \dfrac{Cn(n-1)x^{n-2}}{[1+(cnx^{n-1})^2]^{3/2}}$

At $x = 0$, $k = 0$ as desired. At $x = 1$, want $k = \dfrac{Cn(n-1)}{(1+C^2n^2)^{3/2}} = \dfrac{96}{125}$

$y = \dfrac{1}{4}x^3$ works.

CHAPTER 14

SECTION 14.1

1. dom (f) = the first and third quadrants, including the axes;   ran $(f) = [0, \infty)$

3. dom (f) = the set of all points (x, y) except those on the line $y = -x$;   ran $(f) = (-\infty, 0) \cup (0, \infty)$

5. dom (f) = the entire plane;   ran $(f) = (-1, 1)$ since

$$\frac{e^x - e^y}{e^x + e^y} = \frac{e^x + e^y - 2e^y}{e^x + e^y} = 1 - \frac{2}{e^{x-y} + 1}$$

and the last quotient takes on all values between 0 and 2.

7. dom (f) = the first and third quadrants, excluding the axes;   ran $(f) = (-\infty, \infty)$

9. dom (f) = the set of all points (x, y) with $x^2 < y$ —in other words, the set of all points of the plane above the parabola $y = x^2$;   ran $(f) = (0, \infty)$

11. dom (f) = the set of all points (x, y) with $-3 \le x \le 3$, $-2 \le y \le 2$ (a rectangle);   ran $(f) = [-2, 3]$

13. dom (f) = the set of all points (x, y, z) not on the plane $x + y + z = 0$;   ran $(f) = \{-1, 1\}$

15. dom (f) = the set of all points (x, y, z) with $|y| < |x|$;   ran $(f) = (-\infty, 0]$

17. dom (f) = the set of all points (x, y) with $x^2 + y^2 < 9$ —in other words, the set of all points of the plane inside the circle $x^2 + y^2 = 9$;   ran $(f) = [2/3, \infty)$

19. dom (f) = the set of all points (x, y, z) with $x + 2y + 3z > 0$ —in other words, the set of all points in space that lie on the same side of the plane $x + 2y + 3z = 0$ as the point $(1, 1, 1)$;   ran $(f) = (-\infty, \infty)$

21. dom (f) = all of space;   ran $(f) = (0, \infty)$

23. dom $(f) = \{x : x \ge 0\}$; range $(f) = [0, \infty)$

dom $(g) = \{(x, y) : x \ge 0, \ y \ \text{real}\}$; range $(g) = [0, \infty)$

dom $(h) = \{(x, y, z) : x \ge 0, \ y, z \ \text{real}\}$; range $(h) = [0, \infty)$

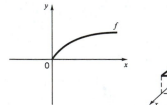
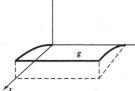

25. $\displaystyle \lim_{h \to 0} \frac{f(x + h, y) - f(x, y)}{h} = \lim_{h \to 0} \frac{2(x + h)^2 - y - (2x^2 - y)}{h} = \lim_{h \to 0} \frac{4xh + 2h^2}{h} = 4x$

$\displaystyle \lim_{h \to 0} \frac{f(x, y + h) - f(x, y)}{h} = \lim_{h \to 0} \frac{2x^2 - (y + h) - (2x^2 - y)}{h} = -1$

27. $\displaystyle \lim_{h \to 0} \frac{f(x + h, y) - f(x, y)}{h} = \lim_{h \to 0} \frac{3(x + h) - (x + h)y + 2y^2 - (3x - xy + 2y^2)}{h} = \lim_{h \to 0} \frac{3h - hy}{h} = 3 - y$

$$\lim_{h \to 0} \frac{f(x, y+h) - f(x, y)}{h} = \lim_{h \to 0} \frac{3x - x(y+h) + 2(y+h)^2 - (3x - xy + 2y^2)}{h}$$

$$= \lim_{h \to 0} \frac{-xh + 4yh + 2h^2}{h} = -x + 4y$$

29.

$$\lim_{h \to 0} \frac{f(x+h, y) - f(x, y)}{h} = \lim_{h \to 0} \frac{\cos[(x+h)y] - \cos[xy]}{h}$$

$$= \lim_{h \to 0} \frac{\cos[xy] \cos[hy] - \sin[xy] \sin[hy] - \cos[xy]}{h}$$

$$= \cos[xy] \left(\lim_{h \to 0} \frac{\cos[hy] - 1}{h} \right) - \sin[xy] \lim_{h \to 0} \frac{\sin hy}{h}$$

$$= y \cos[xy] \left(\lim_{h \to 0} \frac{\cos[hy] - 1}{hy} \right) - y \sin[xy] \lim_{h \to 0} \frac{\sin hy}{hy}$$

$$= -y \sin[xy]$$

$$\lim_{h \to 0} \frac{f(x, y+h) - f(x, y)}{h} = \lim_{h \to 0} \frac{\cos[x(y+h)] - \cos[xy]}{h}$$

$$= \lim_{h \to 0} \frac{\cos[xy] \cos[hx] - \sin[xy] \sin[hx] - \cos[xy]}{h}$$

$$= \cos[xy] \left(\lim_{h \to 0} \frac{\cos[hx] - 1}{h} \right) - \sin[xy] \lim_{h \to 0} \frac{\sin hx}{h}$$

$$= x \cos[xy] \left(\lim_{h \to 0} \frac{\cos[hx] - 1}{hx} \right) - x \sin[xy] \lim_{h \to 0} \frac{\sin hx}{hx}$$

$$= -x \sin[xy]$$

31. (a) $f(x, y) = x^2 y$ (b) $f(x, y) = \pi x^2 y$ (c) $f(x, y) = |2\mathbf{i} \times (x\mathbf{i} + y\mathbf{j})| = 2|y|$

33. Surface area: $S = 2lw + 2lh + 2hw = 20 \implies w = \frac{20 - 2lh}{2l + 2h} = \frac{10 - lh}{l + h}$

Volume: $V = lwh = \frac{lh(10 - lh)}{l + h}$

35. $V = \pi r^2 h + \frac{4}{3} \pi r^3$

SECTION 14.2

1. a quadric cone

3. a parabolic cylinder

5. a hyperboloid of one sheet

7. a sphere

9. an elliptic paraboloid

11. a hyperbolic paraboloid

13.

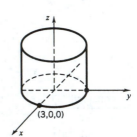

15.

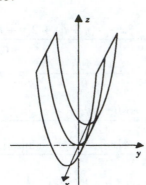

17.

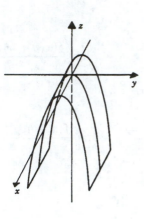

19.

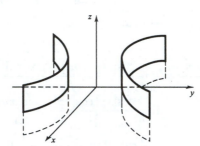

21.

23.

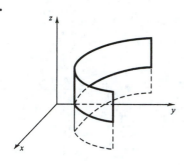

25. elliptic paraboloid
xy-trace: the origin
xz-trace: the parabola $x^2 = 4z$
yz-trace: the parabola $y^2 = 9z$
surface has the form of Figure 14.2.5

27. quadric cone
xy-trace: the origin
xz-trace: the lines $x = \pm 2z$
yz-trace: the lines $y = \pm 3z$
surface has the form of Figure 14.2.4

29. hyperboloid of two sheets
xy-trace: none
xz-trace: the hyperbola $4z^2 - x^2 = 4$
yz-trace: the hyperbola $9z^2 - y^2 = 9$
surface has the form of Figure 14.2.3

31. hyperboloid of two sheets
xy-trace: the hyperbola $4x^2 - 9y^2 = 36$
xz-trace: the hyperbola $x^2 - 4z^2 = 4$
yz-trace: none
see Figure 14.2.3

33. elliptic paraboloid
xy-trace: the origin
xz-trace: the parabola $x^2 = 4z$
yz-trace: the parabola $y^2 = 9z$
surface has the form of Figure 14.2.5

35. hyperboloid of two sheets
xy-trace: the hyperbola $9y^2 - 4x^2 = 36$
xz-trace: none
yz-trace: the hyperbola $y^2 - 4z^2 = 4y$
see Figure 14.2.3

37. paraboloid of revolution
xy-trace: the origin
xz-trace: the parabola $x^2 = 4z$
yz-trace: the parabola $y^2 = 4z$
surface has the form of Figure 14.2.5

39. (a) an elliptic paraboloid (vertex down if A and B are both positive, vertex up if A and B are both negative)

   (b) a hyperbolic paraboloid

   (c) the xy-plane if A and B are both zero; otherwise a parabolic cylinder

41. $x^2 + y^2 - 4z = 0$ (paraboloid of revolution)

43. (a) a circle

   (b) (i) $\sqrt{x^2 + y^2} = -3z$ (ii) $\sqrt{x^2 + z^2} = \frac{1}{3}y$

45. $x + 2y + 3\left(\dfrac{x + y - 6}{2}\right) = 6$ or $5x + 7y = 30,$ a line

47. $\left.\begin{array}{l} x^2 + y^2 + (z-1)^2 = \frac{3}{2} \\ x^2 + y^2 - z^2 = 1 \end{array}\right\}$ $(z^2 + 1) + (z-1)^2 = \dfrac{3}{2};$ $(2z-1)^2 = 0,$ $z = \dfrac{1}{2}$ so that $x^2 + y^2 = \dfrac{5}{4}$

49. $x^2 + y^2 + (x^2 + 3y^2) = 4$ or $x^2 + 2y^2 = 2,$ an ellipse

51. $x^2 + y^2 = (2-y)^2$ or $x^2 = -4(y-1),$ a parabola

SECTION 14.3

1. lines of slope 1: $y = x - c$

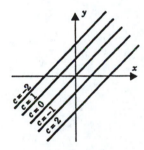

3. parabolas: $y = x^2 - c$

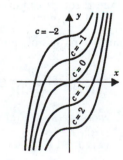

5. the y-axis and the lines $y = \left(\dfrac{1-c}{c}\right)x$

with the origin omitted throughout

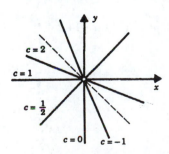

7. the cubics $y = x^3 - c$

9. the lines $y = \pm x$ and the hyperbolas $x^2 - y^2 = c$

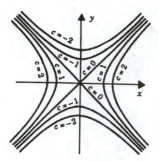

11. pairs of horizontal lines $y = \pm\sqrt{c}$ and the x-axis

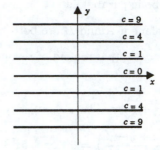

13. the circles $x^2 + y^2 = e^c$, c real

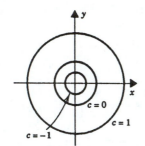

15. the curves $y = e^{cx^2}$ with the point $(0,1)$ omitted

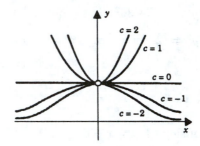

17. the coordinate axes and pairs of lines

$$y = \pm\frac{\sqrt{1-c}}{\sqrt{c}}\,x, \text{ the origin}$$

omitted throughout

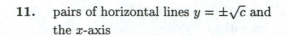

19. $x + 2y + 3z = 0$,   plane through the origin

21. $z = \sqrt{x^2 + y^2}$,   the upper nappe of the circular cone $z^2 = x^2 + y^2$ (Figure 14.2.4)

23. the elliptic paraboloid $\dfrac{x^2}{(1/2)^2} + \dfrac{y^2}{(1/3)^2} = 72z$ (Figure 14.2.5)

25.   (i)   hyperboloid of two sheets   (Figure 14.2.3)

   (ii)   circular cone   (Figure 14.2.4)

   (iii)   hyperboloid of one sheet   (Figure 14.2.2)

27. The level curves of f are: $1 - 4x^2 - y^2 = c$. Substituting $P(0,1)$ into this equation,

we have

$$1 - 4(0)^2 - (1)^2 = c \implies c = 0$$

The level curve that contains P is: $1 - 4x^2 - y^2 = 0$, or $4x^2 + y^2 = 1$.

29. The level curves of f are: $y^2 \tan^{-1} x = c$. Substituting $P(1,2)$ into this equation,

we have

$$4^2 \tan^{-1} 1 = c \implies c = \pi$$

The level curve that contains P is: $y^2 \tan^{-1} x = \pi$.

31. The level surfaces of f are: $x^2 + 2y^2 - 2xyz = c$. Substituting $P(-1,2,1)$ into this equation,

we have

$$(-1)^2 + 2(2)^2 - 2(-1)(2)(1) = c \implies c = 13$$

The level surface that contains P is: $x^2 + 2y^2 - 2xyz = 13$.

33. $\dfrac{GmM}{x^2 + y^2 + z^2} = c \implies x^2 + y^2 + z^2 = \dfrac{GmM}{c}$; the surfaces of constant gravitational force
are concentric spheres.

35. (a) $T(x,y) = \dfrac{k}{x^2 + y^2}$, where k is a constant.

(b) $\dfrac{k}{x^2 + y^2} = c \implies x^2 + y^2 = \dfrac{k}{c}$; the level curves are concentric circles.

(c) $T(1,2) = \dfrac{k}{1^2 + 2^2} = 50 \implies k = 250 \implies T(x,y) = \dfrac{250}{x^2 + y^2}$

Now, $T(3,4) = \dfrac{250}{3^2 + 4^2} = 10°$

37. $f(x,y) = y^2 - y^3$; F   **39.** $f(x,y) = \cos \sqrt{x^2 + y^2}$; A   **41.** $f(x,y) = xye^{-(x^2+y^2)/2}$; E

PROJECT 14.3

1.

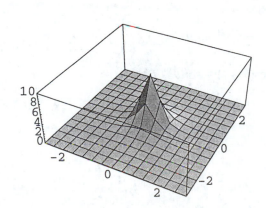

3.

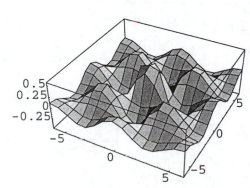

SECTION 14.4

1. $\dfrac{\partial f}{\partial x} = 6x - y, \quad \dfrac{\partial f}{\partial y} = 1 - x$
   3. $\dfrac{\partial \rho}{\partial \phi} = \cos\phi\cos\theta, \quad \dfrac{\partial \rho}{\partial \theta} = -\sin\phi\sin\theta$

5. $\dfrac{\partial f}{\partial x} = e^{x-y} + e^{y-x}, \quad \dfrac{\partial f}{\partial y} = -e^{x-y} - e^{y-x}$
   7. $\dfrac{\partial g}{\partial x} = \dfrac{(AD - BC)y}{(Cx + Dy)^2}, \quad \dfrac{\partial g}{\partial y} = \dfrac{(BC - AD)x}{(Cx + Dy)^2}$

9. $\dfrac{\partial u}{\partial x} = y + z, \quad \dfrac{\partial u}{\partial y} = x + z, \quad \dfrac{\partial u}{\partial z} = x + y$

11. $\dfrac{\partial f}{\partial x} = z\cos(x - y), \quad \dfrac{\partial f}{\partial y} = -z\cos(x - y), \quad \dfrac{\partial f}{\partial z} = \sin(x - y)$

13. $\dfrac{\partial \rho}{\partial \theta} = e^{\theta + \phi}[\cos(\theta - \phi) - \sin(\theta - \phi)], \quad \dfrac{\partial \rho}{\partial \phi} = e^{\theta + \phi}[\cos(\theta - \phi) + \sin(\theta - \phi)]$

15. $\dfrac{\partial f}{\partial x} = 2xy\sec xy + x^2 y(\sec xy)(\tan xy)y = 2xy\sec xy + x^2 y^2 \sec xy \ \tan xy$

$\dfrac{\partial f}{\partial y} = x^2 \sec xy + x^2 y(\sec xy)(\tan xy)x = x^2 \sec xy + x^3 y \sec xy \ \tan xy$

17. $\dfrac{\partial h}{\partial x} = \dfrac{x^2 + y^2 - x(2x)}{(x^2 + y^2)^2} = \dfrac{y^2 - x^2}{(x^2 + y^2)^2}$

$\dfrac{\partial h}{\partial y} = \dfrac{-2xy}{(x^2 + y^2)^2}$

19. $\dfrac{\partial f}{\partial x} = \dfrac{(y\cos x)\sin y - (x\sin y)(-y\sin x)}{(y\cos x)^2} = \dfrac{\sin y(\cos x + x\sin x)}{y\cos^2 x}$

$\dfrac{\partial f}{\partial y} = \dfrac{(y\cos x)(x\cos y) - (x\sin y)\cos x}{(y\cos x)^2} = \dfrac{x(y\cos y - \sin y)}{y^2 \cos x}$

21. $\dfrac{\partial h}{\partial x} = 2f(x)f'(x)g(y), \quad \dfrac{\partial h}{\partial y} = [f(x)]^2 g'(y)$

23. $\dfrac{\partial f}{\partial x} = (y^2 \ln z)z^{xy^2}, \quad \dfrac{\partial f}{\partial y} = (2xy\ln z)z^{xy^2}, \quad \dfrac{\partial f}{\partial z} = xy^2 z^{xy^2 - 1}$

25. $\dfrac{\partial h}{\partial r} = 2re^{2t}\cos(\theta - t) \qquad\qquad \dfrac{\partial h}{\partial \theta} = -r^2 e^{2t}\sin(\theta - t)$

$\dfrac{\partial h}{\partial t} = 2r^2 e^{2t}\cos(\theta - t) + r^2 e^{2t}\sin(\theta - t) = r^2 e^{2t}[2\cos(\theta - t) + \sin(\theta - t)]$

27. $\dfrac{\partial f}{\partial x} = z\dfrac{1}{1 + (y/x)^2}\left(\dfrac{-y}{x^2}\right) = -\dfrac{yz}{x^2 + y^2}$

$\dfrac{\partial f}{\partial y} = z\dfrac{1}{1 + (y/x)^2}\left(\dfrac{1}{x}\right) = \dfrac{xz}{x^2 + y^2}$

$\dfrac{\partial f}{\partial x} = \tan^{-1}(y/x)$

29. $f_x(x,y) = e^x \ln y,$ $f_x(0,e) = 1;$ $f_y(x,y) = \dfrac{1}{y}e^x,$ $f_y(0,e) = e^{-1}$

31. $f_x(x,y) = \dfrac{y}{(x+y)^2},$ $f_x(1,2) = \dfrac{2}{9};$ $f_y(x,y) = \dfrac{-x}{(x+y)^2},$ $f_y(1,2) = -\dfrac{1}{9}$

33. $f_x(x,y) = \lim\limits_{h\to 0} \dfrac{(x+h)^2 y - x^2 y}{h} = \lim\limits_{h\to 0} y\left(\dfrac{2xh + h^2}{h}\right) = y \lim\limits_{h\to 0}(2x + h) = 2xy$

$f_x(x,y) = \lim\limits_{h\to 0} \dfrac{x^2(y+h) - x^2 y}{h} = \lim\limits_{h\to 0} \dfrac{x^2 h}{h} = \lim\limits_{h\to 0} x^2 = x^2$

35. $f_x(x,y) = \lim\limits_{h\to 0} \dfrac{\ln\left(y(x+h)^2\right) - \ln x^2 y}{h} = \lim\limits_{h\to 0} \dfrac{\ln y + 2\ln(x+h) - 2\ln x - \ln y}{h}$

$\qquad = 2 \lim\limits_{h\to 0} \dfrac{\ln(x+h) - \ln x}{h} = 2\dfrac{d}{dx}(\ln x) = \dfrac{2}{x}$

$f_y(x,y) = \lim\limits_{h\to 0} \dfrac{\ln\left(x^2(y+h)\right) - \ln x^2 y}{h} = \lim\limits_{h\to 0} \dfrac{\ln x^2 + \ln(y+h) - \ln x^2 - \ln y}{h}$

$\qquad = \lim\limits_{h\to 0} \dfrac{\ln(y+h) - \ln y}{h} = \dfrac{d}{dy}(\ln y) = \dfrac{1}{y}$

37. $f_x(x,y) = \lim\limits_{h\to 0} \dfrac{1}{h}\left\{\dfrac{1}{(x+h) - y} - \dfrac{1}{x-y}\right\} = \lim\limits_{h\to 0} \dfrac{1}{h}\left\{\dfrac{-h}{(x+h-y)(x-y)}\right\}$

$\qquad = \lim\limits_{h\to 0} \dfrac{-1}{(x+h-y)(x-y)} = \dfrac{-1}{(x-y)^2}$

$f_x(x,y) = \lim\limits_{h\to 0} \dfrac{1}{h}\left\{\dfrac{1}{x-(y+h)} - \dfrac{1}{x-y}\right\} = \lim\limits_{h\to 0} \dfrac{1}{h}\left\{\dfrac{h}{(x-y-h)(x-y)}\right\}$

$\qquad = \lim\limits_{h\to 0} \dfrac{1}{(x-y-h)(x-y)} = \dfrac{1}{(x-y)^2}$

39. $f_x(x,y,z) = \lim\limits_{h\to 0} \dfrac{(x+h)y^2 z - xy^2 z}{h} = \lim\limits_{h\to 0} y^2 z = y^2 z$

$f_y(x,y,z) = \lim\limits_{h\to 0} \dfrac{x(y+h)^2 z - xy^2 z}{h} = \lim\limits_{h\to 0} \dfrac{xz(2yh + h^2)}{h}$

$\qquad = \lim\limits_{h\to 0} xz(2y + h) = 2xyz$

$f_z(x,y,z) = \lim\limits_{h\to 0} \dfrac{xy^2(x+h) - xy^2 z}{h} = \lim\limits_{h\to 0} xy^2 = xy^2$

41.   (b) The slope of the tangent line to C at the point $P(x_0, y_0, f(x_0, y_0))$ is $f_y(x_0, y_0)$.

Thus, equations for the tangent line are:

$$y = y_0, \quad z - z_0 = f_y(x_0, y_0)(y - y_0)$$

43. Let $z = f(x,y) = x^2 + y^2$. Then $f(2,1) = 5$, $f_y(x,y) = 2y$ and $f_y(2,1) = 2$

By Exercise 41, equations for the tangent line are:

$$x = 2, \quad z - 5 = 2(y - 1)$$

45. Let $z = f(x,y) = \dfrac{x^2}{y^2 - 3}$. Then $f(3,2) = 9$, $f_x(x,y) = \dfrac{2x}{y^2 - 3}$ and $f_x(3,2) = 6$

By Exercise 41, equations for the tangent line are:

$$y = 2, \quad z - 9 = 6(x - 3)$$

47. $u_x(x,y) = 2x = v_y(x,y)$; $\qquad u_y(x,y) = -2y = -v_x(x,y)$

49. $u_x(x,y) = \dfrac{1}{2} \dfrac{1}{x^2 + y^2} 2x = \dfrac{x}{x^2 + y^2}$; $\quad v_y(x,y) = \dfrac{1}{1 + (y/x)^2}\left(\dfrac{1}{x}\right) = \dfrac{x}{x^2 + y^2}$

Thus, $u_x(x,y) = v_y(x,y)$.

$u_y(x,y) = \dfrac{1}{2} \dfrac{1}{x^2 + y^2} 2y = \dfrac{y}{x^2 + y^2}$; $\quad v_x(x,y) = \dfrac{1}{1 + (y/x)^2}\left(\dfrac{-y}{x^2}\right) = \dfrac{-y}{x^2 + y^2}$

Thus, $u_y(x,y) = -v_x(x,y)$.

51. (a) f depends only on y. $\qquad\qquad\qquad$ (b) f depends only on x.

53. (a) $50\sqrt{3}$ in.^{2}

(b) $\dfrac{\partial A}{\partial b} = \dfrac{1}{2} c \sin\theta$; at time t_0, $\dfrac{\partial A}{\partial b} = 5\sqrt{3}$

(c) $\dfrac{\partial A}{\partial \theta} = \dfrac{1}{2} bc \cos\theta$; at time t_0, $\dfrac{\partial A}{\partial \theta} = 50$

(d) with $h = \dfrac{\pi}{180}$, $A(b, c, \theta + h) - A(b, c, \theta) \cong h\dfrac{\partial A}{\partial \theta} = \dfrac{\pi}{180}(50) = \dfrac{5\pi}{18}$ in.^{2}

(e) $0 = \dfrac{1}{2} \sin\theta \left(b\dfrac{\partial c}{\partial b} + c\right)$; at time t_0, $\dfrac{\partial c}{\partial b} = \dfrac{-c}{b} = -2$

55. (a) y_0-section: $\mathbf{r}(x) = x\mathbf{i} + y_0\mathbf{j} + f(x, y_0)\mathbf{k}$

tangent line: $\mathbf{R}(t) = [x_0\mathbf{i} + y_0\mathbf{j} + f(x_0, y_0)\mathbf{k}] + t\left[\mathbf{i} + \dfrac{\partial f}{\partial x}(x_0, y_0)\mathbf{k}\right]$

(b) x_0-section: $\mathbf{r}(y) = x_0\mathbf{i} + y\mathbf{j} + f(x_0, y)\mathbf{k}$

tangent line: $\mathbf{R}(t) = [x_0\mathbf{i} + y_0\mathbf{j} + f(x_0, y_0)\mathbf{k}] + t\left[\mathbf{j} + \dfrac{\partial f}{\partial y}(x_0, y_0)\mathbf{k}\right]$

(c) For (x, y, z) in the plane

$$[(x - x_0)\mathbf{i} + (y - y_0)\mathbf{j} + (z - f(x_0, y_0))\mathbf{k}] \cdot \left[\left(\mathbf{i} + \dfrac{\partial f}{\partial x}(x_0, y_0)\mathbf{k}\right) \times \left(\mathbf{j} + \dfrac{\partial f}{\partial y}(x_0, y_0)\mathbf{k}\right)\right] = 0.$$

From this it follows that

$$z - f(x_0, y_0) = (x - x_0)\dfrac{\partial f}{\partial x}(x_0, y_0) + (y - y_0)\dfrac{\partial f}{\partial y}(x_0, y_0).$$

57. (a) Set $u = ax + by$. Then

$$b \frac{\partial w}{\partial x} - a \frac{\partial w}{\partial y} = b(a\,g'(u)) - a(b\,g'(u)) = 0.$$

(b) Set $u = x^m y^n$. Then

$$nx \frac{\partial w}{\partial x} - my \frac{\partial w}{\partial y} = nx \left[mx^{m-1}y^n g'(u) \right] - my \left[nx^m y^{n-1} g'(u) \right] = 0.$$

59.
$$x \frac{\partial u}{\partial x} + y \frac{\partial u}{\partial y} = x \left(4Ax^3 + 4Bxy^2 \right) + y \left(4Bx^2 y + 4Cy^3 \right)$$
$$= 4 \left(Ax^4 + 2Bx^2 y^2 + Cy^4 \right) = 4u$$

61. $\dfrac{\partial x}{\partial r} \dfrac{\partial y}{\partial \theta} - \dfrac{\partial x}{\partial \theta} \dfrac{\partial y}{\partial r} = (\cos \theta)\,(r\,\cos \theta) - (-r\,\sin \theta)\,(\sin \theta) = r$

SECTION 14.5

1. interior $= \{(x,y) : 2 < x < 4, \quad 1 < y < 3\}$ (the inside of the rectangle), boundary $=$ the union of the four boundary line segments; set is closed.

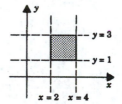

3. interior $=$ the entire set (region between the two concentric circles), boundary $=$ the two circles, one of radius 1, the other of radius 2; set is open.

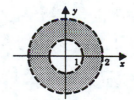

5. interior $= \{(x,y) : 1 < x^2 < 4\} =$
$\{(x,y) : -2 < x < -1\} \cup \{(x,y) : 1 < x < 2\}$
(two vertical strips without the boundary lines),
boundary $= \{(x,y) : x = -2,\ x = -1,\ x = 1,$
or $x = 2\}$ (four vertical lines); set is neither open
nor closed.

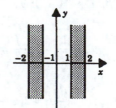

7.   interior = region below the parabola $y = x^2$,
boundary = the parabola $y = x^2$;   the set is closed.

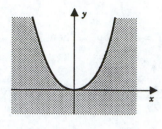

9.   interior = $\{(x, y, z) : x^2 + y^2 < 1, \ 0 < z \le 4\}$
(the inside of the cylinder), boundary = the total
surface of the cylinder (the curved part, the top, and
the bottom);   the set is closed.

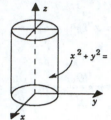

$x^2 + y^2 = 1$

11.   (a) ϕ (b) S (c)   closed

13.   interior = $\{x : 1 < x < 3\}$, boundary = $\{1, 3\}$;   set is closed.

15.   interior = the entire set, boundary = $\{1\}$;   set is open.

17.   interior = $\{x : |x| > 1\}$, boundary = $\{1, -1\}$;   set is neither open nor closed.

19.   interior = ϕ, boundary = $\{$the entire set$\} \cup \{0\}$;   the set is neither open nor closed.

SECTION 14.6

1. $\dfrac{\partial^2 f}{\partial x^2} = 2A, \quad \dfrac{\partial^2 f}{\partial y^2} = 2C, \quad \dfrac{\partial^2 f}{\partial y \partial x} = \dfrac{\partial^2 f}{\partial x \partial y} = 2B$

3. $\dfrac{\partial^2 f}{\partial x^2} = Cy^2 e^{xy}, \quad \dfrac{\partial^2 f}{\partial y^2} = Cx^2 e^{xy}, \quad \dfrac{\partial^2 f}{\partial y \partial x} = \dfrac{\partial^2 f}{\partial x \partial y} = Ce^{xy}(xy + 1)$

5. $\dfrac{\partial^2 f}{\partial x^2} = 2, \quad \dfrac{\partial^2 f}{\partial y^2} = 4(x + 3y^2 + z^3), \quad \dfrac{\partial^2 f}{\partial z^2} = 6z(2x + 2y^2 + 5z^3)$

$\dfrac{\partial^2 f}{\partial x \partial y} = \dfrac{\partial^2 f}{\partial y \partial x} = 4y, \quad \dfrac{\partial^2 f}{\partial z \partial x} = \dfrac{\partial^2 f}{\partial x \partial z} = 6z^2, \quad \dfrac{\partial^2 f}{\partial z \partial y} = \dfrac{\partial^2 f}{\partial y \partial z} = 12yz^2$

7. $\dfrac{\partial^2 f}{\partial x^2} = \dfrac{1}{(x + y)^2} - \dfrac{1}{x^2}, \quad \dfrac{\partial^2 f}{\partial y^2} = \dfrac{1}{(x + y)^2}, \quad \dfrac{\partial^2 f}{\partial y \partial x} = \dfrac{\partial^2 f}{\partial x \partial y} = \dfrac{1}{(x + y)^2}$

9. $\dfrac{\partial^2 f}{\partial x^2} = 2(y + z), \quad \dfrac{\partial^2 f}{\partial y^2} = 2(x + z), \quad \dfrac{\partial^2 f}{\partial z^2} = 2(x + y)$

all the second mixed partials are $2(x + y + z)$

11. $\quad \dfrac{\partial^2 f}{\partial x^2} = y(y-1)x^{y-2}, \quad \dfrac{\partial^2 f}{\partial y^2} = (\ln x)^2 x^y, \quad \dfrac{\partial^2 f}{\partial y \partial x} = \dfrac{\partial^2 f}{\partial x \partial y} = x^{y-1}(1 + y \ln x)$

13. $\quad \dfrac{\partial^2 f}{\partial x^2} = ye^x, \quad \dfrac{\partial^2 f}{\partial y^2} = xe^y, \quad \dfrac{\partial^2 f}{\partial y \partial x} = \dfrac{\partial^2 f}{\partial x \partial y} = e^y + e^x$

15. $\quad \dfrac{\partial^2 f}{\partial x^2} = \dfrac{y^2 - x^2}{(x^2 + y^2)^2}, \quad \dfrac{\partial^2 f}{\partial y^2} = \dfrac{x^2 - y^2}{(x^2 + y^2)^2}, \quad \dfrac{\partial^2 f}{\partial y \partial x} = \dfrac{\partial^2 f}{\partial x \partial y} = -\dfrac{2xy}{(x^2 + y^2)^2}$

17. $\quad \dfrac{\partial^2 f}{\partial x^2} = -2\,y^2 \cos 2xy, \quad \dfrac{\partial^2 f}{\partial y^2} = -2\,x^2 \cos 2xy, \quad \dfrac{\partial^2 f}{\partial y \partial x} = \dfrac{\partial^2 f}{\partial x \partial y} = -[\sin 2xy + 2xy \,\cos 2xy]$

19. $\quad \dfrac{\partial^2 f}{\partial x^2} = 0, \quad \dfrac{\partial^2 f}{\partial y^2} = xz \sin y, \quad \dfrac{\partial^2 f}{\partial z^2} = -\,xy \sin z,$

$$\dfrac{\partial^2 f}{\partial y \partial x} = \dfrac{\partial^2 f}{\partial x \partial y} = \sin z - z\,\cos y, \quad \dfrac{\partial^2 f}{\partial x \partial z} = \dfrac{\partial^2 f}{\partial z \partial x} = y \cos z - \sin y, \quad \dfrac{\partial^2 f}{\partial y \partial z} = \dfrac{\partial^2 u}{\partial z \partial y} = x \cos z - x \cos y$$

21. $\quad x^2 \dfrac{\partial^2 u}{\partial x^2} + 2xy \dfrac{\partial^2 u}{\partial x \partial y} + y^2 \dfrac{\partial^2 u}{\partial y^2} = x^2 \left(\dfrac{-2y^2}{(x+y)^3} \right) + 2xy \left(\dfrac{2xy}{(x+y)^3} \right) + y^2 \left(\dfrac{-2x^2}{(x+y)^3} \right) = 0$

23. (a)   no, since $\quad \dfrac{\partial^2 f}{\partial y \partial x} \ne \dfrac{\partial^2 f}{\partial x \partial y} \qquad$ (b)   no, since $\quad \dfrac{\partial^2 f}{\partial y \partial x} \ne \dfrac{\partial^2 f}{\partial x \partial y} \quad$ for $\quad x \ne y$

25. $\quad \dfrac{\partial^3 f}{\partial x^2 \partial y} = \dfrac{\partial}{\partial x} \left(\dfrac{\partial^2 f}{\partial x \partial y} \right) = \dfrac{\partial}{\partial x} \left(\dfrac{\partial^2 f}{\partial y \partial x} \right) = \dfrac{\partial^2 f}{\partial x \partial y} \left(\dfrac{\partial f}{\partial x} \right) = \dfrac{\partial^2}{\partial y \partial x} \left(\dfrac{\partial f}{\partial x} \right) = \dfrac{\partial}{\partial y} \left(\dfrac{\partial^2 f}{\partial x^2} \right) = \dfrac{\partial^3 f}{\partial y \partial x^2}$

$\qquad\qquad$ by definition $\qquad$ (14.6.5) $\qquad\qquad$ by definition $\qquad$ (14.6.5) $\qquad\qquad$ by def. $\qquad\qquad$ by def.

27. (a) $\quad \lim\limits_{x \to 0} \dfrac{(x)(0)}{x^2 + 0} = \lim\limits_{x \to 0} 0 = 0 \qquad$ (b) $\quad \lim\limits_{y \to 0} \dfrac{(0)(y)}{0 + y^2} = \lim\limits_{y \to 0} 0 = 0$

(c) $\quad \lim\limits_{x \to 0} \dfrac{(x)(mx)}{x^2 + (mx)^2} = \lim\limits_{x \to 0} \dfrac{m}{1 + m^2} = \dfrac{m}{1 + m^2}$

(d) $\quad \lim\limits_{\theta \to 0+} \dfrac{(\theta \cos \theta)(\theta \sin \theta)}{(\theta \cos \theta)^2 + (\theta \sin \theta)^2} = \lim\limits_{\theta \to 0+} \cos \theta \sin \theta = 0$

(e)   By L'Hospital's rule $\quad \lim\limits_{x \to 0} \dfrac{f(x)}{x} = \lim\limits_{x \to 0} f'(x) = f'(0).$ Thus

$$\lim\limits_{x \to 0} \dfrac{x f(x)}{x^2 + [f(x)]^2} = \lim\limits_{x \to 0} \dfrac{f(x)/x}{1 + [f(x)/x]^2} = \dfrac{f'(0)}{1 + [f'(0)]^2}.$$

(f) $\quad \lim\limits_{\theta \to (\pi/3)-} = \dfrac{(\cos \theta \sin 3\theta)(\sin \theta \sin 3\theta)}{(\cos \theta \sin 3\theta)^2 + (\sin \theta \sin 3\theta)^2} = \lim\limits_{\theta \to (\pi/3)-} \cos \theta \sin \theta = \dfrac{1}{4}\sqrt{3}$

(g) $\quad \lim\limits_{t \to \infty} \dfrac{(1/t)(\sin t)/t}{1/t^2 + \left(\sin^2 t \right)/t^2} = \lim\limits_{t \to \infty} \dfrac{\sin t}{1 + \sin^2 t}; \quad$ does not exist

29.   **(a)** $\dfrac{\partial g}{\partial x}(0,0) = \lim\limits_{h\to 0} \dfrac{g(h,0) - g(0,0)}{h} = \lim\limits_{h\to 0} 0 = 0,$

$\dfrac{\partial g}{\partial y}(0,0) = \lim\limits_{h\to 0} \dfrac{g(0,h) - g(0,0)}{h} = \lim\limits_{h\to 0} 0 = 0$

(b)   as (x,y) tends to $(0,0)$ along the x-axis, $g(x,y) = g(x,0) = 0$ tends to 0;

as (x,y) tends to $(0,0)$ along the line $y = x$, $g(x,y) = g(x,x) = \frac{1}{2}$ tends to $\frac{1}{2}$

31.   For $y \neq 0$, $\dfrac{\partial f}{\partial x}(0,y) = \lim\limits_{h\to 0} \dfrac{f(h,y) - f(0,y)}{h} = \lim\limits_{h\to 0} \dfrac{y(y^2 - h^2)}{h^2 + y^2} = y.$

Since $\dfrac{\partial f}{\partial x}(0,0) = \lim\limits_{h\to 0} \dfrac{f(h,0) - f(0,0)}{h} = \lim\limits_{h\to 0} 0 = 0,$

we have $\dfrac{\partial f}{\partial x}(0,y) = y$ for all y.

For $x \neq 0$, $\dfrac{\partial f}{\partial y}(x,0) = \lim\limits_{h\to 0} \dfrac{f(x,h) - f(x,0)}{h} = \lim\limits_{h\to 0} \dfrac{x(h^2 - x^2)}{x^2 + h^2} = -x.$

Since $\dfrac{\partial f}{\partial y}(0,0) = \lim\limits_{h\to 0} \dfrac{f(0,h) - f(0,0)}{h} = \lim\limits_{h\to 0} 0 = 0,$

we have $\dfrac{\partial f}{\partial y}(x,0) = -x$ for all x.

Therefore $\dfrac{\partial^2 f}{\partial y \partial x}(0,y) = 1$ for all y and $\dfrac{\partial^2 f}{\partial x \partial y}(x,0) = -1$ for all x.

In particular $\dfrac{\partial^2 f}{\partial y \partial x}(0,0) = 1$ while $\dfrac{\partial^2 f}{\partial x \partial y}(0,0) = -1.$

33.   Since $f_{xy}(x,y) = 0$, $f_x(x,y)$ must be a function of x alone, and $f_y(x,y)$ must be a function of y alone. Then f must be of the form

$$f(x,y) = g(x) + h(y).$$

PROJECT 14.6

1.   **(a)** $\dfrac{\partial u}{\partial x} = \dfrac{x^2 y^2 + 2xy^3}{(x+y)^2}, \quad \dfrac{\partial u}{\partial y} = \dfrac{x^2 y^2 + 2x^3 y}{(x+y)^2}$

$x\dfrac{\partial u}{\partial x} + y\dfrac{\partial u}{\partial y} = \dfrac{3x^2 y^2 (x+y)}{(x+y)^2} = 3u$

(b) $x\dfrac{\partial u}{\partial x} + y\dfrac{\partial u}{\partial y} = x\left(4Ax^3 + 4Bxy^2\right) + y\left(4Bx^2 y + 4Cy^3\right)$

$= 4\left(Ax^4 + 2Bx^2\, y^2 + Cy^4\right) = 4u$

3. (i) $\dfrac{\partial^2 f}{\partial t^2} = \dfrac{\partial^2 f}{\partial y^2}x = 0 \implies \dfrac{\partial^2 f}{\partial t^2} - c^2\dfrac{\partial^2 f}{\partial y^2}x = 0$

(ii) $\dfrac{\partial^2 f}{\partial t^2} = -5c^2\sin(x+ct)\cos(2x+2ct) - 4c^2\cos(x+ct)\sin(2x+2ct)$

$\dfrac{\partial^2 f}{\partial x^2} = -5\sin(x+ct)\cos(2x+2ct) - 4\cos(x+ct)\sin(2x+2ct)$

It now follows that $\dfrac{\partial^2 f}{\partial t^2} - c^2\dfrac{\partial^2 f}{\partial x^2} = 0$

(iii) $\dfrac{\partial^2 f}{\partial t^2} = -\dfrac{c^2}{(x+ct)^2},\quad \dfrac{\partial^2 f}{\partial y^2}x = -\dfrac{1}{(x+ct)^2} \implies \dfrac{\partial^2 f}{\partial t^2} - c^2\dfrac{\partial^2 f}{\partial y^2}x = 0$

(iv) $\dfrac{\partial^2 f}{\partial t^2} = c^2k^2\left(Ae^{kx} + Be^{-kx}\right)\left(Ce^{ckt} + De^{-ckt}\right),\quad \dfrac{\partial^2 f}{\partial x^2} = k^2\left(Ae^{kx} + Be^{-kx}\right)\left(Ce^{ckt} + De^{-ckt}\right)$

It now follows that $\dfrac{\partial^2 f}{\partial t^2} - c^2\dfrac{\partial^2 f}{\partial x^2} = 0$

CHAPTER 15

SECTION 15.1

1. $\nabla f = e^{xy}[(xy+1)\mathbf{i}+x^2\mathbf{j}]$ **3.** $\nabla f = (6x-y)\mathbf{i}+(1-x)\mathbf{j}$

5. $\nabla f = 2xy^{-2}\mathbf{i}-2x^2y^{-3}\mathbf{j}$

7. $\nabla f = z\cos(x-y)\mathbf{i}-z\cos(x-y)\mathbf{j}+\sin(x-y)\mathbf{k}$

9. $\nabla f = (y+z)\mathbf{i}+(x+z)\mathbf{j}+(x+y)\mathbf{k}$ **11.** $\nabla f = e^{x-y}[(1+x+y)\mathbf{i}+(1-x-y)\mathbf{j}]$

13. $\nabla f = e^x[\ln y\,\mathbf{i}+y^{-1}\mathbf{j}]$ **15.** $\nabla f = \dfrac{AD-BC}{(Cx+Dy)^2}[y\mathbf{i}-x\mathbf{j}]$

17. $\nabla f = (ye^x+xye^x-ze^y\cos xz)\,\mathbf{i}+(xe^x+e^z-e^y\sin xz)\,\mathbf{j}+(ye^z-xe^y\cos xz)\,\mathbf{k}$

19. $\nabla f = e^{x+2y}\cos(z^2+1)\,\mathbf{i}+2e^{x+2y}\cos(z^2+1)\,\mathbf{j}-2ze^{x+2y}\sin(z^2+1)\,\mathbf{k}$

21. $\nabla f = (4x-3y)\mathbf{i}+(8y-3x)\mathbf{j}$;   at $(2,3)$, $\nabla f = -\mathbf{i}+18\mathbf{j}$

23. $\nabla f = \dfrac{2x}{x^2+y^2}\mathbf{i}+\dfrac{2y}{x^2+y^2}\mathbf{j}$;   at $(2,1)$, $\nabla f = \frac{4}{5}\mathbf{i}+\frac{2}{5}\mathbf{j}$

25. $\nabla f = (\sin xy+xy\cos xy)\mathbf{i}+x^2\cos xy\,\mathbf{j}$;   at $(1,\pi/2)$, $\nabla f = \mathbf{i}$

27. $\nabla f = -e^{-x}\sin(z+2y)\mathbf{i}+2e^{-x}\cos(z+2y)\mathbf{j}+e^{-x}\cos(z+2y)\mathbf{k}$;

at $(0,\pi/4,\pi/4)$, $\nabla f = -\frac{1}{2}\sqrt{2}\,(\mathbf{i}+2\mathbf{j}+\mathbf{k})$

29. $\nabla f = \mathbf{i}-\dfrac{y}{\sqrt{y^2+z^2}}\mathbf{j}-\dfrac{z}{\sqrt{y^2+z^2}}\mathbf{k}$;   at $(2,-3,4)$, $\nabla f = \mathbf{i}+\frac{3}{5}\mathbf{j}-\frac{4}{5}\mathbf{k}$

31. For the function $f(x,y)=3x^2-xy+y$, we have

$$f(\mathbf{x}+\mathbf{h})-f(\mathbf{x})=f(x+h_1,y+h_2)-f(x,y)$$

$$=3(x+h_1)^2-(x+h_1)(y+h_2)+(y+h_2)-\left[3x^3-xy+y\right]$$

$$=[(6x-y)\mathbf{i}+(1-x)\mathbf{j}]\cdot(h_1\mathbf{i}+h_2\mathbf{j})+3h_1^2-h_1h_2$$

$$=[(6x-y)\mathbf{i}+(1-x)\mathbf{j}]\cdot\mathbf{h}+3h_1^2-h_1h_2$$

The remainder $g(\mathbf{h})=3h_1^2-h_1h_2=(3h_1\mathbf{i}-h_1\mathbf{j})\cdot(h_1\mathbf{i}+h_2\mathbf{j})$,   and

$$\frac{|g(\mathbf{h})|}{\|\mathbf{h}\|}=\frac{\|3h_1\mathbf{i}-h_1\mathbf{j}\|\cdot\|\mathbf{h}\|\cdot\cos\theta}{\|\mathbf{h}\|}\le\|3h_1\mathbf{i}-h_1\mathbf{j}\|$$

Since $\|3h_1\mathbf{i}-h_1\mathbf{j}\|\to0$ as $\mathbf{h}\to\mathbf{0}$ it follows that

$$\nabla f=(6x-y)\mathbf{i}+(1-x)\mathbf{j}$$

33.   For the function $f(x, y, z) = x^2 y + y^2 z + z^2 x,$ we have

$$f(\mathbf{x} + \mathbf{h}) - f(\mathbf{x}) = f(x + h_1, y + h_2, z + h_3) - f(x, y, z)$$

$$= (x + h_1)^2 (y + h_2) + (y + h_2)^2 (z + h_3) + (z + h_3)^2 (x + h_1) - x^2 y + y^2 z + z^2 x$$

$$= \left(2xy + z^2\right) h_1 + \left(2yz + x^2\right) h_2 + \left(2xz + y^2\right) h_3 + \left(2xh_2 + yh_1 + h_1 h_2\right) h_1 +$$

$$\left(2yh_3 + zh_2 + h_2 h_3\right) h_2 + \left(2zh_1 + xh_3 + h_1 h_3\right) h_3$$

$$= \left[\left(2xy + z^2\right) \mathbf{i} + \left(2yz + x^2\right) \mathbf{j} + \left(2xz + y^2\right) \mathbf{k}\right] \cdot \mathbf{h} + g(\mathbf{h}) \cdot \mathbf{h},$$

where $g(\mathbf{h}) = (2xh_2 + yh_1 + h_1 h_2) \mathbf{i} + (2yh_3 + zh_2 + h_2 h_3) \mathbf{j} + (2zh_1 + xh_3 + h_1 h_3) \mathbf{k}$

Since $\dfrac{|g(\mathbf{h})|}{\|\mathbf{h}\|} \to 0$ as $\mathbf{h} \to 0$ it follows that

$$\nabla f = \left(2xy + z^2\right) \mathbf{i} + \left(2yz + x^2\right) \mathbf{j} + \left(2xz + y^2\right) \mathbf{k}$$

35. $\nabla f = \mathbf{F}(x, y) = 2xy \, \mathbf{i} + \left(1 + x^2\right) \mathbf{j} \quad \Rightarrow \quad \dfrac{\partial f}{\partial x} = 2xy \quad \Rightarrow \quad f(x, y) = x^2 y + g(y)$ for some function g.

Now, $\dfrac{\partial f}{\partial y} = x^2 + g'(y) = 1 + x^2 \quad \Rightarrow \quad g'(y) = 1 \quad \Rightarrow \quad g(y) = y + C, \ \ C$ a constant.

Thus, $f(x, y) = x^2 y + y$ (take $C = 0$) is a function whose gradient is $\mathbf{F}$.

37. $\nabla f = \mathbf{F}(x, y) = (x + \sin y) \, \mathbf{i} + (x \cos y - 2y) \, \mathbf{j} \quad \Rightarrow \quad \dfrac{\partial f}{\partial x} = x + \sin y \quad \Rightarrow \quad f(x, y) = \tfrac{1}{2} x^2 + x \sin y + g(y)$

for some function g.

Now, $\dfrac{\partial f}{\partial y} = x \cos y + g'(y) = x \cos y - 2y \quad \Rightarrow \quad g'(y) = -2y \quad \Rightarrow \quad g(y) = -y^2 + C, \ \ C$ a constant.

Thus, $f(x, y) = \tfrac{1}{2} x^2 + x \sin y - y^2$ (take $C = 0$) is a function whose gradient is $\mathbf{F}$.

39.   With $r = (x^2 + y^2 + z^2)^{1/2}$ we have

$$\frac{\partial r}{\partial x} = \frac{x}{r}, \quad \frac{\partial r}{\partial y} = \frac{y}{r}, \quad \frac{\partial r}{\partial z} = \frac{z}{r}.$$

(a)

$$\nabla (\ln r) = \frac{\partial}{\partial x} (\ln r) \, \mathbf{i} + \frac{\partial}{\partial y} (\ln r) \, \mathbf{j} + \frac{\partial}{\partial z} (\ln r) \mathbf{k}$$

$$= \frac{1}{r} \frac{\partial r}{\partial x} \mathbf{i} + \frac{1}{r} \frac{\partial r}{\partial y} \mathbf{j} + \frac{1}{r} \frac{\partial r}{\partial z} \mathbf{k}$$

$$= \frac{x}{r^2} \mathbf{i} + \frac{y}{r^2} \mathbf{j} + \frac{z}{r^2} \mathbf{k}$$

$$= \frac{\mathbf{r}}{r^2}$$

(b)
$$\nabla(\sin r) = \frac{\partial}{\partial x}(\sin r)\,\mathbf{i} + \frac{\partial}{\partial y}(\sin r)\,\mathbf{j} + \frac{\partial}{\partial z}(\sin r)\mathbf{k}$$

$$= \cos r\,\frac{\partial r}{\partial x}\,\mathbf{i} + \cos r\,\frac{\partial r}{\partial y}\,\mathbf{j} + \cos r\,\frac{\partial r}{\partial z}\,\mathbf{k}$$

$$= (\cos r)\frac{x}{r}\,\mathbf{i} + (\cos r)\frac{y}{r}\,\mathbf{j} + (\cos r)\frac{z}{r}\,\mathbf{k}$$

$$= \left(\frac{\cos r}{r}\right)\mathbf{r}$$

(c) $\nabla e^r = \left(\dfrac{e^r}{r}\right)\mathbf{r}$ [same method as in (a) and (b)]

41.   (a) $\nabla f = 2x\,\mathbf{i} + 2y\,\mathbf{j} = \mathbf{0} \implies x = y = 0;\quad \nabla f = \mathbf{0}$ at $(0,0)$.

(b)   (c) f has an absolute minimum at $(0,0)$

(0,0,1)

43.   (a)   Let $\mathbf{c} = c_1\mathbf{i} + c_2\mathbf{j} + c_3\mathbf{k}$. First, we take $\mathbf{h} = h\mathbf{i}$. Since $\mathbf{c}\cdot\mathbf{h}$ is $o(\mathbf{h})$,

$$0 = \lim_{h\to 0}\frac{\mathbf{c}\cdot\mathbf{h}}{\|\mathbf{h}\|} = \lim_{h\to 0}\frac{c_1 h}{h} = c_1.$$

Similarly, $c_2 = 0$ and $c_3 = 0$.

(b) $(\mathbf{y} - \mathbf{z})\cdot\mathbf{h} = [f(\mathbf{x}+\mathbf{h}) - f(\mathbf{x}) - \mathbf{z}\cdot\mathbf{h}] + [\mathbf{y}\cdot\mathbf{h} - f(\mathbf{x}+\mathbf{h}) + f(\mathbf{x})] = o(\mathbf{h}) + o(\mathbf{h}) = o(\mathbf{h})$,
so that, by part (a), $\mathbf{y} - \mathbf{z} = \mathbf{0}$.

45.   (a)   In Section 14.6 we showed that f was not continuous at $(0,0)$. It is therefore not differentiable at $(0,0)$.

(b)   For $(x,y) \neq (0,0)$, $\dfrac{\partial f}{\partial x} = \dfrac{2y(y^2 - x^2)}{(x^2 + y^2)^2}$.   As (x,y) tends to $(0,0)$ along the positive y-axis,

$\dfrac{\partial f}{\partial x} = \dfrac{2y^3}{y^4} = \dfrac{2}{y}$ tends to ∞.

PROJECT 15.1

1. $f(x,y) = 2x^2 - y^2 - x^4 + 2$

(a) $\nabla f = (4x - 4x^3)\mathbf{i} + 2y\mathbf{j}$

Set $\nabla f = 0$:

$$4x - 4x^3 = 0 \qquad 2y = 0$$

$$x(1 - x^2) = 0$$

$$x = 0, 1, -1$$

$$\nabla f = 0 \text{ at } (0,0), (1,0), (-1,0)$$

(b)

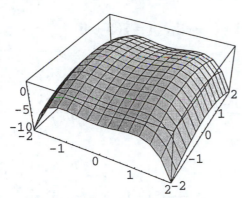

3. $f(x,y) = (x^2 + 4y^2)e^{1-(x^2+y^2)}$

(a) $\nabla f = e^{1-(x^2+y^2)}[2x - 2x(x^2 + 4y^2)]\mathbf{i} + e^{1-(x^2+y^2)}[8y - 2y(x^2 + 4y^2)]\mathbf{j}$

    Set $\nabla f = 0$:

    $2x(1 - x^2 - 4y^2) = 0 \qquad 2y(4 - x^2 - 4y^2) = 0$

    $x = 0 \text{ or } x^2 + 4y^2 = 1$

    $x = 0: \quad 2y(4 - 4y^2) = 0 \Longrightarrow y = 0, \pm 1$

    $x^2 + 4y^2 = 1: \quad 2y(3) = 0 \Longrightarrow y = 0, \ x = \pm 1$

    $\nabla f = 0 \text{ at } (0,0), (0,\pm 1), (\pm 1, 0)$

(b)

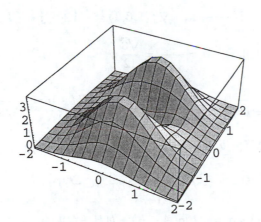

SECTION 15.2

1. $\nabla f = 2x\mathbf{i} + 6y\mathbf{j}$, $\nabla f(1,1) = 2\mathbf{i} + 6\mathbf{j}$, $\mathbf{u} = \frac{1}{2}\sqrt{2}\,(\mathbf{i} - \mathbf{j})$,

   $f'_{\mathbf{u}}(1,1) = \nabla f(1,1) \cdot \mathbf{u} = -2\sqrt{2}$

3. $\nabla f\,(e^y - ye^x)\,\mathbf{i} + (xe^y - e^x)\,\mathbf{j}$, $\nabla f(1,0) = \mathbf{i} + (1-e)\mathbf{j}$, $\mathbf{u} = \frac{1}{5}(3\mathbf{i} + 4\mathbf{j})$,

   $f'_{\mathbf{u}}(1,0) = \nabla f(1,0) \cdot \mathbf{u} = \frac{1}{5}(7 - 4e)$

5. $\nabla f = \dfrac{(a-b)y}{(x+y)^2}\,\mathbf{i} + \dfrac{(b-a)x}{(x+y)^2}\,\mathbf{j}$, $\nabla f(1,1) = \dfrac{a-b}{4}(\mathbf{i} - \mathbf{j})$, $\mathbf{u} = \frac{1}{2}\sqrt{2}\,(\mathbf{i} - \mathbf{j})$,

   $f'_{\mathbf{u}}(1,1) = \nabla f(1,1) \cdot \mathbf{u} = \frac{1}{4}\sqrt{2}\,(a - b)$

7. $\nabla f = \dfrac{2x}{x^2 + y^2}\,\mathbf{i} + \dfrac{2y}{x^2 + y^2}\,\mathbf{j}$, $\nabla f(0,1) = 2\mathbf{j}$, $\mathbf{u} = \dfrac{1}{\sqrt{65}}(8\mathbf{i} + \mathbf{j})$,

   $f'_{\mathbf{u}}(0,1) = \nabla f \cdot \mathbf{u} = \dfrac{2}{\sqrt{65}}$

9. $\nabla f = (y+z)\mathbf{i} + (x+z)\mathbf{j} + (y+x)\mathbf{k}$, $\nabla f(1,-1,1) = 2\mathbf{j}$, $\mathbf{u} = \frac{1}{6}\sqrt{6}\,(\mathbf{i} + 2\mathbf{j} + \mathbf{k})$,

   $f'_{\mathbf{u}}(1,-1,1) = \nabla f(1,-1,1) \cdot \mathbf{u} = \frac{2}{3}\sqrt{6}$

11. $\nabla f = 2\left(x + y^2 + z^2\right)\left(\mathbf{i} + 2y\mathbf{j} + 3z^2\mathbf{k}\right)$, $\nabla f(1,-1,1) = 6(\mathbf{i} - 2\mathbf{j} + 3\mathbf{k})$, $\mathbf{u} = \frac{1}{2}\sqrt{2}\,(\mathbf{i} + \mathbf{j})$,

   $f'_{\mathbf{u}}(1,-1,1) = \nabla f(1,-1,1) \cdot \mathbf{u} = -3\sqrt{2}$

13. $\nabla f = \tan^{-1}(y+z)\,\mathbf{i} + \dfrac{x}{1+(y+z)^2}\,\mathbf{j} + \dfrac{x}{1+(y+z)^2}\,\mathbf{k}$, $\nabla f(1,0,1) = \dfrac{\pi}{4}\mathbf{i} + \dfrac{1}{2}\mathbf{j} + \dfrac{1}{2}\mathbf{k}$,

   $\mathbf{u} = \dfrac{1}{\sqrt{3}}(\mathbf{i} + \mathbf{j} - \mathbf{k})$, $f'_{\mathbf{u}}(1,0,1) = \nabla f(1,0,1) \cdot \mathbf{u} = \dfrac{\pi}{4\sqrt{3}} = \dfrac{\sqrt{3}}{12}\pi$

15. $\nabla f = \dfrac{x}{x^2 + y^2}\,\mathbf{i} + \dfrac{y}{x^2 + y^2}\,\mathbf{j}$, $\mathbf{u} = \dfrac{1}{\sqrt{x^2 + y^2}}\,(-x\mathbf{i} - y\mathbf{j})$, $f'_{\mathbf{u}}(x,y) = \nabla f \cdot \mathbf{u} = -\dfrac{1}{\sqrt{x^2 + y^2}}$

17. $\nabla f = (2Ax + 2By)\,\mathbf{i} + (2Bx + 2Cy)\,\mathbf{j}$, $\nabla f(a,b) = (2aA + 2bB)\mathbf{i} + (2aB + 2bC)\mathbf{j}$

   (a) $\mathbf{u} = \frac{1}{2}\sqrt{2}\,(-\mathbf{i} + \mathbf{j})$, $f'_{\mathbf{u}}(a,b) = \nabla f(a,b) \cdot \mathbf{u} = \sqrt{2}\,[a(B - A) + b(C - B)]$

   (b) $\mathbf{u} = \frac{1}{2}\sqrt{2}\,(\mathbf{i} - \mathbf{j})$, $f'_{\mathbf{u}}(a,b) = \nabla f(a,b) \cdot \mathbf{u} = \sqrt{2}\,[a(A - B) + b(B - C)]$

19. $\nabla f = e^{y^2 - z^2}(\mathbf{i} + 2xy\mathbf{j} - 2xz\mathbf{k})$, $\nabla f(1,2,-2) = \mathbf{i} + 4\mathbf{j} + 4\mathbf{k}$, $\mathbf{r}'(t) = \mathbf{i} - 2\sin(t-1)\mathbf{j} - 2e^{t-1}\mathbf{k}$,

   at $(1,2,-2)$ $t = 1$, $\mathbf{r}'(1) = \mathbf{i} - 2\mathbf{k}$, $\mathbf{u} = \frac{1}{5}\sqrt{5}\,(\mathbf{i} - 2\mathbf{k})$, $f'_{\mathbf{u}}(1,2,-2) = \nabla f(1,2,-2) \cdot \mathbf{u} = -\frac{7}{5}\sqrt{5}$

21. $\nabla f = (2x + 2yz)\,\mathbf{i} + (2xz - z^2)\,\mathbf{j} + (2xy - 2yz)\,\mathbf{k}, \quad \nabla f(1,1,2) = 6\,\mathbf{i} - 2\,\mathbf{k}$

The vector $\mathbf{v} = 2\,\mathbf{i} + \mathbf{j} - 3\,\mathbf{k}$ is a direction vector for the given line; $\mathbf{u} = \dfrac{1}{\sqrt{14}}(2\,\mathbf{i} + \mathbf{j} - 3\,\mathbf{k})$

is a corresponding unit vector; $f'_{\mathbf{u}}(1,1,2) = \nabla f(1,1,2) \cdot \mathbf{u} = \dfrac{18}{\sqrt{14}}$

Note: we could have used $\mathbf{u} = -\dfrac{1}{\sqrt{14}}(2\,\mathbf{i} + \mathbf{j} - 3\,\mathbf{k})$ as the unit vector. In this case,

$f'_{\mathbf{u}}(1,1,2) = -\dfrac{18}{\sqrt{14}}$

23. $\nabla f = 2y^2 e^{2x}\,\mathbf{i} + 2y e^{2x}\,\mathbf{j}, \quad \nabla f(0,1) = 2\,\mathbf{i} + 2\,\mathbf{j}, \quad \|\nabla f\| = 2\sqrt{2}, \quad \dfrac{\nabla f}{\|\nabla f\|} = \dfrac{1}{\sqrt{2}}(\mathbf{i} + \mathbf{j})$

f increases most rapidly in the direction $\mathbf{u} = \dfrac{1}{\sqrt{2}}(\mathbf{i} + \mathbf{j})$; the rate of change is $2\sqrt{2}$.

f decreases most rapidly in the direction $\mathbf{v} = -\dfrac{1}{\sqrt{2}}(\mathbf{i} + \mathbf{j})$; the rate of change is $-2\sqrt{2}$.

25. $\nabla f = \dfrac{x}{\sqrt{x^2 + y^2 + z^2}}\,\mathbf{i} + \dfrac{y}{\sqrt{x^2 + y^2 + z^2}}\,\mathbf{j} + \dfrac{z}{\sqrt{x^2 + y^2 + z^2}}\,\mathbf{k},$

$\nabla f(1,-2,1) = \dfrac{1}{\sqrt{6}}(\mathbf{i} - 2\,\mathbf{j} + \mathbf{k}), \quad \|\nabla f\| = 1$

f increases most rapidly in the direction $\mathbf{u} = \dfrac{1}{\sqrt{6}}(\mathbf{i} - 2\,\mathbf{j} + \mathbf{k})$; the rate of change is 1.

f decreases most rapidly in the direction $\mathbf{v} = -\dfrac{1}{\sqrt{6}}(\mathbf{i} - 2\,\mathbf{j} + \mathbf{k})$; the rate of change is -1.

27. $\nabla f = f'(x_0)\,\mathbf{i}$. If $f'(x_0) \neq 0$, the gradient points in the direction in which f increases: to the right if $f'(x_0) > 0$, to the left if $f'(x_0) < 0$.

29. (a) $\displaystyle\lim_{h\to 0}\dfrac{f(h,0) - f(0,0)}{h} = \lim_{h\to 0}\dfrac{\sqrt{h^2}}{h} = \lim_{h\to 0}\dfrac{|h|}{h}$ does not exist

(b) no; by Theorem 15.2.5 f cannot be differentiable at $(0,0)$

31. $\nabla \lambda(x,y) = -\tfrac{8}{3}x\,\mathbf{i} - 6y\,\mathbf{j}$

(a) $\nabla \lambda(1,-1) = -\tfrac{8}{3}\mathbf{i} = 6\mathbf{j}, \quad \mathbf{u} = \dfrac{-\nabla\lambda(1,-1)}{\|\nabla\lambda(1,-1)\|} = \dfrac{\tfrac{8}{3}\mathbf{i} - 6\mathbf{j}}{\tfrac{2}{3}\sqrt{97}}, \quad \lambda'_{\mathbf{u}}(1,-1) = \nabla\lambda(1,-1)\cdot\mathbf{u} = -\tfrac{2}{3}\sqrt{97}$

(b) $\mathbf{u} = \mathbf{i}, \quad \lambda'_{\mathbf{u}}(1,2) = \nabla\lambda(1,2)\cdot\mathbf{u} = \left(-\tfrac{8}{3}\mathbf{i} - 12\mathbf{j}\right)\cdot\mathbf{i} = -\tfrac{8}{3}$

(c) $\mathbf{u} = \tfrac{1}{2}\sqrt{2}\,(\mathbf{i}+\mathbf{j}), \quad \lambda'_{\mathbf{u}}(2,2) = \nabla\lambda(2,2)\cdot\mathbf{u} = \left(-\tfrac{16}{3}\mathbf{i} - 12\mathbf{j}\right)\cdot\left[\tfrac{1}{2}\sqrt{2}\,(\mathbf{i}+\mathbf{j})\right] = -\tfrac{26}{3}\sqrt{2}$

33. (a) The projection of the path onto the xy-plane is the curve

$$C: \ \mathbf{r}(t) = x(t)\mathbf{i} + y(t)\mathbf{j}$$

which begins at $(1,1)$ and at each point has its tangent vector in the direction of $-\nabla f$. Since

$$\nabla f = 2x\mathbf{i} + 6y\mathbf{j},$$

we have the initial-value problems

$$x'(t) = -2x(t), \quad x(0) = 1 \qquad \text{and} \qquad y'(t) = -6y(t), \quad y(0) = 1.$$

From Theorem 7.6.1 we find that

$$x(t) = e^{-2t} \qquad \text{and} \qquad y(t) = e^{-6t}.$$

Eliminating the parameter t, we find that C is the curve $y = x^3$ from $(1,1)$ to $(0,0)$.

(b) Here

$$x'(t) = -2x(t), \quad x(0) = 1 \qquad \text{and} \qquad y'(t) = -6y(t), \quad y(0) = -2$$

so that

$$x(t) = e^{-2t} \qquad \text{and} \qquad y(t) = -2e^{-6t}.$$

Eliminating the parameter t, we find that the projection of the path onto the xy-plane is the curve $y = -2x^3$ from $(1,-2)$ to $(0,0)$.

35. The projection of the path onto the xy-plane is the curve

$$C: \ \mathbf{r}(t) = x(t)\mathbf{i} + y(t)\mathbf{j}$$

which begins at $\left(a^2, b^2\right)$ and at each point has its tangent vector in the direction of $-\nabla f = -\left(2a^2 x\mathbf{i} + 2b^2 y\mathbf{j}\right)$. Thus,

$$x'(t) = -2a^2 x(t), \quad x(0) = a^2 \qquad \text{and} \qquad y'(t) = -2b^2 y(t), \quad y(0) = b^2$$

so that

$$x(t) = a^2 e^{-2a^2 t} \qquad \text{and} \qquad y(t) = b^2 e^{-2b^2 t}.$$

Since

$$\left[\frac{x}{a^2}\right]^{b^2} = \left(e^{-2a^2 t}\right)^{b^2} = \left[\frac{y}{b^2}\right]^{a^2},$$

C is the curve $\left(b^2\right)^{a^2} x^{b^2} = \left(a^2\right)^{b^2} y^{a^2}$ from $\left(a^2, b^2\right)$ to $(0,0)$.

37.   We want the curve

$$C:\ \mathbf{r}(t) = x(t)\mathbf{i} + y(t)\mathbf{j}$$

which begins at $(\pi/4, 0)$ and at each point has its tangent vector in the direction of

$$\nabla T = -\sqrt{2}\,e^{-y}\sin x\,\mathbf{i} - \sqrt{2}\,e^{-y}\cos x\,\mathbf{j}.$$

From

$$x'(t) = -\sqrt{2}\,e^{-y}\sin x \qquad \text{and} \qquad y'(t) = -\sqrt{2}\,e^{-y}\cos x$$

we obtain

$$\frac{dy}{dx} = \frac{y'(t)}{x'(t)} = \cot x$$

so that

$$y = \ln|\sin x| + C.$$

Since $y = 0$ when $x = \pi/4$, we get $C = \ln\sqrt{2}$ and $y = \ln|\sqrt{2}\sin x|$. As $\nabla T(\pi/4, 0) = -\mathbf{i} - \mathbf{j}$, the curve $y = \ln|\sqrt{2}\,\sin x|$ is followed in the direction of decreasing x.

39. (a) $$\lim_{h \to 0} \frac{f\left(2+h,\,(2+h)^2\right) - f(2,4)}{h} = \lim_{h \to 0} \frac{3(2+h)^2 + (2+h)^2 - 16}{h}$$

$$= \lim_{h \to 0} 4\left[\frac{4h + h^2}{h}\right] = \lim_{h \to 0} 4(4+h) = 16$$

(b) $$\lim_{h \to 0} \frac{f\left(\dfrac{h+8}{4},\, 4+h\right) - f(2,4)}{h} = \lim_{h \to 0} \frac{3\left(\dfrac{h+8}{4}\right)^2 + (4+h) - 16}{h}$$

$$= \lim_{h \to 0} \frac{\frac{3}{16}h^2 + 3h + 12 + 4 + h - 16}{h}$$

$$= \lim_{h \to 0} \left(\tfrac{3}{16}h + 4\right) = 4$$

(c) $\mathbf{u} = \tfrac{1}{17}\sqrt{17}\,(\mathbf{i} + 4\mathbf{j})$, $\nabla f(2,4) = 12\mathbf{i} + \mathbf{j}$; $f'_{\mathbf{u}}(2,4) = \nabla f(2,4) \cdot \mathbf{u} = \tfrac{16}{17}\sqrt{17}$

(d)   The limits computed in (a) and (b) are not directional derivatives. In (a) and (b) we have, in essence, computed $\nabla f(2,4) \cdot \mathbf{r_0}$ taking $\mathbf{r_0} = \mathbf{i} + 4\mathbf{j}$ in (a) and $\mathbf{r_0} = \tfrac{1}{4}\mathbf{i} + \mathbf{j}$ in (b). In neither case is $\mathbf{r_0}$ a unit vector.

41. (a) $\mathbf{u} = \cos\theta\,\mathbf{i} + \sin\theta\,\mathbf{j}$, $\nabla f(x,y) = \dfrac{\partial f}{\partial x}\mathbf{i} + \dfrac{\partial f}{\partial y}\mathbf{j}$;

$$f'_{\mathbf{u}}(x,y) = \nabla f \cdot \mathbf{u} = \left(\frac{\partial f}{\partial x}\mathbf{i} + \frac{\partial f}{\partial y}\mathbf{j}\right) \cdot (\cos\theta\,\mathbf{i} + \sin\theta\,\mathbf{j}) = \frac{\partial f}{\partial x}\cos\theta + \frac{\partial f}{\partial y}\sin\theta$$

(b) $\nabla f = \left(3x^2 + 2y - y^2\right)\mathbf{i} + (2x - 2xy)\mathbf{j}$, $\nabla f(-1,2) = 3\mathbf{i} + 2\mathbf{j}$

$$f'_u(-1, 2) = 3\cos(2\pi/3) + 2\sin(2\pi/3) = \frac{2\sqrt{3} - 3}{2}$$

43.
$$\nabla(fg) = \frac{\partial fg}{\partial x}\mathbf{i} + \frac{\partial fg}{\partial y}\mathbf{j} = \left(f\frac{\partial g}{\partial x} + g\frac{\partial f}{\partial x}\right)\mathbf{i} + \left(f\frac{\partial g}{\partial y} + g\frac{\partial f}{\partial y}\right)\mathbf{j}$$
$$= f\left(\frac{\partial g}{\partial x}\mathbf{i} + \frac{\partial g}{\partial y}\mathbf{j}\right) + g\left(\frac{\partial f}{\partial x}\mathbf{i} + \frac{\partial f}{\partial y}\mathbf{j}\right) = f\nabla g + g\nabla f$$

45. $\nabla f^n = \dfrac{\partial f^n}{\partial x}\mathbf{i} + \dfrac{\partial f^n}{\partial y}\mathbf{j} = nf^{n-1}\dfrac{\partial f}{\partial x}\mathbf{i} + nf^{n-1}\dfrac{\partial f}{\partial y}\mathbf{j} = nf^{n-1}\nabla f$

SECTION 15.3

1. $f(\mathbf{b}) = f(1, 3) = -2; \quad f(\mathbf{a}) = f(0, 1) = 0; \quad f(\mathbf{b}) - f(\mathbf{a}) = -2$

$\nabla f = (3x^2 - y)\,\mathbf{i} - x\mathbf{j}; \quad \mathbf{b} - \mathbf{a} = \mathbf{i} + 2\mathbf{j} \quad \text{and} \quad \nabla f \cdot (\mathbf{b} - \mathbf{a}) = 3x^2 - y - 2x$

The line segment joining $\mathbf{a}$ and $\mathbf{b}$ is parametrized by

$$x = t, \quad y = 1 + 2t, \quad 0 \le t \le 1$$

Thus, we need to solve the equation

$$3t^2 - (1 + 2t) - 2t = -2, \quad \text{which is the same as} \quad 3t^2 - 4t + 1 = 0, \;\; 0 \le t \le 1$$

The solutions are: $t = \frac{1}{3}, t = 1$. Thus, $\mathbf{c} = (\frac{1}{3}, \frac{5}{3})$ satisfies the equation.

Note that the endpoint $\mathbf{b}$ also satisfies the equation.

3. (a) $f(x, y, z) = a_1 x + a_2 y + a_3 z + C$ (b) $f(x, y, z) = g(x, y, z) + a_1 x + a_2 y + a_3 z + C$

5. (a) U is not connected

(b) (i) $g(\mathbf{x}) = f(\mathbf{x}) - 1$ (ii) $g(\mathbf{x}) = -f(\mathbf{x})$

7. $\nabla f = 2xy\mathbf{i} + x^2\mathbf{j};$

$\nabla f(\mathbf{r}(t)) \cdot \mathbf{r}'(t) = (2\mathbf{i} + e^{2t}\mathbf{j}) \cdot (e^t\mathbf{i} - e^{-t}\mathbf{j}) = e^t$

9. $\nabla f = \dfrac{-2x}{1 + (y^2 - x^2)^2}\mathbf{i} + \dfrac{2y}{1 + (y^2 - x^2)^2}\mathbf{j}, \quad \nabla f(\mathbf{r}(t)) = \dfrac{-2\sin t}{1 + \cos^2 2t}\mathbf{i} + \dfrac{2\cos t}{1 + \cos^2 2t}\mathbf{j}$

$\nabla f(\mathbf{r}(t)) \cdot \mathbf{r}'(t) = \left(\dfrac{-2\sin t}{1 + \cos^2 2t}\mathbf{i} + \dfrac{2\cos t}{1 + \cos^2 2t}\mathbf{j}\right) \cdot (\cos t\,\mathbf{i} - \sin t\,\mathbf{j}) = \dfrac{-4\sin t\cos t}{1 + \cos^2 2t} = \dfrac{-2\sin 2t}{1 + \cos^2 2t}$

11. $\nabla f = (e^y - ye^{-x})\,\mathbf{i} + (xe^y + e^{-x})\,\mathbf{j}; \quad \nabla f(\mathbf{r}(t)) = (t^t - \ln t)\,\mathbf{i} + \left(t^t\ln t + \dfrac{1}{t}\right)\mathbf{j}$

$\nabla f(\mathbf{r}(t)) \cdot \mathbf{r}'(t) = \left((t^t - \ln t)\,\mathbf{i} + \left(t^t\ln t + \dfrac{1}{t}\right)\mathbf{j}\right) \cdot \left(\dfrac{1}{t}\mathbf{i} + [1 + \ln t]\mathbf{j}\right) = t^t\left(\dfrac{1}{t} + \ln t + [\ln t]^2\right) + \dfrac{1}{t}$

13. $\nabla f = y\mathbf{i} + (x - z)\mathbf{j} - y\mathbf{k};$

$\nabla f(\mathbf{r}(t)) \cdot \mathbf{r}'(t) = (t^2\mathbf{i} + (t - t^3)\,\mathbf{j} - t^2\mathbf{k}) \cdot (\mathbf{i} + 2t\mathbf{j} + 3t^2\mathbf{k}) = 3t^2 - 5t^4$

15. $\nabla f = 2x\mathbf{i} + 2y\mathbf{j} + \mathbf{k}$;

$$\nabla f(\mathbf{r}(t)) \cdot \mathbf{r}'(t) = (2a\cos \omega t\,\mathbf{i} + 2b\sin \omega t\,\mathbf{j} + \mathbf{k}) \cdot (-a\omega \sin \omega t\,\mathbf{i} + b\omega \cos \omega t\,\mathbf{j} + b\omega \mathbf{k})$$

$$= 2\omega\left(b^2 - a^2\right)\sin \omega t\,\cos \omega t + b\omega$$

17. $\dfrac{du}{dt} = \dfrac{\partial u}{\partial x}\dfrac{dx}{dt} + \dfrac{\partial u}{\partial y}\dfrac{dy}{dt} = (2x - 3y)(-\sin t) + (4y - 3x)(\cos t)$

$$= 2\cos t\sin t + 3\sin^2 t - 3\cos^2 t = \sin 2t - 3\cos 2t$$

19. $\dfrac{du}{dt} = \dfrac{\partial u}{\partial x}\dfrac{dx}{dt} + \dfrac{\partial u}{\partial y}\dfrac{dy}{dt}$

$$= (e^x \sin y + e^y \cos x)\left(\tfrac{1}{2}\right) + (e^x \cos y + e^y \sin x)\,(2)$$

$$= e^{t/2}\left(\tfrac{1}{2}\sin 2t + 2\cos 2t\right) + e^{2t}\left(\tfrac{1}{2}\cos\tfrac{1}{2}t + 2\sin\tfrac{1}{2}t\right)$$

21. $\dfrac{du}{dt} = \dfrac{\partial u}{\partial x}\dfrac{dx}{dt} + \dfrac{\partial u}{\partial y}\dfrac{dy}{dt} = (e^x \sin y)\,(2t) + (e^x \cos y)\,(\pi)$

$$= e^{t^2}[2t\,\sin(\pi t) + \pi\,\cos(\pi t)]$$

23. $\dfrac{du}{dt} = \dfrac{\partial u}{\partial x}\dfrac{dx}{dt} + \dfrac{\partial u}{\partial y}\dfrac{dy}{dt} + \dfrac{\partial u}{\partial z}\dfrac{dz}{dt}$

$$= (y + z)(2t) + (x + z)(1 - 2t) + (y + x)(2t - 2)$$

$$= (1 - t)(2t) + (2t^2 - 2t + 1)(1 - 2t) + t(2t - 2)$$

$$= 1 - 4t + 6t^2 - 4t^3$$

25. $V = \dfrac{1}{3}\pi r^2 h, \quad \dfrac{dV}{dt} = \dfrac{\partial V}{\partial r}\dfrac{dr}{dt} + \dfrac{\partial V}{\partial h}\dfrac{dh}{dt} = \left(\dfrac{2}{3}\pi rh\right)\dfrac{dr}{dt} + \left(\dfrac{1}{3}\pi r^2\right)\dfrac{dh}{dt}.$

At the given instant,

$$\dfrac{dV}{dt} = \dfrac{2}{3}\pi(280)(3) + \dfrac{1}{3}\pi(196)(-2) = \dfrac{1288}{3}\pi.$$

The volume is increasing at the rate of $\dfrac{1288}{3}\pi$ in.$^3/$ sec.

27. $A = \tfrac{1}{2}xy \sin \theta; \quad \dfrac{dA}{dt} = \dfrac{\partial A}{\partial x}\dfrac{dx}{dt} + \dfrac{\partial A}{\partial y}\dfrac{dy}{dt} + \dfrac{\partial A}{\partial \theta}\dfrac{d\theta}{dt} = \tfrac{1}{2}\left[(y\sin\theta)\dfrac{dx}{dt} + (x\sin\theta)\dfrac{dy}{dt} + (xy\cos\theta)\dfrac{d\theta}{dt}\right].$

At the given instant

$$\dfrac{dA}{dt} = \dfrac{1}{2}\left[(2\sin 1)\,(0.25) + (1.5\sin 1)\,(0.25) + (2(1.5)\cos 1)\,(0.1)\right] \cong 0.2871 \text{ ft}^2/s \cong 41.34\,\text{in}^2/s$$

29. $\dfrac{\partial u}{\partial s} = \dfrac{\partial u}{\partial x}\dfrac{\partial x}{\partial s} + \dfrac{\partial u}{\partial y}\dfrac{\partial y}{\partial s} = (2x - y)(\cos t) + (-x)(t\cos s)$

$$= 2s\cos^2 t - t\sin s\cos t - st\cos s\cos t$$

$$\frac{\partial u}{\partial t} = \frac{\partial u}{\partial x}\frac{\partial x}{\partial t} + \frac{\partial u}{\partial y}\frac{\partial y}{\partial t} = (2x - y)(-s\sin t) + (-x)(\sin s)$$

$$= -2s^2\cos t\sin t + st\sin s\sin t - s\cos t\sin s$$

31. $\dfrac{\partial u}{\partial s} = \dfrac{\partial u}{\partial x}\dfrac{\partial x}{\partial s} + \dfrac{\partial u}{\partial y}\dfrac{\partial y}{\partial s} = (2x\tan y)(2st) + \left(x^2\sec^2 y\right)(1)$

$$= 4s^3t^2\tan\left(s + t^2\right) + s^4t^2\sec^2\left(s + t^2\right)$$

$$\frac{\partial u}{\partial t} = \frac{\partial u}{\partial x}\frac{\partial x}{\partial t} + \frac{\partial u}{\partial y}\frac{\partial y}{\partial t} = (2x\tan y)\left(s^2\right) + \left(x^2\sec^2 y\right)(2t)$$

$$= 2s^4t\tan\left(s + t^2\right) + 2s^4t^3\sec^2\left(s + t^2\right)$$

33. $\dfrac{\partial u}{\partial s} = \dfrac{\partial u}{\partial x}\dfrac{\partial x}{\partial s} + \dfrac{\partial u}{\partial y}\dfrac{\partial y}{\partial s} + \dfrac{\partial u}{\partial z}\dfrac{\partial z}{\partial s}$

$$= (2x - y)(\cos t) + (-x)(-\cos(t - s)) + 2z(t\cos s)$$

$$= 2s\cos^2 t - \sin(t - s)\cos t + s\cos t\cos(t - s) + 2t^2\sin s\cos s$$

$$\frac{\partial u}{\partial t} = \frac{\partial u}{\partial x}\frac{\partial x}{\partial t} + \frac{\partial u}{\partial y}\frac{\partial y}{\partial t} + \frac{\partial u}{\partial z}\frac{\partial z}{\partial t}$$

$$= (2x - y)(-s\sin t) + (-x)(\cos(t - s)) + 2z(\sin s)$$
$$= -2s^2\cos t\sin t + s\sin(t - s)\sin t - s\cos t\cos(t - s) + 2t\sin^2 s$$

35.
$$\frac{d}{dt}[f(\mathbf{r}(t))] = \left[\nabla f(\mathbf{r}(t)) \cdot \frac{\mathbf{r}'(t)}{\|\mathbf{r}'(t)\|}\right]\|\mathbf{r}'(t)\|$$

$$= f'_{\mathbf{u}(t)}(\mathbf{r}(t))\|\mathbf{r}'(t)\| \quad \text{where} \quad \mathbf{u}(t) = \frac{\mathbf{r}'(t)}{\|\mathbf{r}'(t)\|}$$

37. (a) $(\cos r)\dfrac{\mathbf{r}}{r}$ (b) $(r\cos r + \sin r)\dfrac{\mathbf{r}}{r}$

39. (a) $(r\cos r - \sin r)\dfrac{\mathbf{r}}{r^3}$ (b) $\left(\dfrac{\sin r - r\cos r}{\sin^2 r}\right)\dfrac{\mathbf{r}}{r}$

41. (a)

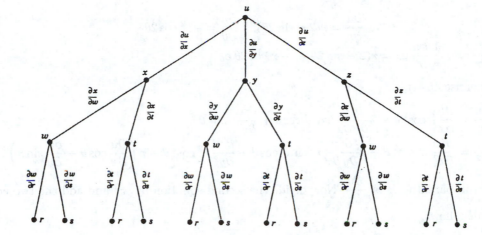

(b) $\dfrac{\partial u}{\partial r} = \dfrac{\partial u}{\partial x}\left(\dfrac{\partial x}{\partial w}\dfrac{\partial w}{\partial r} + \dfrac{\partial x}{\partial t}\dfrac{\partial t}{\partial r}\right) + \dfrac{\partial u}{\partial y}\left(\dfrac{\partial y}{\partial w}\dfrac{\partial w}{\partial r} + \dfrac{\partial y}{\partial t}\dfrac{\partial t}{\partial r}\right) + \dfrac{\partial u}{\partial z}\left(\dfrac{\partial z}{\partial w}\dfrac{\partial w}{\partial r} + \dfrac{\partial z}{\partial t}\dfrac{\partial t}{\partial r}\right).$

To obtain $\partial u/\partial s$, replace each r by s.

43. $\dfrac{du}{dt} = \dfrac{\partial u}{\partial x}\dfrac{dx}{dt} + \dfrac{\partial u}{\partial y}\dfrac{dy}{dt}$

$\dfrac{d^2u}{dt^2} = \dfrac{\partial u}{\partial x}\dfrac{d^2x}{dt^2} + \dfrac{dx}{dt}\left[\dfrac{\partial^2 u}{\partial x^2}\dfrac{dx}{dt} + \dfrac{\partial^2 u}{\partial y\partial x}\dfrac{dy}{dt}\right] + \dfrac{\partial u}{\partial y}\dfrac{d^2y}{dt^2} + \dfrac{dy}{dt}\left[\dfrac{\partial^2 u}{\partial x\,\partial y}\dfrac{dx}{dt} + \dfrac{\partial^2 u}{\partial y^2}\dfrac{dy}{dt}\right]$

and the result follows.

45. (a) $\dfrac{\partial u}{\partial r} = \dfrac{\partial u}{\partial x}\dfrac{\partial x}{\partial r} + \dfrac{\partial u}{\partial y}\dfrac{\partial y}{\partial r} = \dfrac{\partial u}{\partial x}\cos\theta + \dfrac{\partial u}{\partial y}\sin\theta$

$\dfrac{\partial u}{\partial \theta} = \dfrac{\partial u}{\partial x}\dfrac{\partial x}{\partial \theta} + \dfrac{\partial u}{\partial y}\dfrac{\partial y}{\partial \theta} = \dfrac{\partial u}{\partial x}(-r\sin\theta) + \dfrac{\partial u}{\partial y}(r\cos\theta)$

(b) $\left(\dfrac{\partial u}{\partial r}\right)^2 = \left(\dfrac{\partial u}{\partial x}\right)^2\cos^2\theta + 2\dfrac{\partial u}{\partial x}\dfrac{\partial u}{\partial y}\cos\theta\sin\theta + \left(\dfrac{\partial u}{\partial y}\right)^2\sin^2\theta,$

$\dfrac{1}{r^2}\left(\dfrac{\partial u}{\partial \theta}\right)^2 = \left(\dfrac{\partial u}{\partial x}\right)^2\sin^2\theta - 2\dfrac{\partial u}{\partial x}\dfrac{\partial u}{\partial y}\cos\theta\sin\theta + \left(\dfrac{\partial u}{\partial y}\right)^2\cos^2\theta,$

$\left(\dfrac{\partial u}{\partial r}\right)^2 + \dfrac{1}{r^2}\left(\dfrac{\partial u}{\partial \theta}\right)^2 = \left(\dfrac{\partial u}{\partial x}\right)^2\left(\cos^2\theta + \sin^2\theta\right) + \left(\dfrac{\partial u}{\partial y}\right)^2\left(\sin^2\theta + \cos^2\theta\right) = \left(\dfrac{\partial u}{\partial x}\right)^2 + \left(\dfrac{\partial u}{\partial y}\right)^2$

47. Solve the equations in Exercise 45 (a) for $\dfrac{\partial u}{\partial x}$ and $\dfrac{\partial u}{\partial y}$:

$$\dfrac{\partial u}{\partial x} = \dfrac{\partial u}{\partial r}\cos\theta - \dfrac{1}{r}\dfrac{\partial u}{\partial \theta}\sin\theta, \quad \dfrac{\partial u}{\partial y} = \dfrac{\partial u}{\partial r}\sin\theta + \dfrac{1}{r}\dfrac{\partial u}{\partial \theta}\cos\theta$$

Then $\nabla u = \dfrac{\partial u}{\partial x}\mathbf{i} + \dfrac{\partial u}{\partial y}\mathbf{j} = \dfrac{\partial u}{\partial r}(\cos\theta\,\mathbf{i} + \sin\theta\,\mathbf{j}) + \dfrac{1}{r}\dfrac{\partial u}{\partial \theta}(-\sin\theta\,\mathbf{i} + \cos\theta\,\mathbf{j})$

49. $u(x,y) = x^2 - xy + y^2 = r^2 - r^2\cos\theta\sin\theta = r^2\left(1 - \tfrac{1}{2}\sin 2\theta\right)$

$$\frac{\partial u}{\partial r} = r(2 - \sin 2\theta), \qquad \frac{\partial u}{\partial \theta} = -r^2 \cos 2\theta$$

$$\nabla u = \frac{\partial u}{\partial r}\, \mathbf{e_r} + \frac{1}{r}\frac{\partial u}{\partial \theta}\, \mathbf{e_\theta} = r(2 - \sin 2\theta)\mathbf{e_r} - r\cos 2\theta\, \mathbf{e_\theta}$$

51.   From Exercise 45 (a),

$$\frac{\partial^2 u}{\partial r^2} = \frac{\partial^2 u}{\partial x^2}\cos^2\theta + 2\frac{\partial^2 u}{\partial y\,\partial x}\sin\theta\cos\theta + \frac{\partial^2 u}{\partial y^2}\sin^2\theta$$

$$\frac{\partial^2 u}{\partial \theta^2} = \frac{\partial^2 u}{\partial x^2}r^2\sin^2\theta - 2\frac{\partial^2 u}{\partial y\,\partial x}r^2\sin\theta\cos\theta + \frac{\partial^2 u}{\partial y^2}r^2\cos^2\theta - r\left(\frac{\partial u}{\partial x}\cos\theta + \frac{\partial u}{\partial y}\sin\theta\right).$$

The term in parentheses is $\dfrac{\partial u}{\partial r}$. Now divide the second equation by r^2 and add the two equations.

The result follows.

53.   Set $u = xe^y + ye^x - 2x^2y$.   Then

$$\frac{\partial u}{\partial x} = e^y + ye^x - 4xy, \qquad \frac{\partial u}{\partial y} = xe^y + e^x - 2x^2$$

$$\frac{dy}{dx} = -\frac{\partial u/\partial x}{\partial u/\partial y} = -\frac{e^y + ye^x - 4xy}{xe^y + e^x - 2x^2}.$$

55.   Set $u = x\cos xy + y\cos x - 2$.   Then

$$\frac{\partial u}{\partial x} = \cos xy - xy\sin xy - y\sin x, \qquad \frac{\partial u}{\partial y} = -x^2\sin xy + \cos x$$

$$\frac{dy}{dx} = -\frac{\partial u/\partial x}{\partial u/\partial y} = \frac{\cos xy - xy\sin xy - y\sin x}{x^2\sin xy - \cos x}.$$

57.   Set $u = \cos xyz + \ln\left(x^2 + y^2 + z^2\right)$.   Then

$$\frac{\partial u}{\partial x} = -yz\sin xyz + \frac{2x}{x^2 + y^2 + z^2}, \qquad \frac{\partial u}{\partial y} = -xz\sin xyz + \frac{2y}{x^2 + y^2 + z^2}, \qquad \text{and}$$

$$\frac{\partial u}{\partial z} = -xy\sin xyz + \frac{2z}{x^2 + y^2 + z^2}.$$

$$\frac{\partial z}{\partial x} = -\frac{\partial u/\partial x}{\partial u/\partial z} = -\frac{2x - yz\left(x^2 + y^2 + z^2\right)\sin xyz}{2z - xy\left(x^2 + y^2 + z^2\right)\sin xyz},$$

$$\frac{\partial z}{\partial y} = -\frac{\partial u/\partial y}{\partial u/\partial z} = -\frac{2y - xz\left(x^2 + y^2 + z^2\right)\sin xyz}{2z - xy\left(x^2 + y^2 + z^2\right)\sin xyz}.$$

59. $\dfrac{\partial \mathbf{u}}{\partial s} = \dfrac{\partial \mathbf{u}}{\partial x}\dfrac{\partial x}{\partial s} + \dfrac{\partial \mathbf{u}}{\partial y}\dfrac{\partial y}{\partial s}, \qquad \dfrac{\partial \mathbf{u}}{\partial t} = \dfrac{\partial \mathbf{u}}{\partial x}\dfrac{\partial x}{\partial t} + \dfrac{\partial \mathbf{u}}{\partial y}\dfrac{\partial y}{\partial t}$

SECTION 15.4

1.   Set $f(x,y) = x^2 + xy + y^2$.   Then,

$$\nabla f = (2x + y)\mathbf{i} + (x + 2y)\mathbf{j}, \quad \nabla f(-1,-1) = -3\mathbf{i} - 3\mathbf{j}.$$

normal vector $\mathbf{i} + \mathbf{j}$; tangent vector $\mathbf{i} - \mathbf{j}$

tangent line $x + y + 2 = 0$; normal line $x - y = 0$

3. Set $f(x,y) = (x^2 + y^2)^2 - 9(x^2 - y^2)$.   Then,

$$\nabla f = [4x(x^2 + y^2) - 18x]\mathbf{i} + [4y(x^2 + y^2) + 18y]\mathbf{j}, \quad \nabla f\left(\sqrt{2}, 1\right) = -6\sqrt{2}\,\mathbf{i} + 30\mathbf{j}.$$

normal vector $\sqrt{2}\,\mathbf{i} - 5\mathbf{j}$;   tangent vector $5\mathbf{i} + \sqrt{2}\,\mathbf{j}$

tangent line $\sqrt{2}x - 5y + 3 = 0$;   normal line $5x + \sqrt{2}\,y - 6\sqrt{2} = 0$

5. Set $f(x,y) = xy^2 - 2x^2 + y + 5x$.   Then,

$$\nabla f = (y^2 - 4x + 5)\mathbf{i} + (2xy + 1)\mathbf{j}, \quad \nabla f(4, 2) = -7\mathbf{i} + 17\mathbf{j}.$$

normal vector $7\mathbf{i} - 17\mathbf{j}$;   tangent vector $17\mathbf{i} + 7\mathbf{j}$

tangent line $7x - 17y + 6 = 0$;   normal line $17x + 7y - 82 = 0$

7. Set $f(x,y) = 2x^3 - x^2y^2 - 3x + y$.   Then,

$$\nabla f = (6x^2 - 2xy^2 - 3)\mathbf{i} + (-2x^2y + 1)\mathbf{j}, \quad \nabla f(1, -2) = -5\mathbf{i} + 5\mathbf{j}.$$

normal vector $\mathbf{i} - \mathbf{j}$;   tangent vector $\mathbf{i} + \mathbf{j}$

tangent line $x - y - 3 = 0$;   normal line $x + y + 1 = 0$

9. Set $f(x,y) = x^2y + a^2y$.   By (15.4.4)

$$m = -\frac{\partial f/\partial x}{\partial f/\partial y} = -\frac{2xy}{x^2 + a^2}.$$

At $(0, a)$ the slope is 0.

11. Set $f(x,y,z) = x^3 + y^3 - 3xyz$.   Then,

$$\nabla f = (3x^2 - 3yz)\mathbf{i} + (3y^2 - 3xz)\mathbf{j} - 3xy\mathbf{k}, \quad \nabla f\left(1, 2, \tfrac{3}{2}\right) = -6\mathbf{i} + \tfrac{15}{2}\mathbf{j} - 6\mathbf{k};$$

tangent plane at $\left(1, 2, \tfrac{3}{2}\right)$: $-6(x - 1) + \tfrac{15}{2}(y - 2) - 6\left(z - \tfrac{3}{2}\right) = 0$, which reduces to $4x - 5y + 4z = 0$.

Normal: $x = 1 + 4t, \quad y = 2 - 5t, \quad z = \tfrac{3}{2} + 4t$

13. Set $z = g(x,y) = axy$.   Then, $\nabla g = ay\mathbf{i} + ax\mathbf{j}, \quad \nabla g\left(1, \dfrac{1}{a}\right) = \mathbf{i} + a\mathbf{j}$.

tangent plane at $\left(1, \dfrac{1}{a}, 1\right)$: $z - 1 = 1(x - 1) + a\left(y - \dfrac{1}{a}\right)$, which reduces to $x + ay - z - 1 = 0$

Normal: $x = 1 + t, \quad y = \tfrac{1}{a} + at, \quad z = 1 - t$

15. Set $z = g(x,y) = \sin x + \sin y + \sin(x + y)$.   Then,

$$\nabla g = [\cos x + \cos(x + y)]\mathbf{i} + [\cos y + \cos(x + y)]\mathbf{j}, \quad \nabla g(0,0) = 2\mathbf{i} + 2\mathbf{j};$$

tangent plane at $(0,0,0)$: $z - 0 = 2(x - 0) + 2(y - 0), \quad 2x + 2y - z = 0$.

Normal: $x = 2t, \quad y = 2t, \quad z = -t$

17. Set $f(x,y,z) = b^2c^2x^2 - a^2c^2y^2 - a^2b^2z^2$.   Then,

$$\nabla f(x_0, y_0, z_0) = 2b^2c^2x_0\mathbf{i} - 2a^2c^2y_0\mathbf{j} - 2a^2b^2z_0\mathbf{k};$$

tangent plane at (x_0, y_0, z_0):

$$2b^2c^2x_0(x - x_0) - 2a^2c^2y_0(y - y_0) - 2a^2b^2z_0(z - z_0) = 0,$$

which can be rewritten as follows:

$$b^2c^2x_0x - a^2c^2y_0y - a^2b^2z_0z = b^2c^2x_0{}^2 - a^2c^2y_0{}^2 - a^2b^2z_0{}^2$$
$$= f(x_0, y_0, z_0) = a^2 + b^2 + c^2.$$

Normal: $x = x_0 + 2b^2c^2x_0t, \quad y = y_0 - 2a^2c^2y_0t, \quad z = z_0 - 2a^2b^2z_0t$

19.   Set $z = g(x, y) = xy + a^3x^{-1} + b^3y^{-1}$.

$$\nabla g = (y - a^3x^{-2})\mathbf{i} + (x - b^3y^{-2})\mathbf{j}, \quad \nabla g = 0 \implies y = a^3x^{-2} \text{ and } x = b^3y^{-2}.$$

Thus,

$$y = a^3b^{-6}y^4, \quad y^3 = b^6a^{-3}, \quad y = b^2/a, \quad x = b^3y^{-2} = a^2/b \text{ and } g(a^2/b, b^2/a) = 3ab.$$

The tangent plane is horizontal at $(a^2/b, b^2/a, 3ab)$.

21.   Set $z = g(x, y) = xy$.   Then, $\nabla g = y\mathbf{i} + x\mathbf{j}$.

$$\nabla g = 0 \implies x = y = 0.$$

The tangent plane is horizontal at $(0, 0, 0)$.

23.   Set $z = g(x, y) = 2x^2 + 2xy - y^2 - 5x + 3y - 2$.   Then,

$$\nabla g = (4x + 2y - 5)\mathbf{i} + (2x - 2y + 3)\mathbf{j}.$$

$$\nabla g = 0 \implies 4x + 2y - 5 = 0 = 2x - 2y + 3 \implies x = \tfrac{1}{3}, \quad y = \tfrac{11}{6}.$$

The tangent plane is horizontal at $\left(\tfrac{1}{3}, \tfrac{11}{6}, -\tfrac{1}{12}\right)$.

25. $\dfrac{x - x_0}{(\partial f/\partial x)(x_0, y_0, z_0)} = \dfrac{y - y_0}{(\partial f/\partial y)(x_0, y_0, z_0)} = \dfrac{z - z_0}{(\partial f/\partial z)(x_0, y_0, z_0)}$

27.   Since the tangent planes meet at right angles, the normals ∇F and ∇G meet at right angles:

$$\frac{\partial F}{\partial x}\frac{\partial G}{\partial x} + \frac{\partial F}{\partial y}\frac{\partial G}{\partial y} + \frac{\partial F}{\partial z}\frac{\partial G}{\partial z} = 0.$$

29.   The tangent plane at an arbitrary point (x_0, y_0, z_0) has equation

$$y_0z_0(x - x_0) + x_0z_0(y - y_0) + x_0y_0(z - z_0) = 0,$$

which simplifies to

$$y_0z_0x + x_0z_0y + x_0y_0z = 3x_0y_0z_0 \text{ and thus to } \frac{x}{3x_0} + \frac{y}{3y_0} + \frac{z}{3z_0} = 1.$$

The volume of the pyramid is

$$V = \frac{1}{3}Bh = \frac{1}{3}\left[\frac{(3x_0)(3y_0)}{2}\right](3z_0) = \frac{9}{2}x_0y_0z_0 = \frac{9}{2}a^3.$$

31.   The point $(2, 3, -2)$ is the tip of $\mathbf{r}(1)$.

Since $\mathbf{r}'(t) = 2\mathbf{i} - \dfrac{3}{t^2}\mathbf{j} - 4t\mathbf{k}$,   we have $\mathbf{r}'(1) = 2\mathbf{i} - 3\mathbf{j} - 4\mathbf{k}$.

Now set $f(x, y, z) = x^2 + y^2 + 3z^2 - 25$.   The function has gradient $2x\mathbf{i} + 2y\mathbf{j} + 6z\mathbf{k}$.

At the point $(2, 3, -2)$,

$$\nabla f = 2(2\mathbf{i} + 3\mathbf{j} - 6\mathbf{k}).$$

The angle θ between $\mathbf{r}'(1)$ and the gradient gives

$$\cos\theta = \frac{(2\mathbf{i} - 3\mathbf{j} - 4\mathbf{k})}{\sqrt{29}} \cdot \frac{(2\mathbf{i} + 3\mathbf{j} - 6\mathbf{k})}{7} = \frac{19}{7\sqrt{29}} \cong 0.504.$$

Therefore $\theta \cong 1.043$ radians. The angle between the curve and the plane is

$$\frac{\pi}{2} - \theta \cong 1.571 - 1.043 \cong 0.528 \text{ radians.}$$

33. Set $f(x, y, z) = x^2 y^2 + 2x + z^3.$ Then,

$$\nabla f = (2xy^2 + 2)\,\mathbf{i} + 2x^2 y\mathbf{j} + 3z^2\mathbf{k}, \quad \nabla f(2, 1, 2) = 6\mathbf{i} + 8\mathbf{j} + 12\mathbf{k}.$$

The plane tangent to $f(x, y, z) = 16$ at $(2, 1, 2)$ has equation

$$6(x - 2) + 8(y - 1) + 12(z - 2) = 0, \quad \text{or} \quad 3x + 4y + 6z = 22.$$

Next, set $g(x, y, z) = 3x^2 + y^2 - 2z.$ Then,

$$\nabla g = 6x\mathbf{i} + 2y\mathbf{j} - 2\mathbf{k}, \quad \nabla g(2, 1, 2) = 12\mathbf{i} + 2\mathbf{j} - 2\mathbf{k}.$$

The plane tangent to $g(x, y, z) = 9$ at $(2, 1, 2)$ is

$$12(x - 2) + 2(y - 1) - 2(z - 2) = 0, \quad \text{or} \quad 6x + y - z = 11.$$

35. The gradient to the sphere at $(1, 1, 2)$ is

$$2x\mathbf{i} + (2y - 4)\,\mathbf{j} + (2z - 2)\mathbf{k} = 2\mathbf{i} - 2\mathbf{j} + 2\mathbf{k}.$$

The gradient to the paraboloid at $(1, 1, 2)$ is

$$6x\mathbf{i} + 4y\mathbf{j} - 2\mathbf{k} = 6\mathbf{i} + 4\mathbf{j} - 2\mathbf{k}.$$

Since

$$(2\mathbf{i} - 2\mathbf{j} + 2\mathbf{k}) \cdot (6\mathbf{i} + 4\mathbf{j} - 2\mathbf{k}) = 0,$$

the surfaces intersect at right angles.

37. (a) $3x + 4y + 6 = 0$ since plane p is vertical.

(b) $y = -\frac{1}{4}(3x + 6) = -\frac{1}{4}[3(4t - 2) + 6] = -3t$

$z = x^2 + 3y^2 + 2 = (4t - 2)^2 + 3(-3t)^2 + 2 = 43t^2 - 16t + 6$

$\mathbf{r}(t) = (4t - 2)\mathbf{i} - 3t\mathbf{j} + (43t^2 - 16t + 6)\mathbf{k}$

(c)   From part (b) the tip of $\mathbf{r}(1)$ is $(2, -3, 33)$.   We take

$\mathbf{r}'(1) = 4\mathbf{i} - 3\mathbf{j} + 70\mathbf{j}$ as $\mathbf{d}$ to write

$$\mathbf{R}(s) = (2\mathbf{i} - 3\mathbf{j} + 33\mathbf{k}) + s(4\mathbf{i} - 3\mathbf{j} + 70\mathbf{k}).$$

(d)   Set $g(x, y) = x^2 + 3y^2 + 2.$ Then,

$$\nabla g = 2x\mathbf{i} + 6y\mathbf{j} \quad \text{and} \quad \nabla g(2, -3) = 4\mathbf{i} - 18\mathbf{j}.$$

An equation for the plane tangent to $z = g(x, y)$ at $(2, -3, 33)$ is

$$z - 33 = 4(x - 2) - 18(y + 3) \quad \text{which reduces to} \quad 4x - 18y - z = 29.$$

(e) Substituting t for x in the equations for p and p_1, we obtain

$$3t + 4y + 6 = 0 \quad \text{and} \quad 4t - 18y - z = 29.$$

From the first equation

$$y = -\tfrac{3}{4}(t + 2)$$

and then from the second equation

$$z = 4t - 18\left[-\tfrac{3}{4}(t + 2)\right] - 29 = \tfrac{35}{2}t - 2.$$

Thus,

$$(*) \quad \mathbf{r}(t) = t\mathbf{i} - \left(\tfrac{3}{4}t + \tfrac{3}{2}\right)\mathbf{j} + \left(\tfrac{35}{2}t - 2\right)\mathbf{k}.$$

Lines l and l' are the same. To see this, consider how l and l' are formed; to assure yourself, replace t in $(*)$ by $4s + 2$ to obtain $\mathbf{R}(s)$ found in part (c).

SECTION 15.5

1. $\nabla f(x, y) = (2 - 2x)\mathbf{i} - 2y\mathbf{j} = \mathbf{0}$ only at $(1, 0)$.

The difference

$$f(1 + h, k) - f(1, 0) = \left[2(1 + h) - (1 + h)^2 - k^2\right] - 1 = -h^2 - k^2$$

is negative for all small h and k; there is a local maximum of 1 at $(1, 0)$.

3. $\nabla f(x, y) = (2x + y + 3)\mathbf{i} + (x + 2y)\mathbf{j} = \mathbf{0}$ only at $(-2, 1)$.

The difference

$$f(-2 + h,\, 1 + k) - f(-2, 1)$$
$$= \left[(-2 + h)^2 + (-2 + h)(1 + k) + (1 + k)^2 + 3(-2 + h) + 1\right] - (-2) = h^2 + hk + k^2$$

is positive for all small h and k. To see this, note that

$$h^2 + hk + k^2 \geq h^2 + k^2 - |h|\,|k| > 0;$$

there is a local minimum of -2 at $(-2, 1)$.

5. $\nabla f = (2x + y - 6)\mathbf{i} + (x + 2y)\mathbf{j} = \mathbf{0}$ only at $(4, -2)$.

$f_{xx} = 2, \quad f_{xy} = 1, \quad f_{yy} = 2.$

At $(4, -2)$, $D = -3 < 0$ and $A = 2 > 0$ so we have a local min; the value is -10.

7. $\nabla f = (3x^2 - 6y)\mathbf{i} + \left(3y^2 - 6x\right)\mathbf{j} = \mathbf{0}$ at $(2, 2)$ and $(0, 0)$.

$f_{xx} = 6x, \quad f_{xy} = -6, \quad f_{yy} = 6y, \quad D = 36 - 36xy.$

At $(2, 2)$, $D = -108 < 0$ and $A = 12 > 0$ so we have a local min; the value is -8.

At $(0, 0)$, $D = 36 > 0$ so we have a saddle point.

9. $\nabla f = (3x^2 - 6y + 6)\mathbf{i} + (2y - 6x + 3)\mathbf{j} = 0$ at $\left(5, \frac{27}{2}\right)$ and $\left(1, \frac{3}{2}\right)$.

$f_{xx} = 6x, \quad f_{xy} = -6, \quad f_{yy} = 2, \quad D = 36 - 12x.$

At $\left(5, \frac{27}{2}\right)$, $D = -24 < 0$ and $A = 30 > 0$ so we have a local min; the value is $-\frac{117}{4}$.

At $\left(1, \frac{3}{2}\right)$, $D = 24 > 0$ so we have a saddle point.

11. $\nabla f = \sin y\,\mathbf{i} + x \cos y\,\mathbf{j} = 0$ at $(0, n\pi)$ for all integral n.

$f_{xx} = 0, \quad f_{xy} = \cos y, \quad f_{yy} = -x \sin y, \quad D = \cos^2 y.$

Since $D = \cos^2 n\pi = 1 > 0$, each stationary point is a saddle point.

13. $\nabla f = (2xy + 1 + y^2)\mathbf{i} + (x^2 + 2xy + 1)\mathbf{j} = 0$ at $(1, -1)$ and $(-1, 1)$.

$f_{xx} = 2y, \quad f_{xy} = 2x + 2y, \quad f_{yy} = 2x, \quad D = 4(x + y)^2 - 4xy.$

At both $(1, -1)$ and $(-1, 1)$ we have saddle points since $D = 4 > 0$.

15. $\nabla f = (y - x^{-2})\mathbf{i} + (x - 8y^{-2})\mathbf{j} = 0$ only at $\left(\frac{1}{2}, 4\right)$.

$f_{xx} = 2x^{-3}, \quad f_{xy} = 1, \quad f_{yy} = 16y^{-3}, \quad D = 1 - 32x^{-3}y^{-3}.$

At $\left(\frac{1}{2}, 4\right)$, $D = -3 < 0$ and $A = 16 > 0$ so we have a local min; the value is 6.

17. $\nabla f = (y - x^{-2})\mathbf{i} + (x - y^{-2})\mathbf{j} = 0$ only at $(1, 1)$.

$f_{xx} = 2x^{-3}, \quad f_{xy} = 1, \quad f_{yy} = 2y^{-3}, \quad D = 1 - 4x^{-3}y^{-3}.$

At $(1, 1)$, $D = -3 < 0$ and $A = 2 > 0$ so we have a local min; the value is 3.

19. $\nabla f = \dfrac{2\left(x^2 - y^2 - 1\right)}{\left(x^2 + y^2 + 1\right)^2}\mathbf{i} + \dfrac{4xy}{\left(x^2 + y^2 + 1\right)^2}\mathbf{j} = 0$ at $(1, 0)$ and $(-1, 0)$.

$f_{xx} = \dfrac{-4x^3 + 12xy^2 + 12x}{\left(x^2 + y^2 + 1\right)^3}, \quad f_{xy} = \dfrac{4y^3 + 4y - 12x^2y}{\left(x^2 + y^2 + 1\right)^3}, \quad f_{yy} = \dfrac{4x^3 + 4xy^2 + 4x - 16xy^2}{\left(x^2 + y^2 + 1\right)^3}.$

At $(1, 0)$,

$$A = f_{xx}(1, 0) = 1 > 0, \quad B = f_{xy}(1, 0) = 0, \quad C = f_{yy}(1, 0) = 1, \quad D = -1 < 0.$$

Thus, $(1, 0)$ is a local min; $f(1, 0) = -1$.

At $(-1, 0)$,

$$A = f_{xx}(-1, 0) = -1 < 0, \quad B = f_{xy}(-1, 0) = 0, \quad C = f_{yy}(-1, 0) = -1, \quad D = -1 < 0.$$

Thus, $(1, 0)$ is a local max; $f(-1, 0) = 1$.

21. $\nabla f = \left(4x^3 - 4x\right)\mathbf{i} + 2y\,\mathbf{j} = 0$ at $(0, 0)$, $(1, 0)$, and $(-1, 0)$.

$f_{xx} = 12x^2 - 4, \quad f_{xy} = 0, \quad f_{yy} = 2, \quad D = 8 - 24x^2.$

At $(0, 0)$, $D = 8 > 0$. Thus, $(0, 0)$ is a saddle point.

At $(\pm 1, 0)$, $D = -16 < 0$ and $A = 8 > 0$. Thus, the points $(1, 0)$ and $(-1, 0)$ are local minima;

$f(\pm 1, 0) = -3$.

23.   (a) $\nabla f = (2x + ky)\,\mathbf{i} + (2y + kx)\,\mathbf{j}$ and $\nabla f(0,0) = \mathbf{0}$ independent of the value of k.

(b) $f_{xx} = 2$, $f_{xy} = k$, $f_{yy} = 2$, $D = k^2 - 4$. Thus, $D > 0$ for $|k| > 2$ and $(0,0)$ is a saddle point

(c) $D = k^2 - 4 < 0$ for $|k| < 2$. Since $A = f_{xx} = 2 > 0$, $(0,0)$ is a local minimum.

(d) The test is inconclusive when $D = k^2 - 4 = 0$ i.e., for $k = \pm 2$.

25.   Let $P(x, y, z)$ be a point in the plane. We want to find the minimum of $f(x, y, z) = \sqrt{x^2 + y^2 + z^2}$.

However, it is sufficient to minimize the square of the distance: $F(x, y, z) = x^2 + y^2 + z^2$. It is clear

that F has a minimum value, but no maximum value. Since P lies in the plane, $2x - y + 2z = 16$

which implies $y = 2x + 2z - 16 = 2(x + z - 8)$. Thus, we want to find the minimum value of

$$F(x, z) = x^2 + 4(x + z - 8)^2 + z^2$$

Now,

$$\nabla F = [2x + 8(x + z - 8)]\,w\mathbf{i} + [8(x + z - 8)\,\mathbf{k}$$

The gradient is $\mathbf{0}$ when

$$2x + 8(x + z - 8) = 0 \quad \text{and} \quad 8(x + z - 8) + 2z = 0$$

The only solution to this pair of equations is: $x = z = \dfrac{32}{9}$, from which it follows that $y = -\dfrac{16}{9}$.

The point in the plane that is closest to the origin is $P\left(\frac{32}{9}, -\frac{16}{9}, \frac{32}{9}\right)$.

The distance from the origin to the plane is: $F(P) = \frac{16}{3}$.

Check using (12.6.7): $d(P, 0) = \dfrac{|2 \cdot 0 - 0 + 2 \cdot 0 - 16|}{\sqrt{2^2 + (-1)^2 + 2^2}} = \dfrac{16}{3}$.

27.   Using the hint, we want to find the maximum value of $f(x, y) = 18xy - x^2 y - xy^2$.

The gradient of f is:

$$\nabla D = \left(18y - 2xy - y^2\right)\,\mathbf{i} + \left(18x - x^2 - 2xy\right)\,\mathbf{j}$$

The gradient is $\mathbf{0}$ when

$$18y - 2xy - y^2 = 0 \quad \text{and} \quad 18x - x^2 - 2xy = 0$$

The solution set of this pair of equations is: $(0, 0)$, $(18, 0)$, $(0, 18)$, $(6, 6)$.

It is easy to verify that f is a maximum when $x = y = 6$. The three numbers that satisfy $x + y + z = 18$

and maximize the product xyz are: $x = 6$, $y = 6$, $z = 6$.

29. $\nabla f = \dfrac{1}{(x^2 + y^2)^{3/2}}(-x\mathbf{i} - y\mathbf{j})$ is never $\mathbf{0}$ on D. Note that $f(x, y)$ is the reciprocal of the distance

of (x, y) from the origin. The point of D closest to the origin (draw a figure) is $(1, 1)$. Therefore

$f(1,1) = 1/\sqrt{2}$ is the maximum value of f. The point of D furthest from the origin is $(3,4)$. Therefore $f(3,4) = 1/5$ is the least value taken on by f.

31. $\nabla f = 2(x-1)\mathbf{i} + 2(y-1)\mathbf{j} = \mathbf{0}$ only at $(1,1)$. As the sum of two squares, $f(x,y) \geq 0$. Thus, $f(1,1) = 0$ is a minimum. To examine the behavior of f on the boundary of D, we note that f represents the square of the distance between (x,y) and $(1,1)$. Thus, f is maximal at the point of the boundary furthest from $(1,1)$. This is the point $(-\sqrt{2}, -\sqrt{2})$; the maximum value of f is $f(-\sqrt{2}, -\sqrt{2}) = 6 + 4\sqrt{2}$.

33. $\nabla f = 2(x-y)\mathbf{i} - 2(x-y)\mathbf{j} = \mathbf{0}$ at each point of the line segment $y = x$ from $(0,0)$ to $(4,4)$. Since $f(x,x) = 0$ and $f(x,y) \geq 0$, f takes on its minimum of 0 at each of these points.

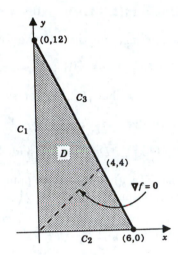

Next we consider the boundary of D. We parametrize each side of the triangle:

$$C_1: \; \mathbf{r}_1(t) = t\mathbf{j}, \quad t \in [0,12]$$
$$C_2: \; \mathbf{r}_2(t) = t\mathbf{i}, \quad t \in [0,6]$$
$$C_3: \; \mathbf{r}_3(t) = t\mathbf{i} + (12-2t)\mathbf{j}, \quad t \in [0,6]$$

and observe from

$$f(\mathbf{r}_1(t)) = t^2, \quad t \in [0,12]$$
$$f(\mathbf{r}_2(t)) = t^2, \quad t \in [0,6]$$
$$f(\mathbf{r}_3(t)) = (3t-12)^2, \quad t \in [0,6]$$

that f takes on its maximum of 144 at the point $(0,12)$.

35. $\nabla f = 2(x-4)\mathbf{i} + 2y\mathbf{j}$ is never $\mathbf{0}$ at an interior point of D. Next we examine f on the boundary of D:

$$C_1: \; \mathbf{r}_1(t) = t\mathbf{i} + 4t\mathbf{j}, \quad t \in [0,2,],$$
$$C_2: \; \mathbf{r}_2(t) = t\mathbf{i} + t^3\mathbf{j}, \quad t \in [0,2].$$

Note that

$$f_1(t) = f(\mathbf{r}_1(t)) = 17t^2 - 8t + 16,$$
$$f_2(t) = f(\mathbf{r}_2(t)) = (t-4)^2 + t^6.$$

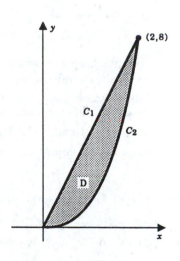

Next

$$f_1'(t) = 34t - 8 = 0 \quad \Longrightarrow \quad t = 4/17 \quad \text{and gives} \quad x = 4/17, \ y = 16/17$$

and

$$f_2'(t) = 6t^5 + 2t - 8 = 0 \quad \Longrightarrow \quad t = 1 \quad \text{and gives} \quad x = 1, \ y = 1.$$

The extreme values of f can be culled from the following list:

$$f(0,0) = 16, \quad f(2,8) = 68, \quad f\left(\tfrac{4}{17}, \tfrac{16}{17}\right) = \tfrac{256}{17}, \quad f(1,1) = 10.$$

We see that $f(1,1) = 10$ is the absolute minimum and $f(2,8)$ is the absolute maximum.

37. $f(x,y) = xy(1 - x - y), \quad 0 \le x \le 1, \quad 0 \le y \le 1 - x.$

[dom (f) is the triangle with vertices $(0,0)$, $(1,0)$, $(0,1)$.]

$$\nabla f = (y - 2xy - y^2)\mathbf{i} + (x - 2xy - x^2)\mathbf{j} = 0 \quad \Longrightarrow \quad x = y = \tfrac{1}{3}.$$

(Note that $[0,0]$ is not an interior point of the domain of f.)

$$f_{xx} = -2x, \quad f_{xy} = 1 - 2x - 2y, \quad f_{yy} = -2x, \quad D = (1 - 2x - 2y)^2 - 4xy.$$

At $\left(\tfrac{1}{3}, \tfrac{1}{3}\right)$, $D = -\tfrac{1}{3} < 0$ and $A > 0$ so we have a local max; the value is $1/27$.

Since $f(x,y) = 0$ at each point on the boundary of the domain, the local max of $1/27$ is also the absolute max.

39. $f(x,y) = (x-1)^2 + (y-2)^2 + z^2 = (x-1)^2 + (y-2)^2 + x^2 + 2y^2$ $\left[\text{since } z = \sqrt{x^2 + 2y^2}\right]$

$$\nabla f = [2(x-1) + 2x]\mathbf{i} + [2(y-2) + 4y]\mathbf{j} = 0 \quad \Longrightarrow \quad x = \frac{1}{2}, \ y = \frac{2}{3}.$$

$f_{xx} = 4 > 0, \quad f_{xy} = 0, \quad f_{yy} = 6, \quad D = -24 < 0.$ Thus, f has a local minimum at $(1/2, 2/3)$.

The shortest distance from $(1,2,0)$ to the cone is $f\left(\tfrac{1}{2}, \tfrac{2}{3}\right) = \tfrac{1}{6}\sqrt{114}$

41. (a) $\nabla f = \tfrac{1}{2}x\mathbf{i} - \tfrac{2}{9}y\mathbf{j} = 0$ only at $(0,0)$.

(b) The difference
$$f(h,k) - f(0,0) = \tfrac{1}{4}h^2 - \tfrac{1}{9}k^2$$

does not keep a constant sign for all small h and k; $(0,0)$ is a saddle point. The function has no local extreme values.

(c) Being the difference of two squares, f can be maximized by maximizing $\tfrac{1}{4}x^2$ and minimizing $\tfrac{1}{9}y^2$; $(1,0)$ and $(-1,0)$ give absolute maximum value $\tfrac{1}{4}$. Similarly, $(0,1)$ and $(0,-1)$ give absolute minimum value $-\tfrac{1}{9}$.

43. (a) $f(x,y) = 0$ along the plane curve $y = x^{2/3}$. Since $f(x,y)$ is positive for points below the curve and is negative for points above the curve, there is a saddle point at the origin.

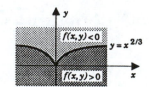

(b) $\nabla f = (ye^{xy} - 2\sin{(x+y)})\mathbf{i} + (xe^{xy} - 2\sin{(x+y)})\mathbf{j}$

$$f_{xx} = -2\cos{(x+y)} + y^2 e^{xy},$$

$$f_{xy} = -2\cos{(x+y)} + e^{xy}(1+xy),$$

$$f_{yy} = -2\cos{(x+y)} + x^2 e^{xy}.$$

At the origin $A = -2$, $B = -1$, $C = -2$, and $D = -3$. We have a local max; the value is 3.

45.
$$f(x,y) = \sum_{i=1}^{3}\left[(x - x_i)^2 + (y - y_i)^2\right]$$

$$\nabla f(x,y) = 2\left[(3x - x_1 - x_2 - x_3)\,\mathbf{i} + (3y - y_1 - y_2 - y_3)\,\mathbf{j}\right]$$

$$\nabla f = \mathbf{0} \quad \text{only at} \quad \left(\frac{x_1 + x_2 + x_3}{3}, \frac{y_1 + y_2 + y_3}{3}\right) = (x_0, y_0).$$

The difference $\quad f(x_0 + h,\, y_0 + k) - f(x_0, y_0)$

$$= \sum_{i=1}^{3}\left[(x_0 + h - x_i)^2 + (y_0 + k - y_i)^2 - (x_0 - x_i)^2 - (y_0 - y_i)^2\right]$$

$$= \sum_{i=1}^{3}\left[2h(x_0 - x_i) + h^2 + 2k(y_0 - y_i) + k^2\right]$$

$$= 2h(3x_0 - x_1 - x_2 - x_3) + 2k(3y_0 - y_1 - y_2 - y_3) + h^2 + k^2$$

$$= h^2 + h^2$$

is positive for all small h and k. Thus, f has its absolute minimum at (x_0, y_0).

47.
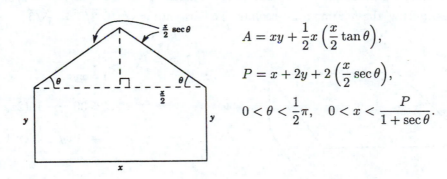

$$A = xy + \frac{1}{2}x\left(\frac{x}{2}\tan\theta\right),$$

$$P = x + 2y + 2\left(\frac{x}{2}\sec\theta\right),$$

$$0 < \theta < \frac{1}{2}\pi, \quad 0 < x < \frac{P}{1 + \sec\theta}.$$

$$A(x, \theta) = \tfrac{1}{2}x(P - x - x\sec\theta) + \tfrac{1}{4}x^2\tan\theta,$$

$$\nabla A = \left(\frac{P}{2} - x - x\sec\theta + \frac{x}{2}\tan\theta\right)\mathbf{i} + \left(\frac{x^2}{4}\sec^2\theta - \frac{x^2}{2}\sec\theta\tan\theta\right)\mathbf{j},$$

$$\nabla A = \frac{1}{2}[P + x(\tan\theta - 2\sec\theta - 2)]\mathbf{i} + \frac{x^2}{4}\sec\theta\,(\sec\theta - 2\tan\theta)\mathbf{j}.$$

From $\dfrac{\partial A}{\partial \theta} = 0$ we get $\theta = \tfrac{1}{6}\pi$ and then from $\dfrac{\partial A}{\partial x} = 0$ we get

$$P + x\left(\tfrac{1}{3}\sqrt{3} - \tfrac{4}{3}\sqrt{3} - 2\right) = 0 \quad \text{so that} \quad x = (2 - \sqrt{3})P.$$

Next,

$$A_{xx} = \tfrac{1}{2}(\tan\theta - 2\sec\theta - 2),$$

$$A_{x\theta} = \frac{x}{2}\sec\theta\,(\sec\theta - 2\tan\theta),$$

$$A_{\theta\theta} = \frac{x^2}{2}\sec\theta\,\left(\sec\theta\tan\theta - \sec^2\theta - \tan^2\theta\right).$$

By the second-partials test

$$A = -\tfrac{1}{2}(2 + \sqrt{3}), \quad B = 0, \quad C = -\tfrac{1}{3}P^2\sqrt{3}\,(2 - \sqrt{3})^2, \quad D < 0.$$

The area is a maximum when $\theta = \tfrac{1}{6}\pi$, $x = (2 - \sqrt{3})\,P$ and $y = \tfrac{1}{6}(3 - \sqrt{3})P$.

49.   From $x = \tfrac{1}{2}y = \tfrac{1}{3}z = t$ and $x = y - 2 = z = s$

we take $(t, 2t, 3t)$ and $(s, 2 + s, s)$

as arbitrary points on the lines. It suffices to minimize the square of the distance between these points:

$$f(t, s) = (t - s)^2 + (2t - 2 - s)^2 + (3t - s)^2$$

$$= 14t^2 - 12ts + 3s^2 - 8t + 4s + 4, \qquad t, s \text{ real.}$$

$$\nabla f = (28t - 12s - 8)\mathbf{i} + (-12t + 6s + 4)\mathbf{j}; \qquad \nabla f = 0 \implies t = 0,\ s = -2/3.$$

$$f_{tt} = 28, \quad f_{ts} = -12, \quad f_{ss} = 6, \quad D = (-12)^2 - 6(28) = -24 < 0.$$

By the second-partials test, the distance is a minimum when $t = 0$, $s = -2/3$; the nature of the problem tells us the minimum is absolute. The distance is $\sqrt{f(0, 2/3)} = \tfrac{2}{3}\sqrt{6}$.

51.

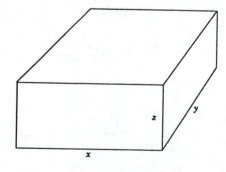

$$96 = xyz,$$

$$C = 30xy + 10(2xz + 2yz)$$

$$= 30xy + 20(x + y)\frac{96}{xy}.$$

$$C(x, y) = 30\left[xy + \frac{64}{x} + \frac{64}{y}\right],$$

$$\nabla C = 30(y - 64x^{-2})\mathbf{i} + 30(x - 64y^{-2})\mathbf{j} = 0 \implies x = y = 4.$$

$$C_{xx} = 128x^{-3}, \quad C_{xy} = 1, \quad C_{yy} = 128y^{-3}.$$

When $x = y = 4$, we have $D = -3 < 0$ and $A = 2 > 0$ so the cost is minimized by making the dimensions of the crate $4 \times 4 \times 6$ meters.

53. Let x, y and z be the length, width and height of the box. The surface area is given by

$$S = 2xy + 2xz + 2yz, \quad \text{so} \quad z = \frac{S - 2xy}{2(x + y)}, \quad \text{where } S \text{ is a constant, and } x, y, z > 0.$$

Now, the volume $V = xyz$ is given by:

$$V(x, y) = xy \left[\frac{S - 2xy}{2(x + y)} \right]$$

and

$$\nabla V = y \left\{ \left[\frac{S - 2xy}{2(x + y)} \right] + xy \frac{2(x + y)(-2y) - (S - 2xy)(2)}{4(x + y)^2} \right\} \mathbf{i}$$

$$+ \left\{ x \left[\frac{S - 2xy}{2(x + y)} \right] + xy \frac{2(x + y)(-2x) - (S - 2xy)(2)}{4(x + y)^2} \right\} \mathbf{j}$$

Setting $\dfrac{\partial V}{\partial x} = \dfrac{\partial V}{\partial y} = 0$ and simplifying, we get the pair of equations

$$2S - 4x^2 - 8xy = 0$$

$$2S - 4y^2 - 8xy = 0$$

from which it follows that $x = y = \sqrt{S/6}$. From practical considerations, we conclude that V has a maximum value at $(\sqrt{S/6}, \sqrt{S/6})$. Substituting these values into the equation for z, we get $z = \sqrt{S/6}$ and so the box of maximum volume is a cube.

55. (a) $f(m, b) = [2 - b]^2 + [-5 - (m + b)]^2 + [4 - (2m + b)]^2$.

$f_m = 10m + 6b - 6, \quad f_b = 6m + 6b - 2; \qquad f_m = f_b = 0 \implies m = 1, \quad b = -\frac{2}{3}$.

$f_{mm} = 10, \quad f_{mb} = 6, \quad f_{bb} = 6, \quad D = -24 < 0 \implies$ a min.

Answer: the line $y = x - \frac{2}{3}$.

(b) $f(\alpha, \beta) = [2 - \beta]^2 + [-5 - (\alpha + \beta)]^2 + [4 - (4\alpha + \beta)]^2$.

$f_\alpha = 34\alpha + 10\beta - 22, \quad f_\beta = 10\alpha + 6\beta - 2; \qquad f_\alpha = f_\beta = 0 \implies \left[\begin{array}{l} \alpha = \frac{14}{13} \\ \beta = -\frac{19}{13} \end{array} \right].$

$f_{\alpha\alpha} = 34, \quad f_{\alpha\beta} = 10, \quad f_{\beta\beta} = 6, \quad D = -104 < 0 \implies$ a min.

Answer: the parabola $y = \frac{1}{13} \left(14x^2 - 19 \right)$.

57. (a) Let x and y be the cross-sectional measurements of the box, and let l be its length.

Then

$$V = xyl, \quad \text{where} \quad 2x + 2y + l \leq 108, \quad x, y > 0$$

To maximize V we will obviously take $2x+2y+l = 108$. Therefore, $V(x,y) = xy(108-2x-2y)$ and

$$\nabla V = [y(108 - 2x - 2y) - 2xy]\,\mathbf{i} + [x(108 - 2x - 2y) - 2xy]\,\mathbf{j}$$

Setting $\dfrac{\partial V}{\partial x} = \dfrac{\partial V}{\partial y} = 0$, we get the pair of equations

$$\frac{\partial V}{\partial x} = 108y - 4xy - 2y^2 = 0$$

$$\frac{\partial V}{\partial y} = 108x - 4xy - 2x^2 = 0$$

from which it follows that $x = y = 18 \implies l = 36$.

Now, at $(18, 18)$, we have

$$A = V_{xx} = -4y = -72 < 0, \quad B = V_{xy} = 108 - 4x - 4y = -36,$$

$$C = V_{yy} = -4x = -72, \quad \text{and} \quad D = (36)^2 - (72)^2 < 0.$$

Thus, V is a maximum when $x = y = 18$ and $l = 36$.

(b)   Let r be the radius of the tube and let l be its length.

Then

$$V = \pi r^2 l, \quad \text{where} \quad 2\pi r + l \le 108, \quad r > 0$$

To maximize V we take $2\pi r + l = 108$. Then $V(r) = \pi r^2(108 - 2\pi r) = 108\pi r^2 - 2\pi^2 r^3$. Now

$$\frac{dV}{dr} = 216\pi r - 6\pi^2 r^2$$

Setting $\dfrac{dV}{dr} = 0$, we get

$$216\pi r - 6\pi^2 r^2 = 0 \implies r = \frac{36}{\pi} \implies l = 36$$

Now, at $r = 36/\pi$, we have

$$\frac{d^2V}{dr^2} = 216\pi - 12\pi^2 \frac{36}{\pi} = -216\pi < 0$$

Thus, V is a maximum when $r = 36/\pi$ and $l = 36$.

59.   Let S denote the cross-sectional area. Then

$$S = \frac{1}{2}\,(12 - 2x + 12 - 2x + 2x\cos\theta)\,x\sin\theta = 12x\sin\theta - 2x^2\sin\theta + \frac{1}{2}x^2\sin 2\theta,$$

where $0 < x < 6, \quad 0 < \theta < \pi/2$

Now,

$$\nabla S = (12\sin\theta - 4x\sin\theta + x\sin 2\theta)\,\mathbf{i} + (12x\cos\theta - 2x^2\cos\theta + x^2\cos 2\theta)\,\mathbf{j}$$

Setting $\dfrac{\partial S}{\partial x} = \dfrac{\partial S}{\partial \theta} = 0$, we get the pair of equations

$$12\sin\theta - 4x\sin\theta + x\sin 2\theta = 0$$

$$12x\cos\theta - 2x^2\cos\theta + x^2\cos 2\theta = 0$$

from which it follows that $x = 4, \theta = \pi/3$.

Now, at $(4, \pi/3)$, we have

$$A = S_{xx} = -4\sin\theta + \sin 2\theta = -\frac{3}{2}\sqrt{3}, \quad B = S_{x\theta} = 12\cos\theta - 4x\cos\theta + 2x\cos 2\theta = -6,$$

$$C = S_{\theta\theta} = -12x\sin\theta + 2x^2\sin\theta - 2x^2\sin 2\theta = -24\sqrt{3} \quad \text{and} \quad D = 36 - 108 < 0.$$

Thus, S is a maximum when $x = 4$ and $\theta = \pi/3$.

61. $(0,0)$ saddle point; $(1,1)$ local maximum

63. $(0,0)$ local min., $(0,-1)$, $(0,1)$ local max.

saddle points at $(1,0)$, $(-1,0)$.

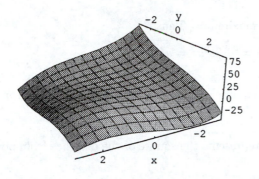

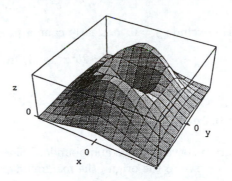

SECTION 15.6

1.

$$f(x,y) = x^2 + y^2, \qquad g(x,y) = xy - 1$$

$$\nabla f = 2x\mathbf{i} + 2y\mathbf{j}, \qquad \nabla g = y\mathbf{i} + x\mathbf{j}.$$

$$\nabla f = \lambda \nabla g \implies 2x = \lambda y \text{ and } 2y = \lambda x.$$

Multiplying the first equation by x and the second equation by y, we get

$$2x^2 = \lambda xy = 2y^2.$$

Thus, $x = \pm y$. From $g(x,y) = 0$ we conclude that $x = y = \pm 1$. The points $(1,1)$ and $(-1,-1)$ clearly give a minimum, since f represents the square of the distance of a point on the hyperbola from the origin. The minimum is 2.

3.

$$f(x,y) = xy, \qquad g(x,y) = b^2x^2 + a^2y^2 - a^2b^2$$

$$\nabla f = y\mathbf{i} + x\mathbf{j}, \qquad \nabla g = 2b^2x\mathbf{i} + 2a^2y\mathbf{j}.$$

$$\nabla f = \lambda \nabla g \implies y = 2\lambda b^2 x \text{ and } x = 2\lambda a^2 y.$$

Multiplying the first equation by a^2y and the second equation by b^2x, we get

$$a^2y^2 = 2\lambda a^2 b^2 xy = b^2 x^2.$$

Thus, $ay = \pm bx$. From $g(x,y) = 0$ we conclude that $x = \pm\frac{1}{2}a\sqrt{2}$ and $y = \pm\frac{1}{2}b\sqrt{2}$.

Since f is continuous and the ellipse is closed and bounded, the minimum exists. It occurs at $\left(\frac{1}{2}a\sqrt{2}, -\frac{1}{2}b\sqrt{2}\right)$ and $\left(-\frac{1}{2}a\sqrt{2}, \frac{1}{2}b\sqrt{2}\right)$; the minimum is $-\frac{1}{2}ab$.

5.   Since f is continuous and the ellipse is closed and bounded, the maximum exists.

$$f(x,y) = xy^2, \qquad\qquad g(x,y) = b^2x^2 + a^2y^2 - a^2b^2$$

$$\nabla f = y^2\mathbf{i} + 2xy\mathbf{j}, \qquad\qquad \nabla g = 2b^2x\mathbf{i} + 2a^2y\mathbf{j}.$$

$$\nabla f = \lambda\nabla g \implies y^2 = 2\lambda b^2 x \quad\text{and}\quad 2xy = 2\lambda a^2 y.$$

Multiplying the first equation by a^2y and the second equation by b^2x, we get

$$a^2y^3 = 2\lambda a^2 b^2 xy = 2b^2x^2y.$$

We can exclude $y = 0$; it clearly cannot produce the maximum. Thus,

$$a^2y^2 = 2b^2x^2 \quad\text{and, from}\quad g(x,y) = 0, \quad 3b^2x^2 = a^2b^2.$$

This gives us $x = \pm\frac{1}{3}\sqrt{3}\,a$ and $y = \pm\frac{1}{3}\sqrt{6}\,b$. This maximum occurs at $x = \frac{1}{3}\sqrt{3}\,a$, $y = \pm\frac{1}{3}\sqrt{6}\,b$; the value there is $\frac{2}{9}\sqrt{3}\,ab^2$.

7.   The given curve is closed and bounded. Since $x^2 + y^2$ represents the square of the distance from points on this curve to the origin, the maximum exists.

$$f(x,y) = x^2 + y^2, \qquad\qquad g(x,y) = x^4 + 7x^2y^2 + y^4 - 1$$

$$\nabla f = 2x\mathbf{i} + 2y\mathbf{j}, \qquad\qquad \nabla g = \left(4x^3 + 14xy^2\right)\mathbf{i} + \left(4y^3 + 14x^2y\right)\mathbf{j}.$$

We use the cross-product equation (16.8.4):

$$2x(4y^3 + 14x^2y) - 2y(4x^3 + 14xy^2) = 0,$$

$$20x^3y - 20xy^3 = 0,$$

$$xy(x^2 - y^2) = 0.$$

Thus, $x = 0$, $y = 0$, or $x = \pm y$. From $g(x,y) = 0$ we conclude that the points to examine are

$$(0, \pm 1), \quad (\pm 1, 0), \quad \left(\pm\tfrac{1}{3}\sqrt{3}, \pm\tfrac{1}{3}\sqrt{3}\right).$$

The value of f at each of the first four points is 1; the value at the last four points is 2/3. The maximum is 1.

9.   The maximum exists since xyz is continuous and the ellipsoid is closed and bounded.

$$f(x,y,z) = xyz, \qquad\qquad g(x,y,z) = \frac{x^2}{a^2} + \frac{y^2}{b^2} + \frac{z^2}{c^2} - 1$$

$$\nabla f = yz\mathbf{i} + xz\mathbf{j} + xy\mathbf{k}, \qquad\qquad \nabla g = \frac{2x}{a^2}\mathbf{i} + \frac{2y}{b^2}\mathbf{j} + \frac{2z}{c^2}\mathbf{k}.$$

$$\nabla f = \lambda\nabla g \implies yz = \frac{2x}{a^2}\lambda, \quad xz = \frac{2y}{b^2}\lambda, \quad xy = \frac{2z}{c^2}\lambda.$$

We can assume x, y, z are non-zero, for otherwise $f(x, y, z) = 0$, which is clearly not a maximum. Then from the first two equations

$$\frac{yza^2}{x} = 2\lambda = \frac{xzb^2}{y} \quad \text{so that} \quad a^2y^2 = b^2x^2 \quad \text{or} \quad x^2 = \frac{a^2y^2}{b^2}.$$

Similarly from the second and third equations we get

$$b^2z^2 = c^2y^2 \quad \text{or} \quad z^2 = \frac{c^2y^2}{b^2}.$$

Substituting these expressions for x^2 and z^2 in $g(x, y, z) = 0$, we obtain

$$\frac{1}{a^2}\left[\frac{a^2y^2}{b^2}\right] + \frac{y^2}{b^2} + \frac{1}{c^2}\left[\frac{c^2y^2}{b^2}\right] - 1 = 0, \quad \frac{3y^2}{b^2} = 1, \quad y = \pm\frac{1}{3}b\sqrt{3}.$$

Then, $x = \pm\frac{1}{3}a\sqrt{3}$ and $z = \pm\frac{1}{3}c\sqrt{3}$. The maximum value is $\frac{1}{9}\sqrt{3}\,abc$.

11.   Since the sphere is closed and bounded and $2x + 3y + 5z$ is continuous, the maximum exists.

$$f(x, y, z) = 2x + 3y + 5z, \qquad g(x, y, z) = x^2 + y^2 + z^2 - 19$$

$$\nabla f = 2\mathbf{i} + 3\mathbf{j} + 5\mathbf{k}, \qquad \nabla g = 2x\mathbf{i} + 2y\mathbf{j} + 2z\mathbf{k}.$$

$$\nabla f = \lambda \nabla g \quad \Longrightarrow \quad 2 = 2\lambda x, \quad 3 = 2\lambda y, \quad 5 = 2\lambda z.$$

Since $\lambda \neq 0$ here, we solve the equations for x, y and z:

$$x = \frac{1}{\lambda}, \quad y = \frac{3}{2\lambda}, \quad z = \frac{5}{2\lambda},$$

and substitute these results in $g(x, y, z) = 0$ to obtain

$$\frac{1}{\lambda^2} + \frac{9}{4\lambda^2} + \frac{25}{4\lambda^2} - 19 = 0, \quad \frac{38}{4\lambda^2} - 19 = 0, \quad \lambda = \pm\frac{1}{2}\sqrt{2}.$$

The positive value of λ will produce positive values for x, y, z and thus the maximum for f. We get $x = \sqrt{2}$, $y = \frac{3}{2}\sqrt{2}$, $z = \frac{5}{2}\sqrt{2}$, and $2x + 3y + 5z = 19\sqrt{2}$.

13. $$f(x, y, z) = xyz, \qquad\qquad g(x, y, z) = \frac{x}{a} + \frac{y}{b} + \frac{z}{c} - 1$$

$$\nabla f = yz\mathbf{i} + xz\mathbf{j} + xy\mathbf{k}, \qquad \nabla g = \frac{1}{a}\mathbf{i} + \frac{1}{b}\mathbf{j} + \frac{1}{c}\mathbf{k}.$$

$$\nabla f = \lambda \nabla g \quad \Longrightarrow \quad yz = \frac{\lambda}{a}, \quad xz = \frac{\lambda}{b}, \quad xy = \frac{\lambda}{c}.$$

Multiplying these equations by x, y, z respectively, we obtain

$$xyz = \frac{\lambda x}{a}, \quad xyz = \frac{\lambda y}{b}, \quad xyz = \frac{\lambda z}{c}.$$

Adding these equations and using the fact that $g(x, y, z) = 0$, we have

$$3xyz = \lambda\left(\frac{x}{a} + \frac{y}{b} + \frac{z}{c}\right) = \lambda.$$

Since x, y, z are non-zero,

$$yz = \frac{\lambda}{a} = \frac{3xyz}{a}, \quad 1 = \frac{3x}{a}, \quad x = \frac{a}{3}.$$

Similarly, $y = \dfrac{b}{3}$ and $z = \dfrac{c}{3}$. The maximum is $\dfrac{1}{27}abc$.

15. It suffices to minimize the square of the distance from $(0, 1)$ to the parabola. Clearly, the minimum exists.

$$f(x, y) = x^2 + (y-1)^2, \qquad\qquad g(x, y) = x^2 - 4y$$

$$\nabla f = 2x\mathbf{i} + 2(y-1)\mathbf{j}, \qquad\qquad \nabla g = 2x\mathbf{i} - 4\mathbf{j}.$$

We use the cross-product equation (15.6.4):

$$2x(-4) - 2x(2y - 2) = 0, \quad 4x + 4xy = 0, \quad x(y+1) = 0.$$

Since $y \geq 0$, we have $x = 0$ and thus $y = 0$. The minimum is 1.

17. It suffices to maximize and minimize the square of the distance from $(2, 1, 2)$ to the sphere. Clearly, these extreme values exist.

$$f(x, y, z) = (x - 2)^2 + (y - 1)^2 + (z - 2)^2 \qquad\qquad g(x, y, z) = x^2 + y^2 + z^2 - 1$$

$$\nabla f = 2(x - 2)\,\mathbf{i} + 2(y - 1)\,\mathbf{j} + 2(z - 2)\,\mathbf{k}, \qquad\qquad \nabla g = 2x\,\mathbf{i} + 2y\,\mathbf{j} + 2z\,\mathbf{k}.$$

$$\nabla f = \lambda = \nabla g \quad\Longrightarrow\quad 2(x - 2) = 2x\lambda, \quad 2(y - 1) = 2y\lambda, \quad 2(z - 2) = 2z\lambda$$

Thus,

$$x = \frac{2}{1 - \lambda}, \quad y = \frac{1}{1 - \lambda}, \quad z = \frac{2}{1 - \lambda}.$$

Using the fact that $x^2 + y^2 + z^2 = 1$, we have

$$\left(\frac{2}{1 - \lambda}\right)^2 + \left(\frac{1}{1 - \lambda}\right)^2 + \left(\frac{2}{1 - \lambda}\right)^2 = 1 \quad\Longrightarrow\quad \lambda = -2, 4$$

At $\lambda = -2$, $(x, y, z) = (2/3, 1/3, 2/3)$ and $f(2/3, 1/3/2/3) = 4$

At $\lambda = 4$, $(x, y, z) = (-2/3, -1/3, -2/3)$ and $f(-2/3, -1/3/ - 2/3) = 16$

Thus, $(2/3, 1/3, 2/3)$ is the closest point and $(-2/3, -1/3, -2/3)$ is the furthest point.

19. $$f(x, y, z) = 3x - 2y + z, \qquad\qquad g(x, y, z) = x^2 + y^2 + z^2 - 14$$

$$\nabla f = 3\mathbf{i} - 2\mathbf{j} + \mathbf{k}, \qquad\qquad \nabla g = 2x\mathbf{i} + 2y\mathbf{j} + 2z\mathbf{k}.$$

$$\nabla f = \lambda\nabla g \quad\Longrightarrow\quad 3 = 2x\lambda, \quad -2 = 2y\lambda, \quad 1 = 2z\lambda.$$

Thus,

$$x = \frac{3}{2\lambda}, \quad y = -\frac{1}{\lambda}, \quad z = \frac{1}{2\lambda}.$$

Using the fact that $x^2 + y^2 + z^2 = 14$, we have

$$\left(\frac{3}{2\lambda}\right)^2 + \left(-\frac{1}{\lambda}\right)^2 + \left(\frac{1}{2\lambda}\right)^2 = 14 \implies \lambda = \pm\frac{1}{2}.$$

At $\lambda = -\frac{1}{2}$, $(x,y,z) = (3,-2,1)$ and $f(3,-2,1) = 14$

At $\lambda = \frac{1}{2}$, $(x,y,z) = (-3,2,-1)$ and $f(-3,2,-1) = -14$

Thus, the maximum value of f on the sphere is 14.

21. It's easier to work with the square of the distance; the minimum certainly exists.

$$f(x,y,z) = x^2 + y^2 + z^2, \qquad g(x,y,z) = Ax + By + Cz + D$$

$$\nabla f = 2x\mathbf{i} + 2y\mathbf{j} + 2z\mathbf{k}, \qquad \nabla g = A\mathbf{i} + B\mathbf{j} + C\mathbf{k}.$$

$$\nabla f = \lambda \nabla g \implies 2x = A\lambda, \ 2y = B\lambda, \ 2z = C\lambda.$$

Substituting these equations in $g(x,y,z) = 0$, we have

$$\frac{1}{2}\lambda\left(A^2 + B^2 + C^2\right) + D = 0, \quad \lambda\frac{-2D}{A^2 + B^2 + C^2}.$$

Thus, in turn,

$$x = \frac{-DA}{A^2 + B^2 + C^2}, \quad y = \frac{-DB}{A^2 + B^2 + C^2}, \quad z = \frac{-DC}{A^2 + B^2 + C^2}$$

so the minimum value of $\sqrt{x^2 + y^2 + z^2}$ is $|D|\left(A^2 + B^2 + C^2\right)^{-1/2}$.

23.

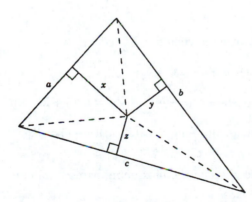

area $A = \frac{1}{2}ax + \frac{1}{2}by + \frac{1}{2}cz$.

The geometry suggests that

$$x^2 + y^2 + z^2$$

has a minimum.

$$f(x,y,z) = x^2 + y^2 + z^2, \qquad g(x,y,z) = ax + by + cz - 2A$$

$$\nabla f = 2x\mathbf{i} + 2y\mathbf{j} + 2z\mathbf{k}, \qquad \nabla g = a\mathbf{i} + b\mathbf{j} + c\mathbf{k}.$$

$$\nabla f = \lambda \nabla g \implies 2x = a\lambda, \ 2y = b\lambda, \ 2z = c\lambda.$$

Solving these equations for x, y, z and substituting the results in $g(x,y,z) = 0$, we have

$$\frac{a^2\lambda}{2} + \frac{b^2\lambda}{2} + \frac{c^2\lambda}{2} - 2A = 0, \quad \lambda = \frac{4A}{a^2 + b^2 + c^2}$$

and thus

$$x = \frac{2aA}{a^2+b^2+c^2}, \quad y = \frac{2bA}{a^2+b^2+c^2}, \quad z = \frac{2cA}{a^2+b^2+c^2}.$$

The minimum is $4A^2(a^2+b^2+c^2)^{-1}$.

25. Since the curve is asymptotic to the line $y = x$ as $x \to -\infty$ and as $x \to \infty$, the maximum exists. The distance between the point (x, y) and the line $y - x = 0$ is given by

$$\frac{|y-x|}{\sqrt{1+1}} = \frac{1}{2}\sqrt{2}\,|y-x|. \qquad \text{(by Theorem 9.2.2)}$$

Since the points on the curve are below the line $y = x$, we can replace $|y - x|$ by $x - y$. To simplify the work we drop the constant factor $\frac{1}{2}\sqrt{2}$.

$$f(x, y) = x - y, \qquad g(x, y) = x^3 - y^3 - 1$$

$$\nabla f = \mathbf{i} - \mathbf{j}, \qquad \nabla g = 3x^2\mathbf{i} - 3y^2\mathbf{j}.$$

We use the cross-product equation (16.8.4):

$$1\left(-3y^2\right) - \left(3x^2\right)(-1) = 0, \quad 3x^2 - 3y^2 = 0, \quad x = -y \ (x \neq y).$$

Now $g(x, y) = 0$ gives us

$$x^3 - (-x)^3 - 1 = 0, \quad 2x^3 = 1, \quad x = 2^{-1/3}.$$

The point is $\left(2^{-1/3}, -2^{-1/3}\right)$.

27. It suffices to show that the square of the area is a maximum when $a = b = c$.

$$f(a, b, c) = s(s - a)(s - b)(s - c), \quad g(a, b, c) = a + b + c - 2s$$

$$\nabla f = -s(s-b)(s-c)\mathbf{i} - s(s-a)(s-c)\mathbf{j} - s(s-a)(s-b)\mathbf{k}, \quad \nabla g = \mathbf{i} + \mathbf{j} + \mathbf{k}.$$

$$\nabla f = \lambda \nabla g \implies -s(s-b)(s-c) = -s(s-a)(s-c) = -s(s-a)(s-b) = \lambda.$$

Thus, $s - b = s - a = s - c$ so that $a = b = c$. This gives us the maximum, as no minimum exists. [The area can be made arbitrarily small by taking a close to s.]

29. (a)

$$f(x, y) = (xy)^{1/2}, \qquad g(x, y) = x + y - k, (x, y \geq 0, \ k \text{ a nonnegative constant}$$

$$\nabla f = \frac{y^{1/2}}{2x^{1/2}}\mathbf{i} + \frac{x^{1/2}}{2y^{1/2}}\mathbf{j} \qquad \nabla g = \mathbf{i} + \mathbf{j}.$$

$$\nabla f = \lambda \nabla g \implies \frac{y^{1/2}}{2x^{1/2}} = \lambda = \frac{x^{1/2}}{2y^{1/2}} \implies x = y = \frac{k}{2}.$$

Thus, the maximum value of f is: $f(k/2, k/2) = \dfrac{k}{2}$.

(b) For all x, y $(x, y \geq 0)$ we have

$$(xy)^{1/2} = f(x,y) \leq f(k/2, k/2) = \frac{k}{2} = \frac{x+y}{2}.$$

31. Simply extend the arguments used in Exercises 29 and 30.

33. $\qquad S(r,h) = 2\pi r^2 + 2\pi rh, \qquad\qquad\qquad g(r,h) = \pi r^2 h, \quad (V \text{ constant}$

$$\nabla S = (4\pi r + 2\pi h)\,\mathbf{i} + 2\pi r\,\mathbf{j}, \qquad\qquad \nabla g = 2\pi rh\,\mathbf{i} + \pi r^2\,\mathbf{j}.$$

$$\nabla S = \lambda \nabla g \implies 4\pi r + 2\pi h = 2\pi rh\lambda, \quad 2\pi r = \pi r^2 \lambda \implies r = \frac{2}{\lambda}, \quad h = \frac{4}{\lambda}.$$

Now $\pi r^2 h = V, \implies \lambda = \sqrt[3]{\dfrac{16\pi}{V}} \implies r = \sqrt[3]{\dfrac{V}{2\pi}}, \quad h = \sqrt[3]{\dfrac{4V}{\pi}}.$

To minimize the surface area, take $r = \sqrt[3]{\dfrac{V}{2\pi}}, \quad$ and $h = \sqrt[3]{\dfrac{4V}{\pi}}.$

35. $\qquad S(l,w,h) = lw + 2(lh + wh), \qquad\qquad\qquad g(l,w,h) = lwh - 12, \quad (V \text{ constant})$

$$\nabla S = (w + 2h)\,\mathbf{i} + (l + 2h)\,\mathbf{j} + 2(l + w)\,\mathbf{k}, \qquad\qquad \nabla g = wh\,\mathbf{i} + lh\,\mathbf{j} + lw\,\mathbf{k}.$$

$$\nabla S = \lambda \nabla g \implies w + 2h = \lambda wh, \quad l + 2h = \lambda lh, \quad 2(l + w) = \lambda lw \implies w = l = 2h.$$

Now $\quad lwh = 12, \implies h = \sqrt[3]{3}, \quad l = w = 2\sqrt[3]{3}.$

To minimize the surface area, take $l = w = 2\sqrt[3]{3}$ ft. and $h = \sqrt[3]{3}$ ft.

37. (a) $\qquad f(x,y,l) = xyl, \qquad\qquad\qquad g(x,y,l) = 2x + 2y + l - 108,$

$$\nabla f = yl\,\mathbf{i} + xl\,\mathbf{j} + xy\,w\mathbf{k}, \qquad\qquad \nabla g = 2\,\mathbf{i} + 2\,\mathbf{j} + \mathbf{k}.$$

$$\nabla f = \lambda \nabla g \implies yl = 2\lambda, \quad xl = 2\lambda, \quad xy = \lambda \implies y = x \text{ and } l = 2x.$$

Now $2x + 2y + l = 108, \implies x = 18$ and $l = 36.$

To maximize the volume, take $x = y = 18$ in. and $l = 36$ in.

(b) $\qquad f(r,l) = \pi r^2 l, \qquad\qquad\qquad g(r,l) = 2\pi r + l - 108,$

$$\nabla f = 2\pi rl\,\mathbf{i} + \pi r^2\,\mathbf{j}, \qquad\qquad \nabla g = 2\pi\,\mathbf{i} + \mathbf{j}.$$

$$\nabla f = \lambda \nabla g \implies 2\pi rl = 2\pi\lambda, \quad \pi r^2 = \lambda, \quad l = \pi r.$$

Now $2\pi r + l = 108, \implies r = \dfrac{36}{\pi}$ and $l = 36.$

To maximize the volume, take $r = 36/\pi$ in. and $l = 36$ in.

39. To simplify notation we set $x = Q_1, \quad y = Q_2, \quad z = Q_3.$

$$f(x,y,z) = 2x + 8y + 24z, \qquad\qquad g(x,y,z) = x^2 + 2y^2 + 4z^2 - 4{,}500{,}000{,}000$$

$$\nabla f = 2\mathbf{i} + 8\mathbf{j} + 24\mathbf{k}, \qquad\qquad \nabla g = 2x\mathbf{i} + 4y\mathbf{j} + 8z\mathbf{k}.$$

$$\nabla f = \lambda \nabla g \quad \Longrightarrow \quad 2 = 2\lambda x, \quad 8 = 4\lambda y, \quad 24 = 8\lambda z.$$

Since $\lambda \neq 0$ here, we solve the equations for x, y, z:

$$x = \frac{1}{\lambda}, \quad y = \frac{2}{\lambda}, \quad z = \frac{3}{\lambda},$$

and substitute these results in $g(x, y, z) = 0$ to obtain

$$\frac{1}{\lambda^2} + 2\left(\frac{4}{\lambda^2}\right) + 4\left(\frac{9}{\lambda^2}\right) - 45 \times 10^8 = 0, \quad \frac{45}{\lambda^2} = 45 \times 10^8, \quad \lambda = \pm 10^{-4}.$$

Since x, y, z are non-negative, $\lambda = 10^{-4}$ and

$$x = 10^4 = Q_1, \quad y = 2 \times 10^4 = Q_2, \quad z = 3 \times 10^4 = Q_3.$$

PROJECT 15.6

1. $f(x, y, z) = xy^2z - x^2yz, \quad g(x, y, z) = x^2 + y^2 - 1, \quad h(x, y, z) = z^2 - x^2 - y^2, \ z \geq 0$

$\nabla f = (y^2z - 2xyz)\,\mathbf{i} + (2xyz - x^2z)\,\mathbf{j} + (xy^2 - x^2y)\,\mathbf{k}, \quad \nabla g = 2x\,\mathbf{i} + 2y\,\mathbf{j}, \quad \nabla h = -2x\,\mathbf{i} + 2y\,\mathbf{j} + 2z\,\mathbf{k}.$

$\nabla f = \lambda \nabla g + \mu \nabla h \quad \Longrightarrow$

$$y^2z - 2xyz = 2x\lambda - 2x\mu 2x$$

$$2xyz - x^2z = 2y\lambda - 2y\mu$$

$$xy^2 - yx^2 = 2z\mu$$

Note first that $z = 1$. Setting $z = 1$ in the first two equations and adding, we get

$$y^2 - x^2 = 2x(\lambda - \mu) + 2y(\lambda - \mu) \quad \text{or} \quad (y + x)(y - x) = 2(y + x)(\lambda - \mu)$$

Therefore, either $y = -x$ or $\lambda - \mu = \dfrac{y - x}{2}$.

First, let $y = -x$. Then it follows that $x = \dfrac{1}{\sqrt{2}}, \quad y = -\dfrac{1}{\sqrt{2}}, \quad z = 1.$

If $y \neq -x$, then $\lambda - \mu = \dfrac{y - x}{2}$ and

$$y^2 - 2xy = x(y - x)$$

$$2xy - x^2 = y(y - x)$$

Solving these equations simultaneously, using the fact that $x^2 + y^2 = 1$, we obtain

$$9x^4 - 9x^2 + 1 = 0 \quad \Longrightarrow \quad x^2 = \frac{3 \pm \sqrt{5}}{6} \cong 0.873, \ 0.127$$

Now,

$$x^2 = 0.873 \quad \Longrightarrow \quad y^2 = 0.127 \quad \Longrightarrow \quad x = \pm 0.934, \ y = \pm 0.356;$$

$$x^2 = 0.127 \quad \Longrightarrow \quad y^2 = 0.873 \quad \Longrightarrow \quad x = \pm 0.356, \ y = \pm 0.934.$$

Clearly, the maximum value of f will occur when $x > 0$ and $y < 0$:

$$f\left(\frac{1}{\sqrt{2}}, -\frac{1}{\sqrt{2}}, 1\right) = \frac{\sqrt{2}}{2} \cong 0.707,$$

$$f(0.934, -0.356, 1) = f(0.356, -0.934, 1) \cong 0.429.$$

Therefore, the maximum value of f subject to the constraints is $\dfrac{\sqrt{2}}{2}$.

3. $f(x, y, z) = x^2 + y^2 + z^2,$ $g(x, y, z) = x + y - z + 1 = 0,$ $h(x, y, z) = x^2 + y^2 - z^2$

$$\nabla f = 2x\,\mathbf{i} + 2y\,\mathbf{j} + 2z\,w\mathbf{k}, \qquad \nabla g = \mathbf{i} + \mathbf{j} - \mathbf{k}, \qquad \nabla h = 2x\,\mathbf{i} + 2y\,\mathbf{j} - 2z\,\mathbf{k}.$$

$$\nabla f = \lambda \nabla g + \mu \nabla h \quad \Longrightarrow \quad 2x = \lambda + 2x\mu, \quad 2y = \lambda + 2y\mu, \quad 2z = -\lambda - 2z\mu$$

Multiplying the first equation by y, the second equation by x and subtracting, yields

$$\lambda(y - x) = 0.$$

Now $\lambda = 0 \quad \Longrightarrow \quad \mu = 1 \quad \Longrightarrow x = y = z = 0$. This is impossible since $x + y - z = -1$.

Therefore, we must have $y = x \quad \Longrightarrow \quad z = \pm\sqrt{2}\,x$.

Substituting $y = x$, $z = \sqrt{2}\,x$ into the equation $x + y - z + 1 = 0$, we get

$$x = -1 - \frac{\sqrt{2}}{2} \quad \Longrightarrow \quad y = -1 - \frac{\sqrt{2}}{2}, \; z = -1 - \sqrt{2}$$

Substituting $y = x$, $z = -\sqrt{2}\,x$ into the equation $x + y - z + 1 = 0$, we get

$$x = -1 + \frac{\sqrt{2}}{2} \quad \Longrightarrow \quad y = -1 + \frac{\sqrt{2}}{2}, \; z = -1 + \sqrt{2}$$

Since

$$f\left(-1 - \frac{\sqrt{2}}{2}, -1 - \frac{\sqrt{2}}{2}, -1 - \sqrt{2}\right) = 6 + 4\sqrt{2} \;\; \text{and}$$

$$f\left(-1 + \frac{\sqrt{2}}{2}, -1 + \frac{\sqrt{2}}{2}, -1 + \sqrt{2}\right) = 6 - 4\sqrt{2},$$

it follows that $\left(-1 + \dfrac{\sqrt{2}}{2}, -1 + \dfrac{\sqrt{2}}{2}, -1 + \sqrt{2}\right)$ is closest to the origin and

$\left(-1 - \dfrac{\sqrt{2}}{2}, -1 - \dfrac{\sqrt{2}}{2}, -1 - \sqrt{2}\right)$ is furthest from the origin.

SECTION 15.7

1. $df = \left(3x^2 y - 2xy^2\right) \Delta x + \left(x^3 - 2x^2 y\right) \Delta y$

3. $df = \left(\cos y + y \sin x\right) \Delta x - \left(x \sin y + \cos x\right) \Delta y$

5. $df = \Delta x - \left(\tan z\right) \Delta y - \left(y \sec^2 z\right) \Delta z$

7. $df = \dfrac{y(y^2 + z^2 - x^2)}{(x^2 + y^2 + z^2)^2} \Delta x + \dfrac{x(x^2 + z^2 - y^2)}{(x^2 + y^2 + z^2)^2} \Delta y - \dfrac{2xyz}{(x^2 + y^2 + z^2)^2} \Delta z$

9. $df = [\cos(x + y) + \cos(x - y)] \Delta x + [\cos(x + y) - \cos(x - y)] \Delta y$

11. $df = (y^2 z e^{xz} + \ln z)\, \Delta x + 2y e^{xz}\, \Delta y + \left(xy^2 e^{xz} + \dfrac{x}{z}\right) \Delta z$

13.
$$\Delta u = \left[(x + \Delta x)^2 - 3(x + \Delta x)(y + \Delta y) + 2(y + \Delta y)^2\right] - (x^2 - 3xy + 2y^2)$$

$$= \left[(1.7)^2 - 3(1.7)(-2.8) + 2(-2.8)^2\right] - \left(2^2 - 3(2)(-3) + 2(-3)^2\right)$$

$$= (2.89 + 14.28 + 15.68) - 40 = -7.15$$

$$du = (2x - 3y)\, \Delta x + (-3x + 4y)\, \Delta y$$

$$= (4 + 9)(-0.3) + (-18)(0.2) = -7.50$$

15. $\Delta u = \left[(x + \Delta x)^2(z + \Delta z) - 2(y + \Delta y)(z + \Delta z)^2 + 3(x + \Delta x)(y + \Delta y)(z + \Delta z)\right]$

$$- \left(x^2 z - 2yz^2 + 3xyz\right)$$

$$= \left[(2.1)^2(2.8) - 2(1.3)(2.8)^2 + 3(2.1)(1.3)(2.8)\right] - \left[(2)^2 3 - 2(1)(3)^2 + 3(2)(1)(3)\right] = 2.896$$

$$du = (2xz + 3yz)\, \Delta x + \left(-2z^2 + 3xz\right)\Delta y + \left(x^2 - 4yz + 3xy\right)\Delta z$$

$$= [2(2)(3) + 3(1)(3)](0.1) + [-2(3)^2 + 3(2)(3)](0.3) + [2^2 - 4(1)(3) + 3(2)(1)](-0.2) = 2.5$$

17. $f(x, y) = x^{1/2} y^{1/4}; \quad x = 121, \ \ y = 16, \ \ \Delta x = 4, \ \ \Delta y = 1$

$$f(x + \Delta x, y + \Delta y) \cong f(x, y) + df$$

$$= x^{1/2}\, y^{1/4} + \tfrac{1}{2} x^{-1/2}\, y^{1/4}\, \Delta x + \tfrac{1}{4} x^{1/2}\, y^{-3/4}\, \Delta y$$

$$\sqrt{125}\, \sqrt[4]{17} \cong \sqrt{121}\, \sqrt[4]{16} + \tfrac{1}{2}(121)^{-1/2}\, (16)^{1/4}\, (4) + \tfrac{1}{4}(121)^{1/2}\, (16)^{-3/4}\, (1)$$

$$= 11(2) + \tfrac{1}{2}\left(\tfrac{1}{11}\right)(2)(4) + \tfrac{1}{4}(11)\left(\tfrac{1}{8}\right)$$

$$= 22 + \tfrac{4}{11} + \tfrac{11}{32} = 22\,\tfrac{249}{352} \cong 22.71$$

19. $$f(x, y) = \sin x \cos y; \quad x = \pi, \ \ y = \dfrac{\pi}{4}, \ \ \Delta x = -\dfrac{\pi}{7}, \ \ \Delta y = -\dfrac{\pi}{20}$$

$$df = \cos x \cos y\, \Delta x - \sin x \sin y\, \Delta y$$

$$f(x + \Delta x, y + \Delta y) \cong f(x, y) + df$$

$$\sin \dfrac{6}{7}\pi \cos \dfrac{1}{5}\pi \cong \sin \pi \cos \dfrac{\pi}{4} + \left(\cos \pi \cos \dfrac{\pi}{4}\right)\left(-\dfrac{\pi}{7}\right) - \left(\sin \pi \sin \dfrac{\pi}{4}\right)\left(-\dfrac{\pi}{20}\right)$$

$$= 0 + \left(\dfrac{1}{2}\sqrt{2}\right)\left(\dfrac{\pi}{7}\right) + 0 = \dfrac{\pi\sqrt{2}}{14} \cong 0.32$$

21. $f(2.9, 0.01) \cong f(3, 0) + df$,   where df is to be evaluated at $x = 3$, $y = 0$, $\Delta x = -0.1$, $\Delta y = 0.01$.

$$df = \left(2xe^{xy} + x^2ye^{xy}\right)\Delta x + x^3e^{xy}\Delta y = \left[2(3)e^0 + (2)^2(0)e^0\right](-0.1) + 3^3e^0(0.01) = -0.33$$

Thus, $f(2.9, .01) \cong 3^2e^0 - 0.33 = 8.67$.

23. $f(2.94, 1.1, 0.92) \cong f(3, 1, 1) + df$,   where df is to be evaluated at $x = 3$, $y = 1$, $z = 1$,

$\Delta x = -0.06$, $\Delta y = 0.1$, $\Delta z = -0.08$

$$df = \tan^{-1}yz\,\Delta x + \frac{xz}{1 + y^2z^2}\Delta y + \frac{xy}{1 + y^2z^2}\Delta z = \frac{\pi}{4}(-0.06) + (1.5)(0.1) + (1.5)(-0.08) \cong -0.441$$

Thus, $f(2.94, 1.1, 0.92) \cong 3 - 0.441 = 2.559$

25. $df = \dfrac{\partial z}{\partial x}\Delta x + \dfrac{\partial z}{\partial y}\Delta y = \dfrac{2y}{(x+y)^2}\Delta x - \dfrac{2x}{(x+y)^2}\Delta y$

With $x = 4$, $y = 2$, $\Delta x = 0.1$, $\Delta y = 0.1$, we get

$$df = \tfrac{4}{36}(0.1) - \tfrac{8}{36}(0.1) = -\tfrac{1}{90}.$$

The exact change is $\dfrac{4.1 - 2.1}{4.1 + 2.1} - \dfrac{4 - 2}{4 + 2} = \dfrac{2}{6.2} - \dfrac{1}{3} = -\dfrac{1}{93}.$

27. $S = 2\pi r^2 + 2\pi rh$; $r = 8$, $h = 12$, $\Delta r = -0.3$, $\Delta h = 0.2$

$$dS = \frac{\partial S}{\partial r}\Delta r + \frac{\partial S}{\partial h}\Delta h = (4\pi r + 2\pi h)\,\Delta r + (2\pi r)\,\Delta h$$

$$= 56\pi(-0.3) + 16\pi(0.2) = -13.6\pi.$$

The area decreases about 13.6π in.^{2}.

29. $S(9.98, 5.88, 4.08) \cong S(10, 6, 4) + dS = 248 + dS$, where

$dS = (2w + 2h)\,\Delta l + (2l + 2h)\,\Delta w + (2l + 2w)\,\Delta h = 20(-0.02) + 28(-0.12) + 32(0.08) = -1.20$

Thus, $S(9.98, 5.88, 4.08) \cong 248 - 1.20 = 246.80$.

31. (a) $dV = yx\,\Delta x + xz\,\Delta y + xy\,\Delta z = (8)(6)(0.02) + (12)(6)(-0.05) + (12)(8)(0.03) = 0.24$

(b) $\Delta V = (12.02)(7.95)(6.03) - (12)(8)(6) = 0.22077$

33. $T(P) - T(Q) \cong dT = (-2x + 2yz)\,\Delta x + (-2y + 2xz)\,\Delta y + (-2z + 2xy)\,\Delta z$

Letting $x = 1$, $y = 3$, $z = 4$, $\Delta x = 0.15$, $\Delta y = -0.10$, $\Delta z = 0.10$, we have

$$dT = (22)(0.15) + (2)(-0.10) + (-2)(0.10) = 2.9$$

35. The area is given by $A = \tfrac{1}{2}x^2\tan\theta$. The change in area is approximated by:

$$dA = x\tan\theta\,\Delta x + \tfrac{1}{2}x^2\sec^2\theta\,\Delta\theta \cong 4(0.75)\,\Delta x + 8(1.5625)\,\Delta\theta = 3\,\Delta x + 12.5\,\Delta\theta$$

The area is more sensitive to a change in θ.

37.   (a) $\pi r^2 h = \pi(r + \Delta r)^2(h + \Delta h) \implies \Delta h \dfrac{r^2 h}{(r + \Delta r)^2} - h = -\dfrac{(2r + \Delta r)h}{(r + \Delta r)^2}\Delta r.$

$$df = (2\pi r h)\,\Delta r + \pi r^2\,\Delta h, \qquad df = 0 \implies \Delta h = \dfrac{-2h}{r}\Delta r.$$

(b) $2\pi r^2 + 2\pi rh = 2\pi(r + \Delta r)^2 + 2\pi(r + \Delta r)(h + \Delta h).$

Solving for Δh,

$$\Delta h = \dfrac{r^2 + rh - (r + \Delta r)^2}{r + \Delta r} - h = -\dfrac{(2r + h + \Delta r)}{r + \Delta r}\Delta r.$$

$$df = (4\pi r + 2\pi h)\,\Delta r + 2\pi r\,\Delta h, \qquad df = 0 \implies \Delta h = -\left(\dfrac{2r + h}{r}\right)\Delta r.$$

39.   (a) $c(x, y) = \sqrt{x^2 + y^2};\ x = 5, y = 12, \Delta x = \pm 1.5, \Delta y = \pm 1.5$

$$dc = \dfrac{\partial c}{\partial x}\Delta x + \dfrac{\partial c}{\partial y}\Delta y = \dfrac{x}{\sqrt{x^2 + y^2}}\Delta x + \dfrac{y}{\sqrt{x^2 + y^2}}\Delta y$$

$$= \dfrac{5}{13}(\pm 1.5) + \dfrac{12}{13}(\pm 1.5) \cong \pm 1.962$$

The maximum possible error in the value of the hypotenuse is 1.962 cm.

(b) $A(x, y) = \frac{1}{2}xy;\ x = 5, y = 12, \Delta x = \pm 1.5, \Delta y = \pm 1.5$

$$dA = \dfrac{\partial A}{\partial x}\Delta x + \dfrac{\partial A}{\partial y}\Delta y = \tfrac{1}{2}y\,\Delta x + \tfrac{1}{2}y\,\Delta y$$

$$= \tfrac{1}{2}(12)(\pm 1.5) + \tfrac{1}{2}(5)(\pm 1.5) \cong \pm 12.75$$

The maximum possible error in the value of the area is 12.75 cm^2.

41. $s = \dfrac{A}{A - W};\ A = 9,\ W = 5,\ \Delta A = \pm 0.01,\ \Delta W = \pm 0.02$

$$ds = \dfrac{\partial s}{\partial A}\Delta A + \dfrac{\partial s}{\partial W}\Delta W = \dfrac{-W}{(A - W)^2}\Delta A + \dfrac{A}{(A - W)^2}\Delta W$$

$$= -\dfrac{5}{16}(\pm 0.01) + \dfrac{9}{16}(\pm 0.02) \cong \pm 0.014$$

The maximum possible error in the value of s is 0.014 lbs; $2.23 \le s + \Delta s \le 2.27$

SECTION 15.8

1. $\dfrac{\partial f}{\partial x} = xy^2, \quad f(x, y) = \tfrac{1}{2}x^2y^2 + \phi(y), \quad \dfrac{\partial f}{\partial y} = x^2 y + \phi'(y) = x^2 y.$

Thus, $\phi'(y) = 0,\ \phi(y) = C,$ and $f(x, y) = \tfrac{1}{2}x^2y^2 + C.$

3. $\dfrac{\partial f}{\partial x} = y$, $f(x,y) = xy + \phi(y)$, $\dfrac{\partial f}{\partial y} = x + \phi'(y) = x$.

   Thus, $\phi'(y) = 0$, $\phi(y) = C$,   and $f(x,y) = xy + C$.

5. No; $\dfrac{\partial}{\partial y}\left(y^3 + x\right) = 3y^2$ whereas $\dfrac{\partial}{\partial x}\left(x^2 + y\right) = 2x$.

7. $\dfrac{\partial f}{\partial x} = \cos x - y\sin x$, $f(x,y) = \sin x + y\cos x + \phi(y)$, $\dfrac{\partial f}{\partial y} = \cos x + \phi'(y) = \cos x$.

   Thus, $\phi'(y) = 0$, $\phi(y) = C$,   and $f(x,y) = \sin x + y\cos x + C$.

9. $\dfrac{\partial f}{\partial x} = e^x \cos y^2$, $f(x,y) = e^x \cos y^2 + \phi(y)$, $\dfrac{\partial f}{\partial y} = -2ye^x \sin y^2 + \phi'(y) = -2ye^x \sin y^2$.

   Thus, $\phi'(y) = 0$, $\phi(y) = C$,   and $f(x,y) = e^x \cos y^2 + C$.

11. $\dfrac{\partial f}{\partial y} = xe^x - e^{-y}$, $f(x,y) = xye^x + e^{-y} + \phi(x)$, $\dfrac{\partial f}{\partial x} = ye^x + xye^x + \phi'(x) = ye^x(1 + x)$.

   Thus, $\phi'(x) = 0$, $\phi(x) = C$,   and $f(x,y) = xye^x + e^{-y} + C$.

13. No; $\dfrac{\partial}{\partial y}\left(xe^{xy} + x^2\right) = x^2 e^{xy}$ whereas $\dfrac{\partial}{\partial x}\left(ye^{xy} - 2y\right) = y^2 e^{xy}$

15. $\dfrac{\partial}{\partial x} = 1 + y^2 + xy^2$, $f(x,y) = x + xy^2 + \frac{1}{2}x^2y^2 + \phi(y)$, $\dfrac{\partial}{\partial y} = 2xy + x^2y + \phi'(y) = x^2y + y + 2xy + 1$.

   Thus, $\phi'(y) = y + 1$, $\phi(y) = \frac{1}{2}y^2 + y + C$ and $f(x,y) = x + xy^2 + \frac{1}{2}x^2y^2 + \frac{1}{2}y^2 + y + C$.

17. $\dfrac{\partial f}{\partial x} = \dfrac{x}{\sqrt{x^2 + y^2}}$, $f(x,y) = \sqrt{x^2 + y^2} + \phi(y)$, $\dfrac{\partial f}{\partial y} = \dfrac{y}{\sqrt{x^2 + y^2}} + \phi'(y) = \dfrac{y}{\sqrt{x^2 + y^2}}$.

   Thus, $\phi'(y) = 0$, $\phi(y) = C$,   and $f(x,y) = \sqrt{x^2 + y^2} + C$.

19. $\dfrac{\partial f}{\partial x} = x^2 \sin^{-1} y$, $f(x,y) = \frac{1}{3}x^3 \sin^{-1} y + \phi(y)$, $\dfrac{\partial f}{\partial y} = \dfrac{x^3}{3\sqrt{1 - y^2}} + \phi'(y) = \dfrac{x^3}{3\sqrt{1 - y^2}} - \ln y$.

   Thus, $\phi'(y) = -\ln y$, $\phi(y) = y - y\ln y + C$,   and

$$f(x,y) = \frac{1}{3}\sin^{-1} y + y - y\ln y + C.$$

21. $\dfrac{\partial f}{\partial x} = f(x,y)$, $\dfrac{\partial f/\partial x}{f(x,y)} = 1$, $\ln f(x,y) = x + \phi(y)$, $\dfrac{\partial f/\partial y}{f(x,y)} = 0 + \phi'(y)$, $\dfrac{\partial f}{\partial y} = f(x,y)$.

   Thus, $\phi'(y) = 1$, $\phi(y) = y + K$,   and $f(x,y) = e^{x+y+K} = Ce^{x+y}$.

23. (a) $P = 2x$, $Q = z$, $R = y$; $\dfrac{\partial P}{\partial y} = 0 = \dfrac{\partial Q}{\partial x}$, $\dfrac{\partial P}{\partial z} = 0 = \dfrac{\partial R}{\partial x}$, $\dfrac{\partial Q}{\partial z} = 1 = \dfrac{\partial R}{\partial y}$

(b), (c), and (d)

$$\frac{\partial f}{\partial x} = 2x, \quad f(x,y,z) = x^2 + g(y,z).$$

$$\frac{\partial f}{\partial y} = 0 + \frac{\partial g}{\partial y} \quad \text{with} \quad \frac{\partial f}{\partial y} = z \implies \frac{\partial g}{\partial y} = z.$$

Then,

$$g(y,z) = yz + h(z),$$

$$f(x,y,z) = x^2 + yz + h(z),$$

$$\frac{\partial f}{\partial z} = 0 + y + h'(z) \quad \text{and} \quad \frac{\partial f}{\partial z} = y \implies h'(z) = 0.$$

Thus, $h(z) = C$ and $f(x,y,z) = x^2 + yz + C$.

25. The function is a gradient by the test stated before Exercise 23.

Take $P = 2x + y$, $Q = 2y + x + z$, $R = y - 2z$. Then

$$\frac{\partial P}{\partial y} = 1 = \frac{\partial Q}{\partial x}, \quad \frac{\partial P}{\partial z} = 0 = \frac{\partial R}{\partial x}, \quad \frac{\partial Q}{\partial z} = 1 = \frac{\partial R}{\partial y}.$$

Next, we find f where $\nabla f = P\mathbf{i} + Q\mathbf{j} + R\mathbf{k}$.

$$\frac{\partial f}{\partial x} = 2x + y,$$

$$f(x,y,z) = x^2 + xy + g(y,z).$$

$$\frac{\partial f}{\partial y} = x + \frac{\partial g}{\partial y} \quad \text{with} \quad \frac{\partial f}{\partial y} = 2y + x + z \implies \frac{\partial g}{\partial y} = 2y + z.$$

Then,

$$g(y,z) = y^2 + yz + h(z),$$

$$f(x,y,z) = x^2 + xy + y^2 + yz + h(z).$$

$$\frac{\partial f}{\partial z} = y + h'(z) = y - 2z \implies h'(z) = -2z.$$

Thus, $h(z) = -z^2 + C$ and $f(x,y,z) = x^2 + xy + y^2 + yz - z^2 + C$.

27. The function is a gradient by the test stated before Exercise 23.

Take $P = y^2z^3 + 1$, $Q = 2xyz^3 + y$, $R = 3xy^2z^2 + 1$. Then

$$\frac{\partial P}{\partial y} = 2yz^3 = \frac{\partial Q}{\partial x}, \quad \frac{\partial P}{\partial z} = 3y^2z^2 = \frac{\partial R}{\partial x}, \quad \frac{\partial Q}{\partial z} = 6xyz^2 = \frac{\partial R}{\partial y}.$$

Next, we find f where $\quad \nabla f = P\mathbf{i} + Q\mathbf{j} + R\mathbf{k}$.

$$\frac{\partial f}{\partial x} = y^2 z^3 + 1,$$

$$f(x, y, z) = xy^2 z^3 + x + g(y, z).$$

$$\frac{\partial f}{\partial y} = 2xyz^3 \frac{\partial g}{\partial y} \quad \text{with} \quad \frac{\partial f}{\partial y} = 2xyz^3 + y \quad \Longrightarrow \quad \frac{\partial g}{\partial y} = y.$$

Then,

$$g(y, z) = \tfrac{1}{2} y^2 + h(z),$$

$$f(x, y, z) = xy^2 z^3 + x + \tfrac{1}{2} y^2 + h(z).$$

$$\frac{\partial f}{\partial z} = 3xy^2 z^2 + h'(z) = 3xy^2 z^2 + 1 \quad \Longrightarrow \quad h'(z) = 1.$$

Thus, $h(z) = z + C \quad$ and $\quad f(x, y, z) = xy^2 z^3 + x + \tfrac{1}{2} y^2 + z + C.$

29. $\mathbf{F(r)} = \nabla \left(\dfrac{GmM}{r} \right)$

SECTION 15.9

1. $\dfrac{\partial P}{\partial y} = 2xy - 1 = \dfrac{\partial Q}{\partial x};\qquad$ the equation is exact.

$$\frac{\partial f}{\partial x} = xy^2 - y \quad \Longrightarrow \quad f(x, y) = \tfrac{1}{2} x^2 y^2 - xy + \varphi(y)$$

$$\frac{\partial f}{\partial y} = x^2 y - x + \varphi'(y) = x^2 y - x \quad \Longrightarrow \quad \varphi'(y) = 0 \quad \Longrightarrow \quad \varphi(y) = 0 \ \text{(omit the constant)} * \cdot$$

Therefore $f(x, y) = \tfrac{1}{2} x^2 y^2 - xy, \quad$ and a one-parameter family of solutions is:

$$\tfrac{1}{2} x^2 y^2 - xy = C$$

* We will omit the constant at this step throughout this section.

3. $\dfrac{\partial P}{\partial y} = e^y - e^x = \dfrac{\partial Q}{\partial x};\qquad$ the equation is exact.

$$\frac{\partial f}{\partial x} = e^y - ye^x \quad \Longrightarrow \quad f(x, y) = xe^y - ye^x + \varphi(y)$$

$$\frac{\partial f}{\partial y} = xe^y - e^x + \varphi'(y) = xe^y - e^x \quad \Longrightarrow \quad \varphi'(y) = 0 \quad \Longrightarrow \quad \varphi(y) = 0$$

Therefore $f(x, y) = xe^y - ye^x, \quad$ and a one-parameter family of solutions is:

$$xe^y - ye^x = C$$

5. $\dfrac{\partial P}{\partial y} = \dfrac{1}{y} + 2x = \dfrac{\partial Q}{\partial x};$ the equation is exact.

$\dfrac{\partial f}{\partial x} = \ln y + 2xy \implies f(x,y) = x \ln y + x^2 y + \varphi(y)$

$\dfrac{\partial f}{\partial y} = \dfrac{x}{y} + x^2 + \varphi'(y) = \dfrac{x}{y} + x^2 \implies \varphi'(y) = 0 \implies \varphi(y) = 0$

Therefore $f(x,y) = x \ln y + x^2 y,$ and a one-parameter family of solutions is:

$$x \ln y + x^2 y = C$$

7. $\dfrac{\partial P}{\partial y} = \dfrac{1}{x} = \dfrac{\partial Q}{\partial x};$ the equation is exact.

$\dfrac{\partial f}{\partial x} = \dfrac{y}{x} + 6x \implies f(x,y) = y \ln x + 3x^2 + \varphi(y)$

$\dfrac{\partial f}{\partial y} = \ln x + \varphi'(y) = \ln x - 2 \implies \varphi'(y) = -2 \implies \varphi(y) = -2y$

Therefore $f(x,y) = y \ln x + 3x^2 - 2y,$ and a one-parameter family of solutions is:

$$y \ln x + 3x^2 - 2y = C$$

9. $\dfrac{\partial P}{\partial y} = 3y^2 - 2y \sin x = \dfrac{\partial Q}{\partial x};$ the equation is exact.

$\dfrac{\partial f}{\partial x} = y^3 - y^2 \sin x - x \implies f(x,y) = xy^3 + y^2 \cos x - \tfrac{1}{2} x^2 + \varphi(y)$

$\dfrac{\partial f}{\partial y} = 3xy^2 + 2y \cos x + \varphi'(y) = 3xy^2 + 2y \cos x + e^{2y} \implies \varphi'(y) = e^{2y} \implies \varphi(y) = \tfrac{1}{2} e^{2y}$

Therefore $f(x,y) = xy^3 + y^2 \cos x - \tfrac{1}{2} x^2 + \tfrac{1}{2} e^{2y},$ and a one-parameter family of solutions is:

$$xy^3 + y^2 \cos x - \tfrac{1}{2} x^2 + \tfrac{1}{2} e^{2y} = C$$

11. (a) Yes: $\dfrac{\partial}{\partial y}[p(x)] = 0 = \dfrac{\partial}{\partial x}[q(y)].$

(b) For all x, y such that $p(y)q(x) \neq 0,$ $\dfrac{1}{p(y)q(x)}$ is an integrating factor.

Multiplying the differential equation by $\dfrac{1}{p(y)q(x)},$ we get

$$\dfrac{1}{q(x)} + \dfrac{1}{p(y)} y' = 0$$

which has the form of the differential equation in part (a).

13. $\dfrac{\partial P}{\partial y} = e^{y-x} - 1$ and $\dfrac{\partial Q}{\partial x} = e^{y-x} - xe^{y-x};$ the equation is not exact.

Since $\dfrac{1}{Q}\left(\dfrac{\partial P}{\partial y} - \dfrac{\partial Q}{\partial x} \right) = \dfrac{1}{xe^{y-x} - 1} \left(xe^{y-x} - 1 \right) = 1,$ $\mu(x) = e^{\int dx} = e^x$ is an an integrating factor. Multiplying the given equation by e^x, we get

$$(e^y - ye^x) + (xe^y - e^x)\, y' = 0$$

This is the equation given in Exercise 3. A one-parameter family of solutions is:

$$xe^y - ye^x = C$$

15. $\dfrac{\partial P}{\partial y} = 6x^2 y + e^y = \dfrac{\partial Q}{\partial x};$ the equation is exact.

$$\frac{\partial f}{\partial x} = 3x^2 y^2 + x + e^y \implies f(x,y) = x^3 y^2 + \tfrac{1}{2}x^2 + xe^y + \varphi(y)$$

$$\frac{\partial f}{\partial y} = 2x^3 y + xe^y + \varphi'(y) = 2x^3 y + y + xe^y \implies \varphi'(y) = y \implies \varphi(y) = \tfrac{1}{2}y^2$$

Therefore $f(x,y) = x^3 y^2 + \tfrac{1}{2}x^2 + xe^y + \tfrac{1}{2}y^2,$ and a one-parameter family of solutions is:

$$x^3 y^2 + \tfrac{1}{2}x^2 + xe^y + \tfrac{1}{2}y^2 = C$$

17. $\dfrac{\partial P}{\partial y} = 3y^2$ and $\dfrac{\partial Q}{\partial x} = 0;$ the equation is not exact.

Since $\dfrac{1}{Q}\left(\dfrac{\partial P}{\partial y} - \dfrac{\partial Q}{\partial x}\right) = \dfrac{1}{3y^2}\,(3y^2) = 1,$ $\mu(x) = e^{\int dx} = e^x$ is an

an integrating factor. Multiplying the given equation by e^x, we get

$$\left(y^3 e^x + xe^x + e^x\right) + \left(3y^2 xe^x\right)y' = 0$$

$$\frac{\partial f}{\partial x} = y^3 e^x + xe^x + e^x \implies f(x,y) = y^3 e^x + xe^x + \varphi(y)$$

$$\frac{\partial f}{\partial y} = 3y^2 e^x + \varphi'(y) = 3y^2 e^x \implies \varphi'(y) = 0 \implies \varphi(y) = 0$$

Therefore $f(x,y) = y^3 e^x + xe^x,$ and a one-parameter family of solutions is:

$$y^3 e^x + xe^x = C$$

19. $\dfrac{\partial P}{\partial y} = 1 = \dfrac{\partial Q}{\partial x};$ the equation is exact.

$$\frac{\partial f}{\partial x} = x^2 + y \implies f(x,y) = \tfrac{1}{3}x^3 + xy + \varphi(y)$$

$$\frac{\partial f}{\partial y} = x + \varphi'(y) = x + e^y \implies \varphi'(y) = e^y \implies \varphi(y) = e^y$$

Therefore $f(x,y) = \tfrac{1}{3}x^3 + xy + e^y,$ and a one-parameter family of solutions is:

$$\tfrac{1}{3}x^3 + xy + e^y = C$$

Setting $x = 1,\ y = 0$, we get $C = \tfrac{4}{3}$ and

$$\tfrac{1}{3} x^3 + xy + e^y = \tfrac{4}{3} \quad \text{or} \quad x^3 + 3xy + 3e^y = 4$$

21. $\dfrac{\partial P}{\partial y} = 4y$ and $\dfrac{\partial Q}{\partial x} = 2y;$ the equation is not exact.

Since $\quad \dfrac{1}{Q}\left(\dfrac{\partial P}{\partial y} - \dfrac{\partial Q}{\partial x}\right) = \dfrac{1}{2xy}(2y) = \dfrac{1}{x}, \quad \mu(x) = e^{\int (1/x)\, dx} = e^{\ln x} = x \quad$ is an

integrating factor. Multiplying the given equation by x, we get

$$(2xy^2 + x^3 + 2x) + (2x^2 y)\, y' = 0$$

$\dfrac{\partial f}{\partial y} = 2x^2 y \quad \Longrightarrow \quad f(x,y) = x^2 y^2 + \varphi(x)$

$\dfrac{\partial f}{\partial x} = 2xy^2 + \varphi'(x) = 2xy^2 + x^3 + 2x \quad \Longrightarrow \quad \varphi'(x) = x^3 + 2x \quad \Longrightarrow \quad \varphi = \tfrac{1}{4} x^4 + x^2$

Therefore $\quad f(x,y) = x^2 y^2 + \tfrac{1}{4} x^4 + x^2,\quad$ and a one-parameter family of solutions is:

$$x^2 y^2 + \tfrac{1}{4} x^4 + x^2 = C$$

Setting $x = 1,\ y = 0,$ we get $C = \tfrac{5}{4}$ and

$$x^2 y^2 + \tfrac{1}{4} x^4 + x^2 = \tfrac{5}{4} \quad \text{or} \quad 4x^2 y^2 + x^4 + 4x^2 = 5$$

23. $\dfrac{\partial P}{\partial y} = 3y^2$ and $\dfrac{\partial Q}{\partial x} = y^2;$ the equation is not exact.

Since $\quad \dfrac{1}{P}\left(\dfrac{\partial P}{\partial y} - \dfrac{\partial Q}{\partial x}\right) = \dfrac{1}{y^3}(2y^2) = \dfrac{2}{y}, \quad \mu(y) = e^{-\int (2/y)\, dy} = e^{-2\ln y} = y^{-2} \quad$ is an

integrating factor. Multiplying the given equation by y^{-2}, we get

$$y + (y^{-2} + x)\, y' = 0$$

$\dfrac{\partial f}{\partial x} = y \quad \Longrightarrow \quad f(x,y) = xy + \varphi(y)$

$\dfrac{\partial f}{\partial y} = x + \varphi'(y) = y^{-2} + x \quad \Longrightarrow \quad \varphi'(y) = y^{-2} \quad \Longrightarrow \quad \varphi(y) = -\dfrac{1}{y}$

Therefore $\quad f(x,y) = xy - \dfrac{1}{y},\quad$ and a one-parameter family of solutions is:

$$xy - \dfrac{1}{y} = C$$

Setting $x = -2,\ y = -1,$ we get $C = 3$ and the solution $xy - \dfrac{1}{y} = 3.$

25. $\dfrac{\partial P}{\partial y} = -2y\sinh(x - y^2) = \dfrac{\partial Q}{\partial x};$ the equation is exact.

$\dfrac{\partial f}{\partial x} = \cosh(x - 2y^2) + e^{2x} \quad \Longrightarrow \quad f(x,y) = \sinh(x - y^2) + \tfrac{1}{2} e^{2x} + \varphi(y)$

$$\frac{\partial f}{\partial y} = -2y\,\cosh(x - y^2) + \varphi'(y) = y - 2y\,\cosh(x - y^2) \quad \Longrightarrow \quad \varphi'(y) = y \quad \Longrightarrow \quad \varphi(y) = \tfrac{1}{2}\,y^2$$

Therefore $f(x, y) = \sinh(x - y^2) + \tfrac{1}{2}\,e^{2x} + \tfrac{1}{2}\,y^2,$ and a one-parameter family of solutions is:

$$\sinh(x - y^2) + \tfrac{1}{2}\,e^{2x} + \tfrac{1}{2}\,y^2 = C$$

Setting $x = 2$, $y = \sqrt{2}$, we get $C = \tfrac{1}{2}e^4 + 1$ and the solution

$$\sinh(x - y^2) + \tfrac{1}{2}\,e^2 x + \tfrac{1}{2}\,y^2 = \tfrac{1}{2}e^4 + 1$$

27. (a) $\dfrac{\partial P}{\partial y} = 2xy + kx^2$ and $\dfrac{\partial Q}{\partial x} = 2xy + 3x^2 \quad \Longrightarrow \quad k = 3.$

   (b) $\dfrac{\partial P}{\partial y} = e^{2xy} + 2xye^{2xy}$ and $\dfrac{\partial Q}{\partial x} = ke^{2xy} + 2kxye^{2xy} \quad \Longrightarrow \quad k = 1.$

29. $y' = y^2 x^3$;   the equation is separable.

$$y^{-2}\,dy = x^3\,dx \quad \Longrightarrow \quad -\frac{1}{y} = \tfrac{1}{4}\,x^4 + C \quad \Longrightarrow \quad y = \frac{-4}{x^4 + C}$$

31. $y' + \dfrac{4}{x}\,y = x^4$;   the equation is linear.

$$H(x) = \int (4/x)\,dx = 4\ln x = \ln x^4, \quad \text{integrating factor:} \quad e^{\ln x^4} = x^4$$

$$x^4 y' + 4x^3 y = x^8$$

$$\frac{d}{dx}\left[x^4 y\right] = x^8$$

$$x^4 y = \tfrac{1}{9}\,x^9 + C$$

$$y = \tfrac{1}{9}\,x^5 + Cx^{-4}$$

33. $\dfrac{\partial P}{\partial y} = e^{xy} + xye^{xy} = \dfrac{\partial Q}{\partial x};$ the equation is exact.

$$\frac{\partial f}{\partial x} = ye^{xy} - 2x \quad \Longrightarrow \quad f(x, y) = e^{xy} - x^2 + \varphi(y)$$

$$\frac{\partial f}{\partial y} = xe^{xy} + \varphi'(y) = \frac{2}{y} + xe^{xy} \quad \Longrightarrow \quad \varphi'(y) = \frac{2}{y} \quad \Longrightarrow \quad \varphi(y) = 2\ln|y|$$

Therefore $f(x, y) = e^{xy} - x^2 + 2\ln|y|,$ and a one-parameter family of solutions is:

$$e^{xy} - x^2 + 2\ln|y| = C$$

CHAPTER 16

SECTION 16.1

1. $\displaystyle\sum_{i=1}^{3}\sum_{j=1}^{3} 2^{i-1}3^{j+1} = \left(\sum_{i=1}^{3} 2^{i-1}\right)\left(\sum_{j=1}^{3} 3^{j+1}\right) = (1+2+4)(9+27+81) = 819$

3. $\displaystyle\sum_{i=1}^{4}\sum_{j=1}^{3}(i^2+3i)(j-2) = \left[\sum_{i=1}^{4}(i^2+31)\right]\left[\sum_{j=1}^{3}(j-2)\right] = (4+10+18+28)(-1+0+1) = 0$

5. $\displaystyle\sum_{i=1}^{m}\Delta x_i = \Delta x_1 + \Delta x_2 + \cdots + \Delta x_n = (x_1 - x_0) + (x_2 - x_1) + \cdots + (x_n - x_{n-1})$

$$= x_n - x_0 = a_2 - a_1$$

7. $\displaystyle\sum_{i=1}^{m}\sum_{j=1}^{n}\Delta x_i\,\Delta y_j = \left(\sum_{i=1}^{m}\Delta x_i\right)\left(\sum_{j=1}^{n}\Delta y_j\right) = (a_2 - a_1)(b_2 - b_1)$

9. $\displaystyle\sum_{i=1}^{m}(x_i + x_{i-1})\,\Delta x_i = \sum_{i=1}^{m}(x_i + x_{i-1})(x_i - x_{i-1}) = \sum_{i=1}^{m}\left(x_i{}^2 - x_{i-1}^2\right)$

$$= x_m{}^2 - x_0{}^2 = a_2{}^2 - a_1{}^2$$

11. $\displaystyle\sum_{i=1}^{m}\sum_{j=1}^{n}(x_i + x_{i-1})\,\Delta x_i\,\Delta y_j = \left(\sum_{i=1}^{m}(x_i + x_{i-1})\,\Delta x_i\right)\left(\sum_{j=1}^{n}\Delta y_j\right)$

    (Exercise 9)

$$= \left(a_2{}^2 - a_1{}^2\right)(b_2 - b_1)$$

13. $\displaystyle\sum_{i=1}^{m}\sum_{j=1}^{n}(2\Delta x_i - 3\Delta y_j) = 2\left(\sum_{i=1}^{m}\Delta x_i\right)\left(\sum_{j=1}^{n}1\right) - 3\left(\sum_{i=1}^{m}1\right)\left(\sum_{j=1}^{n}\Delta y_j\right)$

$$= 2n(a_2 - a_1) - 3m(b_2 - b_1)$$

15. $\displaystyle\sum_{i=1}^{m}\sum_{j=1}^{n}\sum_{k=1}^{q}\Delta x_i\,\Delta y_j\,\Delta z_k = \left(\sum_{i=1}^{m}\Delta x_i\right)\left(\sum_{j=1}^{n}\Delta y_j\right)\left(\sum_{k=1}^{q}\Delta z_k\right)$

$$= (a_2 - a_1)(b_2 - b_1)(c_2 - c_1)$$

17. $\displaystyle\sum_{i=1}^{n}\sum_{j=1}^{n}\sum_{k=1}^{n}\delta_{ijk}a_{ijk} = a_{111} + a_{222} + \cdots + a_{nnn} = \sum_{p=1}^{n} a_{ppp}$

19. $\displaystyle\sum_{i=1}^{m}\sum_{j=1}^{n}\alpha a_{ij} = \sum_{i=1}^{m}\alpha\left(\sum_{j=1}^{n}a_{ij}\right) = \alpha\sum_{i=1}^{m}\sum_{j=1}^{n}a_{ij}$

SECTION 16.2

1. $L_f(P) = 2\frac{1}{4}, \quad U_f(P) = 5\frac{3}{4}$

3. (a) $\displaystyle L_f(P) = \sum_{i=1}^{m}\sum_{j=1}^{n}(x_{i-1} + 2y_{j-1})\,\Delta x_i\,\Delta y_j, \quad U_f(P) = \sum_{i=1}^{m}\sum_{j=1}^{n}(x_i + 2y_j)\,\Delta x_i\,\Delta y_j$

(b) $\displaystyle L_f(P) \leq \sum_{i=1}^{m}\sum_{j=1}^{n}\left[\frac{x_{i-1}+x_i}{2} + 2\left(\frac{y_{j-1}+y_j}{2}\right)\right]\Delta x_i\,\Delta y_j \leq U_f(P).$

The middle expression can be written

$$\sum_{i=1}^{m}\sum_{j=1}^{n}\frac{1}{2}\left(x_i^2 - x_{i-1}^2\right)\Delta y_j + \sum_{i=1}^{m}\sum_{j=1}^{n}\left(y_j^2 - y_{j-1}^2\right)\Delta x_i.$$

The first double sum reduces to

$$\sum_{i=1}^{m}\sum_{j=1}^{n}\frac{1}{2}\left(x_i^2 - x_{i-1}^2\right)\Delta y_j = \frac{1}{2}\left(\sum_{i=1}^{m}\left(x_i^2 - x_{i-1}^2\right)\right)\left(\sum_{j=1}^{n}\Delta y_j\right) = \frac{1}{2}\left(4-0\right)\left(1-0\right) = 2.$$

In like manner the second double sum also reduces to 2. Thus, $I = 4$; the volume of the prism bounded above by the plane $z = x + 2y$ and below by R.

5. $L_f(P) = -7/24, \quad U_f(P) = 7/24$

7. (a) $\displaystyle L_f(P) = \sum_{i=1}^{m}\sum_{j=1}^{n}(4x_{i-1}\,y_{j-1})\,\Delta x_i\,\Delta y_j, \quad U_f(P) = \sum_{i=1}^{m}\sum_{j=1}^{n}(4x_i\,y_j)\,\Delta x_i\,\Delta y_j$

(b) $\displaystyle L_f(P) \leq \sum_{i=1}^{m}\sum_{j=1}^{n}(x_i + x_{i-1})(y_j + y_{j-1})\,\Delta x_1\,\Delta y_j \leq U_f(P).$

The middle expression can be written

$$\sum_{i=1}^{m}\sum_{j=1}^{n}\left(x_i{}^2 - x_{i-1}^2\right)\left(y_j{}^2 - y_{j-1}^2\right) = \left(\sum_{i=1}^{m} x_i{}^2 - x_{i-1}^2\right)\left(\sum_{j=1}^{n} y_j{}^2 - y_{j-1}^2\right)$$

by (16.1.5)

$$= \left(b^2 - 0^2\right)\left(d^2 - 0^2\right) = b^2 d^2.$$

It follows that $I = b^2 d^2$.

9. (a) $\displaystyle L_f(P) = \sum_{i=1}^{m}\sum_{j=1}^{n} 3\left(x_{i-1}^2 - y_j{}^2\right)\Delta x_i\,\Delta y_j, \quad U_f(P) = \sum_{i=1}^{m}\sum_{j=1}^{n} 3\left(x_i{}^2 - y_{j-1}^2\right)\Delta x_i\,\Delta y_j$

(b) $\displaystyle L_f(P) \le \sum_{i=1}^{m}\sum_{j=1}^{n}\left[\left(x_i{}^2 + x_i x_{i-1} + x_{i-1}^2\right) - \left(y_j{}^2 + y_j y_{j-1} + y_{j-1}^2\right)\right]\Delta x_i\,\Delta y_j \le U_f(P).$

Since in general $\left(A^2 + AB + B^2\right)(A - B) = A^3 - B^3,$ the middle expression can be written

$$\sum_{i=1}^{m}\sum_{j=1}^{n}\left(x_i{}^3 - x_{i-1}^3\right)\Delta y_j - \sum_{i=1}^{m}\sum_{j=1}^{n}\left(y_j{}^3 - y_{j-1}^3\right)\Delta x_i,$$

which reduces to

$$\left(\sum_{i=1}^{m} x_i{}^3 - x_{i-1}^3\right)\left(\sum_{j=1}^{n}\Delta y_j\right) - \left(\sum_{i=1}^{m}\Delta x_i\right)\left(\sum_{j=1}^{n} y_j{}^3 - y_{j-1}^3\right).$$

This can be evaluated as $b^3 d - b d^3 = b d\left(b^2 - d^2\right).$ It follows that $I = b d\left(b^2 - d^2\right).$

11. $\displaystyle \iint_{\Omega} dx\,dy = \int_a^b \phi(x)\,dx$

13. Suppose $f(x_0, y_0) \ne 0$. Assume $f(x_0, y_0) > 0$. Since f is continuous, there exists a disc Ω_ϵ with radius ϵ centered at (x_0, y_0) such that $f(x, y) > 0$ on Ω_ϵ. Let R be a rectangle contained in Ω_ϵ. Then $\displaystyle \iint_R f(x, y)\,dx\,dy > 0$, which contradicts the hypothesis.

15. By Exercise 7, $\displaystyle \iint_R 4xy\,dx\,dy = 2^2 3^2 = 36.$ Thus

$$f_{avg} = \frac{1}{\text{area}(R)}\iint_R 4xy\,dx\,dy = \frac{1}{6}(36) = 6$$

17. By Theorem 16.2.10, there exists a point $(x_1, y_1) \in D_r$ such that

$$\iint_{D_r} f(x, y)\,dx\,dy = f(x_1, y_1)\iint_R dx\,dy = f(x_1, y_1)\pi r^2 \quad\Longrightarrow\quad f(x_1, y_1) = \frac{1}{\pi r^2}\iint_{D_r} f(x, y)\,dx\,dy$$

As $r \to 0$, $(x_1, y_1) \to (x_0, y_0)$ and $f(x_1, y_1) \to f(x_0, y_0)$ since f is continuous. The result follows.

SECTION 16.3

1. $\displaystyle\int_0^1\int_0^3 x^2\,dy\,dx = \int_0^1 3x^2\,dx = 1$

3. $\displaystyle\int_0^1\int_0^3 xy^2\,dy\,dx = \int_0^1 x\left[\frac{1}{3}y^3\right]_0^3 dx = \int_0^1 9x\,dx = \frac{9}{2}$

5. $\displaystyle\int_0^1\int_0^x xy^3\,dy\,dx = \int_0^1 x\left[\frac{1}{4}y^4\right]_0^x dx = \int_0^1 \frac{1}{4}x^5\,dx = \frac{1}{24}$

7. $\displaystyle\int_0^{\pi/2}\int_0^{\pi/2}\sin(x+y)\,dy\,dx = \int_0^{\pi/2}[-\cos(x+y)]_0^{\pi/2}\,dx = \int_0^{\pi/2}\left[\cos x - \cos\left(x+\frac{\pi}{2}\right)\right]dx = 2$

9. $\displaystyle\int_0^{\pi/2}\int_0^{\pi/2}(1+xy)\,dy\,dx = \int_0^{\pi/2}\left[y+\frac{1}{2}xy^2\right]_0^{\pi/2}dx = \int_0^{\pi/2}\left(\frac{1}{2}\pi+\frac{1}{8}\pi^2 x\right)dx = \frac{1}{4}\pi^2+\frac{1}{64}\pi^4$

11. $\displaystyle\int_0^1\int_{y^2}^y \sqrt{xy}\,dx\,dy = \int_0^1\sqrt{y}\left[\frac{2}{3}x^{3/2}\right]_{y^2}^y dy = \int_0^1\frac{2}{3}\left(y^2-y^{7/2}\right)dy = \frac{2}{27}$

13. $\displaystyle\int_{-2}^2\int_{\frac{1}{2}y^2}^{4-\frac{1}{2}y^2}\left(4-y^2\right)dx\,dy = \int_{-2}^2\left(4-y^2\right)\left[\left(4-\frac{1}{2}y^2\right)-\left(\frac{1}{2}y^2\right)\right]dy$

$$= 2\int_0^2\left(16-8y^2+y^4\right)dy = \frac{512}{15}$$

15. 0 by symmetry (integrand odd in y, Ω symmetric about x-axis)

17. $\displaystyle\int_0^2\int_0^{x/2}e^{x^2}\,dy\,dx = \int_0^2\frac{1}{2}xe^{x^2}\,dx = \left[\frac{1}{4}e^{x^2}\right]_0^2 = \frac{1}{4}\left(e^4-1\right)$

19.

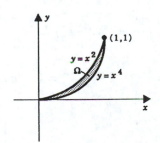

21.

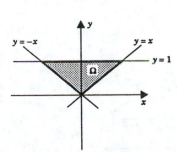

$\displaystyle\int_0^1\int_{y^{1/2}}^{y^{1/4}} f(x,y)\,dx\,dy$

$\displaystyle\int_{-1}^0\int_{-x}^1 f(x,y)\,dy\,dx + \int_0^1\int_x^1 f(x,y)\,dy\,dx$

23.

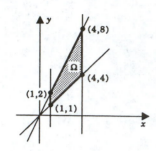

$$\int_1^2 \int_1^y f(x,y)\,dx\,dy + \int_2^4 \int_{y/2}^y f(x,y)\,dx\,dy$$

$$+ \int_4^8 \int_{y/2}^4 f(x,y)\,dx\,dy$$

25. $\displaystyle \int_{-2}^4 \int_{1/4x^2}^{\frac{1}{2}x+2} dy\,dx = \int_{-2}^4 \left[\frac{1}{2}x + 2 - \frac{1}{4}x^2\right] dx = 9$

27. $\displaystyle \int_0^{1/4} \int_{2y^{3/2}}^y dx\,dy = \int_0^{1/4} \left[y - 2y^{3/2}\right] dy = \frac{1}{160}$

29.

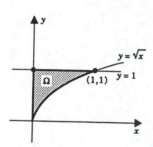

$$\int_0^1 \int_0^{y^2} \sin\left(\frac{y^3+1}{2}\right) dx\,dy = \int_0^1 y^2 \sin\left(\frac{y^3+1}{2}\right) dy$$

$$= \left[-\frac{2}{3}\cos\left(\frac{y^3+1}{2}\right)\right]_0^1$$

$$= \tfrac{2}{3}\left(\cos\tfrac{1}{2} - \cos 1\right)$$

31.

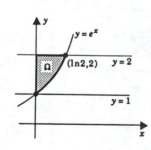

$$\int_0^{\ln 2} \int_{e^x}^2 e^{-x}\,dy\,dx = \int_0^{\ln 2} e^{-x}(2 - e^x)\,dx$$

$$= \left[-2e^{-x} - x\right]_0^{\ln 2} = 1 - \ln 2$$

33. $\displaystyle \int_1^2 \int_{y-1}^{2/y} dx\,dy = \int_1^2 \left[\frac{2}{y} - (y-1)\right] dy = \ln 4 - \frac{1}{2}$

35. $\displaystyle \int_0^2 \int_0^{3-\frac{3}{2}x} \left(4 - 2x - \frac{4}{3}y\right) dy\,dx = \int_0^3 \int_0^{2-\frac{2}{3}y} \left(4 - 2x - \frac{4}{3}y\right) dx\,dy = 4$

37. $\displaystyle \int_0^2 \int_0^{1-\frac{1}{2}x} x^3 y\,dy\,dx = \int_0^2 \int_0^{2-2y} x^3 y\,dx\,dy = \frac{2}{15}$

39.
$$\int_0^2 \int_{-\sqrt{2x-x^2}}^{\sqrt{2x-x^2}} (2x+1) \, dy \, dx = \int_{-1}^1 \int_{1-\sqrt{1-y^2}}^{1+\sqrt{1-y^2}} (2x+1) \, dx \, dy$$

$$= \int_{-1}^1 \left[x^2 + x \right]_{1-\sqrt{1-y^2}}^{1+\sqrt{1-y^2}} dy$$

$$= 6 \int_{-1}^1 \sqrt{1-y^2} \, dy = 6 \left(\frac{\pi}{2} \right) = 3\pi$$

41.
$$\int_0^1 \int_0^{1-x} (x^2 + y^2) \, dy \, dx = \int_0^1 \left(2x^2 - \frac{4}{3}x^3 - x + \frac{1}{3} \right) dx = \frac{1}{6}$$

43.
$$\int_0^1 \int_{x^2}^x (x^2 + 3y^2) \, dy \, dx = \int_0^1 (2x^3 - x^4 - x^6) \, dx = \frac{11}{70}$$

45.
$$\int_0^a \int_0^{\sqrt{a^2-x^2}} \sqrt{a^2 - x^2} \, dy \, dx = \int_0^a (a^2 - x^2) \, dx \, dy = \frac{2}{3} a^2$$

47.
$$\int_0^1 \int_y^1 e^{y/x} \, dx \, dy = \int_0^1 \int_0^x e^{y/x} \, dy \, dx = \int_0^1 \left[xe^{y/x} \right]_0^x dx = \int_0^1 x(e-1) \, dx = \frac{1}{2}(e-1)$$

49.
$$\int_0^1 \int_x^1 x^2 e^{y^4} \, dy \, dx = \int_0^1 \int_0^y x^2 e^{y^4} \, dx \, dy = \int_0^1 \left[\frac{1}{3} x^3 e^{y^4} \right]_0^y dy = \frac{1}{3} \int_0^1 y^3 e^{y^4} \, dy = \frac{1}{12}(e-1)$$

51.
$$f_{avg} = \frac{1}{8} \int_{-1}^1 \int_0^4 x^2 y \, dy \, dx = \frac{1}{8} \int_{-1}^1 8x^2 \, dx = \int_{-1}^1 x^2 \, dx = \frac{2}{3}$$

53.
$$f_{avg} = \frac{1}{(\ln 2)^2} \int_{\ln 2}^{2\ln 2} \int_{\ln 2}^{2\ln 2} \frac{1}{xy} \, dy \, dx = \frac{1}{(\ln 2)^2} \int_{\ln 2}^{2\ln 2} \frac{1}{x} \ln 2 \, dx = 1$$

55.
$$\iint_R f(x)g(y) \, dx dy = \int_c^d \int_a^b f(x)g(y) \, dx \, dy = \int_c^d \left(\int_a^b f(x)g(y) \, dx \right) dy$$

$$= \int_c^d g(y) \left(\int_a^b f(x) \, dx \right) dy = \left(\int_a^b f(x) \, dx \right) \left(\int_c^d g(y) \, dy \right)$$

57. We have $R: \ -a \le x \le a, \quad c \le y \le d.$ Set $f(x,y) = g_y(x).$ For each fixed $y \in [c,d]$, g_y is an odd function. Thus

$$\int_{-a}^a g_y(x) \, dx = 0. \qquad (5.8.8)$$

Therefore

$$\iint_R f(x,y)\,dxdy = \int_c^d \int_{-a}^a f(x,y)\,dx\,dy$$

$$= \int_c^d \int_{-a}^a g_y(x)\,dx\,dy = \int_c^d 0\,dy = 0.$$

59. Note that $\Omega = \{(x,y): 0 \le x \le y, \quad 0 \le y \le 1\}.$

Set $\Omega' = \{(x,y): 0 \le y \le x, \quad 0 \le x \le 1\}.$

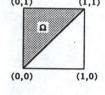 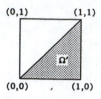

$$\iint_\Omega f(x)f(y)\,dxdy = \int_0^1 \int_0^y f(x)f(y)\,dx\,dy$$

$$= \int_0^1 \int_0^x f(y)f(x)\,dy\,dx$$

x and y are dummy variables

$$= \int_0^1 \int_0^x f(x)f(y)\,dy\,dx = \iint_{\Omega'} f(x)f(y)\,dxdy.$$

Note that Ω and Ω' don't overlap and their union is the unit square

$$R = \{(x,y): 0 \le x \le 1, \quad 0 \le y \le 1\}.$$

If $\int_0^1 f(x)\,dx = 0,$ then

$$0 = \left(\int_0^1 f(x)\,dx\right)\left(\int_0^1 f(y)\,dy\right) = \iint_R f(x)f(y)\,dxdy$$

by Exercise 55

$$= \iint_\Omega f(x)f(y)\,dxdy + \iint_{\Omega'} f(x)f(y)\,dxdy$$

$$= 2\iint_\Omega f(x)f(y)\,dxdy$$

and therefore $\iint_\Omega f(x)f(y)\,dxdy = 0.$

61. Let M be the maximum value of $|f(x,y)|$ on Ω.

$$\int_{\phi_1(x+h)}^{\phi_2(x+h)} = \int_{\phi_1(x+h)}^{\phi_1(x)} + \int_{\phi_1(x)}^{\phi_2(x)} + \int_{\phi_2(x)}^{\phi_2(x+h)}$$

$$|F(x+h) - F(x)| = \left| \int_{\phi_1(x+h)}^{\phi_2(x+h)} f(x,y)\,dy - \int_{\phi_1(x)}^{\phi_2(x)} f(x,y)\,dy \right|$$

$$= \left| \int_{\phi_1(x+h)}^{\phi_1(x)} f(x,y)\,dy + \int_{\phi_2(x)}^{\phi_2(x+h)} f(x,y)\,dy \right|$$

$$\leq \left| \int_{\phi_1(x+h)}^{\phi_1(x)} f(x,y)\,dy \right| + \left| \int_{\phi_2(x)}^{\phi_2(x+h)} f(x,y)\,dy \right|$$

$$\leq |\phi_1(x) - \phi_1(x+h)|\,M + |\phi_2(x+h) - \phi_2(x)|\,M.$$

The expression on the right tends to 0 as h tends to 0 since ϕ_1 and ϕ_2 are continuous.

PROJECT 16.3

1.   (a) $M_{32} = \frac{2-0}{3} \cdot \frac{3-0}{2} \left[f(\frac{1}{3}, \frac{3}{4}) + f(1, \frac{3}{4}) + f(\frac{5}{3}, \frac{3}{4}) + f(\frac{1}{3}, \frac{9}{4}) + f(1, \frac{9}{4}) + f(\frac{5}{3}, \frac{9}{4}) \right] = \frac{144}{6} = 24.$

   (c) $\displaystyle\int_0^3 \int_0^2 (x + 2y)\,dx\,dy = 24$

3.   (a) $M_{32} = \frac{\frac{\pi}{2}-0}{2} \cdot \frac{\frac{\pi}{2}-0}{2} \left[f(\frac{\pi}{8}, \frac{\pi}{8}) + f(\frac{3\pi}{8}, \frac{\pi}{8}) + f(\frac{\pi}{8}, \frac{3\pi}{8}) + f(\frac{3\pi}{8}, \frac{3\pi}{8}) \right] = \frac{\pi^2}{16}(1.70710) \simeq 1.05303.$

   (c) $\displaystyle\int_0^{\frac{\pi}{2}} \int_0^{\frac{\pi}{2}} \sin x \sin y \, dx\,dy = 1$

5.   (a) Applying the formula in Problem 4, we get

$$T_{32} = \frac{2-0}{3} \cdot \frac{3-0}{2} \cdot \frac{1}{4} \left[f(0,0) + 2f(\tfrac{2}{3},0) + 2f(\tfrac{4}{3},0) + f(2,0) + 2f(0,\tfrac{3}{2}) + 4f(\tfrac{2}{3},\tfrac{3}{2}) + 4f(\tfrac{4}{3},\tfrac{3}{2}) + 2f(2,\tfrac{3}{2}) + \right.$$

$$\left. f(0,3) + 2f(\tfrac{2}{3},3) + 2f(\tfrac{4}{3},3) + f(2,3) \right] = \tfrac{1}{4}(96) = 24$$

SECTION 16.4

1. $\displaystyle\int_0^{\pi/2} \int_0^{\sin\theta} r \cos\theta \, dr\,d\theta = \int_0^{\pi/2} \frac{1}{2} \sin^2\theta \cos\theta \, d\theta = \left[\frac{1}{6} \sin^3\theta \right]_0^{\pi/2} = \frac{1}{6}$

3. $\displaystyle\int_0^{\pi/2} \int_0^{3\sin\theta} r^2 \, dr\,d\theta = \int_0^{\pi/2} 9\sin^3\theta \, d\theta = 9\int_0^{\pi/2}(1 - \cos^2\theta)\sin\theta \, d\theta = 9\left[-\cos\theta + \frac{1}{3}\cos^3\theta \right]_0^{\pi/2} = 6$

5.   (a) $\Gamma: \; 0 \leq \theta \leq 2\pi, \quad 0 \leq r \leq 1$

$$\iint\limits_{\Gamma} (\cos r^2)r\,dr d\theta = \int_0^{2\pi}\int_0^1 (\cos r^2)r\,dr\,d\theta$$

$$= 2\pi \int_0^1 r\cos r^2\,dr = \pi\,\sin 1 \cong 0.84\,\pi$$

(b) $\Gamma:\ 0 \le \theta \le 2\pi,\quad 1 \le r \le 2$

$$\iint\limits_{\Gamma} (\cos r^2)r\,dr d\theta = \int_0^{2\pi}\int_1^2 (\cos r^2)r\,dr\,d\theta$$

$$= 2\pi \int_1^2 r\cos r^2\,dr = \pi(\sin 4 - \sin 1) \cong -1.60\pi$$

7.   (a) $\Gamma:\ 0 \le \theta \le \pi/2,\quad 0 \le r \le 1$

$$\iint\limits_{\Gamma} (r\cos\theta + r\sin\theta)r\,dr d\theta = \int_0^{\pi/2}\int_0^1 r^2(\cos\theta + \sin\theta)\,dr\,d\theta$$

$$= \left(\int_0^{\pi/2}(\cos\theta + \sin\theta)\,d\theta\right)\left(\int_0^1 r^2\,dr\right) = 2\left(\frac{1}{3}\right) = \frac{2}{3}$$

(b) $\Gamma:\ 0 \le \theta \le \pi/2,\quad 1 \le r \le 2$

$$\iint\limits_{\Gamma} (r\cos\theta + r\sin\theta)r\,dr d\theta = \int_0^{\pi/2}\int_1^2 r^2(\cos\theta + \sin\theta)\,dr\,d\theta$$

$$= \left(\int_0^{\pi/2}(\cos\theta + \sin\theta)\,d\theta\right)\left(\int_1^2 r^2\,dr\right) = 2\left(\frac{7}{3}\right) = \frac{14}{3}$$

9. $\displaystyle\int_{-\pi/2}^{\pi/2}\int_0^1 r^2\,dr\,d\theta = \frac{1}{3}\,\pi$

11. $\displaystyle\int_0^1\int_0^{\sqrt{1-x^2}} (x^2 + y^2)^{\frac{3}{2}}\,dy\,dx = \int_0^{\frac{\pi}{2}}\int_0^1 (r^3)r\,dr\,d\theta = \int_0^{\frac{\pi}{2}}\int_0^1 r^4\,dr\,d\theta = \frac{\pi}{10}$

13. $\displaystyle\int_0^1\int_0^{\sqrt{1-x^2}} \sin(x^2 + y^2)\,dx\,dy = \int_0^{\pi/2}\int_0^1 \sin(r^2)\,r\,dr\,d\theta = \int_0^{\pi/2}\frac{1}{2}(1 - \cos 1)\,d\theta = \frac{\pi}{4}(1 - \cos 1)$

15. $\displaystyle\int_0^2\int_0^{\sqrt{2x-x^2}} x\,dx\,dy = \int_0^{\pi/2}\int_0^{2\cos\theta} r\cos\theta\,r\,dr\,d\theta = \frac{8}{3}\int_0^{\pi/2}\cos^4\theta\,d\theta = \frac{8}{3}\cdot\frac{3}{4}\cdot\frac{1}{2}\cdot\frac{\pi}{2} = \frac{\pi}{2}$

(See Exercise 62, Section 8.3)

17. $\displaystyle A = \int_0^{\pi/3}\int_0^{3\sin 3\theta} r\,dr\,d\theta = \frac{9}{2}\int_0^{\pi/3}\sin^2 3\theta\,d\theta = \frac{9}{4}\int_0^{\pi/3}(1 - 6\cos\theta)\,d\theta = \frac{3\pi}{4}$

19. First we find the points of intersection:

$$r = 4\cos\theta = 2 \quad \Longrightarrow \quad \cos\theta = \frac{1}{2}$$

$$\Longrightarrow \quad \theta = \pm\frac{\pi}{3}.$$

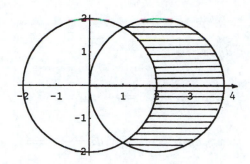

$$A = \int_{-\pi/3}^{\pi/3} \int_2^{4\cos\theta} r\, dr\, d\theta = \int_{-\pi/3}^{\pi/3} (8 - \cos^2\theta - 2)\, d\theta = \int_{-\pi/3}^{\pi/3} (2 + 4\cos 2\theta)\, d\theta = \frac{4\pi}{3} + 2\sqrt{3}$$

21. $\displaystyle A = 4\int_0^{\pi/4} \int_0^{2\sqrt{\cos 2\theta}} r\, dr\, d\theta = 8\int_0^{\pi/4} \cos 2\theta\, d\theta = 4$

23.
$$\int_0^{2\pi} \int_0^b (r^2\sin\theta + br)\, dr\, d\theta = \int_0^{2\pi} \left[\frac{1}{3}r^3\sin\theta + \frac{b}{2}r^2\right]_0^b d\theta$$

$$= b^3\int_0^{2\pi} \left(\frac{1}{3}\sin\theta + \frac{1}{2}\right)\, d\theta = b^3\pi$$

25.
$$8\int_0^{\pi/2} \int_0^2 \frac{r}{2}\sqrt{12 - 3r^2}\, dr\, d\theta = 8\int_0^{\pi/2} \left[-\frac{1}{18}\left(12 - 3r^2\right)^{3/2}\right]_0^2 d\theta$$

$$= 8\int_0^{\pi/2} \frac{4}{3}\sqrt{3}\, d\theta = \frac{16}{3}\sqrt{3}\,\pi$$

27.
$$\int_0^{2\pi} \int_0^1 r\sqrt{4 - r^2}\, dr\, d\theta = \int_0^{2\pi} \left[-\frac{1}{3}\left(4 - r^2\right)^{3/2}\right]_0^1 d\theta$$

$$= \int_0^{2\pi} \left(\frac{8}{3} - \sqrt{3}\right)\, d\theta = \frac{2}{3}(8 - 3\sqrt{3})\pi$$

29.
$$\int_{-\pi/2}^{\pi/2} \int_0^{2\cos\theta} 2r^2\cos\theta\, dr\, d\theta = \int_{-\pi/2}^{\pi/2} \left[\frac{2}{3}r^3\cos\theta\right]_0^{2\cos\theta} d\theta$$

$$= \int_{-\pi/2}^{\pi/2} \frac{16}{3}\cos^4\theta\, d\theta = \frac{32}{3}\int_0^{\pi/2} \cos^4\theta\, d\theta = \frac{32}{3}\left(\frac{3}{16}\pi\right) = 2\pi$$

Ex. 46, Sect. 8.3

31.
$$\frac{b}{a}\int_0^{\pi} \int_0^{a\sin\theta} r\sqrt{a^2 - r^2}\, dr\, d\theta = \frac{b}{a}\int_0^{\pi} \left[-\frac{1}{3}\left(a^2 - r^2\right)^{3/2}\right]_0^{a\sin\theta} d\theta$$

$$= \frac{1}{3}a^2 b\int_0^{\pi} (1 - \cos^3\theta)\, d\theta = \frac{1}{3}\pi a^2 b$$

SECTION 16.5

1. $M = \int_{-1}^{1} \int_0^1 x^2 \, dy \, dx = \dfrac{2}{3}$

$x_M M = \int_{-1}^{1} \int_0^1 x^3 \, dy \, dx = 0 \implies x_M = 0$

$y_M M = \int_{-1}^{1} \int_0^1 x^2 y \, dy \, dx = \int_{-1}^{1} \dfrac{1}{2} x^2 \, dx = \dfrac{1}{3} \implies y_M = \dfrac{1/3}{1/2} = \dfrac{1}{2}$

3. $M = \int_0^1 \int_{x^2}^1 xy \, dy \, dx = \dfrac{1}{2} \int_0^1 (x - x^5) \, dx = \dfrac{1}{6}$

$x_M M = \int_0^1 \int_{x^2}^1 x^2 y \, dy \, dx \, \dfrac{1}{2} \int_0^1 (x^2 - x^6) \, dx = \dfrac{2}{21} \implies x_M = \dfrac{2/21}{1/6} = \dfrac{12}{21}$

$y_M M = \int_0^1 \int_{x^2}^1 xy^2 \, dy \, dx = \dfrac{1}{3} \int_0^1 (x - x^7) \, dx = \dfrac{1}{8} \implies y_M = \dfrac{1/8}{1/6} = \dfrac{3}{4}$

5. $M = \int_0^8 \int_0^{x^{1/3}} y^2 \, dy \, dx = \dfrac{1}{3} \int_0^8 x \, dx = \dfrac{32}{3}$

$x_M M = \int_0^8 \int_0^{x^{1/3}} xy^2 \, dy \, dx \, \dfrac{1}{3} \int_0^8 x^2 \, dx = \dfrac{512}{9} \implies x_M = \dfrac{512/9}{32/3} = \dfrac{16}{3}$

$y_M M = \int_0^8 \int_0^{x^{1/3}} y^3 \, dy \, dx = \dfrac{1}{4} \int_0^8 x^{4/3} \, dx = \dfrac{96}{7} \implies y_M = \dfrac{96/7}{32/3} = \dfrac{9}{7}$

7. $M = \int_0^1 \int_{2x}^{3x} xy \, dy \, dx = \dfrac{5}{2} \int_0^1 x^3 \, dx = \dfrac{5}{8}$

$x_M M = \int_0^1 \int_{2x}^{3x} x^2 y \, dy \, dx = \dfrac{5}{2} \int_0^1 x^4 \, dx = \dfrac{1}{2} \implies x_M = \dfrac{1/2}{5/8} = \dfrac{4}{5}$

$y_M M = \int_0^1 \int_{2x}^{3x} xy^2 \, dy \, dx = \dfrac{19}{3} \int_0^1 x^4 \, dx = \dfrac{19}{15} \implies y_M = \dfrac{19/15}{5/8} = \dfrac{152}{75}$

9. $M = \int_0^{2\pi} \int_0^{1+\cos\theta} r^2 \, dr \, d\theta = \dfrac{1}{3} \int_0^{2\pi} (1 + 3\cos\theta + 3\cos^2\theta + \cos^3\theta) \, d\theta = \dfrac{5\pi}{3}$

$x_M M = \int_0^{2\pi} \int_0^{1+\cos\theta} r^3 \cos\theta \, dr \, d\theta = \dfrac{1}{4} \int_0^{2\pi} (1 + \cos\theta)^4 \cos\theta \, d\theta$

$\qquad\qquad = \dfrac{1}{4} \int_0^{2\pi} \left[\cos\theta + 4\cos^2\theta + 6\cos^3\theta + 4\cos^4\theta + \cos^5\theta \right] d\theta$

$\qquad\qquad = \dfrac{7\pi}{4}$

Therefore, $x_M = \dfrac{7\pi/4}{5\pi/3} = \dfrac{21}{20}$.

$$y_M\, M = \int_0^{2\pi}\int_0^{1+\cos\theta} r^3 \sin\theta\, dr\, d\theta = \frac{1}{4}\int_0^{2\pi}(1+\cos\theta)^4 \sin\theta\, d\theta = \frac{1}{4}\left[\frac{1}{5}(1+\cos\theta)^5\right]_0^{2\pi} = 0$$

Therefore, $y_M = 0$.

11. $\quad \Omega:\ -L/2 \le x \le L/2, \quad -W/2 \le y \le W/2$

$$I_x = \iint_\Omega \frac{M}{LW} y^2\, dxdy = \frac{4M}{LW}\int_0^{W/2}\int_0^{L/2} y^2\, dx\, dy = \frac{1}{12}MW^2$$

symmetry

$$I_y = \iint_\Omega \frac{M}{LW} x^2\, dxdy = \frac{1}{12}ML^2, \quad I_z = \iint_\Omega \frac{M}{LW}(x^2+y^2)\, dxdy = \frac{1}{12}M(L^2+W^2)$$

$$K_x = \sqrt{I_x/M} = W/2\sqrt{3}, \quad K_y = \sqrt{I_y/M} = L/2\sqrt{3}$$

$$K_z = \sqrt{I_z/M} = \sqrt{L^2+W^2}\Big/2\sqrt{3}$$

13.
$$M = \iint_\Omega k\left(x+\frac{L}{2}\right) dxdy = \iint_\Omega \frac{1}{2}kL\, dxdy = \frac{1}{2}kL(\text{ area of }\Omega) = \frac{1}{2}kL^2W$$

symmetry

$$x_M\, M = \iint_\Omega x\left[k\left(x+\frac{L}{2}\right)\right] dxdy = \iint_\Omega \left(kx^2+\frac{1}{2}Lx\right) dxdy$$

$$= \iint_\Omega kx^2\, dxdy = 4k\int_0^{W/2}\int_0^{L/2} x^2\, dx\, dy = \frac{1}{12}kWL^3$$

symmetry symmetry

$$= \tfrac{1}{6}\left(\tfrac{1}{2}kL^2W\right)L = \tfrac{1}{6}ML; \quad x_M = \tfrac{1}{6}L$$

$$y_M\, M = \iint_\Omega y\left[k\left(x+\frac{L}{2}\right)\right] dxdy = 0; \quad y_M = 0$$

by symmetry

15.
$$I_x = \iint_\Omega \frac{4M}{\pi R^2} y^2 \, dx dy = \frac{4M}{\pi R^2} \int_0^{\pi/2} \int_0^R r^3 \sin^2 \theta \, dr \, d\theta$$

$$= \frac{4M}{\pi R^2} \left(\int_0^{\pi/2} \sin^2 \theta \, d\theta \right) \left(\int_0^R r^3 \, dr \right) = \frac{4M}{\pi R^2} \left(\frac{\pi}{4} \right) \left(\frac{1}{4} R^4 \right) = \frac{1}{4} M R^2$$

$$I_y = \frac{1}{4} M R^2, \quad I_z = \frac{1}{2} M R^2$$

$$K_x = K_y = \frac{1}{2} R, \quad K_z = R/\sqrt{2}$$

17. I_M, the moment of inertia about the vertical line through the center of mass, is

$$\iint_\Omega \frac{M}{\pi R^2} \left(x^2 + y^2 \right) dx dy$$

where Ω is the disc of radius R centered at the origin. Therefore

$$I_M = \frac{M}{\pi R^2} \int_0^{2\pi} \int_0^R r^3 \, dr \, d\theta = \frac{1}{2} M R^2.$$

We need $I_0 = \frac{1}{2} M R^2 + d^2 M$ where d is the distance from the center of the disc to the origin. Solving this equation for d, we have $d = \sqrt{I_0 - \frac{1}{2} M R^2} \Big/ \sqrt{M}$.

19. $\Omega: 0 \le x \le a, \quad 0 \le y \le b$

$$I_x = \iint_\Omega \frac{4M}{\pi ab} y^2 \, dx dy = \frac{4M}{\pi ab} \int_0^a \int_0^{\frac{b}{a}\sqrt{a^2-x^2}} y^2 \, dy \, dx = \frac{1}{4} M b^2$$

$$I_y = \iint_\Omega \frac{4M}{\pi ab} x^2 \, dx dy = \frac{4M}{\pi ab} \int_0^a \int_0^{\frac{b}{a}\sqrt{a^2-x^2}} x^2 \, dy \, dx = \frac{1}{4} M a^2$$

$$I_z = \frac{1}{4} M \left(a^2 + b^2 \right)$$

21. $I_x = \int_0^1 \int_{x^2}^1 xy^3 \, dy \, dx = \frac{1}{4} \int_0^1 (x - x^9) \, dx = \frac{1}{10}$

$$I_y = \int_0^1 \int_{x^2}^1 x^3 y \, dy \, dx \frac{1}{2} \int_0^1 (x^3 - x^7) \, dx = \frac{1}{16}$$

$$I_z = \int_0^1 \int_{x^2}^1 xy(x^2 + y^2) \, dy \, dx = I_x + I_y = \frac{13}{80}$$

23. $I_x = \int_0^{2\pi} \int_0^{1+\cos\theta} r^4 \sin^2\theta \, dr \, d\theta = \frac{1}{5} \int_0^{2\pi} (1 + \cos\theta)^5 \sin^2\theta \, d\theta = \frac{33\pi}{40}$

$$I_y = \int_0^{2\pi} \int_0^{1+\cos\theta} r^4 \cos^2\theta \, dr \, d\theta = \frac{1}{5} \int_0^{2\pi} (1+\cos\theta)^5 \cos^2\theta \, d\theta = \frac{93\pi}{40}$$

$$I_z = \int_0^{2\pi} \int_0^{1+\cos\theta} r^4 \, dr \, d\theta = I_x + I_y = \frac{63\pi}{20}$$

25. $\Omega: \quad r_1{}^2 \le x^2 + y^2 \le r_2{}^2, \quad A = \pi\left(r_2{}^2 - r_1{}^2\right)$ ´

(a) Place the diameter on the x-axis.

$$I_x = \iint_\Omega \frac{M}{A} y^2 \, dx \, dy = \frac{M}{A} \int_0^{2\pi} \int_{r_1}^{r_2} \left(r^2 \sin^2\theta\right) r \, dr \, d\theta = \frac{1}{4} M \left(r_2{}^2 + r_1{}^2\right)$$

(b) $\frac{1}{4}M\left(r_2{}^2 + r_1{}^2\right) + Mr_1{}^2 = \frac{1}{4}M\left(r_2{}^2 + 5r_1{}^2\right)$ (parallel axis theorem)

(c) $\frac{1}{4}M\left(r_2{}^2 + r_1{}^2\right) + Mr_2{}^2 = \frac{1}{4}M\left(5r_2{}^2 + r_1{}^2\right)$

27. $\Omega: \quad r_1{}^2 \le x^2 + y^2 \le r_2{}^2, \quad A = \pi\left(r_2{}^2 - r_1{}^2\right)$

$$I = \iint_\Omega \frac{M}{A}\left(x^2 + y^2\right) dx \, dy = \frac{M}{A} \int_0^{2\pi} \int_{r_1}^{r_2} r^3 \, dr \, d\theta = \frac{1}{2}M\left(r_2{}^2 + r_1{}^2\right)$$

29.

$$M = \iint_\Omega k\left(R - \sqrt{x^2 + y^2}\right) dx \, dy = k \int_0^\pi \int_0^R \left(Rr - r^2\right) dr \, d\theta = \frac{1}{6}k\pi R^3$$

$x_M = 0$ by symmetry

$$y_M M = \iint_\Omega y\left[k\left(R - \sqrt{x^2 + y^2}\right)\right] dx \, dy = k \int_0^\pi \int_0^R \left(Rr^2 - r^3\right) \sin\theta \, dr \, d\theta = \frac{1}{6}kR^4$$

$y_M = R/\pi$

31. Place P at the origin.

$$M = \iint\limits_{\Omega} k\sqrt{x^2 + y^2}\; dxdy$$

$$= k \int_0^{\pi} \int_0^{2R\sin\theta} r^2\; dr\; d\theta = \frac{32}{9}kR^3$$

$x_M = 0$ by symmetry

$$y_M M = \iint\limits_{\Omega} y\left(k\sqrt{x^2+y^2}\right) dxdy = k \int_0^{\pi} \int_0^{2R\sin\theta} r^3\sin\theta\; dr\; d\theta = \frac{64}{15}kR^4$$

$$y_M = 6R/5$$

Answer: the center of mass lies on the diameter through P at a distance $6R/5$ from P.

33. Suppose Ω, a basic region of area A, is broken up into n basic regions $\Omega_1, \cdots, \Omega_n$ with areas $A_1, \cdots, A_n$. Then

$$\overline{x}A = \iint\limits_{\Omega} x\; dxdy = \sum_{i=1}^{n}\left(\iint\limits_{\Omega_i} x\; dxdy\right) = \sum_{i=1}^{n} \overline{x}_i\, A_i = \overline{x}_1\, A_1 + \cdots + \overline{x}_n\, A_n.$$

The second formula can be derived in a similar manner.

SECTION 16.6

1. They are equal; they both give the volume of T.

3. $$\iiint\limits_{\Pi} \alpha\; dx\, dy\, dz = \alpha \iiint\limits_{\Pi} dx\, dy\, dz = \alpha\,[\text{volume}\,(\Pi)] = \alpha(a_2 - a_1)(b_2 - b_1)(c_2 - c_1)$$

5. Let $P_1 = \{x_0, \cdots, x_m\}$, $P_2 = \{y_0, \cdots, y_n\}$, $P_3 = \{z_0, \cdots, z_q\}$ be partitions of $[0, a]$, $[0, b]$, $[0, c]$ respectively and let $P = P_1 \times P_2 \times P_3$. Note that

$$x_{i-1}y_{j-1} \leq \left(\frac{x_i + x_{i-1}}{2}\right)\left(\frac{y_j + y_{j-1}}{2}\right) \leq x_i y_j$$

and therefore

$$x_{i-1}y_{j-1}\,\Delta x_i\,\Delta y_j\,\Delta z_k \leq \tfrac{1}{4}\left(x_i^2 - x_{i-1}^2\right)\left(y_j^2 - y_{j-1}^2\right)\Delta z_k \leq x_i y_j\,\Delta x_i\,\Delta y_j\Delta z_k.$$

It follows that

$$L_f(P) \leq \frac{1}{4}\sum_{i=1}^{m}\sum_{j=1}^{n}\sum_{k=1}^{q}\left(x_i^2 - x_{i-1}^2\right)\left(y_j^2 - y_{j-1}^2\right)\Delta z_k \leq U_f(P).$$

The middle term can be written

$$\frac{1}{4}\left(\sum_{i=1}^{m} x_i^2 - x_{i-1}^2\right)\left(\sum_{j=1}^{n} y_j^2 - y_{j-1}^2\right)\left(\sum_{k=1}^{q}\Delta z_k\right) = \frac{1}{4}a^2b^2c.$$

7. $\overline{x}_1 = a, \quad \overline{y}_1 = b, \quad \overline{z}_1 = c; \qquad \overline{x}_0 = A, \quad \overline{y}_0 = B, \quad \overline{z}_0 = C$

$$\overline{x}_1 V_1 + \overline{x} V = \overline{x}_0 V_0 \quad \Longrightarrow \quad a^2 bc + (ABC - abc)\,\overline{x} = A^2 BC$$

$$\Longrightarrow \quad \overline{x} = \frac{A^2 BC - a^2 bc}{ABC - abc}$$

similarly

$$\overline{y} = \frac{AB^2 C - ab^2 c}{ABC - abc}, \quad \overline{z} = \frac{ABC^2 - abc^2}{ABC - abc}$$

9. $M = \iiint\limits_{\Pi} Kz\,dxdydz$

Let $P_1 = \{x_0, \cdots, x_m\}, \quad P_2 = \{y_0, \cdots, y_n\}, \quad P_3 = \{z_0, \cdots, z_q\}$ be partitions of $[0, a]$ and let $P = P_1 \times P_2 \times P_3$.   Note that

$$z_{k-1} \le \tfrac{1}{2}(z_k + z_{k-1}) \le z_k$$

and therefore

$$Kz_{k-1}\,\Delta x_i\,\Delta y_j\,\Delta z_k \le \tfrac{1}{2} K\,\Delta x_i\,\Delta y_j\,(z_k^2 - z_{k-1}^2) \le Kz_k\,\Delta x_i\,\Delta y_j \Delta z_k.$$

It follows that

$$L_f(P) \le \frac{1}{2} K \sum_{i=1}^{m} \sum_{j=1}^{n} \sum_{k=1}^{q} \Delta x_i\,\Delta y_j\,(z_k^2 - z_{k-1}^2) \le U_f(P).$$

The middle term can be written

$$\frac{1}{2} K \left(\sum_{i=1}^{m} \Delta x_i \right) \left(\sum_{j=1}^{n} \Delta y_j \right) \left(\sum_{k=1}^{q} z_k^2 - z_{k-1}^2 \right) = \frac{1}{2} K(a)(a)(a^2) = \frac{1}{2} Ka^4.$$

$M = \tfrac{1}{2} Ka^4$ where K is the constant of proportionality for the density function.

11. $$I_z = \iiint\limits_{\Pi} Kz\left(x^2 + y^2\right)\,dxdydz$$

$$= \underbrace{\iiint\limits_{\Pi} Kx^2 z\,dxdydz}_{I_1} + \underbrace{\iiint\limits_{\Pi} Ky^2 z\,dxdydz}_{I_2}\,.$$

We will calculate I_1 using the partitions we used in doing Exercise 9.   Note that

$$x_{i-1}^2 z_{k-1} \le \left(\frac{x_i^2 + x_i x_{i-1} + x_{i-1}^2}{3} \right) \left(\frac{z_k + z_{k-1}}{2} \right) \le x_i^2 z_k$$

and therefore

$$Kx_{i-1}^2 z_{k-1}\,\Delta x_i\,\Delta y_j\,\Delta z_k \le \tfrac{1}{6} K\left(x_i^3 - x_{i-1}^3\right)\,\Delta y_j\,(z_k^2 - z_{k-1}^2) \le Kx_i^2 z_k^2\,\Delta x_i\,\Delta y_j\,\Delta z_k.$$

It follows that

$$L_f(P) \le \frac{1}{6} K \sum_{i=1}^{m} \sum_{j=1}^{n} \sum_{k=1}^{q} \left(x_i{}^3 - x_{i-1}^3\right) \Delta y_j \left(z_k{}^2 - z_{k-1}^2\right) \le U_f(P).$$

The middle term can be written

$$\frac{1}{6} K \left(\sum_{i=1}^{m} x_i{}^3 - x_{i-1}^3\right)\left(\sum_{j=1}^{n} \Delta y_j\right)\left(\sum_{k=1}^{q} z_k{}^2 - z_{k-1}^2\right) = \frac{1}{6} K a^3 \, (a)(a^2) = \frac{1}{6} K a^6.$$

Similarly $I_2 = \frac{1}{6} K a^6$ \quad and therefore

by Exercise 9

$$I_z = \frac{1}{3} K a^6 = \frac{2}{3}\left(\frac{1}{2} K a^4\right)a^2 = \frac{2}{3} M a^2.$$

SECTION 16.7

1. $\displaystyle \int_0^a \int_0^b \int_0^c dx\, dy\, dz = \int_0^a \int_0^b c\, dy\, dz = \int_0^a bc\, dz = abc$

3. $\displaystyle \int_0^1 \int_1^{2y} \int_0^x (x + 2z)\, dz\, dx\, dy = \int_0^1 \int_1^{2y} \left[xz + z^2\right]_0^x dx\, dy = \int_0^1 \int_1^{2y} 2x^2\, dx\, dy$

$$= \int_0^1 \left[\frac{2}{3} x^3\right]_1^{2y} dy = \int_0^1 \left(\frac{16}{3} y^3 - \frac{2}{3}\right) dy = \frac{2}{3}$$

5. $\displaystyle \int_0^2 \int_{-1}^1 \int_0^3 (z - xy)\, dz\, dy\, dx = \int_0^2 \int_{-1}^1 \left[\frac{1}{2} z^2 - xyz\right]_1^3 dy\, dx$

$$= \int_0^2 \int_{-1}^1 (4 - 2xy)\, dy\, dx = \int_0^2 \left[2y - xy^2\right]_{-1}^1 dx = \int_0^2 8\, dy = 16$$

7. $\displaystyle \int_0^{\pi/2} \int_0^1 \int_0^{\sqrt{1-x^2}} x \cos z\, dy\, dx\, dz = \int_0^{\pi/2} \int_0^1 \left[xy \cos z\right]_0^{\sqrt{1-x^2}} dx\, dz$

$$= \int_0^{\pi/2} \int_0^1 x\sqrt{1 - x^2} \cos z\, dx\, dz = \int_0^{\pi/2} \left[-\frac{1}{3}(1 - x^2)^{3/2} \cos z\right]_0^1 dz = \frac{1}{3} \int_0^{\pi/2} \cos z\, dz = \frac{1}{3}$$

9. $\displaystyle \int_1^2 \int_y^{y^2} \int_0^{\ln x} ye^z\, dz\, dx\, dy = \int_1^2 \int_y^{y^2} \left[ye^z\right]_0^{\ln x} dx\, dy$

$$= \int_1^2 \int_y^{y^2} y(x - 1)\, dx\, dy = \int_1^2 \left[\frac{1}{2} x^2 y - xy\right]_y^{y^2} dy = \int_1^2 \left(\frac{1}{2} y^5 - \frac{3}{2} y^3 + y^2\right) dy = \frac{47}{24}$$

11.
$$\iiint_\Pi f(x)g(y)h(z)\,dxdydz = \int_{c_1}^{c_2}\left[\int_{b_1}^{b_2}\left(\int_{a_1}^{a_2}f(x)g(y)h(z)\,dx\right)dy\right]dz$$

$$= \int_{c_1}^{c_2}\left[\int_{b_1}^{b_2}g(y)h(z)\left(\int_{a_1}^{a_2}f(x)\,dx\right)dy\right]dz$$

$$= \int_{c_1}^{c_2}\left[h(z)\left(\int_{a_1}^{a_2}f(x)\,dx\right)\left(\int_{b_1}^{b_2}g(y)\,dy\right)dz\right]$$

$$= \left(\int_{a_1}^{a_2}f(x)\,dx\right)\left(\int_{b_1}^{b^2}g(y)\,dy\right)\left(\int_{c_1}^{c_2}h(z)\,dz\right)$$

13. $\left(\int_0^1 x^2\,dx\right)\left(\int_0^2 y^2\,dy\right)\left(\int_0^3 z^2\,dz\right) = \left(\frac{1}{3}\right)\left(\frac{8}{3}\right)\left(\frac{27}{3}\right) = 8$

15.
$$x_M M = \iiint_\Pi kx^2yz\,dxdydz = k\left(\int_0^a x^2\,dx\right)\left(\int_0^b y\,dy\right)\left(\int_0^c z\,dz\right)$$

$$= k\left(\tfrac{1}{3}a^3\right)\left(\tfrac{1}{2}b^2\right)\left(\tfrac{1}{2}c^2\right) = \tfrac{1}{12}ka^3b^2c^2.$$

By Exercise 14, $M = \frac{1}{8}ka^2b^2c^2$. Therefore $\bar{x} = \frac{2}{3}a$. Similarly, $\bar{y} = \frac{2}{3}b$ and $\bar{z} = \frac{2}{3}c$.

17.

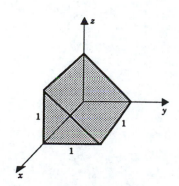

19. center of mass is the centroid

$$\bar{x} = \tfrac{1}{2} \quad \text{by symmetry}$$

$$\bar{y}V = \iiint_T y\,dxdydz = \int_0^1\int_0^1\int_0^{1-y} y\,dz\,dy\,dx = \int_0^1\int_0^1 (y - y^2)\,dy\,dx$$

$$= \int_0^1\left[\frac{1}{2}y^2 - \frac{1}{3}y^3\right]_0^1 dx = \int_0^1 \frac{1}{6}\,dx = \frac{1}{6}$$

$$\bar{z}V = \iiint_T z\,dxdydz = \int_0^1\int_0^1\int_0^{1-y} z\,dz\,dy\,dx = \int_0^1\int_0^1 \frac{1}{2}(1-y)^2\,dy\,dx$$

$$= \frac{1}{2}\int_0^1\int_0^1 (1 - 2y + y^2)\,dy\,dx = \frac{1}{2}\int_0^1\left[y - y^2\frac{1}{3}y^3\right]_0^1 dx = \frac{1}{2}\int_0^1 \frac{1}{3}\,dx = \frac{1}{6}$$

$V = \frac{1}{2}$ (by Exercise 18); $\bar{y} = \frac{1}{3}$, $\bar{z} = \frac{1}{3}$

21. $\displaystyle\int_{-r}^{r}\int_{-\phi(x)}^{\phi(x)}\int_{-\psi(x,y)}^{\psi(x,y)} k\left(r - \sqrt{x^2 + y^2 + z^2}\right)\, dz\, dy\, dx$ with $\phi(x) = \sqrt{r^2 - x^2}$,

$\psi(x,y) = \sqrt{r^2 - (x^2 + y^2)}$, k the constant of proportionality

23. $\displaystyle\int_{0}^{1}\int_{-\sqrt{x-x^2}}^{\sqrt{x-x^2}}\int_{-2x-3y-10}^{1-y^2} dz\, dy\, dx$

25. $\displaystyle\int_{-1}^{1}\int_{-2\sqrt{2-2x^2}}^{2\sqrt{2-2x^2}}\int_{3x^2+y^2/4}^{4-x^2-y^2/4} k\left(z - 3x^2 - \frac{1}{4}y^2\right)\, dz\, dy\, dx$

27. $\displaystyle\iiint_{T}(x^2 z + y)\, dx\, dy\, dz = \int_{0}^{2}\int_{1}^{3}\int_{0}^{1}(x^2 z + y)\, dx\, dy\, dz = \int_{0}^{2}\int_{1}^{3}\left[\frac{1}{3}x^3 z + xy\right]_{0}^{1} dy\, dz$

$\displaystyle = \int_{0}^{2}\int_{1}^{3}\left(\frac{1}{3}z + y\right) dy\, dz = \int_{0}^{2}\left[\frac{1}{3}zy + \frac{1}{2}y^2\right]_{1}^{3} dz = \int_{0}^{2}\left(\frac{2}{3}z + 4\right) dz = \frac{28}{3}$

29. $\displaystyle\iiint_{T} x^2 y^2 z^2\, dx\, dy\, dz = \int_{-1}^{0}\int_{0}^{y+1}\int_{0}^{1} x^2 y^2 z^2\, dx\, dz\, dy + \int_{0}^{1}\int_{0}^{1-y}\int_{0}^{1} x^2 y^2 z^2\, dx\, dz\, dy$

$\displaystyle = \int_{-1}^{0}\int_{0}^{y+1}\left[\frac{1}{3}x^3 y^2 z^2\right]_{0}^{1} dz\, dy + \int_{0}^{1}\int_{0}^{1-y}\left[\frac{1}{3}x^3 y^2 z^2\right]_{0}^{1} dz\, dy$

$\displaystyle = \frac{1}{3}\int_{-1}^{0}\int_{0}^{y+1} y^2 z^2\, dz\, dy + \frac{1}{3}\int_{0}^{1}\int_{0}^{1-y}[y^2 z^2]_{0}^{1}\, dz\, dy$

$\displaystyle = \frac{1}{3}\int_{-1}^{0}\left[\frac{1}{3}y^2 z^3\right]_{0}^{y+1} dy + \frac{1}{3}\int_{0}^{1}\left[\frac{1}{3}y^2 z^3\right]_{0}^{1-y} dy$

$\displaystyle = \frac{1}{9}\int_{-1}^{0}(y^5 + 3y^4 + 3y^3 + y^2)\, dy + \frac{1}{9}\int_{0}^{1}(y^2 - 3y^3 + 3y^4 - y^5)\, dy = \frac{1}{270}$

31. $\displaystyle\iiint_{T} y^2\, dx\, dy\, dz = \int_{0}^{3}\int_{0}^{2-2x/3}\int_{0}^{6-2x-3y} y^2\, dz\, dy\, dx = \int_{0}^{3}\int_{0}^{2-2x/3}[y^2 z]_{0}^{6-2x-3y}\, dy\, dx$

$\displaystyle = \int_{0}^{3}\int_{0}^{2-2x/3}(6y^2 - 2xy^2 - 3y^3)\, dy\, dx$

$\displaystyle = \int_{0}^{3}\left[2y^3 - \frac{2}{3}xy^3 - \frac{3}{4}y^4\right]_{0}^{2-2x/3} dx$

$\displaystyle = \frac{1}{4}\int_{0}^{3}\left(2 - \frac{2}{3}x\right) dx = \frac{12}{5}$

33.
$$V = \int_0^2 \int_{x^2}^{x+2} \int_0^x dz\, dy\, dx = \int_0^2 \int_{x^2}^{x+2} x\, dy\, dx = \int_0^2 (x^2 + 2x - x^3)\, dx = \frac{8}{3}$$

$$\bar{x}V = \int_0^2 \int_{x^2}^{x+2} \int_0^x x\, dz\, dy\, dx = \int_0^2 \int_{x^2}^{x+2} x^2\, dy\, dx = \int_0^2 (x^3 + 2x^2 - x^4)\, dx = \frac{44}{15}$$

$$\bar{y}V = \int_0^2 \int_{x^2}^{x+2} \int_0^x y\, dz\, dy\, dx = \int_0^2 \int_{x^2}^{x+2} xy\, dy\, dx = \int_0^2 \frac{1}{2}(x^3 + 4x^2 + 4x - x^5)\, dx = 6$$

$$\bar{z}V = \int_0^2 \int_{x^2}^{x+2} \int_0^x z\, dz\, dy\, dx = \int_0^2 \int_{x^2}^{x+2} \frac{1}{2}x^2\, dy\, dx = \int_0^2 \frac{1}{2}(x^3 + 2x^2 - x^4)\, dx = \frac{22}{15}$$

$$\bar{x} = \frac{11}{10}, \quad \bar{y} = \frac{9}{4}, \quad \bar{z} = \frac{11}{20}$$

35. $V = \int_{-1}^2 \int_0^3 \int_{2-x}^{4-x^2} dz\, dy\, dx = \frac{27}{2}; \quad (\bar{x}, \bar{y}, \bar{z}) = \left(\frac{1}{2}, \frac{3}{2}, \frac{12}{5}\right)$

37. $V = \int_0^a \int_0^{\phi(x)} \int_0^{\psi(x,y)} dz\, dy\, dx = \frac{1}{6}abc \text{ with } \phi(x) = b\left(1 - \frac{x}{a}\right), \quad \psi(x,y) = c\left(1 - \frac{x}{a} - \frac{y}{b}\right)$

$(\bar{x}, \bar{y}, \bar{z}) = \left(\frac{1}{4}a, \frac{1}{4}b, \frac{1}{4}c\right)$

39. $\Pi: 0 \le x \le a, \quad 0 \le y \le b, \quad 0 \le z \le c$

(a) $I_z = \int_0^a \int_0^b \int_0^c \frac{M}{abc}(x^2 + y^2)\, dz\, dy\, dx = \frac{1}{3}M(a^2 + b^2)$

(b) $I_M = I_z - d^2 M = \frac{1}{3}M(a^2 + b^2) - \frac{1}{4}(a^2 + b^2)M = \frac{1}{12}M(a^2 + b^2)$

             parallel axis theorem (16.5.7)

(c) $I = I_M + d^2 M = \frac{1}{12}M(a^2 + b^2) + \frac{1}{4}a^2 M = \frac{1}{3}Ma^2 + \frac{1}{12}Mb^2$

             parallel axis theorem (16.5.7)

41.
$$M = \int_0^1 \int_0^1 \int_0^y k(x^2 + y^2 + z^2)\, dz\, dy\, dx = \int_0^1 \int_0^1 k\left(x^2 y + y^3 + \frac{1}{3}y^3\right)\, dy\, dx$$

$$= \int_0^1 k\left(\frac{1}{2}x^2 + \frac{1}{3}\right)\, dx = \frac{1}{2}k$$

$(x_M, y_M, z_M) = \left(\frac{7}{12}, \frac{34}{45}, \frac{37}{90}\right)$

43. (a)   0 by symmetry

(b) $\iiint\limits_{T} (a_1\,x + a_2\,y + a_3 z + a_4)\,dxdydz = \iiint\limits_{T} a_4\,dxdydz = a_4$ (volume of ball) $= \frac{4}{3}\,\pi a_4$

by symmetry

45. $V = 8\int_0^a \int_0^{\sqrt{a^2-x^2}} \int_0^{\sqrt{a^2-x^2-y^2}} dz\,dy\,dx = 8\int_0^a \int_0^{\sqrt{a^2-x^2}} \sqrt{a^2-x^2-y^2}\,dy\,dx$

polar coordinates

$$= 8\int_0^{\pi/2} \int_0^a \sqrt{a^2-r^2}\,r\,dr\,d\theta$$

$$= -4\int_0^{\pi/2} \left[\frac{2}{3}(a^2-r^2)^{3/2}\right]_0^a d\theta$$

$$= \frac{8}{3}\int_0^{\pi/2} d\theta = \frac{4}{3}\,\pi\,a^3$$

47. $M = \int_{-2}^{2} \int_{-\sqrt{4-x^2}/2}^{\sqrt{4-x^2}/2} \int_{x^2+3y^2}^{4-y^2} k|x|\,dz\,dy\,dx = 4\int_0^2 \int_0^{\sqrt{4-x^2}/2} \int_{x^2+3y^2}^{4-y^2} kx\,dz\,dy\,dx$

$$= 4\,k\int_0^2 \int_0^{\sqrt{4-x^2}/2} \left(4x - x^3 - 4xy^2\right)dy\,dx = \frac{4}{3}\,k\int_0^2 x\,(4-x^2)^{3/2}\,dx = \frac{128}{15}\,k$$

49. $M = \int_{-1}^{2} \int_0^3 \int_{2-x}^{4-x^2} k(1+y)\,dz\,dy\,dx = \frac{135}{4}\,k; \quad (x_M, y_M, z_M) = \left(\frac{1}{2}, \frac{9}{5}, \frac{12}{5}\right)$

51. (a) $V = \int_0^6 \int_{z/2}^3 \int_x^{6-x} dy\,dx\,dz$

(b) $V = \int_0^3 \int_0^{2x} \int_x^{6-x} dy\,dz\,dx$

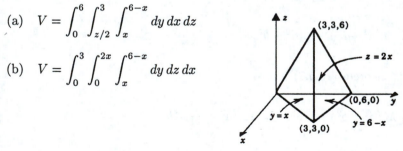

(c) $V = \int_0^6 \int_{z/2}^3 \int_{z/2}^y dx\,dy\,dz + \int_0^6 \int_3^{(12-z)/2} \int_{z/2}^{6-y} dx\,dy\,dz$

53. (a) $V = \iint\limits_{\Omega_{yz}} 2y\,dydz$

(b) $V = \iint\limits_{\Omega_{yz}} \left(\int_{-y}^y dx\right) dydz$

(c) $V = \int_0^4 \int_{-\sqrt{4-y}}^{\sqrt{4-y}} \int_{-y}^y dx\,dz\,dy$

(d) $V = \int_{-2}^2 \int_0^{4-z^2} \int_{-y}^y dx\,dy\,dz$

SECTION 16.8

1. $r^2 + z^2 = 9$ **3.** $z = 2r$ **5.** $4r^2 = z^2$

7. $\displaystyle\int_0^\pi \int_0^2 \int_0^{4-r^2} r\, dz\, dr\, d\theta = \int_0^\pi \int_0^2 \left(4r - r^2\right) dr\, d\theta = \int_0^\pi 4\, d\theta = 4\pi$

9. $\displaystyle\int_0^{\pi/4} \int_0^1 \int_0^{r\cos\theta} r\sec^3\theta\, dz\, dr\, d\theta = \int_0^{\pi/4} \int_0^1 r^2 \sec^2\theta\, dr\, d\theta = \frac{1}{3}\int_0^{\pi/4} \sec^2\theta\, d\theta = \frac{1}{3}$

11.   Set the lower base of the cylinder on the xy-plane so that the axis of the cylinder coincides with the z-axis. Assume that the density varies directly as the distance from the lower base.

$$M = \int_0^{2\pi} \int_0^R \int_0^h kzr\, dz\, dr\, d\theta = \frac{1}{2}\, k\pi R^2 h^2$$

13.

$$I = I_z = k\int_0^{2\pi} \int_0^R \int_0^h zr^3\, dr\, d\theta\, dz$$

$$= \tfrac{1}{4}\, k\pi R^4 h^2 = \tfrac{1}{2}\left(\tfrac{1}{2}\, k\pi R^2 h^2\right) R^2 = \tfrac{1}{2}\, MR^2$$

from Exercise 11

15.   Inverting the cone and placing the vertex at the origin, we have

$$V = \int_0^h \int_0^{2\pi} \int_0^{(R/h)z} r\, dr\, d\theta\, dz = \frac{1}{3}\pi R^2 h.$$

17. $\displaystyle I = \frac{M}{V}\int_0^h \int_0^{2\pi} \int_0^{(R/h)z} r^3\, dr\, d\theta\, dz = \frac{3}{10}\, MR^2$

19. $\displaystyle V = \int_0^{2\pi} \int_0^1 \int_0^{1-r^2} r\, dz\, dr\, d\theta = \frac{1}{2}\pi$

21. $\displaystyle M = \int_0^{2\pi} \int_0^1 \int_0^{1-r^2} k\left(r^2 + z^2\right) r\, dz\, dr\, d\theta = \frac{1}{4}k\pi$

23. $\displaystyle\int_0^{\pi/2} \int_0^1 \int_0^{\sqrt{4-r^2}} r\, dz\, dr\, d\theta = \int_0^{\pi/2} \int_0^1 r\sqrt{4-r^2}\, dr\, d\theta = \int_0^{\pi/2} \left(\frac{8}{3} - \sqrt{3}\right) d\theta = \frac{1}{6}\left(8 - 3\sqrt{3}\right)\pi$

25.

$$\int_0^3 \int_0^{\sqrt{9-y^2}} \int_0^{\sqrt{9-x^2y^2}} \frac{1}{\sqrt{x^2+y^2}} \, dz \, dx \, dy = \int_0^{\pi/2} \int_0^3 \int_0^{\sqrt{9-r^2}} \frac{1}{r} \cdot r \, dz \, dr \, d\theta$$

$$= \int_0^{\pi/2} \int_0^3 \sqrt{9-r^2} \, dr \, d\theta$$

$$= \int_0^{\pi/2} \left[\frac{r}{2}\sqrt{9-r^2} + \frac{9}{2}\sin^{-1}\frac{r}{3} \right]_0^3 \, d\theta$$

$$= \frac{9\pi}{4} \int_0^{\pi/2} d\theta = \frac{9}{8}\pi^2$$

27.

$$\int_0^1 \int_0^{\sqrt{1-x^2}} \int_0^2 \sin(x^2+y^2) \, dz \, dy \, dx = \int_0^{\pi/2} \int_0^1 \int_0^2 \sin(r^2)r \, dz \, dr \, d\theta = 2\int_0^{\pi/2} \int_0^1 r\sin(r^2) \, dr \, d\theta$$

$$= 2\int_0^{\pi/2} \left[-\frac{1}{2}\cos(r^2) \right]_0^1 \, d\theta = (1-\cos 1)\int_0^{\pi/2} d\theta = \frac{\pi}{2}(1-\cos 1) \cong 0.7221$$

29.

$$V = \int_{-\pi/2}^{\pi/2} \int_0^{2a\cos\theta} \int_0^r r \, dz \, dr \, d\theta = \int_{-\pi/2}^{\pi/2} \int_0^{2a\cos\theta} r^2 \, dr \, d\theta$$

$$= \int_{-\pi/2}^{\pi/2} \frac{8}{3}a^3\cos^3\theta \, d\theta = \frac{32}{9}a^3$$

31.

$$V = \int_{-\pi/2}^{\pi/2} \int_0^{a\cos\theta} \int_0^{a-r} r \, dz \, dr \, d\theta = \int_{-\pi/2}^{\pi/2} \int_0^{a\cos\theta} r(a-r) \, dr \, d\theta$$

$$= \int_{-\pi/2}^{\pi/2} a^3 \left(\frac{1}{2}\cos^2\theta - \frac{1}{3}\cos^3\theta \right) d\theta = \frac{1}{36}a^2(9\pi - 16)$$

33.

$$V = \int_{-\pi/2}^{\pi/2} \int_0^{\cos\theta} \int_{r^2}^{r\cos\theta} r \, dz \, dr \, d\theta = \int_{-\pi/2}^{\pi/2} \int_0^{\cos\theta} (r^2\cos\theta - r^3) \, dr \, d\theta$$

$$= \int_{-\pi/2}^{\pi/2} \frac{1}{12}\cos^4\theta \, d\theta = \frac{1}{32}\pi$$

35. $$V = \int_0^{2\pi} \int_0^{1/2} \int_{r\sqrt{3}}^{\sqrt{1-r^2}} r \, dz \, dr \, d\theta = \int_0^{2\pi} \int_0^{1/2} \left(r\sqrt{1-r^2} - r^2\sqrt{3} \right) dr \, d\theta = \frac{1}{3}\pi \left(2 - \sqrt{3} \right)$$

SECTION 16.9

1. $\left(\sqrt{3},\ \frac{1}{4}\pi,\ \cos^{-1}\left[\frac{1}{3}\sqrt{3}\right]\right)$

3. $\left(\frac{3}{4},\ \frac{3}{4}\sqrt{3},\ \frac{3}{2}\sqrt{3}\right)$

5. $\rho = \sqrt{2^2 + 2^2 + (2\sqrt{6}/3)^2} = \dfrac{4\sqrt{6}}{3}$

$\phi = \cos^{-1}\left(\dfrac{2\sqrt{6}/3}{4\sqrt{6}/3}\right) = \cos^{-1}(1/2) = \dfrac{\pi}{3}$

$\theta = \tan^{-1}(1) = \dfrac{\pi}{4}$

$(\rho, \theta, \phi) = \left(\dfrac{4\sqrt{6}}{3},\ \dfrac{\pi}{4},\ \dfrac{\pi}{3}\right)$

7. $x = \rho \sin\phi \cos\theta = 3\sin 0 \cos(\pi/2) = 0$

$y = \rho \sin\phi \sin\theta = 3\sin 0 \sin(\pi/2) = 0$

$z = \rho \cos\phi = 3\cos 0 = 3$

$(x, y, z) = (0, 0, 3)$

9. The circular cylinder $x^2 + y^2 = 1$; the radius of the cylinder is 1 and the axis is the z-axis.

11. The lower nappe of the circular cone $z^2 = x^2 + y^2$.

13. Horizontal plane one unit above the xy-plane.

15. $\displaystyle\int_0^{\pi/3}\int_0^{2\pi}\int_0^1 \rho^2 \sin\phi\, d\rho\, d\theta\, d\phi = \frac{1}{3}\int_0^{\pi/3}\int_0^{2\pi} \sin\phi\, d\theta\, d\phi = \frac{2\pi}{3}\int_0^{\pi/3}\sin\phi\, d\phi = \frac{\pi}{3}$

17. $\displaystyle\int_0^{\pi/4}\int_0^{\pi}\int_0^{2\cos\phi} \rho^2 \sin\phi\, d\rho\, d\theta\, d\phi = \frac{8}{3}\int_0^{\pi/4}\int_0^{\pi} \cos^3\phi \sin\phi\, d\theta\, d\phi$

$\displaystyle = \frac{8}{3}\pi \int_0^{\pi/4} \cos^3\phi \sin\phi\, d\phi$

$\displaystyle = \frac{8}{3}\pi \left[-\frac{1}{4}\cos^4\phi\right]_0^{\pi/4} = \frac{\pi}{2}$

19. $\displaystyle\int_0^1\int_0^{\sqrt{1-x^2}}\int_{\sqrt{x^2+y^2}}^{\sqrt{2-x^2-y^2}} dz\, dy\, dx = \int_0^{\pi/4}\int_0^{\pi/2}\int_0^{\sqrt{2}} \rho^2 \sin\phi\, d\rho\, d\theta\, d\phi$

$\displaystyle = \frac{2}{3}\sqrt{2}\int_0^{\pi/4}\int_0^{\pi/2} \sin\phi\, d\theta\, d\phi$

$\displaystyle = \frac{\sqrt{2}}{3}\pi \int_0^{\pi/4} \sin\phi\, d\phi = \frac{\pi}{3}\left[\sqrt{2} - 1\right]$

21. $\displaystyle\int_0^3\int_0^{\sqrt{9-y^2}}\int_0^{\sqrt{9-x^2-y^2}} z\sqrt{x^2+y^2+x^2}\, dz\, dx\, dy$

$\displaystyle = \int_0^{\pi/2}\int_0^{\pi/2}\int_0^3 \rho\cos\phi \cdot \rho \cdot \rho^2 \sin\phi\, d\rho\, d\theta\, d\phi$

$\displaystyle = \int_0^{\pi/2}\frac{1}{2}\sin 2\phi\, d\phi \int_0^{\pi/2} d\theta \int_0^3 \rho^4\, d\rho = \left[-\frac{1}{4}\cos 2\phi\right]_0^{\pi/2} \left(\frac{\pi}{2}\right)\left[\frac{1}{5}\rho^5\right]_0^3$

$\displaystyle = \frac{243\pi}{20}$

23. $\displaystyle V = \int_0^{2\pi}\int_0^{\pi}\int_0^R \rho^2 \sin\phi\, d\rho\, d\phi\, d\theta = \frac{4}{3}\pi R^3$

23. $\quad V = \int_0^{2\pi} \int_0^{\pi} \int_0^R \rho^2 \sin\phi \, d\rho \, d\phi \, d\theta = \dfrac{4}{3}\pi R^3$

25. $\quad V = \int_0^{\alpha} \int_0^{\pi} \int_0^R \rho^2 \sin\phi \, d\rho \, d\phi \, d\theta = \dfrac{2}{3}\alpha R^3$

27.
$$M = \int_0^{2\pi} \int_0^{\tan^{-1}(r/h)} \int_0^{h\sec\phi} k\rho^3 \sin\phi \, d\rho \, d\phi \, d\theta$$

$$= \int_0^{2\pi} \int_0^{\tan^{-1}(r/h)} \frac{kh^4}{4} \tan\phi \sec^3\phi \, d\phi \, d\theta$$

$$= \frac{kh^4}{4} \int_0^{2\pi} \frac{1}{3}\left[\sec^3\phi\right]_0^{\tan^{-1}(r/h)} d\theta = \frac{kh^4}{4} \int_0^{2\pi} \frac{1}{3}\left[\left(\frac{\sqrt{r^2+h^2}}{h}\right)^3 - 1\right] d\theta$$

$$= \frac{1}{6} k\pi h \left[(r^2+h^2)^{3/2} - h^3\right]$$

29. center ball at origin; $\quad$ density $= \dfrac{M}{V} = \dfrac{3M}{4\pi R^3}$

(a) $\quad I = \dfrac{3M}{4\pi R^3} \int_0^{2\pi} \int_0^{\pi} \int_0^R \rho^4 \sin^3\phi \, d\rho \, d\phi \, d\theta = \dfrac{2}{5}MR^2$

(b) $\quad I = \frac{2}{5}MR^2 + R^2 M = \frac{7}{5}MR^2 \quad$ (parallel axis theorem)

31. center balls at origin; $\quad$ density $= \dfrac{M}{V} = \dfrac{3M}{4\pi\left(R_2{}^3 - R_1{}^3\right)}$

(a) $\quad I = \dfrac{3M}{4\pi\left(R_2{}^3 - R_1{}^3\right)} \int_0^{2\pi} \int_0^{\pi} \int_{R_1}^{R_2} \rho^4 \sin^3\phi \, d\rho \, d\phi \, d\theta = \dfrac{2}{5}M\left(\dfrac{R_2{}^5 - R_1{}^5}{R_2{}^3 - R_1{}^3}\right)$

This result can be derived from Exercise 29 without further integration. View the solid as a ball of mass M_2 from which is cut out a core of mass M_1.

$$M_2 = \frac{M}{V}V_2 = \frac{3M}{4\pi\left(R_2{}^3 - R_1{}^3\right)}\left(\frac{4}{3}\pi R_2{}^3\right) = \frac{MR_2{}^3}{R_2{}^3 - R_1{}^3}; \quad \text{similarly} \quad M_1 = \frac{MR_1{}^3}{R_2{}^3 - R_1{}^3}.$$

Then

$$I = I_2 - I_1 = \frac{2}{5}M_2 R_2{}^2 - \frac{2}{5}M_1 R_1{}^2 = \frac{2}{5}\left(\frac{MR_2{}^3}{R_2{}^3 - R_1{}^3}\right)R_2{}^2 - \frac{2}{5}\left(\frac{MR_1{}^3}{R_2{}^3 - R_1{}^3}\right)R_1{}^2$$

$$= \frac{2}{5}M\left(\frac{R_2{}^5 - R_1{}^5}{R_2{}^3 - R_1{}^3}\right).$$

(b) Outer radius R and inner radius R_1 gives

$$\text{moment of inertia} = \frac{2}{5}M\left(\frac{R^5 - R_1{}^5}{R^3 - R_1{}^3}\right). \qquad \text{[part (a)]}$$

As $R_1 \to R$,

$$\frac{R^5 - R_1{}^5}{R^3 - R_1{}^3} = \frac{R^4 + R^3 R_1 + R^2 R_1{}^2 + RR_1{}^3 + R_1{}^4}{R^2 + RR_1 + R_1{}^2} \longrightarrow \frac{5R^4}{3R^2} = \frac{5}{3} R^2.$$

Thus the moment of inertia of spherical shell of radius R is

$$\tfrac{2}{5} M \left(\tfrac{5}{3} R^2 \right) = \tfrac{2}{3} MR^2.$$

(c) $I = \tfrac{2}{3} MR^2 + R^2 M = \tfrac{5}{3} MR^2$ (parallel axis theorem)

33. $V = \int_0^{2\pi} \int_0^\alpha \int_0^a \rho^2 \sin\phi \, d\rho \, d\phi \, d\theta = \frac{2}{3} \pi \left(1 - \cos\alpha \right) a^3$

35.   (a)   Substituting $x = \rho \sin\phi \cos\theta, \quad y = \rho \sin\phi \sin\theta, \quad z = \rho \cos\phi$

into $x^2 + y^2 + (z - R)^2 = R^2$

we have $\rho^2 \sin^2 \phi + (\rho \cos\phi - R)^2 = R^2,$

which simplifies to $\rho = 2R \cos\phi.$

(b) $0 \le \theta \le 2\pi, \quad 0 \le \phi \le \pi/4, \quad R \sec\phi \le \rho \le 2R \cos\phi$

37. $V = \int_0^{2\pi} \int_0^{\pi/4} \int_0^2 \rho^2 \sin\phi \, d\rho \, d\phi \, d\theta + \int_0^{2\pi} \int_{\pi/4}^{\pi/2} \int_0^{2\sqrt{2} \cos\phi} \rho^2 \sin\phi \, d\rho \, d\phi \, d\theta$

$$= \frac{1}{3} \left(16 - 6\sqrt{2} \right) \pi$$

39.   Encase T in a spherical wedge W. W has spherical coordinates in a box Π that contains S.   Define f to be zero outside of T.   Then

$$F(\rho, \theta, \phi) = f \left(\rho \sin\phi \cos\theta, \ \rho \sin\phi \sin\theta, \ \rho \cos\phi \right)$$

is zero outside of S and

$$\iiint_T f(x, y, z) \, dxdydz = \iiint_W f(x, y, z) \, dxdydz$$

$$= \iiint_\Pi F(\rho, \theta, \phi) \, \rho^2 \sin\phi \, d\rho d\theta d\phi$$

$$= \iiint_S F(\rho, \theta, \phi) \, \rho^2 \sin\phi \, d\rho d\theta d\phi.$$

41. T is the set of all (x, y, z) with spherical coordinates (ρ, θ, ϕ) in the set

$$S: \quad 0 \le \theta \le 2\pi, \quad 0 \le \phi \le \pi/4, \quad R \sec\phi \le \rho \le 2R \cos\phi.$$

T has volume $V = \tfrac{2}{3} \pi R^3$. By symmetry the $\mathbf{i}, \mathbf{j}$ components of force are zero and

$$\mathbf{F} = \left\{ \frac{3GmM}{2\pi R^3} \iiint_T \frac{z}{(x^2 + y^2 + z^2)^{3/2}} \, dxdydz \right\} \mathbf{k}$$

$$= \left\{ \frac{3GmM}{2\pi R^3} \iiint_S \left(\frac{\rho \cos \phi}{\rho^3} \right) \rho^2 \sin \phi \, d\rho d\theta d\phi \right\} \mathbf{k}$$

$$= \left\{ \frac{3GmM}{2\pi R^3} \int_0^{2\pi} \int_0^{\pi/4} \int_{R \sec \phi}^{2R \cos \phi} \cos \phi \sin \phi \, d\rho \, d\phi \, d\theta \right\} \mathbf{k}$$

$$= \frac{GmM}{R^2} \left(\sqrt{2} - 1 \right) \mathbf{k}.$$

SECTION 16.10

1. $ad - bc$

3. $2\left(v^2 - u^2\right)$

5. $u^2 v^2 - 4uv$

7. abc

9. r

11. $w\left(1 + w \cos v\right)$

13. Set $u = x + y$, $v = x - y$. Then

$$x = \frac{u + v}{2}, \quad y = \frac{u - v}{2} \quad \text{and} \quad J(u, v) = -\frac{1}{2}.$$

Ω is the set of all (x, y) with uv-coordinates in

$$\Gamma: \quad 0 \le u \le 1, \quad 0 \le v \le 2.$$

Then

$$\iint_\Omega \left(x^2 - y^2\right) dxdy = \iint_\Gamma \frac{1}{2} uv \, dudv = \frac{1}{2} \int_0^1 \int_0^2 uv \, dv \, du$$

$$= \frac{1}{2} \left(\int_0^1 u \, du \right) \left(\int_0^2 v \, dv \right) = \frac{1}{2} \left(\frac{1}{2} \right) (2) = \frac{1}{2}.$$

15. $\frac{1}{2} \int_0^1 \int_0^2 u \cos(\pi v) \, dv \, du = \frac{1}{2} \left(\int_0^1 u \, du \right) \left(\int_0^2 \cos(\pi v) \, dv \right) = \frac{1}{2} \left(\frac{1}{2} \right) (0) = 0$

17. Set $u = x - y$, $v = x + 2y$. Then

$$x = \frac{2u + v}{3}, \quad y = \frac{v - u}{3}, \quad \text{and} \quad J(u, v) = \frac{1}{3}.$$

Ω is the set of all (x, y) with uv-coordinates in the set

$$\Gamma: 0 \le u \le \pi, \quad 0 \le v \le \pi/2.$$

Therefore

$$\iint_\Omega \sin(x-y)\cos(x+2y)\,dxdy = \iint_\Gamma \frac{1}{3}\sin u \cos v\,dudv = \frac{1}{3}\int_0^\pi \int_0^{\pi/2} \sin u \cos v \, dv \, du$$

$$= \frac{1}{3}\left(\int_0^\pi \sin u \, du\right)\left(\int_0^{\pi/2} \cos v \, dv\right) = \frac{1}{3}(2)(1) = \frac{2}{3}.$$

19.   Set $u = xy,\quad v = y.$ Then

$$x = u/v,\quad y = v \quad \text{and} \quad J(u,v) = 1/v.$$

$$xy = 1,\qquad xy = 4 \quad\Longrightarrow\quad u = 1,\qquad u = 4$$

$$y = x,\qquad y = 4x \quad\Longrightarrow\quad u/v = v,\qquad 4u/v = v \quad\Longrightarrow\quad v^2 = u,\qquad v^2 = 4u$$

Ω is the set of all (x,y) with uv-coordinates in the set

$$\Gamma: \quad 1 \le u \le 4,\quad \sqrt{u} \le v \le 2\sqrt{u}.$$

(a) $A = \displaystyle\iint_\Gamma \frac{1}{v}\,dudv = \int_1^4 \int_{\sqrt{u}}^{2\sqrt{u}} \frac{1}{v}\,dv\,du = \int_1^4 \ln 2 \, du = 3\ln 2$

(b) $\bar{x}A = \displaystyle\int_1^4 \int_{\sqrt{u}}^{2\sqrt{u}} \frac{u}{v^2}\,dv\,du = \frac{7}{3};\quad \bar{x} = \frac{7}{9\ln 2}$

$\bar{y}A = \displaystyle\int_1^4 \int_{\sqrt{u}}^{2\sqrt{u}} dv\,du = \frac{14}{3};\quad \bar{y} = \frac{14}{9\ln 2}$

21.   Set $u = x+y,\quad v = 3x - 2y.$ Then

$$x = \frac{2u+v}{5},\quad y = \frac{3u-v}{5} \quad \text{and} \quad J(u,v) = -\frac{1}{5}.$$

With $\Gamma: \quad 0 \le u \le 1,\quad 0 \le v \le 2$

$$M = \int_0^1 \int_0^2 \frac{1}{5}\lambda\,dv\,du = \frac{2}{5}\lambda \quad \text{where} \quad \lambda \text{ is the density.}$$

Then

$$I_x = \int_0^1 \int_0^2 \left(\frac{3u-v}{5}\right)^2 \frac{1}{5}\lambda\,dv\,du = \frac{8\lambda}{375} = \frac{4}{75}\left(\frac{2}{5}\lambda\right) = \frac{4}{75}M,$$

$$I_y = \int_0^1 \int_0^2 \left(\frac{2u+v}{5}\right)^2 \frac{1}{5}\lambda\,dv\,du = \frac{28\lambda}{375} = \frac{14}{75}\left(\frac{2}{5}\lambda\right) = \frac{14}{75}M,$$

$$I_z = I_x + I_y = \tfrac{18}{75}M.$$

23.   Set $u = x - 2y,\quad v = 2x + y.$ Then

$$x = \frac{u+2v}{5},\quad y = \frac{v-2u}{5} \quad \text{and} \quad J(u,v) = \frac{1}{5}.$$

Γ is the region between the parabola $v = u^2 - 1$ and the line $v = 2u + 2$.   A sketch of the curves shows that

$$\Gamma: \quad -1 \le u \le 3, \quad u^2 - 1 \le v \le 2u + 2.$$

Then

$$A = \frac{1}{5} \text{ (area of } \Gamma) = \frac{1}{5} \int_{-1}^{3} \left[(2u + 2) - (u^2 - 1) \right] du = \frac{32}{15}.$$

25.   The choice $\theta = \pi/6$ reduces the equation to $13u^2 + 5v^2 = 1$.   This is an ellipse in the uv-plane with area $\pi a b = \pi/\sqrt{65}$.   Since $J(u, v) = 1$,   the area of Ω is also $\pi/\sqrt{65}$.

27. $J = abc\rho^2 \sin \phi; \quad V = \int_0^{2\pi} \int_0^{\pi} \int_0^1 abc\rho^2 \sin \phi \, d\rho \, d\phi \, d\theta = \frac{4}{3} \pi abc$

29. $\qquad\qquad V = \frac{2}{3} \pi abc, \quad \lambda = \frac{M}{V} = \frac{3M}{2\pi abc}$

$\qquad I_x = \frac{3M}{2\pi abc} \int_0^{2\pi} \int_0^{\pi/2} \int_0^1 (b^2\rho^2 \sin^2 \phi \sin^2 \theta + c^2\rho^2 \cos^2 \phi) abc\rho^2 \sin \phi \, d\rho \, d\phi \, d\theta$

$\qquad\quad = \frac{1}{5} M (b^2 + c^2)$

$\qquad I_y = \frac{1}{5} M (a^2 + c^2), \quad I_z = \frac{1}{5} M (a^2 + b^2)$

PROJECT 16.10

1.   (a) $\theta = \tan^{-1}\left[\left(\dfrac{ay}{bx}\right)^{1/\alpha}\right]$, $r = \left[\left(\dfrac{x}{a}\right)^{2/\alpha} + \left(\dfrac{y}{b}\right)^{2/\alpha}\right]^{\alpha/2}$

     (b)

$$\left.\begin{array}{c} ar_1(\cos\theta_1)^\alpha = ar_2(\cos\theta_2)^\alpha \\[2mm] br_1(\sin\theta_1)^\alpha = br_2(\sin\theta_2)^\alpha \\[2mm] r_1 > 0, \quad 0 < \theta < \dfrac{1}{2}\pi \end{array}\right\} \implies r_1 = r_2, \quad \theta_1 = \theta_2$$

3.   (a)

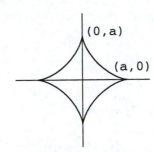

(0,a)

(a,0)

(b) $x = ar\cos^3\theta,\ y = ar\sin^3\theta;\quad x^{\frac{2}{3}} + y^{\frac{2}{3}} = a^{\frac{2}{3}} \implies r = 1$ and $x = a\cos^3\theta,\ y = a\sin^3\theta$

$A = \displaystyle\int_{\frac{\pi}{2}}^{0} y(\theta)x'(\theta)\,d\theta = \int_{\frac{\pi}{2}}^{0} a\sin^3\theta(3a\cos^2\theta[-\sin\theta])\,d\theta$

$\quad = 3a^2 \displaystyle\int_0^{\frac{\pi}{2}} \sin^4\theta\cos^2\theta\,d\theta = 3a^2 \int_0^{\frac{\pi}{2}} (\sin^4\theta - \sin^6\theta)\,d\theta$

$\quad = 3a^2 \left[\dfrac{3\cdot 1}{4\cdot 2}\dfrac{\pi}{2} - \dfrac{5\cdot 3\cdot 1}{6\cdot 4\cdot 2}\dfrac{\pi}{2}\right]$ (See Exercise 62(b) in 8.3)

$\quad = \dfrac{3a^2\pi}{32}$

(c) Entire area enclosed: $4\cdot\dfrac{3a^2\pi}{32} = \dfrac{3a^2\pi}{8}$

CHAPTER 17

SECTION 17.1

1.   (a) $\mathbf{h}(x,y) = y\,\mathbf{i} + x\,\mathbf{j};\quad \mathbf{r}(u) = u\,\mathbf{i} + u^2\,\mathbf{j},\quad u \in [0,1]$

   $x(u) = u,\; y(u) = u^2;\quad x'(u) = 1,\; y'(u) = 2u$

   $\mathbf{h}(\mathbf{r}(u)) \cdot \mathbf{r}'(u) = y(u)\,x'(u) + x(u)\,y'(u) = u^2(1) + u(2u) = 3u^2$

   $\displaystyle \int_C \mathbf{h}(\mathbf{r}) \cdot d\mathbf{r} = \int_0^1 3u^2\,du = 1$

   (b) $h(x,y) = y\,\mathbf{i} + x\,\mathbf{j};\quad \mathbf{r}(u) = u^3\,\mathbf{i} - 2u\,\mathbf{j},\quad u \in [0,1]$

   $x(u) = u^3,\; y(u) = -2u;\quad x'(u) = 3u^2,\; y'(u) = -2$

   $\mathbf{h}(\mathbf{r}(u)) \cdot \mathbf{r}'(u) = y(u)\,x'(u) + x(u)\,y'(u) = (-2u)(3u^2) + u^3(-2) = -8u^3$

   $\displaystyle \int_C \mathbf{h}(\mathbf{r}) \cdot d\mathbf{r} = \int_0^1 -8u^3\,du = -2$

3. $h(x,y) = y\,\mathbf{i} + x\,\mathbf{j};\quad \mathbf{r}(u) = \cos u\,\mathbf{i} - \sin u\,\mathbf{j},\quad u \in [0, 2\pi]$

   $x(u) = \cos u,\; y(u) = -\sin u;\quad x'(u) = -\sin u,\; y'(u) = -\cos u$

   $\mathbf{h}(\mathbf{r}(u)) \cdot \mathbf{r}'(u) = y(u)\,x'(u) + x(u)\,y'(u) = \sin^2 u - \cos^2 u$

   $\displaystyle \int_C \mathbf{h}(\mathbf{r}) \cdot d\mathbf{r} = \int_0^{2\pi} (\sin^2 u - \cos^2 u)\,du = 0$

5.   (a) $\mathbf{r}(u) = (2 - u)\,\mathbf{i} + (3 - u)\,\mathbf{j},\quad u \in [0,1]$

   $\displaystyle \int_C \mathbf{h}(\mathbf{r}) \cdot d\mathbf{r} = \int_0^1 (-5 + 5u - u^2)\,du = -\frac{17}{6}$

   (b) $\mathbf{r}(u) = (1 + u)\,\mathbf{i} + (2 + u)\,\mathbf{j},\quad u \in [0,1]$

   $\displaystyle \int_C \mathbf{h}(\mathbf{r}) \cdot d\mathbf{r} = \int_0^1 (1 + 3u + u^2)\,du = \frac{17}{6}$

7. $C = C_1 \cup C_2 \cup C_3$ where,

   $C_1 : \mathbf{r}(u) = (1 - u)(-2\,\mathbf{i}) + u(2\,\mathbf{i}) = (4u - 2)\,\mathbf{i},\quad u \in [0,1]$

   $C_2 : \mathbf{r}(u) = (1 - u)(2\,\mathbf{i}) + u(2\,wj) = (2 - 2u)\,\mathbf{i} + 2u\,\mathbf{j},\quad u \in [0,1]$

   $C_3 : \mathbf{r}(u) = (1 - u)(2\,\mathbf{j}) + u(-2\,\mathbf{i}) = -2u\,\mathbf{i} + (2 - 2u)\,\mathbf{j},\quad u \in [0,1]$

   $\displaystyle \int_C = \int_{C_1} + \int_{C_2} + \int_{C_3} = 0 + (-4) + (-4) = -8$

9.

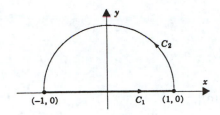

$$C_1 : \mathbf{r}(u) = (-1 + 2u)\,\mathbf{i}, \quad u \in [0,1]$$

$$C_2 : \mathbf{r}(u) = \cos u\,\mathbf{i} + \sin u\,\mathbf{j}, \quad u \in [0, \pi]$$

$$\int_C = \int_{C_1} + \int_{C_2} = 0 + (-\pi) = -\pi$$

11. (a) $\mathbf{r}(u) = u\,\mathbf{i} + u\,\mathbf{j} + u\mathbf{k}, \quad u \in [0,1]$

$$\int_C \mathbf{h}(\mathbf{r}) \cdot d\mathbf{r} = \int_0^1 3u^2 \, du = 1$$

(b) $\displaystyle\int_C \mathbf{h}(\mathbf{r}) \cdot d\mathbf{r} = \int_0^1 (2u^3 + u^5 + 3u^6) \, du = \frac{23}{21}$

13. (a) $\mathbf{r}(u) = 2u\,\mathbf{i} + 3u\,\mathbf{j} - u\,\mathbf{k}, \quad u \in [0,1]$

$$\int_C \mathbf{h}(\mathbf{r}) \cdot d\mathbf{r} = \int_0^1 (2 \cos 2u + 3 \sin 3u + 3u^2) \, du = \left[\sin 2u - \cos 3u + u^3\right]_0^1 = 2 + \sin 2 - \cos 3$$

(b) $\displaystyle\int_C \mathbf{h}(\mathbf{r}) \cdot d\mathbf{r} = \int_0^1 \left(2u \cos u^2 + 3u^2 \sin u^3 - u^4\right) du = \left[\sin u^2 - \cos u^3 - \frac{1}{5}u^5\right]_0^1 = \frac{4}{5} + \sin 1 - \cos 1$

15. $\mathbf{r}(u) = (1 - u)(\mathbf{j} + 4\mathbf{k}) + u(\mathbf{i} - 4\mathbf{k})$

$$= u\mathbf{i} + (1 - u)\mathbf{j} + (4 - 8u)\mathbf{k}, \quad u \in [0,1]$$

$$\int_C \mathbf{F}(\mathbf{r}) \cdot d\mathbf{r} = \int_0^1 (-32u + 97u^2 - 64u^3) \, du = \frac{1}{3}$$

17. (a) $\mathbf{r}(u) = u\,\mathbf{i} + u^2\,\mathbf{j}, \quad u \in [0,2]$

$$\int_C \mathbf{F}(\mathbf{r}) \cdot d\mathbf{r} = \int_0^2 \left[(u + 2)u^2 + (2u + u^2)2u\right] du = \int_0^2 \left(3u^3 + 6u^2\right) du = 28$$

(b) $\displaystyle\int_C \mathbf{F}(\mathbf{r}) \cdot d\mathbf{r} = \int_0^{\pi/2} \left[-(\cos u + 2)\sin^2 u + \cos u(2 \cos u + \sin u)\right] du$

$$= \int_0^{\pi/2} \left[-\sin^2 u \cos u + \cos 2u + \sin u \cos u\right] du = \frac{1}{6}$$

19. $\mathbf{r}(u) = \cos u\,\mathbf{i} + \sin u\,\mathbf{j} + u\mathbf{k}, \quad u \in [0, 2\pi]$

$$\int_C \mathbf{F}(\mathbf{r}) \cdot d\mathbf{r} = \int_0^{2\pi} \left[-\cos^2 u \sin u + \cos^2 u \sin u + u^2\right] du = \int_0^{2\pi} u^2 \, du = \frac{8\pi^3}{3}$$

21.
$$\int_C \mathbf{q} \cdot d\mathbf{r} = \int_a^b [\mathbf{q} \cdot \mathbf{r}'(u)] \, du = \int_a^b \frac{d}{du} [\mathbf{q} \cdot \mathbf{r}(u)] \, du$$

$$= [\mathbf{q} \cdot \mathbf{r}(b)] - [\mathbf{q} \cdot \mathbf{r}(a)]$$

$$= \mathbf{q} \cdot [\mathbf{r}(b) - \mathbf{r}(a)]$$

$$\int_C \mathbf{r} \cdot d\mathbf{r} = \int_a^b [\mathbf{r}(u) \cdot \mathbf{r}'(u)] \, du$$

$$= \frac{1}{2} \int_a^b \|\mathbf{r}\| \, d\|\mathbf{r}\| \quad \text{(see Exercise 53, Section 13.1)}$$

$$= \frac{1}{2} \left(\|\mathbf{r}(b)\|^2 - \|\mathbf{r}(a)\|^2 \right)$$

23. $\displaystyle \int_C \mathbf{f}(\mathbf{r}) \cdot d\mathbf{r} = \int_a^b [\mathbf{f}(\mathbf{r}(u)) \cdot \mathbf{r}'(u)] \, du = \int_a^b [f(u)\mathbf{i} \cdot \mathbf{i}] \, du = \int_a^b f(u) \, du$

25. $E : \mathbf{r}(u) = a \cos u \, \mathbf{i} + b \sin u \, \mathbf{j}, \quad u \in [0, 2\pi]$

$$W = \int_0^{2\pi} \left[\left(-\frac{1}{2} b \sin u \right) (-a \sin u) + \left(\frac{1}{2} a \cos u \right) (b \cos u) \right] \, du = \int_0^{2\pi} ab \, du = \pi ab$$

If the ellipse is traversed in the opposite direction, then $W = -\pi ab$. In both cases $|W| = \pi ab = $ area of the ellipse.

27. $\mathbf{r}(t) = \alpha t \mathbf{i} + \beta t^2 \mathbf{j} + \gamma t^3 \mathbf{k}$

$\mathbf{r}'(t) = \alpha \mathbf{i} + 2\beta t \mathbf{j} + 3\gamma t^2 \mathbf{k}$

force at time $t = m\mathbf{r}''(t) = m(2\beta \mathbf{j} + 6\gamma t \mathbf{k})$

$$W = \int_0^1 [m(2\beta \mathbf{j} + 6\gamma t \mathbf{k}) \cdot (\alpha \mathbf{i} + 2\beta t \mathbf{j} + 3\gamma t^2 \mathbf{k})] \, dt$$

$$= m \int_0^1 (4\beta^2 t + 18\gamma^2 t^3) \, dt = \left(2\beta^2 + \frac{9}{2}\gamma^2 \right) m$$

29. Take $C : \mathbf{r}(t) = r \cos t \, \mathbf{i} + r \sin t \, \mathbf{j}, \quad t \in [0, 2\pi]$

$$\int_C \mathbf{v}(\mathbf{r}) \cdot d\mathbf{r} = \int_0^{2\pi} [\mathbf{v}(\mathbf{r}(t)) \cdot \mathbf{r}'(t)] \, dt$$

$$= \int_0^{2\pi} [f(x(t), y(t)) \, \mathbf{r}(t) \cdot \mathbf{r}'(t)] \, dt$$

$$= \int_0^{2\pi} f(x(t), y(t)) [\mathbf{r}(t) \cdot \mathbf{r}'(t)] \, dt = 0$$

since for the circle $\mathbf{r}(t) \cdot \mathbf{r}'(t) = 0$ identically. The circulation is zero.

31. (a) $\mathbf{r}(u) = (1-u)(\mathbf{i} + 2\mathbf{k}) + u(\mathbf{i} + 3\mathbf{j} + 2\mathbf{k}) = \mathbf{i} + 3u\mathbf{j} + 2\mathbf{k}, \quad u \in [0,1].$

$$\int_C \mathbf{F}(\mathbf{r}) \cdot d\mathbf{r} = \int_0^1 \frac{9u\mathbf{k}}{5 + 9u^2)^{3/2}} \, du = \left[\frac{-\mathbf{k}}{\sqrt{5+9u^2}} \right]_0^1 = \frac{\mathbf{k}}{\sqrt{5}} - \frac{\mathbf{k}}{\sqrt{14}}$$

(b) $C = C_1 \cup C_2,$ where

$C_1 : \mathbf{r}(u) = (1-u)\mathbf{i} + 5u\mathbf{i} = (1+4u)\mathbf{i}, \quad u \in [0,1],$

$C_2 : x^2 + y^2 + z^2 = 25 \implies \quad \|\mathbf{r}\| = 5$

$$\int_{C_1} \mathbf{F}(\mathbf{r}) \cdot d\mathbf{r} = \int_0^1 \frac{4\mathbf{k}(1+4u)}{(1+4u)^3} \, du = \int_0^1 \frac{4\mathbf{k}}{(1+4u)^2} \, du \left[\frac{-\mathbf{k}}{1+4u} \right]_0^1 = \frac{4}{5}\mathbf{k}$$

$$\int_{C_2} \mathbf{F}(\mathbf{r}) \cdot d\mathbf{r} = \int_{C_2} \frac{k\mathbf{r}}{\|\mathbf{r}\|^3} \cdot d\mathbf{r}$$

$$= \frac{k}{5^3} \int_{C_2} \mathbf{r} \cdot d\mathbf{r} = \frac{k}{5^3} \int_{C_2} \|\mathbf{r}\| \, d\|\mathbf{r}\| \quad \text{(see Exercise 53, Section 13.1)}$$

$$= \frac{k}{5^3} \left[\frac{1}{2} \|\mathbf{r}\|^2 \right]_{(5,0,0)}^{(0,5/\sqrt{2},5/\sqrt{2})} = 0$$

Therefore, $\displaystyle\int_C \mathbf{F}(\mathbf{r}) \cdot d\mathbf{r} = \frac{4}{5}\mathbf{k}.$

33. $\mathbf{r}(u) = u\mathbf{i} + \alpha u(1-u)\mathbf{j}, \quad \mathbf{r}'(u) = \mathbf{i} + \alpha(1-2u)\mathbf{j}, \quad u \in [0,1]$

$$W(\alpha) = \int_C \mathbf{F}(\mathbf{r}) \cdot d\mathbf{r} = \int_0^1 \left[(\alpha^2 u^2(1-u)^2 + 1] + [u + \alpha u(1-u)]\alpha(1-2u) \right] dx$$

$$= \int_0^1 \left[1 + (\alpha + \alpha^2)u - (2\alpha + 2\alpha^2)u^2 + \alpha^2 u^4 \right] du = 1 - \frac{1}{6}\alpha + \frac{1}{30}\alpha^2$$

$$W'(\alpha) = -\frac{1}{6} + \frac{1}{15}\alpha \implies \alpha = \frac{15}{6}$$

The work done by $\mathbf{F}$ is a minimum when $\alpha = 15/6.$

SECTION 17.2

1. $\mathbf{h}(x,y) = \nabla f(x,y)$ where $f(x,y) = \frac{1}{2}(x^2 + y^2)$

C is closed $\implies \displaystyle\int_C \mathbf{h}(\mathbf{r}) \cdot d\mathbf{r} = 0$

3. $\mathbf{h}(x,y) = \nabla f(x,y)$ where $f(x,y) = x \cos \pi y; \quad \mathbf{r}(0) = \mathbf{0}, \quad \mathbf{r}(1) = \mathbf{i} - \mathbf{j}$

$$\int_C \mathbf{h}(\mathbf{r}) \cdot d\mathbf{r} = \int_C \nabla f(\mathbf{r}) \cdot d\mathbf{r} = f(\mathbf{r}(1)) - f(\mathbf{r}(0)) = f(1,-1) - f(0,0) = -1$$

5. $\mathbf{h}(x,y) = \nabla f(x,y)$ where $f(x,y) = \frac{1}{2}x^2 y^2; \quad \mathbf{r}(0) = \mathbf{j}, \quad \mathbf{r}(1) = -\mathbf{j}$

$$\int_C \mathbf{h}(\mathbf{r}) \cdot d\mathbf{r} = \int_C \nabla f(\mathbf{r}) \cdot d\mathbf{r} = f(\mathbf{r}(1)) - f(\mathbf{r}(0)) = f(0,-1) - f(0,1) = 0 - 0 = 0$$

7. $\mathbf{h}(x,y) = \nabla f(x,y)$ where $f(x,y) = x^2 y - xy^2$; $\mathbf{r}(0) = \mathbf{i}$, $\mathbf{r}(\pi) = -\mathbf{i}$

$$\int_C \mathbf{h}(\mathbf{r}) \cdot d\mathbf{r} = \int_C \nabla f(\mathbf{r}) \cdot d\mathbf{r} = f(\mathbf{r}(\pi)) - f(\mathbf{r}(0)) = f(-1,0) - f(1,0) = 0 - 0 = 0$$

9. $\mathbf{h}(x,y) = \nabla f(x,y)$ where $f(x,y) = (x^2 + y^4)^{3/2}$

$$\int_C \mathbf{h}(\mathbf{r}) \cdot d\mathbf{r} = \int_C \nabla f(\mathbf{r}) \cdot d\mathbf{r} = f(1,0) - f(-1,0) = 1 - 1 = 0$$

11. $\mathbf{h}(x,y)$ is not a gradient, but part of it,

$$2x \cosh y \, \mathbf{i} + (x^2 \sinh y - y)\mathbf{j},$$

is a gradient. Since we are integrating over a closed curve, the contribution of the gradient part is 0. Thus

$$\int_C \mathbf{h}(\mathbf{r}) \cdot d\mathbf{r} = \int_C (-y\mathbf{i}) \cdot d\mathbf{r}.$$

$$C_1: \mathbf{r}(u) = \mathbf{i} + (-1 + 2u)\mathbf{j}, \quad u \in [0,1]$$

$$C_2: \mathbf{r}(u) = (1 - 2u)\mathbf{i} + \mathbf{j}, \quad u \in [0,1]$$

$$C_3: \mathbf{r}(u) = -\mathbf{i} + (1 - 2u)\mathbf{j}, \quad u \in [0,1]$$

$$C_4: \mathbf{r}(u) = (-1 + 2u)\mathbf{i} - \mathbf{j}, \quad u \in [0,1]$$

$$
\begin{aligned}
\int_C \mathbf{h}(\mathbf{r}) \cdot d\mathbf{r} &= \int_{C_1}(-y\mathbf{i}) \cdot d\mathbf{r} + \int_{C_2}(-y\mathbf{i}) \cdot d\mathbf{r} + \int_{C_3}(-y\mathbf{i}) \cdot d\mathbf{r} + \int_{C_4}(-y\mathbf{i}) \cdot d\mathbf{r} \\
&= \quad 0 \quad + \int_0^1 -\mathbf{i} \cdot (-2\mathbf{i})\, du + \quad 0 \quad + \int_0^1 \mathbf{i} \cdot (2\mathbf{i})\, du \\
&= \quad 0 \quad + \quad \int_0^1 2\, du \quad + \quad 0 \quad + \quad \int_0^1 2\, du \\
&= \quad 4
\end{aligned}
$$

13. $\mathbf{h}(x,y) = (3x^2 y^3 + 2x)\,\mathbf{i} + (3x^3 y^2 - 4y)\,\mathbf{j}$; $\quad \dfrac{\partial P}{\partial y} = 9x^2 y^2 = \dfrac{\partial Q}{\partial x}$. Thus $\mathbf{h}$ is a gradient.

(a) $\mathbf{r}(u) = u\,\mathbf{i} + e^u\,\mathbf{j}$, $\quad \mathbf{r}'(u) = \mathbf{i} + e^u\,\mathbf{j}$, $\quad u \in [0,1]$

$$\int_C \mathbf{h}(\mathbf{r}) \cdot d\mathbf{r} = \int_0^1 \left[(3u^2 e^{3u} + 2u) + 3u^3 e^3 u - 4e^{2u}\right] du = \left[u^3 e^{3u} + u^2 - 2e^{2u}\right]_0^1 = e^3 - 2e^2 + 3$$

(b) $\dfrac{\partial f}{\partial x} = 3x^2 y^3 + 2x \implies f(x,y) = x^3 y^3 + x^2 + g(y)$;

$\dfrac{\partial f}{\partial y} = 3x^3 y^2 + g'(y) = 3x^3 - 4y \implies g'(y) = -4y \implies g(y) = -2y^2$

Therefore, $f(x,y) = x^3 y^3 + x^2 - 2y^2$.

Now, at $u = 0$, $r(0) = 0\,\mathbf{i} + \mathbf{j} = (0,1)$; at $u = 1$, $r(1) = \mathbf{i} + e\,\mathbf{j} = (1, e)$ and

$$\int_C \mathbf{h}(\mathbf{r}) \cdot d\mathbf{r} = \left[x^3 y^3 + x^2 - 2y^2\right]_{(0,1)}^{(1,e)} = e^3 - 2e^2 + 3$$

15. $\mathbf{h}(x, y) = (e^{2y} - 2xy)\,\mathbf{i} + (2xe^{2y} - x^2 + 1)\,\mathbf{j}$; $\quad \dfrac{\partial P}{\partial y} = 2e^{2y} - 2x = \dfrac{\partial Q}{\partial x}$. Thus $\mathbf{h}$ is a gradient.

(a) $\mathbf{r}(u) = ue^u\,\mathbf{i} + (1 + u)\,\mathbf{j}$, $\quad \mathbf{r}'(u) = (1 = u)e^u\,\mathbf{i} + \mathbf{j}$, $\quad u \in [0, 1]$

$$\int_C \mathbf{h}(\mathbf{r}) \cdot d\mathbf{r} = \int_0^1 \left[e^2(3ue^{3u} + e^{3u} - 2u^3 e^{2u} - 5u^2 e^{2u} - 2ue^{2u} + 1\right] du$$

$$= \left[e^2 ue^{3u} - u^3 e^{2u} - u^2 e^{2u} + u\right]_0^1 = e^5 - 2e^2 + 1$$

(b) $\dfrac{\partial f}{\partial x} = e^{2y} - 2xy \implies f(x, y) = xe^{2y} - x^2 y + g(y)$.

$$\dfrac{\partial f}{\partial y} = 2xe^{2y} - x^2 + g'(y) = 3x^3 - 4y \implies g'(y) = 1 \implies g(y) = y$$

Therefore, $f(x, y) = xe^{2y} - x^2 y + y$.

Now, at $u = 0$, $r(0) = 0\,\mathbf{i} + \mathbf{j} = (0, 1)$; at $u = 1$, $r(1) = e\,\mathbf{i} + 2\,\mathbf{j} = (e, 2)$ and

$$\int_C \mathbf{h}(\mathbf{r}) \cdot d\mathbf{r} = \left[xe^{2y} - x^2 y + y\right]_{(0,1)}^{(e,2)} = e^5 - 2e^2 + 1$$

17. $\mathbf{h}(x, y, z) = (2xz + \sin y)\,\mathbf{i} + x\,\cos y\,\mathbf{j} + x^2\,\mathbf{k}$;

$$\dfrac{\partial P}{\partial y} = \cos y = \dfrac{\partial Q}{\partial x}, \quad \dfrac{\partial P}{\partial z} = 2x = \dfrac{\partial R}{\partial x}, \quad \dfrac{\partial Q}{\partial z} = 0 = \dfrac{\partial R}{\partial y}. \quad \text{Thus } \mathbf{h} \text{ is a gradient.}$$

$$\dfrac{\partial f}{\partial x} = 2xz + \sin y, \quad \longrightarrow \quad f(x, y, z) = x^2 z + x\,\sin y + g(y, z)$$

$$\dfrac{\partial f}{\partial y} = x\,\cos y + \dfrac{\partial g}{\partial y} = x\,\cos y, \quad \implies \quad g(y, z) = h(z) \quad \implies f(x, y, z) = x^2 z + x\,\sin y + h(z)$$

$$\dfrac{\partial f}{\partial z} = x^2 + h'(z) = x^2 \quad \implies \quad h'(z) = 0 \quad \implies \quad h(z) = C$$

Therefore, $f(x, y, z) = x^2 z + x\,\sin y \quad (\text{take } C = 0)$

$$\int_C \mathbf{h}(\mathbf{r}) \cdot d\mathbf{r} = \int_C \nabla f \cdot d\mathbf{r} = \left[x^2 z + x\,\sin y\right]_{\mathbf{r}(0)}^{\mathbf{r}(2\pi)} = \left[x^2 z + x\,\sin y\right]_{(1,0,0)}^{(1,0,2\pi)} = 2\pi$$

19. $\mathbf{F}(x, y) = (x + e^{2y})\,\mathbf{i} + (2y + 2xe^{2y})\,\mathbf{j}$; $\quad \dfrac{\partial P}{\partial y} = 2e^{2y} = \dfrac{\partial Q}{\partial x}$. Thus $\mathbf{F}$ is a gradient.

$$\dfrac{\partial f}{\partial x} = x + e^{2y} \quad \implies \quad f(x, y) = \dfrac{1}{2}x^2 + xe^{2y} + g(y);$$

$$\dfrac{\partial f}{\partial y} = 2xe^{2y} + g'(y) = 2y + 2xe^{2y} \quad \implies \quad g'(y) = 2y \quad \implies g(y) = y^2 \ \text{ take } C = 0$$

Therefore, $f(x, y) = \dfrac{1}{2}x^2 + xe^{2y} + y^2 \quad (\text{take } C = 0)$.

$$\int_C \mathbf{F}(\mathbf{r}) \cdot d\mathbf{r} = \int_C \nabla f \cdot d\mathbf{r} = \left[\dfrac{1}{2}x^2 + xe^{2y} + y^2\right]_{\mathbf{r}(0)}^{\mathbf{r}(2\pi)} = \left[\dfrac{1}{2}x^2 + xe^{2y} + y^2\right]_{(3,0)}^{(3,0)} = 0$$

21. Set $f(x, y, z) = g(x)$ and $C : \mathbf{r}(u) = u\mathbf{i}$, $u \in [a, b]$.

In this case

$$\nabla f(\mathbf{r}(u)) = g'(x(u))\mathbf{i} = g'(u)\mathbf{i} \quad \text{and} \quad \mathbf{r}'(u) = \mathbf{i},$$

so that

$$\int_C \nabla f(\mathbf{r}) \cdot d\mathbf{r} = \int_a^b [\nabla f(\mathbf{r}(u)) \cdot \mathbf{r}'(u)] \, du = \int_a^b g'(u) \, du.$$

Since $\quad f(\mathbf{r}(b)) - f(\mathbf{r}(a)) = g(b) - g(a),$

$$\int_C \nabla f(\mathbf{r}) \cdot d\mathbf{r} = f(\mathbf{r}(b)) - f(\mathbf{r}(a)) \quad \text{gives} \quad \int_a^b g'(u) \, du = g(b) - g(a).$$

23. $\mathbf{F}(\mathbf{r}) = \nabla \left(\dfrac{mG}{r} \right);$ $\quad W = \displaystyle\int_C \mathbf{F}(\mathbf{r}) \cdot d\mathbf{r} = mG \left(\dfrac{1}{r_2} - \dfrac{1}{r_1} \right)$

25. $\mathbf{F}(x, y, z) = 0\mathbf{i} + 0\mathbf{j} + \dfrac{-mGr_0^2}{(r_0 + z)^2} \mathbf{k};$ $\quad \dfrac{\partial P}{\partial y} = 0 = \dfrac{\partial Q}{\partial x},$ $\quad \dfrac{\partial P}{\partial z} = 0 = \dfrac{\partial R}{\partial x},$ $\quad \dfrac{\partial Q}{\partial z} = 0 = \dfrac{\partial R}{\partial y}.$

Therefore, $\mathbf{F}(x, y, z)$ is a gradient.

$$\dfrac{\partial f}{\partial x} = 0 \implies f(x, y, z) = g(y, z); \quad \dfrac{\partial f}{\partial y} = \dfrac{\partial g}{\partial y} = 0 \implies g(y, z) = h(z).$$

Therefore $f(x, y, z) = h(z)$.

Now $\dfrac{\partial f}{\partial z} = h'(z) = \dfrac{-mGr_0^2}{(r_0 + z)^2} \implies f(x, y, z) = h(z) = \dfrac{mGr_0^2}{r_0 + z}$

27. By Exercise 25, the work required to lift an object of mass m a distance of h miles above the surface of the earth is:

$$W = \int_0^h -\mathbf{F} \cdot d\mathbf{r} = -f(0, 0, h) + f(0, 0, 0) = \dfrac{mGr_0^2 h}{r_0(r_0 + h)}$$

where $\mathbf{F} = \dfrac{-mGr_0^2}{(r_0 + z)^2} \mathbf{k}$, $\mathbf{r} = 0\mathbf{i} + 0\mathbf{j} + u\mathbf{k}$, $u \in [0, h]$, and $f = \dfrac{mGr_0^2}{r_0 + z}$.

In this particular case, put $r_0 = 4000$, $h = 500/5280$ and $m = 8000/32 = 250$.

Then $W \cong 23.67 \, G$.

SECTION 17.3

1. If f is continuous, then $-f$ is continuous and has antiderivatives u. The scalar fields $U(x, y, z) = u(x)$ are potential functions for $\mathbf{F}$:

$$\nabla U = \dfrac{\partial U}{\partial x} \mathbf{i} + \dfrac{\partial U}{\partial y} \mathbf{j} + \dfrac{\partial U}{\partial z} \mathbf{k} = \dfrac{du}{dx} \mathbf{i} = -f\mathbf{i} = -\mathbf{F}.$$

3. The scalar field $U(x, y, z) = cz + d$ is a potential energy function for **F**. We know that the total mechanical energy remains constant. Thus, for any times t_1 and t_2,

$$\tfrac{1}{2}m[v(t_1)]^2 + U(\mathbf{r}(t_1)) = \tfrac{1}{2}m[v(t_2)]^2 + U(\mathbf{r}(t_2)).$$

This gives

$$\tfrac{1}{2}m[v(t_1)]^2 + cz(t_1) + d = \tfrac{1}{2}m[v(t_2)]^2 + cz(t_2) + d.$$

Solve this equation for $v(t_2)$ and you have the desired formula.

5. (a) We know that $-\nabla U$ points in the direction of maximum decrease of U. Thus $\mathbf{F} = -\nabla U$ attempts to drive objects toward a region where U has lower values.

    (b) At a point where u has a minimum, $\nabla U = 0$ and therefore $\mathbf{F} = 0$.

7. (a) By conservation of energy $\tfrac{1}{2}mv^2 + U = E$. Since E is constant and U is constant, v is constant.

    (b) ∇U is perpendicular to any surface where U is constant. Obviously so is $\mathbf{F} = -\nabla U$.

9. $f(x, y, z) = -\dfrac{k}{\sqrt{x^2 + y^2 + z^2}}$ is a potential function for **F**. The work done by **F** moving an object along C is:

$$W = \int_C \mathbf{F}(\mathbf{r}) \cdot d\mathbf{r} = \int_a^b \nabla f \cdot d\mathbf{r} = f[\mathbf{r}(b)] - f[\mathbf{r}(a)].$$

Since $\mathbf{r}(a) = (x_0, y_0, z_0)$ and $\mathbf{r}(b) = (x_1, y_1, z_1)$ are points on the unit sphere,

$$f[\mathbf{r}(b)] - f[\mathbf{r}(a)] = -k \quad \text{and so} \quad W = 0$$

SECTION 17.4

1. $\mathbf{r}(u) = u\,\mathbf{i} + 2u\,\mathbf{j}, \quad u \in [0, 1]$

$$\int_C (x - 2y)\, dx + 2x\, dy = \int_0^1 \{[x(u) - 2y(u)]x'(u) + 2x(u)\, y'(u)\}\, du = \int_0^1 u\, du = \frac{1}{2}$$

3. $C = C_1 \cup C_2$

$C_1: \mathbf{r}(u) = u\,\mathbf{i}, \quad u \in [0, 1]; \qquad C_2: \mathbf{r}(u) = \mathbf{i} + 2u\mathbf{j}, \quad u \in [0, 1]$

$$\int_{C_1} (x - 2y)\, dx + 2x\, dy = \int_{C_1} x\, dx = \int_0^1 x(u)\, x'(u)\, du = \int_0^1 u\, du = \frac{1}{2}$$

$$\int_{C_2} (x - 2y)\, dx + 2x\, dy = \int_{C_2} 2x\, dy = \int_0^1 4\, du = 4$$

$$\int_C = \int_{C_1} + \int_{C_2} = 4\tfrac{1}{2}$$

5. $\mathbf{r}(u) = 2u^2\,\mathbf{i} + u\,\mathbf{j}, \quad u \in [0,1]$

$$\int_C y\,dx + xy\,dy = \int_0^1 [y(u)\,x'(u) + x(u)\,y(u)\,y'(u)]\,du = \int_0^1 (4u^2 + 2u^3)\,du = \frac{11}{6}$$

7. $C = C_1 \cup C_2$

$C_1 : \mathbf{r}(u) = u\,\mathbf{j}, \quad u \in [0,1]; \qquad C_2 : \mathbf{r}(u) = 2u\,\mathbf{i} + \mathbf{j}, \quad u \in [0,1]$

$$\int_{C_1} y\,dx + xy\,dy = 0$$

$$\int_{C_2} y\,dx + xy\,dy = \int_{C_2} y\,dx = \int_0^1 y(u)\,x'(u)\,du = \int_0^1 2\,du = 2$$

$$\int_C = \int_{C_1} + \int_{C_2} = 2$$

9. $\mathbf{r}(u) = 2u\,\mathbf{i} + 4u\,\mathbf{j}, \quad u \in [0,1]$

$$\int_C y^2\,dx + (xy - x^2)\,dy = \int_0^1 \left\{ y^2(u)x'(u) + [x(u)y(u) - x^2(u)]\,y'(u) \right\}\,du$$

$$= \int_0^1 \left[(4u)^2(2) + (8u^2 - 4u^2)(4) \right]\,du = \int_0^1 48u^2\,du = 16$$

11. $\mathbf{r}(u) = \dfrac{1}{8}u^2\,\mathbf{i} + u\,\mathbf{j}, \quad u \in [0,4]$

$$\int_C y^2\,dx + (xy - x^2)\,dy = \int_0^4 \left\{ y^2(u)x'(u) + [x(u)y(u) - x^2(u)]\,y'(u) \right\}\,du$$

$$= \int_0^4 \left[u^2 \left(\frac{u}{4}\right) + \left(\frac{u^2}{8}(u) - \left(\frac{u^2}{8}\right)^2 (1) \right) \right]\,du$$

$$= \int_0^1 \left[\frac{3}{8}u^3 - \frac{1}{64}u^4 \right]\,du = \frac{104}{5}$$

13. $\mathbf{r}(u) = u\,\mathbf{i} + u\,\mathbf{j}, \quad u \in [0,1]$

$$\int_C (y^2 + 2x + 1)\,dx + (2xy + 4y - 1)\,dy$$

$$= \int_0^1 \left\{ [y^2(u) + 2x(u) + 1]x'(u) + [2x(u)y(u) + 4y(u) - 1]y'(u) \right\}\,du$$

$$\int_0^1 \left[(u^2 + 2u + 1) + (2u^2 + 4u - 1) \right]\,du = \int_0^1 (3u^2 + 6u)\,du = 4$$

15. $\mathbf{r}(u) = u\,\mathbf{i} + u^3\,\mathbf{j}, \quad u \in [0,1]$

$$\int_C (y^2 + 2x + 1)\,dx + (2xy + 4y - 1)\,dy$$

$$= \int_0^1 \left\{ [y^2(u) + 2x(u) + 1]x'(u) + [2x(u)y(u) + 4y(u) - 1]y'(u) \right\} \, du$$

$$= \int_0^1 \left[(u^6 + 2u + 1) + (2u^4 + 4u^3 - 1)3u^2 \right] \, du = \int_0^1 \left(7u^6 + 12u^5 - 3u^2 + 2u + 1 \right) \, du = 4$$

17. $\mathbf{r}(u) = u\mathbf{i} + u\mathbf{j} + u\mathbf{k}, \quad u \in [0,1]$

$$\int_C y \, dx + 2z \, dy + x \, dz = \int_0^1 [y(u) \, x'(u) + 2z(u) \, y'(u) + x(u) \, z'(u)] \, du = \int_0^1 4u \, du = 2$$

19. $C = C_1 \cup C_2 \cup C_3$

$C_1 : \mathbf{r}(u) = u\mathbf{k}, \quad u \in [0,1]; \quad C_2 : \mathbf{r}(u) = u\mathbf{j} + \mathbf{k}, \quad u \in [0,1]; \quad C_3 : \mathbf{r}(u) = u\mathbf{i} + \mathbf{j} + \mathbf{k}, \quad u \in [0,1]$

$$\int_{C_1} y \, dx + 2z \, dy + x \, dz = 0$$

$$\int_{C_2} y \, dx + 2z \, dy + x \, dz = \int_{C_2} 2z \, dy = \int_0^1 2z(u) \, y'(u) \, du = \int_0^1 2 \, du = 2$$

$$\int_{C_3} y \, dx + 2z \, dy + x \, dz = \int_{C_3} y \, dx = \int_0^1 y(u) \, x'(u) \, du = \int_0^1 du = 1$$

$$\int_C = \int_{C_1} + \int_{C_2} + \int_{C_3} = 3$$

21. $\mathbf{r}(u) = 2u\,\mathbf{i} + 2u\,\mathbf{j} + 8u\,\mathbf{k}, \quad u \in [0,1]$

$$\int_C xy \, dx + 2z \, dy + (y + z) \, dz$$

$$= \int_0^1 \left\{ x(u)y(u)x'(u) + 2z(u)y'(u) + [y(u) + z(u)]z'(u) \right\} \, du$$

$$= \int_0^1 \left[(2u)(2u)(2) + 2(8u)(2) + (2u + 8u)(8) \right] \, du$$

$$= \int_0^1 \left(8u^2 + 112u \right) \, du = \frac{176}{3}$$

23. $\mathbf{r}(u) = u\,\mathbf{i} + u\,\mathbf{j} + 2u^2\,\mathbf{k}, \quad u \in [0,2]$

$$\int_C xy \, dx + 2z \, dy + (y + z) \, dz$$

$$= \int_0^2 \left\{ x(u)y(u)x'(u) + 2z(u)y'(u) + [y(u) + z(u)]z'(u) \right\} \, du$$

$$= \int_0^2 \left[(u)(u)(1) + 2(2u^2)(1) + (u + 2u^2)(4u) \right] \, du$$

$$= \int_0^2 \left(8u^3 + 9u^2 \right) \, du = 56$$

25. $\mathbf{r}(u) = (u-1)\mathbf{i} + (1+2u^2)\mathbf{j} + u\mathbf{k}, \quad u \in [1,2]$

$$\int_C x^2 y \, dx + y \, dy + xz \, dz$$

$$= \int_1^2 \left[x^2(u)y(u)x'(u) + y(u)y'(u) + x(u)z(u)z'(u) \right] du$$

$$= \int_1^2 \left[(u-1)^2(1+2u^2)(1) + (1+2u^2)(4u) + (u-1)u \right] du$$

$$= \int_1^2 \left(2u^4 + 4u^3 + 4u^2 + u + 1 \right) du = \frac{1177}{30}$$

27.   (a) $\dfrac{\partial P}{\partial y} = 6x - 4y = \dfrac{\partial Q}{\partial x}$

$$\frac{\partial f}{\partial x} = x^2 + 6xy - 2y^2 \quad \Longrightarrow \quad f(x,y) = \frac{1}{3}x^3 + 3x^2 y - 2xy^2 + g(y)$$

$$\frac{\partial f}{\partial y} = 3x^2 - 4xy + g'(y) = 3x^2 - 4xy + 2y \quad \Longrightarrow \quad g'(y) = 2y \quad \Longrightarrow g(y) = y^2 + C$$

Therefore, $f(x,y) = \dfrac{1}{3}x^3 + 3x^2 y - 2xy^2 + y^2$ (take $C = 0$)

(b) $\displaystyle\int_C (x^2 + 6xy - 2y^2)\, dx + (3x^2 - 4xy + 2y)\, dy = [f(x,y)]_{(3,0)}^{(0,4)} = 7$

(c) $\displaystyle\int_C' (x^2 + 6xy - 2y^2)\, dx + (3x^2 - 4xy + 2y)\, dy = [f(x,y)]_{(4,0)}^{(0,3)} = -\frac{37}{3}$

29. $s'(u) = \sqrt{[x'(u)]^2 + [y'(u)]^2} = a$

(a) $M = \displaystyle\int_C k(x+y)\, ds = k \int_0^{\pi/2} [x(u) + y(u)]\, s'(u)\, du = ka^2 \int_0^{\pi/2} (\cos u + \sin u)\, du = 2ka^2$

(b)
$$x_M M = \int_C kx(x+y)\, ds = k \int_0^{\pi/2} x(u)\, [x(u) + y(u)]\, s'(u)\, du$$

$$= ka^3 \int_0^{\pi/2} (\cos^2 u + \cos u \sin u)\, du = \frac{1}{4} ka^3(\pi + 2)$$

$$y_M M = \int_C ky(x+y)\, ds = k \int_0^{\pi/2} y(u)\, [x(u) + y(u)]\, s'(u)\, du$$

$$= ka^3 \int_0^{\pi/2} (\sin u \cos u + \sin^2 u)\, du = \frac{1}{4} ka^3(\pi + 2)$$

$x_M = y_M = \frac{1}{8} a(\pi + 2)$

31.   (a) $I_z = \displaystyle\int_C k(x+y)a^2\, ds = a^2 \int_C k(x+y)\, ds = a^2 M = Ma^2$

(b)   The distance from the point (x,y) to the line $y = x$ is $|y - x|/\sqrt{2}$.   Therefore

$$I = \int_C k(x+y) \left[\frac{1}{2}(y-x)^2\right] ds = \frac{1}{2}k \int_0^{\pi/2} (a\cos u + a\sin u)(a\sin u - a\cos u)^2 a\,du$$

$$= \frac{1}{2}ka^4 \int_0^{\pi/2} (\sin u - \cos u)^2 \frac{d}{du}(\sin u - \cos u)\,du$$

$$= \frac{1}{2}ka^4 \left[\frac{1}{3}(\sin u - \cos u)^3\right]_0^{\pi/2} = \frac{1}{3}ka^4.$$

From Exercise 29, $M = 2ka^2$.   Therefore

$$I = \tfrac{1}{6}(2ka^2)a^2 = \tfrac{1}{6}Ma^2.$$

33. (a) $s'(u) = \sqrt{a^2 + b^2}$

$$L = \int_C ds = \int_0^{2\pi} \sqrt{a^2 + b^2}\,du = 2\pi\sqrt{a^2 + b^2}$$

(b) $x_M = 0, \quad y_M = 0$ (by symmetry)

$$z_M = \frac{1}{L}\int_C z\,ds = \frac{1}{2\pi\sqrt{a^2+b^2}}\int_0^{2\pi} bu\sqrt{a^2+b^2}\,du = b\pi$$

(c) $I_x = \int_C \frac{M}{L}(y^2 + z^2)\,ds = \frac{M}{2\pi}\int_0^{2\pi}(a^2\sin^2 u + b^2 u^2)\,du = \frac{1}{6}M(3a^2\pi + 8b^2\pi^2)$

$I_y = \frac{1}{6}M(3a^2\pi + 8b^2\pi^2)$ similarly

$I_z = Ma^2$ (all the mass is at distance a from the z-axis)

35.
$$M = \int_C k(x^2 + y^2 + z^2)\,ds$$

$$= k\sqrt{a^2 + b^2}\int_0^{2\pi}(a^2 + b^2 u^2)\,du = \frac{2}{3}\pi k\sqrt{a^2 + b^2}\,(3a^2 + 4\pi^2 b^2)$$

SECTION 17.5

1. (a) $\oint_C xy\,dx + x^2\,dy = \int_{C_1} xy\,dx + x^2\,dy + \int_{C_2} xy\,dx + x^2\,dy + \int_{C_3} xy\,dx + x^2\,dy,$ where

$C_1 : \mathbf{r}(u) = u\,\mathbf{i} + u\,\mathbf{j}, \quad u \in [0,1]; \quad C_2 : \mathbf{r}(u) = (1-u\,\mathbf{i} + \mathbf{j}, \quad u \in [0,1]$

$C_3 : \mathbf{r}(u) = (1-u)\,\mathbf{j}, \quad u \in [0,1].$

$$\int_{C_1} xy\,dx + x^2\,dy = \int_0^1 (u^2 + u^2)\,du = \frac{2}{3}$$

$$\int_{C_2} xy\,dx + x^2\,dy = \int_0^1 -(1-u)\,du = -\frac{1}{2}$$

$$\int_{C_3} xy\,dx + x^2\,dy = \int_0^1 0^2(-1)\,du = 0$$

Therefore, $\oint_C xy\,dx + x^2\,dy = \frac{2}{3} - \frac{1}{2} = \frac{1}{6}$.

(b) $\oint_C xy\,dx + x^2\,dy = \iint_\Omega x\,dx\,dy = \int_0^1 \int_0^y x\,dx\,dy = \int_0^1 \left[\frac{1}{2}x^2\right]_0^y du = \frac{1}{2}\int_0^1 y^2\,dy = \frac{1}{6}$

3. (a) $C: \mathbf{r}(u) = 2\cos u\,\mathbf{i} + 3\sin u\,\mathbf{j}, \quad u \in [0, 2\pi]$

$$\oint_C (3x^2 + y)\,dx + (2x + y^3)\,dy = \int_0^{2\pi} [(12\cos^2 u + 3\sin u)(-2\sin u) + (4\cos u + 27\sin^3 u)3\cos u]\,du$$

$$= \int_0^{2\pi} [-24\cos^2 u \sin u - 6\sin^2 u + 12\cos^2 u + 81\sin^3 u \cos u]\,du$$

$$= \left[8\cos^3 u - 3u + \frac{3}{2}\sin 2u + 6u + 3\sin 2u + \frac{81}{4}\sin^4 u\right]_0^{2\pi} = 6\pi$$

(b) $\oint_C (3x^2 + y)\,dx + (2x + y^3)\,dy = \iint_\Omega 1\,dx\,dy = \text{area}\,(\Omega) = \pi(3)(2) = 6\pi$

5. $\oint_C 3y\,dx + 5x\,dy = \iint_\Omega (5-3)\,dx\,dy = 2A = 2\pi$

7. $\oint_C x^2\,dy = \iint_\Omega 2x\,dx\,dy = 2\bar{x}A = 2\left(\frac{a}{2}\right)(ab) = a^2 b$

9. $\oint_C (3xy + y^2)\,dx + (2xy + 5x^2)\,dy = \iint_\Omega [(2y + 10x) - (3x + 2y)]\,dx\,dy$

$$= \iint_\Omega 7x\,dx\,dy = 7\bar{x}A = 7(1)(\pi) = 7\pi$$

11. $\oint_C (2x^2 + xy - y^2)\,dx + (3x^2 - xy + 2y^2)\,dy = \iint_\Omega [(6x - y) - (x - 2y)]\,dx\,dy$

$$= \iint_\Omega (5x + y)\,dx\,dy = (5\bar{x} + \bar{y})A = (5a + 0)(\pi r^2) = 5a\pi r^2$$

13. $\oint_C e^x \sin y\,dx + e^x \cos y\,dy = \iint_\Omega [e^x \cos y - e^x \cos y]\,dx\,dy = 0$

15. $\oint_C 2xy\,dx + x^2\,dy = \iint_\Omega [2x - 2x]\,dx\,dy = 0$

17. $C: \mathbf{r}(u) = a\cos u\,\mathbf{i} + a\sin u\,\mathbf{j}; \quad u \in [0, 2\pi]$

$$A = \oint_C -y\,dx = \int_0^{2\pi} (-a\sin u)(-\sin u)\,du = a^2 \int_0^{2\pi} \sin^2 u\,du = a^2 \left[\frac{1}{2}u - \frac{1}{4}\sin 2u\right]_0^{2\pi} = \pi a^2$$

19. $A = \oint_C x\,dy,$ where $C = C_1 \cup C_2 \cup C_3;$

$$C_1 : \mathbf{r}(u) = au\,\mathbf{i}, \quad u \in [0,1]; \quad C_2 : \mathbf{r}(u) = a(1-u)\,\mathbf{i} + bu\,\mathbf{j}, \quad u \in [0,1];$$

$$C_3 : \mathbf{r}(u) = b(1-u)\,\mathbf{j}, \quad u \in [0,1].$$

$$\int_{C_1} x\,dy = 0; \quad \int_{C_2} x\,dy = \int_0^1 ab(1-u)\,du = \frac{1}{2}ab; \quad \int_{C_3} x\,dy = 0$$

Therefore, $A = \dfrac{1}{2}ab.$

21. $\displaystyle\oint_C (ay+b)\,dx + (cx+d)\,dy = \iint_\Omega (c-a)\,dxdy = (c-a)A$

23.   We take the arch from $x = 0$ to $x = 2\pi R.$ (Figure 9.11.1)   Let C_1 be the line segment from $(0,0)$ to $(2\pi R, 0)$ and let C_2 be the cycloidal arch from $(2\pi R, 0)$ back to $(0,0)$. Letting $C = C_1 \cup C_2$, we have

$$A = \oint_C x\,dy = \int_{C_1} x\,dy + \int_{C_2} x\,dy = 0 + \int_{C_2} x\,dy$$

$$= \int_{2\pi}^0 R(\theta - \sin\theta)(R\sin\theta)\,d\theta$$

$$= R^2 \int_0^{2\pi} (\sin^2\theta - \theta\sin\theta)\,d\theta$$

$$= R^2 \left[\frac{\theta}{2} - \frac{\sin 2\theta}{4} + \theta\cos\theta - \sin\theta\right]_0^{2\pi} = 3\pi R^2.$$

25. $\mathbf{F}(x,y) = (x^2 - y^3)\,\mathbf{i} + (x^2 + y^2)\,\mathbf{j}; \quad C : \mathbf{r} = \cos u\,\mathbf{i} + \sin u\,\mathbf{j}, \quad u \in [0, 2\pi]$

$$W = \oint_C \mathbf{F} \cdot d\mathbf{r} = \iint_\Omega (2x + 3y^2)\,dx\,dy = \int_0^{2\pi} \int_0^1 (2r\cos\theta + 3r^2\sin^2\theta)\,r\,dr\,d\theta$$

$$= \int_0^{2\pi} \int_0^1 (2r^2\cos\theta + 3r^3\sin^2\theta)\,dr\,d\theta$$

$$= \frac{2}{3} \int_0^{2\pi} \cos\theta\,d\theta + \frac{3}{4} \int_0^{2\pi} \sin^2\theta\,d\theta$$

$$= \frac{2}{3} [\sin\theta]_0^{2\pi} + \frac{3}{4} \left[\frac{1}{2}u - \frac{1}{4}\sin 2u\right]_0^{2\pi} = \frac{3}{4}\pi$$

27. Taking Ω to be of type II (see Figure 17.5.2), we have

$$\iint_\Omega \frac{\partial Q}{\partial x}(x,y)\,dxdy = \int_c^d \int_{\phi_3(y)}^{\phi_4(y)} \frac{\partial Q}{\partial x}(x,y)\,dx\,dy$$

$$= \int_c^d \{Q[\phi_4(y),y] - Q[\phi_3(y),y]\}\,dy$$

$$(*) = \int_c^d Q[\phi_4(y),y]\,dy - \int_c^d Q[\phi_3(y),y]\,dy.$$

The graph of $x = \phi_4(y)$ from $x = c$ to $x = d$ is the curve

$$C_4 : \mathbf{r}_4(u) = \phi_4(u)\mathbf{i} + u\mathbf{j}, \qquad u \in [c,d].$$

The graph of $x = \phi_3(y)$ from $x = c$ to $x = d$ is the curve

$$C_3 : \mathbf{r}_3(u) = \phi_3(u)\mathbf{i} + u\mathbf{j}, \qquad u \in [c,d].$$

Then

$$\oint_C Q(x,y)\,dy = \int_{C_4} Q(x,y)\,dy - \int_{C_3} Q(x,y)\,dy$$

$$= \int_c^d Q[\phi_4(u),u]\,du - \int_c^d Q[\phi_3(u),u]\,du.$$

Since u is a dummy variable, it can be replaced by y. Comparison with $(*)$ gives the result.

29. $\oint_{C_1} = \oint_{C_2} + \oint_{C_3}$

31. $\dfrac{\partial P}{\partial y} = \dfrac{-2xy}{(x^2+y^2)^2} = \dfrac{\partial Q}{\partial x}$ except at $(0,0)$

(a) If C does not enclose the origin, and Ω is the region enclosed by C, then

$$\oint_C \frac{x}{x^2+y^2}\,dx + \frac{y}{x^2+y^2}\,dy = \iint_\Omega 0\,dxdy = 0.$$

(b) If C does enclose the origin, then

$$\oint_C = \oint_{C_a}$$

where $C_a : \mathbf{r}(u) = a\cos u\,\mathbf{i} + a\sin u\,\mathbf{j}, \quad u \in [0,2\pi]$ is a small circle in the inner region of C.

In this case

$$\oint_C = \int_0^{2\pi} \left[\frac{a\cos u}{a^2}(-a\sin u) + \frac{a\sin u}{a^2}(a\cos u) \right] du = \int_0^{2\pi} 0\,du = 0.$$

The integral is still 0.

33. If Ω is the region enclosed by C, then

$$\oint_C \mathbf{v} \cdot \mathbf{dr} = \oint_C \frac{\partial \phi}{\partial x}\,dx + \frac{\partial \phi}{\partial y}\,dy = \iint_\Omega \left\{ \frac{\partial}{\partial x}\left(\frac{\partial \phi}{\partial y}\right) - \frac{\partial}{\partial y}\left(\frac{\partial \phi}{\partial x}\right) \right\}\,dxdy$$

$$= \iint_\Omega 0\,dxdy = 0.$$

equality of mixed partials

35. $A = \dfrac{1}{2} \oint_C (-y\,dx + x\,dy)$

$$= \left[\int_{C_1} + \int_{C_2} + \cdots \int_{C_n} \right]$$

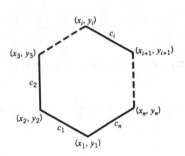

Now

$$\int_{C_i} (-y\,dx + x\,dy) = \int_0^1 \left\{ [y_i + u(y_{i+1} - y_i)](x_{i+1} - x_i) + [x_i + u(x_{i+1} - x_i)](y_{i+1} - y_i) \right\}\,du$$

$$= x_i y_{i+1} - x_{i+1} y_i, \quad i = 1, 2, \ldots, n; \ x_{n+1} = x_1,\ y_{n+1} = y_1$$

Thus, $A = \dfrac{1}{2}\left[(x_1 y_2 - x_2 y_1) + (x_2 y_3 - x_3 y_2) + \cdots + (x_n y_1 - x_1 y_n) \right]$

PROJECT 17.5

1. Replace y by xt and solve for x to get $\quad x = \dfrac{3at}{1+t^3}$

Then $\quad y = xt = \dfrac{3at^2}{1+t^3}$

3. The curve is symmetric with respect to the line $y = x$ since interchanging x and y leaves the equation unchanged.

As t varies from $-\infty$ to -1, the point $(x(t), y(t))$ moves along the curve from $(0,0)$ to "$(\infty, -\infty)$";
as t varies from -1 to ∞, the point $(x(t), y(t))$ moves along the curve from "$(-\infty, \infty)$" to $(0,0)$.
The loop is traced out in the counter-clockwise direction at t increases from 0 to ∞.

5. Rotate the axes through an angle of $\frac{\pi}{4}$ in the clockwise direction and then translate the origin to the point $\left(0, -\frac{a}{\sqrt{2}}\right)$. In the new coordinate system, the parametric equations of the curve are:

$$u = \frac{3a}{\sqrt{2}}\left(\frac{t - t^2}{1 + t^3}\right), \quad v = \frac{3a}{\sqrt{2}}\left(\frac{t + t^2}{1 + t^3}\right) + \frac{a}{\sqrt{2}}$$

and the u-axis is the asymptote. Using the third equation in (17.5.2), it can be shown that the area between the curve and the asymptote is given by:

$$A = \int_{-1}^{0} [v(t)u'(t) - u(t)v'(t)]\, dt = \frac{3}{2}a^2$$

SECTION 17.6

1. $4[(u^2 - v^2)\mathbf{i} - (u^2 + v^2)\mathbf{j} + 2uv\mathbf{k}]$ 3. $2(\mathbf{j} - \mathbf{i})$

5. $\mathbf{r}(u, v) = 3\cos u \cos v\, \mathbf{i} + 2\sin u \cos v\, \mathbf{j} + 6\sin v\, \mathbf{k}, \quad u \in [0\, 2\pi], \ v \in [0, \pi/2]$

7. $\mathbf{r}(u, v) = 2\cos u \cos v\, \mathbf{i} + 2\sin u \cos v\, \mathbf{j} + 2\sin v\, \mathbf{k}, \quad u \in [0\, 2\pi], \ v \in (\pi/4, \pi/2]$

9.   The surface consists of all points of the form $(x, g(x, z), z)$ with $(x, z) \in \Omega$. This set of points

is given by

$$\mathbf{r}(u, v) = u\mathbf{i} + g(u, v)\mathbf{j} + v\mathbf{k}, \quad (u, v) \in \Omega.$$

11. $x^2/a^2 + y^2/b^2 + z^2/c^2 = 1;$ ellipsoid

13. $x^2/a^2 - y^2/b^2 = z;$ hyperbolic paraboloid

15.   For each $v \in [a, b]$, the points on the surface at level $z = f(v)$ form a circle of radius v. That circle can be parametrized

$$\mathbf{R}(u) = v\cos u\, \mathbf{i} + v\sin u\, \mathbf{j} + f(v)\mathbf{k}, \quad u \in [0, 2\pi].$$

Letting v range over $[a, b]$, we obtain the entire surface:

$$\mathbf{r}(u, v) = v\cos u\, \mathbf{i} + v\sin u\, \mathbf{j} + f(v)\mathbf{k}; \quad 0 \le u \le 2\pi, \quad a \le v \le b.$$

17.   Since γ is the angle between p and the xy-plane, γ is the angle between the upper normal to p and $\mathbf{k}$. (Draw a figure.) Therefore, by 17.6.5,

$$\text{area of } \Gamma = \iint_{\Omega} \sec \gamma\, dxdy = (\sec \gamma)A_{\Omega} = A_{\Omega}\sec \gamma.$$

$$\gamma \text{ is constant}$$

19. The surface is the graph of the function

$$f(x,y) = c\left(1 - \frac{x}{a} - \frac{y}{b}\right) = \frac{c}{ab}(ab - bx - ay)$$

defined over the triangle $\Omega : 0 \leq x \leq a, \quad 0 \leq y \leq b(1 - x/a)$. Note that Ω has area $\frac{1}{2}ab$.

$$A = \iint_{\Omega} \sqrt{[f_x'(x,y)]^2 + [f_y'(x,y)]^2 + 1} \; dxdy$$

$$= \iint_{\Omega} \sqrt{c^2/a^2 + c^2/b^2 + 1} \; dxdy$$

$$= \frac{1}{ab}\sqrt{a^2b^2 + a^2c^2 + b^2c^2} \iint_{\Omega} dxdy = \frac{1}{2}\sqrt{a^2b^2 + a^2c^2 + b^2c^2}.$$

21. $f(x,y) = x^2 + y^2, \quad \Omega : 0 \leq x^2 + y^2 \leq 4$

$$A = \iint_{\Omega} \sqrt{4x^2 + 4y^2 + 1} \; dxdy \qquad [\text{change to polar coordinates}]$$

$$= \int_0^{2\pi} \int_0^2 \sqrt{4r^2 + 1}\, r \; dr \; d\theta$$

$$= 2\pi \left[\tfrac{1}{12}(4r^2 + 1)^{3/2}\right]_0^2 = \tfrac{1}{6}\pi(17\sqrt{17} - 1)$$

23. $f(x,y) = a^2 - (x^2 + y^2), \quad \Omega : \tfrac{1}{4}a^2 \leq x^2 + y^2 \leq a^2$

$$A = \iint_{\Omega} \sqrt{4x^2 + 4y^2 + 1} \; dxdy \qquad [\text{change to polar coordinates}]$$

$$= \int_0^{2\pi} \int_{a/2}^a r\sqrt{4r^2 + 1} \; dr \; d\theta = 2\pi \left[\frac{1}{12}(4r^2 + 1)^{3/2}\right]_{a/2}^a$$

$$= \frac{\pi}{6}\left[(4a^2 + 1)^{3/2} - (a^2 + 1)^{3/2}\right]$$

25. $f(x,y) = \tfrac{1}{3}(x^{3/2} + y^{3/2}), \quad \Omega : 0 \leq x \leq 1, \quad 0 \leq y \leq x$

$$A = \iint_{\Omega} \frac{1}{2}\sqrt{x + y + 4} \; dxdy$$

$$= \int_0^1 \int_0^x \frac{1}{2}\sqrt{x + y + 4} \; dy \; dx = \int_0^1 \left[\frac{1}{3}(x + y + 4)^{3/2}\right]_0^x dx$$

$$= \int_0^1 \frac{1}{3}\left[(2x + 4)^{3/2} - (x + 4)^{3/2}\right] dx = \frac{1}{3}\left[\frac{1}{5}(2x + 4)^{5/2} - \frac{2}{5}(x + 4)^{5/2}\right]_0^1$$

$$= \tfrac{1}{15}(36\sqrt{6} - 50\sqrt{5} + 32)$$

27.   The surface $x^2 + y^2 + z^2 - 4z = 0$ is a sphere of radius 2 centered at $(0, 0, 2)$:

$$x^2 + y^2 + z^2 - 4z = 0 \quad \Longleftrightarrow \quad x^2 + y^2 + (z-2)^2 = 4.$$

The quadric cone $z^2 = 3(x^2 + y^2)$ intersects the sphere at height $z = 3$:

$$\left. \begin{array}{r} x^2 + y^2 + z^2 - 4z = 0 \\ z^2 = 3(x^2 + y^2) \end{array} \right\} \quad \Longrightarrow \quad \begin{array}{l} 3(x^2 + y^2) + 3z^2 - 12z = 0 \\ 4z^2 - 12z = 0 \\ z = 3. \quad (\text{since } z \geq 2) \end{array}$$

The surface of which we are asked to find the area is a spherical segment of width 1 (from $z = 3$ to $z = 4$) in a sphere of radius 2. The area of the segment is 4π. (Exercise 27, Section 9.9.)

A more conventional solution. The spherical segment is the graph of the function

$$f(x, y) = 2 + \sqrt{4 - (x^2 + y^2)}, \quad \Omega : 0 \leq x^2 + y^2 \leq 3.$$

Therefore

$$A = \iint_\Omega \sqrt{\left(\frac{-x}{\sqrt{4 - x^2 - y^2}}\right)^2 + \left(\frac{-y}{\sqrt{4 - x^2 - y^2}}\right)^2 + 1} \, dx dy$$

$$= \iint_\Omega \frac{2}{\sqrt{4 - (x^2 + y^2)}} \, dx dy$$

$$= \int_0^{2\pi} \int_0^{\sqrt{3}} \frac{2r}{\sqrt{4 - r^2}} \, dr \, d\theta \qquad [\text{changed to polar coordinates}]$$

$$= 2\pi \left[-2\sqrt{4 - r^2} \right]_0^{\sqrt{3}} = 4\pi$$

29.   **(a)** $\displaystyle \iint_\Omega \sqrt{\left[\frac{\partial g}{\partial y}(y, z)\right]^2 + \left[\frac{\partial g}{\partial z}(y, z)\right]^2 + 1} \, dy dz = \iint_\Omega \sec\left[\alpha(y, z)\right] \, dy dz$

where α is the angle between the unit normal with positive $\mathbf{i}$ component and the positive x-axis

(b) $\displaystyle \iint_\Omega \sqrt{\left[\frac{\partial h}{\partial x}(x, z)\right]^2 + \left[\frac{\partial h}{\partial z}(x, z)\right]^2 + 1} \, dx dz = \iint_\Omega \sec\left[\beta(x, z)\right] \, dx dz$

where β is the angle between the unit normal with positive $\mathbf{j}$ component and the positive y-axis

31.   **(a)** $\mathbf{N}(u, v) = v \cos u \sin \alpha \cos \alpha \, \mathbf{i} + v \sin u \sin \alpha \cos \alpha \, \mathbf{j} - v \sin^2 \alpha \, \mathbf{k}$

(b) $$A = \iint_\Omega \|\mathbf{N}(u, v)\| \, du dv = \iint_\Omega v \sin \alpha \, du dv$$

$$= \int_0^{2\pi} \int_0^s v \sin \alpha \, dv \, du = \pi s^2 \sin \alpha$$

33. $A = \sqrt{A_1{}^2 + A_2{}^2 + A_3{}^2};$ the unit normal to the plane of Ω is a vector of the form

$$\cos\gamma_1\,\mathbf{i} + \cos\gamma_2\,\mathbf{j} + \cos\gamma_3\,\mathbf{k}.$$

Note that

$$A_1 = A\cos\gamma_1, \quad A_2 = A\cos\gamma_2, \quad A_3 = A\cos\gamma_3.$$

Therefore

$$A_1{}^2 + A_2{}^2 + A_3{}^2 = A^2[\cos^2\gamma_1 + \cos^2\gamma_2 + \cos^2\gamma_3] = A^2.$$

35. (a) (We use Exercise 34.) $f(r,\theta) = r + \theta;$ $\Omega : 0 \le r \le 1, \quad 0 \le \theta\pi$

$$A = \iint_\Omega \sqrt{r^2\,[f_r'(r,\theta)]^2 + [f_\theta'(r,\theta)]^2 + r^2}\; drd\theta = \iint_\Omega \sqrt{2r^2 + 1}\; drd\theta$$

$$= \int_0^\pi \int_0^1 \sqrt{2r^2 + 1}\; dr\, d\theta = \tfrac{1}{4}\sqrt{2}\pi\left[\sqrt{6} + \ln\left(\sqrt{2} + \sqrt{3}\right)\right]$$

(b) $f(r,\theta) = re^\theta;$ $\Omega : 0 \le r \le a, \quad 0 \le \theta \le 2\pi$

$$A = \iint_\Omega r\sqrt{2e^{2\theta} + 1}\; drd\theta = \left(\int_0^{2\pi} \sqrt{2e^{2\theta} + 1}\; d\theta\right)\left(\int_0^a r\, dr\right)$$

$$= \tfrac{1}{2}a^2[\sqrt{2e^{4\pi} + 1} - \sqrt{3} + \ln\left(1 + \sqrt{3}\right) - \ln\left(1 + \sqrt{2e^{4\pi} + 1}\right)]$$

SECTION 17.7

For Exercises 1–5 we have $\sec\left[\gamma(x,y)\right] = \sqrt{y^2 + 1}.$

1. $\displaystyle\iint_S d\sigma = \int_0^1 \int_0^1 \sqrt{y^2 + 1}\; dx\, dy = \int_0^1 \sqrt{y^2 + 1}\; dy = \tfrac{1}{2}[\sqrt{2} + \ln\left(1 + \sqrt{2}\right)]$

3. $\displaystyle\iint_S 3y\, d\sigma = \int_0^1 \int_0^1 3y\sqrt{y^2 + 1}\; dy\, dx = \int_0^1 3y\sqrt{y^2 + 1}\; dy = \left[(y^2 + 1)^{3/2}\right]_0^1 = 2\sqrt{2} - 1$

5. $\displaystyle\iint_S \sqrt{2z}\, d\sigma = \iint_S y\, d\sigma = \tfrac{1}{3}(2\sqrt{2} - 1)$ (Exercise 3)

7. $\displaystyle\iint_S xy\, d\sigma;$ $S : \mathbf{r}(u,v) = (6 - 2u - 3v)\,\mathbf{i} + u\,\mathbf{j} + v\,\mathbf{k}, \quad 0 \le u \le 3 - \tfrac{3}{2}v, \quad 0 \le v \le 2$

$$\|\mathbf{N}(u,v)\| = \|(-2\,\mathbf{i} + \mathbf{j}) \times (-3\,\mathbf{i} + \mathbf{k})\| = \sqrt{14}$$

$$\iint_S xy \, d\sigma = \sqrt{14} \iint_\Omega x(u,v)y(u,v) \, du \, dv$$

$$= \sqrt{14} \iint_\Omega (6 - 2u - 3v)u \, du \, dv$$

$$= \sqrt{14} \int_0^2 \int_0^{3-3v/2} (6u - 2u^2 - 3uv) \, du \, dv$$

$$= \sqrt{14} \left[3\left(3 - \tfrac{3}{2}v\right)^2 - \tfrac{2}{3}\left(3 - \tfrac{3}{2}v\right)^3 - \tfrac{3}{2}v\left(3 - \tfrac{3}{2}v\right)^2 \right] dv = \frac{9}{2}\sqrt{14}$$

9. $\iint_S x^2 z \, d\sigma; \quad S : \mathbf{r}(u,v) = (\cos u \, \mathbf{i} + v \, \mathbf{j} + \sin u \, \mathbf{k}, \quad 0 \le u \le \pi, \quad 0 \le v \le 2.$

$$\mathbf{N}(u,v) = \begin{vmatrix} \mathbf{i} & \mathbf{j} & \mathbf{k} \\ -\sin u & 0 & \cos u \\ 0 & 1 & 0 \end{vmatrix} = -\cos u \, \mathbf{i} - \sin u \, \mathbf{k} \quad \text{and} \quad \|\mathbf{N}(u,v)\| = 1.$$

$$\iint_S x^2 z \, d\sigma = \iint_\Omega \cos^2 u \sin u \, du = \int_0^2 \int_0^\pi \cos^2 u \sin u \, du = \frac{4}{3}$$

11. $\iint_S (x^2 + y^2) \, d\sigma; \quad S : \mathbf{r}(u,v) = (\cos u \cos v \, \mathbf{i} + \cos u \sin v \, \mathbf{j} + \sin u \, \mathbf{k}, \quad 0 \le u \le \pi/2, \quad 0 \le v \le 2\pi.$

$$\mathbf{N}(u,v) = \begin{vmatrix} \mathbf{i} & \mathbf{j} & \mathbf{k} \\ -\sin u \cos v & -\sin u \sin v & \cos u \\ -\cos u \cos v & \cos u \cos v 1 & 0 \end{vmatrix} = -\cos^2 u \cos v \, \mathbf{i} + \cos^2 u \sin v \, w\mathbf{j} - \sin u \cos u \, \mathbf{k};$$

$$\|\mathbf{N}(u,v)\| = \cos u.$$

$$\iint_S (x^2 + y^2) \, d\sigma = \iint_\Omega \cos^2 u \cos u \, du = \int_0^{2\pi} \int_0^{\pi/2} \cos^3 u \, du = \frac{4}{3}\pi$$

For Exercises 13–15 the surface S is given by

$$f(x,y) = a - x - y; \quad 0 \le x \le a, \quad 0 \le y \le a - x.$$

Then $\sec[\gamma(x,y)] = \sqrt{3}.$

13. $M = \iint_S \lambda(x,y,x) \, d\sigma = \int_0^a \int_0^{a-x} k\sqrt{3} \, dy \, dx = \int_0^a k\sqrt{3}(a-x) \, dx = \frac{1}{2}a^2 k\sqrt{3}$

15. $M = \iint_S \lambda(x,y,z) \, d\sigma = \int_0^a \int_0^{a-x} kx^2\sqrt{3} \, dy \, dx = \int_0^a k\sqrt{3}x^2(a-x) \, dx = \frac{1}{12}a^4 k\sqrt{3}$

17. $S : \mathbf{r}(u,v) = a\cos u\cos v\,\mathbf{i} + a\sin u\cos v\,\mathbf{j} + a\sin v\,\mathbf{k}$ with $0 \le u \le 2\pi,\quad 0 \le v \le \frac{1}{2}\pi.$ By a previous calculation $\|\mathbf{N}(u,v)\| = a^2\cos v.$

$\bar{x} = 0,\quad \bar{y} = 0$ (by symmetry)

$$\bar{z}A = \iint_S z\,d\sigma = \iint_\Omega z(u,v)\,\|\mathbf{N}(u,v)\|\,dudv = \int_0^{2\pi}\int_0^{\pi/2} a^3\sin v\cos v\,dv\,du = \pi a^3$$

$\bar{z} = \frac{1}{2}a$ since $A = 2\pi a^2$

19. $\mathbf{N}(u,v) = (\mathbf{i}+\mathbf{j}+2\mathbf{k}) \times (\mathbf{i}-\mathbf{j}) = 2\mathbf{i}+2\mathbf{j}-2\mathbf{k}$

$$\text{flux in the direction of } \mathbf{N} = \iint_S \left(\mathbf{v} \cdot \frac{\mathbf{N}}{\|\mathbf{N}\|}\right) d\sigma = \iint_\Omega [\mathbf{v}(x(u),\,y(u),\,z(u)) \cdot \mathbf{N}(u,v)]\,dudv$$

$$= \iint_\Omega [(u+v)\mathbf{i} - (u-v)\mathbf{j}] \cdot [2\mathbf{i}+2\mathbf{j}-2\mathbf{k}]\,dudv.$$

$$= \iint_\Omega 4v\,dudv = 4\int_0^1\int_0^1 v\,dv\,du = 2$$

For Exercises 21–23 $\mathbf{n} = \dfrac{1}{a}(x\mathbf{i}+y\mathbf{j}+z\mathbf{k})$

$S : \mathbf{r}(u,v) = a\cos u\cos v\,\mathbf{i} + a\sin u\cos v\,\mathbf{j} + a\sin v\,\mathbf{k}$ with $0 \le u \le 2\pi,\quad -\frac{1}{2}\pi \le v \le \frac{1}{2}\pi$

$\|\mathbf{N}(u,v)\| = a^2\cos v$

21.   with $\mathbf{v} = z\mathbf{k}$

$$\text{flux} = \iint_S (\mathbf{v} \cdot \mathbf{n})\,d\sigma = \frac{1}{a}\iint_S z^2\,d\sigma = \frac{1}{a}\iint_\Omega (a^2\sin^2 v)(a^2\cos v)\,dudv$$

$$= a^3\int_0^{2\pi}\int_{-\pi/2}^{\pi/2}(\sin^2 v\cos v)\,d\sigma = \frac{4}{3}\pi a^3$$

23.   with $\mathbf{v} = y\mathbf{i} - x\mathbf{j}$

$$\text{flux} = \iint_S (\mathbf{v} \cdot \mathbf{n})\,d\sigma = \frac{1}{a}\iint_S \underbrace{(yx - xy)}_{0}\,d\sigma = 0$$

For Exercises 25–27 the triangle S is the graph of the function

$$f(x,y) = a - x - y \quad \text{on} \quad \Omega : 0 \le x \le a,\quad 0 \le y \le a - x.$$

The triangle has area $A = \frac{1}{2}\sqrt{3}a^2.$

25. with $\mathbf{v} = x\mathbf{i} + y\mathbf{j} + z\mathbf{k}$

$$\text{flux} = \iint_S (\mathbf{v} \cdot \mathbf{n})\, d\sigma = \iint_\Omega (-v_1 f_x' - v_2 f_y' + v_3)\, dxdy$$

$$= \iint_\Omega [-x(-1) - y(-1) + (a - x - y)]\, dxdy = a \iint_\Omega dxdy = aA = \frac{1}{2}\sqrt{3}a^3$$

27. with $v = x^2\mathbf{i} - y^2\mathbf{j}$

$$\text{flux} = \iint_S (\mathbf{v} \cdot \mathbf{n})\, d\sigma = \iint_\Omega (-v_1 f_x' - v_2 f_y' + v_3)\, dxdy$$

$$= \iint_\Omega [-x^2(-1) - (-y^2)(-1) + 0]\, dxdy = \int_0^a \int_0^{a-x} (x^2 - y^2)\, dy\, dx$$

$$= \int_0^a \left[ax^2 - x^3 - \frac{1}{3}(a - x)^3 \right] dx = \left[\frac{1}{3}ax^3 - \frac{1}{4}x^4 + \frac{1}{12}(a - x)^4 \right]_0^a = 0$$

29.

$$\text{flux} = \iint_S (\mathbf{v} \cdot \mathbf{n})\, d\sigma = \iint_\Omega (-v_1 f_x' - v_2 f_y' + v_3)\, dxdy$$

$$= \iint_\Omega (-x^3 y - xy)\, dxdy = \int_0^1 \int_0^2 (-x^3 y - xy)\, dy\, dx$$

$$= \int_0^1 -2(x^3 + x)\, dx = -\frac{3}{2}$$

31. $\mathbf{n} = \dfrac{1}{a}(x\mathbf{i} + y\mathbf{j})$

$$\text{flux} = \iint_S (\mathbf{v} \cdot \mathbf{n})\, d\sigma = \frac{1}{a} \iint_S [(x\mathbf{i} + y\mathbf{j} + z\mathbf{k}) \cdot (x\mathbf{i} + y\mathbf{j})]\, d\sigma$$

$$= \frac{1}{a} \iint_S (x^2 + y^2)\, d\sigma = a \iint_S d\sigma = a\,(\text{area of } S) = a\,(2\pi al) = 2\pi a^2 l$$

33.

$$\text{flux} = \iint_S (\mathbf{v} \cdot \mathbf{n})\, d\sigma = \iint_\Omega (-v_1 f_x' - v_2 f_y' + v_3)\, dxdy = \iint_\Omega 2y^{3/2}\, dxdy$$

$$= \int_0^1 \int_0^{1-x} 2y^{3/2}\, dy\, dx = \int_0^1 \frac{4}{5}(1 - x)^{5/2}\, dx = \frac{8}{35}$$

35.
$$\text{flux} = \iint_S (\mathbf{v} \cdot \mathbf{n}) \, d\sigma = \iint_\Omega (-v_1 f_x' - v_2 f_y' + v_3) \, dx \, dy = \iint_\Omega -y^{5/2} \, d\sigma$$

$$= \int_0^1 \int_0^{1-x} -y^{5/2} \, dy \, dx = \int_0^1 -\frac{2}{7}(1-x)^{7/2} \, dx = -\frac{4}{63}$$

37. $\bar{x} = 0, \quad \bar{y} = 0 \qquad$ by symmetry

verify that $\quad \|\mathbf{N}(u, v)\| = v \sin \alpha$

$$\bar{z}A = \iint_S z \, d\sigma = \iint_\Omega (v \cos \alpha)(v \sin \alpha) \, du \, dv = \sin \alpha \cos \alpha \int_0^{2\pi} \int_0^s v^2 \, dv \, du = \frac{2}{3}\pi \sin \alpha \cos \alpha \, s^3$$

$\bar{z} = \frac{2}{3}s \cos \alpha \quad$ since $\quad A = \pi s^2 \sin \alpha$

39. $f(x, y) = \sqrt{x^2 + y^2} \quad$ on $\quad \Omega : 0 \le x^2 + y^2 \le 1; \quad \lambda(x, y, z) = k\sqrt{x^2 + y^2}$

$x_M = 0, \quad y_M = 0 \qquad$ (by symmetry)

$$z_M M = \iint_S z\lambda(x, y, z) \, d\sigma = \iint_\Omega k(x^2 + y^2) \sec\left[\gamma(x, y)\right] \, dx \, dy$$

$$= k\sqrt{2} \iint_\Omega (x^2 + y^2) \, dx \, dy$$

$$= k\sqrt{2} \int_0^{2\pi} \int_0^1 r^3 \, dr \, d\theta = \frac{1}{2}\sqrt{2}\pi k$$

$z_M = \frac{3}{4} \quad$ since $\quad M = \frac{2}{3}\sqrt{2}\pi k \qquad$ (Exercise 38)

41. no answer required

43.
$$x_M M = \iint_S x\lambda(x, y, z) \, d\sigma = \iint_S kx(x^2 + y^2) \, d\sigma$$

$$= 2\sqrt{3}k \iint_\Omega (u + v)\left[(u - v)^2 + 4u^2\right] \, du \, dv$$

$$= 2\sqrt{3}k \int_0^1 \int_0^1 (5u^3 - 2u^2 v + uv^2 + 5u^2 v - 2uv^2 + v^3) \, dv \, du$$

$$= 2\sqrt{3} \int_0^1 \left(5u^3 - u^2 + \frac{1}{3}u + \frac{5}{2}u^2 - \frac{2}{3}u + \frac{1}{4}\right) \, du = \frac{11}{3}\sqrt{3}k$$

$x_M = \frac{11}{9} \quad$ since $\quad M = 3\sqrt{3}k \qquad$ (Exercise 42)

45. Total flux out of the solid is 0. It is clear from a diagram that the outer unit normal to the cylindrical side of the solid is given by $\mathbf{n} = x\mathbf{i} + y\mathbf{j}$ in which case $\mathbf{v} \cdot \mathbf{n} = 0$. The outer unit normals to the top and bottom of the solid are $\mathbf{k}$ and $-\mathbf{k}$ respectively. So, here as well, $\mathbf{v} \cdot \mathbf{n} = 0$ and the total flux is 0.

47. The surface $z = \sqrt{2 - (x^2 + y^2)}$ is the upper half of the sphere $x^2 + y^2 + z^2 = 2$. The surface intersects the surface $z = x^2 + y^2$ in a circle of radius 1 at height $z = 1$. Thus the upper boundary of the solid, call it S_1, is a segment of width $\sqrt{2} - 1$ on a sphere of radius $\sqrt{2}$. The area of S_1 is therefore $2\pi\sqrt{2}(\sqrt{2} - 1)$. (Exercise 25, Section 10.10.) The upper unit normal to S_1 is the vector

$$\mathbf{n} = \frac{1}{\sqrt{2}}(x\mathbf{i} + y\mathbf{j} + z\mathbf{k}).$$

Therefore

$$\text{flux through } S_1 = \iint_{S_1} (\mathbf{v} \cdot \mathbf{n})\, d\sigma = \frac{1}{\sqrt{2}} \iint_{S_1} \overbrace{(x^2 + y^2 + z^2)}^{2}\, d\sigma$$

$$= \sqrt{2} \iint_{S_1} d\sigma = \sqrt{2}\,(\text{area of } S_1) = 4\pi(\sqrt{2} - 1).$$

The lower boundary of the solid, call it S_2, is the graph of the function

$$f(x, y) = x^2 + y^2 \quad \text{on} \quad \Omega : 0 \le x^2 + y^2 \le 1.$$

Taking $\mathbf{n}$ as the lower unit normal, we have

$$\text{flux through } S_2 = \iint_{S_2} (\mathbf{v} \cdot \mathbf{n})\, d\sigma = \iint_{\Omega} \left(v_1 f_x' + v_2 f_y' - v^3 \right) dx\,dy$$

$$= \iint_{\Omega} (x^2 + y^2)\, dx\,dy = \int_0^{2\pi} \int_0^1 r^3\, dr\, d\theta = \frac{1}{2}\pi.$$

The total flux out of the solid is $4\pi(\sqrt{2} - 1) + \frac{1}{2}\pi = (4\sqrt{2} - \frac{7}{2})\pi$.

SECTION 17.8

1. $\nabla \cdot \mathbf{v} = 2, \quad \nabla \times \mathbf{v} = 0$ 　　　　　　**3.** $\nabla \cdot \mathbf{v} = 0, \quad \nabla \times \mathbf{v} = 0$

5. $\nabla \cdot \mathbf{v} = 6, \quad \nabla \times \mathbf{v} = 0$ 　　　　　　**7.** $\nabla \cdot \mathbf{v} = yz + 1, \quad \nabla \times \mathbf{v} = -x\mathbf{i} + xy\mathbf{j} + (1 - x)z\mathbf{k}$

9. $\nabla \cdot \mathbf{v} = 1/r^2, \quad \nabla \times \mathbf{v} = 0$

11. $\nabla \cdot \mathbf{v} = 2(x + y + z)e^{r^2}, \quad \nabla \times \mathbf{v} = 2e^{r^2}\left[(y - z)\mathbf{i} - (x - z)\mathbf{j} + (x - y)\mathbf{k}\right]$

13. $\nabla \cdot \mathbf{v} = f'(x), \quad \nabla \times \mathbf{v} = 0$ 　　　　**15.** use components

17. $\nabla \cdot \mathbf{F} = \dfrac{\partial P}{\partial x} + \dfrac{\partial Q}{\partial y} + \dfrac{\partial R}{\partial z} = 2 + 4 - 6 = 0$

19. $\nabla \times \mathbf{F} = \begin{vmatrix} \mathbf{i} & \mathbf{j} & \mathbf{k} \\ \dfrac{\partial}{\partial x} & \dfrac{\partial}{\partial y} & \dfrac{\partial}{\partial z} \\ x & y & -2z \end{vmatrix} = 0$

21. $\nabla^2 f = 12(x^2 + y^2 + z^2)$

23. $\nabla^2 f = 2y^3 z^4 + 6x^2 y z^4 + 12x^2 y^3 z^2$

25. $\nabla^2 f = e^r(1 + 2r^{-1})$

27. (a) $2r^2$ (b) $-1/r$

29. $\nabla^2 f = \nabla^2 g(r) = \nabla \cdot (\nabla g(r)) = \nabla \cdot (g'(r) r^{-1} \mathbf{r})$

$$= \left[(\nabla g'(r)) \cdot r^{-1} \mathbf{r} \right] + g'(r) \left(\nabla \cdot r^{-1} \mathbf{r} \right)$$

$$= \left\{ \left[g''(r) r^{-1} \mathbf{r} \right] \cdot r^{-1} \mathbf{r} \right\} + g'(r)(2r^{-1})$$

$$= g''(r) + 2r^{-1} g'(r)$$

31. $\dfrac{\partial f}{\partial x} = 2x + y + 2z, \quad \dfrac{\partial^2 f}{\partial x^2} = 2; \quad \dfrac{\partial f}{\partial y} = 4y + x - 3z, \quad \dfrac{\partial^2 f}{\partial y^2} = 4;$

$\dfrac{\partial f}{\partial z} = -6z + 2x - 3y, \quad \dfrac{\partial^2 f}{\partial z^2} = -6;$

$$\dfrac{\partial^2 f}{\partial x^2} + \dfrac{\partial^2 f}{\partial y^2} + \dfrac{\partial^2 f}{\partial z^2} = 2 + 4 - 6 = 0$$

33. $n = -1$

SECTION 17.9

1. $\displaystyle\iint_S (\mathbf{v} \cdot \mathbf{n})\, d\sigma = \iiint_T (\nabla \cdot \mathbf{v})\, dx\,dy\,dz = \iiint_T 3\, dx\,dy\,dz = 3V = 4\pi$

3. $\displaystyle\iint_S (\mathbf{v} \cdot \mathbf{n})\, d\sigma = \iiint_T (\nabla \cdot \mathbf{v})\, dx\,dy\,dz = \iiint_T 2(x + y + z)\, dx\,dy\,dz.$

The flux is zero since the function $f(x,y,z) = 2(x + y + z)$ satisfies the relation $f(-x, -y, -z) = -f(x,y,z)$ and T is symmetric about the origin.

5.

face	n	v · n	flux
$x = 0$	$-\mathbf{i}$	0	0
$x = 1$	$\mathbf{i}$	1	1
$y = 0$	$-\mathbf{j}$	0	0
$y = 1$	$\mathbf{j}$	1	1
$z = 0$	$-\mathbf{k}$	0	0
$z = 1$	$\mathbf{k}$	1	1

total flux = 3

$$\iiint_T (\nabla \cdot \mathbf{v})\, dxdydz = \iiint_T 3\, dxdydz = 3V = 3$$

7.

face	n	v · n	flux
$x = 0$	$-\mathbf{i}$	0	0
$x = 1$	$\mathbf{i}$	1	1
$y = 0$	$-\mathbf{j}$	xz	
$y = 1$	$\mathbf{j}$	$-xz$	
$z = 0$	$-\mathbf{k}$	0	0
$z = 1$	$\mathbf{k}$	1	1

fluxes add up to 0   total flux = 2

$$\iiint_T (\nabla \cdot \mathbf{v})\, dxdydz = \iiint_T 2\,(x + z)\, dxdydz = 2\,(\bar{x} + \bar{z})V = 2\,(\tfrac{1}{2} + \tfrac{1}{2})1 = 2$$

9. $\text{flux} = \displaystyle\iiint_T (1 + 4y + 6z)\, dxdydz = (1 + 4\bar{y} + 6\bar{z})V = (1 + 0 + 3)\, 9\pi = 36\pi$

11.

$$\text{flux} = \iiint_T (2x + x - 2x)\, dxdydz \iiint_T x\, dxdydz$$

$$= \int_0^1 \int_0^{1-x} \int_0^{1-x-y} x\, dz\, dy\, dx$$

$$= \int_0^1 \int_0^{1-x} (x - x^2 - xy)\, dy\, dx$$

$$= \int_0^1 \left[xy - x^2 y - \frac{1}{2} xy^2 \right]_0^{1-x} dx$$

$$= \int_0^1 \left(\frac{1}{2} x - x^2 + \frac{1}{2} x^3 \right) dx = \frac{1}{24}$$

13.

$$\text{flux} = \iiint_T 2(x + y + z)\, dxdydz = \int_0^4 \int_0^2 \int_0^{2\pi} 2(r\cos\theta + r\sin\theta + z)r\, dr\, d\theta\, dz$$

$$= \int_0^4 \int_0^2 4\pi\, rz\, dr\, dz$$

$$= \int_0^4 8\pi\, z\, dz = 64\pi$$

15. $\text{flux} = \displaystyle\iiint_T (2y + 2y + 3y)\, dxdydz = 7\bar{y}V = 0$

17. $\text{flux} = \displaystyle\iiint_T (A + B + C)\, dxdydz = (A + B + C)V$

19.   Let T be the solid enclosed by S and set $\mathbf{n} = n_1\mathbf{i} + n_2\mathbf{j} + n_3\mathbf{k}$.

$$\iint_S n_1\, d\sigma = \iint_S (\mathbf{i} \cdot \mathbf{n})\, d\sigma = \iiint_T (\nabla \cdot \mathbf{i})\, dxdydz = \iiint_T 0\, dxdydz = 0.$$

Similarly

$$\iint_S n_2\, d\sigma = 0 \quad \text{and} \quad \iint_S n_3\, d\sigma = 0.$$

21.   A routine computation shows that $\nabla \cdot (\nabla f \times \nabla g) = 0$.   Therefore

$$\iint_S [(\nabla f \times \nabla g) \cdot \mathbf{n}]\, d\sigma = \iiint_T [\nabla \cdot (\nabla f \times \nabla g)]\, dxdydz = 0.$$

23.   Set $\mathbf{F} = F_1\mathbf{i} + F_2\mathbf{j} + F_3\mathbf{k}$.

$$F_1 = \iint_S [\rho(z - c)\mathbf{i} \cdot \mathbf{n}]\, d\sigma = \iiint_T [\nabla \cdot \rho(z - c)\mathbf{i}]\, dxdydz$$

$$= \iiint_T \underbrace{\frac{\partial}{\partial x}[\rho(z - c)]}_{0}\, dxdydz = 0.$$

Similarly $F_2 = 0$.

$$F_3 = \iint_S [\rho(z - c)\mathbf{k} \cdot \mathbf{n}]\, d\sigma = \iiint_T [\nabla \cdot \rho(z - c)\mathbf{k}]\, dxdydz$$

$$= \iiint_T \frac{\partial}{\partial z}[\rho(z - c)]\, dxdydz$$

$$= \iiint_T \rho\, dxdydz = W.$$

PROJECT 17.9

1. For $\mathbf{r} \neq 0$, $\nabla \cdot \mathbf{E} = \nabla \cdot qr^{-3}\mathbf{r} = q(-3+3)r^{-3} = 0$ by (17.8.8)

3. On $S_a, \mathbf{n} = \dfrac{\mathbf{r}}{r}$, and thus $\mathbf{E} \cdot \mathbf{n} = q\dfrac{\mathbf{r}}{r^3} \cdot \dfrac{\mathbf{r}}{r} = \dfrac{q}{r^2} = \dfrac{q}{a^2}$

Thus flux of $\mathbf{E}$ out of $S_a = \displaystyle\iint\limits_{S_a} (\mathbf{E} \cdot \mathbf{n})\, d\sigma = \iint\limits_{S_a} \dfrac{q}{a^2}\, d\sigma = \dfrac{q}{a^2}(\text{area of } S_a) = \dfrac{q}{a^2}(4\pi a^2) = 4\pi q.$

SECTION 17.10

For Exercises 1–3: $\mathbf{n} = x\mathbf{i} + y\mathbf{j} + z\mathbf{k}$ and $C : \mathbf{r}(u) = \cos u\,\mathbf{i} + \sin u\,\mathbf{j}, \quad u \in [0, 2\pi].$

1. (a) $\displaystyle\iint\limits_{S} [(\nabla \times \mathbf{v}) \cdot \mathbf{n}]\, d\sigma = \iint\limits_{S} (\mathbf{0} \cdot \mathbf{n})\, d\sigma = 0$

(b) S is bounded by the unit circle $C : \mathbf{r}(u) = \cos u\,\mathbf{i} + \sin u\,\mathbf{j}, \quad u \in [0, 2\pi].$

$\displaystyle\oint_{C} \mathbf{v}(\mathbf{r}) \cdot d\mathbf{r} = 0$ since $\mathbf{v}$ is a gradient.

3. (a) $\displaystyle\iint\limits_{S} [(\nabla \times \mathbf{v}) \cdot \mathbf{n}]\, d\sigma = \iint\limits_{S} [(-3y^2\mathbf{i} + 2z\mathbf{j} + 2\mathbf{k}) \cdot \mathbf{n}]\, d\sigma$

$\displaystyle = \iint\limits_{S} (-3xy^2 + 2yz + 2z)\, d\sigma$

$\displaystyle = \underbrace{\iint\limits_{S} (-3xy^2)\, d\sigma}_{0} + \underbrace{\iint\limits_{S} 2yz\, d\sigma}_{0} + 2\iint\limits_{S} z\, d\sigma = 2\bar{z}V = 2(\tfrac{1}{2})2\pi = 2\pi$

Exercise 17, Section 17.7

(b) $\displaystyle\oint_{C} \mathbf{v}(\mathbf{r}) \cdot d\mathbf{r} = \oint_{C} z^2\, dx + 2x\, dy = \oint_{C} 2x\, dy = \int_{0}^{2\pi} 2\cos^2 u\, du = 2\pi$

For Exercises 5–7 take $S: z = 2 - x - y$ with $0 \leq x \leq 2, \quad 0 \leq y \leq 2 - x$
and C as the triangle $(2,0,0), (0,2,0), (0,0,2)$.   Then $C = C_1 \cup C_2 \cup C_3$ with

$$C_1 : \mathbf{r}_1(u) = 2(1-u)\,\mathbf{i} + 2u\mathbf{j}, \quad u \in [0,1],$$

$$C_2 : \mathbf{r}_2(u) = 2(1-u)\,\mathbf{j} + 2u\mathbf{k}, \quad u \in [0,1],$$

$$C_3 : \mathbf{r}_3(u) = 2(1-u)\,\mathbf{k} + 2u\mathbf{i}, \quad u \in [0,1].$$

$\mathbf{n} = \tfrac{1}{3}\sqrt{3}(\mathbf{i} + \mathbf{j} + \mathbf{k})$ area of $S: A = 2\sqrt{3}$ centroid: $(\tfrac{2}{3}, \tfrac{2}{3}, \tfrac{2}{3})$

5. (a) $\displaystyle\iint\limits_{S} [(\nabla \times \mathbf{v}) \cdot \mathbf{n}]\, d\sigma = \iint\limits_{S} \tfrac{1}{3}\sqrt{3}\, d\sigma = \tfrac{1}{3}\sqrt{3}A = 2$

(b) $\oint_C \mathbf{v(r)} \cdot \mathbf{dr} = \left(\int_{C_1} + \int_{C_2} + \int_{C_3} \right) \mathbf{v(r)} \cdot \mathbf{dr} = -2 + 2 + 2 = 2$

7.   (a) $\iint\limits_S [(\nabla \times \mathbf{v}) \cdot \mathbf{n}] \, d\sigma = \iint\limits_S (y\mathbf{k} \cdot \mathbf{n}) \, d\sigma = \frac{1}{3}\sqrt{3} \iint\limits_S y \, d\sigma = \frac{1}{3}\sqrt{3}\bar{y}A = \frac{4}{3}$

     (b) $\oint_C \mathbf{v(r)} \cdot \mathbf{dr} = \left(\int_{C_1} + \int_{C_2} + \int_{C_3} \right) \mathbf{v(r)} \cdot \mathbf{dr} = \left(\frac{4}{3} - \frac{32}{5} \right) + \frac{32}{5} + 0 = \frac{4}{3}$

9.   The bounding curve is the set of all (x, y, z) with

$$x^2 + y^2 = 4 \quad \text{and} \quad z = 4.$$

Traversed in the positive sense with respect to $\mathbf{n}$, it is the curve $-C$ where

$$C : \mathbf{r}(u) = 2\cos u\, \mathbf{i} + 2\sin u\, \mathbf{j} + 4\mathbf{k}, \qquad u \in [0, 2\pi].$$

By Stokes's theorem the flux we want is

$$-\int_C \mathbf{v(r)} \cdot \mathbf{dr} = -\int_C y\, dx + z\, dy + x^2 z^2\, dz$$

$$= -\int_0^{2\pi} \left(-4\sin^2 u + 8\cos u \right) du = 4\pi.$$

11.   The bounding curve C for S is the bounding curve of the elliptical region $\Omega : \frac{1}{4}x^2 + \frac{1}{9}y^2 = 1$. Since

$$\nabla \times \mathbf{v} = 2x^2 yz^2 \mathbf{i} - 2xy^2 z^2 \mathbf{j}$$

is zero on the xy-plane, the flux of $\nabla \times \mathbf{v}$ through Ω is zero, the circulation of $\mathbf{v}$ about C is zero, and therefore the flux of $\nabla \times \mathbf{v}$ through S is zero.

13. C bounds the surface

$$S: z = \sqrt{1 - \tfrac{1}{2}(x^2 + y^2)}, \qquad (x, y) \in \Omega$$

with $\Omega : x^2 + (y - \frac{1}{2})^2 \leq \frac{1}{4}$. Routine calculation shows that $\nabla \times \mathbf{v} = y\mathbf{k}$. The circulation of $\mathbf{v}$ with respect to the upper unit normal $\mathbf{n}$ is given by

$$\iint\limits_S (y\mathbf{k} \cdot \mathbf{n}) \, d\sigma = \iint\limits_\Omega y \, dxdy = \bar{y}A = \frac{1}{2}\left(\frac{\pi}{4} \right) = \frac{1}{8}\pi.$$

(17.7.9)

If $-\mathbf{n}$ is used, the circulation is $-\frac{1}{8}\pi$.   Answer: $\pm\frac{1}{8}\pi$.

15. $\nabla \times \mathbf{v} = \mathbf{i} + 2\mathbf{j} + \mathbf{k}$. The paraboloid intersects the plane in a curve C that bounds a flat surface S that projects onto the disc $x^2 + (y - \frac{1}{2})^2 = \frac{1}{4}$ in the xy-plane. The upper unit normal to S is the vector

$\mathbf{n} = \frac{1}{2}\sqrt{2}\,(-\mathbf{j}+\mathbf{k})$. The area of the base disc is $\frac{1}{4}\pi$. Letting γ be the angle between $\mathbf{n}$ and $\mathbf{k}$, we have $\cos\gamma = \mathbf{n}\cdot\mathbf{k} = \frac{1}{2}\sqrt{2}$ and $\sec\gamma = \sqrt{2}$. Therefore the area of S is $\frac{1}{4}\sqrt{2}\pi$. The circulation of $\mathbf{v}$ with respect to $\mathbf{n}$ is given by

$$\iint\limits_{S} [(\nabla\times\mathbf{v})\cdot\mathbf{n}]\,d\sigma = \iint\limits_{S} -\frac{1}{2}\sqrt{2}\,d\sigma = \left(-\frac{1}{2}\sqrt{2}\right)(\text{area of } S) = -\frac{1}{4}\pi.$$

If $-\mathbf{n}$ is used, the circulation is $\frac{1}{4}\pi$.   Answer: $\pm\frac{1}{4}\pi$.

17. Straightforward calculation shows that

$$\nabla\times(\mathbf{a}\times\mathbf{r}) = \nabla\times[(a_2 z - a_3 y)\,\mathbf{i} + (a_3 x - a_1 z)\,\mathbf{j} + (a_1 y - a_2 x)\mathbf{k}] = 2\mathbf{a}.$$

19. In the plane of C, the curve C bounds some Jordan region that we call Ω. The surface $S\cup\Omega$ is a piecewise–smooth surface that bounds a solid T. Note that $\nabla\times\mathbf{v}$ is continuously differentiable on T. Thus, by the divergence theorem,

$$\iiint\limits_{T} [\nabla\cdot(\nabla\times\mathbf{v})]\,dx\,dy\,dz = \iint\limits_{S\cup\Omega} [(\nabla\times\mathbf{v})\cdot\mathbf{n}]\,d\sigma$$

where $\mathbf{n}$ is the outer unit normal. Since the divergence of a curl is identically zero, we have

$$\iint\limits_{S\cup\Omega} [(\nabla\times\mathbf{v})\cdot\mathbf{n}]\,d\sigma = 0.$$

Now $\mathbf{n}$ is $\mathbf{n}_1$ on S and $\mathbf{n}_2$ on Ω. Thus

$$\iint\limits_{S} [(\nabla\times\mathbf{v})\cdot\mathbf{n}_1]\,d\sigma + \iint\limits_{\Omega} [(\nabla\times\mathbf{v})\cdot\mathbf{n}_2]\,d\sigma = 0.$$

This gives

$$\iint\limits_{S} [(\nabla\times\mathbf{v})\cdot\mathbf{n}_1]\,d\sigma = \iint\limits_{\Omega} [(\nabla\times\mathbf{v})\cdot(-\mathbf{n}_2)]\,d\sigma = \oint_{C} \mathbf{v}(\mathbf{r})\cdot d\mathbf{r}$$

where C is traversed in a positive sense with respect to $-\mathbf{n}_2$ and therefore in a positive sense with respect to $\mathbf{n}_1$. ($-\mathbf{n}_2$ points toward S.)

CHAPTER 18

SECTION 18.1

1. first order, ordinary   **3.** first order, partial   **5.** second order, ordinary

7. second order, partial

9. $y_1'(x) = \frac{1}{2} e^{x/2}$; $2y_1' - y_1 = 2\left(\frac{1}{2}\right) e^{x/2} - e^{x/2} = 0$; y_1 is a solution.

$y_2'(x) = 2x + e^{x/2}$; $2y_2' - y_2 = 2\left(2x + e^{x/2}\right) - \left(x^2 + 2e^{x/2}\right) = 4x - x^2 \neq 0$;

y_2 is not a solution.

11. $y_1'(x) = \dfrac{-e^x}{(e^x + 1)^2}$; $y_1' + y_1 = \dfrac{-e^x}{(e^x + 1)^2} + \dfrac{1}{e^x + 1} = \dfrac{1}{(e^x + 1)^2} = y_1^2$; y_1 is a solution.

$y_2'(x) = \dfrac{-Ce^x}{(Ce^x + 1)^2}$; $y_2' + y_2 = \dfrac{-Ce^x}{(Ce^x + 1)^2} + \dfrac{1}{Ce^x + 1} = \dfrac{1}{(Ce^x + 1)^2} = y_2^2$;

y_2 is a solution.

13. $y_1'(x) = 2e^{2x}$, $y_1'' = 4e^{2x}$; $y_1'' - 4y_1 = 4e^{2x} - 4e^{2x} = 0$; y_1 is a solution.

$y_2'(x) = 2C \cosh 2x$, $y_2'' = 4C \sinh 2x$; $y_2'' - 4y_2 = 4C \sinh 2x - 4C \sinh 2x = 0$;

y_2 is a solution.

15. $\dfrac{\partial u_1}{\partial x} = -\lambda \sin \lambda x \sin \lambda at$, $\dfrac{\partial^2 u_1}{\partial x^2} = -\lambda^2 \cos \lambda x \sin \lambda at$

$\dfrac{\partial u_1}{\partial t} = -\lambda a \cos \lambda x \cos \lambda at$, $\dfrac{\partial^2 u_1}{\partial t^2} = -(\lambda a)^2 \cos \lambda x \sin \lambda at$

$$a^2 \frac{\partial^2 u_1}{\partial x^2} = -a^2 \lambda^2 \cos \lambda x \sin \lambda at = \frac{\partial^2 u_1}{\partial t^2};   u_1 \text{ is a solution.}$$

$\dfrac{\partial u_2}{\partial x} = \cos(x - at)$, $\dfrac{\partial^2 u_2}{\partial x^2} = -\sin(x - at)$

$\dfrac{\partial u_2}{\partial t} = -a \cos(x - at)$, $\dfrac{\partial^2 u_2}{\partial t^2} = a^2 \sin(x - at)$

$$a^2 \frac{\partial^2 u_2}{\partial x^2} = a^2 \sin(x - at) = \frac{\partial^2 u_2}{\partial t^2};   u_2 \text{ is a solution.}$$

17. $y_1'(x) = -\frac{1}{2} x^{-3/2}$, $y_1'' = \frac{3}{4} x^{-5/2}$;

$$4x^2 y_1'' - 12x\, y_1' - 9y_1 = 4x^2 \left(\tfrac{3}{4} x^{-5/2}\right) - 12x \left(-\tfrac{1}{2} x^{-3/2}\right) - 9x^{-1/2}$$

$$= 3x^{-1/2} + 6x^{-1/2} - 9x^{-1/2} = 0;$$

y_1 is a solution.

$y_2'(x) = -\frac{1}{2} C_1 x^{-3/2} + \frac{9}{2} C_2 x^{7/2}$, $y_2'' = \frac{3}{4} C_1 x^{-5/2} + \frac{63}{4} C_2 x^{5/2}$;

$$4x^2 y_2'' - 12x\, y_2' - 9y_2$$

$$= 4x^2 \left(\tfrac{3}{4} C_1 x^{-5/2} + \tfrac{63}{4} C_2 x^{5/2} \right) - 12x \left(-\tfrac{1}{2} C_1 x^{-3/2} + \tfrac{9}{2} C_2 x^{7/2} \right) - 9 \left(C_1 x^{-1/2} + C_2 x^{9/2} \right)$$

$$= C_1 \left(3x^{-1/2} + 6x^{-1/2} - 9x^{-1/2} \right) + C_2 \left(63x^{9/2} - 54x^{9/2} - 9x^{9/2} \right) = 0;$$

y_2 is a solution.

19. $y'(x) = 5Ce^{5x} = 5y$ Thus, $y = Ce^{5x}$ is a solution.

$y(0) = Ce^{5 \cdot 0} = 2 \implies C = 2;$ $y = 2e^{5x}$ satisfies the side condition.

21. It was shown in Exercise 11 that $y = \dfrac{1}{Ce^x + 1}$ is a one-parameter family of solutions of $y' + y = y^2$.

$y(1) = \dfrac{1}{Ce + 1} = -1 \implies C = -\dfrac{2}{e};$ $y = \dfrac{1}{-2e^{x-1} + 1}$ satisfies the side condition.

23. $y' = C_1 + \tfrac{1}{2} C_2 x^{-1/2}, \quad y'' = -\tfrac{1}{4} C_2 x^{-3/2};$

$$2x^2 y'' - xy' + y = 2x^2 \left(-\tfrac{1}{4} C_2 x^{-3/2} \right) - x \left(C_1 + \tfrac{1}{2} C_2 x^{-1/2} \right) + C_1 x + C_2 x^{1/2}$$

$$= -\tfrac{1}{2} C_2 x^{1/2} - C_1 x - \tfrac{1}{2} C_2 x^{1/2} + C_1 x + C_2 x^{1/2} = 0$$

Therefore, $y = C_1 x + C_2 x^{1/2}$ is a two-parameter family of solutions.

$$y(4) = C_1(4) + C_2(4)^{1/2} = 1 = 4C_1 + 2C_2$$

$$y'(4) = C_1 + \tfrac{1}{2} C_2(4)^{-1/2} = -2 = C_1 + \tfrac{1}{4} C_2$$

implies $C_1 = -\tfrac{17}{4}, \quad C_2 = 9;$ $y = -\tfrac{17}{4} x + 9x^{1/2}$ satisfies the side conditions.

25. $y' = 2C_1 x + 2C_2 x \ln x + C_2 x, \quad y'' = 2C_1 + 3C_2 + 2C_2 \ln x;$

$$x^2 y'' - 3xy' + 4y$$

$$= x^2 (2C_1 + 3C_2 + 2C_2 \ln x) - 3x (2C_1 x + 2C_2 x \ln x + C_2 x) + 4 \left(C_1 x^2 + C_2 x^2 \ln x \right)$$

$$= x^2 (2C_1 + 3C_2 - 6C_1 - 3C_2 + 4C_1) + x^2 \ln x (2C_2 - 6C_2 + 4C_2) = 0$$

Therefore, $y = C_1 x^2 + C_2 x^2 \ln x$ is a two parameter family of solutions.

$y(1) = C_1 = 0, \quad y'(1) = 2C_1 + C_2 = 1 \implies C_2 = 1;$ $y = x^2 \ln x$ satisfies the side conditions.

27. $y = e^{rx}, \quad y' = re^{rx};$ $y' + 3y = re^{rx} + 3e^{rx} = 0 \implies r = -3.$

29. $y = e^{rx}, \quad y' = re^{rx}, \quad y'' = r^2 e^{rx};$

$$y'' + 6y' + 9y = r^2 e^{rx} + 6re^{rx} + 9e^{rx} = e^{rx}(r+3)^2 = 0 \implies r = -3$$

31. $y = x^r$, $y' = rx^{r-1}$, $y'' = r(r-1)x^{r-2}$;

$$xy'' + y' = x\left[r(r-1)x^{r-2}\right] + rx^{r-1} = \left[r^2 - r + r\right]x^{r-1} = 0 \implies r = 0$$

33. $y = x^r$, $y' = rx^{r-1}$, $y'' = r(r-1)x^{r-2}$;

$$4x^2y'' - 4xy' + 3y = 4x^2r(r-1)x^{r-2} - 4x\,r\,x^{r-1} + 3\,x^r$$

$$= \left(4r^2 - 8r + 3\right)x^r = 0 \implies 4r^2 - 8r + 3 = 0$$

$$\implies (2r - 1)(2r - 3) = 0 \implies r = \tfrac{1}{2},\ \tfrac{3}{2}$$

35. (a) $y(0) = C_1 \sin(0) + C_2 \cos(0) = 0 \implies C_2 = 0$;

Since $y = C_1 \sin(4 \cdot \pi/2) = C_1 \sin(2\pi) = 0$, $y = C_1 \sin 4x$ satisfies the boundary conditions

$$y(0) = 0, \qquad y(\pi/2)$$

for all values of C_1.

(b) As shown above, $y(0) = 0 \implies C_2 = 0$.

Now $y(\pi/8) = C_1 \sin(4 \cdot \pi/8) = C_1 \sin(\pi/2) = 0 \implies C_1 = 0$. Therefore, $y = 0$

is the only member of the family that satisfies the boundary conditions

$$y(0) = 0, \qquad y(\pi/8) = 0.$$

SECTION 18.2

1. $y' + xy = xy^3 \implies y^{-3}y' + xy^{-2} = x$. Let $v = y^{-2}$, $v' = -2y^{-3}y'$.

$$-\frac{1}{2}v' + xv = x$$

$$v' - 2xv = -2x$$

$$e^{-x^2}v' - 2xe^{-x^2}v = -2xe^{-x^2}$$

$$e^{-x^2}v = e^{-x^2} + C$$

$$v = 1 + Ce^{x^2}$$

$$y^2 = \frac{1}{1 + Ce^{x^2}}.$$

3. $y' - 4y = 2e^x y^{\frac{1}{2}} \implies y^{-\frac{1}{2}}y' - 4y^{\frac{1}{2}} = 2e^x.$ Let $v = y^{\frac{1}{2}}, \quad v' = \frac{1}{2}y^{-\frac{1}{2}}y'.$

$$2v' - 4v = 2e^x$$

$$v' - 2v = e^x$$

$$e^{-2x}v' - 2e^{-2x}v = e^{-x}$$

$$e^{-2x}v = -e^{-x} + C$$

$$v = -e^x + Ce^{2x}$$

$$y = (Ce^{2x} - e^x)^2.$$

5. $(x-2)y' + y = 5(x-2)^2 y^{\frac{1}{2}} \implies y^{-\frac{1}{2}}y' + \frac{1}{x-2}y^{\frac{1}{2}} = 5(x-2).$ Let $v = y^{\frac{1}{2}}, quad v' = \frac{1}{2}y^{-\frac{1}{2}}y'.$

$$2v' + \frac{1}{x-2}v = 5(x-2)$$

$$v' + \frac{1}{2(x-2)}v = \frac{5}{2}(x-2)$$

$$\sqrt{x-2}\,v' + \frac{1}{2\sqrt{x-2}}v = \frac{5}{2}(x-2)^{\frac{3}{2}}$$

$$\sqrt{x-2}\,v = (x-2)^{\frac{5}{2}} + C$$

$$v = (x-2)^2 + \frac{C}{\sqrt{x-2}}$$

$$y = \left[(x-2)^2 + \frac{C}{\sqrt{x-2}}\right]^2.$$

7. $y' + xy = y^3 e^{x^2} \implies y^{-3}y' + xy^{-2} = e^{x^2}.$ Let $v = y^{-2}, \quad v' = -2y^{-3}y'.$

$$-\frac{1}{2}v' + xv = e^{x^2}$$

$$v' - 2xv = -2e^{x^2}$$

$$e^{-x^2}v' - 2xe^{-x^2}v = -2$$

$$e^{-x^2}v = -2x + C$$

$$v = -2xe^{x^2} + Ce^{x^2}$$

$$y^{-2} = Ce^{x^2} - 2xe^{x^2}.$$

$C = 4 \implies y^{-2} = 4e^{x^2} - 2xe^{x^2}.$

9. $2x^3 y' - 3x^2 y = y^3 \implies y^{-3} y' - \dfrac{3}{2x} y^{-2} = \dfrac{1}{2x^3}$.   Let $v = y^{-2}$, $v' = -2y^{-3} y'$.

$$-\frac{1}{2} v' - \frac{3}{2x} v = \frac{1}{2x^3}$$

$$v' + \frac{3}{x} v = -\frac{1}{x^3}$$

$$x^3 v' + 3x^2 v = -1$$

$$x^3 v = -x + C$$

$$v = \frac{C - x}{x^3}$$

$$y^2 = \frac{x^3}{C - x}$$

$$1 = \frac{1}{C - x} \implies C = 2 \implies y^2 = \frac{x^3}{2 - x}.$$

11. $y' - \dfrac{y}{x} \ln y = xy \implies \dfrac{y'}{y} - \dfrac{1}{X} \ln y = x$.   Let $\mu = \ln y$, $\mu' = \dfrac{y'}{y}$.

$$\mu' - \frac{1}{x} \mu = x$$

$$\frac{1}{x} \mu' - \frac{1}{x^2} \mu = 1$$

$$\frac{1}{x} \mu = x + C$$

$$\mu = x^2 + Cx$$

$$\ln y = x^2 + Cx.$$

13. $f(x, y) = \dfrac{x^2 + y^2}{2xy}$; $f(tx, ty) = \dfrac{(tx)^2 + (ty)^2}{2(tx)(ty)} = \dfrac{t^2(x^2 + y^2)}{t^2(2xy)} = \dfrac{x^2 + y^2}{2xy} = f(x, y)$

Set $vx = y$. Then, $v + xv' = y'$ and

$$v + xv' = \frac{x^2 + v^2 x^2}{2vx^2} = \frac{1 + v^2}{2v}$$

$$v - \frac{1 + v^2}{2v} + xv' = 0$$

$$v^2 - 1 + 2xvv' = 0$$

$$\frac{1}{x} dx + \frac{2v}{v^2 - 1} dv = 0$$

$$\int \frac{1}{x} dx + \int \frac{2v}{v^2 - 1} dv = C$$

$$\ln |x| + \ln |v^2 - 1| = K \qquad \text{or} \qquad x(v^2 - 1) = C$$

Replacing v by y/x, we get

$$x \left(\frac{y^2}{x^2} - 1 \right) = C \qquad \text{or} \qquad y^2 - x^2 = Cx$$

15. $f(x,y) = \dfrac{x-y}{x+y};$ $f(tx,ty) = \dfrac{(tx)-(ty)}{tx+ty} = \dfrac{t(x-y)}{t(x+y)} = \dfrac{x-y}{x+y} = f(x,y)$

Set $vx = y$. Then, $v + xv' = y'$ and

$$v + xv' = \frac{x - vx}{x + vx} = \frac{1-v}{1+v}$$

$$v^2 + 2v - 1 + x(1+v)v' = 0$$

$$\frac{1}{x}\,dx + \frac{1+v}{v^2 + 2v - 1}\,dv = 0$$

$$\int \frac{1}{x}\,dx + \int \frac{1+v}{v^2 + 2v - 1}\,dv = C$$

$$\ln|x| + \tfrac{1}{2}\ln|v^2 + 2v - 1| = K \qquad \text{or} \qquad x\sqrt{v^2 + 2v - 1} = C$$

Replacing v by y/x, we get

$$x\sqrt{\frac{y^2}{x^2} + 2\frac{y}{x} - 1} = C \qquad \text{or} \qquad y^2 + 2xy - x^2 = C$$

17. $f(x,y) = \dfrac{x^2 e^{y/x} + y^2}{xy};$ $f(tx,ty) = \dfrac{(tx)^2 - e^{(ty)/(tx)} + (ty)^2}{(tx)(ty)} = \dfrac{t^2\left(x^2 e^{y/x} + y^2\right)}{t^2(xy)} = f(x,y)$

Set $vx = y$. Then, $v + xv' = y'$ and

$$v + xv' = \frac{x^2 e^v + v^2 x^2}{vx^2} = \frac{e^v + v^2}{v}$$

$$v^2 + 2v - 1 + x(1+v)v' = 0$$

$$v^2 + xvv' = e^v + v^2$$

$$-e^v + xvv' = 0$$

$$\frac{1}{x}\,dx = ve^{-v}\,dv$$

$$\int \frac{1}{x}\,dx = \int ve^{-v}\,dv$$

$$\ln|x| = -ve^{-v} - e^{-v} + C$$

Replacing v by y/x, and simplifying, we get

$$y + x = xe^{y/x}(C - \ln|x|)$$

19. $f(x,y) = \dfrac{y}{x} + \sin(y/x);$ $f(tx,ty) = \dfrac{(ty)}{tx} + \sin[(ty/tx)] = \dfrac{y}{x} + \sin(y/x) = f(x,y)$

Set $vx = y$. Then, $v + xv' = y'$ and

$$v + xv' = \frac{vx}{x} + \sin[(vx)/x] = v + \sin v$$

$$xv' = \sin v$$

$$\csc v\,dv = \frac{1}{x}\,dx$$

$$\int \csc v\,dv = \int \frac{1}{x}\,dx$$

$$\ln|\csc v - \cot v| = \ln|x| + K \qquad \text{or} \qquad \csc v - \cot v = Cx$$

Replacing v by y/x, and simplifying, we get

$$1 - \cos(y/x) = Cx \sin(y/x)$$

21. The differential equation is homogeneous since

$$f(x,y) = \frac{y^3 - x^3}{xy^2}; \qquad f(tx, ty) = \frac{(ty)^3 - (tx)^3}{(tx)(ty)^2} = \frac{t^3(y^3 - x^3)}{t^3(xy^2)} = \frac{y^3 - x^3}{xy^2} = f(x,y)$$

Set $vx = y$. Then, $v + xv' = y'$ and

$$v + xv' = \frac{(vx)^3 - x^3}{v^2 x^3} = \frac{v^3 - 1}{v^2}$$

$$1 + xv^2 v' = 0$$

$$\frac{1}{x} dx + v^2 dv = 0$$

$$\int \frac{1}{x} dx + \int v^2 dv = 0$$

$$\ln|x| + \frac{1}{3} v^3 = C$$

Replacing v by y/x, we get

$$y^3 + 3x^3 \ln|x| = Cx^3$$

Applying the side condition $y(1) = 2$, we have

$$8 + 3 \ln 1 = C \quad \Longrightarrow \quad C = 8 \quad \text{and} \quad y^3 + 3x^3 \ln|x| = 8x^3$$

23. $y' = y \Longrightarrow y = Ce^x$.   Also, $y(0) = 1 \Longrightarrow C = 1$

Thus $y = e^x$ and $y(1) = 2.71828$

(a) 2.48832, relative error= 8.46%.

(b) 2.71825, relative error= 0.001%.

25. (a) 2.59374, relative error= 4.58%.

(b) 2.71828, relative error= 0%.

27. $y' = 2x \Longrightarrow y = x^2 + C$.   Also, $y(2) = 5 \Longrightarrow C = 1$

Thus $y = x^2 + 1$ and $y(1) = 2$.

(a) 1.9, relative error= 5.0%.

(b) 2.0, relative error= 0%.

29. $y' = \dfrac{1}{2y}$

Thus $y = \sqrt{x}$ and $y(2) = \sqrt{2} \simeq 1.41421$.

(a) 1.42052, relative error= -0.45%.

    (b) 1.41421, relative error= 0%.

31.   (a) 2.65330, relative error= 2.39%.

    (b) 2.71828, relative error= 0%.

PROJECT 18.2

1.   (a) and (b)

$$y' = y, \ y(0) = 1$$

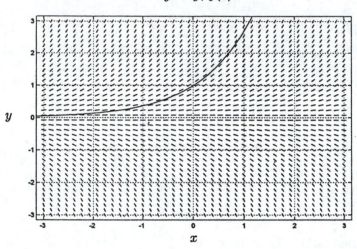

(c) $y - y' = 0$ $H(x) = \int -dx = -x$;   integrating factor: e^{-x}

$$e^{-x}y' - e^{-x}y = 0$$

$$\frac{d}{dx}(e^{-x}y) = 0$$

$$e^{-x}y = C$$

$$y = Ce^{x}$$

$y(0) = 1 \Longrightarrow C = 1.$ Thus $y = e^{x}$.

3. (a) and (b)

$$y' = -4x/y$$

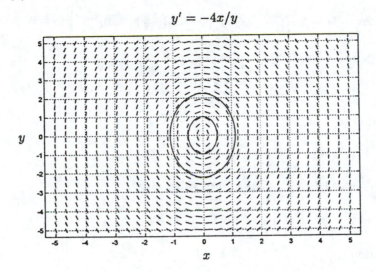

(c) $yy' = -\dfrac{4x}{y}$

$$\frac{1}{2}y^2 = -2x^2 + C \quad \text{or} \quad x^2 + \frac{1}{4}y^2 = C$$

$$y(0) = 1 \Longrightarrow C = \frac{1}{4}. \qquad \text{Thus } 4x^2 + y^2 = 1.$$

SECTION 18.3

1. The characteristic equation is:

$$r^2 + 2r - 15 = 0 \qquad \text{or} \qquad (r+5)(r-3) = 0.$$

The roots are: $r = 3,\ -5.$ The general solution is:

$$y = C_1 e^{3x} + C_2 e^{-5x}.$$

3. The characteristic equation is:

$$r^2 + 8r + 16 = 0 \qquad \text{or} \qquad (r+4)^2 = 0.$$

There is only one root: $r = -4.$ By Theorem 18.5.6 II, the general solution is:

$$y = C_1 e^{-4x} + C_2 x e^{-4x}.$$

5. The characteristic equation is:

$$r^2 + 2r + 5 = 0.$$

The roots are complex: $r = -1 \pm 2i.$ By Theorem 18.5.6 III, the general solution is:

$$y = e^{-x}\left(C_1 \cos 2x + C_2 \sin 2x\right).$$

7. The characteristic equation is:

$$2r^2 + 5r - 3 = 0 \qquad \text{or} \qquad (2r - 1)(r + 3) = 0.$$

The roots are: $r = \frac{1}{2}, -3.$ The general solution is:

$$y = C_1 e^{x/2} + C_2 e^{-3x}.$$

9. The characteristic equation is:

$$r^2 + 12 = 0.$$

The roots are complex: $r = \pm 2\sqrt{3}\, i.$ The general solution is:

$$y = C_1 \cos 2\sqrt{3}\, x + C_2 \sin 2\sqrt{3}\, x.$$

11. The characteristic equation is:

$$5r^2 + \tfrac{11}{4}r - \tfrac{3}{4} = 0 \qquad \text{or} \qquad 20r^2 + 11r - 3 = (5r - 1)(4r + 3) = 0.$$

The roots are: $r = \frac{1}{5}, -\frac{3}{4}.$ The general solution is:

$$y = C_1 e^{x/5} + C_2 e^{-3x/4}.$$

13. The characteristic equation is:

$$r^2 + 9 = 0.$$

The roots are complex: $r = \pm 3i.$ The general solution is:

$$y = C_1 \cos 3x + C_2 \sin 3x.$$

15. The characteristic equation is:

$$2r^2 + 2r + 1 = 0.$$

The roots are complex: $r = -\frac{1}{2} \pm \frac{1}{2} i.$ The general solution is:

$$y = e^{-x/2} \left[C_1 \cos(x/2) + C_2 \sin(x/2) \right].$$

17. The characteristic equation is:

$$8r^2 + 2r - 1 = 0 \qquad \text{or} \qquad (4r - 1)(2r + 1) = 0.$$

The roots are: $r = \frac{1}{4}, -\frac{1}{2}.$ The general solution is:

$$y = C_1 e^{x/4} + C_2 e^{-x/2}.$$

19. The characteristic equation is:

$$r^2 - 5r + 6 = 0 \qquad \text{or} \qquad (r - 3)(r - 2) = 0.$$

The roots are: $r = 3, 2$.   The general solution and its derivative are:

$$y = C_1 e^{3x} + C_2 e^{2x}, \qquad y' = 3C_1 e^{3x} + 2C_2 e^{2x}.$$

The conditions: $y(0) = 1$, $y'(0) = 1$ require that

$$C_1 + C_2 = 1 \qquad \text{and} \qquad 3C_1 + 2C_2 = 1.$$

Solving these equations simultaneously gives $C_1 = -1$, $C_2 = 2$.

The solution of the initial value problem is: $y = 2e^{2x} - e^{3x}$.

21.   The characteristic equation is:

$$r^2 + \tfrac{1}{4} = 0.$$

The roots are: $r = \pm \tfrac{1}{2} i$.   The general solution and its derivative are:

$$y = C_1 \cos(x/2) + C_2 \sin(x/2) \qquad y' = -\tfrac{1}{2} C_1 \sin(x/2) + \tfrac{1}{2} C_2 \cos(x/2).$$

The conditions: $y(\pi) = 1$, $y'(\pi) = -1$ require that

$$C_2 = 1 \qquad \text{and} \qquad C_1 = 2.$$

The solution of the initial value problem is: $y = 2\cos(x/2) + \sin(x/2)$.

23.   The characteristic equation is:

$$r^2 + 4r + 4 = 0 \qquad \text{or} \qquad (r+2)^2 = 0.$$

There is only one root: $r = -2$.   The general solution and its derivative are:

$$y = C_1 e^{-2x} + C_2 x e^{-2x} \qquad y' = -2C_1 e^{-2x} + C_2 e^{-2x} - 2C_2 x e^{-2x}.$$

The conditions: $y(-1) = 2$, $y'(-1) = 1$ require that

$$C_1 e^2 - C_2 e^2 = 2 \qquad \text{and} \qquad -2C_1 e^2 + 3C_2 e^2 = 1.$$

Solving these equations simultaneously gives $C_1 = 7e^{-2}$, $C_2 = 5e^{-2}$.

The solution of the initial value problem is: $y = 7e^{-2}e^{-2x} + 5e^{-2}xe^{-2x} = 7e^{-2(x+1)} + 5xe^{-2(x+1)}$.

25.   The characteristic equation is:

$$r^2 - r - 2 = 0 \qquad \text{or} \qquad (r-2)(r+1) = 0.$$

The roots are: $r = 2, -1$.   The general solution and its derivative are:

$$y = C_1 e^{2x} + C_2 e^{-x} \qquad y' = 2C_1 e^{2x} - C_2 e^{-x}.$$

(a) $y(0) = 1 \implies C_1 + C_2 = 1 \implies C_2 = 1 - C_1$.

Thus, the solutions that satisfy $y(0) = 1$ are: $y = Ce^{2x} + (1 - C)e^{-x}$.

(b) $y'(0) = 1 \Longrightarrow 2C_1 - C_2 = 1 \Longrightarrow C_2 = 2C_1 - 1.$

Thus, the solutions that satisfy $y'(0) = 1$ are: $y = Ce^{2x} + (2C - 1)e^{-x}.$

(c) To satisfy both conditions, we must have $2C - 1 = 1 - C \Longrightarrow C = \frac{2}{3}.$

The solution that satisfies $y(0) = 1,\ \ y'(0) = 1$ is:

$$y = \tfrac{2}{3}\,e^{2x} + \tfrac{1}{3}\,e^{-x}.$$

27. $\alpha = \dfrac{r_1 + r_2}{2}, \qquad \beta = \dfrac{r_1 - r_2}{2};$

$y = k_1 e^{r_1 x} + k_2 e^{r_2 x} = e^{\alpha x}\left(C_1 \cosh \beta x + C_2 \sinh \beta x\right),$ where $k_1 = \dfrac{C_1 + C_2}{2},\quad k_2 = \dfrac{C_1 - C_2}{2}.$

29.   (a) Let $y_1 = e^{rx},\ \ y_2 = xe^{rx}.$ Then

$$W(x) = y_1 y_2' - y_2 y_1' = e^{rx}\left[e^{rx} + rxe^{rx}\right] - xe^{rx}\left[re^{rx}\right] = e^{2rx} \neq 0$$

(b) Let $y_1 = e^{\alpha x}\cos \beta x,\ \ y_2 = e^{\alpha x}\sin \beta x,\ \ \beta \neq 0.$ Then

$W(x) = y_1 y_2' - y_2 y_1'$

$\qquad = e^{\alpha x}\cos \beta x\left[\alpha e^{\alpha x}\sin \beta x + \beta e^{\alpha x}\cos \beta x\right] - e^{\alpha x}\sin \beta x\left[\alpha e^{\alpha x}\cos \beta x - \beta e^{\alpha x}\sin \beta x\right]$

$\qquad = \beta e^{2\alpha x} \neq 0$

31.   (a) The solutions $y_1 = e^{2x},\ \ y_2 = e^{-4x}$ imply that the roots of the characteristic equation are $r_1 = 2\ \ r_2 = -4.$ Therefore, the characteristic equation is:

$$(r - 2)(r + 4) = r^2 + 2r - 8 = 0$$

and the differential equation is: $y'' + 2y' - 8y = 0.$

(b) The solutions $y_1 = 3e^{-x},\ \ y_2 = 4e^{5x}$ imply that the roots of the characteristic equation are $r_1 = -1\ \ r_2 = 5.$ Therefore, the characteristic equation is

$$(r + 1)(r - 5) = r^2 - 4r - 5 = 0$$

and the differential equation is: $y'' - 4y' - 5y = 0.$

(c) The solutions $y_1 = 2e^{3x},\ \ y_2 = xe^{3x}$ imply that 3 is the only root of the characteristic equation. Therefore, the characteristic equation is

$$(r - 3)^2 = r^2 - 6r + 9 = 0$$

and the differential equation is: $y'' - 6y' + 9y = 0.$

33.   Suppose that $y_1(x) = ky_2(x)$, where k is a constant. Then

$$W(x) = y_1 y_2' - y_2 y_1' = ky_2 y_2' - ky_2 y_2' = 0.$$

Now suppose that $y_1 y_2' - y_2 y_1' = 0.$ Then

$$\left[\frac{y_1(x)}{y_2(x)} \right]' = \frac{y_2 y_1' - y_1 y_2'}{y_2^2} = -\frac{y_1 y_2' - y_2 y_1'}{y_2^2} = 0.$$

This implies that $\dfrac{y_1}{y_2} = k,$ for some constant $k,$ that is $y_1 = ky_2.$

35.   (a) If $a = 0,\ b > 0,$ then the general solution of the differential equation is:

$$y = C_1 \cos \sqrt{b}\, x + C_2 \sin \sqrt{b}\, x = A \cos \left(\sqrt{b}\, x + \phi \right)$$

where A and ϕ are constants. Clearly $|y(x)| \le |A|$ for all $x.$

(b) If $a > 0,\ b = 0,$ then the general solution of the differential equation is:

$$y = C_1 + C_2 e^{-ax} \qquad \text{and} \qquad \lim_{x \to \infty} y(x) = C_1.$$

The solution which satisfies the conditions: $y(0) = y_0,\ \ y'(0) = y_1$ is:

$$y = y_0 + \frac{y_1}{a} - \frac{y_1}{a} e^{-ax} \quad \text{and} \quad \lim_{x \to \infty} y(x) = y_0 + \frac{y_1}{a}; \qquad k = y_0 + \frac{y_1}{a}.$$

37.   From Exercise 36, the change of variable $z = \ln x$ transforms the equation

$$x^2 y'' - xy' - 8y = 0$$

into the differential equation with constant coefficients

$$\frac{d^2 y}{dz^2} - 2\frac{dy}{dz} - 8y = 0.$$

The characteristic equation is:

$$r^2 - 2r - 8 = 0 \qquad \text{or} \qquad (r - 4)(r + 2) = 0$$

The roots are: $r = 4,\ r = -2,$ and the general solution (in terms of z) is:

$$y = C_1 e^{4z} + C_2 e^{-2z}.$$

Replacing z by $\ln x$ we get

$$y = C_1 e^{4 \ln x} + C_2 e^{-2 \ln x} = C_1 x^4 + C_2 x^{-2}.$$

39.   From Exercise 36, the change of variable $z = \ln x$ transforms the equation

$$x^2 y'' - 3xy' + 4y = 0$$

into the differential equation with constant coefficients

$$\frac{d^2 y}{dz^2} - 4\frac{dy}{dz} + 4y = 0.$$

The characteristic equation is:

$$r^2 - 4r + 4 = 0 \quad \text{or} \quad (r-2)^2 = 0.$$

The only root is: $r = 2$,   and the general solution (in terms of z) is:

$$y = C_1 e^{2z} + C_2 z e^{-2z}.$$

Replacing z by $\ln x$ we get

$$y = C_1 e^{2\ln x} + C_2 \ln x\, e^{2\ln x} = C_1 x^2 + C_2 x^2 \ln x.$$

41.   (a) $e^x = 1 + x + \dfrac{x^2}{2!} + \dfrac{x^3}{3!} + \cdots + \dfrac{x^n}{n!} \cdots = \displaystyle\sum_{k=0}^{\inf} \dfrac{1}{k!} x^k$

(b) $e^{i\theta} = 1 + (i\theta) + \dfrac{(i\theta)^2}{2!} + \dfrac{(i\theta)^3}{3!} + \cdots + \dfrac{(i\theta)^n}{n!} + \cdots$

$\qquad = 1 + i\theta - \dfrac{1}{2!}\theta^2 - i\dfrac{1}{3!}\theta^3 + \cdots + (i)^n \dfrac{1}{n!}\theta^n + \cdots = \cos\theta + i\sin\theta$

(c) $e^{-i\theta} = 1 + (-i\theta) + \dfrac{(-i\theta)^2}{2!} + \dfrac{(-i\theta)^3}{3!} + \cdots + \dfrac{(-i\theta)^n}{n!} + \cdots$

$\qquad = 1 - i\theta - \dfrac{1}{2!}\theta^2 + i\dfrac{1}{3!}\theta^3 + \cdots + (-i)^n \dfrac{1}{n!}\theta^n + \cdots = \cos\theta - i\sin\theta$

SECTION 18.4

1.   First consider the reduced equation. The characteristic equation is:

$$r^2 + 5r + 6 = (r+2)(r+3) = 0$$

and $u_1(x) = e^{-2x}$, $u_2(x) = e^{-3x}$ are fundamental solutions. Therefore, a particular solution of the given equation has the form

$$y = Ax + B.$$

The derivatives of y are: $y' = A, \quad y'' = 0.$

Substituting y and its derivatives into the given equation gives

$$0 + 5A + 6(Ax + B) = 3x + 4.$$

Thus,

$$6A = 3$$

$$5A + 6B = 4$$

The solution of this pair of equations is: $A = \frac{1}{2}$, $B = \frac{1}{4}$,   and $y = \frac{1}{2}x + \frac{1}{4}$.

3.   First consider the reduced equation. The characteristic equation is:

$$r^2 + 2r + 5 = 0$$

and $u_1(x) = e^{-x}\cos 2x$, $u_2(x) = e^{-x}\sin 2x$ are fundamental solutions. Therefore, a particular solution of the given equation has the form

$$y = Ax^2 + Bx + C.$$

The derivatives of y are: $y' = 2Ax + B$, $y'' = 2A$.

Substituting y and its derivatives into the given equation gives

$$2A + 2(2Ax + B) + 5(Ax^2 + Bx + C) = x^2 - 1.$$

Thus,

$$5A = 1$$

$$4A + 5B = 0$$

$$2A + 2B + 5C = -1$$

The solution of this system of equations is: $A = \tfrac{1}{5}$, $B = -\tfrac{4}{25}$, $C = -\tfrac{27}{125}$,   and

$$y = \tfrac{1}{5}x^2 - \tfrac{4}{25}x - \tfrac{27}{125}.$$

5.   First consider the reduced equation. The characteristic equation is:

$$r^2 + 6r + 9 = (r+3)^2 = 0$$

and $u_1(x) = e^{-3x}$, $u_2(x) = xe^{-3x}$ are fundamental solutions. Therefore, a particular solution of the given equation has the form

$$y = Ae^{3x}.$$

The derivatives of y are: $y' = 3Ae^{3x}$, $y'' = 9Ae^{3x}0$.

Substituting y and its derivatives into the given equation gives

$$9Ae^{3x} + 18Ae^{3x} + 9Ae^{3x} = e^{3x}.$$

Thus, $36A = 1 \Longrightarrow A = \tfrac{1}{36}$,   and $y = \tfrac{1}{36}e^{3x}.$

7.   First consider the reduced equation. The characteristic equation is:

$$r^2 + 2r + 2 = 0$$

and $u_1(x) = e^{-x}\cos x$, $u_2(x) = e^{-x}\sin x$ are fundamental solutions. Therefore, a particular solution of the given equation has the form

$$y = Ae^x.$$

The derivatives of y are: $y' = Ae^x$, $y'' = Ae^x$.

Substituting y and its derivatives into the given equation gives

$$Ae^x + 2Ae^x + 2Ae^x = e^x.$$

Thus, $5A = 1 \Longrightarrow A = \frac{1}{5}$ and $y = \frac{1}{5} e^x$.

9.   First consider the reduced equation. The characteristic equation is:

$$r^2 - r - 12 = (r - 4)(r + 3) = 0$$

and $u_1(x) = e^{4x}$, $u_2(x) = e^{-3x}$ are fundamental solutions. Therefore, a particular solution of the given equation has the form

$$y = A \cos x + B \sin x.$$

The derivatives of y are: $y' = -A \sin x + B \cos x$, $y'' = -A \cos x - B \sin x$.

Substituting y and its derivatives into the given equation gives

$$-A \cos x - B \sin x - (-A \sin x + B \cos x) - 12(A \cos x + B \sin x) = \cos x.$$

Thus,

$$-13A - B = 1$$

$$A - 13B = 0$$

The solution of this system of equations is: $A = -\frac{13}{170}$, $B = -\frac{1}{170}$,   and

$$y = -\frac{13}{170} \cos x - \frac{1}{170} \sin x.$$

11.   First consider the reduced equation. The characteristic equation is:

$$r^2 + 7r + 6 = (r + 6)(r + 1) = 0$$

and $u_1(x) = e^{-6x}$, $u_2(x) = e^{-x}$ are fundamental solutions. Therefore, a particular solution of the given equation has the form

$$y = A \cos 2x + B \sin 2x.$$

The derivatives of y are: $y' = -2A \sin 2x + 2B \cos 2x$, $y'' = -4A \cos 2x - 4B \sin 2x$.

Substituting y and its derivatives into the given equation gives

$$-4A \cos 2x - 4B \sin 2x + 7(-2A \sin 2x + 2B \cos 2x) + 6(A \cos 2x + B \sin 2x) = 3 \cos 2x.$$

Thus,

$$2A + 14B = 3$$

$$-14A + 2B = 0$$

The solution of this system of equations is: $A = \frac{3}{100}$, $B = \frac{21}{100}$ and

$$y = \frac{3}{100} \cos 2x + \frac{21}{100} \sin 2x.$$

13.   First consider the reduced equation. The characteristic equation is:

$$r^2 - 2r + 5 = 0$$

and $u_1(x) = e^x \cos 2x$, $u_2(x) = e^x \sin 2x$ are fundamental solutions. Therefore, a particular solution of the given equation has the form

$$y = Ae^{-x} \cos 2x + Be^{-x} \sin 2x$$

The derivatives of y are: $y' = -Ae^{-x} \cos 2x - 2Ae^{-x} \sin 2x - Be^{-x} \sin 2x + 2be^{-x} \cos 2x$,

$y'' = 4Ae^{-x} \sin -3Ae^{-x} \cos 2x - 4be^{-x} \cos 2x - 3Be^{-x} \sin 2x.$

Substituting y and its derivatives into the given equation gives

$4Ae^{-x} \sin -3Ae^{-x} \cos 2x - 4be^{-x} \cos 2x - 3Be^{-x} \sin 2x -$

$2\left(-Ae^{-x} \cos 2x - 2Ae^{-x} \sin 2x - Be^{-x} \sin 2x + 2be^{-x} \cos 2x\right) +$

$5\left(Ae^{-x} \cos 2x + Be^{-x} \sin 2x\right) = e^{-x} \sin 2x.$

Equating the coefficients of $e^{-x} \cos 2x$ and $e^{-x} \sin 2x$ we get,

$$8A + 4B = 1$$

$$4A - 8B = 0$$

The solution of this system of equations is: $A = \frac{1}{10}$, $B = \frac{1}{20}$ and

$$y = \frac{1}{10} e^{-x} \cos 2x + \frac{1}{20} e^{-x} \sin 2x.$$

15.   First consider the reduced equation. The characteristic equation is:

$$r^2 + 6r + 8 = (r + 4)(r + 2) = 0$$

and $u_1(x) = e^{-4x}$, $u_2(x) = e^{-2x}$ are fundamental solutions. Therefore, a particular solution of the given equation has the form

$$y = Axe^{-2x}.$$

The derivatives of y are: $y' = Ae^{-2x} - 2Axe^{-2x}, \quad y'' = -4Ae^{-2x} + 4Axe^{-2x}.$

Substituting y and its derivatives into the given equation gives

$$-4Ae^{-2x} + 4Axe^{-2x} + 6\left(Ae^{-2x} - 2Axe^{-2x}\right) + 8Axe^{-2x} = 3e^{-2x}$$

Thus, $2A = 3 \Longrightarrow A = \frac{3}{2}$ and $y = \frac{3}{2}xe^{-2x}.$

17.   First consider the reduced equation: $y'' + y = 0.$ The characteristic equation is:

$$r^2 + 1 = 0$$

and $u_1(x) = \cos x, \quad u_2(x) = \sin x$ are fundamental solutions. A particular solution of the given equation has the form

$$y = Ae^x.$$

The derivatives of y are: $y' = y'' = Ae^x.$

Substitute y and its derivatives into the given equation:

$$Ae^x + Ae^x = e^x \Longrightarrow A = \frac{1}{2} \quad \text{and} \quad y = \frac{1}{2}e^x.$$

The general solution of the given equation is: $y = C_1 \cos x + C_2 \sin x + \frac{1}{2}e^x.$

19.   First consider the reduced equation: $y'' - 3y' - 10y = 0.$ The characteristic equation is:

$$r^2 - 3r - 10 = (r - 5)(r + 2) = 0$$

and $u_1(x) = e^{5x}, \quad u_2(x) = e^{-2x}$ are fundamental solutions. A particular solution of the given equation has the form

$$y = Ax + B.$$

The derivatives of y are: $y' = A, \quad y'' = 0.$

Substitute y and its derivatives into the given equation:

$$-3A - 10(Ax + B) = -x - 1 \Longrightarrow A = \frac{1}{10}, \quad B = \frac{7}{100} \text{and} \quad y = \frac{1}{10}x + \frac{7}{100}$$

The general solution of the given equation is:

$$y = C_1e^{5x} + C_2e^{-2x} + \cos x + C_2 \sin x + \frac{1}{10}x + \frac{7}{100}$$

21.   First consider the reduced equation: $y'' + 3y' - 4y = 0.$ The characteristic equation is:

$$r^2 + 3r - 4 = (r + 4)(r - 1) = 0$$

and $u_1(x) = e^x, \quad u_2(x) = e^{-4x}$ are fundamental solutions. A particular solution of the given equation has the form

$$y = Axe^{-4x}.$$

The derivatives of y are: $y' = Ae^{-4x} - 4Axe^{-4x}, \quad y'' = -8Ae^{-4x} + 16Axe^{-4x}.$

Substitute y and its derivatives into the given equation:

$$-8Ae^{-4x} + 16Axe^{-4x} + 3\left(Ae^{-4x} - 4Axe^{-4x}\right) - 4Axe^{-4x} = e^{-4x}.$$

This implies $-5A = 1$, so $A = -\frac{1}{5}$ and $y = -\frac{1}{5}xe^{-4x}.$

The general solution of the given equation is: $y = C_1e^x + C_2e^{-4x} - \frac{1}{5}xe^{-4x}.$

23.   First consider the reduced equation: $y'' + y' - 2y = 0.$ The characteristic equation is:

$$r^2 + r - 2 = (r + 2)(r - 1) = 0$$

and $u_1(x) = e^{-2x}, \quad u_2(x) = e^x$ are fundamental solutions. A particular solution of the given equation has the form

$$y = (Ax^2 + Bx)e^x = Ax^2e^x + Bxe^x.$$

The derivatives of y are:

$$y' = 2Axe^x + Ax^2e^x + Be^x + bxe^x, \quad y'' = 2Ae^x + 4Axe^x + Ax^2e^x + 2Be^x + Bxe^x.$$

Substitute y and its derivatives into the given equation:

$$2Ae^x + 4Axe^x + Ax^2e^x + 2Be^x + Bxe^x +$$

$$\left(2Axe^x + Ax^2e^x + Be^x + Bxe^x\right) - 2\left(Ax^2e^x + Bxe^x\right) = 3xe^x.$$

Equating coefficients, we get

$$6A = 3$$

$$2A + 3B = 0$$

The solution of this system of equations is: $A = \frac{1}{2}, \quad B = -\frac{1}{3}$ and $y = \frac{1}{2}x^2e^x - \frac{1}{3}xe^x.$

The general solution of the given equation is: $y = C_1e^{-2x}x + C_2e^x + \frac{1}{2}x^2e^x - \frac{1}{3}xe^x.$

25.   Let $y_1(x)$ be a solution of $y'' + ay' + by = \phi_1(x),$ let $y_2(x)$ be a solution of $y'' + ay' + by = \phi_2(x),$ and let $z = y_1 + y_2.$ Then

$$z'' + az' + bz = (y_1'' + y_2'') + a(y_1' + y_2') + b(y_1 + y_2)$$

$$= (y_1'' + ay_1' + by_1) + (y_2'' + ay_2' + y_2) = \phi_1 + \phi_2.$$

27.   First consider the reduced equation: $y'' + 4y' + 3y = 0.$ The characteristic equation is:

$$r^2 + 4r + 3 = (r + 3)(r + 1) = 0$$

and $u_1(x) = e^{-3x}$, $u_2(x) = e^{-x}$ are fundamental solutions. Since $\cosh x = \frac{1}{2}(e^x + e^{-x})$,   a particular solution of the given equation has the form

$$y = Ae^x + Bxe^{-x}$$

The derivatives of y are: $y' = Ae^x + Be^{-x} - Bxe^{-x}$ $y'' = Ae^x - 2Be^{-x} + Bxe^{-x}$.

Substitute y and its derivatives into the given equation:

$$Ae^x - 2Be^{-x} + Bxe^{-x} + 4\left(Ae^x + Be^{-x} - Bxe^{-x}\right) + 3\left(Ae^x + Bxe^{-x}\right) = \frac{1}{2}\left(e^x + e^{-x}\right).$$

Equating coefficients, we get $A = \frac{1}{16}$, $B = \frac{1}{4}$,   and so $y = \frac{1}{16}e^x + \frac{1}{4}xe^{-x}$.

The general solution of the given equation is: $y = C_1 e^{-3x} + C_2 e^{-x} + \frac{1}{16}e^x + \frac{1}{4}xe^{-x}$.

29.   First consider the reduced equation $y'' - 2y' + y = 0$.   The characteristic equation is:

$$r^2 - 2r + 1 = (r - 1)^2 = 0$$

and $u_1(x) = e^x$, $u_2(x) = xe^x$ are fundamental solutions. Their Wronskian is given by

$$W = u_1 u_2' - u_2 u_1' = e^x(e^x + xe^x) - xe^x(e^x) = e^{2x}$$

Using variation of parameters, a particular solution of the given equation will have the form

$$y = u_1 z_1 + u_2 z_2,$$

where

$$z_1 = -\int \frac{xe^x(xe^x \cos x)}{e^{2x}}\, dx = -\int x^2 \cos x\, dx = -x^2 \sin x - 2x \cos x + 2 \sin x,$$

$$z_2 = \int \frac{e^x(xe^x \cos x)}{e^{2x}}\, dx = \int x \cos x\, dx = x \sin x + \cos x$$

Therefore,

$$y = e^x\left(-x^2 \sin x - 2x \cos x + 2 \sin x\right) + xe^x\left(x \sin x + \cos x\right) = 2e^x \sin x - xe^x \cos x.$$

31.   First consider the reduced equation $y'' - 4y' + 4y = 0$.   The characteristic equation is:

$$r^2 - 4r + 4 = (r - 2)^2 = 0$$

and $u_1(x) = e^{2x}$, $u_2(x) = xe^{2x}$ are fundamental solutions. Their Wronskian is given by

$$W = u_1 u_2' - u_2 u_1' = e^{2x}\left(e^{2x} + 2xe^{2x}\right) - xe^{2x}(2e^{2x}) = e^{4x}.$$

Using variation of parameters, a particular solution of the given equation will have the form

$$y = u_1 z_1 + u_2 z_2,$$

where

$$z_1 = -\int \frac{xe^{2x}\left(\frac{1}{3}x^{-1}e^{2x}\right)}{e^{4x}}\, dx = -\frac{1}{3}\int dx = -\frac{1}{3}x,$$

$$z_2 = \int \frac{e^{2x}\left(\frac{1}{3}x^{-1}e^{2x}\right)}{e^{4x}}\,dx = -\frac{1}{3}\int \frac{1}{x}\,dx = \frac{1}{3}\ln|x|.$$

Therefore,

$$y = e^{2x}\left(-\frac{1}{3}x\right) + xe^{2x}\left(\frac{1}{3}\ln|x|\right) = -\frac{1}{3}xe^{2x} + \frac{1}{3}x\ln|x|\,e^{2x}.$$

Note: Since $u = -\frac{1}{3}xe^{2x}$ is a solution of the reduced equation,

$$y = \frac{1}{3}x\ln|x|\,e^{2x}$$

is also a particular solution of the given equation.

33. First consider the reduced equation $y'' + 4y' + 4y = 0$. The characteristic equation is:

$$r^2 + 4r + 4 = (r+2)^2 = 0$$

and $u_1(x) = e^{-2x}$, $u_2(x) = xe^{-2x}$ are fundamental solutions. Their Wronskian is given by

$$W = u_1 u_2' - u_2 u_1' = e^{-2x}\left(e^{-2x} - 2xe^{2x}\right) - xe^{-2x}(-2e^{-2x}) = e^{-4x}.$$

Using variation of parameters, a particular solution of the given equation will have the form

$$y = u_1 z_1 + u_2 z_2,$$

where

$$z_1 = -\int \frac{xe^{-2x}\left(x^{-2}e^{-2x}\right)}{e^{-4x}}\,dx = -\int \frac{1}{x}\,dx = -\ln|x|$$

$$z_2 = \int \frac{e^{-2x}\left(x^{-2}e^{-2x}\right)}{e^{-4x}}\,dx = \int \frac{1}{x^2}\,dx = -\frac{1}{x}$$

Therefore,

$$y = e^{-2x}\left(-\ln|x|\right) + xe^{-2x}\left(-\frac{1}{x}\right) = -e^{-2x}\ln|x| - e^{-2x}.$$

Note: Since $u = -e^{-2x}$ is a solution of the reduced equation, we can take

$$y = -e^{-2x}\ln|x|$$

35. First consider the reduced equation $y'' - 2y' + 2y = 0$. The characteristic equation is:

$$r^2 - 2r + 2 = 0$$

and $u_1(x) = e^x \cos x$, $u_2(x) = e^x \sin x$ are fundamental solutions. Their Wronskian is given by

$$W = e^x \cos x \left[e^x \sin x + e^x \cos x\right] - e^x \sin x \left[e^x \cos x - e^x \sin x\right] = e^{2x}$$

Using variation of parameters, a particular solution of the given equation will have the form

$$y = u_1 z_1 + u_2 z_2,$$

where

$$z_1 = -\int \frac{e^x \sin x \cdot e^x \sec x}{e^{2x}}\,dx = -\int \tan x\,dx = -\ln|\sec x| = \ln|\cos x|$$

$$z_2 = \int \frac{e^x \cos x \cdot e^x \sec x}{e^{2x}} \, dx = \int dx = x$$

Therefore,

$$y = e^x \cos x \, (\ln|\cos x|) + e^x \sin x(x) = e^x \cos x \ln|\cos x| + x e^x \sin x.$$

37. Assume that the forcing function $F(t) = F_0$ (constant). Then the differential equation has a particular solution of the form $i = A$. The derivatives of i are: $i' = i'' = 0$. Substituting i and its derivatives into the equation, we get

$$\frac{1}{C} A = F_0 \implies A = CF_0 \implies i = CF_0.$$

The characteristic equation for the reduced equation is:

$$Lr^2 + Rr + \frac{1}{C} = 0 \implies r_1, r_2 = \frac{-R \pm \sqrt{R^2 - 4L/C}}{2L} = \frac{-R \pm \sqrt{CR^2 - 4L}}{2L\sqrt{C}}$$

(a) If $CR^2 = 4L$, then the characteristic equation has only one root: $r = -R/2L$,

and $u_1 = e^{-(R/2L)t}$, $u_2 = t e^{-(R/2L)t}$ are fundamental solutions.

The general solution of the given equation is:

$$i(t) = C_1 e^{-(R/2L)t} + C_2 t e^{-(R/2L)t} + CF_0$$

and its derivative is:

$$i'(t) = -C_1 (R/2L) e^{-(R/2L)t} + C_2 e^{-(R/2L)t} - C_2 (R/2L) t e^{-(R/2L)t}.$$

Applying the side conditions $i(0) = 0$, $i'(0) = F_0/L$, we get

$$C_1 + CF_0 = 0$$

$$(-R/2L)C_1 + C_2 = F_0/L$$

The solution is $C_1 = -CF_0$, $C_2 = \frac{F_0}{2L}(2 - RC)$.

The current in this case is:

$$i(t) = -CF_0 e^{-(R/2L)t} + \frac{F_0}{2L}(2 - RC) \, t e^{-(R/2L)t} + CF_0.$$

(b) If $CR^2 - 4L < 0$ then the characteristic equation has complex roots:

$$r_1 = -R/2L \pm i\beta, \quad \text{where} \quad \beta = \sqrt{\frac{4L - CR^2}{4CL^2}} \quad (\text{here } i^2 = -1)$$

and fundamental solutions are: $u_1 = e^{-(R/2L)t} \cos \beta t$, $u_2 = e^{-(R/2L)t} \sin \beta t$.

The general solution of the given differential equation is:

$$i(t) = e^{-(R/2L)t} (C_1 \cos \beta t + C_2 \sin \beta t) + CF_0$$

and its derivative is:

$$i'(t) = (-R/2L) e^{-(R/2L)t} (C_1 \cos \beta t + C_2 \sin \beta t) + \beta e^{-(R/2L)t} (-C_1 \sin \beta t + C_2 \cos \beta t).$$

Applying the side conditions $i(0) = 0, \quad i'(0) = F_0/L, \quad$ we get

$$C_1 + CF_0 = 0$$

$$(-R/2L)C_1 + \beta C_2 = F_0/L$$

The solution is $\quad C_1 = -CF_0, \quad C_2 = \dfrac{F_0}{2L\beta}(2 - RC).$

The current in this case is:

$$i(t) = e^{-(R/2L)t}\left(\frac{F_0}{2L\beta}(2 - RC)\sin\beta t - CF_0\cos\beta t\right) + CF_0.$$

39.   (a) Let $\ y_1(x) = \sin\left(\ln x^2\right).$ Then

$$y_1' = \left(\frac{2}{x}\right)\cos\left(\ln x^2\right) \quad \text{and} \quad y_1'' = -\left(\frac{4}{x^2}\right)\sin\left(\ln x^2\right) - \left(\frac{2}{x^2}\right)\cos\left(\ln x^2\right)$$

Substituting y_1 and its derivatives into the differential equation, we have

$$x^2\left[-\left(\frac{4}{x^2}\right)\sin\left(\ln x^2\right) - \left(\frac{2}{x^2}\right)\cos\left(\ln x^2\right)\right] - x\left[\left(\frac{2}{x}\right)\cos\left(\ln x^2\right)\right] + 4\sin\left(\ln x^2\right) = 0$$

The verification that y_2 is a solution is done in exactly the same way.

The Wronskian of y_1 and y_2 is:

$$\begin{aligned}W(x) &= y_1 y_2' - y_2 y_1' \\ &= \sin\left(\ln x^2\right)\left[-\left(\frac{2}{x}\right)\sin\left(\ln x^2\right)\right] - \cos\left(\ln x^2\right)\left[\left(\frac{2}{x}\right)\cos\left(\ln x^2\right)\right] \\ &= -\frac{2}{x} \neq 0 \text{ on } (0, \infty)\end{aligned}$$

(b) To use the method of variation of parameters as described in the text, we first re-write the equation in the form

$$y'' + x^{-1}y' + 4x^{-2}y = x^{-2}\sin(\ln x).$$

Then, a particular solution of the equation will have the form $\quad y = y_1 z_1 + y_2 z_2, \quad$ where

$$\begin{aligned}z_1 &= -\int \frac{\cos(\ln x^2)x^{-2}\sin(\ln x)}{-2/x}\,dx \\ &= \tfrac{1}{2}\int \cos(2\ln x)x^{-1}\sin(\ln x)\,dx \\ &= \tfrac{1}{2}\int \cos 2u\,\sin u\,du \quad (u = \ln x) \\ &= \tfrac{1}{2}\int(2\cos^2 u - 1)\sin u\,du \\ &= -\tfrac{1}{3}\cos^3 u + \tfrac{1}{2}\sin u\end{aligned}$$

and

$$z_2 = \int \frac{\sin(\ln x^2)x^{-2}\sin(\ln x)}{-2/x}\,dx$$
$$= -\tfrac{1}{2}\int \sin(2\ln x)x^{-1}\sin(\ln x)\,dx$$
$$= -\tfrac{1}{2}\int \sin 2u \, \sin u \, du \qquad (u = \ln x)$$
$$= -\int \sin^2 u \, \cos u \, du$$
$$= -\tfrac{1}{3}\sin^3 u$$

Thus, $y = \sin 2u\left(-\tfrac{1}{3}\cos^3 u + \tfrac{1}{2}\sin u\right) - \cos 2u\left(\tfrac{1}{3}\sin^3 u\right)$ which simplifies to:

$$y = \tfrac{1}{3}\sin u = \tfrac{1}{3}\sin(\ln x).$$

SECTION 18.5

1. The equation of motion is of the form

$$x(t) = A\sin(\omega t + \phi_0).$$

The period is $T = 2\pi/\omega = \pi/4$. Therefore $\omega = 8$. Thus

$$x(t) = A\sin(8t + \phi_0) \text{ and } v(t) = 8A\cos(8t + \phi_0).$$

Since $x(0) = 1$ and $v(0) = 0$, we have

$$1 = A\sin\phi_0 \quad \text{and} \quad 0 = 8A\cos\phi_0.$$

These equations are satisfied by taking $A = 1$ and $\phi_0 = \pi/2$.

Therefore the equation of motion reads

$$x(t) = \sin\left(8t + \tfrac{1}{2}\pi\right).$$

The amplitude is 1 and the frequency is $8/2\pi = 4/\pi$.

3. We can write the equation of motion as

$$x(t) = A\sin\left(\frac{2\pi}{T}t\right).$$

Differentiation gives

$$v(t) = \frac{2\pi A}{T}\cos\left(\frac{2\pi}{T}t\right).$$

The object passes through the origin whenever $\sin[(2\pi/T)] = 0$.

Then $\cos[(2\pi/T)t] = \pm 1$ and $v = \pm 2\pi A/T$.

5. In this case $\phi_0 = 0$ and, measuring t_2 in seconds, $T = 6$.

Therefore $\omega = 2\pi/6 = \pi/3$ and we have

$$x(t) = A\sin\left(\frac{\pi}{3}t\right), \quad v(t) = \frac{\pi A}{3}\cos\left(\frac{\pi}{3}t\right).$$

Since $v(0) = 5$, we have $\pi A/3 = 5$ and therefore $A = 15/\pi$.

The equation of motion can be written

$$x(t) = (15/\pi)\sin\left(\tfrac{1}{3}\pi t\right)$$

7. $x(t) = x_0 \sin\left(\sqrt{k/m}\, t + \tfrac{1}{2}\pi\right)$

9.   The equation of motion for the bob reads

$$x(t) = x_0 \sin\left(\sqrt{k/m}\, t + \tfrac{1}{2}\pi\right). \qquad \text{(Exercise 7)}$$

Since $v(t) = \sqrt{k/m}\, x_0 \cos\left(\sqrt{k/m}\, t + \tfrac{1}{2}\pi\right)$, the maximum speed is $\sqrt{k/m}\, x_0$.

The bob takes on half of that speed where $\left|\cos\left(\sqrt{k/m}\, t + \tfrac{1}{2}\pi\right)\right| = \tfrac{1}{2}$. Therefore

$$\left|\sin\left(\sqrt{k/m}\, t + \tfrac{1}{2}\pi\right)\right| = \sqrt{1 - \tfrac{1}{4}} = \tfrac{1}{2}\sqrt{3} \quad \text{and} \quad x(t) = \pm\tfrac{1}{2}\sqrt{3}\, x_0.$$

11. $\text{KE} = \tfrac{1}{2}m[v(t)]^2 = \tfrac{1}{2}m(k/m)x_0^2 \cos^2\left(\sqrt{k/m}\, t + \tfrac{1}{2}\pi\right)$

$$= \tfrac{1}{4}kx_0^2 \left[1 + \cos\left(2\sqrt{k/m}\, t + \tfrac{1}{2}\pi\right)\right].$$

$$\text{Average KE} = \frac{1}{2\pi\sqrt{m/k}}\int_0^{2\pi\sqrt{m/k}} \tfrac{1}{4}kx_0^2 \left[1 + \cos\left(2\sqrt{k/m}\, t + \frac{1}{2}\pi\right)\right]\, dt$$

$$= \tfrac{1}{4}kx_0^2.$$

13.   Setting $y(t) = x(t) - 2,$ we can write $x''(t) = 8 - 4x(t)$ as $y''(t) + 4y(t) = 0.$

This is simple harmonic motion about the point $y = 0$; that is, about the point $x = 2$. The equation of motion is of the form

$$y(t) = A\sin(2t + \phi_0).$$

Since $y(0) = x(0) - 2 = -2$, the amplitude A is 2. Since $\omega = 2$, the period T is $2\pi/2 = \pi$.

15.   (a)   Take the downward direction as positive. We begin by analyzing the forces on the buoy at a general position x cm beyond equilibrium. First there is the weight of the buoy: $F_1 = mg$. This is a downward force. Next there is the buoyancy force equal to the weight of the fluid displaced; this force is in the opposite direction: $F_2 = -\pi r^2 (L + x)\rho$. We are neglecting friction so the total force is

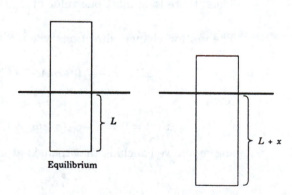

Equilibrium

$$F = F_1 + F_2 = mg - \pi r^2 (L + x) \rho = (mg - \pi r^2 L\rho) - \pi r^2 x\rho.$$

We are assuming at the equilibrium point that the forces (weight of buoy and buoyant force of fluid) are in balance:

$$mg - \pi r^2 L\rho = 0.$$

Thus,

$$F = -\pi r^2 x\rho.$$

By Newton's

$$F = ma \qquad \text{(force} = \text{mass} \times \text{acceleration)}$$

we have

$$ma = -\pi r^2 x\rho \qquad \text{and thus} \qquad a + \frac{\pi r^2 \rho}{m} x = 0.$$

Thus, at each time t,

$$x''(t) + \frac{\pi r^2 \rho}{m} x(t) = 0.$$

(b)   The usual procedure shows that

$$x(t) = x_0 \sin \left(r\sqrt{\pi\rho/m}\, t + \tfrac{1}{2}\pi \right).$$

The amplitude A is x_0 and the period T is $(2/r)\sqrt{m\pi/\rho}$.

17.   From (18.5.4), we have

$$x(t) = Ae^{(-c/2m)t} \sin(\omega t + \phi_0) = \frac{A}{e^{(c/2m)t}} \sin(\omega t + \phi_0) \quad \text{where} \quad \omega = \frac{\sqrt{4km - c^2}}{2m}$$

If c increases, then both the amplitude, $\left| \dfrac{A}{e^{(c/2m)t}} \right|$ and the frequency $\dfrac{\omega}{2\pi}$ decrease.

19.   Set $x(t) = 0$ in (18.5.6). The result is:

$$C_1 e^{(-c/2m)t} + C_2 t e^{(-c/2m)t} = 0 \Longrightarrow C_1 + C_2 t = 0 \Longrightarrow t = -C_1/C_2$$

Thus, there is at most one value of t at which $x(t) = 0$.

The motion changes directions when $x'(t) = 0$:

$$x'(t) = -C_1(c/2m)e^{(-c/2m)t} + C_2 e^{(-c/2m)t} - C_2(c/2m)t e^{(-c/2m)t}.$$

Now,

$$x'(t) = 0 \Longrightarrow -C_1(c/2m) + C_2 - C_2 t(c/2m) = 0 \Longrightarrow t = \frac{C_2 - C_1(c/2m)}{C_2(c/2m)}$$

and again we conclude that there is at most one value of t at which $x'(t) = 0$.

21. $$x(t) = A\sin(\omega t + \phi_0) + \frac{F_0/m}{\omega^2 - \gamma^2}\cos(\gamma t)$$

If $\omega/\gamma = m/n$ is rational, then $m/\omega = n/\gamma$ is a period.

23. The characteristic equation is

$$r^2 + 2\alpha r + \omega^2 = 0; \quad \text{the roots are} \quad r_1, r_2 = -\alpha \pm \sqrt{\alpha^2 - \omega^2}$$

Since $0 < \alpha < \omega$, $\alpha^2 < \omega^2$ and the roots are complex. Thus, $u_1(t) = e^{-\alpha t}\cos\beta t$, $u_2(t) = e^{-\alpha t}\sin\beta t$, where $\beta = \sqrt{\alpha^2 - \omega^2}$ are fundamental solutions, and the general solution is:

$$x(t) = e^{-\alpha t}(C_1\cos\beta t + C_2\sin\beta t) = Ae^{-\alpha t}\sin(\beta t + \phi_0); \quad \beta = \sqrt{\alpha^2 - \omega^2}$$

25. Set $\omega = \gamma$ in the particular solution x_p given in Exercise 24. Then we have

$$x_p = \frac{F_0}{2\alpha\gamma m}\sin\gamma t$$

As $c = 2\alpha m \to 0^+$, the amplitude $\left|\dfrac{F_0}{2\alpha\gamma m}\right| \to \infty$

27. $\left(\omega^2 - \gamma^2\right)^2 + 4\alpha^2\gamma^2 = \omega^4 + \gamma^4 + 2\gamma^2(2\alpha^2 - \omega^2)$ increases as γ increases.